Université Joseph Fourier – NATO Advanced Study Institute

*Les Houches*

Session LXXX

2003

Methods and Models in Neurophysics

Méthodes et Modèles en Neurophysique

# Contributors to this volume

P. Bressloff

E. Brown

N. Brunel

D. Golomb

E. Marder

G. Mato

C. Pouzat

J. Rinzel

M. Shelley

H. Sompolinsky

D. Terman

T. Tishby

A. Treves

M. Tsodyks

C. van Vreeswijk

F. Wolf

ÉCOLE D'ÉTÉ DE PHYSIQUE DES HOUCHES

SESSION LXXX, 28 JULY – 29 AUGUST 2003

NATO ADVANCED STUDY INSTITUTE
ÉCOLE THÉMATIQUE DU CNRS

METHODS AND MODELS IN NEUROPHYSICS

MÉTHODES ET MODÈLES EN NEUROPHYSIQUE

Edited by
C. C. Chow, B. Gutkin, D. Hansel, C. Meunier, and J. Dalibard

ELSEVIER

2005

Amsterdam – Boston – Heidelberg – London – New York – Oxford
Paris – San Diego – San Francisco – Singapore – Sydney – Tokyo

| ELSEVIER B.V. | ELSEVIER Inc. | ELSEVIER Ltd | ELSEVIER Ltd |
|---|---|---|---|
| Sara Burgerhartstraat 25 | 525 B Street, Suite 1900 | The Boulevard, Langford Lane | 84 Theobalds Road |
| P.O. Box 211, 1000 AE | San Diego, CA 92101-4495 | Kidlington, Oxford OX5 1GB | London WC1X 8RR |
| Amsterdam, | USA | UK | UK |
| The Netherlands | | | |

First edition 2005

Library of Congress Cataloging in Publication Data
A catalog record is available from the Library of Congress.

British Library Cataloguing in Publication Data
A catalogue record is available from the British Library.

ISBN: 0-444-51792-8
ISSN: 0924-8099

Transferred to digital print 2007

Printed and bound by CPI Antony Rowe, Eastbourne

## Working together to grow libraries in developing countries

www.elsevier.com | www.bookaid.org | www.sabre.org

ELSEVIER      BOOK AID
              International      Sabre Foundation

# ÉCOLE DE PHYSIQUE DES HOUCHES

Service inter-universitaire commun
à l'Université Joseph Fourier de Grenoble
et à l'Institut National Polytechnique de Grenoble

Subventionné par le Ministère de l'Éducation Nationale,
de l'Enseignement Supérieur et de la Recherche,
le Centre National de la Recherche Scientifique,
le Commissariat à l'Énergie Atomique

# Previous sessions

**Publishers:**
- Session VIII: Dunod, Wiley, Methuen
- Sessions IX and X: Herman, Wiley
- Session XI: Gordon and Breach, Presses Universitaires
- Sessions XII–XXV: Gordon and Breach
- Sessions XXVI–LVIII: North-Holland
- Session LXIX–LXXVIII: EDP Sciences, Springer
- Session LXXIX–LXXX: Elsevier

# Lecturers

Larry ABBOTT, Volen Center for Complex Systems and Department of Biology, Brandeis University, Waltham, MA 02454-9110, USA

Philippe ASCHER, Laboratoire de Physiologie Cérébrale, UMR 8118, Université René Descartes, 45 Rue des Saints-Pères, 75270, Paris, France

Paul BRESSLOFF, Department of Mathematics, 155 South 1400 East 233 JWB, University of Utah, Salt Lake City, UT 84112, USA

Emery N. BROWN, Neuroscience Statistics Research Laboratory, Massachusetts General Hospital, 55 Fruit Street, Clincs 3, Boston, MA 02114, USA

Nicolas BRUNEL, Laboratoire de Neurophysique et Physiologie du Système Moteur, CNRS-UMR8119, Universite René Descartes, 45 rue des Saints Pères, 75270 Paris Cedex 06, France

Wulfram GERSTNER, Laboratory of Computational Neuroscience, EPFL - Batiment AA-B, CH-1015 Lausanne, Switzerland

David GOLOMB, Department of Physics, University of California, San Diego, 9500 Gilman Drive, La Jolla, CA 92093-0319, USA

Eve MARDER, Volen Center for Complex Systems and Department of Biology, Brandeis University, Waltham, MA 02454-9110, USA

German MATO, Centro Atómico Bariloche e Instituto Balseiro, 8400 S.C. de Bariloche, Río Negro, Argentina

Christophe POUZAT, Laboratoire de Physiologie Cérébrale, CNRS UMR 8118, UFR biomédicale de l'Université Paris V, 45 rue des Saints Pères 75006, Paris, France

John RINZEL, Center for Neural Science, New York University, 4 Washington Place, New York, NY 10003, USA

Mike SHELLEY, Courant Institute, 251 Mercer St. New York, NY 10012, USA

Haim SOMPOLINSKY, Racah Institute of Physics, The Hebrew University, E. Safra Campus, Givat Ram 91904, Jerusalem Israel

David TERMAN, Department of Mathematics, 231 W. 18th Ave., The Ohio State University, Columbus, Ohio 43210-1174, USA

Naftali TISHBY, Institute of Computer Science, The Hebrew University, The E. Safra Campus, 91904 Jerusalem, Israel

Alessandro TREVES, SISSA - Cognitive Neuroscience, Via Beirut 2-4, I-34014, Italy

Misha TSODYKS, Brain Research Building Room 133, Department of Neurobiology, Weizmann, Institute of Science, Rehovot 76100, Israel

Carl van VREESWIJK, Neurophysique et Physiologie du Système Moteur, CNRS-UMR8119, Universite René Descartes, 45 rue des Saints Pères, 75270 Paris Cedex 06, France

Fred WOLF, MPI für Strömungsforschung, Dept. of Nonlinear Dynamics, Postfach 2853 Building 2, Room 02207, Germany

# Teaching assistants

Mickey LONDON, Department of Neurobiology, Inst. Life Science, The Hebrew University of Jerusalem, Jerusalem 91904, Israel

Oren SHRIKI, Racah Institute of Physics, The Hebrew University, Jerusalem 91904, Israel

# Organizers

Carson C. CHOW, Department of Mathematics, University of Pittsburgh, Pittsburgh, PA 15260, USA

Jean DALIBARD, Laboratoire Kastler Brossel, Ecole normale supérieure, 24 rue Lhomond, 75231 Paris cedex 05, France

Boris GUTKIN, Gatsby Computational Neuroscience Unit, Alexandra House, 17 Queen Square, London WC1N 3AN, UK

David HANSEL, Laboratoire de Neurophysique et Physiologie du Système Moteur, UMR-8119, 45 Rue des Saints-Pères, 75270 Paris, France

Claude MEUNIER, Laboratoire de Neurophysique et Physiologie du Système Moteur, UMR-8119, 45 Rue des Saints-Pères, 75270, Paris, France

Idan SEGEV, Institute of Life Sciences, Department of Neurobiology, The Hebrew University, Edmond Safra Campus, Givat Ram, Jerusalem, 91904, Israel.

# Speakers at the workshops

## Workshop 1: The neuron in the network

Alexander BORST, Department of Systems and Computational Neurobiology, Max-Planck-Institute of Neurobiology, Am Klopferspitz 18a, D-82152 Martinsried, Germany

Lyle GRAHAM, Laboratoire de Neurophysique et Physiologie du Système Moteur, CNRS UMR 8119, UFR Biomédicale de l'Université René Descartes, 45 rue des Saint-Pères 75006 Paris, France

Peter JONAS, Physiologisches Institut, Universität Freiburg, Hermann-Herder-Strasse 7, D-79104, Freiburg, Germany

Matthew LARKUM, Max Planck Institute for Medical Research, Jahnstr. 29, D-69120 Heidelberg, Germany

## Workshop 2: Dynamics in the somatosensory system

Ehud AHISSAR, Department of Neurobiology, The Weizmann Institute, Rehovot 76100, Israel

Michael BRECHT, Max-Planck Institut für medizinische Forschung, Abteilung Zellphysiologie, Jahnstraße 29, D-69120 Heidelberg, Germany

Rasmus PETERSEN, SISSA - Cognitive Neuroscience, Via Beirut 2-4, I-34014, Italy

## Workshop 3: Learning and Memory

Merav AHISSAR, Department of Psychology, The Hebrew University, Mount Scopus, Jerusalem 91905, Israel

Guo-Qiang BI, Department of Neurobiology, School of Medicine, E1451 Biomedical Science Tower, University of Pittsburgh, 3500 Terrace Street, Pittsburgh, PA 15261, USA

Yadin DUDAI, Department of Neurobiology, Weizmann Institute of Science, Rehovot 76100, Israel

Dr. Mayank MEHTA, E18-366, M.I.T., 50 Ames St., Cambridge, MA 02139, USA

# Participants

BAKER Tanya, Department of Physics, University of Chicago, 5720 South Ellis Avenue, IL 60637, USA

BERNACCHIA Alberto, INFM, Gruppo NAL, Dipartimento di Fisica E.Fermi, Universita La Sapienza, Ple Aldo Moro 5, 00185 Roma, Italy

BIBICHKOV Dmitry, Department of Neurobiology, Weizmann Insitute of Science, 76100 Rehovot, Israel

BORISHYUK Alla, Mathematical Biosciences Institute, Ohio State University, Columbus, OH, USA

BRUUN Georg, Niels Bohr Institute, Blegdamsvej 17, 2100 Copenhagen, Denmark

BUI Tuan, Departement of Physiology, Botterell Hall, Queen's University, Kinston, Ontario, Canada

COCCO Simona, Laboratoire de dynamique des fluides complexes, ULP-CNRS, 3 rue de l'Université, 67000 Strasbourg, France

DA SILVEIRA Rava, 6 ch. Beau-Soleil, 1206 Genève, Switzerland

DELESCLUSE Matthieu, Laboratoire de Physiologie Cérébrale, CNRS UMR 8118, 45 rue des Saints Pères, 75006 Paris, France

DREMENCOV Eliyahu, Life Sciences Faculty, Bar Ilan University, Ramat-gan 52900, Israel

FEDANOV Alexander, Institute of Mathematical Problems in Biology, Russian Academy of Sciences, Puschino, Moscow region 142 290, Russia

FLANAGIN Virginia, MPI of Neurobiology, Systems and Computational Neurobiology, Am Klopferspitz 18a, 82152 Martinsried, Germany

FOLIAS Stefanos, Department of Mathematics, University of Utah, 155 South, 1400 East, JWB 233, Salt Lake City, Utah 84112-0090, USA

GOLLISCH Tim, Institute for Theoretical Biology, Humboldt University Berlin, Invalidenstrasse 43, 10115 Berlin, Germany

GÜTIG Robert, Institute for Theoretical Biology, Humboldt University Berlin, Invalidenstraße 43, 10115 Berlin, Germany

HISTED Mark, MIT E25-236, 77 Massachusetts Avenue, Cambridge MA 02139, USA

HSU Anne, Tolman Hall, University of California, Berkeley CA 94720, USA

KANG Kukjin, Center for Neural Science, New York University, 4 Washington place, Room 809, New York, NY 10003, USA

KASCHUBE Matthias, MPI Strömungsforschung, Bunsenstr.10, 37073 Göttingen, Germany

KATKOV Mikhail, Weizmann Institute of Science, Neurophysiology Department, Rehovot, 76100 Israel

KOEPSELL Kilian, Redwood Neuroscience Institute, 1010 El Camino Real, Suite 380, Menlo Park, CA 94025, USA

KUMAR Arvind, Laboratory for Neurobiology and Biophysics, Albert Ludwigis University of Freiburg, 79104 Freiburg, Germany

LEBLOIS Arthur, Laboratoire de Neurophysique et Physiologie du Système Moteur, Université René Descartes, 45 rue des Saints Pères, 75270 Paris cedex 06, France

LEE Eunjeong, PFC Workgroup, Netherlands Institute for Brain Research, Meibergdreef 33, 1105 AZ, The Netherlands

LERCHNER Alexander, NORDITA, Blegdamsvej 17, 2100 Copenhagen, Denmark

LESNIK Dmitry, Heinrich Heine Universität, Institut für Theoretische Physik, Universitätsstr.1, Gebaüde 25.32, 40225 Düsseldorf, Germany

LOEBEL Alex, The Weizmann Institute of Science, Neurobiology Department, Rehovot, 76100, Israel

MIURA Keiji, Kyoto University, Department of Physics, Kyoto 606-8502, Japan

MOLDAKARIMOV Samat, Department of Mathematics, 301 Thackeray Hall, University of Pittsburgh, Pittsburgh, PA 15260, USA

MONASSON Rémi, LPT/ENS, 24 rue Lhomond, 75231 Paris cedex, France

PERSI Erez, Laboratoire de Neurophysique et Physiologie du Système Moteur, Université René Descartes, 45 rue des Saints Pères, 75270 Paris, France

PFEUTY Benjamin, Laboratoire de Neurophysique et Neurophysiologie du Système Moteur, Université René Descartes, 45 rue des Saints Pères, 75270 Paris cedex 06, France

PFISTER Jean-Pascal, Laboratory of Computational Neuroscience, Swiss Federal Institute of Technology Lausanne EPFL, 1015 Lausanne, Switzerland

PITKOW Xaq, Harvard University, 16 Divinity Avenue, Biological Laboratories room 4033, Cambridge MA 02138, USA

PLEWCZYNSKI Dariusz, Plac Przymierza 3m1, 03-944, Warsaw, Poland

POPOVA Irina, Institute of Theoretical and Experimental Biophysics, Russian Academy of Sciences, Puschino, Moscow district, 142290, Russia

PREMINGER Son, Weizmann Institute of Science, Department of Neurobiology, Rehovot 76100, Israel

ROUDI RASHTABADI Yasser, Cognitive Neurosciences, SISSA, via Beirut 2-4, 34014 Trieste, Italy

RUSSELL Noah, Department of Neurophysiology, National Institute for Medical Research, The Ridgeway, Mill Hill, London NW7 1AA, U.K.

SCHNABEL Michael, MPI Strömungsforschung, Bunsenstr. 10, 37073 Göttingen, Germany

SCHULZKE Erich, Institute of Theoretical Neurophysics, P.O. Box 330440, 28334 Bremen, Germany

SHAHREZAEI Vahid, Physics Department, Simon Fraser University, 8888 University Drive, Burnaby, V5A 154BC, Canada

STEPHENS Greg, Biophysics Group, MS-D454, Los Alamos National Laboratory, Los Alamos, New Mexico, 81545, USA

TCHIJOV Anton, Computational Physics Laboratory, A.F. Ioffe Physico-Technical Institute of RAS, Polytekhnicheskaya str. 26, 194021 St Petersburg, Russia

WANG Jennifer, MIT, Department of Brain and Cognitive Sciences, Seung lab., 77 Massachusetts Avenue, room E25-425, Cambridge, MA 02139, USA

WHITE Olivia, Harvard University, Department of Physics, Jefferson Laboratories, 17 Oxford Str., Cambridge, MA 02140 USA

ZHIGULIN Valentin, California Institute of Technology, Physics Department, MC 103-33, Caltech Pasadena, CA 91125, USA

# PREFACE

Neuroscience is an inherently interdisciplinary field that strives to understand the functioning of neural systems at levels ranging from molecules and cells to behaviour and cognition. Theoretical neuroscience has flourished over the past three decades, becoming an indelible part of neuroscience, and has arguably entered its maturity. It encompasses a vast array of approaches stemming from theoretical physics, computer science, and applied mathematics.

Historically, the application of theoretical techniques to unravelling the operating principles of neurons started with the seminal work of Hodgkin and Huxley, who explained the generation of action potentials and their propagation. A decade later, Rall introduced an appropriate formal framework for investigating synaptic integration. Theoretical physics inspired new theories of network behaviour and organization, following the early work by Wilson and Cowan. The seminal work of Hopfield and of Amit, Gutfreund, and Sompolinsky demonstrated that statistical mechanics provided a powerful framework to analyse memory and learning. Progressively, mathematicians and physicists have elaborated a theoretical framework for analysing neural systems, creating a subfield one may term "neurophysics". At the same time, the emphasis has moved beyond models loosely inspired by biology to ones that directly reflect organisational and biological principles of neural systems.

Methods from statistical mechanics, dynamical systems theory, singular perturbation theory, theory of stochastic processes, theory of signal processing and information theory have enriched neurophysics over the years. For example, geometrical methods for nonlinear differential equations considerably improved our understanding of the dynamics of individual neurons and small circuits. Concepts from oscillation theory, such as averaging, phase reductions and discrete maps have elucidated how cell and synaptic properties determine the dynamical states of systems of interacting neurons. Mean field theories have been successfully used to describe and analyse collective dynamics of large scale networks. Recently, non-trivial extensions of mean field theory have arisen together with the use of symmetric bifurcation theory and group theoretic approaches to study pattern formation in cortical areas. Information theory is now commonly employed

xvii

to analyse spike trains, to uncover neural coding strategies and to examine their optimality.

This volume comes out of the five weeks course of the Summer School in Physics that took place at Les Houches in July–August 2003. The course was composed of a series of theoretical lectures on neurophysics. These lectures comprise this volume. To provide a functional counterpart to some of the theoretical issues addressed, there were three workshops dedicated to selected issues in experimental neuroscience on the topics of: 1) The neuron in the network, 2) The dynamics of the somatosensory system, and 3) Learning and Memory. There were also special seminars by Philippe Ascher and Idan Segev.

A number of excellent textbooks in computational and theoretical neuroscience are now available. However, many of these books lack detailed coverage of the approaches originating from mathematics and physics. This book fills this need by providing complete derivations of the theoretical techniques now used in neurophysics. The lecturers have strived to present in their contributions coherent frameworks that include the important theorems and axioms upon which the methods are based and to delineate the limitations of the formalism. We hope this volume will provide the reader with the necessary tools and background to master theoretical neuroscience and to explore new tracks.

We address this volume to an audience of advanced graduate students and post-doctoral fellows in mathematics and physics, practicing researchers who are moving into theoretical neuroscience, and to neuroscientists who would like to gain deeper understanding of theory. We expect the book to be both a desktop reference as well as a textbook used in advanced courses and seminars on neurophysics.

We would like to thank the faculty for their invaluable contributions to the summer school and for the contents of this volume. We express our gratitude to all the experimentalists who invigorated the topical workshops through their inspired talks. We are indebted to Professor Philippe Ascher for his public lecture intended for a general audience. We would like to acknowledge CNRS, IBRO, NATO, and NSF for their generous support as well as the Les Houches Physics School, and its administrative team. Last but not least we would like to thank the attendees of the course who were instrumental in enlivening the school for five weeks.

Carson C. Chow
Boris Gutkin
David Hansel
Claude Meunier
Jean Dalibard

# CONTENTS

*Course 5.* *Some useful numerical techniques for simulating integrate-and-fire networks, by M. Shelley*    *179*

*Course 6.* *Propagation of pulses in cortical networks: the single-spike approximation, by D. Golomb*    *197*

## Course 7. *Activity-dependent transmission in neocortical synapses, by M. Tsodyks*      245

## Course 8. *Theory of large recurrent networks: from spikes to behavior, by H. Sompolinsky and O.L. White*      267

## *Course 9.    Irregular activity in large networks of neurons, by C. van Vreeswijk and H. Sompolinsky    341*

## *Course 12. Symmetry breaking and pattern selection in visual cortical development, by F. Wolf* 575

*Course 13. Of the evolution of the brain, by A. Treves and Y. Roudi*    *641*

Contents

Course 1

# EXPERIMENTING WITH THEORY

## Eve Marder

*Volen Center MS 013*
*Brandeis University*
*Waltham, MA 02454-9110*
*U.S.A.*

*C.C. Chow, B. Gutkin, D. Hansel, C. Meunier and J. Dalibard, eds.*
*Les Houches, Session LXXX, 2003*
*Methods and Models in Neurophysics*
*Méthodes et Modèles en Neurophysique*
© *2005 Elsevier B.V. All rights reserved*

1

# Contents

The summer course at Les Houches was designed to provide additional training in methods of computational neuroscience to physicists and mathematicians interested in working on problems fundamental to our understanding of the brain. Although a biologist by training and an experimentalist by practice, I have been privileged for almost 20 years to work and interact with a number of theorists. I was initially drawn to collaboration with theorists 20 years ago because it was already clear to me that understanding the dynamics of the rhythmic motor patterns produced by central pattern generating networks on the basis of the cellular mechanisms underlying the network behavior would require more than the reductionist methods we were then using.

Starting about 15 years ago the number of young people who were initially trained in physics, computer science, and mathematics who have been drawn to neuroscience has increased dramatically. The advent of gene microarray technology in the post-genomic era has sensitized many biologists to the need for new quantitative methods of all kinds as part of the analysis of complex biological systems in general, and neuroscience in specific. Thus, both the need for formal models to capture the dynamics of networks of molecules and neurons and the need for sophisticated statistical measures to extract meaning from multi-dimensional data sets have made experimental biologists more interested than ever before in the tools that can be provided by theory of all kinds. That said, there are significant cultural differences in the training of physicists and mathematicians and that of experimental biologists that may lead to different assumptions about how knowledge is created and transmitted that can be barriers to communication between theorists and experimentalists. In the first section of this essay I will make some general comments that may help those moving from physics and math into biology understand some of the culture of experimental biology that may at first mystify them. In the second section of this essay I will discuss some "traps" or "confounds" that are inherent as we try to build useful models of neuronal and network function as well as some modeling methods that I have been involved in developing. Finally, I will describe a specific case in which theory has been extremely instructive, and perhaps essential in illuminating a fundamental biological problem.

5

## 1. Overcoming communication barriers

Like many of my colleagues who did not benefit from advanced training in mathematics or physics, I was initially intimidated by many theory papers and talks. Nonetheless, over the years I have developed a strategy that helps me in approaching many theoretical studies. I ask the following questions: 1) What is the problem that the investigator is posing? Is it a problem that I as an experimental neuroscientist think important or interesting? 2) What are the assumptions and simplifications that are made in the model construction? Do any of these fly in the face of something that I consider a fundamental biological principle relevant to the problem at hand? 3) Does the final result make sense to me? If not, is it because the model was inappropriately chosen or is it because the model has revealed an essential inadequacy of my previous knowledge and set of assumptions? Obviously, these questions are not very different from those that I ask of experimental papers.

Good theory, like good experimental work, has a single definition: at the end of the work the investigator has created new knowledge. New knowledge can sometimes be created with a very simple calculation or a very simple piece of equipment. For me, some of the most beautiful or useful papers, even in today's world of high technology, come from experiments done with a single tension transducer (Morris and Hooper, 1997; Morris et al., 2000) or a trivial simulation (Kepler et al., 1990), if they result in new insights. Nonetheless one of the problems endemic to working at the juncture between two fields is that a given piece of work may provide new insight for one part of the community but fail to do so for another. For example, if you had asked the experimental neuroscience community in 1989 what would result if you depolarized an oscillatory neuron, almost everyone would have said the period would invariably decrease. I and many of my colleagues were very surprised that merely changing the relative conductance densities of the currents in the oscillator could alter this result. Indeed, depending on the specifics of the conductances found in a neuronal oscillator, it may either increase or decrease in period when depolarized (Kepler et al., 1990; Skinner et al., 1993). This result was hardly surprising to anyone with any mathematical sophistication or experience, required no special computational skills to demonstrate, but was illuminating to many experimentalists trying to understand what the consequences of altering a specific set of currents in a specific cell are likely to be.

The converse also occurs: there are times when elegant mathematics that was stimulated by a problem in neuroscience may fail to bring insight to the experimental community. Some work may be perceived by the experimental community as nothing more than the translation of knowledge from the language of biophysics to mathematics. For example, most experimental neuroscientists believe that they "understand" the squid axon action potential, and most would not find much additional new knowledge in the analysis of many simplified models of the Hodgkin-Huxley equations. That said, there are many mathematicians who will not feel they "understand" the squid axon action potential until it has been translated into a mathematical model that describes its threshold and firing properties in geometric or analytical terms. So are mathematical models that reduce the Hodgkin-Huxley equations to forms that are more suited to mathematics useful? The answer to this is certainly yes for the mathematicians who will not find understanding in their absence. Less so for many experimentalists who will find it difficult to extract any new understanding from these translations from the language of biology to the language of mathematics.

Thus there is a constant and unresolved tension that results from the fact that some of the most useful theory for the neurobiological community may not challenge the cleverness of the theorist, while large components of the neuroscience community may not be able to appreciate all of the theory inspired by neuroscience. Nevertheless, there continue to be numerous examples of theory papers that address important neuroscience problems and that speak to both the experimental and theoretical communities. Many of these studies are successful because they explore the computational consequences of biological phenomena that have been well-characterized but whose implications are incompletely understood (Abbott et al., 1997).

Not infrequently theorists are mystified that many of the experimental papers that they read do not contain the data that they need to adequately constrain their models. Theorists are often very frustrated by "incomplete" data sets in experimental papers, and must often say to themselves, "Why didn't they measure x while they were measuring y?" I am never surprised by these complaints, as I think it would be remarkable if the data needed to constrain a formal model would exist in the literature unless it was collected explicitly for that purpose. The key to understanding this is to appreciate that experimentalists will publish papers when they feel that they have a first pass "understanding" of a new finding. Such papers may often conclude with a "word model" or a "cartoon" that captures the experimentalist's intuition of what the findings may mean. Papers that lead to new insights are easy to publish and satisfying to read. Papers that go over old ground and merely provide more quantitative data or more rigorous analyses are difficult for an experimentalist to publish, as they lack "novelty" unless these more quantitatively collected and analyzed data are explicitly coupled to the testing of a formal model that requires rigorous data collection and analysis. Moreover, motivating an experimentalist to spend months or even years going over what may be perceived as old ground will only be possible when the experimentalist himself or herself is actively involved in the implicit or explicit development and testing of more rigorous and formal models. Theorists who ask experimentalists to collect data for them are unlikely to often meet with much success. Theorists who ask experimentalists which problems intrigue and puzzle them are likely to find collaborators willing to do difficult experiments to test models.

Many failures of communication between some theorists and experimentalists surround the issue of "details." Biologists are trained to look for and exploit the differences among cells, synapses, channels, animals, and behaviors as clues to understanding what factors may be important in understanding a physiological or behavioral process. Therefore biologists will naturally assume that the details of synaptic properties, neuronal structure, channel subtypes, neurotransmitter content, etc will eventually be critical to understanding how the networks in which these neurons and synapses function operate. In the past, some theorists have been known to be dismissive of the differences among cells and synapses and look for global simplifications, suggesting that the large numbers of neurons in the brain make the diversity among cells and synapses irrelevant for understanding the system properties by which the brain computes. Most biologists understand fully well the necessity for simplification in the search for understanding, and have a highly developed intuition for what details are likely to be important for which questions and which can be ignored in a first approximation explanation of a phenomenon. Successful collaboration among experimental biologists and theorists requires achieving consensus about how to make decisions about the importance of specific details in the explanation of system properties when constructing models.

## 2. Modeling with biological neurons-the dynamic clamp

All model neurons by definition are incomplete representations of the biological neurons whose properties they are intended to capture. Even when attempts are made to construct models that are highly realistic representations of the conductances and geometry of biological neurons, and even when attempts are made to incorporate some pieces of intracellular signaling pathways, models will at best be missing some important features that contribute to the dynamics of neurons, and at worst, will be in a parameter regime that drastically fails to capture something important or essential. The dynamic clamp is a method that allows simulations to be done with biological neurons (Prinz et al., 2004a), thus serving as an intermediary between theoretical studies and experimental analyses. In fact, it is likely that many experimentalists who use the dynamic clamp will become more comfortable with theory in general as they experience the kinds of intuitions about system dynamics that can be obtained when one can systematically alter properties of currents and synapses in biological neurons and see their impact on cell and network behavior in living tissue.

The dynamic clamp allows the investigator to ask a number of questions about the role of a current or a synapse, or a neuron in controlling the excitability of a single neuron or a network (Sharp et al., 1993a, b). In a dynamic clamp experiment intracellular electrodes are used to record membrane potential and inject current into one or several neurons (Sharp et al., 1993a, b). The injected current is calculated from a modeled conductance, which allows the investigator to set the maximal conductance and dynamic variables of the conductance. Thus the injected current varies as a function of the neuron's membrane potential, essentially creating an artificial conductance in parallel with the other conductances in the neuron's membrane. The dynamic clamp can be used in a number of different configurations: 1) The dynamic clamp allows the investigator to add or subtract conductances from a neuron, and to ask how the cell's excitability is altered (Ma and Koester, 1996; Turrigiano et al., 1996). Unlike the case in a conventional simulation, it is not necessary to specify the densities and properties of all of the other conductances in the neuron, because the neuron itself provides these to the simulation. 2) The dynamic clamp can be used to create artificial synaptic connections among neurons, as the membrane potential of one neuron can be used to control an artificial synaptic conductance of a second neuron (Sharp et al., 1996). 3) The dynamic clamp can be used to construct hybrid circuits, in which model neurons are coupled to biological neurons (Bal et al., 2000; Le Masson et al., 2002). 4) The dynamic clamp can be used to apply realistic patterns of synaptic inputs to neurons in slice preparations to evaluate their dynamics under "natural" synaptic drive conditions (Chance et al., 2002; Reyes, 2003). All of these configurations have become invaluable for understanding how varying the parameters of one or more membrane or synaptic conductances influences system properties (Prinz et al., 2004a).

Today the dynamic clamp is widely used in laboratories around the world (Prinz et al., 2004a). A number of different implementations have been developed, and these are continuing to evolve to take advantage of improved computer speed and board performance. Because the dynamic clamp allows simulations with biological neurons it provides a mechanism to easily test many predictions of both "word" and "formal" models without requiring the investigator to make too many assumptions or oversimplications. As such, it is likely to become more and more used as a modeling tool at the cellular level.

## 3. The traps inherent in building conductance-based models

For some, the holy grail of modeling neurons and ultimately networks, is to build a "realistic neuron" model that incorporates known measurements from biological neurons in a detailed compartmental model that captures the anatomical structure of the neuron. While this may appear to be more "realistic", there are some fundamental assumptions that go into the construction of such models that are often neglected, but that should be taken into account as one evaluates which questions are must suited for this kind of approach.

1) How much does the modeler have to "make-up"? The premise underlying building more and more detailed models is that as one incorporates more currents, signal transduction pathways and anatomical structure, that one is coming closer to the "real thing". However, the unfortunate reality is that because of the nature of the data that we obtain on biological neurons adding increased complexity to a model often means increasing the number of parameters that have not been measured, or cannot be measured properly. For example, there are often benefits to building multi-compartment models that segregate different conductances into regions of the neuron (Traub et al., 1991). However, adding compartments to a model requires that the modeler decide which conductances to put where in what amounts. This is particularly difficult because there are relatively few cases in which conductance densities have been measured in different regions of the neuron (Magee, 1999; Magee and Carruth, 1999; Magee and Cook, 2000; Frick et al., 2004). While it would be certainly interesting and important to model explicitly and correctly the intracellular Ca2+ concentrations and dynamics at all portions of the neuron, this is often unknown, and only rarely correlated with direct biophysical measurements (Frick et al., 2004). Likewise, the first models of intracellular signal transduction are being developed (Bhalla and Iyengar, 2001; Bhalla, 2002, 2003a, b; Yu et al., 2004) but again, it is rare that the data on signal transduction pathways are collected on the same neuron type that is modeled, although the hippocampal CA1 neuron is a fairly notable exception.

2) Lack of uniqueness in models: is this a problem or a solution? The construction of conductance-based models with many different currents immediately brings up the issue of uniqueness. To what extent is some combination of maximal conductances a unique solution to modeling a specific neuron's intrinsic electrical excitability? To what extent is it the job of the modeler to "tune" the model to a specific set of conductances that "adequately" captures the firing properties of the neuron? And what criteria does one use to decide how good a fit to expect? These are difficult questions, and obviously must be answered on a case by case basis. Nonetheless, several things are clear: a) In models with more than 5 or so different currents, it is almost certainly the case that there will be many different combinations of conductance densities that give quite similar behavior (Goldman et al., 2001; Prinz et al., 2003). b) In some models there will be minor changes in a single parameter that will produce qualitative changes in behavior (Guckenheimer et al., 1993; Guckenheimer et al., 1997), and most importantly, it is the correlated values of many conductances that determine the neuron's ac-

tivity, not the density of a single conductance (Goldman et al., 2001; Golowasch et al., 2002; Prinz et al., 2003). This raises an important experimental difficulty: it is often very difficult to measure more than one conductance properly in a single neuron, and if averaged values are not ideally suited to constrain a model (Golowasch et al., 2002), then it becomes extremely difficult to collect the data necessary to build a detailed, conductance-based model. But the most important difficulty is that the range of variation of the properties of the individual neurons of that class is not often measured or known. This, most models are tuned to reflect "mean" behavior, although it is not clear many neurons of a given type actually conform to the modeled mean.

As single neuron and network models become larger, the practical difficulties of hand-tuning models become more onerous. A new solution is now possible with the advent of cheap Beowulf clusters that allow the construction of data bases of model neurons (Prinz et al., 2003) in which multiple forms of models are simulated by "brute force" and then searched for individual models with certain desired properties. This approach avoids some of the potential pitfalls of hand-tuning (Prinz et al., 2003), as it is more likely to find all of the likely configurations of a given model by sampling over a large region of parameter space.

The crustacean stomatogastric nervous system is a central pattern generating circuit that produces several rhythmic motor patterns, including the fast pyloric rhythm and the slower gastric mill rhythm (Harris-Warrick et al., 1992). The stomatogastric ganglion (STG) has only 30 neurons, all of which are individually identifiable. This has facilitated the study of the intrinsic membrane properties of the neurons of the STG and the synaptic connections among them. Consequently, this preparation allows a whole series of investigations into the mechanisms by which its system properties emerge from the interactions among its component parts. Towards this end we have used the data base approach to ask how variations in synaptic strength and intrinsic properties together cooperate to produce the triphasic pyloric rhythm. We first constructed a data base of 1.7 million single neuron models (Prinz et al., 2003). We then selected from this data base candidate neuron models to represent three different cell types in the pyloric rhythm of the lobster stomatogastric ganglion. Using these, we constructed more than 20,000,000 versions of the pyloric network of the crustacean stomatogastric ganglion (Prinz et al., 2004b), thus allowing us to ask how the strengths of the synaptic connections and the properties of the individual neurons lead to the production of a well-constrained pyloric rhythm. We see that some synaptic strengths are inconsistent with the production of a pyloric rhythm, while other synapses can vary dramatically in strength, as long as these changes are accompanied by other changes in network parameters. These findings lead to the conclusion that there are multiple solutions consistent with the production of very similar network output patterns (Prinz et al., 2004a).

## 4. Theory can drive new experiments

One of the most important uses of theory is to allow the investigator to step back from a puzzling problem and imagine a possible solution to that problem, and make specific predictions that can stimulate new lines of investigation. It is extremely gratifying when the underlying premise of the model turns out to have experimental validity. An example of this is our work on the control of intrinsic neuronal excitability. In the early 1990's we were attempting to construct a realistic model of one of the neurons, the LP neuron, in the crab stomatogastric ganglion based on our voltage-clamp measurements of the currents in that neuron (Buchholtz et al., 1992; Golowasch et al., 1992; Golowasch and Marder, 1992). While we were able to hand-tune this model to approximate much of the behavior of the LP neuron, this procedure was inherently unsatisfying, because like others before us, we were unable to measure all of the currents in the LP neuron. More critically, in the process of tuning this model, it became clear that the model was very sensitive to some parameters, including some that we were not able to measure, and others that showed considerable variance in their measurements. This caused us to pose the question of how individual neurons regulated their conductance densities to maintain their intrinsic membrane properties over the life-time of the animal, while the individual membrane channels turn-over in the membrane in hours, days, or weeks (LeMasson et al., 1993; Marder et al., 1996; Liu et al., 1998; Marder and Prinz, 2002).

We have constructed a class of self-tuning models in which activity sensors are used as a feedback signal to slowly alter the density of the membrane channels so that constant neuronal and network activities are maintained (LeMasson et al., 1993; Siegel et al., 1994; Marder et al., 1996; Liu et al., 1998; Golowasch et al., 1999b; Marder and Prinz, 2002). In conventional conductance-based models the maximal conductance of each current is a fixed parameter and a neuron's activity is a consequence of the number and distribution of its ion channels. This assumption presumes that the number of each kind of membrane channel is independently controlled. Alternatively, in these self-tuning models we assume that early in development a neuron's target activity levels are specified, and then homeostatically regulated. The paradigm shift comes from the assumption that it is the neuron's output that is regulated rather than the number of ion channels in the membrane. In the first generation models of this sort the activity sensor was a measure of the bulk intracellular $Ca2+$ concentration (LeMasson et al., 1993). In response to elevated activity, the inward currents would be decreased and the outward currents increased to make the neuron less excitable. In later models, multiple sensors were used: a fast, slow and DC filter of the Ca2+ current (Liu et al., 1998). In this class of models, each membrane current was individually controlled to a varying degree by all three sensors. In all of these models the change in conductance density must occur slowly relative to the firing dynamics of the neuron. Specifically, the change in channel density must be orders of magnitude slower than the firing of the neuron or the activation time constants of its currents.

These models make several predictions: A) Individual neurons of the same class might have similar activity profiles but different current densities. B) The same neuron might have different current densities at different times in its life. C) Genetic knock-outs of some channels may be compensated for by alterations in the densities of other channels. D) Strong perturbations of activity might result in substantial alterations in channel densities. Experimental data obtained over the years, much of it motivated by these models, are consistent with all of these predictions (Turrigiano et al., 1994; Turrigiano et al., 1995; Thoby-Brisson and Simmers, 1998; Desai et al., 1999; Golowasch et al., 1999a; Thoby-Brisson and Simmers, 2000; Goldman et al., 2001; Thoby-Brisson and Simmers, 2002; Luther et al., 2003; MacLean et al., 2003). However, much more work will be needed to understand how neurons and networks wire up in development as a consequence of genetic programming and activity. Much more experimental work is needed to discover whether some conductances are obligatorily coregulated (MacLean et al., 2003). At the same time, a great deal of theoretical work remains before we understand how local and global tuning signals interact in the construction and maintenance of complex networks.

## 5. Conclusions

Neuroscience is today ideally poised to profit optimally from the influx of talented theorists. Today more than ever it is clear that theory is necessary to catalyze paradigm shifts in the way we pose problems about the nervous system. These paradigm shifts will occur when smart experimentalists and smart theorists find common language and common ground to reveal how the glorious richness and detailed idiosyncrasies of neurobiological systems contribute to their ability to be at the same time plastic and stable. After all, it is not yet obvious how networks can learn and develop without losing their ability to function as they are changed.

# References

[1] Abbott LF, Sen K, Varela J, Nelson SB (1997) Synaptic depression and cortical gain control. Science 275:220-224.

[2] Bal T, Debay D, Destexhe A (2000) Cortical feedback controls the frequency and synchrony of oscillations in the visual thalamus. J Neurosci 20:7478-7488.

[3] Bhalla US (2002) Biochemical signaling networks decode temporal patterns of synaptic input. J Comput Neurosci 13:49-62.

[4] Bhalla US (2003a) Temporal computation by synaptic signaling pathways. J Chem Neuroanat 26:81-86.

[5] Bhalla US (2003b) Understanding complex signaling networks through models and metaphors. Prog Biophys Mol Biol 81:45-65.

[6] Bhalla US, Iyengar R (2001) Robustness of the bistable behavior of a biological signaling feedback loop. Chaos 11:221-226.

[7] Buchholtz F, Golowasch J, Epstein IR, Marder E (1992) Mathematical model of an identified stomatogastric ganglion neuron. J Neurophysiol 67:332-340.

[8] Chance FS, Abbott LF, Reyes AD (2002) Gain modulation from background synaptic input. Neuron 35:773-782.

[9] Desai NS, Rutherford LC, Turrigiano GG (1999) Plasticity in the intrinsic excitability of cortical pyramidal neurons. Nature Neuroscience 2:515-520.

[10] Frick A, Magee J, Johnston D (2004) LTP is accompanied by an enhanced local excitability of pyramidal neuron dendrites. Nat Neurosci 7:126-135.

[11] Goldman MS, Golowasch J, Marder E, Abbott LF (2001) Global structure, robustness, and modulation of neuronal models. J Neurosci 21:5229-5238.

[12] Golowasch J, Marder E (1992) Ionic currents of the lateral pyloric neuron of the stomatogastric ganglion of the crab. J Neurophysiol 67:318-331.

[13] Golowasch J, Abbott LF, Marder E (1999a) Activity-dependent regulation of potassium currents in an identified neuron of the stomatogastric ganglion of the crab Cancer borealis. J Neurosci 19:RC33.

[14] Golowasch J, Buchholtz F, Epstein IR, Marder E (1992) Contribution of individual ionic currents to activity of a model stomatogastric ganglion neuron. J Neurophysiol 67:341-349.

[15] Golowasch J, Casey M, Abbott LF, Marder E (1999b) Network stability from activity-dependent regulation of neuronal conductances. Neural Comput 11:1079-1096.

[16] Golowasch J, Goldman MS, Abbott LF, Marder E (2002) Failure of averaging in the construction of a conductance-based neuron model. J Neurophysiol 87:1129-1131.

[17] Guckenheimer J, Gueron S, Harris-Warrick RM (1993) Mapping the dynamics of a bursting neuron. Philos Trans R Soc Lond B 341:345-359.

[18] Guckenheimer J, Harris-Warrick R, Peck J, Willms A (1997) Bifurcation, bursting, and spike frequency adaptation. J Computat Neurosci 4:257-277.

[19] Harris-Warrick RM, Marder E, Selverston AI, Moulins M (1992) Dynamic Biological Networks. The Stomatogastric Nervous System. Cambridge: MIT Press.

[20] Kepler TB, Marder E, Abbott LF (1990) The effect of electrical coupling on the frequency of model neuronal oscillators. Science 248:83-85.

[21] Le Masson G, Renaud-Le Masson S, Debay D, Bal T (2002) Feedback inhibition controls spike transfer in hybrid thalamic circuits. Nature 417:854-858.

[22] LeMasson G, Marder E, Abbott LF (1993) Activity-dependent regulation of conductances in model neurons. Science 259:1915-1917.

[23] Liu Z, Golowasch J, Marder E, Abbott LF (1998) A model neuron with activity-dependent conductances regulated by multiple calcium sensors. J Neurosci 18:2309-2320.

[24] Luther JA, Robie AA, Yarotsky J, Reina C, Marder E, Golowasch J (2003) Episodic bouts of activity accompany recovery of rhythmic output by a neuromodulator- and activity-deprived adult neural network. J Neurophysiol 90:2720-2730.

[25] Ma M, Koester J (1996) The role of potassium currents in frequency-dependent spike broadening in Aplysia R20 neurons: a dynamic clamp analysis. J Neurosci 16:4089-4101.

[26] MacLean JN, Zhang Y, Johnson BR, Harris-Warrick RM (2003) Activity-independent homeostasis in rhythmically active neurons. Neuron 37:109-120.

[27] Magee JC (1999) Dendritic Ih normalizes temporal summation in hippocampal CA1 neurons. Nat Neurosci 2:848.

[28] Magee JC, Carruth M (1999) Dendritic voltage-gated ion channels regulate the action potential firing mode of hippocampal CA1 pyramidal neurons. J Neurophysiol 82:1895-1901.

[29] Magee JC, Cook EP (2000) Somatic EPSP amplitude is independent of synapse location in hippocampal pyramidal neurons. Nat Neurosci 3:895-903.

[30] Marder E, Prinz AA (2002) Modeling stability in neuron and network function: the role of activity in homeostasis. Bioessays 24:1145-1154.

[31] Marder E, Abbott LF, Turrigiano GG, Liu Z, Golowasch J (1996) Memory from the dynamics of intrinsic membrane currents. Proc Natl Acad Sci (USA) 93:13481-13486.

[32] Morris LG, Hooper SL (1997) Muscle response to changing neuronal input in the lobster (Panulirus interruptus) stomatogastric system: spike number-versus spike frequency-dependent domains. J Neurosci 17:5956-5971.

[33] Morris LG, Thuma JB, Hooper SL (2000) Muscles express motor patterns of non-innervating neural networks by filtering broad-band input. Nat Neurosci 3:245-250.

[34] Prinz AA, Billimoria CP, Marder E (2003) Alternative to hand-tuning conductance-based models: construction and analysis of databases of model neurons. J Neurophysiol 90:3998-4015.

[35] Prinz AA, Abbott LF, Marder E (2004a) The dynamic clamp comes of age. Trends Neurosci, 27:218-224.

[36] Prinz AA, Bucher D, Marder E (2004b) Multiple combinations of intrinsic properties and synaptic strengths produce similar network activity. submitted.

[37] Reyes AD (2003) Synchrony-dependent propagation of firing rate in iteratively constructed networks in vitro. Nat Neurosci 6:593-599.

[38] Sharp AA, Skinner FK, Marder E (1996) Mechanisms of oscillation in dynamic clamp constructed two-cell half-center circuits. J Neurophysiol 76:867-883.

[39] Sharp AA, O'Neil MB, Abbott LF, Marder E (1993a) The dynamic clamp: artificial conductances in biological neurons. Trends Neurosci 16:389-394.

[40] Sharp AA, O'Neil MB, Abbott LF, Marder E (1993b) Dynamic clamp: computer-generated conductances in real neurons. J Neurophysiol 69:992-995.

[41] Siegel M, Marder E, Abbott LF (1994) Activity-dependent current distributions in model neurons. Proc Natl Acad Sci USA 91:11308-11312.

[42] Skinner FK, Turrigiano GG, Marder E (1993) Frequency and burst duration in oscillating neurons and two-cell networks. Biol Cybern 69:375-383.

[43] Thoby-Brisson M, Simmers J (1998) Neuromodulatory inputs maintain expression of a lobster motor pattern-generating network in a modulation-dependent state: evidence from long-term decentralization In Vitro. J Neurosci 18:212-2225.

[44] Thoby-Brisson M, Simmers J (2000) Transition to endogenous bursting after long-term decentralization requires de novo transcription in a critical time window. J Neurophysiol 84:596-599.

[45] Thoby-Brisson M, Simmers J (2002) Long-term neuromodulatory regulation of a motor pattern-generating network: maintenance of synaptic efficacy and oscillatory properties. J Neurophysiol 88:2942-2953.

[46] Traub RD, Wong RK, Miles R, Michelson H (1991) A model of a CA3 hippocampal pyramidal neuron incorporating voltage-clamp data on intrinsic conductances. J Neurophysiol 66:635-650.

[47] Turrigiano G, Abbott LF, Marder E (1994) Activity-dependent changes in the intrinsic properties of cultured neurons. Science 264:974-977.

[48] Turrigiano GG, LeMasson G, Marder E (1995) Selective regulation of current densities underlies spontaneous changes in the activity of cultured neurons. J Neurosci 15:3640-3652.

[49] Turrigiano GG, Marder E, Abbott LF (1996) Cellular short-term memory from a slow potassium conductance. J Neurophysiol 75:963-966.

[50] Yu X, Byrne JH, Baxter DA (2004) Modeling Interactions between Electrical Activity and Second-Messenger Cascades in Aplysia Neuron R15. J Neurophysiol 91:2297-2311.

Course 2

# UNDERSTANDING NEURONAL DYNAMICS BY GEOMETRICAL DISSECTION OF MINIMAL MODELS

## A. Borisyuk[1], J. Rinzel[2]

[1]*Mathematical Biosciences Institute,*
*Ohio State University,*
*Columbus, OH, USA*
[2]*Center for Neural Science,*
*New York University,*
*New York, NY, USA*

*C.C. Chow, B. Gutkin, D. Hansel, C. Meunier and J. Dalibard, eds.*
*Les Houches, Session LXXX, 2003*
*Methods and Models in Neurophysics*
*Méthodes et Modèles en Neurophysique*

# Contents

# 1. Introduction

## 1.1. Nonlinear behaviors, time scales, our approach

It has been said that the currency of the nervous system is spikes. Indeed, at some level it is important to understand how neurons generate spikes and patterns of spikes. What is their language and how do they convert stimuli into spike patterns? Actually, these are two different questions. The first is about processing and storing information, and a neuron's role in a neural computation. The second is more mechanistic, about the "how" of converting inputs into spike output. With regard to the first, it is rare that we know what neural computation(s) a given neuron carries out, especially since computations more typically involve the collective interaction of many cells. However, we can, as do many cellular neurophysiologists, approach the second question, asking from a more reductionist viewpoint what are the biophysical mechanisms that underlie spike generation and transmission. How do the properties of different ionic channels and their distributions over the cell's dendritic, somatic, axonal membrane determine the neuron's firing modes? How might the various mechanisms be modulated or recruited if there are changes in the cell or circuitry in which it is embedded or in the brain state or in the read-out targets? We usually imagine that the typical time scales for action potential generation are milliseconds (msecs), but there are examples of where even a brief (msecs) stimulus can evoke a long duration transient spike pattern or where pre-conditioning can delay a spike's onset by 100s of msecs. Some neurons fire repetitively (tonically) for steady or slowly changing stimuli, some fire with complex temporal patterns (e.g., bursts of spikes), but some only respond (phasically) to the rapidly changing features of a stimulus. These behaviors reflect a neuron's biophysical makeup.

In these lectures we attempt to describe how different response properties and firing patterns arise. We seek especially to provide insight into the underlying mathematical structure that might be common to classes of firing behaviors. Indeed, the mathematical structure is more general and the physiological implementation could involve different biophysical components. Our approach will be to use concepts from nonlinear dynamics, especially geometrical methods like phase planes or phase space projections from higher dimensional systems. A key feature of our viewpoint is to exploit time scale differences to reduce dimen-

21

sionality by dissecting the dynamics using fast-slow analysis, i.e., to separately understand the behaviors on the different time scales and then patch the behaviors together. We will begin by dissecting the classical Hodgkin-Huxley model in this way to distinguish the rapid upstroke and downstroke of the spike from the slower behavior during the action potential's depolarized plateau phase and hyperpolarized recovery phase. Analogously we will segregate a burst pattern's active and silent phases from the transitions between these phases. Geometrically, the trajectories during the slow phases are restricted to lower dimensional manifolds and the transitions correspond to reaching folds or bifurcations and jumping to a different manifold where slow flow resumes. Our phase plane treatments will be highlighted in the sections that describe the rich dynamic repertoire of the two-variable Morris-Lecar model, as its biophysical parameters are varied and as we allow them to become slow variables, say, for the generation of bursting behaviors.

For the most part here we will exploit the idealization of a point (i.e., electrically compact) neuron, focusing on the nonlinearities of spiking dynamics, and using biophysically minimal but plausible models. While most of our examples are for single-cell dynamics, the qualitative mathematical structures are also applicable to network dynamics, especially in the mean-field approximations. We work through one such example for network-generated rhythms, as seen in developing neural systems.

A take-home message that will be repeated several times is that the essentials of neural excitability and oscillations are relatively rapid autocatalysis (a regenerative process) and slow negative feedback. At the level of spike generation, say in the Hodgkin-Huxley model: autocatalysis is due to the sodium current's rapid voltage-gated activation while negative feedback comes from sodium inactivation and potassium current activation, both relatively slower. In a network setting autocatalysis could be fast recurrent excitation and negative feedback might be due to intrinsic cellular adaptation or slower synaptic inhibition or depression of excitatory synapses.

## 1.2. Electrical activity of cells

Electrical activity of a cell is commonly described by the cell's membrane potential (voltage) which can vary between different parts of the cell and also with time. The voltage $V = V(x, t)$ satisfies the current-balance equation:

$$C_m \frac{\partial V}{\partial t} + I_{ion}(V) + I_{coupling} = \frac{d}{4R_i} \frac{\partial^2 V}{\partial x^2} + I_{app}.$$

Here $C_m \frac{\partial V}{\partial t}$ is current due to the membrane's capacitive property, $I_{ion}(V)$ represents the cell's intrinsic ionic currents, $I_{coupling}$ represents the inputs and interac-

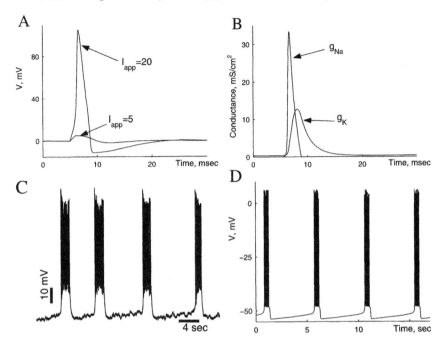

Fig. 1. Modes of neuronal activity. *A-B:* Excitability in response to a brief current pulse (Hodgkin–Huxley model, same as in section 2, pulse duration 1 msec). *A:* Pulse of amplitude $I_{app} = 5\ \mu A/cm^2$ fails to induce a spike, voltage returns to rest. Pulse of amplitude $I_{app} = 20\ \mu A/cm^2$ elicits a single spike. *B:* time courses of $Na^+$ and $K^+$ conductances during the spike from A. *C-D:* Single neuron bursting in brain stem circuit involved with respiration. *A:* Voltage recording in rat. Data courtesy of Christopher A. Del Negro and Jack L. Feldman. See also, Figure 1 of [5]. *B:* Voltage trace in a model. Equations used are the same as in [5], with $I_{app} = 0\ \mu A/cm^2$, $g_{syn} = 0\ mS/cm^2$, $g_{ton} = 0.3$ $mS/cm^2$.

tion currents from coupling with other cells (with their voltages, $V_j$), while $I_{app}$ is the current supplied by the experimentalist's electrode. The term $\frac{\partial^2 V}{\partial x^2}$ represents current spread along dendritic or axonal segments due to spatial gradients in voltage ($d$ is diameter, $R_i$ is cytoplasmic resistivity). We will neglect this term by considering the case of a "point" neuron, i.e. all the membrane currents and inputs are lumped into a single "compartment" with $V$ independent of $x$. This can be a good approximation when the cell is electrically compact.

Coupling to other cells can be via chemical synapses. Neurotransmitter released from the presynaptic cell "$j$" activates receptors on the postsynaptic membrane that in turn open channels, allowing for the flow of some types of ions:

$$\sum_j g_{syn,j}(V_j(t))(V - \bar{V}_{syn}).$$

In fact, $g_{syn,j}$ is not really an instantaneous function of $V_j$ and its dynamics can be quite important, although we will not be addressing these issues here. Coupling may also be "electrical" mediated by gap junctions (formed by local clusters of ionic channels that span the abutting membranes of both cells), that act effectively as resistors:

$$\sum_j g_{elec,j}(V - V_j).$$

The term $I_{ion}$ includes all of the intrinsic ionic currents present in the cell,

$$I_{ion} = \sum_k g_k(V - \bar{V}_k).$$

Although approximately Ohmic instantaneously, these currents provide significant nonlinearities. Their conductances $g_k$ are voltage-dependent and dynamic, expressed in terms of gating variables with a variety of time scales from msecs to 10s or 100s of msecs. Sometimes conductances are also affected by the presence of various substances, for example, by concentrations of other ions, second messengers, etc. The reversal potential $\bar{V}_k$ for current flow depends on the ionic channels' (of "$k$" type) selectivity for ions. The most common ionic species contributing to the electrical activity are $Ca^{2+}$, $Na^+$, $K^+$, and $Cl^-$. There are several different types of channels associated with each of these various ions, and some channels pass more than one type of ion.

The available constellation of different channel types leads to a large variety of nonlinear properties and electrical activity patterns amongst cells, even in the absence of coupling. The simplest example of a cell's nonlinear responsiveness is the generation of a spike or action potential. After stimulation by a small brief pulse of $I_{app}$, $V$ returns back to rest, quite directly (Fig. 1A). However, if the stimulus exceeds a threshold value, $V$ executes a large and characteristic excursion (action potential) and then eventually returns to rest (Fig. 1A). Briefly, the events are as follows (Fig. 1B). Note first that according to the current balance equation, $V$ will tend toward the $\bar{V}_k$ associated with the momentarily dominant $g_k$. Thus, the spike's regenerative upstroke (for example, in the Hodgkin–Huxley model, see section 2) is due to the rapid V-dependent increase in $g_{Na}$ pushing $V$ toward $\bar{V}_{Na}$. Then also driven by $V$ but on a slower time scale, $g_{Na}$ turns off while $g_K$ activates, pushing $V$ toward $\bar{V}_K$. The fall in $V$ causes $g_K$ to eventually die down and the system returns to rest. Excitability involves fast autocatalysis ($V$ rises opening $Na^+$ channels causing $V$ to increase further, etc) and slower negative feedback ($g_{Na}$ shuts down and $g_K$ turns on). Non-linearity is manifested also in the fact that the spike's amplitude is approximately independent of stimulus strength, provided it is superthreshold. Many models and cells respond to a step of $I_{app}$ by firing repetitively. The firing frequency typically exhibits a transient phase and may then adapt to a steady level. The adapted firing frequency ($f$) versus $I_{app}$ is a typical characterization of the cell or model's input-output relation ($f - I$ curve) (e.g., Fig. 5B). How the $f - I$ curve's shape, position, and frequency range depend on state parameters or background activity are of interest. In cells that show bursting behavior (Fig. 1C,D) a much slower negative feedback (for example, a slowly activating $K^+$ current) can bring the cell out of firing mode. Then, during the long quiescent phase, while $V$ is low, the negative feedback process recovers and re-entry into the firing mode occurs eventually. Other instances of nonlinear behavior involve various types of bistability, exhibited by some neurons. A cell might be either quiescent or firing repetitively for a steady stimulus or maybe capable of firing at two different frequencies (a multi-valued $f - I$ curve). Without intervention each state might persist for 100s of msecs. The cell can be switched from one state to the other by brief stimuli.

In some experimental situations $V$ can be measured directly with an electrode, by penetrating or attaching it to the cell (this is much easier to do *in vitro* than *in vivo*). This yields $V(t)$ at one site (typically, the soma) that may or may not reflect what is happening in other parts of the cell, in particular along the axon, which carries the output signal to other areas. While much theoretical research has been done on spatial characteristics, such as action potential propagation (see, e.g. [54, 55]) and the role of dendrites (for reviews see, e.g., [47, 48, 60, 61]), in this chapter we will focus on the point neuron.

## 2. Revisiting the Hodgkin–Huxley equations

### 2.1. Background and formulation

Before we analyze mathematically action potential generation, let's review the experimental basis and determination of the equations. The recipe for experimentally describing the currents that dictate neuronal electric properties comes from the work of Hodgkin and Huxley ( [35] and, for reviews, [33, 38, 52] ). Two major hurdles were overcome. First, in order to isolate $I_{ion}$, the confounding and unknown contribution from the spatial spread of current had to be eliminated. Hodgkin and Huxley chose to use the squid's giant axon [35], extracted and isolated in a dish. It is so big that one can insert a silver wire along its length. Because silver is a good conductor, it equalizes voltage values along the observable segment, constituting the so-called "space-clamp". Second, to dissect the contributions of individual ionic currents one can eliminate some of the ionic species from the bathing solution, thereby revealing the membrane current contributed by other ions. Finally, a tour de force: the voltage-clamp technique involves a feedback circuit to deliver the appropriate current to the axon so that $V$ is held fixed to a commanded level. By systematically using different command $V$'s the dynamics and $V$-dependence of the isolated current can be found. Another use of the voltage-clamp technique is to zero out the contribution of the $k$ type current by clamping $V$ to $\bar{V}_k$, recalling that $\bar{V}_k$ (the Nernst potential) can be altered by changing ion concentrations in bath or axon. After the ionic current time courses are measured they can be empirically fitted with solutions of differential equations.

With the $V$-dependent kinetics of different contributing ionic currents in hand, the test phase involves combining them along with the capacitive membrane current to thus synthesize the current-balance equation. Then, by numerical integration, confirm that the constituted equations describe the evolution of $V$ as a function of $I_{app}$ (i.e., under current-clamp).

Hodgkin and Huxley shared a Nobel prize for their description of $I_{ion}$, accounting for the action potential in squid giant axon and for providing the framework for other excitable membrane systems. Fortunately, there were only two voltage-gated currents, for $Na^+$ and for $K^+$, the delayed-rectifier $K^+$ current, and a constant-conductance leak current. The equations (space-clamped configuration) are:

$$
\begin{aligned}
C_m \dot{V} &= -I_{ion}(V, m, h, n) + I_{app} \\
&= -\bar{g}_{Na} m^3 h (V - \bar{V}_{Na}) - \bar{g}_K n^4 (V - \bar{V}_K) - g_L (V - \bar{V}_L) + I_{app}, \\
\dot{m} &= \phi \left[ m_\infty(V) - m \right] / \tau_m(V), \\
\dot{h} &= \phi \left[ h_\infty(V) - h \right] / \tau_h(V), \\
\dot{n} &= \phi \left[ n_\infty(V) - n \right] / \tau_n(V),
\end{aligned}
\tag{2.1}
$$

where membrane potential $V$ is in mV, and expressed relative to rest; $t$ is in msec; $m$, $h$ and $n$ are the dimensionless phenomenological gating variables (with values between 0 and 1): sodium activation, sodium inactivation, and potassium activation. The applied current $I_{app}$ ($\mu$A/cm$^2$) will be taken as time-independent in this section. The functions $\tau_x(V)$ and $x_\infty(V)$ can be interpreted as, respectively, the "time constant" and the "steady-state" functions for $m, h, n$. Their graphs are shown in Figure 2. The activation variables have steady-state functions that increase with $V$ and asymptote to 1, while for the inactivation variable $h_\infty(V)$ decreases with $V$ and asymptotes to 0. Note also in Fig. 2 that the time

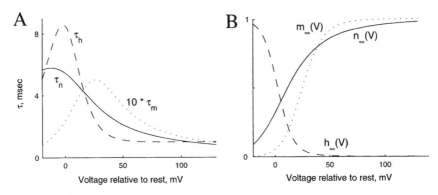

Fig. 2. Functions $\tau_x(V)$ (in A) and $x_\infty(V)$ (in B) in Hodgkin–Huxley equations 2.1 for $m$ (dotted), $n$(solid) and $h$ (dashed). Notice that $\tau_m(V)$ was multiplied by 10 for plotting on the same scale.

constant scale for $m$ is about $1/10$ that for $h$ and $n$, so that $m$ is relatively fast. The temperature factor $\phi$ speeds up the rates for $m, h, n$ for increasing temperature with $Q_{10}$ of 3:

$$\phi = 3^{(\text{Temp}-6.3)/10}.$$

Here, we fix the temperature at $18.5°$ C, unless noted otherwise. Values for the other parameters are as in the original model (see, e.g. [38]).

After these many years, the conceptual approach of Hodgkin and Huxley (voltage-clamp, segregate currents, synthesize and confirm) and general form of the mathematical expressions for conductances (products of gating variables) are still being widely applied. Many cell types, with different types of currents, have been successfully studied experimentally, and described in models using the same framework. Some examples include [33, 38, 39] $Ca^{2+}$ currents ($I_{Ca}$; T,L,N types), A-current ($I_A$; $K^+$ current with inactivation), $Ca^{2+}$-dependent $K^+$ current ($I_{K-Ca}$; $K^+$ current activated by voltage and intracellular $Ca^{2+}$ concentration). These advances are aided by development of new experimental techniques [33]. For example, to reduce the complexity of the system the ionic channels can be selectively blocked by pharmacological agents or disabled by genetic modification; the patch-clamp recording technique gives accurate access to currents and channels in electrically compact neurons or even a small patch of membrane, sometimes containing only a single channel whose opening/closing statistics are then obtained.

From the modeling perspective, it is important to keep in mind that even though many of the ionic currents can be described in the Hodgkin-Huxley formalism, and written in the form of equations 2.1, the parameters can be difficult to measure experimentally. In fact, it is rare that the voltage-clamp data are available for all channel types that are present in a given neuronal system. For example, the time constants of activation and inactivation of currents are often hard to measure. Therefore, the parameters in the model have to be estimated from empirical observations and indirect measurements. Also, in developing a model, the parameters of a particular channel are sometimes borrowed from other experimental systems where the channel has been studied in more detail. This latter technique has to be used with caution — there are many different types of channels for each ionic species, and their dynamics often differs significantly, so one model cannot always be substituted to describe other types of channels for the same ion.

## 2.2. Hodgkin–Huxley gating equations as idealized kinetic models

In this section we outline an interpretation of the Hodgkin-Huxley gating variables in terms of idealized kinetic schemes, but these schemes are not meant as realistic representations of a channel's molecular dynamics. In particular, the "gating subunits" in the Hodgkin–Huxley model (see below) are not the molecular subunits of the channel conformation. For an introduction to the molecular biology of ionic channels see, e.g., [33]. The representation here treats activation and inactivation as independent, and each gating subunit obeys a one-step kinetic scheme.

Each of the quantities $m$, $h$, $n$ can be interpreted as a probability for a specific gating subunit to be "available". In this interpretation a $Na^+$ channel is said to consist of 3 "$m$-subunits" and 1 "$h$-subunit", and a $K^+$ channel has 4 "$n$-subunits". If all gating subunits must be available to open a channel, and the subunits are independent, then the probability of the channel to be open is the product of probabilities of the subunits to be available. Further, the total conductance of the channels of a given type is proportional to the fraction of channels open, which, in turn, is proportional to the probability of the channel being open if the number of channels is large. Hence, say, for the $K^+$ current the instantaneous conductance is $n^4$, and the coefficient of proportionality, $\bar{g}_K$ (say, mS/cm$^2$) is the maximal conductance if all channels are open, i.e. $\bar{g}_K$ equals the local channel density times the conductance of a single open channel.

For the dynamics of a gating subunit of type $x$ suppose the subunit has only 2 states: "available" and "unavailable", and the rate of going from unavailable to available is $\alpha_x(V)$ and from available to unavailable is $\beta_x(V)$. Notice that the rates are voltage-dependent. Then the probability (or fraction) of available subunits, $x$, satisfies

$$\dot{x} = \alpha_x(V)(1 - x) - \beta_x(V)x.$$

Dividing both sides of this equation by $\alpha_x(V) + \beta_x(V)$, and introducing notations

$\tau_x(V) = 1/(\alpha_x(V) + \beta_x(V))$ and $x_\infty(V) = \alpha_x(V)/(\alpha_x(V) + \beta_x(V))$, we get

$$\tau_x(V)\dot{x} = x_\infty(V) - x,$$

which has the same form as in Hodgkin-Huxley equations 2.1.

## 2.3. Dissection of the action potential

To understand the dynamics of action potential generation (figure 3A) we will use the fact that there are two different time scales in the system; $\tau_m$ is much smaller than $\tau_h$ and $\tau_n$ (see Figure 2A). Using the methods of fast-slow analysis we dissect the system into fast (for $V$, $m$) and slow (for $n$, $h$) subsystems. In fact, $m$ is so fast that we will treat it as instantaneous for now, setting $m = m_\infty(V)$, and thus the fast subsystem is one-dimensional. Here, we describe the process in mixed mathematical and biophysical terms; a more formal description of the framework is in Appendix A.

### 2.3.1. Current-voltage relations

Motivated by the biophysicist's interest in the membrane's current-voltage relations we will examine these on two different time scales. First, we consider the steady current that is needed to maintain a constant voltage (as in voltage-clamp). This steady state current, $I_{ss}(V)$, equals $I_{ion}(V, m, h, n)$ with all gating variables set to their steady state values:

$$I_{ss}(V) = I_{ion}(V, m_\infty(V), h_\infty(V), n_\infty(V))$$
$$= \bar{g}_{Na} m_\infty^3(V) h_\infty(V)(V - \bar{V}_{Na}) + \bar{g}_K n_\infty^4(V)(V - \bar{V}_K) + \bar{g}_L(V - \bar{V}_L).$$

Figure 3B shows the steady-state current for each type of ion and the total $I_{ss}$ for equations 2.1. Notice that for the Hodgkin–Huxley equations this current is monotonically increasing. It is dominated by the steadily-activated $K^+$-current. $I_{Na}$ plays little role in $I_{ss}(V)$ because of inactivation; $m_\infty(V)$ and $h_\infty(V)$ "overlap" only slightly.

Second, on a very short time scale the instantaneous current, $I_{inst}(V)$, describes the $V$-dependence of $I_{ion}(V, m, h, n)$ with $m = m_\infty(V)$ and with $n$ and $h$ frozen, since they are so slow (on this time scale):

$$I_{inst}(V; n_0, h_0) = I(V, m_\infty(V), h_0, n_0)$$
$$= \bar{g}_{Na} m_\infty^3(V) h_0(V - \bar{V}_{Na}) + \bar{g}_K n_0^4(V - \bar{V}_K) + \bar{g}_L(V - \bar{V}_L).$$

When $n$ and $h$ are fixed at their rest values $I_{inst}(V)$ approximates the net ionic current that will be produced if the membrane potential is quickly perturbed from rest to the value $V$. Figure 3C shows this current, as a function of $V$, for equations 2.1 with $n$ and $h$ fixed at rest values.

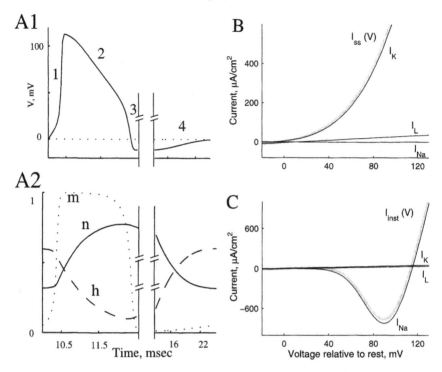

Fig. 3. Illustration of the Hodgkin–Huxley equations. *A:* Time courses of voltage (*A1*) and gating variables (*A2*) during an action potential. Notice that time scale is compressed at the second part of the graph to include the recovery phase. Numbers in the upper panel denote different phases of the action potential (see text). Stimulus is a step of current delivered at $t = 10$ msec of duration 1 msec and with amplitude $I_{app} = 20\mu A/cm^2$. *C:* Steady state currents for each ionic type (black) and total $I_{ss}(V)$ (grey). *D:* Instantaneous currents each ionic type (black) and total $I_{inst}(V)$ (grey).

### 2.3.2. Qualitative view of fast-slow dissection

Here, using the current-voltage relations defined in the previous section, we can describe a Hodgkin-Huxley action potential (Fig. 3A). Except for a brief initial transient pulse we will assume that $I_{app} = 0$ in this section.

We idealize the action potential as consisting of 4 phases (see Figs. 3A and 4): (1) upstroke, (2) plateau, (3) downstroke, and (4) recovery. The upstroke and downstroke happen on the fast time scale. Therefore we will use the approximation that the slow variables $h$ and $n$ are constant during these phases, so that $V$ satisfies

$$C_m \frac{dV}{dt} = -I_{inst}(V; n_0, h_0).$$   (2.2)

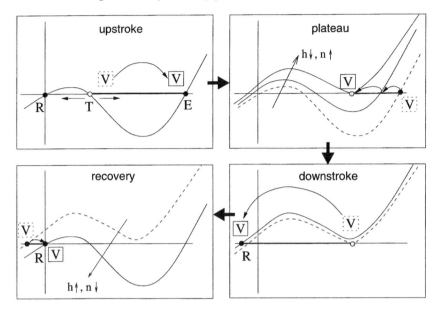

Fig. 4. Phases of Hodgkin–Huxley action potential shown in the $V - I_{inst}$ plane (see text). In each panel the dashed curve is the position of $I_{inst}(V)$ at the beginning of the phase, and $V$ in dashed square marks the initial value of $V$. Solid curve and label are the curve and $V$-value at the end of the phase. Arrows above the x-axis show the motion of $V$, and arrows below axis in the first panel show the direction of flow for $V$. R, T, E denote rest, threshold and excited steady states, respectively.

The plateau and recovery, on the other hand, happen on the slow time scale. During these two phases the dynamics is determined by $h$ and $n$, whereas $V$ and $m$ are "slaved", or "equilibrated": their values follow the dynamics of $h, n$ in such a way that the right-hand sides of their equations remain zero:

$$0 = -I(V, m_\infty(V), h, n) = -I_{inst}(V, h_0, n_0).$$

Note, this does not mean that $dV/dt = 0$ but rather that $V$ is so fast that it can be treated as tracking instantaneously a zero of $I_{inst}(V, h, n)$.

*Upstroke* (phase 1, characterized by the very rapid activation of $Na^+$ channels, Figure 4 *top left*): assume $h, n$ are slow (fixed at their rest values); $m$ instantaneous ($m = m_\infty(V)$), $V$ satisfies 2.2. This a one-dimensional dynamical system and its dynamics is determined by the zeros and signs of the right hand side of the equation 2.2. The graph of $I_{inst}(V)$ is N-shaped, therefore there are 3 steady states (R,T,E) (see Fig. 4), corresponding to *Rest, Threshold* and *Excited* states. We show in the section 2.3.3 below that their stability is determined by the sign of the derivative $\frac{dI_{inst}}{dV}$, and that R and E are stable, and T is unstable . If $V$ is depolarized above T, it increases on the fast time scale toward E (Figure 4 *top left*).

*Plateau* (phase 2, Figure 4 *top right*): as $V$ reaches E (depolarizes), the slow dynamics for $h$ and $n$ take place. The solution remains at $I(V, m_\infty(V), h, n) = 0$, but this curve as a function of $V$ is parameterized by $h$ and $n$ and these parameters are changing according to their dynamics 2.1. For depolarized $V$: $h$ decreases, $n$ increases, i.e. the (positive) contribution of K$^+$ current increases and the negative feedback contribution of Na$^+$ inactivation decreases $I_{Na}$. Therefore, the N-shaped curve of $I(V; h, n)$ drifts upward. As a result the steady states T and E move closer together until they coalesce and disappear, ending phase 2 (Figure 4 *top right*).

*Downstroke* (phase 3, Figure 4 *bottom right*): now there is only one steady state (hyperpolarized R) and $V$ decreases toward it (on the fast time scale).

*Recovery* (phase 4, Figure 4 *bottom left*): In this phase $V$ is hyperpolarized. Therefore, $h$ increases and $n$ decreases. As a result $I_{inst}$ slowly moves downward; R returns to the original value and $V$ drifts with it.

### 2.3.3. Stability of the fast subsystem's steady states

Now, for the analysis of the action potential it remains to show that the stability of any steady state $V_{ss}$ of the equation

$$C_m \frac{dV}{dt} = -I_{inst}(V, m_\infty(V), h_0, n_0) + I_{app} = -I_{inst}(V; n_0, h_0) + I_{app}$$

is determined by the derivative of the right-hand side.

Consider the effect of a small perturbation $v$ from the steady state $V(t) = V_{ss} + v(t)$. Then

$$C_m \frac{dV}{dt} = C_m \frac{dv}{dt} = -I_{inst}(V_{ss} + v; n_0, h_0) + I_{app}$$

$$= -I_{inst}(V_{ss}; n_0, h_0) - \frac{dI_{inst}}{dV}\bigg|_{V_{ss}} v + O(v^2) + I_{app}.$$

The first and last terms sum to 0, because $V_{ss}$ is a steady state. As a linear approximation, we have:

$$C_m \frac{dv}{dt} = -\frac{dI_{inst}}{dV}\bigg|_{V_{ss}} v.$$

The steady state is

$$\text{stable if } \frac{dI_{inst}}{dV}\bigg|_{V_{ss}} > 0,$$

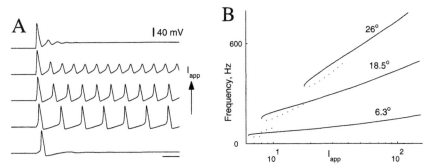

Fig. 5. Repetitive firing through Hopf bifurcation in Hodgkin–Huxley model. *A:* Voltage time courses in response to a step of constant depolarizing current (several levels of current: from bottom to top: $I_{app}$= 5, 15, 50, 100, 200 in $\mu A/cm^2$). Scale bar is 10 msec. *B:* f-I curves for temperatures of 6.3 18.5, 26°C, as marked. Dotted curves show frequency of the unstable periodic orbits.

and

$$\text{unstable if } \left. \frac{dI_{inst}}{dV} \right|_{V_{ss}} < 0.$$

A biophysical interpretation for this instability condition is that negative resistance, due to the rapidly activating $I_{Na}$, is destabilizing.

## 2.4. *Repetitive firing*

Numerical simulations show that the Hodgkin–Huxley model exhibits repetitive firing in response to steady $I_{app}$ within a certain range of values, $I_{v} < I_{app} < I_2$ (Fig. 5 A) [15, 57]. Generally, the firing rate increases and the amplitude of spikes decreases with increasing current. If $I_{app}$ is too large $V$ settles to a stable depolarized level. This is called depolarization block. As the temperature is increased the frequency range moves upward (Fig. 5 B). This is because the recovery processes $h$ and $n$ become faster and the membrane's refractory period decreases. However, if the temperature is increased too much then excitability and repetitive firing is lost, i.e. the negative feedback is too fast. In order to study the emergence and properties of rhythmic behavior we use stability and bifurcation theory.

### 2.4.1. *Stability of the four-variable model's steady state*
The Hodgkin–Huxley model has a unique steady state voltage $V_{ss}$ for each value of $I_{app}$, because $I_{ss}$ is monotonic. Yet we see that for some levels of $I_{app}$ the membrane oscillates and does not remain stably at $V_{ss}$. To find the conditions for

stability of the steady state, we linearize the full system 2.1 around

$$(V_{ss}, m_\infty(V_{ss}), h_\infty(V_{ss}), n_\infty(V_{ss})).$$

This leads to the constant coefficient 4th order system for evolution of the vector $y(t)$ of perturbations of $(V, m, h, n)$:

$$\frac{dy}{dt} = J_{ss} y,$$

where $J_{ss}$ is the 4x4 Jacobian matrix of 2.1, evaluated at the steady state. The four eigenvalues $\lambda_i$ of $J_{ss}$ determine stability. Stability requires that each $\lambda_i$ have negative real part. If any of them has positive real part then the steady state is unstable. In Fig. 6A we plot the leading eigenvalues as functions of $I_{app}$. Indeed, we have stability for low and high values of $I_{app}$. However, there is an intermediate range of $I_{app}$ : $I_1 < I_{app} < I_2$ where the steady state point is unstable. The leading eigenvalues form a complex pair and the steady state loses stability via a Hopf bifurcation [30,62] as $I_{app}$ increases through the critical value $I_1$ and regains stability via Hopf bifurcation as $I_{app}$ increases through $I_2$. Figure 7A shows $V_{ss}$ as a function of $I_{app}$ (thin lines). (Note, this is just a replotting of $I_{ss}$ from Fig. 3B.) The values $I_1$ and $I_2$ depend on temperature and other parameters of the system. Figure 7B shows how the region of instability shrinks with temperature. As the numerical simulations suggest the model oscillates stably for $I_1 < I_{app} < I_2$ and rhythmicity is lost at high temperature [49,57]. The branches for $I_1(Temp)$ and $I_2(Temp)$ coalesce at $Temp = 28.85°C$. Notice, there are no Hopf bifurcations, and the steady state does not lose stability above this temperature.

### 2.4.2. Stability of periodic solutions

The theory of Hopf bifurcations [30, 62] guarantees that small amplitude oscillations emerge at the critical $I_{app}$ values. In this Hodgkin–Huxley case, the bifurcation is subcritical at $I_1$ (unstable oscillations on a branch directed into the region where the steady state is stable) and supercritical at $I_2$. One could compute these local properties by evaluating complicated expansion formulae [30]. Alternatively, the emergence, extension to large amplitude and stability of these periodic solutions, both stable and unstable orbits, can be traced across a range of parameters, using a variety of numerical methods (e.g. [18, 57]). Such branch tracking software [18] was used to compute the periodic solutions as a function of $I_{app}$, shown in Figure 7A (thick lines).

The stability of a periodic solution, in general, is determined according to the Floquet theory [10, 30, 71]. We describe this formally in Appendix B and here just sketch the idea and give the numerical results from the stability analysis. In

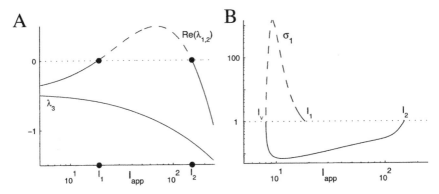

Fig. 6. Stability of steady states and periodic orbits in the Hodgkin–Huxley model. *A:* Real part of the complex pair of eigenvalues $Re(\lambda_{1,2})$ and third negative eigenvalue $\lambda_3$. $\lambda_4$ is even more negative and is off the scale of this plot. For $I_1 < I_{app} < I_2$, $Re(\lambda_{1,2})$ is positive (dashed), showing instability of the steady state. *B:* Values of the leading non-trivial Floquet multiplier along the branch of periodic solutions in log-log axes. The part of the curve with $\sigma_1 > 1$ (dashed) indicates instability of the periodic solutions (see text). In both panels $I_{app}$ is in $\mu A/cm^2$.

analogy to the steady state case, we linearize the equations 2.1 about the periodic solution (period $T$),

$$\frac{dy}{dt} = J_{osc}y,$$

where $J_{osc}$, the 4x4 Jacobian matrix, now has entries that are periodic. This equation has solutions of the form $exp(\lambda t)q(t)$ where $q$ is periodic. The cycle-to-cycle growth or decay of perturbations are governed by the numbers $\sigma = exp(\lambda T)$, the Floquet multipliers. If any $\sigma$ (of the 4) has $|\sigma| > 1$ then the periodic solution is unstable. The (nontrivial) leading multiplier $\sigma_1$ is plotted vs. $I_{app}$ in Fig. 6B along the branch of periodic solutions. (Note, there is always one $\sigma$ equal to unity since the derivative of the periodic solution satisfies the above linear equation, with $\lambda$=0.) The curve is multivalued because for $I_v < I_{app} < I_1$ two periodic solutions exist, one stable (corresponding to repetitive firing) and one unstable. The leading Floquet multiplier is greater than 1 for the unstable orbit (thick dashes), in fact much larger than 1, indicating that this orbit would not be seen in forward integration of 2.1 and, even more unlikely, in experiments.

### 2.4.3. Bistability
Notice that for a range of applied current, $I_v < I_{app} < I_1$, the stable steady state and the stable limit cycle coexist. This means that the system is bistable, i.e. that for the same values of parameters depending on initial conditions the long-term state of the system can be different. In this section we discuss predictions of the

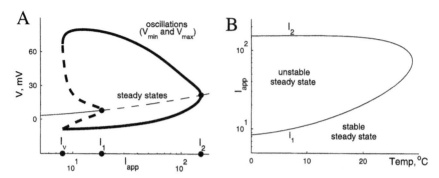

Fig. 7. *A:* Bifurcation diagram showing the possible behaviors for the Hodgkin–Huxley equations for different values of $I_{app}$ at temperature 18.5° C. $I_1$ and $I_2$ denote Hopf bifurcations of the steady states, $I_v$ bifurcation of the periodic solutions. *Thin solid curves:* stable steady state, *thin dashed curves:* unstable steady state, *thick solid curves:* maximum and minimum $V$ of the stable limit cycle, *thick dashed curves:* maximum and minimum $V$ of the unstable limit cycle. *B:* Range of $I_{app}$ where the steady state is unstable plotted as a function of temperature. In both panels $I_{app}$ is in $\mu A/cm^2$.

model derived from the existence of bistability and their experimental verification for the squid giant axon.

One prediction concerns the onset of repetitive firing. The model predicts two critical values of current intensity for the onset of repetitive firing (Fig. 7A): a lower one ($I_v$) for a suddenly applied stimulus, and an upper one ($I_1$) for a slowly increasing ramp. The reason for this is that if the system starts at the steady state and the applied current is below $I_v$, then gradual increase of current will keep the solution at the stable steady state until it loses stability at $I_1$. On the other hand, if a current is turned on abruptly above $I_v$, then the phase space abruptly changes to include the stable limit cycle and the system may find itself in the domain of attraction of the limit cycle. Experimentally determining both critical values for the onset of repetitive firing also indicates the range of currents (between the critical values) where bistability is expected.

A second prediction is that if $I_{app}$ is tuned into the range for bistability, then brief perturbations of appropriate strength and phase will stop the repetitive firing. Moreover, the model predicts that such annihilation can be evoked by depolarizing as well as hyperpolarizing perturbations. This prediction is based on the fact that the domains of attraction are separated by a surface (which presumably contains the unstable limit cycles mentioned above (see [49, 57])). This structure suggests that the trajectory could be forced across the separatrix surface by a brief perturbation. In Hodgkin–Huxley simulations by Cooley et al. [15] and in experiments with stretch receptors of Gregory et al. [29] it was shown that a neural system can be shocked out of the steady state into repetitive firing. Later

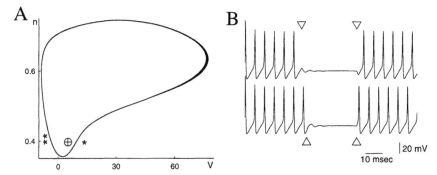

Fig. 8. Bistability in the Hodgkin–Huxley equations ($I_{app}$ = 9 $\mu$A/cm$^2$) A: Projection from 4-dimensional phase space onto the $V - n$ plane. Projections of stable periodic orbit and stable fixed point are shown. Also marked are the phases of the limit cycle where hyperpolarizing (asterisk) or depolarizing (double asterisk) perturbations may switch the system to the domain of attraction of the steady state. B: Time courses of $V$ with continuous spiking that is stopped by a delivery of a short pulse of current. Upper: hyperpolarizing pulses of current delivered at $t$ = 33 msec (strength -5 $\mu$A/cm$^2$), and $t$ = 65 msec (strength -10 $\mu$A/cm$^2$); lower: depolarizing pulses at $t$ = 36 msec (strength 5 $\mu$A/cm$^2$) and $t$ = 65 msec (strength 10 $\mu$A/cm$^2$). All pulses are square shaped current injections of duration 1. Initial conditions: $V_0 = 20mV$, $m = 0.6$, $h = 0.6$, $n = 0.45$.

Guttman et al. [32] demonstrated both theoretically and experimentally that the Hodgkin–Huxley system and squid giant axon can be shocked out of the periodically spiking state. To choose the appropriate phase and amplitude of the perturbation it is useful to look at the projection of the 4-dimensional phase space to a 2-dimensional plane. Figure 8A shows the stable limit cycle and the stable fixed point in the $V - n$ plane. This graph indicates that if we have a trajectory running along the limit cycle, and we deliver a hyperpolarizing perturbation just before the upstroke of the action potential (marked with an asterisk) or a depolarizing perturbation at the end of the recovery period (marked with two asterisks), then it is possible to switch the trajectory into the domain of attraction of the fixed point. The system will stop firing and will exhibit damped oscillations to the steady state (see examples in Fig. 8B). It is important to remember that the two-dimensional projection of the phase space in Figure 8A does not give an accurate picture of the behavior of the solutions. In the $V - n$ plane trajectories can cross while this is ruled out for the actual four-dimensional trajectories. Moreover, the perturbations that bring trajectories close to one of the attractors in the plane do not necessarily bring the actual trajectory across the separating surface. Therefore we treat two-dimensional projections as merely a useful indication of the behavior of the full system.

The model also indicates that it may be easier to stop periodic firing if the bias current is just above $I_\nu$, and it is easier to start the oscillation if the bias current is closer to $I_1$, because the domain of attraction of the fixed point is reduced near $I_1$.

Most of these predictions were qualitatively confirmed experimentally in the squid axon preparation [32]. Squid giant axon bathed in low $Ca^{2+}$ can exhibit co-existent stable states, one oscillatory (repetitive firing) and one time-independent (steady-state), for just superthreshold values of the applied current step. For a slow up and then down current ramp the critical current intensities for switching from one stable state to the other depend on whether current was increasing or decreasing. That is, the experimental system manifests a hysteresis behavior in this near threshold region. Moreover, repetitive firing in response to a just-superthrehsold step of current was annihilated by a brief perturbation provided its amplitude and phase were in an appropriate range of values.

The existence of a range of bistability has important implications for overall dynamics of the system. First, it allows greater stability in the near-threshold regime in the presence of noise. This is the same principle as used in thermostats — once the system is settled to one of the stable states small perturbations have little effect. Bistability in the system can also underlie more complex firing patterns in the autonomous system — for example, it can provide the basis for bursting, observed in many cell types (see section 4). To achieve this, the bifurcation parameter ($I_{app}$ in our case) becomes a dynamic variable whose typical time scale is much longer than the typical oscillation period. Finally, we note that similar phenomena can occur if other quantities are used instead of $I_{app}$ as bifurcation parameters (or as slow dynamic variables), for example the concentration of intracellular $Ca^{2+}$.

## 3. Morris-Lecar model

In order to more fully exploit the power of geometric phase-plane analysis and bifurcation theory in understanding neuronal dynamics we focus on a two-variable model of cellular electrical activity. The model was developed by Morris and Lecar in 1981 [40] in their study of barnacle muscle electrical activity and then popularized as a reduced model for neuronal excitability [53]. Our presentation parallels some of that in [53]. The Morris-Lecar model is formulated in the Hodgkin–Huxley framework, with biophysically meaningful parameters and structure, yet it is simple enough to analyze and it exhibits a large repertoire of interesting behaviors (see e.g. [53]). Of course, it is rather idealized compared to many other neuronal models that are designed to investigate the details of inter-action of many known ionic currents, or the spatial propagation of activity, etc. Such minimal models however are invaluable when the problem or questions at hand require only qualitative or semi-quantitative characterizations of spiking activity. This is especially important in studies of large networks of cells. We note that the geometrical approach via phase plane analysis was pioneered in the study of neuronal excitability by Fitzhugh [22, 23].

The Morris-Lecar model incorporates a delayed-rectifier $K^+$ current similar to the Hodgkin–Huxley $I_K$ and a fast non-inactivating $Ca^{2+}$ current (depolarizing, and regenerative, like the Hodgkin–Huxley $I_{Na}$). The activation of $Ca^{2+}$ is assumed to be so fast that it is modeled as instantaneous. The model equations are

$$C\frac{dV}{dt} = -I_{ion}(V, w) + I_{app} \tag{3.1}$$
$$= -(\bar{g}_{Ca}m_\infty(V)(V - \bar{V}_{Ca}) + \bar{g}_K w(V - \bar{V}_K) + \bar{g}_L(V - \bar{V}_L))$$
$$+ I_{app},$$
$$\frac{dw}{dt} = \phi\left[w_\infty(V) - w\right]/\tau_w(V).$$

The activation variable $w$ is the fraction of $K^+$ channels open, and provides the slow voltage-dependent negative feedback as required for excitability. The V-dependence of $\tau_w$, $m_\infty$ and $w_\infty$ is shown in Figure 9A. We use the same values of model parameters as in [53]. They are summarized in the caption to figure 9 and are the same throughout the section unless noted otherwise.

This two-variable system's behaviors can be fully revealed in the phase plane. We start by finding the nullclines — the two curves defined by setting $dV/dt$ and $dw/dt$ individually equal to zero. These nullclines define some key features of the dynamics. A solution trajectory that crosses a nullcline does so either vertically or horizontally. The nullclines also segregate the phase plane into regions with different directions for a trajectory's vector flow. The nullclines' intersections are the steady states or fixed points of the system. Much can be concluded just by looking at the nullclines, and how they change with parameters. The $V$-nullcline is defined by

$$-I_{ion}(V, w) + I_{app} = 0.$$

To see what this curve looks like, let us say that $I_{app} = 0$, and look at the current $I_{ion}(V, w_0)$ versus $V$ with $w$ fixed. This is the just the instantaneous $I - V$ relation $I_{ion}(V, w_0) = I_{inst}(V; w_0)$, depending on the level of $w$, and we want to know its zeros. Similar to the Hodgkin–Huxley model, this curve is N-shaped (see Fig. 9B1), and increasing $w$ approximately moves the curve upward. Dependence of the zero-crossings its $w$ will determine the $V$ nullcline (Fig. 9B). For moderate $w$ there are three zeros, and therefore three branches of the $V$-nullcline (corresponding to R,T and E: rest, threshold and the excited state); for low $w$

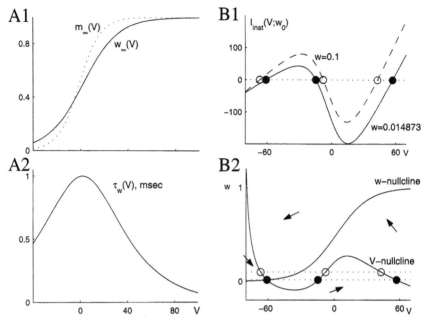

Fig. 9. *A:* Time constants and the steady state functions for the Morris-Lecar model. Upper: $m_\infty(V) = .5(1 + \tanh((V - V_1)/V_2))$ (dotted), $w_\infty(V) = .5(1 + \tanh((V - V_3)/V_4))$ (solid); lower: $\tau_w(V) = 1/\cosh((V - V_3)/(2V_4))$, where $V_1 = -1.2, V_2 = 18, V_3 = 2, V_4 = 30, \phi = .04$. Other parameters: $C = 20\,\mu F/cm^2, \bar{g}_{Ca} = 4, V_{Ca} = 120, \bar{g}_K = 8, V_K = -84, \bar{g}_L = 2, V_L = -60$. *B:* Construction of the nullclines. Upper: $I_{inst}(V; w)$ for $w = 0.014873$ (at rest, solid curve) and $w = 0.1$; lower: $V$ and $w$ nullclines; arrows show direction of the flow in the plane. All parameter values are as in [53]. Voltages are in mV, conductances are in mS/cm$^2$, currents in $\mu A/cm^2$. They are the same throughout the section unless noted otherwise.

there is only one high-$V$ branch; and for large $w$ only a low-$V$ branch. Below the $V$-nullcline in the $V - w$ plane $dV/dt > 0$, i.e. $V$ is increasing, while above the nullcline $V$ is decreasing. The $w$-nullcline is simply the activation curve $w = w_\infty(V)$. Left of this curve $w$ decreases and to the right $w$ increases (Fig. 9B2).

The steady state solution $(\bar{V}, \bar{w})$ of the system is the point where the nullclines intersect. It must satisfy $I_{ss}(\bar{V}) = I_{app}$, where $I_{ss}(V)$ is the steady state $I - V$ relation of the model given by

$$I_{ss}(V) = I_{ion}(V, w_\infty(V)).$$

If $I_{ss}$ is N-shaped, then there can be three steady states for some range of $I_{app}$. However, if $I_{ss}$ is monotone (as is the case for the parameters in figure 9), then there is a unique steady state, for any $I_{app}$.

## 3.1. Excitable regime

We say that the system is in the excitable regime when it has just one stable steady state and the action potential (large regenerative excursion) is evoked following a large enough brief stimulus. Figure 10A,B shows the responses to brief $I_{app}$ pulses of different amplitude. After a small pulse the solution returns directly back to rest (subthreshold). If the pulse is large enough, autocatalysis starts, the solution's trajectory heads rightward toward the $V$-nullcline. After the trajectory crosses the $V$-nullcline (vertically, of course) it follows upward along the nullcline's right branch. After passing above the knee it heads rapidly leftward (downstroke). The number of $K^+$ channels open reaches a maximum during the downstroke, as the $w$-nullcline is crossed. Then the trajectory crosses the $V$-nullcline's left branch (minimum of $V$) and heads downward, returning to rest (recovery).

Notice that if a superthreshold pulse initiates the action potential with different initial conditions (say, closer to the $V$-nullcline's middle branch), then the solution does not go as far rightward in $V$, resulting in an intermediate amplitude (graded) response. This contradicts the traditional view that the action potential is an all-or-none event with a fixed amplitude. This possibility of graded responsiveness in an excitable model was first observed by FitzHugh [23], in studying an idealized analytically tractable two-variable model. His model is often considered as a prototype for excitable systems in many biological and chemical contexts. But in that model, as well as in the Hodgkin–Huxley model and in the Morris-Lecar example of Fig. 10A, there is not a strict threshold. If we plot the peak $V$ vs. the size of the pulse or initial condition $V_0$, we get a continuous curve (Fig. 10C). The steepness of this curve depends on how slow is the negative feedback. If $w$, for example, is very slow then the flow in the phase plane is close to horizontal, $V$ is relatively much faster than $w$, and it takes fine tuning of the initial conditions to evoke the graded responses – the curve is steep. How close to horizontal the flow is, is determined by the size of $\phi$:

$$\frac{dw}{dV} = \left(\frac{dw}{dt}\right) / \left(\frac{dV}{dt}\right) = O(\phi).$$

If $\phi$ is very small ($w$ slow) then the trajectory of the action potential looks like a relaxation oscillator and the plateau, upstroke and the downstroke are more pronounced, like in a cardiac action potential, or an envelope of a burst pattern. This suggests that if the experimentally observed action potentials look like all-or-none events, they may become graded if recordings are made at higher temperatures. This experiment was suggested by FitzHugh and carried out by

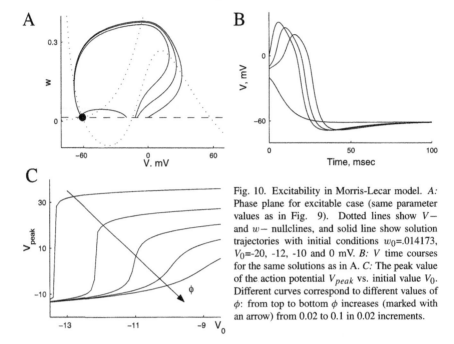

Fig. 10. Excitability in Morris-Lecar model. A:
Phase plane for excitable case (same parameter
values as in Fig. 9). Dotted lines show $V-$
and $w-$ nullclines, and solid line show solution
trajectories with initial conditions $w_0$=.014173,
$V_0$=-20, -12, -10 and 0 mV. B: $V$ time courses
for the same solutions as in A. C: The peak value
of the action potential $V_{peak}$ vs. initial value $V_0$.
Different curves correspond to different values of
$\phi$: from top to bottom $\phi$ increases (marked with
an arrow) from 0.02 to 0.1 in 0.02 increments.

Cole et al. [11]. It was found that if recordings in squid giant axon are made at
38°C instead of, say, 15°C then action potentials do not behave in an all-or none
manner.

### 3.2. Post-inhibitory rebound

Many neurons can fire an action potential when released from hyperpolarization. Namely, if a step with $I_{app} < 0$ is applied for a prolonged period of time
and then switched off, the neuron may respond with a single spike upon the release (Fig. 11A). This phenomenon is called post-inhibitory rebound (PIR), or
in the classical literature, anodal break excitation. To explain PIR, let us look
at the phase-plane (Fig. 11B). Before $I_{app}$ is turned on the resting point is on
the left branch of the $V$-nullcline. When $I_{app}$ comes on, it pulls the $V$-nullcline
down. The steady state moves, accordingly, down and to the left (i.e. it becomes
more hyperpolarized and with $K^+$ further deactivated). If the current is held long
enough, then the solution settles to the new steady state. Next, when the current
is released abruptly the $V$-nullcline snaps back up. The solution location is now
below the nullcline, and if $\phi$ is sufficiently small, the solution will fly all the way
to the right branch and then return to rest through the full action potential.

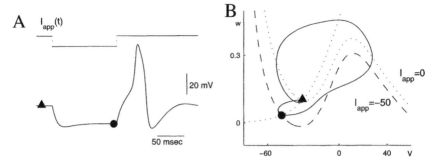

Fig. 11. Post-inhibitory rebound in Morris-Lecar model. *A:* Time courses of hyperpolarizing current and corresponding voltage. *B:* Same response trajectory as in A shown in the $V - w$ phase plane. Also shown are nullclines for $I_{app}$=0 $\mu$A/cm$^2$(dotted) and $V$-nullcline for hyperpolarized current $I_{app}$=-50 $\mu$A/cm$^2$(dashed). Triangle and filled circle mark steady state points for each of these cases.

Physiologically, the rebound is possible because the $K^+$channels that are usually open at rest are closed by the hyperpolarization. This makes the membrane hyperexcitable, i.e. lowers its threshold for firing. When the cell is released it takes awhile for $w$ to activate and during this delay the autocatalytic $I_{Ca}$ is less opposed. Anodal break excitation is also observed in the Hodgkin–Huxley model [24]. There, the hyperpolarization also causes removal of inactivation of $I_{Na}$ (increase in $h$), which contributes to facilitating the rebound.

### 3.3. Single steady state. Onset of repetitive firing, type II

Let us ask now whether and how repetitive firing can arise in this model with $I_{app}$ as a control parameter. As before, we will look for parameter values where the steady state is unstable, using linear stability theory. Let us say the steady state is $(\bar{V}(I_{app}), \bar{w}(I_{app}))$. Now, consider the effects of a perturbation from the steady state: $(V(t), w(t)) = (\bar{V} + x(t), \bar{w} + y(t))$, and ask whether $x$ and $y$ can grow with $t$ (unstable solution) or decay (stable solution). For the first equation:

$$C\frac{dV}{dt} = C\frac{dx}{dt} = I_{app} - I_{ion}(\bar{V} + x, \bar{w} + y) =$$

$$= I_{app} - I_{ion}(\bar{V}, \bar{w}) - \frac{\partial I_{ion}}{\partial V}\Big|_{(\bar{V},\bar{w})} x - \frac{\partial I_{ion}}{\partial w}\Big|_{(\bar{V},\bar{w})} y + h.o.t.,$$

where *h.o.t.* means higher order terms. The first 2 terms sum to zero, leaving (after neglecting h.o.t.)

$$C\frac{dx}{dt} = -\frac{\partial I_{inst}}{\partial V}\Big|_{(\bar{V},\bar{w})} x - \frac{\partial I_{inst}}{\partial w}\Big|_{(\bar{V},\bar{w})} y.$$

For the second equation:

$$\frac{dy}{dt} = \phi \frac{w'_\infty}{\tau_w} x - \frac{\phi}{\tau_w} y.$$

To summarize:

$$\frac{d}{dt}\begin{pmatrix} x \\ y \end{pmatrix} = J \begin{pmatrix} x \\ y \end{pmatrix}, \tag{3.2}$$

where $J$ is the Jacobian matrix, evaluated at the steady state:

$$J = \begin{pmatrix} -\frac{1}{C}\frac{\partial I_{inst}}{\partial V} & -\frac{1}{C}\frac{\partial I_{inst}}{\partial w} \\ \phi \frac{w'_\infty}{\tau_w} & -\phi \frac{1}{\tau_w} \end{pmatrix}\Bigg|_{(\bar{V},\bar{w})}.$$

Stability of the system 3.2 is determined by the eigenvalues $\lambda_1$ and $\lambda_2$ of $J$:

$$\det J = \lambda_1 \cdot \lambda_2,$$

$$\operatorname{tr} J = \lambda_1 + \lambda_2.$$

For the steady state to be stable the real parts of both eigenvalues must be negative. In order for the fixed point to lose stability one of the following two things must happen:
(1) $\lambda_1$ or $\lambda_2$ is equal to 0, i.e. $\det J = 0$;
(2) $Re(\lambda_1) = Re(\lambda_2) = 0$, $Im(\lambda_{1,2}) \neq 0$, i.e. $\operatorname{tr} J = 0$ — Hopf bifurcation.
Case (1) can only happen if $\bar{V}$ is at the "knee" of $I_{ss}(V)$, because $\det J = 0 = \frac{\phi}{\tau_w} \cdot \frac{1}{C}\left[\frac{dI_{ss}}{dV}\right]$. Therefore, if $I_{ss}$ is monotonic, the loss of stability can only happen via case (2), Hopf bifurcation.

For the parameters in figure 9 both eigenvalues are real and negative, i.e. the steady state is stable. Moreover, $I_{ss}(V)$ in this case is monotonic (as in the Hodgkin–Huxley equations) and the loss of stability can happen only through Hopf bifurcation. We have:

$$\operatorname{tr} J = 0 = -\frac{1}{C}\frac{\partial I_{inst}}{\partial V}\Bigg|_{(\bar{V},\bar{w})} - \frac{\phi}{\tau_w}.$$

Instability means that this expression is positive, i.e.

$$-\frac{1}{C}\frac{\partial I_{inst}}{\partial V} > \frac{\phi}{\tau_w}. \tag{3.3}$$

This says that in order to get instability $\frac{\partial I_{inst}}{\partial V}$ must be sufficiently negative, which is equivalent to having the fixed point on the middle branch (and sufficiently away

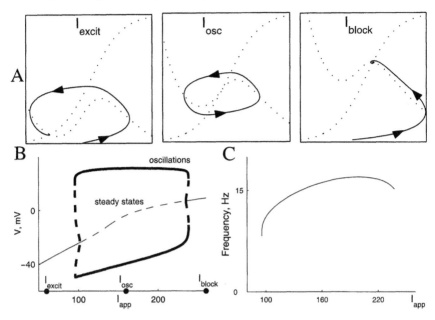

Fig. 12. Onset of repetitive firing with one steady state (Type II). *A:* Nullclines for different values of $I_{app}$ (60, 160 and 260 $\mu$A/cm$^2$), corresponding to excitable, oscillatory and nerve-block states of the system. *B, C:* Bifurcation diagram (*B*) and f-I curve (*C*). *B: Thin solid curves:* stable steady state, *thin dashed curves:* unstable steady state, *thick solid curves:* maximum and minimum $V$ of the stable limit cycle, *thick dashed curves:* maximum and minimum $V$ of the unstable limit cycle. Parameter values are the same as in figure 9.

from the "knees") of the $V$-nullcline. Also the rate of the negative feedback $\frac{\phi}{\tau_w}$ should be slow enough. If the temperature is too high ($\phi$ is too large), then destabilization will not happen. Condition 3.3 can also be interpreted as the autocatalysis rate being faster than the negative feedback's rate. When this condition is met, the steady state loses stability through a Hopf bifurcation, giving rise to a periodic solution, i.e. leading to the onset of repetitive firing. This periodic solution has a non-zero frequency associated with it, which is proportional to the $Im(\lambda_{1,2})$ ( [62]), i.e. the repetitive firing emerges with non-zero frequency. Figures 12B and C show the bifurcation diagram and the $f - I$ relation, respectively. Notice that this is the same type of repetitive firing onset as we have seen in the Hodgkin–Huxley equations. It is called Type II onset, following the terminology of Hodgkin [34].

Figure 12A shows snapshots of the nullcline positions at different $I_{app}$ (the values are marked $I_{excit}$, $I_{osc}$ and $I_{block}$ in Fig. 12B), and the solid lines show examples of solution trajectories for the same values of $I_{app}$. For $I_{app} = I_{excit}$ the membrane is excitable (the steady state is on the left branch). Increasing $I_{app}$ to $I_{osc}$ shifts the $V$-nullcline up, and the steady state moves to the middle branch and becomes unstable. Existence of a stable periodic orbit can be shown qualitatively for $\phi$ small considering the direction of the flow in the phase plane: the flow is always away from the unstable point, horizontally (say to the right), then following the $V$-nullcline to the knee, shoots left again, etc. The cycle can also be constructed by the geometrical singular perturbation or by the Poincare-Bendixon theorem (see chapter by Terman in this volume). If $I_{app}$ is further increased to $I_{block}$ the steady state moves to the right branch and becomes stable again. In this state the $K^+$ and $Ca^{2+}$ currents are strongly activated, but they are in a stable balance. This is nerve block — there is no firing, but the membrane is depolarized. At this point we have to remind ourselves again that this model is a description of a point neuron. It does not address what is happening in the axon. In principle, the axon may be generating action potentials that we do not see in this description.

Finally, we notice in figure 12B that, as in the Hodgkin–Huxley model (see section 2.4.3) there is a range of bistability, and the solution in that range can be brought into the domain of attraction of either the fixed point or the limit cycle by brief perturbations.

### 3.4. Three steady states

The cases of Morris-Lecar model dynamics that we have so far considered are reminiscent of the Hodgkin–Huxley dynamics in that there is always a unique fixed point. In other parameter regimes $I_{ss}(V)$ may not be monotonic. For example, increasing $\bar{V}_K$ to higher values [50], corrsponding experimentally to increasing the extracellular $K^+$ concentration, can cause $I_{ss}(V)$ to become N-shaped. This means that for some values of $I_{app}$ there are three steady states and the curve of steady states vs. $I_{app}$ is S-shaped (Figs. 13-15). We also choose parameters in such a way that only the lower steady state is on the left branch of the $V$-nullcline, while both middle and upper steady states are on the middle branch (Figs. 13-15). The number and location of steady states do not depend on temperature (i.e. on $\phi$). But the stability and the onset of repetitive firing, as we have seen above, do depend on $\phi$. Therefore, we will further study the dynamics at different values of $\phi$.

### 3.4.1. Large $\phi$. Bistability of steady states

We have computed above in equation 3.3 the condition for destabilization of a steady state through Hopf bifurcation. It shows that the negative feedback from $w$ has to be sufficiently slow compared to $V$ for the bifurcation to happen. For

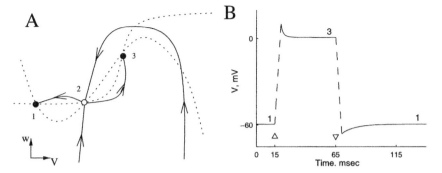

Fig. 13. Large $\phi$ case. *A:* Schematic of $V - w$ phase plane with steady states (filled circles – stable steady states, open circle – unstable), nullclines (dotted curves), and trajectories going in and out the saddle point (solid curves). The curves are slightly modified from the actual computed ones for easier viewing; in particular, the trajectories emanating from the saddle would not have extrema away from nullclines. *B:* Switching the solution from one stable steady state (marked 1 in A) to the other (marked 3 in A) and back with brief current pulses. Parameters are the same as in Fig. 9, except $V_3 = 12$ mV, $V_4 = 17$ mV, $\phi = 1$. Depolarizing pulse of strength 250 $\mu$A/cm$^2$is applied at $t = 15$ msec, hyperpolarizing pulse of strength -250 $\mu$A/cm$^2$is applied at $t = 65$ msec; duration of each pulse 5 msec.

large $\phi$ this condition is not satisfied. Moreover, for large $\phi$ $w$ is so fast that it can be considered as instantaneous $w = w_\infty(V)$. Then the model is reduced to one dynamic variable $V$ and the stability of the steady states is simply determined by the sign of $\frac{dI_{ss}}{dV}$, i.e. the middle steady state is unstable and the upper and lower ones are stable. Figure 13A shows the $V - w$ phase plane with steady states and nullclines for increased $\phi$. There is again a range of bistability, but in this case it is bistability between two steady states — with higher and lower voltage. The voltage can be switched from one constant level to the other by brief pulses (Fig. 13B). This type of behavior is sometimes called "plateau behavior" and it has been used in models describing vertebrate motoneurons [4].

### 3.4.2. Small $\phi$. Onset of repetitive firing, Type I

If $\phi$ is very small (Fig. 14A1), then according to the instability condition 3.3 both middle and upper steady states can be unstable for a range of $I_{app}$. (Notice that when there are three fixed points in the system they are not always "stable-unstable-stable"!) The middle fixed point is a saddle. One of the branches of its unstable ("outgoing") manifold goes directly to the stable steady state, and the other branch goes around the unstable spiral and also comes back to the stable point. These two unstable manifold branches are heteroclinic orbits — connecting two singular points. They effectively form (topologically) a circle that has two fixed points on it (Fig. 14A2, left panel). As $I_{app}$ is increased, the $V$-nullcline moves up and the stable fixed point and the saddle must coalesce and

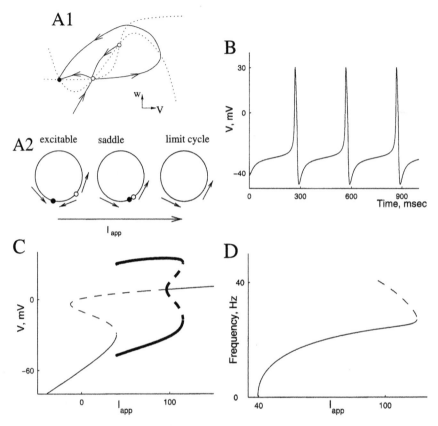

Fig. 14. Small $\phi$ case. Parameters are the same as in Fig. 13, except $\phi = .06666667$. *A:* Schematics of phase plane. *A1:* Schematic of phase plane for $I_{app} = 0$ $\mu$A/cm$^2$. Notations are the same as in Fig. 13A. The curves are slightly modified from the actual computed ones for easier viewing; in particular, actual trajectories would not have extrema away from nullclines. *A2:* Schematic of change in phase plane with change of $I_{app}$ (see text). *B:* Time course of voltage for $I_{app}$=40 $\mu$A/cm$^2$. *C:* Bifurcation diagram. Notation the same as in Fig. 7. *D:* $f - I$ curve. Dashed portion corresponds to unstable periodic orbit.

disappear. Figure 14A2, middle and right panels, show this bifurcation schematically. As the two points coalesce and then disappear the orbits connecting them form a single limit cycle. For $I_{app}$ exactly at the critical value $I_1$ the limit cycle has infinite period, i.e. the closed trajectory is a homoclinic orbit. Such an orbit is called (by some dynamicists) a saddle-node on an invariant circle (SNIC). For $I_{app}$ just above the critical value the frequency is proportional to $\sqrt{I_{app} - I_1}$ (see [62]), i.e. the repetitive firing emerges with zero initial frequency and large

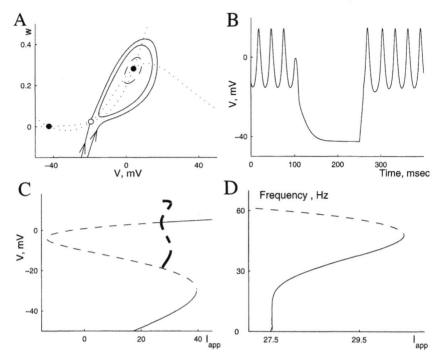

Fig. 15. Intermediate $\phi$ case. Parameters are the same as in Fig. 13, except $\phi = .25$. A: $V - w$ phase plane for $I_{app} = 30 \ \mu A/cm^2$. Dotted curves are nullclines, solid closed curve is the stable periodic orbit, dashed closed curve is the unstable periodic orbit. There are 2 stable steady states (filled circles) and one unstable steady state (open circle). Solid curves with arrows mark the trajectories going into the saddle point. B: Switching of $V$ from the stable periodic orbit to lower steady state and back for $I_{app} = 30 \ \mu A/cm^2$. Hyperpolarizing pulse of amplitude -100 $\mu A/cm^2$ and duration 1 msec arrives at $t = 100$ msec and depolarizing pulse of amplitude 100 $\mu A/cm^2$ and duration 5 msec arrives at $t = 250$ msec. C: Bifurcation diagram for $\phi = 0.25$. *Thin solid curves:* stable steady state, *thin dashed curves:* unstable steady state, *thick solid curves:* maximum and minimum $V$ of the stable limit cycle, *thick dashed curves:* maximum and minimum $V$ of the unstable limit cycle. D: f-I curve. Dashed portion corresponds to unstable periodic orbit.

amplitude. Figure 14B shows an example of the $V$ time-course with $I_{app}$ not far above the critical value, and the period of spiking is very large.

The bifurcation diagram for this case is shown in Fig. 14C and the $f - I$ curve in Fig. 14D. As $I_{app}$ increases beyond the saddle-node bifurcation the frequency of the oscillation increases while the amplitude decreases. The oscillatory solutions terminate via subcritical Hopf bifurcation, generating a small range of bistability.

Emergence of oscillatory behavior with arbitrary low frequencies has been reported in other models as well as in experiments (for example [12, 27, 31, 59]). Sometimes this zero-frequency onset has been attributed to the presence of an inactivating potassium A-type current [12], although as we see here such an $I_K$ is not required [58].

### 3.4.3. Intermediate $\phi$. Bistability of rest state and a depolarized oscillation

At intermediate values of $\phi$ both middle and upper steady states can be unstable, but the upper steady state is surrounded by a stable periodic orbit (Fig. 15A). In this case there is bistability between the low-voltage stationary point and depolarized repetitive firing. As before, brief $I_{app}$ pulses can be used to switch between the states (Fig. 15B). The saddle's stable manifold serves as the separatrix between the domains of attraction of the two states. The model also predicts that if a perturbation from rest is too large, then the system will exhibit one spike and come back to rest, not to the cycle. This happens, mathematically, because the perturbation causes the solution to cross the separatrix twice (see Fig. 15A), and, biophysically, because too much $I_K$ gets recruited. This phenomenon in the model has yet to demonstrated experimentally, to our knowledge.

To see how this phase plane portrait arises, let us look at the bifurcation diagram in Fig. 15C. The lower steady state is stable for all $I_{app}$ and the middle one, a saddle, is always unstable. When $I_{app}$ is very high nerve block occurs. When $I_{app}$ is decreased from large values, the high-$V$ steady state loses stability through a subcritical Hopf bifurcation. The emergent branch of (unstable) periodic orbits bends rightward from the Hopf point, but then turns around, stabilizing at the knee. So for some range of $I_{app}$ there are three steady states and two periodic states. An example from this range of $I_{app}$ is shown in figure 15A. As $I_{app}$ is decreased further the stable periodic orbit's minimum $V$ drops enough to contact the saddle, terminating the periodic branch. At this critical value of $I_{app} = I_1$ when the limit circle collides with the saddle, the unstable manifold of the saddle leaves the fixed point along the cycle and then returns back along the cycle as the stable manifold. This homoclinic loop is called a "saddle loop homoclinic orbit". The firing rate in this case also emerges from zero, but it increases as $1/|\ln(I_{app} - I_1)|$ (Fig. 15D, see also [62]).

This example in which a stable rest state coexists with a depolarized limit cycle provides the basis for square wave bursting (see section 4).

### 3.5. Similar phenomena in the Hodgkin–Huxley model

In section 2 we saw that in the Hodgkin–Huxley model repetitive firing emerges through a Hopf bifurcation with small amplitude and non-zero frequency. In fact, the Hodgkin–Huxley model can be tuned into parameter regimes that yield most of the dynamic behaviors that we described for the Morris-Lecar model [50]. For increased $\bar{V}_K$ $I_{ss}(V)$ becomes N-shaped, because $K^+$ now has a reversal potential that is not below resting $V$. If we also vary the temperature, the model will be tuned into plateauing behavior (coexistence of two steady states) or the type of behavior that we described for intermediate $\phi$ in Morris-Lecar model (resting state and a depolarized stable cycle).

### 3.6. Summary: onset of repetitive firing, Types I and II

Here we summarize the characteristic properties of two generic types of transition from the excitable to oscillatory mode in neuronal models.

**Type II**: (1) $I_{ss}$ monotonic; (2) subthreshold oscillations; (3) excitability without distinct threshold; (4) excitability with finite latency.

**Type I**: (1) $I_{ss}$ N-shaped; (2) no subthreshold oscillations; (3) all-or-none excitability with strict threshold; (4) infinite latencies.

The two cases also have different $(f - I)$ relations. However, if there is noise in the system, the $f - I$ curves for both types will become qualitatively similar. The discontinuity of Type II will disappear with noise since the probability of firing becomes non-zero where it was zero before. Type I's $f - I$ curve will also inherit a smooth foot rather than the infinite slope at zero frequency due to the noise-free case's square root rise (see, for example [31]).

## 4. Bursting, cellular level

Bursting is an oscillation mode with relatively slow rhythmic alternations between an active phase of rapid spiking and a phase of quiescence. A classical case, and well studied (both experimentally and theoretically), is the parabolic bursting $R_{15}$ neuron of the *Aplysia's* abdominal ganglion [2, 63]. Since its early discovery bursting has been found as a primary mode of behavior in many nerve and endocrine cells. Many of the experimentally found examples have been successfully modeled — sometimes with very good agreement between the data and the model. A collection of experimental examples and models was presented by Rinzel and Wang [70], including bursting in pancreatic $\beta$-cells, dopaminergic neurons of the mammalian midbrain, thalamic relay cells, inferior olive cells, and neocortical pyramidal neurons. Several classification schemes of different types of bursting, based on the bifurcation structure of the corresponding mathematical models, have been developed e.g. by Rinzel [51], Bertram et al. [3], de Vries [68], Izhikevich [36], Golubitsky et al. [28]. These classifications characterize the bursting behaviors by describing the topologies of the sequence of invariant sets that the solutions visit (steady states, periodic orbits, etc.) and the ways in which the solutions move between the sets (topological structure of the bursting). They also describe phenomenological properties of different types of bursting, for example, the shape and the amplitude of the spikes, evolution of spiking frequency during the active phase, presence or absence of bistability, etc. These phenomenological properties may be identified experimentally, and used for classification of biological bursters.

Biophysically the rhythmicity in bursts is generated by an autocatalytic depolarization process and slower negative feedback. In some burst mechanisms there is a distinct slow autocatalytic process (in addition to fast autocatalysis of spike generation) with even slower negative feedback, as in the case of $R_{15}$. In this case the spike generation can be disabled and an underlying slow rhythm, on which spike bursts ride, may persist. In other examples the fast autocatalysis of spike generation is adequate to guarantee an active phase and then a single slow (burst time scale) negative feedback process will drive the alternation back and forth over a hysteresis loop (as suggested by Figs. 7 and 15).

For more complete presentations and mathematical details of bursting types the readers are referred to the original papers. In this section we review geometrical methods for fast-slow analysis of bursting and some examples of the different oscillation types. We describe mathematical features of some illustrative examples in the classifications and phenomenological features that can help to identify them in the experiments. For some examples we also describe ionic conductance mechanisms that are sufficient to produce these oscillations and examples of experimental systems where this bursting type has been observed.

### 4.1. Geometrical analysis and fast-slow dissection of bursting dynamics

The general framework described here was first used for analysis of bursting in [50,51], and then extended, for example, in [3,70]. We consider a point neuron model and separate all dynamic variables into two subgroups: "fast" (denoted as $X$) and "slow" (denoted as $Y$). In the context of bursting analysis a variable is considered fast if it changes significantly during an action potential, and a variable is considered slow if it only changes significantly on the time scale of burst duration, not during single spikes. In this formulation both $V$ and $w$ of the above Morris-Lecar examples fall into the "fast" category — they both participate in the spike generation mechanism, even though $w$ may be noticeably slower than $V$. The model equations can now be written as:

$$\dot{X} = F(X, Y),$$

$$\dot{Y} = \epsilon G(X, Y).$$

In the second group of equations $\epsilon$ is a small positive parameter indicating the slower time scale. The slow variables can represent the gating variables with very slow kinetics (e.g., inactivation of a low threshold $I_{Ca}$ or slow $V$-gated activation of some $I_K$), ionic concentrations (for example, intracellular concentration of $Ca^{2+}$), second messenger variables, etc. We focus on cases in which the dynamics of $Y$ depends on $X$, i.e. on activity-dependent feedback from the fast subsystem. In special cases where it does not, i.e., $G(X, Y) = G(Y)$, the fast subsystem is driven non-autonomously by the slow subsystem; the slow burst rhythm cannot be reset by brief perturbations to the fast subsystem variables.

The analysis of the system can be conducted in two steps.

Step 1. First we think of the slow variables as parameters and describe the spike-generating fast subsystem for $X$ as a function of $Y$. This description involves finding steady states, oscillatory orbits and their periods, and transitions between all these solutions (bifurcations) as a function of $Y$:

$$0 = F(X_{ss}, Y) \implies X_{ss} = X_{ss}(Y)$$

or

$$\dot{X}_{osc} = F(X_{osc}, Y) \implies X_{osc}(t) = X_{osc}(t + T), \ T = T(Y).$$

If $Y$ is one-dimensional (there is only one slow variable) then the results can be summarized in a bifurcation diagram of the same type as we have seen above, with $Y$ as the bifurcation parameter. When $Y$ is multi-dimensional it simply adds dimensions to the bifurcation diagram and it may become harder to visualize.

Step 2. To describe the full system we overlay the slow dynamics on the fast system behavior. $Y$ evolves slowly in time according to its equations, while $X$ is tracking its stable states. Therefore, we must understand the direction of change of $Y$ at each part of the bifurcation diagram for $X$.

When the full burst dynamics is projected to the $(Y, V)$ plane, it coincides with portions of the bifurcation diagram. The results of this analysis allow one to make phenomenological descriptions of the bursting behavior and predict effects of parameter changes on behavior.

## 4.2. Examples of bursting behavior

### 4.2.1. Square wave bursting

Square wave bursting (see Fig. 16) is characterized minimally by a single slow variable (i.e., $Y$ is a scalar) and the fast subsystem structure is qualitatively as described in section 3.4.3 above. To recall, in the fast subsystem there is an S-shaped curve of steady states. The lower steady state is stable and corresponds to the silent phase of the bursting solution (Fig. 16B). The upper steady state is surrounded by an oscillatory state which corresponds to the active phase of firing. The oscillatory state terminates when it contacts the middle branch in a saddle-loop homoclinic bifurcation. At the intermediate range of the parameter values there is bistability between the lower steady state and the "upper" periodic orbit (Fig. 16B). Next, the kinetics of the slow parameter $Y$ has to be such that $Y$ decreases when the fast subsystem is at the low steady state and increases when the fast subsystem oscillates around the upper steady state. This dynamics of $Y$ allows fast switching between coexistent stable states, generating bursting, as shown in Fig. 16A. The active phase begins at the saddle-node bifurcation and terminates at the saddle-loop. (Note, it is not essential in Fig. 15 that the Hopf bifurcation on the high-$V$ branch be subcritical.)

Phenomenologically, square wave bursting is characterized by abrupt periodic switching between a silent phase and a state of depolarized repetitive firing. Moreover, the spikes typically ride on a plateau (i.e., do not undershoot), spike frequency decreases towards the end of the active phase (due to proximity to a homoclinic), and the burst rhythm's phase can be reset by brief pulses of $I_{app}$ (due to bistability).

The seminal work of Hodgkin–Huxley -like modeling of square wave bursting [7] was for the electrical activity of pancreatic $\beta$-cells, leading to the subsequent mathematical treatment and characterization of fast-slow analysis ( [50]).

In a number of minimal biophysical models for square wave bursting, spiking is due to high-threshold $Ca^{2+}$and delayed-rectifier $K^+$currents; bursting is due to either a calcium-activated potassium current [7, 50] or due to slow inactivation of the $Ca^{2+}$current by feedback from voltage or from $Ca^{2+}$concentration itself [14]. The $I_{K-Ca}$ bursting mechanism, for example, depends on the slow dynamics of $Ca^{2+}$-handling in the cell. It works in the following way. During the active phase each spike slightly increases $Ca^{2+}$concentration inside the cell, which in turn activates a bit of the $I_{K-Ca}$ current. When the potassium current is large enough, it brings the voltage down, terminating the active phase. During the silent phase $Ca^{2+}$influx is minimal and $Ca^{2+}$concentration slowly decreases turning off $I_{K-Ca}$ until spiking can resume. In these models of square wave bursting intracellular $Ca^{2+}$concentration $Ca$ satisfies

$$\dot{Ca} = f\,[-\alpha I_{Ca} - k \cdot Ca]\,,$$

where $f$ is the buffering constant $f = \frac{[Ca^{2+}_{free}]}{[Ca_{TOT}]}$, and the $Ca^{2+}$concentration is increased by the inward (i.e. negative) membrane calcium current $I_{Ca}$ and calcium is removed from the cytoplasm with rate $k$. The slow time scale is due to the fact that $f$ is small (say, 0.01 or so); most of the $Ca^{2+}$that enters the cell is rapidly buffered by reversible binding to sites on various molecules inside the cell. As the removal rate $k$ is increased parametrically the behavior changes from a low-$V$ steady state, to bursting, to continuous spiking.

Multiple biophysical mechanisms have also been modelled to account for square bursting in brain stem neurons that are involved in neural circuits that

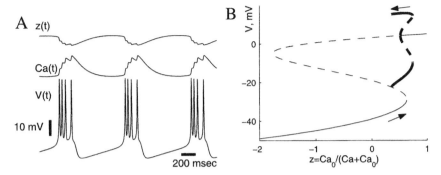

Fig. 16. Square wave burster. Equations and parameters are the same as in Fig. 15 with $I_{app} = 45$ $\mu$A/cm$^2$ and an additional $Ca^{2+}$-dependent $K^+$ current $I_{K-Ca} = g_{K-Ca}(1 - z)(V - V_K)$, where $g_{K-Ca} = .25$ mS/cm$^2$, gating variable $z = Ca_0/(Ca + Ca_0)$, $Ca^{2+}$ concentration is governed by $\dot{Ca} = \epsilon(-\mu \cdot g_{Ca}m_\infty(V)(V - V_{Ca}) - Ca)$, $Ca_0 = 10$, $\epsilon = 0.005$, $\mu = 0.2$. A: Bursting time course. $Ca^{2+}$ is playing the role of a slow variable, accumulating during the burst and slowly decaying during the silent phase. This is entered into the original equation through the $I_{K-Ca}$ term. The quantity that we use as a bifurcation parameter is a function of $Ca$: $z = Ca_0/(Ca + Ca_0)$ (top trace). B: Bifurcation diagram with $z$ as a parameter (compare to Fig 15C). Arrows show direction of change of $z$ during the firing and during the silent phase.

drive respiration (see [5] and Fig. 1C,D in this chapter).

### 4.2.2. Parabolic bursting

Parabolic bursting is generated without bistability in the fast subsystem and it requires at least two-variables in the slow subsystem. Steady states of the fast subsystem are now represented as an S-shaped surface over the two-dimensional plane of slow variables. Oscillatory solutions, similarly, are also represented as surfaces. They terminate as they touch the steady state surface at the (say, low-$V$) knee in a SNIC bifurcation. If the fast subsystem bifurcation surface is projected down to the slow-variable plane, the plane is divided into two non-overlapping regions: one with the resting steady state and the other with the repetitive firing for the fast subsystem. The dynamics of the slow variables is designed in such a way that there is an oscillation in this slow-variable plane that visits both regions (Fig. 17A), creating bursting. An active phase begins and ends with the system's trajectory crossing a SNIC bifurcation.

The monostability of the fast subsystem allows parabolic bursting to have smooth transitions between silent and active phases and to ride on a smooth sub-threshold wave (Fig. 17B). It also precludes resetting of this bursting by brief perturbations. Because the bursting trajectory passes through the homoclinic bi-furcation, the spike frequency at the beginning and at the end of the active phase

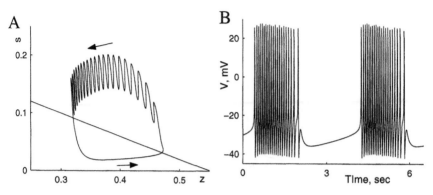

Fig. 17. Parabolic burster. Equations and parameters are the same as in Fig. 14 with $I_{app} = 65$ $\mu$A/cm$^2$ and the same $Ca^{2+}$-dependent $K^+$ current $I_{K-Ca}$ as in Fig. 16. One slow variable will be $z = Ca_0/(Ca + Ca_0)$ as in Fig. 16 above, and the second slow variable will be gating variable $s$ of another, slow, $Ca^{2+}$ current $I_{Ca,s} = g_{Ca,s}s(V - V_{Ca})$. Parameter values: $g_{K-Ca} = .1$ mS/cm$^2$, $Ca_0 = 1$, $\epsilon = 0.0005$, $\mu = 0.25$, $g_{Ca,s} = 1$ mS/cm$^2$, as in [53]. A: Projection of the bursting trajectory to the slow-variable plane. Direction of movement is indicated with arrows. Below the straight-looking curve low-voltage steady state is the only fast-subsystem attractor (silent phase) and above this curve there is an oscillation of the fast variables (spiking). B: Time course of voltage.

is reduced, hence the time course of the spike frequency has a parabolic shape, giving this bursting type its name.

Parabolic bursting was observed, probably first, in the Aplysia R15 neuron [1], and described with minimal biophysical models [46, 51, 56], and with various more detailed models (e.g., [6]). Experimentally, the sodium spikes during a burst can be blocked to reveal an underlying slow periodic wave generated mainly by $Ca^{2+}$ current. In the model [56] the slow activation of this $Ca^{2+}$ current and its even slower inactivation by internal $Ca^{2+}$ concentration provide the two variables for the slow subsystem.

### 4.2.3. Elliptic bursting
In the case of elliptic bursting the fast subsystem has bistability due to a subcritical Hopf bifurcation (as, for example, in the Hodgkin–Huxley model, Fig. 7), and the curve of steady states can be monotonic. As for square wave bursting, the mechanism requires one slow variable and $Y$ alternates back and forth through the region of bistability, allowing the switching between the steady state and the oscillatory solution (Fig. 18A, B). Unlike the above types of bursting, though, the elliptic burster exhibits subthreshold oscillations in the silent phase, and the oscillations grow as the steady state becomes unstable through Hopf bifurcation (Fig. 18C). The active phase starts near the subcritical Hopf (just before or af-

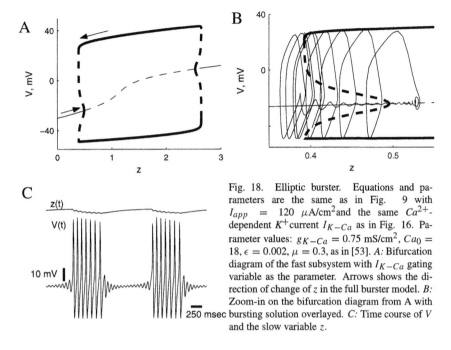

Fig. 18. Elliptic burster. Equations and parameters are the same as in Fig. 9 with $I_{app} = 120$ $\mu$A/cm$^2$ and the same $Ca^{2+}$-dependent $K^+$ current $I_{K-Ca}$ as in Fig. 16. Parameter values: $g_{K-Ca} = 0.75$ mS/cm$^2$, $Ca_0 = 18$, $\epsilon = 0.002$, $\mu = 0.3$, as in [53]. *A:* Bifurcation diagram of the fast subsystem with $I_{K-Ca}$ gating variable as the parameter. Arrows shows the direction of change of $z$ in the full burster model. *B:* Zoom-in on the bifurcation diagram from A with bursting solution overlayed. *C:* Time course of $V$ and the slow variable $z$.

ter, depending on whether or not some noise is present) and terminates at the saddle-node bifurcation of periodic orbits.

In elliptic bursting the envelope of the oscillatory events is modulated at very low frequency (creating an "elliptical" shape), small amplitude just before and after the large amplitude of the burst spikes. The frequency of the "silent" phase's subthreshhold oscillations is comparable to that of the Hopf bifurcation, and may or not be similar to the firing rate during bursts. The burst rhythm's phase can be reset by brief pulses. Elliptic bursting has been described in several computational models [16, 58, 69].

### 4.2.4. Other types of bursting

There are many other types of bursting that have been observed experimentally, and/or constructed theoretically based on the same principles that we have just illustrated. Their classification based on the topology of the underlying structure, rather than on phenomenological description, is not only more accurate, but its predictions are more reliable. The reason for this is that sometimes visually similar bursting patterns can have different ionic and other biophysical mechanisms associated with them. Conversely, sometimes bursting patterns that phenomenologically appear quite different are actually found to belong to the same class, with variations in some parameters. For example, Type Ib and Type IV bursting of Bertram et al. [3] appear to be different from square wave bursting at least in that they have undershooting spikes. However, both of them were revealed to be relatively minor modifications of the square wave bursting.

As in the rest of this chapter, we have focused on bursting mechanisms in isopotential, point neuron models. We would like to note, however, that bursting can also arise due to interactions between different parts of cells, or, in a model, between different compartments (e.g. in pyramidal neocortex cells [13, 45, 66]). In a recent example, ghostbursting [17] of pyramidal cells in the electrosensory organ of weakly electric fish was studied both experimentally and theoretically. It was shown that this type of bursting depends on dendritic properties.

Bursting can also be created via the interaction of many non-bursting cells in a network, as we describe in the next section, rather then by intrinsic cellular mechanisms.

## 5. Bursting, network generated. Episodic rhythms in the developing spinal cord

In this section we present an example of a model which is similar in form to the models described above, and which can be studied with similar methods, but which represents activity of a large network of cells, rather than a single cell.

### 5.1. Experimental background

Many developing neuronal systems exhibit spontaneous activity that is crucial for development. For example, in embryonic rat retina the ganglion cells are spontaneously active [25]. Moreover, the neighboring cells fire synchronous bursts while the activities of cells that are far apart (for example, in different eyes) are uncorrelated. This difference in the degree of synchrony in the spontaneous activity is thought to underlie the formation of ocular dominance regions in the lateral geniculate nucleus, the target of retinal output. In embryonic chick spinal cord the cells exhibit population bursts of activity (called episodes) with episode durations on the order of tens of seconds, and long inter-episode intervals of 2-10 minutes ( [42, 43]). Figure 19 shows an example of such a recording. The recording is made from a bundle of fibers that carry the (motoneuron) output of a segment of the spinal cord. The signal represents the combined activity of a large population of cells. During each episode single cells fire at 20-30 Hz, while the frequency of "population cycles" within an episode is about 0.5 Hz. This pattern of population activity is very robust, and it is generated spontaneously.

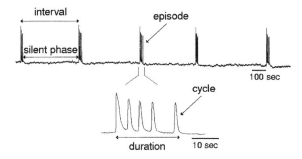

Fig. 19. Population activity of network of cells in a segment of an embryonic chick spinal cord (adapted from [65]).

This means that an isolated section of the spinal cord will produce this activity without any external inputs.

Before introducing the model we identify from the experimental data the key network properties that are essential for such rhythmogenesis. First of all, cells that are intrinsic bursters have not been found in this network. The percentage of cells that are pacemakers (fire periodically in isolation) is estimated to be less than 5%. Therefore, the rhythm is thought not to be a direct consequence of the cells' intrinsic properties alone. Second, if excitatory (glutamatergic) connections are blocked, the rhythm persists [8], i.e. the mechanism of rhythm generation is not dependent on the glutamatergic synapses. However, at this stage in development GABA-A synapses (usually associated with inhibition) can cause depolarization, i.e. be "excitatory" in effect. This is due to the fact that the $Cl^-$ reversal potential is about -30, -40 mV, i.e. close to or even above the threshold for firing. This phenomenon is observed in many developing systems, for example in retina [25].

Experiments also show that the size of the evoked synaptic potentials in the network is transiently reduced after an episode of spontaneous activity [21]. Shorter episodes cause less reduction of evoked synaptic potentials. These experimental observations led to the hypothesis that the episodes of activity can be terminated by the accumulated reduction in the synaptic efficacy and initiated after the synaptic recovery [43]. The primary goal of the modeling is to test this hypothesis.

## 5.2. Firing rate model

This model describes only the average firing rate of the population, rather than the instantaneous membrane potential and spikes of the individual neurons. It

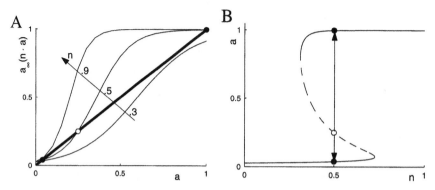

Fig. 20. Basic recurrent network. A: Population input-output function $a_\infty(n \cdot a)$ plotted vs. $a$ for different values of $n$ (thin). For each of these curves points of its intersection with line $y = a$ (thick) determine the steady states of the population at a particular level of $n$. These points are replotted in the $n - a$ plane in B. This is the curve of the steady states and its middle branch is unstable (dashed, and open circles).

assumes a purely excitatory recurrent network in which synaptic coupling is susceptible to both short- and long-term, very slow, activity-dependent depression. The model is described by only three coupled nonlinear differential equations. Therefore, the analysis can be done graphically, step by step, which facilitates an intuitive understanding of the network dynamics.

### 5.2.1. Basic recurrent network
A basic model of the excitatory recurrent network can be written as:

$$\tau_a \dot{a} = a_\infty(input) - a.$$

Here $a$ is the population activity or mean firing rate, and $a_\infty$ is the input-output relation of a neuron, taken to be an increasing sigmoidal function. $\tau_a$ is an effective time constant, reflecting integration time within a cell and recruitment time in the network. Because the network is recurrent the input is proportional to the activity of the network itself:

$$\tau_a \dot{a} = a_\infty(n \cdot a) - a.$$

The coefficient of proportionality $n$ represents the efficacy of coupling. The steady state activity of this network depends on $n$ (see Fig. 20A). For low $n$ there is only one stable low activity steady state; for $n$ high there is only one stable state, of high activity. And for intermediate values of $n$ there are three steady states, the middle of which is unstable and the lower and upper ones are stable.

Next, we allow the effective $n$ to vary dynamically. Based on the assumption that there is synaptic depression the synaptic efficacy $n$ should not only vary with time, but be activity dependent. Figure 20B gives a hint of how the full model will work. When $a$ is low, the synapses are recovering, i.e. $n$ is increasing and the steady state moves along the lower branch of the curve to the right. When the solution reaches the knee of the curve, the steady state switches to the upper branch, which corresponds to high $a$. The population activity is now high which leads to depression of synapses ($n$ decreases), until the solution crashes back to the lower branch.

### 5.2.2. Full model

The full model is described by three differential equations with both "fast" and "slow" depression [65]. Slow depression governs the initiation and termination of the episodes, while fast depression allows cycling within an episode. The equations of the model are:

$$
\begin{aligned}
\tau_a \dot{a} &= a_\infty (s \cdot d \cdot a) - a, \\
\tau_d \dot{d} &= d_\infty (a) - d, \\
\tau_s \dot{s} &= s_\infty (a) - s.
\end{aligned}
$$

The product $s \cdot d$ gives the available synaptic efficacy or fraction of synapses not affected by depression. It has one component ($d$) that changes on a faster time scale, and the other component $s$ that is very slow. The functions $d_\infty$ and $s_\infty$ are depression (turn-off) functions or efficacy recovery functions; they are sigmoidally decreasing with a from 1 to 0. The parameters of the model can be found in [65].

To understand the behavior of the model we first treat the slow variable $s$ as a parameter and study the $a - d$ dynamics. We would like to design the two-dimensional $a - d$ system in such a way that at low values of the parameter $s$ the solution rests at a low-activity steady state (inter-episode interval), and for high $s$ the solution oscillates (episode). Moreover, there needs to be an overlap (bistability) between oscillation and the steady state so that $s$ can provide transition from one to the other.

For a range of values of $s$ there is a limit cycle oscillation in the $a - d$ plane. It originates because the recurrent excitation $a$ provides the autocatalysis and $d$ provides a delayed negative feedback. As activity grows, synapses start to depress, and eventually the depression reduces the activity. As $a$ is reduced, $d$ starts to recover and allows $a$ to grow again, repeating the cycle (see Fig. 21A). If $s$ is small, then there is only one stable steady state in the $a - d$ plane at low values of $a$. However, it is possible to choose parameters in such a way that

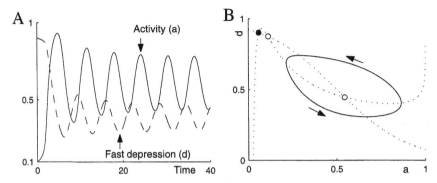

Fig. 21. Fast-subsystem of full model. A: Oscillations in $a - d$ subsystem with $s$ frozen, $s = 0.9$. B: Oscillation can co-exist with a low-activity steady state (filled circle). It is shown here in the $a - d$ phase plane with $s = .78$.

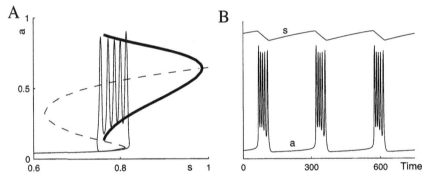

Fig. 22. A: Bifurcation diagram of the $a - d$ subsystem with $s$ as a parameter. There is an S-shaped curve of the steady states (solid – stable, dashed – unstable), periodic orbits are shown with minimal and maximal $a$ values (thick curve). Bursting solution of the full system is overlayed on top. B: Time courses of $a$ and $s$ of the same solution as in A.

there are three steady states, only the one with lowest $a$ is stable, and there is a small stable limit cycle around the upper one (Fig. 21B). Figure 22A shows the bifurcation diagram of the system with $s$ as the bifurcation parameter. We can see that there is a region of bistability, and as $s$ decreases the cycle disappears in a homoclinic (saddle-loop) orbit. Now we can design the s dynamics in such a way that if $a$ is small, $s$ slowly increases, and when $a$ is at higher values (as during the oscillation) $s$ slowly decreases. Figure 22B shows the full model generating the rhythmic spontaneous activity.

## 5.3. Predictions of the model

The model predicts that a brief perturbation should bring the network from one steady state to the other, i.e. prematurely terminate or start an episode. Moreover, the bifurcation diagram in figure 22B predicts that if the perturbation is delivered early in the silent phase, then the following episode will be shorter, if the perturbation is delivered late in the silent phase the episode will be longer. This prediction has been confirmed experimentally [64]: there is a correlation between the time since the previous episode and the length of the next episode. The correlation is the same for both spontaneous and evoked episodes. This confirms that the termination of the episode is well-determined. On the other hand, no correlation has been found between the duration of the episode and the following silent interval, suggesting that stochastic effects influence episode initiation (for example, spontaneous synaptic release may cause some cells to fire, thereby triggering an episode).

The model also predicts the effects of pharmacological agents that alter the effective network connectivity. To do that we re-introduce the synaptic efficacy $n$:

$$\tau_a \dot{a} = a_\infty (n \cdot s \cdot d \cdot a) - a.$$

Synaptic blockers in the model have the effect of reducing $n$. Experimentally it was found that when some of the synapses are blocked, the network exhibits a prolonged silent phase and then settles into a slower rhythm (Fig. 23A). To explain this in the model let us call $\bar{s} = n \cdot s$. The fast subsystem is bistable over a particular range of $\bar{s}$ values. If $n$ is reduced the range of $s$ in which oscillations exist moves to higher values of $s$. Therefore, just after $n$ is decreased, initially the network must collapse to the silent state (as this is the only attractor at that value of $\bar{s}$). Before an episode can start $s$ has to recover to a much higher value than before. This leads to a very prolonged first silent phase. Moreover, when $s$ settles into its new range of operation the silent phase is now closer to the $s$-nullcline, which makes the rhythm's period longer.

The simulation of pharmacological manipulation of synapses was originally done to distinguish between a model that we just presented and one with a spike frequency adaptation instead of a synaptic depression $s$. The spike frequency adaptation model assumes that if a cell is active, then more and more input is needed to keep the activity at the same level. It is modeled by

$$a_\infty (input) = a_\infty (input - \theta),$$

and the threshold $\theta$ slowly increases with $a$:

$$\tau_\theta \dot{\theta} = \theta_\infty (a) - \theta.$$

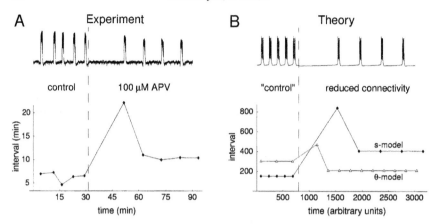

Fig. 23. Effect of a reduction of connectivity. *A:* Experimentally the connectivity is reduced by application of a synaptic blocker, NMDA-receptor antagonist, (adapted from Fig. 11 in [65]), after which there is one very long inter-event interval and then rhythm settles to a new (slower) period. *B:* Simulation results analogous to experiments in A for *s*-model and $\theta$-model.

It is found that for the $\theta$-model, in contrast to the *s*-model the rhythm's period actually becomes shorter as connections are blocked and the length of each episode also significantly decreases (Fig. 23B). Experimental data agreed with predictions of the *s*-model. This favored synaptic depression as the proposed mechanisms for spontaneous rhythm generation in the embryonic spinal cord.

Recent experimental data [9] show that the reversal potential for $Cl^-$ oscillates slowly in phase with the episodic rhythm. This variation contrasts with what is usually assumed, that for firing on short time scales the Nernst potentials for various ion species remain approximately constant. Such a slowly oscillating $V_{Cl}$ could provide a specific mechanism for slowly modulating the GABA-mediated synaptic current, thereby a biophysical realization of the phenomenological *s*-variable in the model above. Thus, $Cl^-$ handling dynamics have been introduced into the model (not shown here) as a specific mechanism for slowly modulating the GABA-mediated synaptic current. Synaptic activity and currents are strong during an episode leading to slow and considerable efflux of $Cl^-$. Consequently $V_{Cl}$ slowly declines, making the GABA-synapses less "functionally excitatory", eventually terminating the episode. In the interepisode interval $Cl^-$ efflux is very small and cloride pumps are able to slowly restore $V_{Cl}$ so that a new episode can eventually begin.

## 6. Chapter summary

We have offered here an introduction to some concepts and techniques of dynamical systems theory that we and others have used to gain insight into neuronal dynamics. Our examples are illustrative of a variety of nonlinear behaviors including bifurcations, multi-stability, oscillations (tonic and bursting), and excitability primarily in the context of individual neurons. We have used reduced modeling formulations, fast-slow dissections, branch-tracking methods, phase plane geometrical treatments, stability analyses, and rapid equilibrium assumptions to gain insight into these behaviors. By applying these approaches to formulate and dissect a model of episodic rhythmicity in developing spinal cord you can see how one could analyze and understand mean field (firing rate) models for networks. We have restricted our attention to point models (non-distributed systems), neglecting the effects of cable properties of axons and dendrites at the cellular level and of spatial extension in networks of neurons. These latter considerations are important and mathematical treatments of the partial differential equations or integro-differential equations (say for networks, where cell coupling is handled by convolution integrals with kernels that represent synaptic weighting/footprints) involve the above and other techniques (some review is provided in [20]).

There are many other dynamical phenomena that we have not addressed including the effects of stochastic fluctuations, intrinsic to a cell or to synapses, or due to network sparseness and complex dynamical states (e.g., see [41, 44, 67]). We have focused on reduced models but yet have by-passed the historical literature having to do with non-continuous models like integrate-and-fire (see [67]) and modern, more general, formulations (e.g. [26]). We have not considered models of synaptic plasticity, in the sense of learning, which seem like natural candidates for fast-slow treatment.

We close by reiterating a primary take-home message: that in many of the continuous models excitability and oscillations are realized mechanistically by fast autocatalytic/regenerative processes and slower negative feedback. The former at the cell level are usually due to rapidly activating inward currents (due to $Na^+$ and/or $Ca^{2+}$) and the latter to slower activating outward currents ($K^+$) or slower inactivation of inward currents. At the network level recurrent excitation may be the autocatalytic process and synaptic depression is just one candidate for negative feedback.

### 6.1. Appendix A. Mathematical formulation of fast-slow dissection

In order to describe the mathematical approach more systematically we imagine exaggerating the time scale differences in the Hodgkin–Huxley equations and introduce a "small" parameter $\epsilon$ to carry out this process. For simplicity we will again assume *a priori* that $m$ is still much faster than $V$ so that $m = m_\infty(V)$ is a valid idealization. Now, restating explicitly that $V$ and $m$ are fast relative to $h$ and $n$ we write the equations as follows:

$$\epsilon dV/dt = [-I_{ion}(V, m_\infty(V), h, n) + I_{app}]/C_m,$$
$$dh/dt = \phi [h_\infty(V) - h]/\tau_h(V),$$
$$dn/dt = \phi [n_\infty(V) - n]/\tau_n(V).$$

In this formulation if $h$ and $n$ vary on the time scale of msec then $V$ ($m$ as well) varies with an effective time constant of order $\epsilon$. The upstroke and downstroke will occur very rapidly. In the limit, with $\epsilon$ very small, these phases appear almost as jumps. In order to describe these 2 fast phases (cf, Fig. 3A) we must stretch the time variable and look at the dynamics "inside" this time window. So we let, $s = t/\epsilon$ be the stretched time variable. Rewriting the equations in terms of $s$ we get:

$$dV/ds = [-I_{ion}(V, m_\infty(V), h, n) + I_{app}]/C_m,$$
$$dh/ds = \epsilon\phi [h_\infty(V) - h]/\tau_h(V),$$
$$dn/ds = \epsilon\phi [n_\infty(V) - n]/\tau_n(V).$$

Now to get this "inner" solution we set $\epsilon = 0$ and find that, on the fast time scale, $h$ and $n$ are constant (say, equal to $h_0, n_0$, respectively), according to the second pair of equations below:

$$dV/ds = [-I_{ion}(V, m_\infty(V), h_0, n_0) + I_{app}]/C_m,$$
$$dh/ds = 0,$$
$$dn/ds = 0.$$

This, of course, is just the mathematical expression of what we did for phases 1 and 3 in Section 2.3.2. With $h_0, n_0$ set to their resting values the equation for $V$ is just the 1st order equation satisfied by the upstroke in this approximation when $\epsilon$ is small. Now to describe the solution during the plateau and recovery phases we look on the unstretched time scale, return to the first equation and set $\epsilon = 0$ there to get:

$$0 = -I_{ion}(V, m_\infty(V), h, n) + I_{app},$$
$$dh/dt = \phi [h_\infty(V) - h]/\tau_h(V),$$
$$dn/dt = \phi [n_\infty(V) - n]/\tau_n(V).$$

Here, $V$ corresponds to an instantaneous crossing by $I_{inst}(V)$ of the level $I_{app}$, i.e. to a zero crossing of $I_{inst}(V)$ if $I_{app} = 0$. Of course there are in general, for the Hodgkin–Huxley equations, up to 3 such crossings (recall, R,T and E) and these $V$-values depend on $h$ and $n$. Meanwhile $h$ and $n$ evolve according the 2nd and 3rd equations with, say during the plateau phase, $V = V_E(h, n)$. Geometrically, the first equation defines a surface in $V, h, n$ space; a multi-valued and folded surface with (if $V$ is the vertical direction) $V_E$ the upper branch, $V_R$ the lower branch, and $V_T$ the middle branch. During the plateau the trajectory drifts along the surface's upper branch, until it reaches the "fold" where $V_E$ and $V_T$ coalesce. This is when the downstroke occurs. Then, because the trajectory leaves the upper branch of the surface, we must revert to the fast time scale, stretch it, and capture the inner solution (the time course of $V$ on the fast time scale of the downstroke) by switching to the fast equations with $h, n$ held at their values when the fold is reached: $h = h_{down}$ and $n = n_{down}$. Then, having described the transition of $V$ from the fold down to $V_R(h_{down}, n_{down})$, the trajectory drifts along the surface's lower branch according to the slow equation as the solution returns to rest, say if $I_{app} = 0$. If $I_{app}$ is positive and large enough then the "rest" state does not lie on the lower surface but rather on the middle surface, in which case the trajectory reaches the knee of the lower surface and another upstroke occurs. This process continues, cycle after cycle, converging towards the periodic orbit of repetitive firing.

The procedure that we have described here is just the lowest order approximation to a series expansion of inner and outer solutions in the technique called matched asymptotic expansions (see, e.g. [37]).

### 6.2. Appendix B. Stability of periodic solutions

Here we discuss how one can determine the stability and characterize bifurcations of the limit cycles. The idea is to reduce the problem to consideration of stability of a fixed point of a special map, called the Poincare map (see, for example [62, 71]). Let us first define this map. Suppose we have a system

$$\dot{x} = f(x) \tag{6.1}$$

and this system has a closed orbit (periodic solution) $p(t)$. Let us choose a plane $S$ transversal to the vector field of the solutions in the neighborhood of a point $p^*$ on the periodic solution $p(t)$, i.e. such that any solution that starts on the plane has to come out of the plane, not along it. Then consider the mapping $P$ in the neighborhood of $p^*$ in the plane defined in the following way. For every point $x_0$ consider the solution that starts at $x_0$ and trace it until it hits the plane again. The point $x_1$ at which the solution hits the plane is the image of $x_0$ under $P$: $x_1 = P(x_0)$. This map is the Poincare map. A closed orbit of the original system corresponds to a fixed point of the Poincare map: $p^* = P(p^*)$. Now we can consider its stability. Let us choose a small perturbation from the fixed point, such that it still lies in the plane $S$: $x = (p^* + v) \in S$, and then look at its image under $P$ and the deviation of the image from $p^*$:

$$p^* + v_1 = P(p^* + v) = P(p^*) + [DP(p^*)]v + O(||v^2||),$$

where $DP(x)$ is the Jacobian matrix of $P$, called the linearized Poincare map. Remembering that $p^* = P(p^*)$, and using linear approximation, we see that

$$v_1 = [DP(p^*)]v.$$

The stability of the solution is now determined by the matrix $DP(p^*)$:
*The closed orbit is linearly stable if and only if all eigenvalues $\sigma_j$ of $DP(p^*)$ satisfy $|\sigma_j| < 1$.*
Intuitively, the condition $|\sigma_j| < 1$ assures that at every iteration of $P$ the deviation from the fixed point decreases. The quantities $\sigma_j$ are called Floquet multipliers. Notice that $P$ operates in the subspace transversal to the solution vector field. If one also considers perturbations along the periodic solution, for the autonomous system the corresponding eigenvalue will always be equal to 1. This is due to the time-invariance of the solutions of the autonomous system: a perturbation of the solution along the periodic orbit does not alter the solution. For more rigorous mathematical discussion the reader is referred to the dynamical systems literature, for example [10, 62, 71].

In practice $P(x)$ and $DP(x)$ are usually hard to compute. To avoid it, instead of considering the perturbed solutions on their return to $S$, one can consider the perturbed solutions at time $T$, where $T$ is the period of $p(t)$ (see e.g. [10]). This gives a new mapping $P'$ that takes a neighborhood of $p^*$ in $S$ to a surface $S'$ which intersects with $S$ at $p^*$. To characterize the linear approximation of $P'$ we linearize the original equation 2.1 around $p(s)$:

$$\dot{x} = \left.\frac{\partial f}{\partial x}\right|_{p(s)} x,$$

then consider a set of solutions with initial conditions $X(0) = I$ (identity matrix), and their image under $P'$, $X(T)$. For stability of the point $p^*$ and consequently the periodic solution $p(s)$, eigenvalues of $X(T)$ and of $DP(p^*)$ provide equivalent information under the same technical conditions that allow $P$ to be well-defined.

Coming back to the Hodgkin–Huxley equations, let us write them in the form

$$\dot{y} = f(y; I_{app}),$$

where $y(t)$ corresponds to the column vector $(V, m, h, n)^t$ and assume that there is a periodic solution $p(t)$, stability of which we would like to compute. Linearize the equations around $p(t)$:

$$\dot{y} = \left.\frac{\partial f}{\partial y}\right|_{p(t)} y. \tag{6.2}$$

The Floquet multipliers can now be found by numerical integration of 6.2 with condition $Y(0) = I$ and computation of the four eigenvalues of $X(T)$. This was first done for the Hodgkin–Huxley equations in [57], where they also studied dependence of periodic solutions on temperature (see Section 2.4.2). It was also shown [57] that for some parameter values (e.g. at temperature $Temp = 6.3°C$) Floquet multipliers cross the unit disk at $\sigma = -1$, which gives birth to secondary (unstable) periodic orbits in a period-doubling bifurcation [71].

## References

[1] Adams W.B and Benson J.A. (1985) The generation and modulation of endogenous rhythmicity in the *Aplysia* bursting pacemaker neurone R15, *Prog. Biophys. Mol. Biol.*, 46: 1-49

[2] Alving B. (1968) Spontaneous activity in isolated somata of Aplysia pacemaker neurons, *J. Gen. Physiol.*, 51: 29-45

[3] Bertram R., Butte M.J., Kiemel T., and Sherman A. (1995) Topological and phenomenological classification of bursting oscillations, *Bull. Math. Bio.*, 57: 413-439

[4] Booth V. and Rinzel J. (1995) A minimal, compartmental model for a dendritic origin of bistability of motoneuron firing patterns, *J. Comp. Neurosci.*, 2: 299-312

[5] Butera R.J., Rinzel J., and Smith J.C. (1999) Models of respiratory rhythm generation in the pre-Botzinger complex: I. Bursting pacemaker neurons, *J. Neurophys.*, 82: 382-397

[6] Canavier C.C., Clark J.W., and Byrne J.H. (1991) Simulation of the bursting activity of neuron R15 in Aplysia: role of ionic currents, calcium balance, and modulatory transmitters. *J. Neurophysiol.*, 66: 2107-2124

[7] Chay T.R. and Keizer J. (1983) Minimal model for membrane oscillations in the pancreatic *beta*-cell, *Biophys. J.*, 42: 181-190

[8] Chub N. and O'Donovan M. (1998) Blockade and recovery of spontaneous rhythmic activity after application of neurotransmitter antagonists to spinal networks of the chick embryo. *J. Neurosci.*, 18: 294 -306

[9] Chub N. and O'Donovan M.J. (2001) Post-episode depression of GABAergic transmission in spinal neurons of the chick embryo. *J. Neurophysiol.*, 85: 2166-2176

[10] Coddington E.A. and Levinson N. Theory of ordinary differential equations, McGraw-Hill, New York, 1955

[11] Cole K.S., Guttman R., and Bezanilla F. (1970) Nerve excitation without threshold, *Proc. Nat. Acad. Sci.*, 65: 884-891

[12] Connor J.A., Walter D. and McKown R. (1977) Neural repetitive firing: modifications of the Hodgkin-Huxley axon suggested by experimental results from crustacean axons, *Biophys. J.*, 18: 81-102

[13] Connors B.W. and Gutnick M.J. (190) Intrinsic firing patterns of diverse neocortical neurons, *Trends Neurosci.*, 13: 99-104

[14]  Cook L.D., Satin L.S., and Hopkins W.F. (1991) Pancreatic $\beta$-cells are bursting, but how?,*Trends Neurosci.*, 14: 411-414

[15]  Cooley J., Dodge F., and Cohen H. (1965) Digital computer solutions for excitable membrane models, *J. Cell Comp. Physiol.*, 66: 99-108

[16]  Del Negro, C. A., Hsiao, C.-F., Chandler, S. H., and Garfinkel, A. (1998) Evidence for a novel mechanism of bursting in rodent trigeminal neurons, *Biophys. J.*, 75: 174-82

[17]  Doiron B., Laing C., Longtin A., and Maler L. (2002) Ghostbursting: a novel neuronal burst mechanism, *J. Comp. Neurosci.*, 12: 5Ũ25

[18]  Ermentrout B. Simulating, Analyzing, and Animating Dynamical Systems: A Guide to XP-PAUT for Researchers and Students. SIAM, Philadelphia, 2002

[19]  Ermentrout B. (1998) Linearization of F-I curves by adaptation, *Neural Computation*, 10: 1721-1729

[20]  Ermentrout B. (1998) Neural networks as spatio-temporal pattern-forming systems, *Reports on progress in physics*, 61: 353-430

[21]  Fedirchuk B., Wenner P., Whelan P., Ho S., Tabak J., and O'Donovan M. (1999) Spontaneous network activity transiently depresses synaptic transmission in the embryonic chick spinal cord, *J. Neurosci.*, 19: 2102-2112

[22]  FitzHugh R. (1960) Thresholds and plateaus in the Hodgkin-Huxley nerve equations, *J. Gen. Physiol.*, 43: 867-896

[23]  FitzHugh R. (1961) Impulses and physiological states in models of nerve membrane, *Biophys. J.*, 1: 445-466

[24]  FitzHugh R. (1976) Anodal excitation in the Hodgkin-Huxley nerve model, *Biophys J.*, 16: 209-226

[25]  Galli L. and Maffei L. (1988) Spontaneous impulse activity of rat retinal ganglion cells in prenatal life, *Science*, 242: 90-91

[26]  Gerstner W. and Kistler W. Spiking Neuron Models: Single Neurons, Populations, Plasticity. Cambridge University Press, Cambridge, 2002.

[27]  Goldstein S.S. and Rall W. (1974) Changes of action potential shape and velocity for changing core conductor geometry, *Biophys. J.*, 14: 731-757

[28]  Golubitsky M., Josic K. and Kaper T.J. (2001) An unfolding theory approach to bursting in fast-slow systems. *In:* Global Analysis of Dynamical Systems: Festschrift dedicated to Floris Takens on the occasion of his 60th birthday. (H.W. Broer, B. Krauskopf and G. Vegter, eds.) Institute of Physics Publ.: 277-308

[29]  Gregory J.E., Harvey R.J., and Proske U. (1977) A late supernormal period in the recovery of excitability following an action potential in muscle spindle and tendon organ receptors and its efefct on their responses near threshold for stretch, *J. Physiol.*, 271: 449-472

[30]  Guckenheimer J. and Holmes P. Nonlinear Oscillations, Dynamical Systems, and Bifurcations of Vector Fields. Springer Verlag, Berlin, 1990

[31]  Gutkin B.S. and Ermentrout G.B. (1998) Dynamics of membrane excitability determine inter-spike interval variability: a link between spike generation mechanisms and cortical spike train statistics, *Neural Computation*, 10: 1047-1065

[32]  Guttman R., Lewis S., and Rinzel J. (1980) Control of repetitive firing in squid axon membrane as a model for a neuroneoscillator, *J. Physiol. (Lond)*, 305: 377-395

[33]  Hille B. Ion Channels of Excitable Membranes (3rd Edition), Sinauer Associates, Sunderland, Mass., 2001

[34]  Hodgkin A.L. (1948) The local electric changes associated with repetitive action in a non-medullated axon, *J. Physiol. (London)*, 107: 165-181

[35] Hodgkin A.L. and Huxley A.F. (1952) A quantitative description of membrane current and its application to conduction and excitation in nerve, *J. Physiol.-London*, 117: 500-544

[36] Izhikevich E. (2000) Neural excitability, spiking and bursting, *Int. J. Bif. Chaos*, 10: 1171-1266

[37] Keener J.P. Principles of Applied Mathematics; Transformation and Approximation , 2nd edition. Perseus Books, 1999.

[38] Koch C. Biophysics of Computation: Information Processing in Single Neurons. Oxford University Press, New York, 1999

[39] Llinas R.R. (1988) The intrinsic electrophysiological properties of mammalian neurons: insights into central nervous system function, *Science*, 242:1654-1664

[40] Morris C. and Lecar H. (1981) Voltage oscillations in the barnacle giant muscle fiber, *Biophys. J.*, 35: 193-213

[41] Nykamp D.Q., Tranchina D. (2000) A population density approach that facilitates large-scale modeling of neural networks: analysis and an application to orientation tuning, *J. Comput. Neurosci.*, 8: 19-50

[42] O'Donovan M. (1999) The origin of spontaneous activity in developing networks of the vertebrate nervous system, *Curr. Opin. Neurobiol.*, 9: 94 -104

[43] O'Donovan M. and Chub N. (1997) Population behavior and self-organization in the genesis of spontaneous rhythmic activity by developing spinal networks, *Semin. Cell Dev. Biol.*, 8: 21-28

[44] Omurtag A., Knight B.W., Sirovich L. (2000) On the simulation of large populations of neurons, *J. Comput. Neurosci.*, 8: 51-63

[45] Pinsky P.F. and Rinzel J. (1994) Intrinsic and network rhythmogenesis in a reduced Traub model for CA3 neurons, *J. Comp. Neurosci.*, 1: 39-60

[46] Plant R.E. (1981) Bifurcation and resonance in a model for bursting nerve cells, *J. Math. Bio.*, 11: 15-32

[47] Rall W. and Agmon-Snir H. (1998) Cable theory for dendritic neurons, *In:* Methods in Neuronal Modeling: From Ions to Networks, 2nd edition. (C. Koch and I. Segev, eds.), MIT Press, 27-92

[48] Reyes A. (2001) Influence of dendritic conductances on the input-output properties of neurons, *Annu. Rev. Neurosci.*, 24: 653-675

[49] Rinzel J. (1978) On repetitive activity in nerve, *Federation Proceedings*, 37: 2793-2802

[50] Rinzel J. (1985) Bursting oscillations in an excitable membrane model. *In:* Ordinary and Partial Differential Equations. (Sleeman B.D. and Jarvis R.D., eds.) *Lecture Notes in Mathematics*, Springer, Berlin, 1151: 304-316

[51] Rinzel J. (1987) A formal classification of bursting mechanisms in excitable systems. *In:* Mathematical Topics in Population Biology, Morphogenesis, and Neurosciences (E.Teramoto and M.Yamaguti, eds.),*Lecture Notes in Biomathematics*, Springer, Berlin, 71: 267-281

[52] Rinzel J. (1990) Electrical excitability of cells, theory and experiment: review of the Hodgkin-Huxley foundation and an update. In: Mangel M, ed. *Classics of Theoretical Biology. Bull Math. Bio.*, 52: 5-23

[53] Rinzel J. and Ermentrout B. (1998) Analysis of neural excitability and oscillations, *In:* Methods in Neuronal Modeling: From Ions to Networks, 2nd edition. (C. Koch and I. Segev, eds.), MIT Press, 251-291

[54] Rinzel J. and Keener J.P. (1983) Hopf bifurcation to repetitive activity in nerve, *SIAM J. Appl. Math.*, 43: 907-922

[55] Rinzel J. and Keller J.B. (1973) Traveling wave solutions of a nerve conduction equation, *Biophys. J.*, 13: 1313-1337

[56] Rinzel J. and Lee Y.S. (1987) Dissection of a model for neuronal parabolic bursting, *J. Math. Biol.*, 25: 653-675

[57] Rinzel J. and Miller R.N. (1980) Numerical calculation of stable and unstable periodic solutions to the Hodgkin–Huxley equations, *Math. Biosci.*, 49: 27-59

[58] Rush M.E. and Rinzel J. (1995) The potassium A-current, low firing rates, and rebound excitation in Hodgkin-Huxley models, *Bull. Math. Biology*, 57: 899-929

[59] Stafstrom C.E., Schwindt P.C., and Crill W.E. (1984) Repetitive firing in layer V neurons from cat neocortex in vitro, *J. Neurophysiol.*, 52: 264-277

[60] Segev I. (1992) Single neurone models: oversimple, complex and reduced., *Trends Neurosci.*, 15: 414-421

[61] Segev I. and London M. (2000) Untangling dendrites with quantitative models, *Science*, 290: 744-750

[62] Strogatz S.H. Nonlinear Dynamics and Chaos: With Applications to Physics, Biology, Chemistry, and Engineering. Addison-Wesley, Reading, Mass., 1994

[63] Strumwasser F. (1967) Types of information stored in single neurons. *In:* Invertebrate Nervous Systems: Their Significance for Mammalian Neurophysiology (C.A.G. Wiersma, ed.), The University of Chicago Press, Chicago, 290-319

[64] Tabak J., Rinzel J., and O'Donovan M.J. (2001) The role of activity-dependent network depression in the expression and self-regulation of spontaneous activity in the developing spinal cord, *J. Neurosci.*, 21: 8966 - 8978

[65] Tabak J., Senn W., O'Donovan M.J., and Rinzel J. (2000) Modeling of spontaneous activity in developing spinal cord using activity-dependent depression in an excitatory network, *J. Neurosci.*, 20: 3041 - 3056

[66] Traub R., Wong. R., Miles. R., and Michelson H. (1991) A model of a CA3 hippocampal pyramidal neuron incorporating voltage-clamp data on itrinsic conductances, *J. Neurophysiol.*, 66: 635-649

[67] Tuckwell H.C. Introduction to Theoretical Neurobiology. Cambridge Univ. Press, Cambridge, 1988, vol. 2.

[68] de Vries G. (1998) Multiple bifurcations in a polynomial model of bursting oscillations, *J. Nonlin. Sci.*, 8: 281-316

[69] Wang X.-J. (1993) Ionic basis for intrinsic 40 Hz neuronal oscillations. *Neuroreport* 5: 221-224

[70] Wang X.-J. and Rinzel J. (1995) Oscillatory and bursting properties of neurons. *In:* Handbook of Brain Theory and Neural Networks, edited by M. Arbib, MIT Press, 686-691

[71] Wiggins S. Introduction to Applied Nonlinear Dynamical Systems and Chaos, Springer Verlag, Berlin, 1990

[72] Wilson H.R. and Cowan J.D. (1972) Excitatory and inhibitory interactions in localized populations of model neurons, *Biophys. J.*, 12: 1-24

Course 3

# GEOMETRIC SINGULAR PERTURBATION ANALYSIS OF NEURONAL DYNAMICS

David Terman

*Department of Mathematics, Ohio State University,*
*Colombus, Ohio, USA*

*C.C. Chow, B. Gutkin, D. Hansel, C. Meunier and J. Dalibard, eds.*
*Les Houches, Session LXXX, 2003*
*Methods and Models in Neurophysics*
*Méthodes et Modèles en Neurophysique*

# Contents

## 1. Introduction

Models for neuronal systems often exhibit a rich structure of dynamic behavior. The firing properties of even a single cell can be quite complicated. An individual cell may, for example, fire repetitive action potentials or bursts of action potentials that are separated by silent phases of near quiescent behavior [9, 14]. Examples of population rhythms include synchronized oscillations, in which every cell in the network fires at the same time and clustering, in which the entire population of cells breaks up into subpopulations or blocks; every cell within a single block fires synchronously and different blocks are desynchronized from each other [6, 19]. Of course, much more complicated population rhythms are possible. The activity may, for example, propagate through the network in a wave-like manner, or exhibit chaotic dynamics [16, 24, 26].

A neuronal network's population rhythm results from interactions between three separate components: the intrinsic properties of individual neurons, the synaptic properties of coupling between neurons, and the architecture of coupling (i.e., which neurons communicate with each other). These components typically involve numerous parameters and multiple time scales. The synaptic coupling, for example, can be excitatory or inhibitory, and its possible turn on and turn off rates can vary widely. Neuronal systems may include several different types of cells as well as different types of coupling. An important and typically very challenging problem is to determine the role each component plays in shaping the emergent network behavior.

In this article, I will discuss models for neuronal systems and dynamical systems methods for analyzing these models. The discussion will focus primarily on models which include a small parameter and results in which geometric singular perturbation methods have been used to analyze the network behavior. I will not consider other types of models which are commonly used in the study of neural systems. The integrate and fire model of a single cell is one such example. A review of these types of models can be found in [7, 11].

An outline of the article is as follows. Chapter 2 presents an informal introduction to the geometric theory of dynamical systems. I introduce the notions of phase space, local and global bifurcation theory, stability theory, oscillations, and geometric singular perturbation theory. All of these techniques are very important in the analysis of models for neuronal systems. In Chapter 3, I show how dynamical systems methods can be used to analyze properties of a single neuron model. We consider how a neuron responds to injected current and briefly discuss traveling wave solutions. These correspond to propagating action potentials. I then discuss the dynamics of small networks of neurons. Conditions will be given for when these networks exhibit either synchronous or desynchronous rhythms. This analysis easily generalizes to larger systems as shown in Chapter 5 where we consider excitatory-inhibitory networks. I conclude by discussing a model for activity patterns in the Basal Ganglia, a part of the brain involved in the generation of movements.

## 2. Introduction to dynamical systems

This chapter provides an informal introduction to the dynamical systems approach for studying nonlinear, ordinary differential equations. A more thorough presentation can be found in [21], for example. Dynamical systems has proved to be very useful framework in which to analyze neuronal systems. This approach associates a picture (the phase space) to each differential equation. Solutions, such as a resting state or oscillations, correspond to geometric objects, such as points or curves, in phase space. Since it is usually impossible to derive an explicit formula for the solution of a nonlinear equation, the phase space provides an extremely useful way for understanding qualitative features of solutions. In fact, even when it is possible to write down a solution in closed form, the geometric phase space approach is often a much easier way to analyze an equation.

### 2.1. First order equations

We illustrate the geometric approach with the following simple example. Consider the first order, nonlinear differential equation

$$\frac{dx}{dt} = x - x^3 \equiv f(x). \tag{2.1}$$

Note that it is possible to solve this equation in closed form by separating variables and then integrating. The resulting formula is so complicated, however, that it is difficult to interpret. Suppose, for example, we are given an initial condition, say $x(0) = \pi$, and we are asked to determine the behavior of the solution $x(t)$ as $t \to \infty$. The answer to this question is not at all obvious by considering the solution formula.

The geometric approach provides a simple solution to this problem and is illustrated in Fig. 1. We think of $x(t)$ as the position of a particle moving along the $x$-axis at some time $t$. The differential equation gives us a formula for the velocity $x'(t)$ of the particle; namely, $x'(t) = f(x)$. Hence, if at time $t$, $f(x(t)) > 0$, then the position of the particle must increase, while if $f(x(t)) < 0$, then the position must be decrease.

Now consider the solution that begins at $x(0) = \pi$. Since $f(\pi) = \pi - \pi^3 < 0$, the solution initially decreases, moving to the left. It continues to move to the left and eventually approaches the fixed point at $x = 1$. A *fixed point* is a value of $x$ where $f(x) = 0$.

This sort of analysis allows us to understand the behavior of every solution, no matter what its initial position. The differential equation tells us what the velocity of a particle is at each position $x$. This defines a vector field; each vector points either to the right or to the left depending on whether $f(x)$ is positive or negative (unless $x$ is a fixed point). By following the position of a particle in the direction of the vector field, one can easily determine the behavior of the solution corresponding to that particle.

This analysis carries over for *every* scalar differential equation of the form $x' = f(x)$, no matter how complicated the nonlinear function $f(x)$ is. Solutions can be thought of as particles moving along the real axis depending on the sign of the velocity $f(x)$. Every solution must either approach a fixed point as $t \to \pm\infty$ or become unbounded. It is not hard to realize that a fixed point $x_0$ is stable if $f'(x_0) < 0$ and is unstable if $f'(x_0) > 0$. If $f'(x_0) = 0$, then one must be careful since $x_0$ may be stable or unstable.

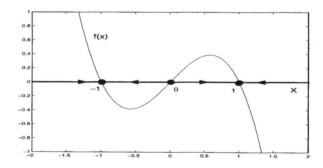

Fig. 1. The phase space for Eq. (2.1).

## 2.2. Bifurcations

Bifurcation theory is concerned with how solutions of a differential equation depend on a parameter. Imagine, for example, that an experimentalist is able to control the level of applied current injected into a neuron. As the level of applied current increases, the neuron may switch its behavior from a resting state to exhibiting sustained oscillations. Here, the level of applied current represents the bifurcation parameter. Bifurcation theory (together with a good model) can explain how the change in dynamics, from a resting state to oscillations, takes place. It can also be used to predict the value of injected current at which the neuron begins to exhibit oscillations. This may be a useful way to test the model.

There are only four major types of so-called local bifurcations and three of them can be explained using one-dimensional equations. We shall illustrate each of these with a simple example. The fourth major type of local bifurcation is the *Hopf bifurcation*. It describes how stable oscillations arise when a fixed point loses its stability. This requires at least a two dimensional system and is discussed later in this chapter.

**Saddle-Node bifurcation**

The following example illustrates the *saddle-node bifurcation*:

$$x' = \lambda + x^2. \tag{2.2}$$

Here, $\lambda$ is a fixed (bifurcation) parameter and may be any real number. We wish to solve this equation for a given value of $\lambda$ and to understand how qualitative features of solutions change as the bifurcation parameter is varied.

Consider the fixed points of (2.2) for different values of the bifurcation parameter. Recall that fixed points are those values of $x$ where the right hand side of (2.2) is zero. If $\lambda < 0$ then (2.2) has two fixed points; these are at $x = \pm\sqrt{-\lambda}$. If $\lambda = 0$ then there is only one fixed point, at $x = 0$, and if $\lambda > 0$ then there are no fixed points of (2.2).

To determine the stability of the fixed points, we let $f_\lambda(x) \equiv \lambda + x^2$ denote the right hand side of (2.2). A fixed point $x_0$ is stable if $f'_\lambda(x_0) < 0$. Here, differentiation is with respect to $x$. Since $f'_\lambda(x) = 2x$, it follows that the fixed point at $-\sqrt{-\lambda}$ is stable and the fixed point at $+\sqrt{-\lambda}$ is unstable.

A very useful way to visualize the bifurcation is shown in Fig 2 (left). This is an example of a *bifurcation diagram*. We plot the fixed points $x = \pm\sqrt{-\lambda}$ as functions of the bifurcation parameter. The upper half of the fixed point curve is drawn with a dashed line since these points correspond to unstable fixed points, and the lower half is drawn with a solid line since these points correspond to stable fixed points. The point $(\lambda, x) = (0, 0)$ is said to be a *bifurcation point*. At a bifurcation point there is a qualitative change in the nature of the fixed point set as the bifurcation parameter varies.

A basic feature of the saddle-node bifurcation is that as the bifurcation parameter changes, two fixed points, one stable and the other unstable, come together and annihilate each other. A closely related example is $x' = -\lambda + x^2$. There are no fixed points for $\lambda < 0$ and two for $\lambda > 0$. Hence, two fixed points are created as $\lambda$ increases through the bifurcation point at $\lambda = 0$. This is also referred to as a saddle-node bifurcation.

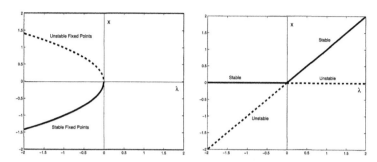

Fig. 2. The saddle-node bifurcation (left) and the transcritical bifurcation (right).

## Transcritical bifurcation

A second type of bifurcation is the *transcritical bifurcation*. Consider the equation

$$x' = \lambda x - x^2. \tag{2.3}$$

Note that $x = 0$ is a fixed point for all values of $\lambda$; moreover, there is a second fixed point at $x = \lambda$.

To determine the stability of the fixed points, we let $f_\lambda(x) \equiv \lambda x - x^2$ denote the right hand side of (2.3). Since $f'_\lambda(x) = \lambda - 2x$, it follows that the fixed point at $x = 0$ is stable if $\lambda < 0$ and is unstable if $\lambda > 0$. The fixed point at $x = \lambda$ is stable if $\lambda > 0$ and is unstable if $\lambda < 0$.

The bifurcation diagram corresponding to this equation is shown in Fig. 2 (right). As before, we plot values of the fixed points versus the bifurcation parameter $\lambda$. Solid curves represent stable fixed points, while dashed curves represent unstable fixed points. Note that there is an *exchange of stability* at the bifurcation point $(\lambda, x) = (0, 0)$ where the two curves cross.

## Pitchfork bifurcation

The third type of bifurcation is the *pitchfork bifurcation*. Consider the equation

$$x' = \lambda x - x^3. \tag{2.4}$$

If $\lambda \leq 0$, then there is one fixed point at $x = 0$. If $\lambda > 0$, then there are three fixed points. One is at $x = 0$ and the other two satisfy $x^2 = \lambda$.

In order to determine the stability of the fixed points, we let $f_\lambda(x) \equiv \lambda x - x^3$. Note that $f_\lambda'(x) = \lambda - 3x^2$. It follows that $x = 0$ is stable for $\lambda < 0$ and unstable for $\lambda > 0$. Moreover, if $\lambda > 0$ then both fixed points $x = \pm\sqrt{\lambda}$ are stable.

The bifurcation diagram corresponding to (2.4) is illustrated in Fig. 3 (left). There are actually two types of pitchfork bifurcations; (2.4) is an example of the *supercritical* case. An example of a *subcritical* pitchfork bifurcation is

$$x' = \lambda x + x^3. \tag{2.5}$$

The bifurcation diagram for this equation is shown in Fig 3 (right). Here, $x_0 = 0$ is a fixed point for all $\lambda$. It is stable for $\lambda < 0$ and unstable for $\lambda > 0$. If $\lambda < 0$, then there are two other fixed points; these are at $x_0 = \pm\sqrt{-\lambda}$. Both of these fixed points are unstable.

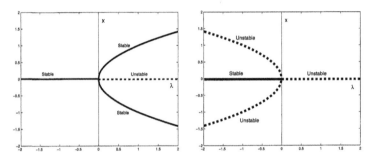

Fig. 3. A supercritical pitchfork bifurcation (left) and a subcritical pitchfork bifurcation (right).

### 2.3. Bistability and hysteresis

Our final example of a scalar ordinary differential equation is:

$$x' = \lambda + 3x - x^3. \tag{2.6}$$

The bifurcation diagram corresponding to (2.6) is shown in Fig. 4. The fixed points lie along the cubic $x^3 - 3x - \lambda = 0$. There are three fixed points for $|\lambda| < 2$ and one fixed point for $|\lambda| > 2$. We note that the upper and lower branches of the cubic correspond to stable fixed points, while the middle branch corresponds to unstable fixed points. Hence, if $|\lambda| < 2$ then there are two stable fixed points and (2.6) is said to be *bistable*.

There are two bifurcation points. These are at $(\lambda, x) = (-2, 1)$ and $(\lambda, x) = (2, -1)$ and both correspond to saddle-node bifurcations.

Suppose we slowly change the parameter $\lambda$, with initially $\lambda = 0$ and $x$ at the stable fixed point $-\sqrt{3}$. As $\lambda$ increases, $(\lambda, x)$ remains close to the lower branch of stable fixed points. (See Fig. 4.) This continues until $\lambda = 2$ when $(\lambda, x)$ crosses the saddle-node bifurcation point at $(\lambda, x) = (2, -1)$. The solution then approaches the stable fixed point along the upper branch. We now decrease $\lambda$ to its initial value $\lambda = 0$. The solution remains on the upper branch. In particular, $x = \sqrt{3}$ when $\lambda = 0$. Note that while $\lambda$ has returned to its initial value, the state variable $x$ has not. This is an example of what is often called a *hysteresis phenomenon*.

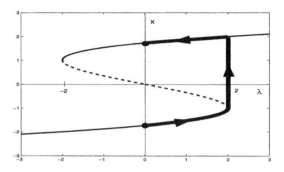

Fig. 4. Example of hysteresis.

### 2.4. The phase plane

We have demonstrated that solutions of first order differential equations can be viewed as particles flowing in a one dimensional phase space. Remarkably, there is a similar geometric interpretation for *every* ordinary differential equation. This is because every ode can be written in the form $x' = f(x)$, $x \in R^n$ for some $n \geq 1$. One can then view solutions as particles flowing in $R^n$ according to the vector field given by $f(x)$. Trajectories in higher dimensions can be very complicated, much more complicated than the one dimensional examples considered above. In one dimension, solutions (other than fixed points) must always flow monotonically to the left or to the right. In higher dimensions, there is a much wider range of possible dynamic behaviors.

A two dimensional system is one of the form

$$
\begin{aligned}
x' &= f(x, y) \\
y' &= g(x, y).
\end{aligned}
\tag{2.7}
$$

Here, $f$ and $g$ are given (smooth) functions. The phase space for this system is simply the $x-y$ plane; this is usually referred to as the *phase plane*. If $(x(t), y(t))$ is a solution of (2.7), then at each time $t_0$, $(x(t_0), y(t_0))$ defines a point in the

phase plane. The point changes with time, so the entire solution $(x(t), y(t))$
traces out a curve, or trajectory, in the phase plane.

Of course, not every arbitrarily drawn curve in the phase plane corresponds
to a solution of (2.7). What is special about solution curves is that the velocity
vector at each point along the curve is given by the right hand side of (2.7).
That is, the velocity vector of the solution curve $(x(t), y(t))$ at a point $(x_0, y_0)$ is
given by $(x'(t), y'(t)) = (f(x_0, y_0), g(x_0, y_0))$. This geometric property – that
the vector $(f(x, y), g(x, y))$ always points in the direction that the solution is
flowing – completely characterizes the solution curves.

For an example, consider the system:

$$x' = y - x^2 + x$$
$$y' = x - y. \qquad\qquad\qquad (2.8)$$

The phase plane associated with this system is shown in Fig. 5. The vectors
correspond to the vector field defined by the right hand side of (2.8). The dashed
curves are the nullclines; these are where either $x' = 0$ or $y' = 0$; that is, the
vector field is either vertical or horizontal. Note that the nullclines intersect at
the fixed points; these are at $(0, 0)$ and $(2, 2)$. One can determine the stability of
the fixed points using the method of linearization. This method is discussed in
any book on ordinary differential equations.

It is now possible to predict the behavior of the solution to (2.8) with some
prescribed initial condition $(x_0, y_0)$. Suppose, for example, that $(x_0, y_0)$ lies in
the intersection of the first quadrant with region (I). Since the vector field points
towards the fourth quadrant, the solution initially flows with $x(t)$ increasing and
$y(t)$ decreasing. There are now three possibilities. The solution must either; (A)
enter region II, (B) enter region V, or (C) remain in region I for all $t > 0$. It is
not hard to see that in cases A or B, the solution must remain in region II or V,
respectively. In each of these three cases, the solution must then approach the
fixed point at $(2, 2)$ as $t \to \infty$.

## 2.5. Oscillations

We say that a solution $(x(t), y(t))$ is *periodic* if $(x(0), y(0)) = (x(T), y(T))$ for
some $T > 0$. A periodic solution corresponds to a closed curve or *limit cycle*
in the phase plane. Periodic solutions can be either stable or unstable. Roughly
speaking, a periodic solution is stable if solutions that begin close to the limit
cycle remain close to the limit cycle for all $t > 0$. We do not give a precise
definition here.

It is usually much more difficult to locate periodic solutions than it is to locate
fixed points. Recall that every ordinary differential equation can be written as
$x' = f(x), \ x \in R^n$ for some $n \geq 1$. A fixed point $x_0$ satisfies the equation

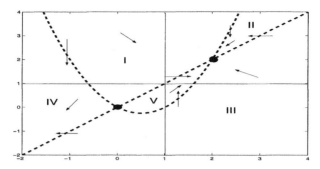

Fig. 5. The phase plane for Eq. (2.8).

$f(x_0) = 0$ and this last equation can usually be solved with straightforward numerical methods. We also note that a fixed point is a local object – it is simply one point in phase space. Oscillations or limit cycles are global objects; they correspond to an entire curve in phase space that retraces itself. This curve may be quite complicated.

One method for demonstrating the existence of a limit cycle for a two dimensional flow is the Poincare-Bendixson theorem [21]. This theorem does not apply for higher dimensional flows, so we shall not discuss it further. Three more general methods for locating limit cycles are the Hopf bifurcation theorem, global bifurcation theory and singular perturbation theory. These methods are discussed in the following sections.

### 2.6. Local bifurcations

Recall that bifurcation theory is concerned with differential equations that depend on a parameter. We saw that one dimensional flows can exhibit saddle-node, transcritical and pitchfork bifurcations. These are all examples of local bifurcations; they describe how the structure of the flow changes near a fixed point as the bifurcation parameter changes. Each of these local bifurcations can arise in higher dimensional flows. In fact, there is only one major new type of bifurcation in dimensions greater than one. This is the so-called Hopf bifurcation. We begin this section by giving a necessary condition for the existence of a local bifurcation point. We then describe the Hopf bifurcation.

Consider a system of the form:

$$u' = F(u, \lambda). \tag{2.9}$$

Suppose that $u_0$ is a fixed point of (2.9) for some value, say $\lambda_0$, of the bifurcation parameter. This simply means that $F(u_0, \lambda_0) = 0$. We will need to consider the Jacobian matrix $J$ of $F$ at $u_0$. Then $u_0$ is stable, for $\lambda = \lambda_0$, if all of the eigenvalues of $J$ have negative real parts and is unstable if at least one eigenvalue has positive real part. We say that $u_0$ is a *hyperbolic fixed point* if $J$ does not have any eigenvalues on the imaginary axis. An important result is that if $u_0$ is hyperbolic, then $(u_0, \lambda_0)$ cannot be a bifurcation point. That is, a necessary condition for $(u_0, \lambda_0)$ to be a bifurcation point is that the Jacobian matrix has purely imaginary eigenvalues. Of course, the converse statement may not be true.

We now describe the Hopf bifurcation using an example. Consider the system

$$
\begin{aligned}
x' &= 3x - x^3 - y \\
y' &= x - \lambda.
\end{aligned}
\tag{2.10}
$$

Note that there is only one fixed point for each value of the bifurcation parameter $\lambda$. This fixed point is at $(x, y) = (\lambda, \ 3\lambda - \lambda^3)$. It lies along the left or right branch of the cubic $x$-nullcline if $|\lambda| > 1$ and lies along the middle branch of this cubic if $|\lambda| < 1$.

We linearize (2.10) about the fixed point and compute the corresponding eigenvalues to find that the fixed point is stable for $|\lambda| > 1$ and unstable for $|\lambda| < 1$. When $|\lambda| = 1$, the fixed points are at the local maximum and local minimum of the cubic; in this case, the eigenvalues are $\pm i$. In particular, the fixed points are not hyperbolic and a bifurcation is possible.

As $\lambda$ increases past $-1$, the fixed point loses its stability. The eigenvalues are complex, so trajectories spiral towards the fixed point for $\lambda < -1$ and trajectories spiral away from the fixed point for $\lambda > -1$. (Here we are assuming that $|\lambda + 1|$ is not too large.) One can show (using a computer) that these unstable trajectories must approach a stable limit cycle. The amplitude of the limit cycle approaches zero as $\lambda \to -1$.

This is an example of a Hopf bifurcation. As the bifurcation parameter varies, a fixed point loses its stability as its corresponding eigenvalues cross the imaginary axis. The Hopf bifurcation Theorem gives precise conditions for when this guarantees the existence of a branch of periodic orbits. Note that there is a second Hopf bifurcation at $(\lambda, x) = (1, 1)$.

Fig. 6 shows a bifurcation diagram corresponding to (2.10). Here we plot the maximum value of the $x$-variable along a solution as a function of the bifurcation parameter $\lambda$. The line $x = \lambda$ corresponds to fixed points. This is drawn as a bold, solid line for $|\lambda| > 1$ since these points correspond to stable fixed points, and as a dashed line for $|\lambda| < 1$ since these points correspond to unstable fixed points. There is a curve corresponding to limit cycles that connects the bifurcation points at $(\lambda, x) = (-1, -1)$ and $(1, 1)$.

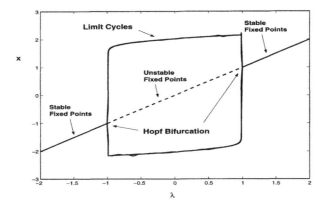

Fig. 6. The bifurcation diagram for Eq. (2.10). There are two Hopf bifurcation points.

A Hopf bifurcation may be subcritical or supercritical. In the supercritical case, the limit cycles are stable and they exist for the same parameter values as the unstable fixed points (near the bifurcation point). In the subcritical case, the limit cycles are unstable and exist for those same parameter values as the stable fixed points.

## 2.7. Global bifurcations

Hopf bifurcations are local phenomena; they describe the creation of limit cycles near a fixed point. As the bifurcation parameter approaches some critical value, the limit cycle approaches the fixed point and the amplitude of the limit cycle approaches zero. There are also global mechanisms by which oscillations can be created or destroyed. It is possible, for example, that the amplitude of oscillations remain bounded away from zero, but the frequency of oscillations approaches zero. This will be referred to as a *homoclinic* bifurcation (for reasons described below). It is also possible for two limit cycles, one stable and the other unstable, to approach and annihilate each other at some critical parameter value. This is referred to as a saddle-node bifurcation of limit cycles and resembles the saddle-node bifurcation of fixed points in which two fixed points come together and annihilate each other.

Fig. 7 illustrates a homoclinic bifurcation. For all values of the bifurcation parameter $\lambda$ there are three fixed points; these are labeled as $l, m$, and $u$, and they are stable, a saddle and unstable, respectively. When $\lambda = \lambda_0$ (shown in the middle panel), there is a homoclinic orbit labeled as $\gamma_h(t)$. This orbit satisfies $\lim_{t \to \pm\infty} \gamma_h(t) = m$ and lies in both the stable and unstable manifolds of the fixed point $m$. If $\gamma < \gamma_0$ (shown in the left panel) then there is no periodic orbit, while if $\gamma > \gamma_0$, then there is a stable limit cycle, labeled as $p(t)$. Note that if $\gamma < \gamma_0$, then the stable manifold of $m$ lies inside of the unstable manifold, while the opposite holds if $\gamma > \gamma_0$.

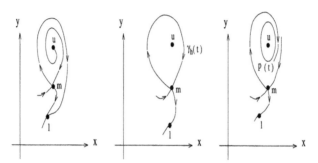

Fig. 7. A homoclinic bifurcation.

Note that solutions move very slowly as they pass near an unstable fixed point. It follows that the period of the periodic solution, for $\lambda > \lambda_0$, must become arbitrarily large as $\lambda$ approaches $\lambda_0$. We also note that the nature of the three fixed points do not change as $\lambda$ is varied. Hence, there is no local bifurcation. The homoclinic orbit is a global object, and the limit cycle disappears via a global bifurcation.

### 2.8. Geometric singular perturbation theory

Models for neuronal systems often involve variables that evolve on very different time scales. The existence of different time scales naturally leads to models that contain small parameters. Geometric singular perturbation theory provides a powerful technique for analyzing these models. The theory gives a systematic way to reduce systems with small parameters to lower dimensional reduced systems that are more easily analyzed. Here we illustrate how this method works with a simple example. The method will be used extensively in the next chapter to study more complicated models arising from neuronal systems.

Consider a general two-dimensional system of the form

$$
\begin{aligned}
v' &= f(v, w) \\
w' &= \epsilon g(v, w).
\end{aligned}
\tag{2.11}
$$

Here, $\epsilon > 0$ is the small, singular perturbation parameter. We assume that the $v$−nullcline is a cubic-shaped curve and the $w$−nullcline is a monotone increas-

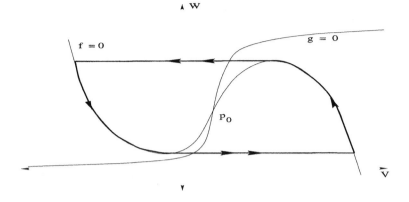

Fig. 8. Nullclines and singular periodic orbit for an oscillatory relaxation oscillator.

ing curve that intersects the cubic at a single fixed point, denoted by $p_0$, that lies along the middle branch of the cubic nullcline. We also need to assume that $v' > 0$ $(< 0)$ below (above) the cubic $v$—nullcline and $w' > 0$ $(< 0)$ below (above) the $w$—nullcline.

One can prove, using the Poincare-Bendixson theorem, that (2.11) has a limit cycle for all $\epsilon$ sufficiently small. Moreover, the limit cycle approaches a singular limit cycle as shown in Fig. 8 as $\epsilon \to 0$. The singular limit cycle consists of four pieces. One of these pieces lies along the left branch of the cubic nullcline. We shall see in the next chapter that this corresponds to the silent phase of an action potential. Another piece of the singular solution lies along the right branch of the cubic nullcline; this corresponds to the active phase of the action potential. The other two pieces are horizontal curves in the phase plane and they connect the left and right branches. The "jump-up" to the active phase occurs at the left knee of the cubic and the "jump-down" occurs at the right knee of the cubic.

We refer to (2.11) as a *singular* perturbation problem because the structure of solutions of (2.11) with $\epsilon > 0$ is very different than the structure of solutions of (2.11) with $\epsilon = 0$. If we set $\epsilon = 0$, then (2.11) reduces to

$$
\begin{aligned}
v' &= f(v, w) \\
w' &= 0.
\end{aligned}
\tag{2.12}
$$

Note that $w$ is constant along every solution of (2.12); that is, every trajectory is horizontal. The fixed point set is the entire cubic-shaped curve $\{f = 0\}$. This is very different from (2.11) with $\epsilon > 0$ in which there is only one fixed point and $w$ is not constant along all other solutions.

The reduced system (2.12) does, however, give a good approximation of solutions away from the cubic $v$−nullcline. In particular, it determines the evolution of the jump-up and jump-down portions of the singular solution. In order to determine the behavior of solutions near the cubic nullcline – that is, during the silent and active phases – we introduce the slow time scale $\tau = \epsilon t$ and then set $\epsilon = 0$ in the resulting system. The leads to the system

$$
\begin{aligned}
0 &= f(v, w) \\
\dot{w} &= g(v, w).
\end{aligned}
\tag{2.13}
$$

The first equation in (2.13) implies that the solution of (2.13) lies along the cubic nullcline. The second equation in (2.13) determines the evolution of the solution along this nullcline. If we the write the left and right branches of the cubic nullcline as $v = H_L(w)$ and $v = H_R(w)$, respectively, then the second equation in (2.13) becomes

$$
\dot{w} = g(H_\alpha(w), w) \equiv G_\alpha(w)
\tag{2.14}
$$

where $\alpha = L$ or $R$.

Note that each piece of the singular solution is determined by a single, scalar differential equation. The silent and active phases correspond to solutions of (2.14). This equation will be referred to as the *slow equation*. The jump-up and jump-down correspond to solutions of the first equation in (2.12); this is referred to as the *fast equation*.

The analysis described here will be used to study more complicated, higher dimensional systems that arise in models for neuronal systems. Using the existence of small parameters, we will construct singular periodic solutions. Each piece of the singular solution will satisfy a reduced system of equations. We note that the order of the slow equations will be equal to the number of slow variables. In particular, if a given model has one variable that evolves at a much slower time-scale than the other variables, then the order of the slow equations will be just one. Hence, we will reduce the analysis of a complicated, high dimensional system to a single, scalar equation.

## 3. Properties of a single neuron

In this chapter we demonstrate how dynamical systems methods can be used to analyze a simple model for a single neuron. The simple model takes the form of a relaxation oscillator. Using phase-plane methods we will describe such concepts as a cell's active, silent and refractory phases and how the cell responds to inputs. This discussion will be very useful when we consider larger networks in the following sections.

## 3.1. Reduced models

The dynamics of even one single neuron can be quite complicated. We are primarily interested in developing techniques to study networks consisting of possibly a large number of coupled neurons. Clearly the analysis of networks may be extremely challenging if each single element exhibits complicated dynamics. For this reason, we often consider simpler, reduced models for single neurons. The insights we gain from analyzing the reduced models are often extremely useful in studying the behavior of more complicated biophysical models.

An example of a reduced model are the Morris-Lecar equations [12]:

$$
\begin{aligned}
v' &= -g_L(v - v_L) - g_K w(v - v_K) - g_{Ca} m_\infty(v)(v - v_{Ca}) + I \\
w' &= \epsilon(w_\infty(v) - w)\cosh((v + .1)/.3)
\end{aligned}
\tag{3.1}
$$

The parameters and nonlinear functions in (3.1) are given by $v_L = -.1$, $g_L = .5$, $g_K = 2$, $v_K = -.7$, $g_{Ca} = 1$, $v_{Ca} = 1$, $\epsilon = .1$, $m_\infty(v) = .5(1+\tanh((v - .01)/.145))$, $w_\infty(v) = .5(1 + \tanh((v + .1)/.15))$.

These equations can be viewed as a simple model for a neuron in which there are potassium and calcium currents along with the leak current. There is also an applied current represented by the constant $I$. Here, $v$ represents the membrane potential, rescaled between $-1$ and $1$, and $w$ is activation of the potassium current. Note that the activation of the calcium current is instantaneous. The small, positive parameter $\epsilon$ is introduced to emphasize that $w$ evolves on a slow time scale. We treat $\epsilon$ as a singular perturbation parameter in the analysis.

Note that this simple model has no spatial dependence; that is, we are viewing the neuron as a single point. This will certainly simplify our analysis of time-dependent oscillatory behavior. Later we add spatial dependence to the model and consider the generation of propagating wave activity.

If $I = 0$, then every solution approaches a stable fixed point. This corresponds to a neuron in its resting state. However, if $I = .4$, then solutions of (3.1) quickly approach a stable limit cycle. The periodic solution alternates between an active phase and a silent phase of near resting behavior. Moreover, there are sharp transitions between the silent and active phases.

These solutions can be easily analyzed using the phase space methods described in Section 2.8. The $v$-nullcline is a cubic-shaped curve, while the $w$-nullcline is a monotone increasing function that intersects the $v$-nullcline at precisely one point; hence, there exists precisely one fixed point of (3.1), denoted by $p_0$. If $I = 0$, then $p_0$ lies along the left branch of the cubic $v$-nullcline and one can easily show that $p_0$ is asymptotically stable. If, on the other hand, $I = .4$, then $p_0$ lies along the middle branch of the cubic nullcline and $p_0$ is unstable.

There must then exist a stable limit cycle for all $\epsilon$ sufficiently small; moreover, the limit cycle approaches a singular limit cycle as shown in Fig. 8 as $\epsilon \to 0$.

The phase plane analysis does not depend on the precise details of the nonlinear functions and other parameters in (3.1). We will consider more general two dimensional systems of the form

$$
\begin{aligned}
v' &= f(v, w) + I \\
w' &= \epsilon g(v, w)
\end{aligned}
\tag{3.2}
$$

where the $v$-nullcline and the $w$-nullcline satisfy the assumptions described in Section 2.8.

The reduced two-dimensional models given by (3.1), or more generally (3.2), exhibit many properties of real neurons. For example, (3.2) exhibits a refractory period: immediately after an action potential it is difficult to generate another one. This is because when the trajectory in phase space corresponding to the action potential returns to the silent phase, it lies along the left branch of the cubic nullcline with an elevated value of the recovery variable $n$. Here, the trajectory is further away from the threshold, corresponding to the middle branch.

Note also that when there is no applied current (that is, $I = 0$), (3.2) exhibits a stable fixed point; this corresponds to the resting state of a neuron. If $I$ is sufficiently large, then (3.2) exhibits sustained oscillations. The simple model also displays *excitability*. That is, a small stimulus will not generate an action potential. In this case the solution returns quickly to rest. In order to produce an action potential, the initial stimulus must be larger than some threshold. Note that the threshold corresponds to the position of the middle branch of the cubic nullcline.

### 3.2. Response to injected current

Here we describe how the geometric singular perturbation approach can be used to understand the response of a neuron to injected current. Consider the system

$$
\begin{aligned}
v' &= f(v, w) + I(t) \\
w' &= \epsilon g(v, w)
\end{aligned}
\tag{3.3}
$$

where $f$ and $g$ are as in (3.2) and $I(t)$ represents the injected current. We assume that when $I(t) = 0$ the system is excitable; that is, the $v$- and $w$-nullclines intersect along the left branch of the cubic. This is a globally stable fixed point and the model neuron is in its resting state. We further assume that there exists $I_0$ and $T_{on} < T_{off}$ such that

$$
I(t) = I_0 \text{ if } T_{on} < t < T_{off} \qquad \text{and} \qquad I(t) = 0 \text{ otherwise.}
$$

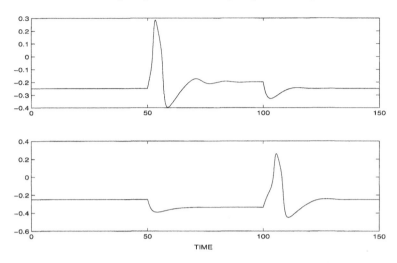

Fig. 9. Response of a model neuron to applied current. Current is applied at time $t = 50$ and turned off at $t = 100$. In the top figure, the current is depolarizing ($I_0 = .1$), while in the bottom figure the current is hyperpolarizing ($I_0 = -.1$) and the neuron exhibits post-inhibitory rebound.

We will consider two cases: either $I_0 > 0$, in which case the injected current is said to be *depolarizing*, or $I_0 < 0$ and the injected current is *hyperpolarizing*. Fig. 8 illustrates the neuron's response when (top) $I_0 = .1$ and (bottom) $I_0 = -.1$. In the depolarizing case, the neuron fires an action potential immediately after the injected current is turned on. The cell then returns to rest. In the hyperpolarizing case, the neuron's membrane potential approaches a more negative steady state until the current is turned off, at which time the neuron fires a single action potential. This last response is often called *post-inhibitory rebound* [5].

The geometric approach is very useful in understanding these responses. As before, we construct singular solutions in which $\epsilon$ is formally set equal to zero. See Fig. 10. The singular solutions lie on the left or right branch of some cubic-shaped nullcline during the silent and active phases. The cubics depend on the values of $I(t)$. We denote the cubic corresponding to $I = 0$ as $C$ and the cubic corresponding to $I_0$ as $C_0$. Note that if $I_0 > 0$, then $C_0$ lies 'above' $C$, while if $I_0 < 0$, then $C_0$ lies 'below' $C$. This is because of our assumption that $f < 0$ ($> 0$) above (below) the $v$-nullcline.

Consider the depolarizing case $I_0 > 0$. This is illustrated in Fig. 10 (left). For $t < T_{on}$, the solution lies at the fixed point $p_0$ along the left branch of $C$. When $t = T_{on}$, $I(t)$ jumps to $I_0$ and the cell's cubic switches to $C_0$. If the left knee of $C_0$ lies above $p_0$ then the cell jumps up to the right branch of $C_0$; this

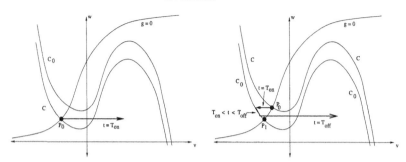

Fig. 10. Phase space representation of the response of a model neuron to applied current. Current is applied at time $t = T_{on}$ and turned off at $t = T_{off}$. (Left) Depolarizing current. The cell jumps up as soon as the current is turned on. (Right) Hyperplorizing current. The cell jumps to the left branch of $C_0$ when the current is turned on and jumps up to the active phase due to post-inhibitory rebound when the current is turned off.

corresponds to the firing of an action potential. If the $w$-nullcline intersects $C_0$ along its left branch, then the cell will approach the stable fixed point along the left branch of $C_0$ until the input is turned off. It is possible that the $w$-nullcline intersects $C_0$ along its middle branch. If this is the case then the cell oscillates, continuing to fire action potentials, until $t = T_{off}$ when the input is turned off. Note that in order for the cell to fire an action potential, the injected current must be sufficiently strong. $I_0$ must be large enough so that the $p_0$ lies below the left knee of $C_0$.

We next consider the hyperpolarizing case $I_0 < 0$, shown in Fig. 10 (right). Then $C_0$ lies below $C$ and the $w$-nullcline intersects $C_0$ at a point denoted by $p_1$. When $t = T_{on}$, the solution jumps to the left branch of $C_0$ and then evolves along this branch approaching $p_1$ for $T_{on} < t < T_{off}$. When $t = T_{off}$, $I$ switches back to 0 and the cell now seeks the left or right branch of $C$. If, at this time, the cell lies below the left knee of $C$, then the cell will jump up to the active phase giving rise to post-inhibitory rebound. In order to have post-inhibitory rebound, the hyperpolarizing input must be sufficiently large and last sufficiently long. $I_0$ must be sufficiently negative so that $p_1$ lies below the left knee of $C$. Moreover, $T_{off} - T_{on}$ must be sufficiently large so that the cell has enough time to evolve along the left branch of $C_0$ so that it lies below the left knee of $C$ when the input is turned off.

### 3.3. Traveling wave solutions

We have so far considered a model for neurons that ignores the spatial dependence. This has allowed us to study how temporal oscillatory behavior arises. One of the most important features of neurons is the propagating nerve impulse and this clearly requires consideration of spatial dynamics. The nerve impulse corresponds to a traveling wave solution and there has been extensive research on the mathematical mechanisms responsible for both the existence and stability of these types of solutions. Here we briefly illustrate how one constructs a traveling wave solution in a simple model, the FitzHugh-Nagumo equations. References related to this important topic are given later.

The FitzHugh-Nagumo equations can be written as:

$$
\begin{aligned}
v_t &= v_{xx} + f(v) - w \\
w_t &= \epsilon(v - \gamma w).
\end{aligned}
\tag{3.4}
$$

Here, $(v, w)$ are functions of $(x, t)$, $x \in R$ and $t \geq 0$. Moreover, $f(v) = v(1 - v)(v - a)$, $0 < a < 1/2$, $\epsilon$ is a small singular perturbation parameter, and $\gamma$ is a positive constant chosen so that the curves $w = f(v)$ and $v = \gamma w$ intersect only at the origin.

A traveling wave solution of (3.4) is a solution of the form $(v(x, t), w(x, t)) = (V(z), W(z)), z = x + ct$; that is, a traveling wave solution corresponds to a solution that propagates with constant shape and velocity. The velocity is $c$ as in not known a priori. We also assume that the traveling wave solution satisfies the boundary conditions $\lim_{z \to \pm\infty}(V(z), W(z)) = (0, 0)$.

Note that a traveling wave solution corresponds to a solution of the first order system of ordinary differential equations

$$
\begin{aligned}
V' &= Y \\
Y' &= cY - f(V) + W \\
W' &= \frac{\epsilon}{c}(V - \gamma W)
\end{aligned}
\tag{3.5}
$$

together with the boundary conditions

$$
\lim_{z \to \pm\infty}(V(z), Y(z), W(z)) = (0, 0, 0)
\tag{3.6}
$$

Hence, a traveling wave solution corresponds to a homoclinic orbit of a first order system. This homoclinic orbit will exist only for special values of the velocity parameter $c$.

One can use geometric singular perturbation methods, as described in Section 2.8, to construct a singular homoclinic orbit in which $\epsilon$ is formally set equal to zero. One needs to then rigorously prove that this singular solution perturbs to an actual homoclinic orbit that lies near the singular orbit for $\epsilon$ sufficiently small.

The singular orbit is constructed as follows. As before, the singular orbit consists of four pieces, as shown in Fig. 11. Two of these pieces correspond to the silent and active phases and the other two pieces correspond to the jump-up and jump-down between these phases. As before, we consider both fast and slow time scales.

The jump-up and jump-down pieces correspond to solutions of the fast equations. These are obtained by simply setting $\epsilon = 0$ in (3.5). The resulting equations are:

$$
\begin{aligned}
V' &= Y \\
Y' &= cY - f(V) + W \\
W' &= 0
\end{aligned}
\tag{3.7}
$$

Note that $W$ must be constant along this solution. For the jump-up (or front), we set $W \equiv 0$ and look for a solution of the first two equations of (3.4) that satisfy

$$
\lim_{z \to -\infty}(V, Y) = (0, 0) \quad \lim_{z \to +\infty}(V, Y) = (1, 0)
\tag{3.8}
$$

It is well known that there exists a unique solution for a unique value of the parameter $c$. We denote this parameter as $c_0$. This is the velocity of the wave in the limit $\epsilon \to 0$.

For the jump-down (or back) we set $W \equiv W_0$, where $W_0$ is chosen so that if $c = c_0$ then there exists a solution of the first two equations in (3.7) such that $\lim_{z \to -\infty}(V, Y, W_0)$ lies along the right branch of the cubic $W = f(V)$ and $\lim_{z \to +\infty}(V, Y, W_0)$ lies along the left branch of this cubic. This is shown in Fig. 11. We note that $W_0$ is indeed uniquely determined.

We now consider the pieces of the singular traveling wave solution corresponding to the silent and active phases. We introduce the slow time scale $\eta = \epsilon z$ and then set $\epsilon = 0$ to obtain the slow equations

$$
\begin{aligned}
Y &= 0 \\
W &= f(V) \\
\dot{W} &= \frac{1}{c_0}(V - \gamma W)
\end{aligned}
\tag{3.9}
$$

Here $\dot{W}$ corresponds to differentiation with respect to $\eta$. These equations demonstrate that during the silent and active phases, the singular solution lies along the cubic curve defined by $W = f(V)$, $Y = 0$. The complete singular homoclinic orbit is shown in Fig. 11.

**Remark 3.1** References to rigorous studies of the existence and stability of traveling wave solutions can be found in [10].

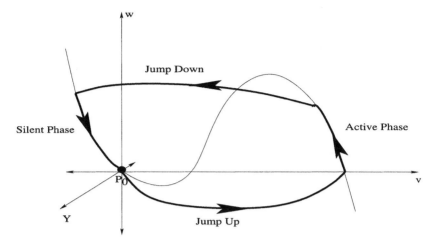

Fig. 11. The singular homoclinic orbit corresponding to a traveling wave solution.

## 4. Two mutually coupled cells

### 4.1. Introduction

In this section, we consider a network consisting simply of two mutually coupled cells. By considering such a simple system, we are able to describe how we model networks of oscillators, the types of behavior that can arise in such systems and the mathematical techniques we use for the analysis of the behavior. For this discussion, we assume that each cell, without any coupling, is modeled as the relaxation oscillator (3.2). In the next section, we describe how we model the two mutually coupled cells. The form of coupling used is referred to as *synaptic coupling* and is meant to correspond to a simple model for chemical synapses. There are many different types of synaptic coupling. For example, it may be excitatory or inhibitory and it may exhibit either fast or slow dynamics. We are particularly interested in how the nature of the synaptic coupling affects the emergent population rhythm. A natural question is whether excitatory or inhibitory coupling leads to either synchronous or desynchronous rhythms. There are four possible combinations and all four may, in fact, be stably realized, depending on the details of the intrinsic and synaptic properties of the cells. Here we discuss conditions for when excitatory coupling leads to synchronous rhythms and inhibitory coupling leads to antiphase behavior.

## 4.2. Synaptic coupling

We model a pair of mutually coupled neurons by the following system of differential equations:

$$
\begin{aligned}
v_1' &= f(v_1, w_1) - s_2 g_{syn}(v_1 - v_{syn}) \\
w_1' &= \epsilon g(v_1, w_1) \\
v_2' &= f(v_2, w_2) - s_1 g_{syn}(v_2 - v_{syn}) \\
w_2' &= \epsilon g(v_2, w_2)
\end{aligned}
\tag{4.1}
$$

Here $(v_1, w_1)$ and $(v_2, w_2)$ correspond to the two cells and the coupling term $s_j g_{syn}(v_i - v_{syn})$ can be viewed as an additional current which may change a cell's membrane potential $v_i$. The parameter $g_{syn} > 0$ corresponds to the maximal conductance of the synapse and is positive, while the reversal potential $v_{syn}$ determines whether the synapse is excitatory or inhibitory. If $v < v_{syn}$ along each bounded singular solution, then the synapse is excitatory, while if $v > v_{syn}$ along each bounded singular solution, then the synapse is inhibitory.

The terms $s_i$, $i = 1, 2$, in (4.1) encode how the postsynaptic conductance depends on the presynaptic potentials $v_i$. There are several possible choices for the $s_i$. The simplest choice is to assume that $s_i = H(v_i - \theta_{syn})$, where $H$ is the Heaviside step function and $\theta_{syn}$ is a threshold above which one cell can influence the other. Note, for example, that if $v_1 < \theta_{syn}$, then $s_1 = H(v_1 - \theta_{syn}) = 0$, so cell 1 has no influence on cell 2. If, on the other hand, $v_1 > \theta_{syn}$, then $s_1 = 1$ and cell 2 is affected by cell 1.

Another choice for the $s_i$ is to assume that they satisfy a first order equation of the form

$$
s_i' = \alpha(1 - s_i) H(v_i - \theta_{syn}) - \beta s_i
\tag{4.2}
$$

where $\alpha$ and $\beta$ are positive constants and $H$ and $\theta_{syn}$ are as before. Note that $\alpha$ and $\beta$ are related to the rates at which the synapses turn on or turn off. For *fast synapses*, we assume that both of these constants are $O(1)$ with respect to $\epsilon$. For a *slow synapse*, we assume that $\alpha = O(1)$ and $\beta = O(\epsilon)$; hence, a slow synapse activates on the fast time scale but turns off on the slow time scale.

The synapses considered so far are referred to as *direct* synapses since they are activated as soon as a membrane potential crosses the threshold $\theta_{syn}$. To more fully represent the range of synapse dynamics observed biologically, it is also necessary to consider more complicated connections. These are referred to as *indirect synapses*, and they are modeled by introducing new dependent variables $x_1$ and $x_2$. Each $(x_i, s_i)$ satisfies the equations

$$
\begin{aligned}
x_i' &= \epsilon \alpha_x (1 - x_i) H(v_i - \theta_v) - \epsilon \beta_x x_i \\
s_i' &= \alpha (1 - s_i) H(x_i - \theta_x) - \beta s_i
\end{aligned}
\tag{4.1}
$$

The constants $\alpha_x$ and $\beta_x$ are assumed to be independent of $\epsilon$. The effect of the indirect synapses is to introduce a delay from the time one oscillator jumps up until the time the other oscillator feels the synaptic input.

### 4.3. Geometric approach

All of the networks in this paper are analyzed by treating $\epsilon$ as a small, singular perturbation parameter. As in the previous section, the first step in the analysis is to identify the fast and slow variables. We then dissect the full system of equations into fast and slow subsystems. The fast subsystem is obtained by simply setting $\epsilon = 0$ in the original equations. This leads to a reduced set of equations for the fast variables with each of the slow variables held constant. The slow subsystems are obtained by first introducing the slow time scale $\tau = \epsilon t$ and then setting $\epsilon = 0$ in the resulting equations. This leads to a reduced system of equations for just the slow variables, after solving for each fast variable in terms of the slow ones. The slow subsystems determine the evolution of the slow variables while the cells are in either the active or the silent phase. During this time, each cell lies on either the left or the right branch of some "cubic" nullcline determined by the total synaptic input which the cell receives. This continues until one of the cells reaches the left or right "knee" of its corresponding cubic. Upon reaching a knee, the cell may either jump up from the silent to the active phase or jump down from the active to the silent phase. The jumping up or down process is governed by the fast equations.

For a concrete example, consider two mutually coupled cells with fast synapses. The dependent variables $(v_i, w_i, s_i)$, $i = 1, 2$, then satisfy (4.1) and (4.2). The slow equations are

$$
\begin{aligned}
0 &= f(v_i, w_i) - s_j g_{syn}(v_i - v_{syn}) \\
\dot{w}_i &= g(v_i, w_i) \\
0 &= \alpha (1 - s_i) H(v_i - \theta_{syn}) - \beta s_i
\end{aligned}
\tag{4.1}
$$

where differentiation is with respect to $\tau$ and $i \neq j$. The first equation in (4.1) states that $(v_i, w_i)$ lies on a curve determined by $s_j$. The third equation states that if cell $i$ is silent ($v_i < \theta_{syn}$), then $s_i = 0$, while if cell $i$ is active, then $s_i = \frac{\alpha}{\alpha + \beta} \equiv s_A$. We demonstrate that it is possible to reduce (4.1) to a single equation for each of the slow variables $w_i$. Before doing this, it will be convenient to introduce some notation.

Let $\Phi(v, w, s) \equiv f(v, w) - g_{syn}s(v - v_{syn})$. If $g_{syn}$ is not too large, then each $C_s \equiv \{\Phi(v, w, s) = 0\}$ defines a cubic-shaped curve. We express the left and right branches of $C_s$ by $\{v = \Phi_L(w, s)\}$ and $\{v = \Phi_R(w, s)\}$, respectively. Finally, let

$$G_L(w, s) = g(\Phi_L(w, s), w) \quad \text{and} \quad G_R(w, s) = g(\Phi_R(w, s), w)$$

Now the first equation in (4.1) can be written as $0 = \Phi(v_i, w_i, s_j)$ with $s_j$ fixed. Hence, $v_i = \Phi_\alpha(w_i, s_j)$ where $\alpha = L$ if cell $i$ is silent and $\alpha = R$ if cell $i$ is active. It then follows that each slow variable $w_i$ satisfies the single equation

$$\dot{w}_i = G_\alpha(w_i, s_j) \tag{4.1}$$

By dissecting the full system into fast and slow subsystems, we are able to construct singular solutions of (4.1), (4.2). In particular, this leads to sufficient conditions for when there exists a singular synchronous solution and when this solution is (formally) asymptotically stable. The second step in the analysis is to rigorously prove that the formal analysis, in which $\epsilon = 0$, is justified for small $\epsilon > 0$. This raises some very subtle issues in the geometric theory of singular perturbations, some of which have not been completely addressed in the literature. For most of the results presented here, we only consider singular solutions.

We note that the geometric approach used here is somewhat different from that used in many dynamical systems studies (see, for example, [15]). All of the networks considered here consist of many differential equations, especially for larger networks. Traditionally, one would interpret the solution of this system as a single trajectory evolving in a very large dimensional phase space. We consider several trajectories, one corresponding to a single cell, moving around in a much lower dimensional phase space (see also [18, 20, 22, 23, 25]). After reducing the full system to a system for just the slow variables, the dimension of the lower dimensional phase space equals the number of slow intrinsic variables and slow synaptic variables corresponding to each cell. In the worst case considered here, there is only one slow intrinsic variable for each cell and one slow synaptic variable; hence, we never have to consider phase spaces with dimension more than two. Of course, the particular phase space we need to consider may change, depending on whether the cells are active or silent and also depending on the synaptic input that a cell receives.

## 4.4. Synchrony with excitatory synapses

Consider two mutually coupled cells with excitatory synapses. Our goal here is to give sufficient conditions for the existence of a synchronous solution and its stability. Note that if the synapses are excitatory, then the curve $C_A \equiv C_{s_A}$ lies 'above' $C_0 \equiv \{f = 0\}$ as shown in Fig. 12. This is because for an excitatory synapse, $v < v_{syn}$ along the synchronous solution. Hence, on $C_A$, $f(v, w) = g_{syn}s_A(v - v_{syn}) < 0$, and we are assuming that $f < 0$ above $C_0$. If $g_{syn}$ is not too large, then both $C_0$ and $C_A$ will be cubic shaped. We assume that the threshold $\theta_{syn}$ lies between the two knees of $C_0$. In the statement of the following result, we denote the left knee of $C_0$ by $(v_{LK}, w_{LK})$.

**Theorem:** Assume that each cell, without any coupling, is oscillatory. Moreover, assume the synapses are fast and excitatory. Then there exists a synchronous periodic solution of (4.1), (4.2). This solution is asymptotically stable if one of the following two conditions is satisfied.

(H1) $\frac{\partial f}{\partial w} < 0$, $\frac{\partial g}{\partial v} > 0$, and $\frac{\partial g}{\partial w} < 0$ near the singular synchronous solution.

(H2) $|g(v_{LK}, w_{LK})|$ is sufficiently small.

**Remark 4.1** We note that the synchronous solution cannot exist if the cells are excitable and the other hypotheses, concerning the synapses, are satisfied. This is because along a synchronous solution, each $(v_i, w_i)$ lies on the left branch of $C_0$ during the silent phase. If the cells are excitable, then each $(v_i, w_i)$ will approach the point where the $w$-nullcline $\{g = 0\}$ intersects the left branch of $C_0$. The cells, therefore, will not be able to jump up to the active phase.

**Remark 4.2** The assumptions concerning the partial derivatives of $f$ and $g$ in (H1) are not very restrictive since we are already assuming that $f > 0 \ (< 0)$ below (above) the $v$-nullcline and $g > 0 \ (< 0)$ below (above) the $w$-nullcline.

**Remark 4.3** A useful way to interpret (H2) is that the silent phases of the cells are much longer than their active phases. This is because $g(v_{LK}, w_{LK})$ gives the rate at which the slow variables $w_i$ evolve near the end of the silent phase. Note that $g(v_{LK}, w_{LK})$ will be small if the left knee of $C_0$ is very close to the $w$-nullcline.

**Proof:** We first consider the existence of the synchronous solution. This is straightforward because along a synchronous solution

$$(v_1, w_1, s_1) = (v_2, w_2, s_2) \equiv (v, w, s)$$

satisfy the reduced system

$$v' = f(v, w) - sg_{syn}(v - v_{syn})$$

$$w' = \epsilon g(v, w)$$

$$s' = \alpha(1 - s)H(v - \theta_{syn}) - \beta s$$

The singular solution consists of four pieces. During the silent phase, $s = 0$ and $(v, w)$ lies on the left branch of $C_0$. During the active phase $s = s_A$ and $(v, w)$ lies on the right branch of $C_A$. The jumps between these two phases occur at the left and right knees of the corresponding cubics.

We next consider the stability of the synchronous solution to small perturbations. We begin with both cells close to each other in the silent phase on the left branch of $C_0$, with cell 1 at the left knee ready to jump up. We follow the cells around in phase space by constructing the singular solution until one of the cells returns to the left knee of $C_0$. As before, the singular solution consists of four pieces. We need to show that the cells are closer to each other after this complete cycle than they were initially.

The first piece of the singular solution begins when cell 1 jumps up. When $v_1(t)$ crosses $\theta_{syn}$, $s_1(t) \rightarrow s_A$. This raises the cubic corresponding to cell 2 from $C_0$ to $C_A$. If $|w_1(0) - w_2(0)|$ is sufficiently small, corresponding to a sufficiently small perturbation, then cell 2 lies below the left knee of $C_A$. The fast equations then force cell 2 to also jump up to the active phase, as shown in Fig. 14. Note that this piece takes place on the fast time scale. Hence, on the slow time scale, both cells jump up at precisely the same time.

During the second piece of the singular solution, both oscillators lie in the active phase on the right branch of $C_A$. Note that the ordering in which the oscillators track along the left and right branches has been reversed. While in the silent phase, cell 1 was ahead of cell 2. In the active phase, cell 2 leads the way. The oscillators remain on the right branch of $C_A$ until cell 2 reaches the right knee.

The oscillators then jump down to the silent phase. Cell 2 is the first to jump down. When $v_2(t)$ crosses $\theta_{syn}$, $s_2$ switches from $s_A$ to 0 on the fast time scale. This lowers the cubic corresponding to cell 1 from $C_A$ to $C_0$. If, at this time, cell 1 lies above the right knee of $C_A$, then cell 1 must jump down to the silent phase. This will certainly be the case if the cells are initially close enough to each other.

During the final piece of the singular solution, both oscillators move down the left branch of $C_0$ until cell 1 reaches the left knee. This completes one full cycle.

To prove that the synchronous solution is stable, we must show that the cells are closer to each other after this cycle; that is, there is compression in the distance between the cells. There are actually several ways to demonstrate this

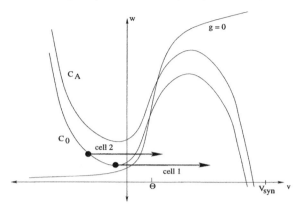

Fig. 12. Nullclines for an oscillatory relaxation oscillator with ($C_A$) and without ($C_0$) excitatory coupling. Note that cell 2 responds to cell 1 through Fast Threshold Modulation.

compression; these correspond to two different ways to define what is meant by the 'distance' between the cells. Here we consider a Euclidean metric, which is defined as follows: Suppose that both cells lie on the same branch of the same cubic and the coordinates of cell $i$ are $(v_i, w_i)$. Then the distance between the cells is defined as simply $|w_1 - w_2|$. Note that during the jump up and the jump down, this metric remains invariant. This is because the jumps are horizontal so the values of $w_i$ do not change. If there is compression, therefore, it must take place as the cells evolve in the silent and active phases. We now show that this is indeed the case if (H1) is satisfied.

Suppose that when $\tau = 0$, both cells lie in the silent phase on $C_0$. We assume, for convenience, that $w_2(0) > w_1(0)$. We need to prove that $w_2(\tau) - w_1(\tau)$ decreases as long as the cells remain in the silent phase. Now each $w_i$ satisfies (4.1) with $\alpha = L$ and $s_j = 0$. Hence,

$$w_i(\tau) = w_i(0) + \int_0^\tau G_L(w_i(\xi), 0)d\xi$$

and, using the Mean Value Theorem,

$$
\begin{aligned}
w_2(\tau) - w_1(\tau) &= w_2(0) - w_1(0) \\
&\quad + \int_0^\tau G_L(w_2(\xi), 0) - G_L(w_1(\xi), 0)\, d\xi \\
&= w_2(0) - w_1(0)
\end{aligned}
$$

(4.1)

$$+ \int_0^\tau \frac{\partial G_L}{\partial w}(w^*, 0)(w_2(\xi) - w_1(\xi))d\xi$$

for some $w^*$. Now $G_L(w, s) = g(\Phi_L(w), w)$. Hence, $\frac{\partial G_L}{\partial w} = g_v \Phi_L'(w) + g_w$. We assume in (H1) that $g_v > 0$ and $g_w < 0$ near the synchronous solution. Moreover, $\Phi_L'(w) < 0$ because $v = \Phi_L(w)$ defines the left branch of the cubic $C_0$ which has negative slope. It follows that $\frac{\partial G_L}{\partial w} < 0$, and therefore, from (34), $w_2(\tau) - w_1(\tau) < w_2(0) - w_1(0)$. This gives the desired compression; a similar computation applies in the active phase. We note that if there exists $\gamma > 0$ such that $\frac{\partial G_L}{\partial w} < -\gamma$ along the left branch, then Gronwall's inequality shows that $w_2(\tau) - w_1(\tau)$ decreases at an exponential rate.

We next consider (H2) and demonstrate why this leads to compression of trajectories. Suppose, for the moment, that $g(v_{LK}, w_{LK}) = 0$; that is, the left knee of $C_0$ touches the $w$-nullcline at some fixed point. Then both cells will approach this fixed point as they evolve along the left branch of $C_0$ in the silent phase. There will then be an infinite amount of compression, since both cells approach the same fixed point. It follows that we can assume that the compression is as large as we please by making $g(v_{LK}, w_{LK})$ sufficiently small. If the compression is sufficiently large, then it will easily dominate any possible expansion over the remainder of the cells' trajectories. This will, in turn, lead to stability of the synchronous solution.

**Remark 4.4** The mechanism by which one cell fires, and thereby raises the cubic of the other cell such that it also fires, was referred to as *Fast Threshold Modulation (FTM)* in [20]. There, a time metric was introduced to establish the compression of trajectories of excitatorily coupled cells, which implies the stability of the synchronous solution. A detailed discussion of the time metric can be found in [11].

We have so far considered a completely homogeneous network with just two cells. The analysis generalizes to larger inhomogeneous networks in a straightforward manner, if the degree of heterogeneity between the cells is not too large. The major difference in the analysis is that, with heterogeneity, the cells may lie on different branches of different cubics during the silent and active phases. The resulting solution cannot be perfectly synchronous; however, as demonstrated in [25], one can often expect synchrony in the jump-up, but not in the jump-down. Related work on heterogeneous networks include [3, 13].

One may also consider, for example, an arbitrarily large network of identical oscillators with nearest neighbor coupling. We do not assume that the strength of coupling is homogeneous. Suppose that we begin the network with each cell in the silent phase. If the cells are identical, then they must all lie on the left branch of $C_0$. Now if one cell jumps up it will excite its neighbors and raise their corresponding cubics. If the cells begin sufficiently close to each other, then these neighbors will jump up due to FTM. In a similar manner, the neighbor's neighbors will also jump due to FTM and so on until every cell jumps up. In this way, every cell jumps up at the same (slow) time. While in the active phase, the cells may receive different input and, therefore, lie on the right branches of different cubics. Once one of the cells jumps down, there is no guarantee that other cells will also jump down at this (slow) time, because the cells to which it is coupled may still receive input from other active cells. Hence, one cannot expect synchrony in the jumping down process. Eventually every cell must jump down. Note that there may be considerable expansion in the distance between the cells in the jumping down process. If $|g(v_{LK}, w_{LK})|$ is sufficiently small, however, as in the previous result, then there will be enough compression in the silent phase so that the cells will still jump up together. Here we assumed that the cells are identical; however, the analysis easily follows if the heterogeneities among the cells are not too large. A detailed analysis of this network is given in [25].

### 4.5. Desynchrony with inhibitory synapses

We now consider two mutually coupled cells with inhibitory synapses. Under this coupling, the curve $C_A$ now lies below $C_0$. As before, we assume that $g_{syn}$ is not too large, such that both $C_0$ and $C_A$ are cubic shaped. We also assume that the right knee of $C_A$ lies above the left knee of $C_0$ as shown in Fig. 13. Some assumptions on the threshold $\theta_{syn}$ are also required. For now, we assume that $\theta_{syn}$ lies between the left knee of $C_0$ and right knee of $C_A$.

We will assume throughout this section that the synapses are fast and inhibitory. In [19] it is shown that if a synchronous solution exists then it must be unstable. This network typically exhibits either out-of-phase oscillations or a completely quiescent state. Here we give sufficient conditions for when either of these arises. We note that the network may exhibit bistability; both the out-of-phase and completely quiescent solutions may exist and be stable for the same parameter values. These results are all for singular solutions. Some rigorous results for $\epsilon > 0$ are given in [23].

The following theorem gives sufficient conditions for when there exists out-of-phase behavior. We first introduce the following notation. Assume that the left and right knees of $C_0$ are at $(v_{LK}, w_{LK})$ and $(v_{RK}, w_{RK})$, respectively. If the $w$-nullcline intersects the left branch of $C_A$, then we denote this point by $(v_A, w_A) = p_A$. We assume that $w_A < w_{LK}$. Let $\tau_L$ be the (slow) time it takes for the solution of (4.1) with $\alpha = L$ and $s = s_A$ to go from $w = w_{RK}$ to $w = w_{LK}$, and let $\tau_R$ be the time it takes for the solution of (4.1) with $\alpha = R$ and $s = 0$ to go from $w = w_{LK}$ to $w = w_{RK}$. Note that $\tau_L$ is related to the time a solution spends in the silent phase, while $\tau_R$ is related to the time a solution spends in the active phase.

**Theorem:** Assume that the cells are excitable for each fixed level of synaptic input and the synapses are fast, direct, and inhibitory. Moreover, assume that $w_A < w_{LK}$ and $\tau_L < \tau_R$. Then the network exhibits stable out-of-phase oscillatory behavior.

**Remark 4.5** We do not claim that the out-of-phase solution is uniquely determined or that it corresponds to antiphase behavior. These results may hold; however, their proofs require more analysis than that given here.

**Remark 4.6** The rest state with each cell at the fixed point on $C_0$ also exists and is stable. Hence, if the hypotheses of Theorem are satisfied, then the network exhibits bistability.

**Proof:** Suppose that we begin with cell 1 at the right knee of $C_0$ and cell 2 on the left branch of $C_A$ with $w_A < w_2(0) < w_{LK}$. Then cell 1 jumps down and, when $v_1$ crosses the threshold $\theta_{syn}$, cell 2's cubic switches from $C_A$ to $C_0$. Since $w_2(0) < w_{LK}$, cell 2 lies below the left knee of $C_0$, so it must jump up to the active phase. After these jumps, cell 1 lies on the left branch of $C_A$, while cell 2 lies on the right branch of $C_0$.

Cell 2 then moves up the right branch of $C_0$ while cell 1 moves down the left branch of $C_A$, approaching $p_A$. This continues until cell 2 reaches the right knee of $C_0$ and jumps down. We claim that at this time, cell 1 lies below the left knee of $C_0$, so it must jump up. We can then keep repeating this argument to obtain the sustained out-of-phase oscillations. The reason why cell 1 lies below the left knee of $C_0$ when cell 2 jumps down is because it spends a sufficiently long amount of time in the silent phase. To estimate this time, note that because cell 2 was initially below the left knee of $C_0$, the time it spends in the active phase before jumping down is greater than $\tau_R$. Hence, the time cell 1 spends in the silent phase from the time it jumps down is greater than $\tau_R > \tau_L$. From the definitions, since cell 1 was initially at the right knee of $C_0$, it follows that cell 1 must be below the left knee of $C_0$ when cell 2 jumps down, which is what we wished to show.

**Remark 4.7** Wang and Rinzel [27] distinguish between "escape" and "release" in producing out-of-phase oscillations. In the proof of the preceding theorem, the silent cell can only jump up to the active phase once the active cell jumps down and releases the silent cell from inhibition. This is referred to as the release mechanism and is often referred to as *post inhibitory rebound* [5]. To describe

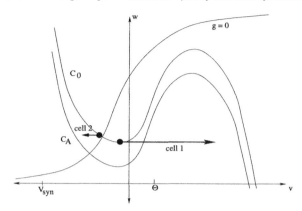

Fig. 13. Instability induced by mutual inhibition. Cell 2 jumps to $C_A$ when cell 1 fires.

the escape mechanism, suppose that each cell is oscillatory for fixed levels of synaptic input. Moreover, one cell is active and the other is inactive. The inactive cell will then be able to escape the silent phase from the left knee of its cubic, despite the inhibition it receives from the active cell. Note that when the silent cell jumps up, it inhibits the active cell. This lowers the cubic of the active cell, so it may be forced to jump down before reaching a right knee.

**Remark 4.8** We have presented rigorous results that demonstrate that excitation can lead to synchrony and inhibition can lead to desynchrony. These results depended on certain assumptions, however. We assumed, for example, that the synapses turned on and off on a fast time scale; moreover, the results hold in some singular limit. In is, in fact, possible for excitatory synapses to generate stable desynchronous oscillations and for inhibitory synapses to generate stable synchronous oscillations. Conditions for when these are possible has been the subject of numerous research articles. References may be found in [22].

**Remark 4.9** The two cell model can generate other rhythms besides those discussed so far. For example, it is possible that one cell periodically fires action potentials, while the other cell is always silent. This is sometimes refereed to as a *suppressed solution*. It arises if the rate at which inhibition turns off is slower than the rate at which a cell recovers in the silent phase. That is, suppose cell 1 fires. This sends inhibition to cell 2, preventing it from firing. Now if cell 1 is able to recover from its silent phase before the inhibition to cell 2 wears off, then cell 1 will fire before cell 2 is able to. This will complete one cycle and cell 2 will continue to be suppressed. If the time for cells to recover in the silent phase is comparable to the time for inhibition to decay, then more exotic solutions are possible (see [22]).

## 5. Excitatory-inhibitory networks

### 5.1. Introduction

The methods discussed in the previous section carry over to larger networks of neuronal oscillators. Here we demonstrate how the geometric singular perturbation approach can be used to analyze excitatory-inhibitory networks. These networks arise in numerous applications including models for thalamic sleep rhythms and Parkinsonian tremor.

We shall consider the network illustrated in Fig.14. This is composed of a population of excitatory ($E$-)cells and a single inhibitory ($J$-)cell. Each $E$-cell sends excitation to the $J$-cell, while the $J$-cell sends inhibition back to every $E$-cell. We assume that all of the $E$-cells are identical, but they may differ from the $J$-cell. This network, analyzed in [17, 18] is motivated by models for thalamic oscillations involved in sleep rhythms. In those networks, there may be a population of $J$-cells with inhibitory coupling among the $J$-cells. Our analysis carries over, in a straightforward manner, to these larger systems.

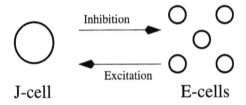

Fig. 14. Globally inhibitory network. The $J$-cell inhibits the $E$-cells, which excite the $J$-cell.

We distinguish two types of rhythms in which the network may engage. In a *synchronous oscillation*, the $E$-cells are completely synchronized. When the $E$-cells fire, or become active, they excite the $J$-cell to fire in response. However, the $J$-cell is not necessarily synchronized with the $E$-cells throughout the entire oscillation; this is because the $J$-cell need not have the same intrinsic properties as the $E$-cells. Alternately, in a *clustered oscillation*, the $E$-cells form subpopulations or clusters; the cells within a cluster fire synchronously but cells from distinct clusters act out of synchrony from each other. The $J$-cell will be induced to fire with each cluster.

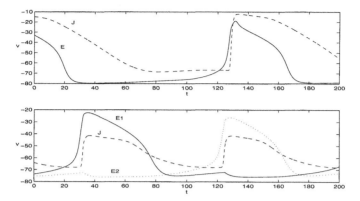

Fig. 15. Solutions for a globally inhibitory network of four $E$-cells and one $J$-cell. A. (top) A synchronous solution; the solid curve is the time course of $E$-cell voltage and the dashed curve is the time course of $J$-cell voltage. B. (bottom) A 2-cluster solution; the solid and dotted curves are the voltage time courses of two different $E$-cell clusters of two cells each. The dashed curve is the time course of $J$-cell voltage.

Two types of solutions for a network with just four $E$-cells are shown in Fig. 15. Fig. 15A shows a synchronous solution, while a 2-cluster oscillation is shown in Fig. 15B. Note that in each of these figures, the $E$-cells within a cluster are perfectly synchronized; moreover, the $J$-cell jumps up together with each cluster. To obtain the different cluster states, we adjusted a parameter in the equations for the $J$-cell; this parameter controls the duration of the $J$-cell's active phase, which, in turn, controls the amount of inhibition sent back to the $E$-cells. The analysis that follows will clarify why the level of inhibition produced by the $J$-cell is an important factor in determining the network behavior of the $E$-cells.

We model the globally inhibitory network as follows (see [17, 18]). Here we assume that all of the synapses are direct. Each $E_i$ satisfies equations of the form

$$
\begin{aligned}
v_i' &= f(v_i, w_i) - g_{inh} s_J (v_i - v_{inh}) \\
w_i' &= \epsilon g(v_i, w_i) \\
s_i' &= \alpha(1 - s_i) H(v_i - \theta) - \beta s_i
\end{aligned}
\tag{5.1}
$$

while the $J$-cell satisfies the equations

$$
\begin{aligned}
v_J' &= f_J(v_J, w_J) - \frac{1}{N} \sum_i s_i g_{exc} (v_J - v_{exc}) \\
w_J' &= \epsilon g_J(v_J, w_J) \\
s_J' &= \alpha_J(1 - s_J) H(v_J - \theta_J) - \epsilon K_J s_J
\end{aligned}
\tag{5.2}
$$

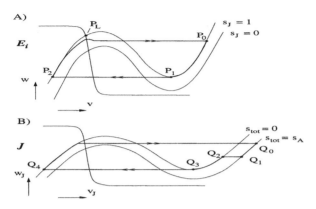

Fig. 16. Nullclines for A) $E$-cells and B) $J$-cells in a globally inhibitory network [18]. The heavy lines and points $P_i$, $Q_i$ correspond to the singular synchronous solution discussed in the text. Note that $s_J$ decays on the slow time scale.

Unlike the previous section, we assume $f > 0$ ($f < 0$) above (below) the $v$-nullcline for $s_J = 0$ and $g > 0$ ($g < 0$) below (above) the $w$-nullcline. The nullclines are illustrated in Fig. 16. Note that the nullclines are "upside-down" relative to the figures in Section 3. This orientation is motivated by the biological model in Section 6.

The variable $s_J$ denotes the inhibitory synaptic input from the $J$-cell, while $s_i$ denotes the excitatory synaptic input from cell $E_i$ to the $J$-cell. The sum in (5.2) is taken over all $N$ $E_i$-cells. Note that turn on of inhibition and excitation both occur on the fast time scale, while the inhibitory variable $s_J$ turns off on the slow time scale. This turn off may be 'fast' or 'slow', depending on whether $K_J$ is large or small. We assume that $\beta = O(1)$, representing fast turn off of excitation, although there is no problem extending the analysis if $\beta = O(\epsilon)$. Note that if $v_i > \theta$, then $s_i \to s_A \equiv \frac{\alpha}{\alpha+\beta}$ on the fast time scale.

Each synapse in (5.1), (5.2) is direct. If the inhibition is indirect, then $s_J$ satisfies system (4.1), with the appropriate adjustment of subscripts and a new variable $x_J$ included, instead of the equation given in (5.2). Indirect synapses will be necessary to obtain stability of synchronous and clustered solutions to (5.1), (5.2).

As in previous section, we analyze this network by constructing singular solutions. The trajectory for each cell lies on the left (right) branch of a cubic nullcline during the silent (active) phase. Which cubic a cell inhabits depends on the total synaptic input that the cell receives. Nullclines for the $E_i$ are shown in Fig. 16A and those for $J$ in Figure 16B. Note in Fig. 16A that the $s_J = 1$ nullcline lies above the $s_J = 0$ nullcline, while in Fig. 16B, the $s_{tot} \equiv \frac{1}{N} \sum s_i = s_A$ nullcline lies below the $s_{tot} = 0$ nullcline. These relations hold because the $E_i$ receive inhibition from $J$ while $J$ receives excitation from the $E_i$. We assume that each cell is excitable for fixed levels of synaptic input.

## 5.2. Synchronous solution

We now give sufficient conditions for the existence of a singular synchronous periodic solution and for its stability. To state the main result, it is necessary to introduce some notation. Let $C_s$ denote the cubic $f(v, w) - g_{inh}s(v - v_{inh}) = 0$. Since the $E$-cells are excitable for fixed levels of the inhibitory input $s_J$, there is a fixed point on the left branch of $C_s$ of the first two equations in (5.1) with $s = s_J$ held constant. We denote this fixed point by $(v_F(s_J), w_F(s_J))$ and the left knee of $C_s$ by $(v_L(s_J), w_L(s_J))$.

**Theorem 5.1.** *A singular synchronous periodic solution exists if* $w_F(1) > w_L(0)$, *$K_J$ is sufficiently large, the active phase of the $J$-cell is sufficiently long, and the recovery of the $J$-cell in the silent phase is sufficiently fast. If the inhibitory synapse $s_J$ is indirect and the active phase of the $J$-cell is long enough, then the synchronous solution is stable.*

**Remark 5.1.** The condition $w_F(1) > w_L(0)$ simply states that the fixed point on the left branch of $C_1$ lies above the left knee of $C_0$. This allows the possibility of the $E$-cells firing upon being released from inhibition from the $J$-cell.

**Remark 5.2.** Recall that $K_J$ corresponds to the rate of decay of the inhibition. We will see that synchronous oscillations cannot exist if this synaptic decay rate is too slow.

**Proof:** We prove the existence by demonstrating how to construct the singular synchronous solution if the hypotheses of Theorem 5.1 are satisfied. We assume throughout this construction that the positions of the $E$-cells are identical. The singular trajectory is shown in Fig. 16.

We begin with each cell in the active phase just after it has jumped up. These are the points labeled $P_0$ and $Q_0$ in Fig. 16. Then each $E_i$ evolves down the right branch of the $s_J = 1$ cubic, while $J$ evolves down the right branch of the $s_{tot} = s_A$ cubic. We assume that the $J$-cell active phase is long, such that the $E_i$ have a shorter active phase than $J$; thus, each $E_i$ reaches the right knee $P_1$ and jumps down to the point $P_2$ before $J$ jumps down. The assumption of a long $J$-cell active phase implies that at this time, $J$ lies above the right knee of the $s_{tot} = 0$ cubic; otherwise, $J$ would jump down as soon as the $E_i$ did. $J$ must therefore jump from the point $Q_1$ to the point $Q_2$ along the $s_{tot} = 0$ cubic when the $E_i$ jump down. On the next piece of the solution, $J$ moves down the right branch of the $s_{tot} = 0$ cubic while the $E_i$ move up the left branch of the $s_J = 1$ cubic. When $J$ reaches the right knee $Q_3$ it jumps down to the point $Q_4$ on the left branch of the $s_{tot} = 0$ cubic.

Now the inhibition $s_J$ to the $E_i$ starts to turn off on the slow time scale; that is, $\dot{s}_J = -K_J s_J$ for $\cdot = \frac{d}{d\tau}$, $\tau = \epsilon t$. Thus, the $E_i$ do not jump immediately to another cubic. Instead, the trajectory for the $E_i$ moves upwards, with increasing $w_i$, until it crosses the $w$ nullcline. Then each $w_i$ starts to decrease. If the $E_i$ are able to reach a left knee, then they jump up to the active phase and this completes one cycle of the synchronous solution. When the $E_i$ jump up, $J$ also jumps up if it lies above the left knee of the $s_{tot} = s_A$ cubic; that is, the recovery of the $J$-cell in its silent phase must be sufficiently fast.

Existence of the synchronous solution requires that the $E$-cells can reach the jump-up curve and escape from the silent phase. We demonstrate that this is indeed the case if the assumptions of Theorem 5.1 are satisfied. It will be convenient to first introduce some notation. This will allow us to obtain simple estimates for when a synchronous oscillation exists and what the period of the oscillation is. This notation will also be useful in the next section.

As in earlier sections, we derive equations for the evolution of the $E$-cells' slow variables; these are $(w_i, s_J)$. Let $v = \Phi_L(w, s)$ denote the left branch of the cubic $f(v, w) - g_{inh}s(v - v_{inh}) = 0$, and let $G_L(w, s) \equiv g(\Phi_L(w, s), s)$. Then each $(w_i, s_J)$ satisfies the slow equations

$$\dot{w} = G_L(w, s_J)$$
$$\dot{s}_J = -K_J s_J$$
(5.3)

Fig. 17 illustrates the phase plane corresponding to this system. Recall that $w = w_L(s_J)$ denotes the jump-up curve or curve of left knees, and the second curve, which is denoted by $w_F(s_J)$, consists of the fixed points of the first two equations in (5.1) with the input $s_J$ held constant. This corresponds to the $w$-nullcline of (5.3).

We need to determine when a solution $(w(\tau), s_J(\tau))$ of (5.3), beginning with $s_J(0) = 1$ and $w(0) < w_F(1)$, can reach the jump-up curve $w_L(s_J)$. This is clearly impossible if $w_F(1) < w_L(0)$. If $w_F(1) > w_L(0)$ and $w(0) > w_L(0)$, with $K_J$ sufficiently large, then the solution will certainly reach the jump-up curve; this holds because the solution will be nearly vertical, as shown in Fig. 17. If, on the other hand, $K_J$ is too small, then the solution will never be able to reach the jump-up curve. Instead, the solution will slowly approach the curve $w_F(s_J)$ and lie very close to this curve as $s_J$ approaches zero. This is also shown in Fig. 17. We conclude that the cells are able to escape the silent phase if the inhibitory synapses turn off sufficiently quickly and the $w$-values of the cells are

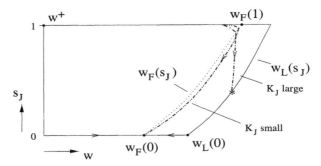

Fig. 17. The slow phase plane for an $E$-cell [18]. The curve $w_L(s_J)$ is the jump-up curve, which trajectories reach if $K_J$ is large enough. The dotted curve $w_F(s_J)$ consists of zeros of $G_L(w, s_J)$ in system (5.3); trajectories tend to $w_F(0)$ as $s_J \to 0$ for small $K_J$. Note that $w' < 0$ for $w > w_F$.

sufficiently large when this turn off begins. Escape is not possible for very slowly deactivating synapses (although it would be possible with slow deactivation if the cells were oscillatory for some levels of synaptic input).

We assume that $K_J$ is large enough so that escape is possible. Choose $w_{esc}$ so that the solution of (5.3) that begins with $s_J(0) = 1$ will be able to reach the jump-up curve if and only if $w(0) > w_{esc}$. The existence of the singular synchronous solution now depends on whether the $E_i$ lie in the region where $w_i > w_{esc}$ when $J$ jumps down to the silent phase. This will be the case if the active phase of $J$ is sufficiently long. One can give a simple estimate on how long this active phase must be as follows.

Let $\tau_E$ ($\tau_J$) denote the duration of the $E$-cell ($J$-cell) active phase. Further, let $w^+$ denote the value of $w$ at the right knee of the $s_J = 1$ cubic (see Figure 15). If $g_{syn}$ is not too large, then $w^+ < w_L(0)$. Finally, let $\tau_{esc}$ denote the time for $w_i$ to increase from $w^+$ to $w_{esc}$ under $\dot{w} = G_L(w, 1)$. Since the $E_i$ spend time $\tau_J - \tau_E$ in the silent phase before they are released from inhibition, the singular synchronous solution exists if $\tau_{esc} < \tau_J - \tau_E$.

For the stability of the synchronous solution, we need to assume that $s_J$ corresponds to an indirect synapse, as in Section 4.2. Suppose we slightly perturb the synchronous solution. If $s_J$ is direct, then when one $E$-cell fires, it will excite the $J$-cell which will, in turn, inhibit the other $E$-cells on the fast time scale. This will prevent the other $E$-cells from firing and desynchrony will result.

The compression of trajectories if $s_J$ is indirect and the $J$ active phase is long enough follows because all the $E$-cells approach exponentially close to the point $w_F(1)$ in the silent phase while the $J$-cell is still active. This exponential compression easily dominates any possible expansion over the remainder of the cells' trajectories.

The domain of attraction of the synchronous solution depends on the ability of the $E$-cells to pass through the 'window of opportunity' provided by the indirect synapse. The size of this domain grows as the $J$-cell's active phase increases: More powerful $J$-cell bursts provide increased inhibition to the $E$-cells and this further compresses the $E$-cells near the point $w_F(1)$ in the silent phase. We note that this source of compression in globally inhibitory networks is considerably more powerful than any compression mechanism in mutually coupled inhibitory networks.

**Remark 5.3.** The analysis leads to simple formulas for the period of the synchronous solution. Let $\tau_J$ be, as above, the time cell $J$ spends in the active phase and let $\tau_S$ be the time for the $E$-cells to reach the jump up curve after the $J$-cell jumps down. Then the period of the synchronous solution is simply $\tau_J + \tau_S$. Now $\tau_J$ is determined by the dynamics of the $J$-cell, while $\tau_S$ is primarily controlled by the rate at which the synapses turn off; this is the parameter $K_J$ in (5.3). Other parameters play a secondary role. Note, for example, that the parameter $g_{syn}$ mildly influences the period by controlling the slope of the jump up curve.

**Remark 5.4.** The domain of attraction of the synchronous solution increases with $K_J$, since a large $K_J$ yields rapid decay of $s_J$. If $K_J$ is too large, however, then the $J$-cells actually cannot recover in time to respond to the excitation from the firing $E$-cells. Hence, this analysis shows that the combination of fast $J$-cell recovery and a large $K_J$ promotes stable synchronization.

### 5.3. Clustered solutions

Here we describe the singular trajectory corresponding to a 2-cluster solution. The number of cells in the network may be arbitrary, but we assume for ease of notation that the clusters have equal numbers of cells. The geometric construction will require certain assumptions on the equations and a precise theorem is stated and proved in the following subsections. As we shown in [19], the construction of a 2-cluster solution easily generalizes to an arbitrary number of clusters.

For the geometric construction of a singular 2-cluster solution, it suffices to consider only one-half of a complete cycle. During this half-cycle, one cluster, call it $E_1$, fires, say at $\tau = 0$, and evolves to the initial position of the other cluster; the non-firing cluster, call it $E_2$, evolves in the silent phase to the initial position of $E_1$. By symmetry, the solution then continues with the roles of the clusters reversed.

When $E_1$ jumps up, it forces $J$ to jump up to the right branch of the $s_{tot} = \frac{1}{2}s_A$ cubic. Then $E_1$ moves down the right branch of the $s_J = 1$ cubic, while $J$ moves down the right branch of the $s_{tot} = \frac{1}{2}s_A$ cubic and $E_2$ moves up the left branch of the $s_J = 1$ cubic. We assume, as before, that the $E$-cells in $E_1$ have shorter active phases than the $J$-cell, so $E_1$ jumps down before $J$ does. The assumption that $J$ has a longer active phase than $E_1$ implies that it lies above the right knee of the $s_{tot} = 0$ cubic at this time, so it moves down the right branch of the $s_{tot} = 0$ cubic until it reaches the right knee and then jumps down. During the time that $J$ remains active, both $E_1$ and $E_2$ move up the left branch of the $s_J = 1$ cubic.

After $J$ jumps down, $s_J(\tau)$ slowly decreases. If $E_2$ is able to reach the jump-up curve, then it fires and this completes the first half cycle of the singular solution. Suppose that $\tau = \tau_F$ when this occurs. By abuse of notation, let $w_i$ denote the $w$-value of all cells in cluster $E_i$. For the trajectories described above to represent one-half of a 2-cluster solution, we need that $w_2(\tau_F) = w_1(0)$, $w_1(\tau_F) = w_2(0)$, and $w_J(\tau_F) = w_J(0)$. The analysis in Subsection 5.2 shows that a 2-cluster solution will exist, with stability between clusters, if the active phase of $J$ is not too long or too short, compared with the active phase of the $E_i$. If $J$'s active phase is too long, then the network exhibits synchronous behavior as described before. If $J$'s active phase is too short, then the system approaches the stable quiescent state.

To conclude that the 2-cluster state is stable, we must also consider stability within each cluster. The stability mechanism here is similar to that for a synchronous solution. Since each cluster experiences a decay of inhibition and subsequent re-inhibition between firings, stability within clusters does not require as long a $J$-cell active phase as does stability of the synchronous solution.

## 6. Activity patterns in the basal ganglia

In this final chapter, we discuss a recent model for neuronal activity patterns in the basal ganglia [24]. This is a part of the brain believed to be involved in the control of movement. Dysfunction of the basal ganglia is associated with movement disorders such as Parkinson's disease and Huntington's disease. Neurons within the basal ganglia are also the target of recent surgical procedures, including deep brain stimulation. An introduction to the basal ganglia can be found in [?].

The issues discussed arise in numerous other neuronal systems. We shall describe how these large neuronal networks are modeled, what population rhythms may arise in these networks and the possible roles of these activity patterns.

### 6.1. The basal ganglia

The basal ganglia consist of several nuclei; these are illustrated in Fig. 18. The primary input nucleus is the striatum; it receives motor information from the cortex. The primary output nuclei are the internal segment of the globus pallidus (GPi) and the substantia nigra par retularis (SNr). Neuronal information passes through the basal ganglia through two routes. The direct pathway passes directly from the striatum to the output nuclei. In the indirect pathway, the information passes from the striatum to the external segment of the globus pallidus (GPe) onto the subthalamic nuclues (STN) and then onto the output nuclei. The final nucleus is the substantia nigra par compacta (SNc). This is the primary source of dopamine.

Fig. 18 illustrates that some of the pathways within the basal ganglia are excitatory and some are inhibitory. Most of the pathways are inhibitory except for those that originate in the STN. Note that the pathways arising from SNc are labeled both inhibitory and excitatory. This is because there are different classes of dopamine receptors within the striatum. We also note that Fig. 18 illustrates only some of the pathways reported in the literature.

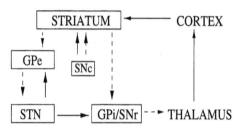

Fig. 18. Nuclei within the basal ganglia. Solid arrows indicate excitatory connections and dashed arrows indicate inhibitory connections

Parkinson's disease is associated with a severe reduction of dopamine. Experiments have also demonstrated that during Parkinson's disease, there is a change in the neuronal activity of the output nucleus GPi. Previous explanations for how a loss of dopamine leads to altered neuronal activity in GPi have been in terms of an average firing rate of neurons; that is, the average number of action potentials in some fixed interval of time. A detailed description of this explanation can be found in [1, 4, 24]. It has been successful in accounting for some features of PD; however, it cannot account for others. For example, it is not at all clear how one can explain tremor. It also cannot account for recent experiments that demonstrate that there is an increased level synchronization among neurons in the STN and GPi during a parkinsonian state [8]. Several authors have suggested that the pattern of neuronal activity, not just the firing rate, is crucially important.

The goal of the modeling study in [24] is to test hypotheses on how the loss of dopamine may lead to tremor-like oscillations and changes in firing patterns. We construct a model for neurons within GPe and STN based on recent experiments [2]. We use computational methods to study the types of activity patterns that arise in this model. In particular, we demonstrate that the model can exhibit irregular uncorrelated patterns, synchronous tremor-like rhythms, and propagating wave-like activity. In the next subsection, we describe the computational model. We then describe the types of activity patterns that arise in the model.

## 6.2. The model

Here we describe the model for the STN and GPe network. The detailed equations are given in [24]. These equations are derived using the Hodgkin-Huxley formalism discussed earlier. The precise equations are different from the Hodgkin-Huxley equations, however. This is because the STN and GPe neurons contain channels different from those in the squid's giant axon. In particular, calcium plays a very important role in generating the action potential of STN and GPe neurons. There are two types of potassium channels, one of which depends on the concentration of intracellular calcium (along with membrane potential). There are also two types of calcium channels in STN neurons.

The membrane potential of each STN neuron obeys the current balance equation:

$$C_m \frac{dV}{dt} = -I_L - I_K - I_{Na} - I_T - I_{Ca} - I_{AHP} - I_{G \to S}.$$

The leak current is given by $I_L = g_L(v - v_L)$, and the other voltage-dependent currents are described by the Hodgkin-Huxley formalism as follows: $I_K = g_K n^4(v - v_K)$, $I_{Na} = g_{Na} m_\infty^3(v) h(v - v_{Na})$, $I_T = g_T a_\infty^3(v) b_\infty^2(r)(v - v_{Ca})$, and $I_{Ca} = g_{Ca} s_\infty^2(v)(v - v_{Ca})$. The slowly-operating gating variables $n$, $h$, and $r$ are treated as functions of both time and voltage, and have first order kinetics governed by differential equations of the form $\frac{dX}{dt} = \phi_X \frac{(X_\infty(v) - X)}{\tau_X(v)}$ (where $X$ can be $n$, $h$, or $r$), with $\tau_X(v) = \tau_X^0 + \frac{\tau_X^1}{1 + \exp[-(v - \theta_X^\tau)/\sigma_X^\tau]}$. Activation gating for the rapidly activating channels ($m$, $a$, and $s$) was treated as instantaneous. For all gating variables $X = n, m, h, a, r$, or $s$, the steady state voltage dependence was determined using $X_\infty(v) = \frac{1}{1 + \exp[-(v - \theta_X)/\sigma_X]}$. The gating variable $b$ was modeled in a similar, but somewhat different, manner; we do not describe this here. As the final intrinsic current, we take $I_{AHP} = g_{AHP}(v - v_K) \frac{[Ca]}{([Ca] + k_1)}$ where $[Ca]$, the intracellular concentration of $Ca^{2+}$ ions, is governed by $[Ca]' = \epsilon(-I_{Ca} - I_T - k_{Ca}[Ca])$.

The current $I_{G \to S}$ that represents synaptic input from the GPe to STN is modeled as $I_{G \to S} = g_{G \to S}(v - v_{G \to S}) \sum s_j$. The summation is taken over the presynaptic GPe neurons, and each synaptic variable $s_j$ solves a first order differential equation $s_j' = \alpha H_\infty(vg_j - \theta_g)(1 - s_j) - \beta s_j$. Here $vg_j$ is the membrane potential of the GPe neuron $j$, and $H_\infty(v) = 1/(1 + \exp[-(v - \theta_g^H)/\sigma_g^H])$.

The precise forms of the nonlinear functions in this model, along with parameter values, are given in [24]. The GPe neurons are modeled in a similar way.
We do not describe these equations here.

The model STN neurons were adjusted to exhibit properties that are characteristic of the firing of STN neurons in experiments [2]. Fig. 19, left column,
shows the firing properties of the model STN neurons. These cells fire intrinsically at approximately 3 Hz and exhibit high frequency sustained firing and
strong rebound bursts after release from hyperpolarizing current. Fig. 19, right
column, illustrates the firing properties of single GPe neurons. These cells can
fire rapid periodic spikes with sufficient applied current. They also display bursts
of activity when subjected to a small constant hyperpolarizing current.

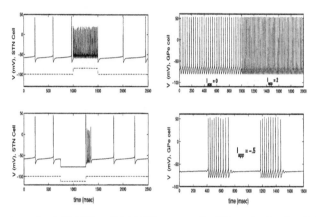

Fig. 19. Voltage traces for (left) STN and (right) GPe neurons for different levels of applied current.
STN cells display high frequency sustained firing with higher input (as shown by the elevated dashed
line) and fire rebound bursts after release from hyperpolarizing current. GPe cells fire rapid periodic
spikes for positive input and fire bursts of spikes for small negative applied current.

Currently the details of connections between STN and GPe cells are poorly
understood. It is known that STN neurons provide one of the largest sources
of excitatory input to the globus pallidus and that the GPe is a major source
of inhibitory afferents to the STN. However, the spatial distribution of axons
in each pathway, as well as the number of cells innervated by single neurons
in each direction, are not known to the precision required for a computer model.
Therefore, in [24] we consider multiple architectures in order to study what types
of activity patterns may arise in a particular class of network architecture.. In the
model networks, each GPe neuron sends inhibition to other GPe neurons as well
as to one or more STN neurons. Each STN neuron sends excitation to one or
more GPe neurons. A prototype network is illustrated in Fig. 20.

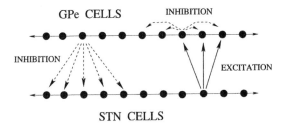

Fig. 20. Architecture of the STN/GPe network.

### 6.3. Activity patterns

Two activity patterns displayed by the model are shown in Fig. 21. The left column displays irregular and weakly correlated firing of each cell. The voltage traces of two STN neurons are shown. shown. Irregular activity arises in sparsely connected, unstructured networks in which each neuron is synaptically connected to only a small number of other neurons chosen at random. It is possible for irregular activity to also arise in structured networks, however.

The right column of Fig. 21 displays clustered activity, in which each structure is divided into subsets of neurons that become highly correlated with each other. The most commonly observed clustered pattern consists of two clusters, with alternating pairs of cells belonging to opposite clusters. Different clusters alternate firing, and in this pattern, cluster membership is persistent over time. Both episodic and continuous clustering are possible. Clustered activity typically arises in networks with a structured, sparsely connected architecture.

A third type of activity pattern is traveling waves (not shown). Propagating wave-like activity can be very robust and exist over a wide range of parameter values. They typically arise in networks with a structured, tightly connected architecture. The speed of the wave depends on both the footprint of network architecture and both intrinsic and synaptic time-scales.

We note that both of the patterns shown in Fig. 21 are generated for a network with exactly the same architecture. In order to switch from the irregular pattern to the synchronous pattern, we increase the applied current to the GPe cells (this corresponds to input from the striatum) and the level of intra-GPe inhibition.

### 6.4. Concluding remarks

We have shown that in a biophysical, conductance-based model that the cellular properties of the STN and GPe cells can give rise to a variety of rhythmic or irregular self-sustained firing patterns, depending on both the arrangement of

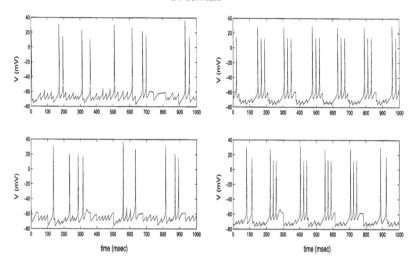

Fig. 21. Irregular and clustered patterns. Each column shows the voltage traces of two STN neurons.

connections among and within the nuclei and the effective strengths of the connections. The dependence on network architecture points out the importance of certain missing pieces of anatomical information. It is important to determine the precision of the spatial organization of connections between the STN and GPe neurons and whether the two nuclei project on each other in a reciprocal or out of register manner.

According to recent studies, correlated oscillatory activity in the GPe and STN neurons is closely related to the generation of the symptoms of Parkinsonism. Previous firing rate models hold that during Parkinsonian states, an increased level of inhibition from the striatum to GPe causes a decrease in the activity of GPe. This in turn would send less inhibition to STN, thus increasing STN activity and ultimately leading to increased inhibitory output from the basal ganglia to the thalamus. In our model network, a more complex picture emerges, in which the STN and GPe are spontaneously oscillatory and synchronous, whereas intra-GPe inhibition and appropriate level of input from the striatum can act to suppress rhythmic behavior.

The analysis described in earlier chapters is extremely useful in understanding the mechanisms responsible for the generation of the different firing patterns arising in the STN/GPe model. The simple two-cell models considered earlier illustrate, for example, that inhibition may play multiple roles in the generation of activity patterns. In the clustered rhythm, for example, active STN neurons need moderate levels of feedback inhibition from GPe to synchronize among themselves. Silent STN neurons, on the other hand, are prevented from firing because they receive more powerful tonic inhibition. For the generation of propagating waves, intra-GPe inhibition is needed to prevent activity from persisting in the wake of the wave. Hence, this inhibition helps to organize the network into a structured activity pattern. If one increases the intra-GPe inhibition, this can desynchronize the GPe oscillations and irregular firing may result.

The STN/GPe model is an example of an excitatory-inhibitory network. This type of model arises in other neuronal systems. For example, recent models for thalamic sleep rhythms share many of the properties of the STN/GPe model. References to papers on the thalamic sleep rhythms can be found in [19].

## References

[1] R.L. Albin, A.B. Young, and J.B. Penney. The functional anatomy of basal ganglia disorders. *Trends in Neurosci.*, **12**:366–375, 1989.

[2] M.D. Bevan and C.J. Wilson. Mechanisms underlying spontaneous oscillation and rhythmic firing in rat subthalamic neurons. *J. Neurosci.*, **19**:7617–7628, 1999.

[3] R.J. Butera, J. Rinzel, and J.C. Smith. Models of respiratory rhythm generation in the pre-Botzinger complex: II. Populations of coupled pacemaker neurons. *J. Neurophysiology*, **82**:398–415, 1999.

[4] M.R. DeLong. Activity of pallidal neurons during movement. *J. Neurophysiol.*, **34**:414–427, 1971.

[5] W. Friesen. Reciprocal inhibition, a mechanism underlying oscillatory animal movements. *Neurosci. Behavior*, **18**:547–553, 1994.

[6] D. Golomb, X.-J. Wang, and J. Rinzel. Synchronization properties of spindle oscillations in a thalamic reticular nucleus model. *J. Neurophysiol.*, **72**:1109–1126, 1994.

[7] F.C. Hoppensteadt and E.M. Izhikevich. *Weakly Connected Neural Networks*. Springer-Verlag, New York, Berlin, and Heidelberg, 1997.

[8] J.M. Hurtado, C.M. Gray, L.B. Tamas, and K.A. Sigvardt. Dynamics of tremor-related oscillations in the human globus pallidus: A single case study. *Proc. Natl. Acad. Sci. USA*, **96**:1674–1679, 1999.

[9] E.M. Izhikevich. Neural excitability, spiking, and bursting. *International Journal of Bifurcation and Chaos*, **10**, 2000.

[10] C.K.R.T. Jones. Stability of the traveling wave solutions of the fitzhugh-nagumo system. *Trans. Amer. Math. Soc.*, 286:431–469, 1984.

[11] N. Kopell and B. Ermentrout. Mechanisms of phase-locking and frequency control in pairs of coupled neural oscillators. In B. Fiedler, G. Iooss, and N. Kopell, editors, *Handbook of Dynamical Systems, vol. 3: Towards Applications*. Elsevier, 2003.

[12] C. Morris and H. Lecar. Voltage oscillations in the barnacle giant muscle fiber. *Biophys. J.*, **35**:193–213, 1981.

[13] P. Pinsky and J. Rinzel. Intrinsic and network rhythmogenesis in a reduced Traub model for CA3 neurons. *J. Comput. Neurosci.*, 1:39–60, 1994.

[14] J. Rinzel. A formal classification of bursting mechanisms in excitable systems. In A.M. Gleason, editor, *Proceedings of the International Congress of Mathematicians*, pages 1578–1594. American Mathematical Society, Providence, RI, 1987.

[15]  J. Rinzel and G.B. Ermentrout. Analysis of neural excitability and oscillations. In C.Koch and
      I. Segev, editors, *Methods in Neuronal Modeling: From Ions to Networks*, pages 251–291. The
      MIT Press, Cambridge, MA, second edition, 1998.

[16]  J. Rinzel, D. Terman, X.-J. Wang, and B. Ermentrout. Propagating activity patterns in large-
      scale inhibitory neuronal networks. *Science*, **279**:1351–1355, 1998.

[17]  J. Rubin and D. Terman. Analysis of clustered firing patterns in synaptically coupled networks
      of oscillators. *J. Math. Biol.*, **41**:513–545, 2000.

[18]  J. Rubin and D. Terman. Geometric analysis of population rhythms in synaptically coupled
      neuronal networks. *Neural Comput.*, **12**:597–645, 2000.

[19]  J. Rubin and D. Terman. Geometric singular perturbation analysis of neuronal dynamics. In
      B. Fiedler, G. Iooss, and N. Kopell, editors, *Handbook of Dynamical Systems, vol. 3: Towards
      Applications*. Elsevier, 2003.

[20]  D. Somers and N. Kopell. Rapid synchronization through fast threshold modulation. *Biol.
      Cybern.*, **68**:393–407, 1993.

[21]  S.H. Strogatz. *Nonlinear Dynamics and Chaos*. Addison-Wesley Publishing Company, Read-
      ing, Ma., 1994.

[22]  D. Terman, N. Kopell, and A. Bose. Dynamics of two mutually coupled inhibitory neurons.
      *Physica D*, **117**:241–275, 1998.

[23]  D. Terman and E. Lee. Partial synchronization in a network of neural oscillators. *SIAM J. Appl.
      Math.*, **57**:252–293, 1997.

[24]  D. Terman, J. Rubin, A.C. Yew, and C.J. Wilson. Activity patterns in a model for the subthala-
      mopallidal network of the basal ganglia. *J. Neuroscience*, **22**:2963–2976, 2002.

[25]  D. Terman and D. L. Wang. Global competition and local cooperation in a network of neural
      oscillators. *Physica D*, **81**:148–176, 1995.

[26]  D.H. Terman, G.B. Ermentrout, and A.C. Yew. Propagating activity patterns in thalamic neu-
      ronal networks. *SIAM J. Appl. Math.*, **61**:1578–1604, 2001.

[27]  X.-J. Wang and J. Rinzel. Spindle rhythmicity in the reticularis thalamic nucleus: synchroniza-
      tion among mutually inhibitory neurons. *Neuroscience*, **53**:899–904, 1993.

Course 4

# THEORY OF NEURAL SYNCHRONY

## G. Mato

*Comisión Nacional de Energía Atómica and CONICET*
*Centro Atómico Bariloche and Instituto Balseiro (CNEA and UNC)*
*8400 S. C. de Bariloche, Argentina*

*C.C. Chow, B. Gutkin, D. Hansel, C. Meunier and J. Dalibard, eds.*
*Les Houches, Session LXXX, 2003*
*Methods and Models in Neurophysics*
*Méthodes et Modèles en Neurophysique*

# Contents

## 1. Introduction

The nervous system controls the behavior of the organism by going through a different number of dynamical states. In some cases the dynamical states have a functional role, such as the synchronized oscillations responsible for the respiratory rhythms, in other cases they are pathological, such as the oscillations observed during the Parkinson disease or their role can be unknown, as in the different oscillations observed during sleep. In this context one objective is to understand which features of the network are relevant to control the dynamics and, if possible, to find some rule that would allow us to predict the effect of some change in the parameters. The dynamics of the network is controlled by a complex system of non-linear equations. One possible approach to the understanding of the collective dynamics is to use extensive numerical simulations to explore the space of parameters. This has the obvious drawback that an exhaustive search is clearly impossible. Moreover, as the various parameters interact in a complex non-linear way, it would be very difficult to extract from simulations general conclusions and predictions about the global behavior.

For these reasons, approximate analytical or semi-analytical methods have been introduced. These methods consider some special cases that allow to simplify considerably the dynamics of the network and to draw general conclusions about which are the more relevant parameters. Subsequently these results can be extended to more realistic situations by performing numerical simulations.

We are mostly interested in the interplay between the intrinsic properties of the neurons (controlled by the strength and kinetics of intrinsic currents) and the interactions between the neurons (type of interaction: excitatory, inhibitory, electrotonic, architecture of the network, kinetics of the interactions). We propose two situations that can be studied in-depth analytically. In the first one the neurons are weakly coupled. In this case we can apply a linearized theory that decouples the intrinsic dynamics from the interactions. In the second we consider the stability of asynchronous states in one-dimensional models. These models can be obtained from more realistic conductance based models near the onset of oscillations. In the asynchronous state the interactions are constant in time. We can evaluate the stability of this state by assuming there are small time dependent perturbations and analyzing its temporal evolution. We study a two-population network and we investigate how the characteristics of the interactions control synchrony. In both cases the results are verified with extensive numerical simulations and extended to situations that are not strictly covered by the theoretical assumptions. In all the cases we restrict ourselves to point-like neurons, we do not consider having delays in the interactions nor architectures with spatial structure. However the methods here presented can be applied to those situations as well.

This work is organized as follows: in Section 2 we introduce the theoretical framework of weakly coupled oscillators. This is first applied to the analysis of the synchronization properties of networks connected with chemical synapses. We study the combined effect of the interactions and the excitability properties of the neurons on the collective dynamics. Networks of neurons connected via gap junctions are also considered. We use the weak coupling approximation to understand the effect of the different intrinsic currents on the synchronization properties. In Section 3 we first introduce the stability analysis of the asynchronous state for a one dimensional model called *quadratic integrate-and-fire* model. This analysis allows us to characterize the generic mechanisms of synchrony in a heterogeneous two-population network. These results are applied to the analysis of the stability of self-sustained states. In Section 4 we summarize the conclusions and future lines of work.

## 2. Weakly coupled oscillators

Let us consider that the state of one neuron is given by the vector $\mathbf{X}$. The number of components of $\mathbf{X}$ will depend on the model we are using for describing the neuron. For instance, for point-like Hodgkin-Huxley neurons [1], the vector is $\mathbf{X} = (V, m, h, n)$, where $V$ is the membrane potential, $m$ and $h$ and the activation and inactivation variables of the sodium current and $n$ is the activation variable of the potassium current.

The state of the neuron evolves according to the equation

$$\frac{d\mathbf{X}}{dt} = \mathbf{F}(\mathbf{X}). \tag{2.1}$$

This set of equations will have a periodic solution with period $T$ if the condition $\mathbf{X}(t + T) = \mathbf{X}(t)$ is verified. This solution will be called a *limit cycle* if it is asymptotically stable, i.e., if there is a neighborhood $\mathcal{G}$ of the periodic solution such that any initial condition in $\mathcal{G}$ will eventually converge to the cycle. For any system with a limit cycle we can define a phase function $\phi(\mathbf{X})$ in the following way [2]: for the points on the limit cycle we choose an arbitrary point $\mathbf{X}_0$ and we assign to it $\phi(\mathbf{X}_0) = 0$. The phase of the other points on the limit cycle is defined via the differential equation:

$$\frac{d\phi(\mathbf{X})}{dt} = 1. \tag{2.2}$$

This definition can be extended to the points in the neighborhood of the limit cycle. Let us consider a point $\mathbf{Y}_0$ in $\mathcal{G}$. This point will evolve according to Eq. 2.1, giving rise to the trajectory $\mathbf{Y}(t)$. This trajectory will converge to the limit cycle. There will we one point $\mathbf{X}(0)$ on the limit cycle, whose trajectory $\mathbf{X}(t)$ will verify that $\lim_{t \to \infty} |\mathbf{Y}(t) - \mathbf{X}(t)| \to 0$. In this case we will define that $\phi(\mathbf{Y}_0) \equiv \phi(\mathbf{X}(0))$. According to this definition the phase evolves with the same velocity for points on the limit cycle or in its neighborhood, i.e.,

$$\frac{d\phi(\mathbf{X})}{dt} = 1 \tag{2.3}$$

for all $\mathbf{X} \in \mathcal{G}$.

The purpose of this extension is to broaden the concept of phase to a situation where the dynamics of the system is perturbed in such a way that, strictly speaking, it never lies exactly on the limit cycle. Let us consider the weakly perturbed system

$$\frac{d\mathbf{X}}{dt} = \mathbf{F}(\mathbf{X}) + \epsilon \mathbf{p}(\mathbf{X}) \tag{2.4}$$

with $\epsilon \ll 1$. The phase will evolve in this case according to:

$$\frac{d\phi(\mathbf{X})}{dt} = \frac{\partial \phi}{\partial \mathbf{X}} \cdot \frac{d\mathbf{X}}{dt} = 1 + \epsilon \frac{\partial \phi}{\partial \mathbf{X}} \cdot \mathbf{p}(\mathbf{X}). \tag{2.5}$$

This equation is exact but useless for determining the time evolution of the phase, because to evaluate the right hand side of the equation we need to know the point $\mathbf{X}$ in which evaluate the functions $\frac{\partial \phi}{\partial \mathbf{X}}$ and $\mathbf{p}(\mathbf{X})$. In general it is not possible to recover that information from the phase $\phi$ alone. However, if the perturbation is small we can assume that the dynamics will depart from the limit cycle a distance proportional to $\epsilon$. In this case we can evaluate the right hand side *on the limit cycle*, i.e.,

$$\frac{d\phi}{dt} = 1 + \epsilon \mathbf{Z}(\phi) \cdot \mathbf{p}(\phi) + O(\epsilon^2) \tag{2.6}$$

with $\mathbf{Z}(\phi) = \frac{\partial \phi(\mathbf{X})}{\partial \mathbf{X}}$ and $\mathbf{p}(\phi) = \mathbf{p}(\mathbf{X})$ for a point $\mathbf{X}$ whose phase is $\phi$.

For any time dependent function $f$ we can define the average over one cycle as

$$< f > (t) = \frac{1}{T} \int_{t-T}^{t} f(t')dt'. \tag{2.7}$$

The evolution of the time averaged phase is given by

$$\frac{d < \phi > (t)}{dt} = 1 + \frac{\epsilon}{T} \int_{t-T}^{t} \mathbf{Z}(\phi(t')) \cdot \mathbf{p}(\phi(t'))dt' + O(\epsilon^2). \tag{2.8}$$

This shows that $\phi$ and $< \phi >$ differ at most by an amount of order $\epsilon$. Therefore we can replace in the right hand side $\phi$ by $< \phi >$:

$$\frac{d < \phi > (t)}{dt} = 1 + \frac{\epsilon}{T} \int_{t-T}^{t} \mathbf{Z}(< \phi(t') >) \cdot \mathbf{p}(< \phi(t') >)dt' + O(\epsilon^2). \tag{2.9}$$

Let us note that $\mathbf{Z}(\phi)$ and $\mathbf{p}(\phi)$ are $T$-periodic functions of $\phi$, therefore we can write

$$\frac{d\phi}{dt} = 1 + \frac{\epsilon}{T} \int_{0}^{T} \mathbf{Z}(\phi).\mathbf{p}(\phi)d\phi + O(\epsilon^2) \tag{2.10}$$

where we have dropped the brackets for simplicity. In the following we will be always writing the dynamics in terms of time averaged quantities.

This result can be easily generalized to the situation where the perturbation of the dynamics comes from the interaction from other neurons. Let us first consider the case of two identical oscillators symmetrically coupled:

$$\frac{d\mathbf{X}_1}{dt} = \mathbf{F}(\mathbf{X}_1) + \epsilon\mathbf{p}(\mathbf{X}_1, \mathbf{X}_2) \tag{2.11}$$

$$\frac{d\mathbf{X}_2}{dt} = \mathbf{F}(\mathbf{X}_2) + \epsilon\mathbf{p}(\mathbf{X}_2, \mathbf{X}_1). \tag{2.12}$$

After using the weak coupling approximation and averaging over one period we find

$$\frac{d\phi_1}{dt} = 1 + \Gamma(\phi_1 - \phi_2) + O(\epsilon^2) \tag{2.13}$$

$$\frac{d\phi_2}{dt} = 1 + \Gamma(\phi_2 - \phi_1) + O(\epsilon^2) \tag{2.14}$$

where

$$\Gamma(\phi) = \frac{\epsilon}{T} \int_{0}^{T} \mathbf{Z}(\phi' + \phi).\mathbf{p}(\phi' + \phi, \phi)d\phi'. \tag{2.15}$$

The function $\Gamma(\phi)$ is the *phase interaction function*. It is a convolution between the $Z(\phi)$ (*phase response function*) and the interaction $p(\phi, \phi')$. It is a periodic function with period $T$ because both $Z(\phi)$ and $p(\phi, \phi')$ are periodic with the same period. The phase response function represents how the phase is advanced or delayed when the oscillator receives an infinitesimal perturbation. The total change of the phase velocity is given by the convolution between this response and the interaction. Let us remark that the first function is a property of the single neuron dynamics and is totally independent on the interaction.

We can also generalize this result to $N$ identical and all-to-all coupled oscillators:

$$\frac{dX_l}{dt} = F(X_l) + \epsilon \sum_{j=1}^{N} p(X_l, X_j) \tag{2.16}$$

with $l = 1, ..., N$, obtaining (up to order $\epsilon^2$):

$$\frac{d\phi_l}{dt} = 1 + \sum_{j=1}^{N} \Gamma(\phi_l - \phi_j). \tag{2.17}$$

## 2.1. Stability of cluster states

### 2.1.1. One-cluster state

The system of equations Eq. 2.17 can display a great variety of solutions. The simplest one is the *one-cluster state, fully synchronized* or *in-phase locked* solution. In this solution all the oscillators have the same phase. For neural system this means that all the neurons have their action potentials at the same time. This solution is described by $\phi_l(t) = \phi_0(t)$ $l = 1, ..., N$. This is obviously a solution of Eq. 2.17 if $\phi_0(t) = (1 + N\Gamma(0))t$. However the existence of this solution does not mean necessarily that it is physically relevant, because it could be unstable. If this were the case any small perturbation would destroy it. To analyze the stability we perform a linearization around the one-cluster state:

$$\phi_l(t) = \phi_0(t) + \delta_l(t) \tag{2.18}$$

where $\delta_l(t) \ll 1$. Inserting this expression in Eq. 2.17 and keeping up to first order in $\delta_l$ we obtain

$$\frac{d\delta_l(t)}{dt} = \sum_{j=1}^{N} A_{lj}\delta_j(t) \tag{2.19}$$

where the matrix elements are given by $A_{ll} = (N-1)\frac{d\Gamma(0)}{d\phi}$ and $A_{lj} = -\frac{d\Gamma(0)}{d\phi}$ for $l \neq j$. The stability will be controlled by the eigenvalues of the matrix $A$. If there is at least one eigenvalue with positive real part the state will be unstable in the direction of the corresponding eigenvector. In our case the eigenvalues of $\mathbf{A}$ are 0 (with multiplicity 1) and $\frac{d\Gamma(0)}{d\phi}$ (with multiplicity $N - 1$). The first eigenvalue corresponds to the constant change of all the phases at the same time. It represents a global shift along the limit cycle and its value is 0 because of the temporal invariance of the dynamical system. The other eigenvalues are always controlled by the same quantity : $\frac{d\Gamma(0)}{d\phi}$. If this is negative the state is stable with respect an arbitrary perturbation.

### 2.1.2. Two-cluster states
Another possibility is that the oscillators split in two groups. Oscillations inside each group are fully synchronized but there is a constant phase difference between he two groups: $\phi_l(t) = \phi_0(t)$ for $l = 1, ..., q$ and $\phi_l(t) = \phi_0(t) + \Delta$ for $l = q + 1, ..., N$. Inserting this ansatz in Eq. 2.17 we find a self-consistency relation between the "mass" of the first cluster $q$ and the dephasing $\Delta$:

$$q = N \frac{\Gamma(0) - \Gamma(-\Delta)}{2\Gamma(0) - \Gamma(\Delta) - \Gamma(-\Delta)}. \tag{2.20}$$

The stability analysis can be performed in a similar way to the previous section. In this case we find that the eigenvalues are

$$\lambda_1 = 0 \tag{2.21}$$

$$\lambda_2 = q\frac{d\Gamma(0)}{d\phi} + (N - q)\frac{d\Gamma(-\Delta)}{d\phi} \tag{2.22}$$

$$\lambda_3 = (N - q)\frac{d\Gamma(0)}{d\phi} + q\frac{d\Gamma(\Delta)}{d\phi} \tag{2.23}$$

$$\lambda_4 = q\frac{d\Gamma(\Delta)}{d\phi} + (N - q)\frac{d\Gamma(-\Delta)}{d\phi} \tag{2.24}$$

with multiplicity 1, $q - 1$, $N - q - 1$ and 1 respectively. The second and third eigenvalues correspond to perturbations inside each one of the clusters. The last one corresponds to perturbations that change the relative distance between the clusters. Let us note that it is possible in principle to have several combinations of the parameters $q$, $\Delta$ for which the two-cluster state is stable. In this case we would have coexistence between several possible dynamical states. Each one of them could be reached by choosing a suitable initial condition.

### 2.1.3. N-cluster state
The results of the previous section can be generalized to higher order cluster states. For instance the general condition for existence and stability of a generic cluster state (in which all the clusters have the same number of oscillators) can

be found in [3]. However, here we will go directly to the opposite case from the fully synchronized state. In this situation each oscillator is in its own cluster, and is separated from the next one by a phase difference equal to $T/N$, i.e., $\phi_j(t) = \phi_0(t) - jT/N$, $j = 1, .., N$. This is the *N-cluster* state, or *splay* state. The self-consistency equation obtained from Eq. 2.17 gives us that

$$\phi_0(t) = \left(1 + \sum_{j=1}^{N} \Gamma(jT/N)\right) t. \tag{2.25}$$

The stability of this state can be easily evaluated. The matrix $\mathbf{A}$ of the linearized system has elements:

$$A_{ll} = \sum_{j=2}^{N} \frac{d\Gamma((j-1)T/N)}{d\phi} \tag{2.26}$$

$$A_{lj} = -\frac{d\Gamma((j-l)T/N)}{d\phi} \quad (j \neq l) \tag{2.27}$$

with $l, j = 1, ..., N$. This is a cyclic matrix. The eigenvalues are simply the Fourier components of the row vector with components $A_{1j}$:

$$\lambda_l = \sum_{j=1}^{N} A_{1j} \exp\left(\frac{2\pi i l(j-1)}{N}\right) \tag{2.28}$$

for $l = 1, ..., N$. In the limit of large $N$ this is simply the Fourier transform of the derivative of $\Gamma$:

$$\lambda_l = \frac{N}{T} \int_0^T -\frac{d\Gamma(\phi)}{d\phi} \exp\left(\frac{2\pi i l\phi}{T}\right) d\phi \tag{2.29}$$

or

$$\lambda_l = \frac{2\pi i l N}{T} \Gamma_l \tag{2.30}$$

where $\Gamma_l = \frac{1}{T} \int_0^T \Gamma(\phi) \exp\left(\frac{2\pi i l\phi}{T}\right) d\phi$ is the Fourier component of the phase interaction function. Therefore the condition for stability of the splay state, $\Re(\lambda_l) \leq 0$, $l = 1, ..., N$ can be written as

$$\Im(\Gamma_l) \geq 0 \tag{2.31}$$

for all $l = 1, ..., N$. If this condition is not satisfied for some value of $l$ then the state is unstable.

As we change some parameter of the dynamics we could have a transition from a stable splay state to a unstable splay state. This occurs when $\Re(\lambda_{l_0}) = 0$ for some index $l_0$. In this case the instability develops in time with a profile that depends on the value of $l_0$. In general, we will find, a *smeared* cluster state. The oscillators tend to generate $l_0$ groups that oscillate with similar phases and there is a constant dephasing between the groups.

### 2.1.4. Stochastic dynamics

In all the previous cases the dynamics we considered is deterministic, i.e., the initial state determines completely the trajectory. It is also interesting to consider the case of a dynamics that includes random forces in order to understand their influence on the synchronization properties. We will consider a system of $N$ fully coupled identical oscillators whose phase evolves according to

$$\frac{d\phi_l}{dt} = 1 + \sum_{j=1}^{N} \Gamma(\phi_l - \phi_j) + \xi_l(t) \tag{2.32}$$

where $\xi_l(t)$ is a Gaussian white noise: $< \xi_l(t)\xi_j(t') >= 2D\delta(t - t')\delta_{lj}$.

In this case the trajectories of the individual oscillators are random variables. However we can define the quantity $n(\phi, t)$ that represents the fraction of oscillators that have a given phase at a given time. This quantity evolves according to a *Fokker-Planck* equation [4]:

$$\frac{\partial n(\phi, t)}{\partial t} + \frac{\partial(v(\phi, t)n(\phi, t))}{\partial \phi} = D\frac{\partial^2 n(\phi, t)}{\partial^2 \phi} \tag{2.33}$$

with

$$v(\phi, t) = 1 + \int_0^T \Gamma(\phi - \phi')n(\phi', t)d\phi'. \tag{2.34}$$

This equation has always a solution $n(\phi, t) = N/T$. This is the *asynchronous state*. This is a generalization of the *splay state* from the previous case. In both cases any network averaged quantity will give a time independent result in the limit of large $N$. In that sense both states are asynchronous. The difference is that in the splay state there is a constant phase difference between any pair of oscillators, that would give rise to a non-zero cross correlation function. Those cross correlations are suppressed by any non-zero value of noise $D$.

As in the previous case we can analyze the stability of the asynchronous state. To do that we expand the density near the asynchronous state solution:

$$n(\phi, t) = \frac{N}{T} + \epsilon(\phi, t). \tag{2.35}$$

Inserting this expression into the Fokker-Planck equation we obtain (up to order $\epsilon^2$):

$$\frac{\partial \epsilon(\phi, t)}{\partial t} + \frac{\partial((1 + N\Gamma_0)\epsilon(\phi, t) + u)}{\partial \phi} = D\frac{\partial^2 \epsilon(\phi, t)}{\partial^2 \phi} \tag{2.36}$$

with $u = \frac{N}{T}\int_0^T \Gamma(\phi - \phi')\epsilon(\phi', t)d\phi'$ and $\Gamma_0 = \frac{1}{T}\int_0^T \Gamma(\phi)d\phi$. Defining the Fourier components as

$$\epsilon_l(t) = \frac{1}{T}\int_0^T \epsilon(\phi, t)\exp\left(\frac{2\pi i l\phi}{T}\right)d\phi \tag{2.37}$$

$$\Gamma_l = \frac{1}{T}\int_0^T \Gamma(\phi)\exp\left(\frac{2\pi i l\phi}{T}\right)d\phi \tag{2.38}$$

we find

$$\frac{d\epsilon_l(t)}{dt} - \frac{2\pi i l}{T}((1 + N\Gamma_0)\epsilon_l(t) + N\epsilon_l(t)\Gamma_l) = -\frac{4\pi^2 l^2 D}{T^2}\epsilon_l(t). \tag{2.39}$$

The Fourier components $\epsilon_l(t)$ will behave as $\epsilon_l(t) \propto \exp(\lambda_l t)$ with

$$\lambda_l = \frac{2\pi i l}{T}((1 + N\Gamma_0) + N\Gamma_l) - \frac{4\pi^2 l^2 D}{T^2}. \tag{2.40}$$

The asynchronous state will be stable iff

$$\Re(\lambda_l) = -\frac{2\pi l N}{T}\Im(\Gamma_l) - \frac{4\pi^2 l^2 D}{T^2} \leq 0 \tag{2.41}$$

for all $l$. In the limit of zero noise we recover the result of the stability of the splay state ($\Im(\Gamma_l) \geq 0$). Let us note that in order to have a well defined limit $N \to \infty$ we must have a finite $N\Gamma_l$. This means that the interaction strength has to go as $1/N$.

## 2.2. Evaluation of the phase interaction function

According to the results of the previous sections the existence and stability of the different dynamical states is totally determined by the phase interaction function $\Gamma$. This function is a convolution between the phase response function $\mathbf{Z}$ and the interaction function $\mathbf{p}$. The last one is given by the model of interaction we are analyzing (chemical synapses, gap junctions, etc.). The phase response function is a single neuron property that can be evaluated in two ways. The first is by using the definition of $\mathbf{Z}$ as the gradient of the phase. The single neuron model is numerically iterated until convergence to the limit cycle is reached. At a given point of the limit cycle each one of the variables $X_l$ is instantaneously perturbed by some small amount $\Delta X_l$ and the system is allowed to relax again to the limit cycle. On the limit cycle we can measure the change on the phase generated by the perturbation with respect to the non perturbed system, $\Delta \phi_l$. A positive value of $\Delta \phi_l$ means that the system has been accelerated. The phase response function can be estimated as $Z_l \approx \frac{\Delta \phi_l}{\Delta X_l}$. Let us note that this procedure has been applied in experimental systems [6] to study the response properties of cortical neurons. Let us also observe that this procedure has to be performed only for the components with index $l$ for which $p_l$ is non zero. If for instance the interaction is only via the membrane potential $V$, we have to evaluate only $Z_V$ and not the other components.

The other method for evaluating $\mathbf{Z}$ is by noting that their components have to be a periodic solution of the system

$$\frac{dZ_i(t)}{dt} = -\sum_{j=1}^{N} \frac{\partial F_j(\mathbf{X}(t))}{\partial X_i} Z_j(t) \tag{2.42}$$

with the normalization condition $\sum_{j=1}^{N} Z_j(0) F_j(\mathbf{X}(0)) = 1$. This can be easily proved by taking the time derivative of the definition of the response function: $\mathbf{Z} = \frac{\partial \phi}{\partial \mathbf{X}}$. To use this method we need to evaluate numerically the Jacobian matrix $\frac{\partial F_j}{\partial X_i}$ along the limit cycle and integrate the *adjoint system* Eq. 2.42. That is done for instance in the package XPP [7].

### 2.3. Synaptic interactions

#### 2.3.1. Conductance based neurons
The neuronal dynamics is controlled by the flux of ions across the membrane. Their equations are given in Appendix A for two models, Hodgkin-Huxley (HH) and Wang-Buszaki (WB). When the neurons are in a network we have to include in the dynamics an additional term for taking into account the interaction. For chemical synapses the interaction term is described by an additional current $I_{syn}(t)$ that has to be included in the right hand side of the evolution equation for the membrane potential $V$ (Eq. A. 1). In this type of interaction, every time the *presynaptic* neuron generates and action potential it opens channels on the *postsynaptic* neuron that give rise to a current. The synaptic current is described by

$$I_{syn}(t) = g_{syn}(t)(V_{syn} - V) \tag{2.43}$$

where $V_{syn}$ is the reversal potential of the interaction. Its value depends on the type of interaction: it will be above the spiking threshold for excitatory interactions and below it for inhibitory interactions. The synaptic conductance is described by

$$g_{syn}(t) = g_{syn}^0 \sum_{spikes} f(t - t_{spike}) \qquad (2.44)$$

where $t_{spikes}$ are the time of the spikes of the presynaptic neurons and $f(t)$ describes the kinetics of the synapses. It is usually written as

$$f(t) = \frac{\exp(-t/\tau_1) - \exp(-t/\tau_2)}{\tau_1 - \tau_2} \Theta(t) \qquad (2.45)$$

where $\tau_1$ and $\tau_2$ are the rise and decay times and $\Theta$ is the Heaviside function. In Fig. 1 we can see the synaptic conductance for a synapses with rise time of 1 msec and decay time of 3 msec.

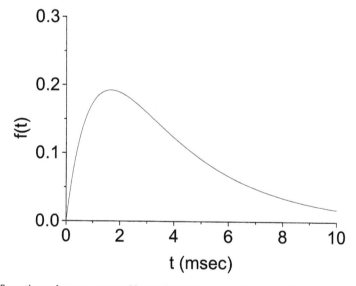

Fig. 1. Synaptic conductance generated by a spike in the presynaptic neuron. Rise time: $\tau_1 = 1$ msec, decay time: $\tau_2 = 3$ msec.

We will now apply the method developed in the previous section to networks of neurons connected with chemical synapses. For applying the phase reduction method we have first to assume that the neurons are spiking periodically with period $T$. As the interaction is not instantaneous we have to take into account the contribution of all the previous spikes:

$$g_{syn}^T(t) = g_{syn}^0 \sum_{n=0}^{\infty} f(t - nT) = g_{syn}^0 \frac{\frac{\exp(-t/\tau_1)}{1-\exp(-T/\tau_1)} - \frac{\exp(-t/\tau_2)}{1-\exp(-T/\tau_2)}}{\tau_1 - \tau_2} \qquad (2.46)$$

for $0 \leq t \leq T$.

We will first apply the method to the analysis of the synchronization properties of a set of HH neurons coupled with excitatory interactions. The firing rate of these neurons depends on the injected current. If the current is below a threshold value $I_c$ of about 9.8 $\mu A/cm^2$ they are inactive. Above that value they undergo an sub-critical Hopf bifurcation [5] and they begin to spike with a firing rate of about 55 Hz. In Fig. 2 we show the membrane potential for one period of oscillation. The injected current is chosen to have an interspike period of 15 msec.

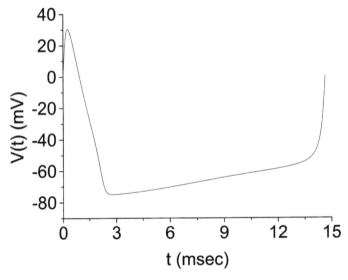

Fig. 2. Membrane potential during one period of oscillation for a HH neuron with a period of oscillation of 15 msec.

For this situation we can evaluate the response function **Z** along the period. As the neurons interact only via the voltage variable we only plot the first component of the response in Fig. 3. We can see that the response function is divided in three main components. Immediately after the spike, its value is very small. This is the *refractory period*. During this part the dynamics is controlled by the large sodium and potassium intrinsic currents and a small perturbation almost has no effect on the timing of the next spike. Then, there is a negative region. This means that a small depolarization during this time window will *delay* the next spike. And finally we have a positive region, where a depolarization will *advance* the next spike. We will see later, that the presence of the negative region has a strong influence on the synchronization properties.

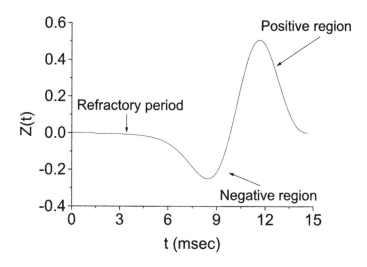

Fig. 3. Phase response function for a HH neuron with a period of oscillation of 15 msec.

We also need to take into account the dependence of the synaptic current on the potential of the postsynaptic neuron (see Eq. 2.43). We will take an excitatory interaction with $V_{syn} = 0$ mV. Therefore the phase interaction function will be the convolution between $-Z(t)V(t)$ and $g_{syn}^T(t)$. Both functions are shown in Fig. 4. From this figure we can see two main effects of the negative region on the response function. First, the product of $-Z(t)V(t)$ and $g_{syn}^T(t)$ will be negative, leading to $\Gamma(0) < 0$. Second, the integral will decrease as we shift the synaptic current curve to the right. But according to Eq. 2.15 this is equivalent to increasing the value of $\phi$. Therefore we have $\frac{d\Gamma(0)}{d\phi} < 0$. This means that the fully synchronized state will be stable but the firing rate of the coupled system will be smaller because of the interaction. The result of the convolution is shown in Fig. 5.

In Fig. 6 we show the firing rate of a system of two HH neurons, identical and symmetrically coupled with an interaction strength $g_{syn}^0$ and the same time constant as the previous figures. The prediction of the phase reduction method is that the firing rate should depend linearly on $g_{syn}^0$ with a slope $\Gamma(0)$. That

G. Mato

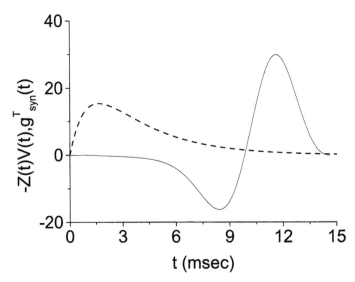

Fig. 4. Solid line: $-V(t)Z(t)$ for a HH neuron with a period of 15 msec. Dashed line: synaptic conductance generated by a spike in the presynaptic neuron. (rise time: $\tau_1 = 1$ msec, decay time: $\tau_2 = 3$ msec.)

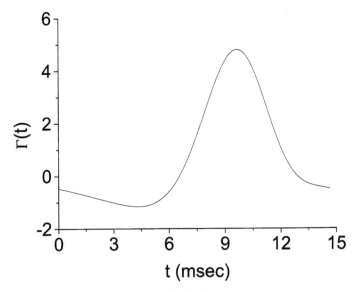

Fig. 5. Phase interaction function $\Gamma$ obtained from the convolution of the curves in the previous figure.

prediction is shown with the solid line. We can see that that the interaction can be strong enough to change the firing rate in about 20 % and still the weak coupling approximation gives a reasonable results. These results will be affected by the time constant of the synaptic interaction or the firing rate of the neuron. If the peak of the synaptic conductance were to the right of the minimum of the response function, then we would have $\frac{d\Gamma(0)}{d\phi} > 0$. Indeed a transition has been found from in-phase locking to out-of-phase locking as synaptic time constants are increased [8].

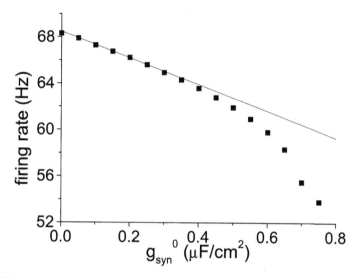

Fig. 6. Firing rate of a pair of identically coupled HH neurons as a function of the synaptic conductance. For all points the two neurons are stably in-phase locked.

This behavior is strongly model dependent. We have performed the same analysis on the WB model [9]. This is a modification of the HH model, but the oscillations appear via a saddle-node bifurcation instead of a Hopf one [5]. The frequency of oscillations $\nu$ goes as $\sqrt{I - I_c}$ for currents near the threshold current $I_c$. In Fig. 7 we show the membrane potential during one period of oscillation and in Fig. 8 the phase response function. We can see that is completely different from the one of the HH model. There is almost no negative region after the spike. This means that $\Gamma(\phi) > 0$ for all $\phi$ and $\frac{d\Gamma(0)}{d\phi} > 0$ (see Figs 9 and 10). Excitatory interactions will lead to out-of-phase locking and the neurons will be always accelerated by the excitatory interaction.

These results show that excitability properties essentially determine synchronization properties of synaptically coupled neurons. Models in which the os-

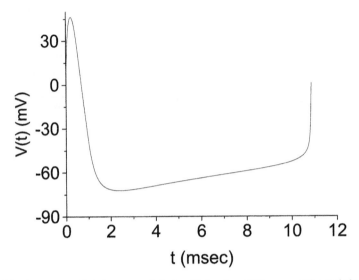

Fig. 7. Membrane potential during one period of oscillation for a WB neuron with a period of oscillation of 11 msec.

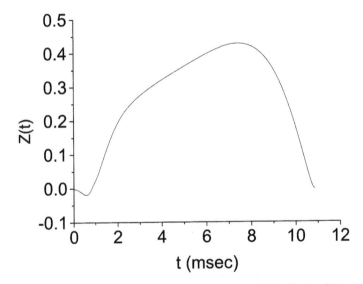

Fig. 8. Phase response function for a WB neuron with a period of oscillation of 11 msec.

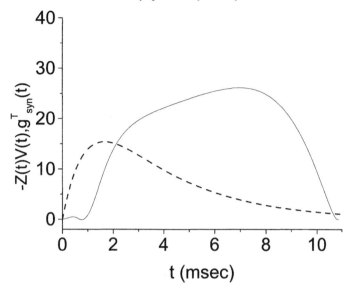

Fig. 9. Solid line: $-V(t)Z(t)$ for a WB neuron with a period of 11 msec. Dashed line: synaptic conductance generated by a spike in the presynaptic neuron. (rise time: $\tau_1 = 1$ msec, decay time: $\tau_2 = 3$ msec.)

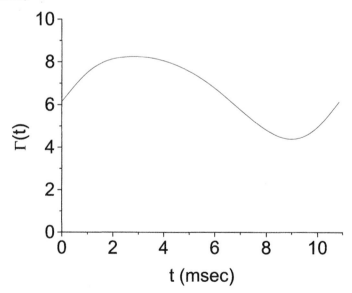

Fig. 10. Phase interaction function $\Gamma$ obtained from the convolution of the curves in the previous figure.

cillations arise after a saddle-node bifurcation are called to be *Type I*. For fast excitatory interactions they have unstable fully synchronized solution and the interaction leads always to higher firing rates. Inversely when oscillations are generated by a Hopf bifurcation we call the neurons to belong to *Type II*. These neurons have significant negative regions in the response function, in-phase locking is stable for fast excitatory interactions and more surprisingly, excitation leads to slower oscillations. All these results are reversed if we interchange excitation by inhibition.

### 2.3.2. Integrate-and-fire neurons

A simplified neuronal model that has been extensively studied is the *Integrate and Fire* (IF) model. In this system a neuron integrates passively its input until the membrane potential reaches a threshold value. At that moment, it is instantaneously reset to a lower value. In this section we will analyze the synchronization properties of a set of IF neurons coupled with synaptic interactions using the phase reduction method. We will see that we can fully understand the stability of the asynchronous state.

The evolution equation for a IF neuron is

$$\frac{dV}{dt} = -V + I \tag{2.47}$$

with $V(t^+) = 0$ if $V(t) = 1$. If the external current $I$ is larger than 1 then the system undergoes oscillations with frequency

$$\nu = -\frac{1}{\log(1 - 1/I)}. \tag{2.48}$$

We will first analyze the response function of this system. For any one-dimensional dynamics the response function is easy to evaluate . This is because the effect of an infinitesimal change of the state variable on the timing will be inversely proportional to the velocity at that point,

$$Z(t) = \frac{1}{dV(t)/dt} = \frac{1}{-V(t) + I}. \tag{2.49}$$

The solution of Eq. 2.47 is given by

$$V(t) = I(1 - \exp(-t)) \tag{2.50}$$

for $0 \leq t \leq T = 1/\nu$, therefore

$$Z(t) = \frac{\exp(t)}{I}. \tag{2.51}$$

We will use as a model for the interaction a synaptic current given by

$$I_{syn}(t) = g_{syn}^0 \sum_{spikes} f(t - t_{spike}) \tag{2.52}$$

where $f(t)$ is given by Eq. 2.45. In this model there is no term $V - V_{syn}$ in the synaptic current. The interaction is excitatory or inhibitory according to the sign of $g_{syn}^0$.

As we have seen in section 2.1.4 the stability of the asynchronous state is controlled by the Fourier components of the phase interaction function $\Gamma_l$. For our model the interaction function can be written as

$$\Gamma(\phi) = \frac{1}{T} \int_0^T Z(\phi + \phi') I_{syn}(\phi') d\phi' \tag{2.53}$$

and the Fourier components are given by

$$\Gamma_l = Z_l I_{syn,l}^* \tag{2.54}$$

where $Z_l$ and $I_{syn,l}$ can be evaluated using Eqs. 2.51, 2.52 and 2.46. We obtain

$$Z_l = \frac{\exp(T) - 1}{IT} \frac{1}{1 + 2\pi i l/T} \tag{2.55}$$

$$I_{syn,l}^* = \frac{g_{syn}^0}{T\tau_1\tau_2} \frac{1}{(1/\tau_1 + 2\pi i l/T)(1/\tau_2 + 2\pi i l/T)}. \tag{2.56}$$

Therefore, up to some positive constant

$$\Im(\Gamma_l) \propto g_{syn}^0 \left( \frac{4\pi^2 \tau_1 \tau_2 l^2}{T^2} - (\tau_1 + \tau_2) - 1 \right). \tag{2.57}$$

Let us first consider excitatory interactions in the absence of noise, ($g_{syn}^0 > 0$, $D = 0$). In that case the sign of $\Re(\lambda_l)$ will be the opposite from the sign of $\frac{4\pi^2 \tau_1 \tau_2 l^2}{T^2} - (\tau_1 + \tau_2) - 1$. At high firing rates the first term will always dominate for any mode and the asynchronous state will be stable. On the other hand if one mode is unstable, all the modes with a lower index $l$ will be also unstable. For inhibitory interactions the behavior is the opposite one: the sign of $\Re(\lambda_l)$ is the same as the sign of $\frac{4\pi^2 \tau_1 \tau_2 l^2}{T^2} - (\tau_1 + \tau_2) - 1$. For any firing rate there will be an index $l$ high enough for which $\Re(\lambda_l) > 0$ and the asynchronous state will be unstable. This will lead to high order clustering for inhibitory IF networks.

We can also see the difference between excitatory and inhibitory coupling as a function of the firing rate. At high firing rate the term proportional to $1/T^2$ will dominate and because of its positive sign, the system will be more easily synchronized by inhibition. At low firing rates this term will be dominated by $-(\tau_1 + \tau_2) - 1$ that has the opposite sign. In this case excitation will be more efficient to destabilize the asynchronous state.

## 2.4. Gap junctions

In this section we apply the theory of phase reduction for analyzing the synchronization properties of *gap junctions* or *electrotonic* interactions. This interactions are present in sites where gap-junctions bridge the membranes of two neurons. There is a flow of current between the two neurons that is proportional to the difference of membrane potential between them. The current that flows from neuron 1 to neuron 2 can be described as

$$I_{gj}(t) = g_{gj}(V_1(t) - V_2(t)). \tag{2.58}$$

This term has to be included in the evolution equation for the membrane potential such as Eq. A. 1. These interactions have long been known to exist in invertebrates [10, 11] but it is only recently that evidence of their ubiquity has been unequivocally found in the mammalian brain. Experiments have revealed that electrical synapses increase the synchronization of neural activity [12–19]. In contrast with the results of these works, it has been recently reported that inspiratory motoneurons may have their activity more strongly synchronized in presence of CBX, a blocker of electrical synapses, than in the control situation [20]. Therefore, in this case, electrical synapses *desynchronize* neural activity.

We analyze here the influence of intrinsic currents on the synchronizing effects of gap-junctions [21]. We first present the results of numerical simulations and then the analytical results.

### 2.4.1. Numerical simulations of conductance based models

In order to study the effect of intrinsic currents we supplement the WB model with two additional currents: a slow potassium current and a persistent sodium current. The equations for these currents are given in Appendix A with a value of $\phi = 3$[1]. Moreover we also change the value of the delayed rectifier potassium current. Therefore, we have three control parameters $g_K$, $g_{Ks}$ and $g_{NaP}$. The effect of these currents on the firing properties of the neurons can be seen in Fig. 11.[2] In this figure each panel corresponds to a different set of values of

---

[1]See Appendix A

[2]In the simulations of the conductance-based model the differential equations were integrated using the second-order Runge-Kutta (RK2) scheme with fixed time step: $\Delta_t = 0.01$ msec. Averaged quantities (firing rate, CV, $\chi$) were computed over a time period of 1 sec after discarding a transient of 500 msec. See Appendix B for the definitions of CV and $\chi$.

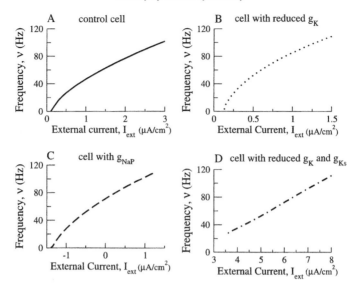

Fig. 11. I-f curves of the conductance-based model neurons. Solid line: $g_K = 9$ mS/cm², $g_{Ks} = g_{NaP} = 0$ (control case). Dotted line: $g_K = 2.5$ mS/cm², $g_{Ks} = g_{NaP} = 0$. Dashed line: $g_K = 9$ mS/cm², $g_{Ks} = 0$, $g_{NaP} = 0.2$ mS/cm². Dash-dotted line: $g_K = 2.5$ mS/cm², $g_{Ks} = 0.2$ mS/cm², $g_{NaP} = 0$.

intrinsic conductances. The firing rate was computed from the steady-state response of the neuron to steps of constant external current of different amplitudes. A, Control case: $g_K = 9$ mS/cm², $g_{Ks} = g_{NaP} = 0$. B, $g_K = 2.5$ mS/cm², $g_{Ks} = 0$. Note that the scale of the x-axis is half the scale used in A. The main effect on the I-f curve of the reduction of $g_K$ is multiplicative (change in the gain). C, $g_{NaP} = 0.2$ mS/cm², $g_K = 9$ mS/cm², $g_{Ks} = g_{NaP} = 0$. Other parameters are as in A. The main effect on the I-f curve of the persistent-sodium is subtractive (change in the rheobase). D, $g_K = 2.5$ mS/cm², $g_{Ks} = 0.2$ mS/cm², $g_{NaP} = 0$. The slow potassium current reduces the gain of the neuron and prevents low frequency firing. It also linearizes the I-f curve (compare with B).

We now show the effect of the intrinsic currents on the size and temporal evolution of the evoked potential generated when one neuron generates an action potential. This evoked potential is called *spikelet*. Fig. 12A: an action potential generated by a short and strong current pulse. The duration $\delta$ and the amplitude $A$ of the pulse are the same for all four traces: $\delta = 1$ msec and $A = 50$ μA/cm². The neuron was hyperpolarized to prevent firing when persistent sodium was present and to obtain the same initial membrane potential for all four parameter sets. The solid line and the dashed line overlap. Fig. 12B: spikelet induced

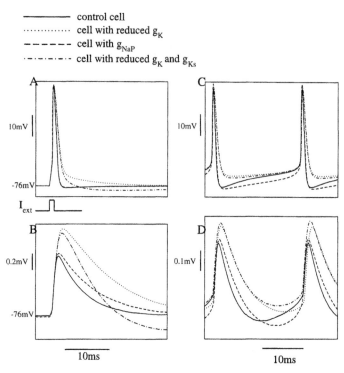

Fig. 12. Traces and spikelets of the conductance-based model neurons. A: spike generated by one pulse of current. B: spikelets generated by this spike. C: constant current induces periodic spikes with a period of about 30 msec. D; spikelets generated by the spike trains. Solid line: $g_K = 9$ mS/cm$^2$, $g_{Ks} = g_{NaP} = 0$ (control case). Dotted line: $g_K = 2.5$ mS/cm$^2$, $g_{Ks} = g_{NaP} = 0$. Dashed line: $g_K = 9$ mS/cm$^2$, $g_{Ks} = 0$, $g_{NaP} = 0.2$ mS/cm$^2$. Dash-dotted line: $g_K = 2.5$ mS/cm$^2$, $g_{Ks} = 0.2$ mS/cm$^2$, $g_{NaP} = 0$.

by a presynaptic neuron firing an action potential as in A. The synaptic conductance is $g_{gj} = 0.005$ mS/cm$^2$ which corresponds to a coupling coefficient [22] $CC \approx 5\%$. The post-synaptic neuron has the same intrinsic properties as the presynaptic neuron. The height and the width of the spikelet increase when $g_K$ is reduced. Increasing the persistent-sodium conductance has a minor effect on the height of the spikelet but changes its width. The slow potassium current decreases slightly the spikelet size but reduces significantly its width and induces hyperpolarization in the postsynaptic neuron. Fig. 12C: a constant external current is injected to make the neuron fire at 50 Hz. The external current is: $I_{ext} = 1.10$ $\mu$A/cm$^2$ for the solid line, $I_{ext} = 0.48$ $\mu$A/cm$^2$ for the dotted line, $I_{ext} = 4.88$ $\mu$A/cm$^2$ for the dash-dotted line, $I_{ext} = -0.55$ $\mu$A/cm$^2$

for the dashed line. The subthreshold voltage of the neuron when the persistent sodium is added (dashed line) is *below* the control (solid line), although the persistent sodium current accelerates the depolarization of the neuron because here we compare the firing patterns for the same firing rate. The external current is therefore smaller when the persistent sodium is present than in the control case. Fig. 12D: spikelets induced by a presynaptic neuron firing tonically, as in C. The postsynaptic and the presynaptic neurons have the same intrinsic properties. The spikelets in the presence (dash-dotted line) and in absence of $I_{Ks}$ (dotted line) are more similar to one another than in B, because the slow potassium current is now saturated. In both cases the amplitude of the spikelets and larger than for the control (solid line). From this figure it seems that increasing potassium conductances tend to reduce the effect of the spikelets and that would tend to reduce synchrony. Inversely increasing slow sodium conductance tends to have the opposite effect. However when we measure the synchrony on the network we find this is not the case. In Fig. 13 we show the synchronization parameter for a network of 1000 neurons randomly connected. The average connectivity is $K = 10$. The dynamics includes white noise with standard deviation $\sigma$. The synaptic conductance is $g_{gj} = 0.005$ mS/cm$^2$. The external current $I_{ext}$, and the standard deviation of the noise $\sigma$, were changed to control the average firing rate and the average CV of the neurons. In all the simulations the average firing rate is 50 Hz $\pm 10\%$, and the CV $\approx 0.1$. In each of the panels, the corresponding values of $I_{ext}$ and $\sigma$ are plotted vs. the conductance that is varied (figures on the right of each panel). In Fig. 13A the conductance $g_K$, varies; $g_{Ks} = g_{NaP} = 0$. The network activity is asynchronous for small $g_K$ and synchronous for $g_K > 4$ mS/cm$^2$. In Fig. 13B, the conductance $g_{Ks}$, varies; $g_K = 2.5$ mS/cm$^2$, $g_{NaP} = 0$. For $g_{Ks} < g_{Ks}^* \approx 0.05$ mS/cm$^2$ the network activity is asynchronous. Above this value, $\chi$ increases monotonously and the network activity is synchronous. In Fig. 13C the conductance $g_{NaP}$, varies; $g_K = 9$ mS/cm$^2$, $g_{Ks} = 0$. The synchrony is reduced when $g_{NaP}$ increases. Transition to asynchronous state occurs for $g_{NaP} \approx 0.15$ mS/cm$^2$.

We can see that in fact potassium currents *promote* synchrony while slow sodium current *depresses* synchrony, even after compensating for all the changes in firing rate and variability.

### 2.4.2. Phase interaction for electrotonic interactions

The phase coupling function, $\Gamma(\phi)$, between two identical neurons, $i$ and $j$, interacting synaptically depends on the phase difference, $\phi_i - \phi_j$, of the two neurons and it is obtained from Eq. 2.15 [2, 8, 23]:

$$\Gamma(\phi) = \frac{1}{T} \int_0^T Z(\phi' + \phi) I_{gj}(\phi' + \phi, \phi') d\phi' \tag{2.59}$$

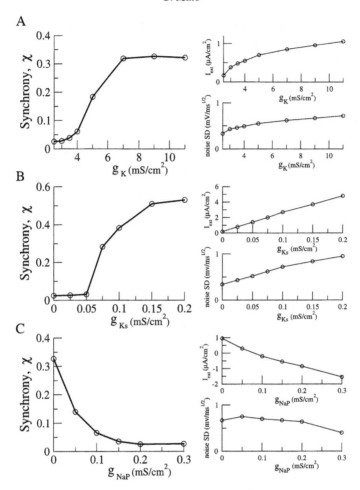

Fig. 13. Dependence of the synchrony level on the intrinsic currents. The firing rate and the coefficient of variability are kept constant (at $f = 50\text{Hz}$ and $CV = 0.1$ respectively) by changing the external current and the noise level. In the right-hand-side figures we show the corresponding values of $I_{ext}$ and $\sigma$. A: $g_{Ks} = g_{NaP} = 0$, B: $g_K = 2.5 \text{ mS/cm}^2$, $g_{NaP} = 0$, C: $g_K = 9 \text{ mS/cm}^2$, $g_{Ks} = 0$.

where $Z$ is the phase response function of the neurons and $I_{gj}(\phi_i, \phi_j)$ is the synaptic current coming from the presynaptic neuron $j$ to the postsynaptic neuron $i$. For electrical synapses, this current is given by Eq. 2.58.

In general, it is not possible to solve Eq. 2.59 analytically for a conductance based model. However, it is possible to analyze the influence of intrinsic currents on the synchronization by reducing the dynamics to a Quadratic Integrate and Fire (QIF) model. This approximation is valid near a saddle node bifurcation (see Appendix C). The intrinsic neural dynamics is given by

$$\tau_0 \frac{dV}{dt} = V^2 + I - I_c \tag{2.60}$$

with $V(t^+) = V_r$ if $V(t) = V_T$. In this reduction all the internal parameters of the model are absorbed in the threshold and reset potentials $V_T$ and $V_r$, the time constant $\tau_0$ and in the threshold current $I_c$. In order to incorporate the effect of the spike on the interaction we add a $\delta$-function to the trace at the time of resetting The integral of this function will be denoted by $\theta$. For this model we can easily solve $V(t)$ and $Z(t)$

$$V(t) = \sqrt{I - I_c} \tan \left( \frac{t\sqrt{I - I_c}}{\tau_0} + \tan^{-1} \frac{V_r}{\sqrt{I - I_c}} \right) \tag{2.61}$$

$$Z(t) = \frac{\tau_0}{V(t)^2 + I - I_c} \tag{2.62}$$

We can now analyze, for instance the stability of the asynchronous state. As we have seen in Section 2.1.4 the stability is controlled by the Fourier components of the phase interaction function. For our model they are given by

$$\Gamma_l = g_{gj} Z_l \left( V_l^* + v\theta \right) \tag{2.63}$$

where $v$ is the firing rate and $Z_l$, $V_l$ are the Fourier components of the phase response function and the sub-threshold potential respectively. We also find that the derivative of $\Gamma$ at a given point can be written as:

$$\frac{d\Gamma(\phi)}{d\phi} = g_{gj} \left( v\theta \frac{dZ(\phi)}{d\phi} + (V_T - V_r) Z(\phi) \right.$$
$$\left. - \frac{1}{T} \int_0^T Z(\phi') \frac{dV(\phi' + \phi)}{d\phi} d\phi' \right) \tag{2.64}$$

We can use this expression to study the stability of locked states, such as in-phase (controlled by $\frac{d\Gamma(0)}{d\phi}$ or anti-phase $\frac{d\Gamma(T/2)}{d\phi}$). In Fig. 14 we show the results from the analytical study of the QIF model. The time constant is $\tau_0 = 10$ msec and the size of the spikes is $\theta = 1$. In panel A se show the phase response function for three different combinations of $V_T$ and $V_r$. In all the cases the difference $V_T - V_r = 3$. In B we show the stability of the asynchronous state for a large network in absence of noise as a function of the firing rate and $|V_r| / V_T$. Note the logarithmic scale in the x-axis. In C we show the stability of the anti-phase state for a pair of identical neurons. Comparing B and C we can see that almost always the region of stability of the asynchronous state corresponds to the stability of the anti-phase state. This region is controlled by the sign of $\frac{d\Gamma(T/2)}{d\phi}$. This is because the dominant contribution comes from the term $\frac{dZ(\phi)}{d\phi}$ in Eq. 2.64 unless the firing rate is very low. If the response function has a peak in the first semi-period of oscillation (as in the first figure of panel A) then the derivative will be negative and the asynchronous state stable. If the peak is in the second half, the derivative will be positive and the asynchronous state unstable. Therefore we can relate in a simple way the response properties of a single neuron with the collective behavior of the network.

Now we can compare these results with the ones from the conductance based models. The phase response function was computed for the same four sets of intrinsic conductances as in Figure 12 using the XPP software [7]. For each set of parameters, the external current is such that the firing frequency of the neuron is 50Hz. The results are shown in Fig. 15. A, Control cases: $g_K = 9$ mS/cm$^2$, $g_{NaP} = g_{Ks} = 0$. External current $I_{ext} = 1.10 \ \mu$A/cm$^2$. B, The conductance of the delayed rectifier is reduced to $g_K = 2.5$ mS/cm$^2$ ($g_{NaP} = 0$, $g_{Ks} = 0$ as in the control case). This skews the phase response function toward the left and shifts its maximum. External current is $I_{ext} = 0.48 \ \mu$A/cm$^2$. C, $g_{NaP} = 0.2$ mS/cm$^2$ (other currents as in the control ($g_K = 9$ mS/cm$^2$, $g_{Ks} = 0$)). The phase response function is more skewed to the left and the maximum of this function is shifted in that direction by the persistent sodium current. External current: $I_{ext} = -1.55 \ \mu$A/cm$^2$. D, The parameters are as in B but $g_{Ks} = 0.2$ mS/cm$^2$. This modifies substantially the phase response function. The refractory period is much larger, the maximum of the response function is shifted to the right and the overall shape of this function is skewed in that direction. External current is $I_{ext} = 4.88 \ \mu$A/cm$^2$.

We are now able to understand the results of our numerical simulations. In Figures 13A,B the network state is asynchronous for small potassium conductances (both delay-rectifier and slow-potassium) and becomes synchronous when these conductances are increased. A similar behavior is found in the QIF network. The asynchronous state is stable when $|V_r|/V_T$ is small and the asynchronous state is destabilized when $|V_r|/V_T$ increases enough. This stems from the fact that the changes in the phase response function of the conductance-based neuron when $g_K$ increases, and the QIF neuron when $|V_r|/V_T$ increases are qualitatively similar. Inversely, increasing the conductance of the persistent sodium current has the same effect on the phase response function of the neuron as a decrease of $|V_r|/V_T$. This explains the results of Figure 13C.

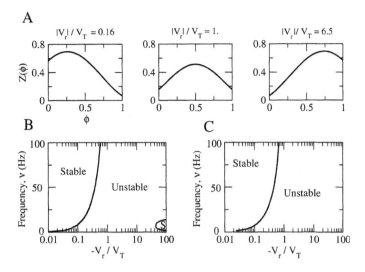

Fig. 14. A: Phase response function for three different values of $|V_r|/V_T$. B: synchronization phase diagrams of QIF networks at weak coupling. S: asynchronous state stable, U: asynchronous state unstable. In all the cases $V_T - V_r = 3$.

## 3. Strongly coupled oscillators: mechanisms of synchrony

When the coupling is not weak the analysis of the dynamical states becomes much more difficult. In principle, there is no simple way to predict if cluster states will exist at all, even less to say if they are stable or not. However there is one special case that is amenable to analysis. In this section we study the stability of the asynchronous state in a two-population network of QIF neurons. In addition to the synaptic input, the neurons receive a tonic external input different from neuron to neuron. We also investigate the different ways the asynchronous state can become unstable.

### 3.1. The two-population QIF model

We consider networks with two populations. One excitatory and the other inhibitory. The dynamics of neuron $i$ in population $\alpha$ is given by

$$\frac{dV_{i\alpha}}{dt} = V_{i\alpha}^2 + I_{i\alpha} + I_{i\alpha}^{syn}(t) \tag{3.1}$$

where $i = 1, ..., N_\alpha$ and $\alpha = E, I$, with the resetting condition: $V_{i\alpha}(t^+) = V_r$ if $V_{i\alpha}(t) = V_T$. The current $I_{i\alpha}$ is constant in time but different from one

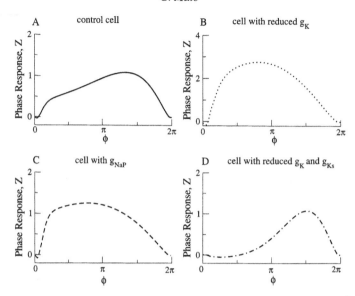

Fig. 15. Phase response functions of conductance based models. A: $g_K = 9$ mS/cm$^2$, $g_{Ks} = g_{NaP} = 0$ (control case). B: $g_K = 2.5$ mS/cm$^2$, $g_{Ks} = g_{NaP} = 0$. C: $g_K = 9$ mS/cm$^2$, $g_{Ks} = 0$, $g_{NaP} = 0.2$ mS/cm$^2$. D: $g_K = 2.5$ mS/cm$^2$, $g_{Ks} = 0.2$ mS/cm$^2$, $g_{NaP} = 0$. The period of 20 msec has been rescaled to $2\pi$.

neuron to another. In the following we will assume that the parameters of the model are obtained from the reduction of the WB neuron [9] near the saddle-node bifurcation. We can fit the I-f curve of this model between 0 and 100 Hz and we obtain $V_T = 4.52$ and $V_r = -0.626$ [24]. The time is measured in units of $\tau_0 = 10$ msec.

The populations are fully connected. We have four parameters of interaction: $G_{EE}$, $G_{EI}$, $G_{IE}$ and $G_{II}$, that represent the strength of interactions between excitatory neurons, from inhibitory neurons to excitatory neurons, from excitatory neurons to inhibitory neurons and between inhibitory neurons respectively. As a model for the interaction we use Eq. 2.52 for each component, therefore

$$I_{i\alpha}^{syn}(t) = \sum_{\beta} G_{\alpha\beta} \sum_{spikes} f(t - t_{spike,\beta}). \tag{3.2}$$

The average number of synaptic inputs per neuron is proportional to the size of the system and the synaptic strengths are varied in inverse proportion to the system size so that

$$G_{\alpha\beta} = \frac{g_{\alpha\beta}}{N_\beta}, \tag{3.3}$$

where the normalized synaptic strengths, $g_{\alpha\beta}$ are independent of the number of neurons. As there is no driving term $V_{syn} - V$ in the synaptic current, the type of the interaction is given by the sign of the coupling parameters, i.e., $G_{EE} > 0$, $G_{IE} > 0$, $G_{EI} < 0$, $G_{II} < 0$.

The distribution of external currents $Q_\alpha(I)$ is assumed to be a Gaussian function with mean value $I_\alpha$ and standard deviation $\sigma_\alpha$

$$Q_\alpha(I) = \frac{1}{\sqrt{2\pi\sigma_\alpha^2}} \exp\left(-\frac{(I - I_\alpha)^2}{2\sigma_\alpha^2}\right). \tag{3.4}$$

In the asynchronous state the synaptic current is constant in time. Neuron $i$ in population $\alpha$ fires periodically with frequency

$$\nu_{i\alpha} = F\left(I_{i\alpha} + \sum_\beta g_{\alpha\beta}\nu_\beta\right), \tag{3.5}$$

where $F(I)$ is the input-output relation of the QIF neuron. For this model it can be easily proven that

$$F(I) = \frac{\sqrt{I}\Theta(I)}{\arctan\frac{V_T}{\sqrt{I}} - \arctan\frac{V_r}{\sqrt{I}}}, \tag{3.6}$$

where $\Theta$ is the Heaviside function. Eqs. 3.4,3.5,3.6 determine the distribution $P_\alpha(\nu)$ of firing rates inside each population. This is given by

$$P_\alpha(\nu) = Q_\alpha(F^{-1}(\nu))|F^{-1'}(\nu)| + \delta(\nu)\int_{-\infty}^{I_\alpha^{min}} dI\, Q_\alpha(I) \tag{3.7}$$

where the last contribution comes from the neurons whose total input is too small to make them fire. As the total current is $I_{i\alpha} + \sum_\beta g_{\alpha\beta}\nu_\beta$, the threshold value is given by $I_\alpha^{min} = -\sum_\beta g_{\alpha\beta}\nu_\beta$. Let us note that we can keep the distribution of firing rates constant even if the coupling strengths change. This can be done simply by adjusting the distribution of external currents with an offset that depends on the coupling and the average firing rate of the population. In this way we can separate the effect of coupling strength and the effect of the average rate on synchronization properties. In Fig. 16 we show an example of the distribution of currents and firing rates.

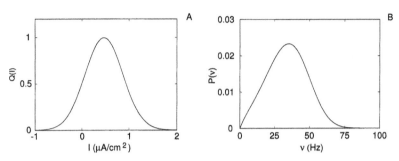

Fig. 16. A: distribution of external currents with $\sigma = 0.2$ The mean value was chosen to have a mean firing rate of 30 Hz. B: distribution of firing rates.

### 3.2. Stability of the asynchronous state

The stability of the asynchronous state in a homogeneous network of leaky integrate-and-fire neurons has been studied in [25]. We present here the generalization of their approach to the case of heterogeneous network of QIF neurons [24]. In this approach we define a change of variable from $V_{i\alpha}$ to $y_{i\alpha}$ given by

$$y_{i\alpha} = v_{i\alpha} \int_{V_r}^{V_{i\alpha}} \frac{dx}{x^2 + I_{i\alpha} + \sum_\beta g_{\alpha\beta} v_\beta} \tag{3.8}$$

where $v_i$ and $v_\alpha$ are the firing rates of neuron $i$ and the average population firing rate in the asynchronous state, respectively and $g_{\alpha\beta}$ are the strengths of the synapses from population $\beta$ to population $\alpha$. This new variable evolves according to

$$\frac{dy_{i\alpha}}{dt} = v_{i\alpha} + \sum_\beta g_{\alpha\beta} K_\alpha(y_{i\alpha}) \epsilon_\beta(t) \tag{3.9}$$

where

$$\epsilon_\beta(t) = \frac{1}{N_\beta} \sum_{j=1}^{N_\beta} (v_{j\beta}(t) - v_\beta) \tag{3.10}$$

is the deviation of the firing rate of neuron $j$ at time $t$ from its value in the asynchronous state, and

$$K_\alpha(y_{i\alpha}) = \frac{v_{i\alpha}}{V_{i\alpha}^2 + I_{i\alpha} + \sum_\beta g_{\alpha\beta} v_\beta} \tag{3.11}$$

The probability density, $\rho_\alpha(y, \nu, t)$ satisfies the continuity equation:

$$\frac{\partial \rho_\alpha(y, \nu, t)}{\partial t} = -\frac{\partial J_\alpha(y, \nu, t)}{\partial y} \tag{3.12}$$

where the flux $J_\alpha(y, \nu, t)$ is

$$J_\alpha(y, \nu, t) = \left( \nu_\alpha + \sum_\beta g_{\alpha\beta} K_\alpha(y) \epsilon_\beta(t) \right) \rho_\alpha(y, \nu, t). \tag{3.13}$$

In the asynchronous state: $J_\alpha(y, \nu, t) = \nu_\alpha$. We analyze the temporal behavior of the deviations from the asynchronous state, $j_\alpha(y, \nu, t) = J_\alpha(y, \nu, t) - \nu_\alpha$. These deviations satisfy:

$$\frac{\partial j_\alpha(y, \nu, t)}{\partial t} = \sum_\beta g_{\alpha\beta} K_\alpha(y) \frac{d\epsilon_\beta(t)}{dt} - \nu_\alpha \frac{\partial j_\alpha(y, \nu, t)}{\partial y}. \tag{3.14}$$

For synaptic interactions modeled with a difference of exponentials given by Eq. 2.45 it is straightforward to show that $\epsilon_\alpha(t)$ evolves according to:

$$\frac{d\epsilon_\alpha(t)}{dt} = -\gamma_{1\alpha} \epsilon_\alpha(t) + h_\alpha(t) \tag{3.15}$$

$$\frac{dh_\alpha(t)}{dt} = -\gamma_{2\alpha} h_\alpha(t) + \gamma_{1\alpha}\gamma_{2\alpha} \sum_\nu j_\alpha(1, \nu, t) \tag{3.16}$$

where $\gamma_{k\alpha} = 1/\tau_{k\alpha}$ for $k = 1, 2$. The perturbation $j_\alpha$ has to be evaluated at $y = 1$ because according to Eq. 3.8 this corresponds to the crossing of the threshold.

The asynchronous state is stable if small perturbations eventually decay. Assuming that $j_\alpha(y, \nu, t)$, $\epsilon_\alpha(t)$, $h_\alpha(t)$ are proportional to $\exp(\lambda t)$, and integrating Eq. 3.14 one finds after a straightforward but tedious calculation that the perturbation rate $\lambda$ satisfies:

$$\prod_{\alpha=E, I} \left[ \frac{(\lambda + \gamma_{1\alpha})(\lambda + \gamma_{2\alpha}) - \gamma_{1\alpha}\gamma_{2\alpha} g_{\alpha\alpha} U_\alpha}{\gamma_{1\alpha}\gamma_{2\alpha}} \right] = g_{EI} g_{IE} U_I U_E \tag{3.17}$$

where

$$U_\alpha(\lambda) = \int_0^\infty d\nu \frac{H_\alpha(\nu)}{1 - \exp(-\lambda/\nu)} \tag{3.18}$$

and

$$H_\alpha(v) = \frac{P_\alpha(v)v}{2F^{-1}(v)}\left[1 - \exp(-\lambda/v) + \frac{\cos(A+\phi) + \sin(A+\phi)\frac{Av}{\lambda}}{1 + (Av/\lambda)^2}\right.$$
$$\left. - \frac{\exp(-\lambda/v)\left(\cos(\phi) + \sin(\phi)\frac{Av}{\lambda}\right)}{1 + (Av/\lambda)^2}\right] \tag{3.19}$$

with

$$A = 2\sqrt{F^{-1}(v)}/v \tag{3.20}$$

$$\phi = 2\arctan(V_r/\sqrt{F^{-1}(v)}). \tag{3.21}$$

The spectral equation, Eq. 3.17, determines the eigenvalues of the dynamical equations linearized around the asynchronous state. Note that this spectral equation also holds if a fraction of the neurons are below threshold and do not fire in the asynchronous state. Indeed an infinitesimal perturbation cannot make these neurons fire. Therefore they do not contribute to the destabilization of the asynchronous state.

The asynchronous state is stable if all the eigenvalues, the solutions to Eq. 3.17, have a negative real part. Therefore, continuous onset of instabilities occurs when at least one of the eigenvalues crosses the imaginary axis when some parameter is changed. At this onset, $\lambda = i\mu$. The onset of instabilities of the asynchronous state are determined by taking the real and imaginary part of Eq. 3.17. The simultaneous solution of these equations determine $\mu$ and the coupling at the onset of instabilities as a function of the other parameters of the model.

At instability onset the synaptic currents in the unstable mode oscillate with a frequency given by the imaginary part $\mu$ of the critical eigenvalue. The dephasing $\delta$ between the oscillations of the two populations is:

$$\exp(i\delta) \propto \epsilon_I/\epsilon_E = \frac{g_{EE}U_E - (\lambda/\gamma_{1E} + 1)(\lambda/\gamma_{2E} + 1)}{|g_{EI}|U_I}. \tag{3.22}$$

A positive (resp. negative) value of the phase lag $\delta$ means that the oscillation of the inhibitory population is in advance (resp. delayed) over the excitatory population.

To study how the stability of the asynchronous states depends on the synaptic properties we construct phase diagrams for a fixed distribution of the firing rates of the two populations, $P_E(v)$ and $P_I(v)$, taking as parameters the strength of the four interactions. This implies that when the interaction strengths are changed, the external average inputs (or the firing thresholds) have to be modified accordingly to keep constant the total input to the two populations. Because of the normalization of the synaptic interactions, changing the synaptic time constants does not affect the firing rate distribution in the asynchronous state. To study their role on the emergence of synchrony, the synaptic time constants can be varied while maintaining constant all the other parameters.

### 3.3. Mechanisms of synchrony

#### 3.3.1. The symmetric case

The stability of the asynchronous state depends on 12 parameters: 4 coupling constants, the mean value and variance of the firing rate of each population and 2 time constants for each interaction. An exhaustive analysis in terms of all these parameters is a very complex task. In this section we focus on the simple case where $\tau_{1E} = \tau_{1I} = \tau_1$, $\tau_{2E} = \tau_{2I} = \tau_2$ and $P_E(\nu) = P_I(\nu)$ . We term this situation the *symmetric* model. It turns out that this case can be, to a large extent, investigated analytically and that the results thus obtained are highly instructive to understand the general properties of the two population network.

In the symmetric case, for all $\alpha$, $U_\alpha(\lambda) = U(\lambda)$ where $U(\lambda)$ is given by Eq. 3.18. Therefore Eq. 3.17 becomes:

$$(i\mu + \gamma_1)^2(i\mu + \gamma_2)^2 - \gamma_1\gamma_2(i\mu + \gamma_1)(i\mu + \gamma_2)UT + \gamma_1^2\gamma_2^2U^2\mathcal{D} = 0$$

$$(3.23)$$

where $U$ is a function of $\mu$, $\mathcal{T} = g_{EE} + g_{II}$ and $\mathcal{D} = g_{EE}g_{II} - g_{EI}g_{IE}$. The latter are the trace and the determinant of the matrix $g_{\alpha\beta}$, respectively. Therefore, the role of the synaptic strength can be completely described in a two dimensional plane spanned by these two effective parameters.

Let us consider point $X_0 = (\mathcal{D} = 0, \mathcal{T} = 0)$. A particular realization of point $X_0$ is obtained in the absence of coupling. Therefore at $X_0$ the asynchronous state is marginally stable if the system is homogeneous, i.e. for $\sigma_E = \sigma_I = \sigma = 0$. If one introduces any level of heterogeneities the asynchronous state becomes stable in some region around this point. The size of this region depends on $\sigma$ (or equivalently on the dispersion of the firing rates in the asynchronous state). This result holds independently of the parameters and the firing rates distribution provided the symmetry is preserved. It shows that there is always some domain in the $\mathcal{D}$, $\mathcal{T}$ plane where the asynchronous state is stable.

What are the boundaries of this region and what are the instabilities on these boundaries? An example of the network phase diagram is shown in Fig. 17. It was obtained by solving Eq. 3.23 for a synaptic rise and decay times of 1 msec and 4 msec respectively, the average firing rate of the two populations, $\nu = 50$ Hz, and the standard deviation of the external input distribution, $\sigma = 0.1$. The region of stability of the asynchronous state is bounded by four lines.

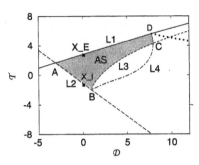

Fig. 17. Phase diagram in the symmetric case for $\sigma = 0.1$. The mean firing rates of both populations are 50 Hz and the synaptic time constants are $\tau_1 = 1$ msec, $\tau_2 = 4$ msec.

Line L1 is deduced from Eq. (3.23) by taking $\mu = 0$. It is straightforward to see that the equation of the line is

$$T = U(0)\mathcal{D} + 1/U(0) \tag{3.24}$$

with:

$$U(0) = \int_0^\infty dv \frac{P(v)v}{2F^{-1}(v)} \left[ 1 + \frac{\sin(A + \phi) - \sin(\phi)}{A} \right] \tag{3.25}$$

For the parameters of Fig. 17, $U(0) = 0.36$. On this line a saddle-node bifurcation occurs [5]. Above the line, at the instability onset, the unstable mode grows exponentially but does not oscillate. This corresponds to a change in the average firing rates of the two populations, $v_E$ and $v_I$, but not to the emergence of synchrony in the network.

Before discussing the general meaning of transition line L2 we consider two particular cases.

**1-** $P \equiv g_{EI}g_{IE} = 0$, $g_{II} = 0$, $g_{EE} > 0$: The two populations are non-interacting and the inhibitory neurons are decoupled. For such parameters the states of the network are represented in the $(\mathcal{D}, T)$ plane by the line segment: $\mathcal{D} = 0, T > 0$. There is one instability with real eigenvalue for the unstable mode on this segment at point $X_E = (0, 1/U(0))$ which belongs to L1. All other instabilities in this segment occur in the region above L1 (not shown in Fig. 17). Therefore, for the parameters of the figure, the excitatory population cannot develop synchrony by itself.

**2-** $P = g_{EI}g_{IE} = 0$, $g_{EE} = 0$, $g_{II} < 0$: As in the previous case the two populations are non-interacting but now the excitatory neurons are decoupled rather than the inhibitory ones. This situation is mapped in the phase diagram on segment $\mathcal{D} = 0, T < 0$. An instability with a pure imaginary eigenvalue $\lambda = i\mu_c$

occurs on this segment at point $X_I = (0, g_c)$ where $g_c$ and $\mu_c$ are solutions to the complex equation:

$$(i\mu_c + \gamma_1)(i\mu_c + \gamma_2) = \gamma_1\gamma_2 g_c U(i\mu_c). \tag{3.26}$$

Solving this equation yields: $g_c = -0.98$, $\mu_c = 3.02$ i.e., $\mu_c/(2\pi\nu_I) \approx 0.96$. Therefore, at $X_I$ the mutual inhibitory interactions I-I become strong enough to destabilize the asynchronous state with a Hopf bifurcation [5] where synchrony emerges on the time scale of the average spiking period. Other solutions of Eq. 3.26 exist but they are not relevant since they occur in the region below $X_I$ where the asynchronous state is already unstable. This mechanism, where synchrony of neural activity emerges in the inhibitory population of neurons, has been studied by several authors [9,27–33]. It has been proposed that hippocampal $\gamma$ rhythms can be driven by populations of inhibitory interneurons whose activity is synchronized through this mechanism [34].

We now consider the general case, $P = g_{EI}g_{IE} \neq 0$ in which the two populations interact. It is straightforward to see that if $T$ and $D$ satisfy the relationship:

$$T = \frac{D}{g_c} + g_c \tag{3.27}$$

then Eq. 3.23 is satisfied with $\mu = \mu_c$.

Eq. 3.27 defines a straight line (L2) in the $(T, D)$ plane on which a Hopf bifurcation occurs. Obviously $X_I$ belongs to this line. The intersection of L1 and L2 defines a point, $A \approx (-2.72, 1.8)$, where the real parts of the two eigenvalues vanish. One corresponds to a saddle-node bifurcation and the other to a Hopf bifurcation. At that point the asynchronous state becomes unstable through a codimension-two bifurcation called a Gavrielov-Guckenheimer bifurcation [35]. The points on L1 and L2 to the left of A are not relevant to instability onset since they are outside the region where the asynchronous state is stable.

The pattern of synchrony into which population $\alpha$ ($\alpha = E, I$) settles depends in general on the ratio:

$$R_\alpha = \frac{\mu}{2\pi\nu_\alpha}. \tag{3.28}$$

Here, $R_\alpha$ is close to one and neurons fire on average about one spike per period in the unstable mode. However the phase relationship between the spikes and the collective oscillations can change from cycle to cycle. Consequently, the phase distribution is broad but unimodal.

It can be easily shown that on L2:

$$\frac{\epsilon_I}{\epsilon_E} = \frac{g_{EE} - g_c}{|g_{EI}|}. \tag{3.29}$$

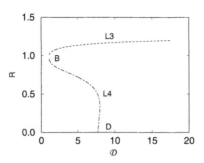

Fig. 18. Frequency of collective oscillations respect to the average frequency of the neurons for the parameters of Fig. 17.

Therefore, on this line, the phase lag is constant, $\delta = 0$, i.e., at the instability onset the activities in the two populations oscillate in-phase. Point B on L2 defined by: $\mathcal{D} = g_c^2 \equiv \mathcal{D}_B$, $\mathcal{T} = 2g_c \equiv \mathcal{T}_B$ is a multicritical point where two eigenvalues have a vanishing real part.

We conclude that on (AB) the synchronous state undergoes an instability to spike-to-spike synchrony. This instability, driven by the I-I interactions is the extension in the framework of two populations of the instability which occurs in one population of inhibitory neurons. The interaction between the two populations makes this mechanism less efficient. Indeed, from Eq. 3.27, one sees that when the feedback between the two populations increases, a stronger mutual inhibitory coupling, $g_{II}$ is required for the synchronous state to be destabilized on L2.

Point D is defined as the intersection of L4 and L1. At that point the asynchronous state is unstable through a codimension two bifurcation called a Takens-Bogdanov bifurcation [35]. The coordinates of D can also be obtained by expanding Eq. 3.17 in the limit $\mu \to 0$. One finds that in D, $\mathcal{D} = U(0)^2$, $\mathcal{T} = 2U(0)$.

The frequency of the unstable mode, $\mu$, varies continuously on L3 and L4, as shown in Fig. 18. On L4, $\mu$ decreases continuously when $\mathcal{D}$ increases, from $\mu = \mu_{II}^c = 3.02$ at B, to $\mu = 0$ at D. On L3, $\mu$ increases slowly when $\mathcal{D}$ increases. In the limit of large $\mathcal{D}$, $\mu$ converges asymptotically to a finite value, $\mu_\infty$ and L3 is asymptotic to the straight line defined by: $\mathcal{T} = \mathcal{D}/g_\infty + g_c$ where $g_\infty$ and $\mu_\infty$ satisfy

$$(i\mu_\infty + \gamma_1)(i\mu_\infty + \gamma_2) = \gamma_1 \gamma_2 g_\infty U(i\mu_\infty). \tag{3.30}$$

Note that Eq. 3.30 is identical to the spectral equation describing the stability of the asynchronous state in one population of neurons. For the parameters of Fig. 17, $\mu_\infty \approx 3.82$ and $g_\infty \approx 6.5$.

The ratio between the frequency of the unstable mode and the average firing rate of the neurons, $R \equiv R_E = R_I$ (see Eq. 3.28) is plotted in Figure 18. This ratio is larger than 1 on L3. This means that the unstable mode oscillates faster than the average frequency of the neurons. However $R$ increases by less than 25% on L3. Therefore the period of the population activity is of the order of the interspike interval. In contrast, on L4, $R$ varies from $R \approx 1$ near B to $R = 0$ in the vicinity of D. This means that along L4 the pattern of synchrony changes continuously from spike-to-spike synchrony to burst synchrony in which the neurons have the time to fire several spikes on average during one period of oscillation of the unstable mode.

The phase lag, $\delta$, on L3 and L4, can be expressed analytically as a function of $\mathcal{D}$, $\mathcal{T}$ and $g_{EE}$. From Eq. 3.23 one finds that:

$$(i\mu + \gamma_1)(i\mu + \gamma_2) = \frac{\gamma_1 \gamma_2 U(i\mu)}{2} \left( \mathcal{T} \pm \sqrt{\mathcal{T}^2 - 4\mathcal{D}} \right). \tag{3.31}$$

Analyzing the numerical solutions of the spectral equation one finds that it is negative on L4 and positive on L3. Substitution in Eq. 3.22 results in

$$\delta = \pm \arctan \frac{\Im \left( \sqrt{\mathcal{T}^2 - 4\mathcal{D}} \right)}{g_{EE} - g_{II}} \tag{3.32}$$

where the positive (resp. negative) sign is for the phase lag on L4 (resp. on L3).

The condition $\mathcal{T}^2 - 4\mathcal{D} = 0$ defines a line in the phase diagram to which B, D and the point $\mathcal{T} = \mathcal{D} = 0$ belong. Therefore, $\mathcal{T}^2 - 4\mathcal{D} < 0$ on L3, except in B and D where $\mathcal{T}^2 - 4\mathcal{D} = 0$ (see definition of points B and D). Since $g_{EE} - g_{II} > 0$

$$\cos \delta > 0. \tag{3.33}$$

The phase lag $\delta$ is a function of the three variables $g_{EE}$, $g_{II}$ and $g_{EI}g_{IE}$ and not only the reduced variables $\mathcal{T}$ and $\mathcal{D}$. In contrast to what happens on L2, on L3 and L4 it changes continuously and non-monotonically. It vanishes in $B$ and $D$.

Eq. 3.33 implies that in general, $|\delta| < \pi/2$. In particular the activity of the two populations can never be in anti-phase. From Eq. 3.32 one sees that the maximum phase lag at the onset of instability, $|\delta| = \pi/2$, is achieved only for $g_{EE} = g_{II} = 0$ ($\delta = -\pi/2$ on L3 and $\delta = \pi/2$ on L4). Finally, on line L3, in the limit of very large $\mathcal{D}$ and $\mathcal{T}$ the phase lag returns to 0, as can be seen in Eq. 3.22 using Eq. 3.30.

In Fig. 19 we show the phase diagram for a value of $\sigma = 0.4$. All the other parameters are as in Fig. 17. We can see that the main qualitative difference is that the line L3 lies entirely beyond line L4. For that reason it becomes irrelevant to the determination of the boundary of the asynchronous state. This result suggests that the different mechanisms of synchrony have very different behaviors when

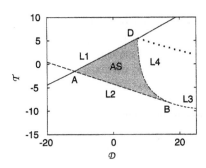

Fig. 19. Phase diagram in the symmetric case for $\sigma = 0.4$. Other parameters as in Fig. 17.

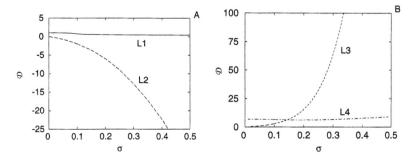

Fig. 20. The determinant of the coupling matrix, $\mathcal{D}$, vs. $\sigma$ on the bifurcation lines L1, L2, L3, L4 for $\mathcal{T} = 1$. The synaptic time constants are as in Fig. 17.

heterogeneities are increased. In order to quantify that effect, we have evaluated the change of the different lines as a function of $\sigma$. This is shown in Fig. 20. We can see that lines L1 and L4 are very weakly dependent on $\sigma$. On the other hand, lines L2 and L3 change very fast. This indicates that spike-to-spike synchrony is very easily destabilized by heterogeneities.

### 3.3.2. The general case

We can now examine the situation in the general case. We examine the more realistic situation where the inhibitory neurons are firing at a higher rate than the excitatory neurons and the inhibitory synaptic time constants are slower than the excitatory ones. Keeping fixed all these parameters the phase diagram will be function of the three variables: $-g_{EI}g_{IE}, g_{EE}, g_{II}$. In Fig. 21 we show the phase diagram of the asynchronous state in the plane $-g_{EI}g_{IE}, g_{EE}$ when $g_{II} = 0$ (A) or $g_{EE} = 0$ (B). Like in the symmetric case, if the recurrent excitation is strong enough, a saddle-node instability occurs. In the plane $-g_{EI}g_{IE}, g_{EE}$ this

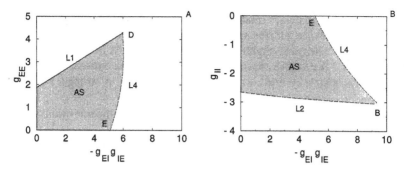

Fig. 21. Phase diagram in the general case. The average firing rate of the excitatory population is 20 Hz and the average rate of the inhibitory population is 40 Hz. The synaptic time constants are $\tau_{1E} = 1$ msec, $\tau_{2E} = 3$ msec, $\tau_{1I} = 1$ msec, $\tau_{2I} = 6$ msec. The standard deviations of the inputs are $\sigma_E = \sigma_I = 0.2$. In A we take $g_{II} = 0$ and in B, $g_{EE} = 0$.

instability occurs on a straight line, whose equation is

$$g_{EE} = \frac{1}{U_E(0)} - g_{EI}g_{IE}U_I(0) \tag{3.34}$$

where $U_\alpha(0)$, $\alpha = E, I$ is given by Eq. 3.25, replacing the distribution $P$ by $P_\alpha$, the firing rate distribution in population $\alpha$. For the parameters of the figure one finds: $U_E(0) = 0.53$, $U_I(0) = 0.4$. When $g_{EE}$ increases, this instability can be prevented by increasing the negative feedback between the two populations measured by $g_{EI}g_{IE} < 0$. When this feedback is strong enough, the asynchronous state becomes unstable through a Hopf bifurcation which occurs on the line to the right of the phase diagram. On this line the period of the population rhythm varies continuously from $\mu \approx 1.25$ at its intersection with the x-axis, E, $(g_{EE}(E) = 0)$ to $\mu = 0$ at D. The value of $\mu$ at E corresponds to a pattern in which the inhibitory (resp. excitatory) neurons fire approximately two spikes (resp. one spike) per period of the population oscillation. Near D it corresponds to a pattern in which neurons tend to fire synchronous bursts with many spikes. The properties of the instability on this line are similar to those found on line L4 in the phase diagrams of the symmetric case. Because of these similarities we have denoted this line by L4 here as well.

In Fig. 21B the region of stability of the asynchronous state is bounded by two lines. One, L2, corresponds to the destabilization of the asynchronous state when $g_{II}$ is strong enough. Like line L2 in Figs. 17 and 19 this line corresponds to the emergence of synchrony through the I-I interaction. In contrast to the symmetric case, here $\mu/2\pi$ and $\delta$ are not constant on L2. However, $\mu/2\pi$ always remains just slightly smaller, by less than 4%, than the average firing rate of the inhibitory population, and $\delta$ is always smaller than 15% of the total period.

Another difference with the symmetric case is that L2 is not straight anymore. However, it deviates only slightly from a straight line. As in the symmetric case the slope of L2 is negative. This means that the feedback between the two populations plays against synchrony through the I-I coupling. A stronger I-I coupling is required to compensate for this feedback.

We can see from these results that the classification of the transitions obtained in the symmetric case still holds in the more general case.

We now focus on the effect of the synaptic time constants. We fix a set of coupling strengths, the rise time constants are also kept fixed at 1 msec and we evaluate the stability as a function of the decay time constants $\tau_{2E}$, $\tau_{2I}$. We consider first the case where $g_{II} = 0$. In the limit $\tau_{2E} \to \infty$ the asynchronous state is always stable. This is because in that limit the excitatory input to the inhibitory population is not modulated in time. Therefore it is equivalent to a positive constant current that excites the inhibitory population. The only way synchrony can occur in this case is through the I-I interactions which we have assumed to be zero. For finite $\tau_{2E}$ the asynchronous state is unstable provided $\tau_{2I}$ is sufficiently large. This defines a transition line $\gamma_{2E} = f(\gamma_{2I})$ where the function $f(\gamma)$ is an increasing function which vanishes linearly when $\gamma \to 0$. On this line, $\mu$ is a continuous decreasing function of $\gamma_{2I}$. It goes to 0 in the limit $\gamma_{2I} \to 0$. For small values of $\gamma_{2I}$ this transition leads to a synchronous bursting state, equivalent to the line L4 in the symmetric case. The bursts are short for fast inhibition and long when the inhibition is slow. In Fig. 22A the asynchronous state becomes stable for fast inhibition and slow enough excitation. In particular, the fastest excitation compatible with a stable asynchronous state (attained for very fast inhibition with $\tau_{1I} = \tau_{2I} = 1$ msec) is $\tau_{2E} \approx 50$ msec, which is in the range of NMDA synaptic time constants.

If $g_{II}$ is large enough, a second instability line appears in the region of the phase diagram corresponding to fast inhibition and slow excitation. This is because for a large enough $g_{II}$ the I-I interaction can induce spike to spike synchrony in the inhibitory population. This situation is shown in Fig. 21B. This is the equivalent of line L2 in the symmetric case. Note that the transition line found for $g_{II} = 0$ is still present but has been shifted to much higher values of $\gamma_{2E}$.

Here, in contrast to what happens for small values of $g_{II}$, when the inhibition is fast ($\tau_{2I} < 9$ msec) the asynchronous state is destabilized for slow enough excitation. For instance, for $\tau_{2I} = 6$ msec, this happens for $\tau_{2E}$ larger than 20 msec. This decay is substantially faster than the typical decay for NMDA synaptic currents. Note that for the parameters of Figure 22B, the asynchronous state is stable if the excitatory and inhibitory synapses are in a range corresponding to AMPA and GABA$_A$ synapses, respectively.

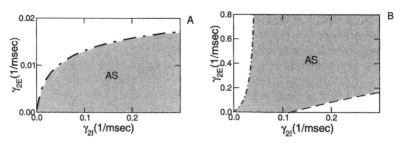

Fig. 22. Phase diagram in the space of the inverse of the decay synaptic time constants $\gamma_{2E} = 1/\tau_{2E}, \gamma_{2I} = 1/\tau_{2I}$. The average firing rate of the excitatory population is 20 Hz and the average rate of the inhibitory population is 30 Hz. The standard deviations of the inputs are $\sigma_E = \sigma_I = 0.2$. The coupling strengths are $g_{EE} = 6$, $g_{EI}g_{IE} = -36$ and $g_{II} = 0$ (A), $g_{II} = -7.4$ (B).

### 3.4. Application: persistent states

If the excitatory couplings are strong enough the possibility of finding *persistent* or *self-sustained* states arises. In these states, the activity of the network is initially generated by some constant external input. However, if this stimulus is removed, the network remains active because of the excitatory recurrent connections. This reverberatory activity has been proposed as a mechanism for memory storage [36]. Experimental evidence of sustained activity during memory tasks has been reported (for instance in [37]).

There are two main problems in this context: the first is the existence of the persistent states. We will see below that this is always solved by having strong enough excitatory couplings. The second is the stability. Persistent states can become unstable in several ways. If the excitatory couplings are too strong then the level of activity of the network can become impossible to control because the network undergoes a saddle-node bifurcation, such as the one of line L1 in Figs. 17 or 19. This can be solved by introducing a suitable inhibitory feedback. But if the inhibition is too strong the system could synchronize via the mechanisms of the lines L2, L3 or L4. Clearly, strong synchrony is not compatible with a persistent state, at least in the excitatory population. Let us suppose that all the excitatory neurons fire at the same time. In that case the excitatory current decays to 0 in a time of order $\tau_{2E}$. Self-sustained activity is impossible, except at rates of order of $1/\tau_{2E}$. For AMPA synapses, this means that self-sustained states should have rates above 100 Hz, which is not the situation observed experimentally [37]. For this reason, a mechanism based on slower NMDA synapses has been also proposed [38]. However, there is still the possibility that even with fast excitatory interactions the asynchronous state be stable in the conditions that allow persistent states. [26]

Let us consider first the conditions for the existence of the persistent state in a homogeneous two-population network of QIF neurons. According to Eqs. 3.5,3.4 the average rate $\nu_\alpha$ can be written as

$$\nu_\alpha = F\left(I_\alpha + \sum_\beta g_{\alpha\beta}\nu_\beta\right). \tag{3.35}$$

If $I_\alpha < 0$ then there will be a solution where $(\nu_\alpha = 0 \ \alpha = E, I)$. On the other hand, the equation

$$I_\alpha = F^{-1}(\nu_\alpha) - \sum_\beta g_{\alpha\beta}\nu_\beta \tag{3.36}$$

will also have a solution with firing rates $\nu_\alpha$ if

$$g_{\alpha E} > \frac{F^{-1}(\nu_\alpha) - g_{\alpha_I}\nu_I}{\nu_E} > 0. \tag{3.37}$$

If the network is heterogeneous, the input-output relation $F(I)$ has to be replaced with $F_{\sigma_\alpha}(I)$, that is its convolution with a Gaussian function with standard deviation $\sigma_\alpha$.

It is very easy to verify that in the symmetric case these constraints are compatible with the stability of the asynchronous state for any value of firing rate. We know that the asynchronous state is always stable if $g_{EE} + g_{II} = 0$, $g_{EE}g_{ii} - g_{EI}g_{IE} = 0$. We first choose any value of $g_{EE}$ that satisfies Eq. 3.37, we put $g_{II} = -g_{EE}$. The second stability constraint is now $g_{EI}g_{IE} = -g_{EE}^2$. Choosing $g_{IE}$ large enough and $g_{EI}$ small enough we can keep the product constant and still satisfy Eq. 3.37. Let us note that this is independent on the synaptic time constants. Of course, if the heterogeneities are small, then the stability region will be also small around this point.

If the degree of heterogeneities is not too small, on one hand we increase the region of stability of the asynchronous state and on the other hand the state with $\nu_\alpha = 0$ now becomes an active state with very low rate. We can interpret this as the background activity of the network. In Fig. 23 we show an example of a bistable network which can jump from one state to the other by the application of transient external inputs.

As we have said before, in order to have a persistent state, the synchrony on the excitatory population must not be too strong. This is of course achieved in the asynchronous state (where the synchrony is strictly 0). However, partially synchronized states are still compatible with persistence. In order to see how to regulate the synchrony for both populations we have simulated a two-population network and measured the synchrony as a function of $-g_{IE}g_{IE}$ for two different

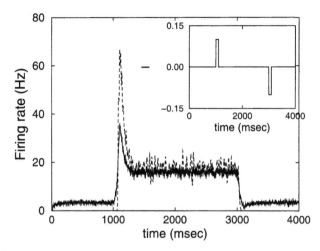

Fig. 23. Switching between the spontaneously active state and the APS in a simulation of a network with $N_E = N_I = 1600$ neurons. The parameters are: $\nu_E = \nu_I = 15$ Hz, $\tau_{1E} = \tau_{1I} = 1$ msec, $\tau_{1E} = 3$ msec, $\tau_{1I} = 6$ msec, $g_{EE} = 4.5$, $g_{EI} = -1$, $g_{IE} = 16$, $g_{II} = -4$, $\sigma_E = \sigma_I = 0.3$. A depolarizing (resp. hyperpolarizing) stimulus is applied at 1000 msec during 100 msec (resp. 3000 msec during 100 msec). Solid line: E population. Dotted line: I population. Inset: stimulus as a function of time.

ratios of $g_{IE}/g_{IE}$. The results are shown in Fig. 24. As expected if $|g_{EI}/g_{IE}|$ is small, then the excitatory population receives a weak input and tends to be less synchronized than the inhibitory population. Fortunately, this is compatible with the condition for having persistence coming from Eq, 3.37. In Fig. 25 we show an example where the asynchronous state is unstable. There are sizeable oscillations in the activity of the inhibitory population, but as the excitatory population is still weakly synchronized we still have persistent activity.

## 4. Conclusion

In these lectures we have analyzed the dynamics of large neural networks. We have proposed analytical methods that, in simplified situations, allow us to understand the relation between the intrinsic dynamics of the neurons, the interactions and the collective dynamics of the system. We have concentrated on fully connected networks. We have not intended to present a complete model of the cortex, but a simple network model that is amenable to analytical treatment. We present here a summary of the results.

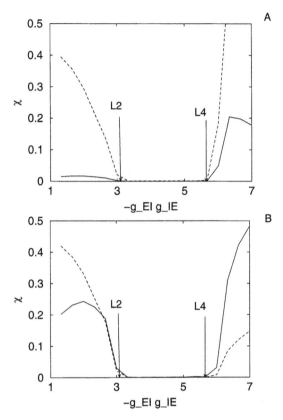

Fig. 24. The synchrony level of the excitatory (solid line) and the inhibitory populations (dashed line) vs. $-g_{EI}g_{IE}$. The strength of the E-E and I-I interactions are fixed: $g_{EE} = 2$, $g_{II} = -2$. Other parameters $\nu_E = \nu_I = 50$ Hz, $\sigma_E = \sigma_I = 0.1$. A network with $N_E = N_I = 1600$ neurons was simulated for different values of $g_{EI}$ and a fixed ratio $g_{EI}/g_{IE}$. A. $g_{EI}/g_{IE} = -1/4$, B. $g_{EI}/g_{IE} = -4$. The arrows show the prediction of the different transitions from the solution of Eq. 3.17.

Using the phase reduction method we have analyzed the dynamics of a network of chemically connected neurons. We have clarified the relation between the interaction and the excitability properties of the neurons:

• Type I excitability: continuous I-f curve (saddle-node bifurcation). Positive phase response function. Desynchronizing for fast excitation. Synchronizing for fast inhibition.

• Type II excitability: discontinuous I-f curve (Hopf bifurcation). Phase response function with sizeable negative region. Synchronizing for fast excitation.

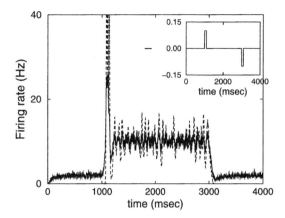

Fig. 25. Switching between the spontaneously active state and the APS in a simulation of a network with $N_E = N_I = 1600$ neurons. The parameters are: $\nu_E = \nu_I = 10$ Hz, $\tau_{1E} = \tau_{1I} = 1$ msec, $\tau_{1E} = 3$ msec, $\tau_{1I} = 6$ msec, $g_{EE} = 5$, $g_{EI} = -1.66$, $g_{IE} = 15$, $g_{II} = -5$, $\sigma_E = \sigma_I = 0.185$. A depolarizing (resp. hyperpolarizing) stimulus is applied at 1000 msec during 100 msec (resp. 3000 msec during 100 msec). Solid line: E population. Dotted line: I population. Inset: stimulus as a function of time. According to the solution of Eq. 3.17 the asynchronous state is unstable.

Desynchronizing for fast inhibition.

We have also analyzed the behavior of networks connected with gap-junctions. By comparing the phase response functions with the theoretical predictions we could conclude that:

• Delayed rectifier or slow potassium currents: shift response function to the right: promotes synchrony.

• Slow sodium current: shift response function to the left: promotes asynchrony.

Regarding the strong coupling situation we have approximated the neural dynamics by a QIF model, and we have analyzed the stability of the asynchronous state for a heterogeneous two population model. We have classified the ways the asynchronous state can become unstable. The generic mechanisms of synchrony are:

• Line L2: synchronization induced by inhibitory-inhibitory interaction: two populations in-phase. The excitatory population goes against synchrony. Collective oscillations with approximately the same frequency as the average rate (spike-to-spike synchrony). Easily disrupted by heterogeneities.

• Line L3: synchronization induced by the feedback loop: present even if $g_{EE} = g_{II} = 0$. Inhibitory oscillations advanced respect to excitatory oscillations. Collective oscillations with approximately the same frequency as the average rate (spike-to-spike synchrony). Easily disrupted by heterogeneities.

• Line L4: synchronization induced by the feedback loop: present even if $g_{EE} = g_{II} = 0$. Excitatory oscillations advanced respect to inhibitory oscillations. Collective oscillations with a lower frequency than the average rate (burst synchrony). Robust against heterogeneities.

We have also seen that if the coupling is strong enough, a stable self-sustained state is possible, even if the synaptic time constants are fast (such as the given by AMPA and GABA$_A$) interactions.

Some natural but non-trivial extension would be to study networks with spatial structure,that can give rise to traveling waves or other spatio-temporal patterns in which hot spots of activity move across the system [47–50], delay effects or multi-compartmental neurons.

## Acknowledgements

The results discussed in this work were obtained in collaboration with D. Hansel from the Laboratoire de Neurophysique et de Physiologie du Système Moteur at Université René Descartes. I thank I. Samengo for a careful reading of the manuscript.

## Appendix A.  Hodgkin–Huxley and Wang–Buszaki models

**The Hodgkin-Huxley model:** The dynamical equations of the HH model read [1]:

$$C\frac{dV}{dt} = I - g_{Na}m^3h(V - V_{Na}) - g_K n^4(V - V_K) - g_l(V - V_l) \quad \text{(A. 1)}$$

$$\frac{dm}{dt} = \frac{m_\infty(V) - m}{\tau_m(V)} \quad \text{(A. 2)}$$

$$\frac{dh}{dt} = \frac{h_\infty(V) - h}{\tau_h(V)} \quad \text{(A. 3)}$$

$$\frac{dn}{dt} = \frac{n_\infty(V) - n}{\tau_n(V)}. \quad \text{(A. 4)}$$

where $I$ is the external current injected into the neuron which determines the neuron's firing rate. The parameters $g_{Na}$, $g_K$ and $g_l$ are the maximum conductances per surface unit for the sodium, potassium and leak currents respectively and $V_{Na}$, $V_K$ and $V_l$ are the corresponding reversal potentials. The capacitance per surface unit is denoted by $C$. For the squid's axon typical values of the parameters (at 6.3 $^0C$) are: $V_{Na} = 50$, $V_K = -77$, $V_l = -54.4$, $g_{Na} = 120$,

$g_K = 36$, $g_l = 0.3$, and $C = 1\mu F/cm^2$. The functions $m_\infty(V)$, $h_\infty(V)$, and $n_\infty(V)$ and the characteristic times (in milliseconds) $\tau_m, \tau_n, \tau_h$, are given by:
$x_\infty(V) = a_x/(a_x + b_x)$, $\tau_x = 1/(a_x + b_x)$
with $x = m, n, h$ and:

$$a_m = 0.1(V + 40)/(1 - \exp((-V - 40)/10)) \tag{A.5}$$
$$b_m = 4\exp((-V - 65)/18) \tag{A.6}$$
$$a_h = 0.07\exp((-V - 65)/20)) \tag{A.7}$$
$$b_h = 1/(1 + \exp((-V - 35)/10) \tag{A.8}$$
$$a_n = 0.01(V + 55)/(1 - \exp((-V - 55)/10)) \tag{A.9}$$
$$b_n = 0.125\exp((-V - 65)/80) \tag{A.10}$$

**The Wang-Buszaki model:** This model, introduced by Wang and Buszaki [9] to describe the firing of hippocampal inhibitory interneurons is a modification of the standard Hodgkin-Huxley model, as follows. The leak current parameters are $g_\ell = 0.1$ and $V_\ell = -65$.
The sodium current has the same form as in the HH model with:

$$a_m = 0.1(V + 35)/(1 - \exp(-0.1(V + 35))) \tag{A.11}$$
$$b_m = 4\exp(-(V + 60)/18) \tag{A.12}$$
$$a_h = 0.07\phi\exp(-(V + 58)/20) \tag{A.13}$$
$$b_h = \phi/(\exp(-0.1(V + 28)) + 1) \tag{A.14}$$

For simplicity the activation variable $m$ instantaneously adjusts to its steady-state value $m_\infty$. The other parameters of the sodium current are: $g_{Na} = 35$ and $V_{Na} = 55$.

The delayed rectifier current is described in a similar way as in the HH model with with:

$$a_n = 0.01\phi(V + 34)/(1 - \exp(-0.1(V + 34))) \tag{A.15}$$
$$b_n = 0.125\phi\exp(-(V + 44)/80) \tag{A.16}$$

and $g_K = 9$ and $V_K = -90$. For $\phi$ we take the value $\phi = 5$ unless other value is stated.
In all cases potentials are measured in millivolts and time in milliseconds.
**Slow potassium and persistent sodium currents:** We also introduce a slow potassium current given by: $I_{Ks} = g_{Ks} s^4 (V - V_K)$ ( [39]) and a non-inactivating (persistent) sodium current ( [40]), $I_{NaP} = g_{NaP} p_\infty (V - V_{Na})$. The kinetics of the gating variable $s$ is controlled by

$$\alpha_s(V) = 0.07(44 + V)/(1 - e^{-(V+44)/4.6}) \tag{A.17}$$

$$\beta_s(V) \quad = \quad 0.008\, e^{(V+44)/68} \tag{A. 18}$$

and $p_\infty(V) = 1/(1 + e^{-(V+50)/6})$.

## Appendix B. Measure of synchrony and variability in numerical simulations

**Measure of synchrony level:** By synchrony of the activity of a *pair* of neurons we mean the tendency of these neurons to fire spikes at the same time. In this sense a pair of neurons will be said to fire in synchrony if the cross-correlation of their spike trains display a central peak which is above chance. The degree of synchrony can be characterized by the amplitude of this peak. According to this definition, if two neurons fire action potentials periodically and in anti-phase they can be considered to fire asynchronously.

In a large network of neurons, the activity is asynchronous if at any time the number of action potentials fired in the network is the same up to some random fluctuations. This implies that in a large network the asynchronous state is unique. When this state is unstable, the network activity is necessarily synchronous. In this case neurons tend to fire preferentially in some windows of time.

One can characterize the degree of synchrony in a population of $N$ neurons by measuring the temporal fluctuations of macroscopic observables, as for example, the membrane potential averaged over the population ( [41–44]). The quantity

$$V(t) = \frac{1}{N} \sum_{i=1}^{N} V_i(t) \tag{B. 1}$$

is evaluated over time and the standard deviation $\sigma_V^2 = \langle [V(t)]^2 \rangle_t - [\langle V(t) \rangle_t]^2$ of its temporal fluctuations is computed, where $\langle \ldots \rangle_t$ denotes time-averaging. After normalization of $\sigma_V$ to the average over the population of the single cell membrane potentials, $\sigma_{V_i}^2 = \langle [V_i(t)]^2 \rangle_t - [\langle V_i(t) \rangle_t]^2$, one defines $\chi(N)$:

$$\chi(N) = \sqrt{\frac{\sigma_V^2}{\frac{1}{N} \sum_{i=1}^{N} \sigma_{V_i}^2}} \tag{B. 2}$$

which varies between 0 and 1. The central limit theorem implies that in the limit $N \to \infty$, $\chi(N)$ behaves as:

$$\chi(N) = \chi(\infty) + \frac{a}{\sqrt{N}} + O(\frac{1}{N}) \tag{B. 3}$$

where $a > 0$ is a constant, and $O(1/N)$ means a term of order $1/N$. In particular, $\chi(N) = 1$, if the activity of the network is fully synchronized (i.e., $V_i(t) = V(t)$ for all $i$), and $\chi(N) = O(1/\sqrt{N})$ if the state of the network activity is asynchronous. In the asynchronous state, $\chi(\infty) = 0$. More generally, the larger $\chi(\infty)$ the more synchronized the population. Note that this measure of synchrony is sensitive not only to the correlations in the spike timing but also to the correlations in the time course of the membrane potentials in the subthreshold range.

**Measure of the irregularity of spike trains:** The irregularity of the spike trains is characterized by the coefficient of variability (CV), i.e., the ratio between the standard deviation of the interspike intervals and their mean values averaged over the whole network. For periodic spike trains, $CV = 0$, whereas $CV = 1$ for spike trains randomly distributed with Poisson statistics.

## Appendix C.  Reduction of a conductance-based model to the QIF model

We consider a Type I conductance-based model. The equations of the single neuron dynamics can be written in compact form:

$$\frac{d\mathbf{X}}{dt} = \mathbf{F}(\mathbf{X}) + \mathbf{G}(\mathbf{X}) \tag{C. 1}$$

In the case of the WB model, $\mathbf{X}$, $\mathbf{F}$ and $\mathbf{G}$ are three dimensional vectors with:

$$
\begin{aligned}
X_1 &= V & \text{(C. 2)} \\
X_2 &= h & \text{(C. 3)} \\
X_3 &= n & \text{(C. 4)} \\
F_1(\mathbf{X}) &= (I_c - g_{Na}m_\infty^3 h(V - V_{Na}) - g_K n^4 (V - V_K) \\
& \quad - g_l(V - V_l))/C & \text{(C. 5)} \\
F_2(\mathbf{X}) &= (h_\infty(V) - h)/\tau_h(V) & \text{(C. 6)} \\
F_3(\mathbf{X}) &= (n_\infty(V) - n)/\tau_n(V) & \text{(C. 7)} \\
G_1 &= I - I_c & \text{(C. 8)}
\end{aligned}
$$

and $G_2 = G_3 = 0$. Following [45] one can study this dynamics in the vicinity of firing onset. One writes:

$$\mathbf{G}(\mathbf{X}) = \epsilon \mathbf{N}(\mathbf{X}) \tag{C. 9}$$

where $\epsilon \ll 1$. For $\epsilon < 0$, the neuron is not firing. It displays a stable fixed point: $\mathbf{X} = \bar{\mathbf{X}}(I)$. For $\epsilon > 0$, the neuron fires action potentials. At $\epsilon = 0$, a bifurcation occurs. The Taylor expansion of $F(\mathbf{X})$ around $\mathbf{X}^\star = \bar{\mathbf{X}}(I_c)$ is:

$$F(\mathbf{X}) = \mathbf{F}(\mathbf{X}^\star) + \mathbf{A}(\mathbf{X} - \mathbf{X}^\star) + \mathbf{Q}(\mathbf{X} - \mathbf{X}^\star, \mathbf{X} - \mathbf{X}^\star) + \dots \tag{C. 10}$$

where **A** and **Q** are respectively the Jacobian and the Hessian matrices of **F** evaluated at $\mathbf{X}^{\star}$. Since a saddle-node bifurcation occurs at $\mathbf{X}^{\star}$, the matrix **A** has one zero eigenvalue. The corresponding normalized eigenvector will be denoted by **e**. Similarly, $\mathbf{A}^{T}$, the transposed of **A**, has a zero eigenvalue. We will denote by **f** the corresponding eigenvector satisfying $\mathbf{f}.\mathbf{e} = 1$. For small $\epsilon > 0$ the solution $\mathbf{X}(t)$ of Equation C. 1 can be expanded:

$$\mathbf{X}(t) = \mathbf{X}^{\star} + \sqrt{\epsilon} z(t) \mathbf{e} \tag{C. 11}$$

where the scalar quantity $z$ satisfies ( [46]):

$$\frac{dz}{dt} = \sqrt{\epsilon}(\eta + qz^{2}) \tag{C. 12}$$

where: $\eta = \mathbf{f}.\mathbf{N}(\mathbf{X}^{\star})$ and $q = \mathbf{f}.\mathbf{Q}(\mathbf{e}, \mathbf{e})$. This equation can easily be solved. Assuming that $\eta$ and q have the same sign one finds:

$$z(t) = \sqrt{\frac{\eta}{q}} \tan \sqrt{\eta q \epsilon}(t - t_0) \tag{C. 13}$$

where $t_0$ is a constant. Function $z$ diverges when the argument of the tangent is a multiple of $\pi/2$. These divergences correspond to the firing of an action potential in the original system ( [46]). Therefore, in the limit $\epsilon \to 0$, the firing frequency of the neuron vanishes as:

$$v = B\sqrt{I - I_c} \tag{C. 14}$$

with:

$$B = \frac{\sqrt{\eta q}}{\pi} \tag{C. 15}$$

Coefficient $B$ can be calculated by diagonalizing matrix A. It turns out that due to the particular structure of this matrix, expressions for **e**, **f**, $\mathbf{Q}(\mathbf{e}, \mathbf{e})$ can be derived analytically in terms of the gating functions (this fact has apparently escaped attention in previous papers). In particular, for the WB model, one finds that

$$\mathbf{e} = K(1, h'_{\infty}, n'_{\infty}) \tag{C. 16}$$

$$\mathbf{f} = \frac{1}{LK}(1, \tau_h \frac{\partial F_1}{\partial h}, \tau_n \frac{\partial F_1}{\partial n}) \tag{C. 17}$$

where:

$$K = \frac{1}{\sqrt{1 + (h'_{\infty})^2 + (n'_{\infty})^2}} \tag{C. 18}$$

$$L = 1 + \tau_h \frac{\partial F_1}{\partial h} h'_{\infty} + \tau_n \frac{\partial F_1}{\partial n} n'_{\infty} \tag{C. 19}$$

The derivative of function $f(V)$ with respect to $V$ has been denoted $f'$. In these expressions, all the functions have to be evaluated for $V, n_\infty, h_\infty$ at the fixed point of the dynamics.

Finally we find:

$$
B = \frac{1}{\pi L} \sqrt{\frac{1}{2} \left( \frac{\partial^2 F_1}{\partial V^2} + 2 \sum_{i=2,3} X_i' \frac{\partial^2 F_1}{\partial V \partial X_i} + (n_\infty')^2 \frac{\partial^2 F_1}{\partial n^2} + \sum_{i=2,3} X_i'' \frac{\partial F_1}{\partial X_i} \right)}
$$

(C. 20)

The Taylor expansion of the dynamics, Equation C. 10, is valid in the limit of low firing rate except during an action potential where the variable $z$ diverges. If $I$ is sufficiently close to threshold an approximate trajectory for the vector $\mathbf{X}(t)$ can be computed from $z(t)$ and the vector $\mathbf{e}$. This approximation is a good description of the behavior of the neuron except during the spikes which are of short duration. It becomes exact in the limit $I \to I_c$. Therefore near the firing onset, one can replace the full model, Equation C. 1, by the reduced model, Equations (C. 11,C. 12) where all the parameters can be computed from the full model. In particular, the shape of the I-f curve near threshold is given by Equation C. 14.

When $I - I_c$ is not small, this reduction is no longer exact. However, one can generalize it heuristically as follows. We consider that action potentials are fired whenever $z$ crosses (from below) some threshold, $z_t$. Then $z$ is immediately reset to a value $z_r$ and it evolves again according to Equation C. 12.

The two phenomenological parameters, $z_t$ and $z_r$ cannot be determined by the exact reduction method detailed above. Instead, we estimate them by requiring that the I-f curve fits the I-f curve of the full model as well as possible.

### References

[1] A.L. Hodgkin and A.F. Huxley, J. Physiol. **117**(4) (1952) 500.

[2] Y. Kuramoto, *Chemical Oscillations, Waves and Turbulence* (Springer, New York, 1984).

[3] K. Okuda, Physica D **63** (1993) 424.

[4] N.G. van Kampen, *Stochastic processes in physics and chemistry* (North Holland, Amsterdam, 1981).

[5] S. Strogatz, *Nonlinear dynamics and chaos* (Perseus, Cambridge, 2000).

[6] E.E. Fetz and B. Gustafsson, J. Physiol. **341** (1983) 387.

[7] B. Ermentrout, *Simulating, analyzing, and animating dynamical systems: a guide to Xppaut for researchers and students (software, environments, tools)* (SIAM, Philadelphia, 2002).

[8] D. Hansel, G. Mato and C. Meunier, Europhys. Lett. **23** (1993) 367.

[9] X.-J. Wang and G. Buzsáki, J. Neurosci. **16** (1996) 6402.

[10] A. Watanabe, Jap. J. Physiol. **8** (1958) 305.

[11] E.J. Furshpan and D.D. Potter, J. Physiol. **145** (1959) 289.

[12] P. Mann-Metzer and Y. Yarom, J. Neurosci. **19** (1999) 3298.

[13] A. Draghun , R.D. Traub , D. Schmitz D. and J.G.R. Jefferys, Nature **394** (1998) 189.

[14] M. Beierlein, J.R. Gibson and B.W. Connors, Nat. neurosci. **3** (2000) 904.

[15] J.L. Perez-Velazquez and P.L. Carlen, Trends Neurosci. **23** (2000) 68.

[16] G. Tamas, E.H. Buhl, A. Lorincz and P. Somogyi, Nat. neurosci. **3** (2000) 366.

[17] M.R. Deans, J.R. Gibson, C. Sellitto, B.W. Connors and D.L. Paul, Neuron **31** (2001) 477.

[18] S.G. Hormuzdi , I. Pais, F.E.N. Lebeau, S.K. Towers, A. Rozov, E. Buhl, M.A. Whittington and H. Monyer, Neuron **31** (2001) 487.

[19] M.S. Jones and D.S. Barth, J. Neurophysiol. **88** (2002) 1016.

[20] C. Bou-Flores and A.J. Berger, J. Neurophysiol. **85** (2001) 1543.

[21] B. Pfeuty, G. Mato, D. Golomb and D. Hansel, J. Neuroscience **23** (2003) 6280.

[22] Y. Amitai, J.R. Gibson, A. Patrick, B. Ho, B.W. Connors and D. Golomb, J. neurosci. **22** (2002) 4142.

[23] D. Hansel, G. Mato and C. Meunier C., Neural Comput. **7** (1995) 307.

[24] D. Hansel and G. Mato, Neural Comp. **15** (2003) 1.

[25] L.F. Abbott and C. van Vreeswijk, Phys. Rev., E**48** (1993) 1483.

[26] D. Hansel and G. Mato, Phys. Rev. Lett. **86** (2001) 4175.

[27] C. van Vreeswijk, Phys. Rev. E **54** (1996) 5522.

[28] J.A. White, C.C. Chow, J. Ritt , C. Soto-Treviño and N. Kopell, J. Comput. Neurosci. **5** (1998) 5.

[29] L. Neltner, D. Hansel, G. Mato and C. Meunier, Neural Comput. **12** (2000) 1607.

[30] D. Golomb and D. Hansel D., Neural Comput. **12** (2001) 1095.

[31] D. Golomb, D. Hansel and G. Mato, *Handbook of Biological Physics"*. Eds.: S. Gielen and F. Moss (Elsevier, 2001).

[32] C.C. Chow, Physica D**118** (1998) 343.

[33] M.A, Whittington, R.D. Traub, N. Kopell, B. Ermentrout and E.H. Buhl, Int. J. of Psychophysiology **38** (2000) 315.

[34] G. Buzsáki and J.J. Chrobak, Curr. Opin. Neurobiol. **5** (1995) 504.

[35] Y. Kuznetsov , *Elements of applied bifurcation theory*, 2nd edition (Springer, New York, 1998).

[36] D. Hebb, *The organization of behavior* (New York, Wiley, 1949).

[37] J. M. Fuster *The prefrontal cortex* (New York, Raven, 1988).

[38] X.-J. Wang , J. Neurosci. **19** (1999) 9587.

[39] A. Erisir, D. Lau, B. Rudy and C.S. Leonard, J. Neurophysiol. **82** (1999) 2476.

[40] C.R. French, P. Sah, K.J. Buckett and P.W. Gage, J. Gen. Physiol. **95** (1990) 1139.

[41] D. Hansel and H. Sompolinsky, Phys. Rev. Lett. **68** (1992) 718.

[42] D. Golomb and J. Rinzel, Phys. Rev. E**48** (1993) 4810.

[43] D. Golomb and J. Rinzel, Physica D**72** (1994) 259.

[44] I. Ginzburg and H. Sompolinsky, Phys. Rev. E**50** (1994) 3171.

[45] G.B. Ermentrout, Neural Comput. **8** (1996) 979.

[46] G.B. Ermentrout and N. Kopell, SIAM J. Appl. Math. **46** (1986) 233.

[47] R. Ben Yshai, D. Hansel H. and Sompolinsky, J. Comput. Neurosci. **4** (1998) 57.

[48] D. Hansel and H. Sompolinsky, *Methods in Neuronal Modeling: From Ions to Networks*. Eds.: C. Koch and I. Segev (The MIT Press, Cambridge, 1998).

[49] D. Golomb, J. Neurophysiol. **79** (1998) 1.

[50] D. Golomb and G.B. Ermentrout, Proc. Natl. Acad. Sci. USA **96** (1999) 13480.

Course 5

# SOME USEFUL NUMERICAL TECHNIQUES FOR SIMULATING INTEGRATE-AND-FIRE NETWORKS

## Michael Shelley

*Courant Institute of Mathematical Sciences and Center for Neural Science, New York university, New York, NY 10012*

*C.C. Chow, B. Gutkin, D. Hansel, C. Meunier and J. Dalibard, eds.*
*Les Houches, Session LXXX, 2003*
*Methods and Models in Neurophysics*
*Méthodes et Modèles en Neurophysique*

# Contents

## 1. Introduction

In my lectures at the Les Houches Summer School I spoke on several topics related to the modeling and computing of neuronal networks, especially as applied to understanding the primary visual cortex. This included material on the architecture and responses of (macaque) V1, the nature and responses of a large-scale V1 model that I and my colleagues at NYU have constructed and are studying, and how such models are efficiently simulated. Here I have decided to focus on the latter as it was that which the students found most useful for their own projects.

Certain scientific disciplines, such as fluid and gas dynamics, have a long and rich history of using computers for simulating dynamical phenomena, and have developed – and are still developing – a large body of sophisticated methods. This is less so for neuroscience. Fluid dynamics has the marked advantage that the relevant dynamical description, the Navier-Stokes equations, has been known since the nineteenth century, and their relevance to military applications has driven much of the development of simulational methods. The use of large-scale computing in neuroscience is much more recent, and it also unlikely that functional activity in the brain could be described usefully by such a single comprehensive set of equations as the Navier-Stokes. Still, one very commonly used approximation for simulating the large-scale dynamics of neuronal networks, such as primary visual cortex, is through systems of integrate-and-fire (I&F) point neurons. Such systems capture the thresholding dynamics of individual neurons, and the modulation of the intracellular potential by network-driven membrane conductances. As the use of I&F networks has become widespread, and are being applied at increasing scale, it is worth thinking about how to solve these systems with both accuracy and efficiency.

Several numerical issues arise in simulating I&F systems. Primary is the treatment of the spike-and-reset mechanism of an I&F neuron, which if handled casually gives a large error that dominates the overall accuracy of the simulation. Second, the order of accuracy of any time-stepping scheme will be limited by smoothness of onset of conductance changes. Hence, to avoid unnecessary work for a given order of accuracy, the time-stepping method should reflect this fact. Finally, efficiency issues arise when accounting for synaptic interactions in the network. In this chapter I will review some useful numerical methods for solving systems of I&F neurons that address these issues, and show the application of these methods to simulating the large-scale dynamics of models of V1 orientation hypercolumns.

## 2. The conductance-based I&F model

Let $v^j$ be the membrane potential of the $j^{th}$ neuron in a coupled integrate-and-fire, conductance-based network. A typical model of the dynamics of $v^j$ is given by

$$C\frac{dv^j}{dt} = -g_{leak}(v^j - V_r) - g_e^j(t)(v^j - V_E) - g_i^j(t)(v^j - V_I), \qquad (2.1)$$

where $C$ is the total capacitance, $g_{leak}$ is the leakage conductance, $V_r$ is the rest potential, $V_E$ and $V_I$ are the excitatory and inhibitory synaptic reversal potentials, and $g_e^j(t)$ and $g_i^j(t)$ are the time-dependent conductances arising from input forcing and from the network activity of excitatory and inhibitory neurons. (See [9,18,20–22] for examples of large-scale conductance-based I&F networks.) Whenever the potential $v^j$ reaches a threshold $\bar{v}$, that neuron fires, and $v^j$ is reset instantaneously to a reset potential $\hat{v}$. (A refractory period can also be included.) We model the time-dependent conductances as arising from external forcing and the network activity of this population of neurons:

$$
\begin{aligned}
g_e^j(t) &= g_{e0}^j(t) + \sum_k \sum_l A_{j,k,l} G_e(t - t_l^k) \\
g_i^j(t) &= g_{i0}^j(t) + \sum_k \sum_l B_{j,k,l} G_i(t - T_l^k),
\end{aligned}
\qquad (2.2)
$$

where $t_l^k$ ($T_l^k$) denotes the time of the $l^{th}$ spike of the $k^{th}$ excitatory (inhibitory) neuron, and $A_{j,k,l}$ and $B_{j,k,l}$ are coupling strengths. The input conductances are $g_{e0}^j(t)$ for excitation and $g_{i0}^j(t)$ for inhibition. In V1 cortical models, the excitatory input conductances would arise, in part, from geniculate excitation in response to visual stimulation, while other contributions (both excitatory and inhibitory) would model other extra-striate sources.

Here we will consider synaptic-induced conductance changes of the form

$$
\begin{aligned}
G(t - t_{spike}) &= \left(\frac{t - t_{spike}}{\tau}\right)^m \exp(-(t - t_{spike})/\tau), \quad t \geq t_{spike} \\
&= \quad 0, \quad t < t_{spike}
\end{aligned}
\qquad (2.3)
$$

where $\tau$ is a time constant and $m$ is an integer. Both the excitatory and inhibitory postsynaptic conductance functions, $G_e$ and $G_i$, are given by Eq. (2.3), but with different time constants. In many computational studies of I&F neuronal networks, the postsynaptic conductance changes are modeled by an $\alpha$ function (setting $m = 1$; see, for example, Somers *et al*, 1995) or as the difference of two exponentials with different time constants (e.g., Hansel *et al*, 1998 and Troyer *et al*, 1998), or again as in Eq. (2.3) but with a different choice for $m$ (e.g. $m = 3$ in Tao *et al* (2004)). These different choices are motivated by the modeling of time-scales of onset, persistence, and relaxation in synaptic currents of differing types (e.g. modulated by AMPA, GABA, NMDA). The main fact we need here is that for each choice the *short time behavior* of $G$ following a spike is $G \approx (t - t_{spike})^m$, for some integer $m$.

It is useful to rewrite the initial value problem in Eq. (2.1) as

$$\frac{dv}{dt} = f(t, v) = -g_T(t)v + I_d(t), \quad \text{with } v(0) = v_0, \tag{2.4}$$

where the superscript $j$ has been dropped for clarity. $g_T(t)$ is the total (non-dimensionalized) conductance, and is the inverse of an effective integration time scale, and $I_d(t)$ is the (non-dimensionalized) "difference current", as it arises as a difference of excitatory and inhibitory currents.

## 3. Modified time-stepping schemes

To simulate Eq. (2.4) a time-stepping method must be chosen. The simplest choice would be a standard explicit, single-step method (e.g., see [19]) such as a Runge-Kutta (RK) scheme. A $2^{nd}$-order RK method, the "trapezoidal rule", applied to Eq. (2.4) gives the iteration:

$$
\begin{aligned}
v_{n+1} &= v_n + \frac{\Delta t}{2}(k_1 + k_2), \quad \text{with } v_0 = v(0) \\
k_1 &= f(t_n, v_n) = -g_{T,n}v_n + I_{d,n} \\
k_2 &= f(t_{n+1}, v_n + \Delta t k_1) \\
&= -g_{T,n+1}[v_n + \Delta t(-g_{T,n}v_n + I_{d,n})] + I_{d,n+1}, \tag{3.1}
\end{aligned}
$$

where $t_n = n\Delta t$, and $v_n \approx v(t_n)$. A commonly used and simply implemented strategy to handle the fire-and-reset part of the dynamics is to append the step:

$$\text{If } v_{n+1} > \bar{v} \text{ then reset } v_{n+1} = \hat{v} \tag{3.2}$$

In the absence of a reset step, classical convergence theory [19] for single-step methods applies and tells us about the accuracy of this method: Assume that $g_T$ and $I_d$ are sufficiently smooth, and fix a final time $T = N\Delta t < \infty$. Then, the convergence is at second-order, or there exists a constant $C$ independent of $\Delta t$ such that

$$|v(T) - v_N| \le C(T, v_0)\Delta t^2 . \tag{3.3}$$

The classical theory goes further and even proves the existence of an asymptotic expansion for the error $v(T) - v_N$ in powers of $\Delta t$. And of course, higher-order methods like the ever popular fourth-order RK method, yield a higher-order error bound. Still, there are a few basic facts to understand about this theoretical statement that help explain what can limit the accuracy of these methods when applied to an I&F system:

1. The theory assumes that the initial data is exact. Still, one can show that for RK2 if the initial data error is instead $O(\Delta t^p)$ for $p \ge 2$, then the error after $N$ steps remains second-order. If instead the error in initial data were $O(\Delta t)$, then the error after $N$ steps is only $O(\Delta t)$. The order of error allowable in initial data so as to retain the order of the scheme increases with the order of the scheme; The fourth-order RK4 method requires initial data that is no more that $O(\Delta t^4)$ in error from the exact data.

2. Roughly speaking, the basic feature that underlays the second-order accuracy of RK2 is that the error committed on each step is of size $O(\Delta t^3)$, so that accumulation of error after $N = T/\Delta t$ steps is $O(\Delta t^2)$. However, this relies on sufficient smoothness in the right-hand-side of Eq. (2.4). If $g_T$ or $I_d$ has an isolated $O(1)$ jump discontinuity in its value, then the final error would degrade to first-order. If instead, the discontinuity were in a first or higher derivative (as might happen with a spike-induced conductance change, with $m \ge 1$), then this yields an isolated $2^{nd}$-order error in $v$ on the step that contains it. By being isolated this does not alter the final error remaining $O(\Delta t^2)$. Again, the higher order the method the more stringent the condition on any discontinuities. To remain globally fourth order in the presence of derivative discontinuities in the right-hand-side of Eq. (2.4) requires that any isolated discontinuity lie only in the third or higher derivative.

The first point explains why it is a bad idea to use the common strategy in (3.2) of resetting $v_{n+1}$ to $\hat{v}$ when the spike threshold is exceeded: The resetting of the potential following a spike is equivalent to initiating a new initial value problem, and $\hat{v}$ is $O(\Delta t)$ in error from the true value $v(t_{n+1})$. It is easy to see this. If the true potential is reset at time $t_s$ with $t_n < t_s \le t_{n+1}$, a simple estimate shows that $v(t_{n+1}) - \hat{v} = (t_{n+1} - t_s)f(t_s, \hat{v}) + O((t_{n+1} - t_s)^2) \le K\Delta t$.

Hence, with the common method the final error at $t = T$ will be only first-order, regardless of the order of the method. So, let's address first how time-stepping schemes might be modified to remove this error, and we will use RK2

(a)                                        (b)

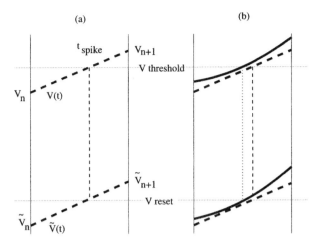

Fig. 1. (a) Recalibration of postsynaptic membrane potential. (b) Local discretization error analysis. We plot $v(t)$ vs. $t$: the solid vertical lines are $t = t_n$ and $t = t_n + \Delta t$, and the dotted horizontal lines are $v = \bar{v}$ (top) and $v = \hat{v}$ (bottom). The numerical interpolation is shown as heavy dashed lines. In (b), the two "true" solutions are in heavy solid lines. (From Shelley & Tao [17], with permission.)

as our working example. Assume that $v_{n+1} > \bar{v}$. Then following Hansel *et al* [7] and Shelley & Tao [17], we use a linear interpolant to estimate the membrane potential between discretized times:

$$v(t) = v_n + \frac{(v_{n+1} - v_n)}{\Delta t}(t - t_n),\tag{3.4}$$

Setting the interpolant equal to $\bar{v}$, we approximate the (relative) spike time by

$$\frac{t_s - t_n}{\Delta t} = \frac{\bar{v} - v_n}{v_{n+1} - v_n} = R_{n,n+1}\ .\tag{3.5}$$

Having found $t_s$, we estimate the value of the post-spike membrane potential by asking the following: What are the values of a new $(\tilde{v}_n, \tilde{v}_{n+1})$ pair so that the (interpolated) solution trajectory passes through the *reset potential* $\hat{v}$ at time $t_s$? The answer corresponds to finding the lower branch in Fig. 1 (a).

Thus we require the post-spike $\tilde{v}_{n+1}$ and a $\tilde{v}_n$ to satisfy two properties: (i) the new membrane potentials are related by a single RK2 step,

$$
\begin{aligned}
\tilde{v}_{n+1} &= \tilde{v}_n + \frac{\Delta t}{2}\left(\tilde{k}_1 + \tilde{k}_2\right),\\
&= (1 - \Delta t A_{n,n+1})\tilde{v}_n + B_{n,n+1}\ ,
\end{aligned}\tag{3.6}
$$

where $\tilde{k}_1 = -g_{T,n}\tilde{v}_n + I_{d,n}$ and $\tilde{k}_2 = -g_{T,n+1}(\tilde{v}_n + \tilde{k}_1 \Delta t) + I_{d,n+1}$ determine $A_{n,n+1}$ and $B_{n,n+1}$; and (ii) at time $t_s$ the new membrane potential is at reset, i.e.,

$$\tilde{v}(t_s) = \tilde{v}_n + (\tilde{v}_{n+1} - \tilde{v}_n)\frac{t_s - t_n}{\Delta t} = \hat{v} . \tag{3.7}$$

(See Fig. 1(a).)

Both requirements are linear relations between $\tilde{v}_n$ and $\tilde{v}_{n+1}$. Eliminating $\tilde{v}_n$ between them yields

$$\tilde{v}_{n+1} = \frac{(1 - R_{n,n+1})\Delta t\, B_{n,n+1} + (1 - \Delta t\, A_{n,n+1})\hat{v}}{1 - \Delta t\, A_{n,n+1} R_{n,n+1}}, \tag{3.8}$$

In this method then, the entire algorithm consists of replacing the potential after reset by the explicit expression above for $\tilde{v}_{n+1}$, with the claim that this will yield a second-order method.

And so it does, as long as a condition on $G$ is satisfied. In deriving expression (3.8), it was assumed that $g_{T,n+1}$ and $I_{d,n+1}$ did not account for any conductance changes induced by spikes occurring in the interval $(t_n, t_{n+1}]$. However, these neglected contributions are still small at time $t = t_n$, and their lack in calculating the post-spike $v_{n+1}$ yields an error of only $O(\Delta t^m)$. As stated in item 2 above, this does not alter the order of the final error if $m \geq 1$. Hence, to see second-order error only requires that the post-synaptic conductance functions be continuous and piece-wise differentiable at the spike time.

We solve Eqs. (2.1) and (2.2) in a one-dimensional ring of $N = 128$ identical neurons. While details are given in Shelley & Tao, each neuron is driven with a sinusoid in time, $g_{e0}^j(t) = 25\sin(t)$, and the effects of inhibition are ignored ($g_i^j(t) = 0$). The spatial couplings are isotropic Gaussians in the angle separation of neurons along the ring (i.e., $A_{j,k,l} = A\exp(-\alpha\Delta\theta^2(j^2+k^2))$) We set $\tau_e = 0.6$ milliseconds, and integrate until $t = 1$ second, after each neuron spiked about 11 times. We use here $m = 5$ (as used in Pugh et al [13] and McLaughlin et al [9]) and examine the errors of our modified scheme and the errors associated with standard RK2 and RK4 schemes without recalibration. (In an earlier paper, Hansel et al [7] developed and applied a scheme similar to the modified RK2 method, and showed the utility of higher accuracy in capturing phase transitions phenomena in neuronal network dynamics.)

In Fig. 2, we plot the error as a function of the time-step. Since the conductances ramp up very smoothly with $m = 5$, the main contribution to the error for the unmodified schemes is the reset mechanism, which dominates their global accuracy. The standard RK2 and RK4 schemes perform no better than a modified Euler scheme, a first-order method. The errors of these schemes are virtually

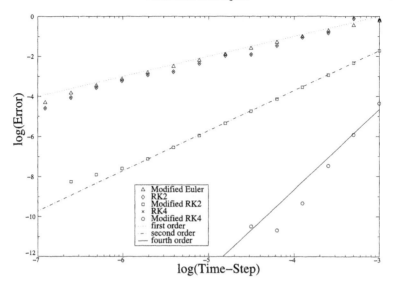

Fig. 2. Errors as a function of time-step for different algorithms. The data points are triangles: Euler with linear interpolant; diamonds: RK2; squares: RK2 with linear interpolant; x's: RK4; and circles: RK4 with cubic interpolant. The various lines are not fits but are only guides to the order of convergence. Dotted line: first order; dash-dotted line: second order; and solid line: fourth order. (From Shelley & Tao [17], with permission.)

indistinguishable and each scheme has errors that decrease linearly with step-size (the dotted line is linear in $\Delta t$).

Higher-order accuracy is restored by our modification; The errors of our modified RK2 method decrease quadratically with $\Delta t$ (as indicated by the dash-dotted line). Moreover, for a time-step of $\Delta t = 10^{-4}$, the modified RK2 method achieves an accuracy over 2 orders of magnitude smaller than that of the unmodified methods. The additional cost was simply an extra line or two of code.

As we show in Shelley & Tao [17], this approach can be generalized fruitfully to higher-order schemes. For example, in our studies of model V1 dynamics we use a modified fourth-order RK method. The main modifications are that instead of the linear (in $t$) interpolant to estimate the spike time (i.e., Eq. (3.4)), we use a cubic Hermite polynomial that interpolates $v_n$, $v_{n+1}$, $\dot{v}_n$, and $\dot{v}_{n+1}$ (these latter two are linear functions of the $v_n$ and $v_{n+1}$, respectively, by the form of the dynamics). This polynomial is used to estimate the spike time to an error of $O(\Delta t^4)$. Relation (3.6) is replaced by the (longer) expression for the RK4 method, and relation (3.7) replaced by the condition a cubic Hermite interpolant to $\tilde{v}$ pass through $\hat{v}$ at the spike time. Again this yields two linear equations for

$\tilde{v}_n$ and $\tilde{v}_{n+1}$, which can be solved for the corrected reset value $\tilde{v}_{n+1}$. It should not be surprising that actually seeing fourth-order accuracy requires a yet greater restriction on the post-synaptic conductance functions. And indeed, we now require that $m \geq 3$.

Fig. 2 also shows the errors of this modified RK4 method (using $m = 5$), and shows that errors now diminish as the fourth power of $\Delta t$ (the solid line). With the fourth-order scheme, we achieve 8-digit accuracy at a time-step of only $1 \times 10^{-4}$ seconds. To realize the same accuracy using suitably modified, lower-order schemes requires small time-steps of $\Delta t \approx 10^{-6}$ for RK2 and $\Delta t \approx 10^{-10}$ for first order schemes.

In general we show in Shelley & Tao [17]) that a $p^{th}$-order time-stepping scheme, coupled with a $q^{th}$-degree spike-time interpolant, has local errors of order $\min(m + 1, p + 1, q + 1)$ and global errors of order $\min(m + 1, p, q + 1)$, where the dependence on $m$ is imposed by the smoothness of $G(t)$. That is, we give the sharp estimate on the global error

$$|V(T) - v_N| \leq O(\Delta t^{\min(m+1,p,q+1)}) \tag{3.9}$$

at $T = N\Delta t$. The more basic result is that with a little care, and a little extra coding, I&F systems can be integrated with a high degree of accuracy using modified, but standard, methods.

Furthermore, for conductance time courses of the form of Eq. (2.3), neglecting the small conductance changes at the next discretized time introduces errors of order $O(\Delta t^{m+1})$, and so is consistent with the overall order of the time-stepping. Thus we can improve the efficiency of these schemes by using the previously calculated conductances, saving an extra cortico-cortical conductance evaluation.

## 4. Synaptic interactions

Another important element when designing a large-scale I&F code is the efficient computation of synaptic interactions. Consider a not uncommon case, when they are of the form:

$$g^j(t) = \sum_{k=1}^{N} A_{j,k} \sum_l G(t - t_l^k), \quad j = 1, \ldots, N \tag{4.1}$$

where $g^j$ is a spike-induced conductance. (Here I ignore, for example, synaptic depression or voltage dependent conductance contributions, though these too can be treated.) A straightforward approach is to write

$$\sum_{k=1}^{N} A_{j,k} \sum_{l} G(t - t_l^k) = \sum_{k=1}^{N} A_{j,k} P_k(t) , \quad j = 1, \dots, N \qquad (4.2)$$

keep track of $P_k(t)$, and evaluate this at every time-step at a cost of $O(N^2)$. Since $k$ denotes the presynaptic neuron that has spiked, this means that information is being saved presynaptically , and then distributed across the network at every time-step, which can easily become prohibitively expensive. The cost of evaluating this form can be considerably reduced if the interactions have special structure such as translation invariance (i.e., $A_{j,k} = A_{j-k}$) so that the Discrete Convolution Theorem can be invoked, and the Fast Fourier Transform used to evaluate the sum. In this particular, non-generic case the cost reduces to $O(N \log N)$ per time-step.

Generally it is much more flexible and cost-effective to employ (if possible) a fire-and-forget strategy, where information is instead saved post-synaptically, and without reference to the presynaptic source of the information. Such an approach requires using a particular form for $G$. As an example, consider using the $G$ given in Eq. (2.3) with $m = 1$. Setting $\tau = 1$ for ease of computation and dropping the $j$ superscript we want to evaluate:

$$g(t) = \sum_{p} A_p G(t - t_p)$$

$$\text{with} \quad G(t) = te^{-t} \text{for } t > 0 \text{ and } 0 \text{ otherwise.} \qquad (4.3)$$

Note that the form of the coupling is now more general than above. This function $g$ satisfies the differential equations

$$\begin{aligned} \dot{g} &= -g + s \\ \dot{s} &= -s + \sum_{p} A_p \delta(t - t_p). \end{aligned} \qquad (4.4)$$

This is most easily seen by writing $G(t) = H(t)te^{-t}$ and setting $S(t) = H(t)e^{-t}$, where $H$ is the Heaviside function. Then a formal calculation gives that $\dot{G}(t) = S(t) - G(t)$ and $\dot{S}(t) = -S(t) + \delta(t)$. Now, consider the solution to system (4.4) between the spike times $t_k$ and $t_{k+1}$, which can be written as:

$$s(t) = s_k e^{-(t-t_k)} \text{ and } g(t) = (g_k + (t - t_k)s_k) e^{-(t-t_k)} \qquad (4.5)$$

At the spike time $t = t_{k+1}$ the value of $s$ must be incremented by $A_{k+1}$, and the expression for $s$ and $g$ found anew. Doing this yields recursive relations for the

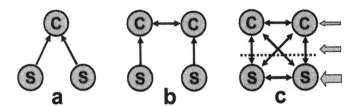

Fig. 3. Schematics of models for Complex cells. Panel **a** represents the hierarchical model of Hubel & Wiesel [8], wherein the summed output of Simple cells drives Complex cells. Panel **b** represents the recurrent excitation model of Chance *et al* [3]. Panel **c** represents the egalitarian model of Tao *et al* [20] where Simple and Complex cells coexist within a common circuit. The Simple cell model of Wielaard *et al* [23] is represented by that portion of the schematic below the dashed line. In the first two models, LGN excitation arrives through the Simple cells, which then drive the Complex cells. In Tao *et al* model, LGN excitation arrives with varying strength throughout the network, with this strength represented by the arrows on the right of the schematic. Weak (strong) LGN excitation is balanced by a stronger (weaker) excitatory coupling to the network. From Tao *et al* [20], with permission.

coefficients of the new solution:

$$s_{k+1} = A_{k+1} + s_k e^{-(t_{k+1}-t_k)} \text{ and } g_{k+1} = (g_k + (t_{k+1} - t_k)s_k)\, e^{-(t_{k+1}-t_k)} \quad (4.6)$$

This approach for synaptic interactions – keeping track of a few coefficients that reconstruct a cell's post-synaptic conductances, and which are only updated with the arrival of a new spike – costs $O(N)$ per spike, where $N$ is the number of post-synaptic neurons coupled to the presynaptic neuron. This approach (first pointed out to me by Dan Tranchina, whom I thank) is especially efficient for sparsely coupled networks, and is easily extended to accomodate effects such as synaptic failure. It can also be easily extended to the general form in Eq. (2.3) and to other forms such as a difference of exponentials.

## 5. Simulating a V1 model

Now I will discuss an application of these numerical methods. I and my colleagues have been studying an "egalitarian" model of macaque primary visual cortex, so-called because it's architecture is (mostly) non-hierarchical (see Fig. 3; also [1]). It is a large-scale, detailed model of a set of orientation hypercolumns, based on an interpretation of the anatomy of macaque V1 layer $4C\alpha$ [20], which receives direct geniculate excitation. This model is an extension of an earlier

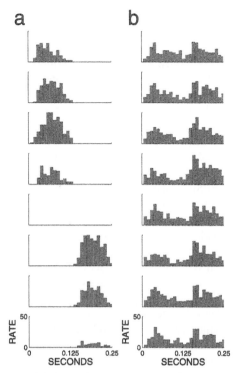

Fig. 4. Responses from model neurons to contrast reversal stimulation [5] at 8 spatial phases (at optimal orientation, and temporal and spatial frequency). **A** and **B** are predicted responses from respectively a Simple and a Complex neuron in the model network. **A**: Model Network Simple cell driven at 4 Hz. The spatial phase is defined so that one spatial cycle of the grating pattern is 360°. At 180°, the response is zero. **B**: Model Network Complex cell driven at 4 Hz. The response is at the second harmonic and is insensitive to spatial phase. From Tao *et al* [20], with permission.

network model that produced the responses of Simple cells [23]. One goal of these studies was to see whether a non-hierarchical model could give a good accounting for recent data on the broad distribution of Simple/Complex response properties found in macaque V1 [15].

Details and a far more thorough exploration of the model can be found in Tao *et al* [20], but I will give a very short description of its main elements: Inputs from visual stimulation are relayed through the LGN to a set of approximately 4000 interacting excitatory (75%) and inhibitory (25%) model V1 neurons. Each model V1 cell "sees" a collection of LGN cells that is probabilistically sampled from a two-dimensional Gabor function [14]. This segregation of convergent On- and

Off-center LGN cells confers orientation and spatial phase preference on individual cortical cells [5]. Orientation preference is laid out in pinwheels patterns, as is found through optical imaging studies (e.g. [2]), while spatial phase preference is distributed randomly [4]. One central assumption is that the number of LGN cells providing afferents varies broadly and randomly from cell to cell in cortex. Intracortical couplings are isotropic with interaction profiles taken to be Gaussians in space, with the lengthscale of excitation greater than that of inhibition (see [9, 23] for a discussion of the relevant anatomical literature). Inhibitory coupling strengths are taken randomly from a Gaussian distribution, while excitatory coupling strengths are drawn from Gaussian distributions whose mean strengths are inversely proportional to the number of LGN afferents [11, 12, 16]. Each cortico-cortical EPSP is taken to be 50% AMPA and 50% NMDA, while an IPSP is divided evenly between $GABA_A$ and a slower inhibition [6]. The simulational code for this model uses a modified RK4 scheme, as described in Shelley & Tao [17], together with a "fire-and-forget" strategy for cortical interactions.

Figure 4 shows the responses of two model cells from the model to contrast reversal stimulation. It demonstrates that its cells reproduce both Simple and Complex responses: As seen in physiological experiments [5], Simple cells show responses at the driving harmonic, with a strong and continuous dependence on spatial phase. Complex cells are frequency doubled in their response, with little dependence on spatial phase. In this model these different responses follow from differing balances of geniculate versus network excitation, with Simple cells having relatively greater geniculate drive than Complex cells. In these cells it is the balance between geniculate excitation and a strong network inhibition that leads to their Simple cell responses [23]. Fig. 5b shows that while many cells in the model are Simple, and many Complex, many are also intermediate, and is in qualitative agreement with cortical measurements (Fig. 5c; [15]). This arises from the fact that the division between sources of excitation varies broadly across the many model cells of the network. Finally, Fig. 5a shows an intracellular metric of Simple versus Complex response, and shows that because of the model's non-hierarchical structure this distribution is unimodal [1, 10], despite the bimodal appearance of the extracellular measure (Figs. 5b & c). This prediction has recently been observed in intracellular measurements of responses of cat V1 cells [24].

## Acknowledgments

I thank my colleagues Bob Shapley, David Cai, Louis Tao, and Dave McLaughlin for their ongoing collaborations with me, and Carson Chow and David Hansel for inviting me to give these lectures.

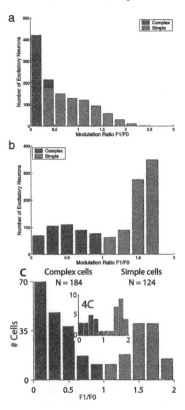

Fig. 5. Comparison of intracellular and extracellular $F_1/F_0$ between the Tao *et al* model (with permission) and experiment. **A**: Distribution of modulation ratio $F_1/F_0$ of membrane potential of excitatory neurons in model network, when stimulated by drifting gratings set at optimal orientation and spatial frequency. The height of each bar indicates the total number of excitatory neurons in each bin, while the blue and red portions correspond to the cells that are classified as "Simple" or "Complex" based on their extracellular responses. **B**: Distribution of the modulation ratio $F_1/F_0$ of the firing rate for excitatory neurons in model network. **C**: Distribution of the modulation ratio $F_1/F_0$ of the firing rate for 308 cells from the experiments of Ringach *et al* [15] (with permission). The detail shows the distribution for the 38 cells identified as being in 4*C*. Here, with a baseline of zero firing rate, the modulation ratio is bounded between zero and two.

# References

[1] Abbott, L & Chance, F. (2002) Rethinking the taxonomy of visual neurons. *Nature Neuroscience* **5**, 391–392.

[2] Blasdel, G. (1992) Differential imaging of ocular dominance and orientation selectivity in monkey striate cortex. *Journal of Neuroscience* **12**, 3115–3138.

[3]  Chance, F, Nelson, S, & Abbott, L. (1999) Complex cells as cortically amplified simple cells. *Nature Neuroscience* **2**, 277–282.

[4]  DeAngelis, G, Ghose, R, Ohzawa, I, & Freeman, R. (1999) Functional Micro-Organization of Primary Visual Cortex: Receptive Field Analysis of Nearby Neurons. *Journal of Neuroscience* **19**, 4046–4064.

[5]  De Valois, R, Albrecht, D, & Thorell, L. (1982) Spatial frequency selectivity of cells in macaque visual cortex. *Vision Res* **22**, 545–559.

[6]  Gibson, JR, Beierlein, M, Connors, BW (1999) Two networks of electrically coupled inhibitory neurons in neocortex. *Nature* 402: 75-79.

[7]  Hansel, D, Mato, G, Meunier, C, Neltner, L (1998) Numerical simulations of integrate-and-fire neural networks. *Neural Comp.* 10: 467-483.

[8]  Hubel, D & Wiesel, T. (1962) Receptive fields, binocular interaction and functional architecture of the cat's visual cortex. *J Physiol (Lond)* **160**, 106–154.

[9]  McLaughlin, D, Shapley, R, Shelley, M, Wielaard, DJ (2000) A neuronal network model of macaque primary visual cortex (V1): Orientation selectivity and dynamics in the input layer $4C\alpha$. *Proc. Natl. Acad. Sci. U. S. A.* 97: 8087-8092.

[10] Mechler, F & Ringach, D. (2002) On the classification of simple and complex cells. *Vis. Res.* **42**, 1017–1033.

[11] Miller, K & MacKay, D. (1994) The role of constraints in Hebbian learning. *Neural Computation* **6**, 100–126.

[12] Miller, K. (1996) Synaptic Economics: Competition and Cooperation in Synaptic Plasticity. *Neuron* **17**, 371–374.

[13] Pugh, MC, Ringach, DL, Shapley, R, Shelley, MJ (2000) *J. Comp. Neurosci.* 8: 143-159.

[14] Reid, R & Alonso, J.-M. (1995) Specificity of monosynaptic connections from thalamus to visual cortex. *Nature* **378**, 281–284.

[15] Ringach, D, Shapley, R, & Hawken, M. (2002) Orientation Selectivity in Macaque V1: Diversity and Laminar Dependence. *J. Neuroscience* **22**, 5639–5651.

[16] Royer, S & Pare, D. (2002) Bidirectional synaptic plasticity in intercalated amygdala neurons and the extinction of conditioned fear responses. *Neuroscience* **115**, 455–462.

[17] Shelley, M & Tao, L. (2001) Efficient and accurate time-stepping schemes for integrate-and-fire neuronal networks. *J. Comput. Neurosci.* **11**, 111–119.

[18] Somers, DC, Nelson, SB, Sur, M (1995) An emergent model of orientation selectivity in cat visual cortical simple cells. *J. Neurosci.* 15: 5448-5465.

[19] Stoer, J, Bulirsch, R (1993) Introduction to Numerical Analysis. Springer-Verlag, New York, NY.

[20] Tao, L, Shelley, M, McLaughlin, D, Shapley, R (2004) An egalitarian network model for the emergence of simple and complex cells in visual cortex, *Proc. Natl. Acad. Sci. USA* 101: 366-371.

[21] Troyer, TW, Krukowski, AE, Priebe, NJ, Miller, KD (1998) Contrast-invariant orientation tuning in cat visual cortex: Thalamocortical input tuning and correlation-based intracortical connectivity. *J. Neurosci.* 18: 5908-5927.

[22] Wang, X-J (1999) Synaptic basis of cortical persistent activity: The importance of NMDA receptors to working memory. *J. Neurosci.* 19: 9587-9603.

[23] Wielaard, J, Shelley, M, Shapley, R, & McLaughlin, D. (2001) How Simple cells are made in a nonlinear network model of the visual cortex. *J. Neuroscience* **21**, 5203–5211.

[24] Priebe, Mechler, Carandini, Ferster (2004) The contribution of spike threshold to the dichotomy of cortical simple and complex cells, *Nature Neuroscience* **7**, 1113-1122 .

Course 6

# PROPAGATION OF PULSES IN CORTICAL NETWORKS: THE SINGLE-SPIKE APPROXIMATION

## David Golomb

*Dept. of Physiology and Zlotowski Center for Neuroscience, Faculty of Health Sciences, Ben Gurion University of the Negev, Be'er-Sheva 84105, Israel*

*C.C. Chow, B. Gutkin, D. Hansel, C. Meunier and J. Dalibard, eds.*
*Les Houches, Session LXXX, 2003*
*Methods and Models in Neurophysics*
*Méthodes et Modèles en Neurophysique*

# Contents

**Abstract**

We study the propagation of traveling solitary pulses in one-dimensional cortical networks with two types of one-dimensional architectures: networks of excitatory neurons, and networks composed of both excitatory and inhibitory neurons. Each neuron is represented by the integrate-and-fire model, and is allowed to fire only one spike. The velocity and stability of propagating, continuous pulses are calculated analytically.

While studying excitatory-only networks, we focus on the effects of synaptic and axonal delays. Two continuous pulses with different velocities exist if the synaptic coupling is larger than a minimal value; the pulse with the lower velocity is always unstable. Above a certain critical value of the constant delay, continuous pulses lose stability via a Hopf bifurcation, and lurching pulses with spatiotemporal periodicity emerge. The parameter regime for which lurching occurs is strongly affected by the synaptic footprint (connectivity) shape. A bistable regime, in which both continuous and lurching pulses can propagate, may occur with square footprint shapes but not with exponential footprint shapes. A perturbation calculation is used in order to calculate the spatial lurching period and the velocity of lurching pulses at large delay values. For strong synaptic coupling, the velocities of both the continuous pulse and the lurching pulse are governed by the tail of the synaptic footprint shape. We find analytically how the axonal propagation velocity reduces the velocity of continuous pulses; it does not affect the critical delay.

We then focus on the propagation of traveling solitary pulses in networks of excitatory and inhibitory neurons. Two types of stable propagating pulses are observed. During fast pulses, inhibitory neurons fire a short time before or after the excitatory neurons. During slow pulses, inhibitory cells fire well before neighboring excitatory cells, and potentials of excitatory cells become negative and then positive before they fire. There is a bistable parameter regime in which both fast and slow pulses can propagate. Fast pulses can propagate at low levels of inhibition, are affected by fast excitation but are almost unaffected by slow excitation, and are easily elicited by stimulating groups of neurons. In contrast, slow pulses can propagate at intermediate levels of inhibition, and are difficult to evoke. They can propagate without slow excitation, but slow excitation makes their propagation substantially more robust. Strong inhibitory-to-inhibitory conductance eliminates the slow pulses and converts the fast traveling pulses into irregular pulses, in which the inhibitory neurons segregate into two groups, which have different firing delays with respect to their neighboring excitatory cells. Lurching pulses may propagate if the excitatory-to-excitatory coupling is slow and very strong. In general, the velocity of the fast pulse increases with the axonal conductance velocity $c$, but there are cases in which it decreases with $c$. We suggest that the fast and slow pulses observed in our model correspond to the fast and slow propagating activity observed in experiments on neocortical slices.

## 1. Introduction

Propagating epileptic-like pulses of neuronal activity appear in disinhibited coronal neocortical slices in response to electrical stimulation above a certain threshold. The average discharge velocity is about 10-15 cm/sec [1]. Neurons are recruited to the wave because of the excitatory, recurrent interactions between neurons. These slice preparations were developed initially as experimental models for epilepsy [2, 3]. Experimental and theoretical investigations [1, 4–7] have tried to relate the dynamics of propagating discharge to the underlying neuronal circuitry. Numerical simulations of a conductance based neuronal model with homogeneous architecture have revealed that the discharge propagates at a constant velocity, as a *continuous* traveling pulse [1, 5]. In both theory and experiment, there was a minimal velocity below which the discharge could not propagate. Propagating discharges with similar properties and velocities have been found in other cortical structures, such as the hippocampus [8, 9] and the piriform cortex [10]. The effects of synaptic and axonal delays were not examined in those studies.

Propagating epileptic-like pulses appear in cortical slices when the strength of inhibition is reduced by only 10%-20% [11–13], but they cannot propagate in healthy cortical slices under physiological conditions when inhibition is intact. Inhibition shapes the form of these pulses, and reduces their velocity [11, 12]. Recent experiments in rodents [7, 14] and ferrets [15, 16] have revealed a different type of propagating pulses in cortical tissue when inhibition is intact or partially reduced. This activity is non-epileptic, with firing rate of individual neurons typically low ($< 10$ Hz). The generation and termination of this propagating state may account for the generation of a subset of cortical rhythm during sleep. In intact ferret slices, the propagating velocity is slow, about 1.1 cm/s [15]. When inhibition is blocked, the activity becomes epileptic-like and the velocity becomes fast, about 9 cm/s. The propagation of slow pulses depends on the existence of slow (NMDA-mediated) excitation. When this excitation is blocked, the slow pulse often, but not always, cannot propagate [15]. In contrast, blocking the slow excitation does not prevent the propagation of the fast pulses [1]. Blocking the fast (AMPA-mediated) excitation prevents the appearance of both fast and slow pulses [1, 15]

In this Book Chapter, we explore the propagation of fronts with and without inhibition. Our goals are to characterize the types of propagating pulses, and to relate the conditions for their appearance, the pulse velocities, and the differences between the firing times of neighboring excitatory and inhibitory neurons with the network architecture, the kinetics of single neurons and synapses, and synaptic and axonal delays. Specifically, we aim to determine the conditions under which the fast, epileptic-like pulses and the slow pulses appear. To achieve these goals, we analyze simple models of a network composed of excitatory (and possibly inhibitory) neurons. The single cell is represented by a simplified version of the integrate-and-fire neuronal model, in which a neuron is allowed to fire only one spike, and then it is silent forever. This model, which is exact in the limit of very long refractory period or very strong synaptic depression, is amenable for analytical treatment. The velocity of traveling pulses in this model, their stability, and the delay between the firing times of neighboring excitatory and inhibitory cells during the pulses, are calculated analytically. Under certain conditions, we also investigate more complicated patterns such as lurching pulses [17, 18]. The content of this Book Chapter is based on several research articles [19–22].

## 2. Propagating pulses in networks of excitatory neurons

### 2.1. The model

We consider neurons along a one-dimensional chain (Fig. 1A), and use the following version of the integrate-and-fire (Lapique) model [5, 23–25]

$$\frac{\partial V(x, t)}{\partial t} = -\frac{V(x, t)}{\tau_0} + I_{\text{syn}}(x, t) + I_{\text{app}}(x, t) \tag{2.1}$$

for $0 < V(x, t) < V_T$, where $V(x, t)$ is the membrane potential of a neuron at a position $x$ and time $t$, $\tau_0$ is the passive membrane time constant of the neuron, $I_{\text{syn}}$ is the normalized synaptic input, and $I_{\text{app}}$ is the normalized applied current; $I_{\text{app}} = 0$ unless stated differently. When $V$ of a neuron reaches the threshold $V_T$ at time $T(x)$, the neuron fires a spike, and cannot fire more spikes afterwords. We assume that the number of neurons within a footprint length is large and therefore use a continuum model and replace the sum over the presynaptic neurons by an integral

$$I_{\text{syn}}(x, t) = g_{\text{syn}} \int_{-\infty}^{\infty} dx' \, w(x - x') \alpha \left[ t - T(x') - \tau_{\text{delay}} \right] \tag{2.2}$$

where $g_{\text{syn}} = \tilde{g}_{\text{syn}} \Delta / C$, $\tilde{g}_{\text{syn}}$ is the synaptic conductance, $C$ is the membrane capacitance, and $\Delta = V - V_{\text{syn}}$ is approximated here to be a constant ("coupling by

## A. Disinhibited Cortical Architecture

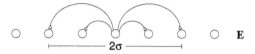

## B. Footprint Shape

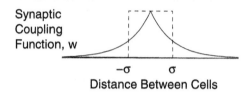

Synaptic
Coupling
Function, w

Distance Between Cells

Fig. 1. A. The model has a one-dimensional architecture, with the coupling between cells decaying with their distance. The footprint length is denoted by $\sigma$. B. Exponential (solid line) and square (dashed line) footprint shapes. Adapted from [19]

currents", see, *e.g.,* [5,26]); $\tau_{\text{delay}}$ is the delay between the post- and presynaptic neurons at positions $x$ and $x'$ respectively. It is given by

$$\tau_{\text{delay}} = \tau_d + \frac{x - x'}{c} \tag{2.3}$$

where $\tau_d$ is the constant delay and $c$ is the axonal conduction velocity.

Eqs. 2.1,2.2 are implicit equations of $T(x)$. The temporal shape of the EPSC that a post-synaptic cell at a position $x$ receives following a spike of a pre-synaptic cell at a position $x'$ is given by the normalized $\alpha$ function $\alpha\left[t - T\left(x'\right)\right]$:

$$\alpha\left(t\right) = \begin{cases} \frac{e^{-t/\tau_1} - e^{-t/\tau_s}}{\tau_1 - \tau_s} & t \geq 0 \\ 0 & \text{otherwise} \end{cases} \tag{2.4}$$

where $\tau_1$ and $\tau_s$ are the synaptic rise and decay time respectively; $\tau_1 \ll \tau_s$. We will assume that $\tau_1 = 0$ unless otherwise stated. The spatial dependence of the synaptic strength on the distance between neurons, $w(x)$, is denoted as the "synaptic footprint shape" [1,17]. We examine two shapes (Fig. 1B):

$$w\left(x\right) = \frac{1}{2\sigma} e^{-|x|/\sigma} \qquad \text{Exponential} \tag{2.5}$$

$$w\left(x\right) = \begin{cases} \frac{1}{2\sigma} & |x| \leq \sigma \\ 0 & |x| > \sigma \end{cases} \qquad \text{Square} \tag{2.6}$$

$\sigma$ is denoted as the "synaptic footprint length". We consider a half-infinite network, *i.e.*, the length of the system is much larger than $\sigma$.

Defining the response (Green) function $G(t)$ for $t > 0$ as

$$\frac{dG}{dt} = -\frac{G}{\tau_0} + \alpha(t) \; ; \quad G(0) = 0 \tag{2.7}$$

and $G(t) = 0$ for $t < 0$. Then, for $\tau_1 = 0$,

$$G(t) = \begin{cases} \frac{\tau_0}{\tau_0 - \tau_s} \left( e^{-t/\tau_0} - e^{-t/\tau_s} \right) & t \geq 0 \\ 0 & \text{otherwise} \end{cases} \tag{2.8}$$

The function $G$ is the normalized excitatory post-synaptic potential (EPSP) developed in the cell as a response to the EPSC (Eq. 2.4). The Volterra representation of Eqs. 2.1,2.2 for neurons that can fire only one spike is

$$\frac{V_T}{g_{syn}} = \int_{-\infty}^{\infty} dx' \, w(x') \, G\left[ T(x) - T(x - x') - \tau_d - \frac{x'}{c} \right] \tag{2.9}$$

together with the condition that $T(x)$ is the first time that the voltage crosses the threshold ("causality criterion"). This condition requires that $V$ increases with time just before the spike, namely

$$\frac{dV[x, T(x)]}{dt} > 0 \tag{2.10}$$

The meaning if Eq. 2.9,2.10 is that the summation of all the contribution to the voltage of one neuron from other neurons is equal to $V_T$ when this neuron fires, and that this neuron does not fire before.

**Numerical methods.** Eqs. 2.1,2.2 are simulated numerically by discretizing space. There are $N$ neurons in the chain, and the density of neurons is $\rho$ per length $\sigma$. The coupled system of ordinary differential equations for the integrate-and-fire neurons is solved using exact integration [27]. To stimulate the network, applied current is "injected" into a group of neurons on the "left" of the system (small $x$ values), that span a length at least equal to the footprint length $\sigma$ ("shock" initial conditions).

### 2.2. Continuous and lurching pulses

A pulse can propagate along the network in response to "shock" initial conditions. For zero or small $\tau_d$ (below a critical value $\tau_{dc}$), the pulse is continuous far from the stimulus region (Fig. 2A), and the firing times of the neurons obey: $T(x) = T_0 + x/v$, where $v$ is the pulse velocity and $T_0$ is an arbitrary time. The neuronal potential fulfills an equation of a traveling pulse as well:

$V(x, t) = \tilde{V}(x - vt)$ [1,5] As $\tau_d$ approaches a critical value $\tau_{dc}$ from below, the convergence of the firing time $T(x)$ to a continuous propagating pulse decelerates. For $\tau_d > \tau_{dc}$, a lurching propagating pulse is observed (Fig. 2B). Space is spontaneously divided into basic spatial units, each with a spatial period length $L$, and the firing time in each unit can be obtained from the spatial period in the previous unit according to

$$T(x + L) = T(x) + T_{\text{per}} \qquad (2.11)$$

where $T_{\text{per}}$ is the time period of a lurching cycle. The average velocity of the pulse is $v = L/T_{\text{per}}$. Supposed that one lurching period starts at $x = 0$ and $T(0) = 0$. The firing time of a neuron at a position $x$ is given by

$$T(x) = nT_{\text{per}} + f(\hat{x}) \qquad (2.12)$$

where $n$ is the integer part of $T(x)/T_{\text{per}}$ (or $x/L$) and $\hat{x} = x - nL$. The function $f$, expressing the firing time within one period relative to the starting point of the period in space and time, is defined on the interval $[0, L)$; $f(0) = 0$. Hence, the function

$$T(x) - x/v = f(\hat{x}) - \hat{x}/v \qquad (2.13)$$

is a periodic function of $x$ with a period $L$. Eq. 2.13 demonstrates the spatiotemporal periodicity of the lurching pulse.

In the following, we calculate first the existence and stability regimes and the velocity of the continuous and lurching pulses for $c \to \infty$. Then, we analyze the effects of finite axonal velocity $c$.

### 2.3. Existence, stability and velocity of continuous pulses

**General formalism** Substituting the condition for a continuous pulse, $T(x) = x/v$, into the evolution equation 2.9, we obtain

$$\int_0^\infty dx'\, w(x' + \tau_d v)\, G(x'/v) = V_T/g_{\text{syn}} \qquad (2.14)$$

We should also confirm that the condition of Eq. 2.10 holds. Stability of the continuous pulse is calculated by considering $T(x) = x/v + s(x)$ and linearizing Eq. 2.9 near the continuous solution, to obtain

$$\int_0^\infty dx'\, w(x' + \tau_d v)\, G'(x'/v)\left[s(x) - s(x - x' - \tau_d v)\right] = 0 \qquad (2.15)$$

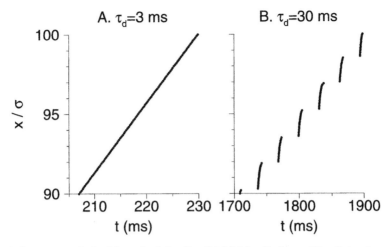

Fig. 2. Rastergrams obtained from simulating Eqs. 2.1,2.2,2.5, with the condition that each neuron can fire only one spike. Parameters: $\tau_0 = 30$ ms, $\tau_s = 2$ ms, $c \to \infty$, $N = 5 \times 10^4$, $\rho = 500$; for these parameters, $\tau_{dc} = 11.15$ ms. The solid circles represent the firing time of neurons as a function of their normalized position $x/\sigma$; spikes of only one out of every 50 neuron are plotted. Together, the groups of solid circles looks almost like one continuous line. A: For $\tau_d < \tau_{dc}$ (3 ms), a continuous pulse is obtained. B. For $\tau_d > \tau_{dc}$ (30 ms), the pulse is lurching. Adapted from [19]

This convolution equation has a general solution $s(x) = \exp(\lambda x)$. Substituting this equation in Eq. 2.15 yields

$$\int_0^\infty dx' \, w\left(x' + \tau_d v\right) G'\left(x'/v\right)\left[1 - e^{-\lambda\left(x' + \tau_d v\right)}\right] = 0 \qquad (2.16)$$

$\lambda = 0$ is always a solution to Eq. 2.16, corresponding to the translation invariance of the continuous pulse. The continuous wave is stable if Re $\lambda < 0$ for all the $\lambda$ values that are solutions of this eigenvalue equation (except for that single zero solution). This means that a small perturbation at a specific, finite $x$ will decay at larger $x$ as the pulse propagate. A similar method to study stability was developed independently by Bressloff [24, 25].

**Exponential footprint shape.** The velocity $v$ is determined using Eqs. 2.5, 2.8, 2.14:

$$\frac{(\tau_0 v + \sigma)(\tau_s v + \sigma)}{\tau_0 v \sigma} \exp\left(\frac{\tau_d v}{\sigma}\right) = \frac{g_{syn}}{2V_T} \qquad (2.17)$$

This is an extension of the equation obtained in [5] for $\tau_d = 0$. From this equation, one can see that:

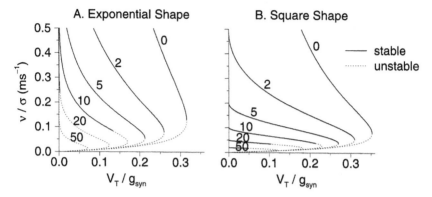

Fig. 3. The velocity of the continuous pulse as a function of $V_T/g_{syn}$ for several values of $\tau_d$. The thick lines represent stable pulses and the thin lines represent unstable pulses. The number above each line, from 0 to 50, denotes the value of $\tau_d$. Parameters: $\tau_0 = 30$ ms, $\tau_s = 2$ ms. A. Exponential footprint shape (Eq. 2.17). B. Square footprint shape (Eq. 2.24). Adapted from [19, 20]

1. For $\tau_d = 0$, the left-hand-side of Eq. 2.17 has a minimum with respect to $v$ at $v_{min} = \sigma/\sqrt{\tau_0\tau_s}$. Continuous pulses cannot propagate below this minimal velocity, which is obtained for a minimal synaptic coupling $g_{syn,min}$. For $g_{syn} > g_{syn,min}$, there are two branches of solutions to Eq. 2.17. At the fast branch, $v$ increases with $g_{syn}$, and at the slow branch $v$ decreases with $g_{syn}$ [5].

2. Because $\exp(\tau_d v/\sigma) > 1$ and increases with $\tau_d$, $v_{min}$ decreases with $\tau_d$ and is obtained for larger $g_{syn,min}$.

3. For $\tau_d > 0$ and at large enough $g_{syn}$, the velocity is determined mainly by the exponential factor in Eq. 2.17, and therefore $v$ depends logarithmically on $g_{syn}$ to the highest order. In contrast, for $\tau_d = 0$, the velocity exhibits a power-law dependence on $g_{syn}$ at at large $g_{syn}$ [5]. Graphs of $v/\sigma$ as a function of $V_T/g_{syn}$ for several values of $\tau_d$ are shown in Fig. 3A.

In order to find the function $V(x, t)$ before the spike (and thus verify that (11) holds), we look without loss of generality at a neuron located at $x = 0$. The Volterra representation of Eqs. 2.1,2.2 for the traveling wave $T(x) = x/v$ for time $t < 0$, taking into account that $G(t) = 0$ for $t < 0$, is

$$V(0, t) = g_{syn} \int_{-\infty}^{\infty} dx'\, w\left(x'\right) G\left[t - T\left(-x'\right) - \tau_d\right] = \qquad (2.18)$$

$$g_{syn} \int_{(\tau_d-t)v}^{\infty} dx'\, w\left(x'\right) G\left(t + \frac{x}{v} - \tau_d\right)$$

Substituting Eq. 2.5,2.8 into Eq. 2.3 yields

$$V(0, t) = \frac{g_{\text{syn}} \tau_0 \nu \sigma}{2 (\tau_0 \nu + \sigma) (\tau_s \nu + \sigma)} \exp\left[\frac{(t - \tau_d) \nu}{\sigma}\right] \tag{2.19}$$

The voltage $V$ rises exponentially from 0 and reaches $V_T$ at $t = 0$ for all $\nu$ values; both the upper and lower branches fulfill the condition of Eq. 2.10.

The stability of the continuous pulse is explored by substituting Eq. 2.5, 2.8 in Eq. 2.16 to obtain

$$e^{\lambda \nu \tau_d} = \frac{(\tau_0 \nu + \sigma) (\tau_s \nu + \sigma) (1 + \lambda \sigma)}{[\tau_0 \nu (1 + \lambda \sigma) + \sigma][\tau_s \nu (1 + \lambda \sigma) + \sigma]} \tag{2.20}$$

The value $\lambda = 0$ is always a root corresponding to translation invariance. In Appendix A we show in general that the lower branch, for which $d\nu/dg_{\text{syn}} < 0$, is unstable. This can be easily demonstrated in the case $\tau_d = 0$, in which there is another solution to Eq. 2.20, $\lambda = \sigma/(\nu^2 \tau_0 \tau_s) - 1/\sigma$. The pulse is stable if $\nu > \sigma/\sqrt{\tau_0 \tau_s}$, and therefore the fast branch is stable and the slow branch is unstable.

In order to examine whether the delay can destabilize pulses that belong to the fast branch, we look for a pair of complex conjugate eigenvalues which cross the imaginary axis. At this Hopf bifurcation $\lambda = i\omega$, and

$$e^{i\omega\nu\tau_{dc}} = \frac{(\tau_0 \nu + \sigma) (\tau_s \nu + \sigma) (1 + i\omega\sigma)}{[\tau_0\nu(1 + i\omega\sigma) + \sigma][\tau_s\nu(1 + i\omega\sigma) + \sigma]} \equiv Z(\omega) \tag{2.21}$$

As $\omega$ varies, the left-hand-side traces out the unit circle. In order to solve this equation, we search for the non-zero $\omega$ value for which $|Z(\omega)| = 1$. This value is given by

$$\omega^2 = \left[\sigma^4 + 2\sigma^3\nu(\tau_0 + \tau_s) + 4\sigma^2\nu^2\tau_0\tau_s - \nu^4\tau_0^2\tau_s^2\right] / \left(\sigma^2\nu^4\tau_0^2\tau_s^2\right) \tag{2.22}$$

For that $\omega$ we find $\tau_{dc}$, the critical value of $\tau_d$ for which the arguments of the complex numbers on the two sides of equation 2.21 are equal

$$\tau_{dc} = \frac{\arg[Z(\omega)]}{\omega\nu} \tag{2.23}$$

Note that $\tau_{dc}$ does not depend explicitly on $g_{\text{syn}}$, but only through $\nu$. Using Eqs. 2.22-2.23, $\tau_{dc}$ is calculated as a function of $\nu$. This solution shows that $\tau_{dc}$ increases with $\nu$, and this increase is steep at small $\nu$ and modest at large $\nu$. As a result, for small values of $\tau_d$, the Hopf bifurcation occurs on the lower, slow branch and does not have an effect on the dynamics. For larger values of $\tau_d$ the Hopf bifurcation occurs on the upper, fast branch, and the continuous pulse is

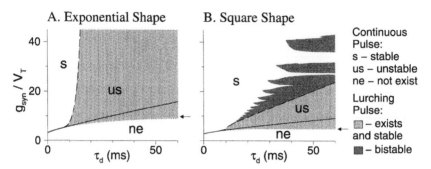

Fig. 4. Regimes of existence and stability of the continuous and lurching pulses in the $\tau_d$-$v$ plane are shown in (A) for exponential footprint shape and in (B) for square footprint shape. Parameters are as in Fig 3. The boundaries of the regime in which the lurching pulse exists and is stable were computed from numerical simulations, in which a pulse was initiated by a "shock" initial stimulus; $N = 20,000$, $\rho = 50$. The solid line denotes the minimal possible velocity as a function of $g_{syn}$; the continuous pulse becomes unstable (via a Hopf bifurcation) on the dashed line. The continuous pulse is therefore stable above both the solid and dashed line, as denoted by "s". It is unstable between the two lines, as denoted by "us", and does not exist below the continuous line, as denoted by "ne". The light-gray shading represents the region for which lurching pulses (and not continuous pulses) are obtained. Bistable regimes, in which the continuous pulse can coexist with the lurching pulse, are denoted by the dark-gray shading. For the square footprint shape (B), but not for the exponential footprint shape, there is such a bistable regime that has a "tongue-like" structure. The arrow at the right of each graph represents the minimal values of $g_{syn}$ for which the lurching pulse is found in simulations for $\tau_d \to \infty$. Adapted from [19]

unstable for low velocities, as shown in Fig. 3A. The various behavioral regimes of the continuous pulse in the $\tau_d$-$g_{syn}$ plane are plotted in Fig. 4A. Below the solid line, the continuous pulse does not exist. Above this line, the continuous pulse is stable if the delay $\tau_d$ is smaller than a critical delay $\tau_{dc}$, and is unstable otherwise. The dashed line denotes the value of $\tau_{dc}$ as a function of $g_{syn}$. At very large $g_{syn}$, $\tau_{dc}$ increases logarithmically with $g_{syn}$.

The boundaries of the regime of existence of lurching pulses have been located using numerical simulations, and are represented by the light-shaded area in Fig. 4A. Lurching pulses are observed in all the parameter regimes in which the continuous pulse is unstable. In addition, lurching pulses are also observed in a parameter regime in which the continuous pulse does not exist at all. At very large delay, lurching pulses are observed above a critical value of $g_{syn}$, denoted by the arrow, that does not depend on $\tau_d$ (see below).

**Square footprint shape.** The velocity $v$ is determined using Eqs. 2.6, 2.8, 2.14:

$$\frac{2V_T}{g_{syn}} = \qquad\qquad\qquad\qquad (2.24)$$

$$\frac{\tau_0 \, \nu}{\sigma} \left\{ 1 - \frac{1}{\tau_0 - \tau_s} \left[ \tau_0 \exp\left(\frac{\tau_d - \sigma/\nu}{\tau_0}\right) - \tau_s \exp\left(\frac{\tau_d - \sigma/\nu}{\tau_s}\right) \right] \right\}$$

Graphs of $\nu/\sigma$ as a function of $V_T/g_{\text{syn}}$ for several values of $\tau_d$ are shown in Fig. 3B. The qualitative results regarding the minimal velocity and its dependence on $\tau_d$ are the same as for the exponential case. The situation is different, however, for large $g_{\text{syn}}$. For $\nu = \sigma/\tau_d$, the right hand side of Eq. 2.24 is zero. Hence, at the limit $g_{\text{syn}} \to \infty$ the velocity of the continuous pulse approaches the finite value $\sigma/\tau_d$.

The stability of the continuous pulse is explored by substituting Eq. 2.6, 2.8 in Eq. 2.16 to obtain

$$(1 + \lambda \tau_0 \, \nu)(1 + \lambda \tau_s \, \nu) \left[ \exp\left(\frac{\tau_d - \sigma/\nu}{\tau_0}\right) - \exp\left(\frac{\tau_d - \sigma/\nu}{\tau_s}\right) \right] =$$

$$\lambda \nu \, (\tau_0 - \tau_s) \exp\left(-\lambda \tau_d \nu\right) + \tag{2.25}$$

$$\left[ (1 + \lambda \tau_s \nu) \exp\left(\frac{\tau_d - \sigma/\nu}{\tau_0}\right) - (1 + \lambda \tau_0 \nu) \exp\left(\frac{\tau_d - \sigma/\nu}{\tau_s}\right) \right] \exp\left(-\sigma \lambda\right)$$

To calculate the $\tau_{dc}$ value for which the continuous pulse becomes unstable, we substitute $\lambda = i\omega$ in Eq. 2.25 and obtain two real transcendental equations for the real variables $\tau_{dc}$ and $\omega$. We solve these equation numerically for one value of large-enough $\nu$ using standard iteration methods [28]. The fact that $\omega$, which is the spatial period of the expanding or decaying fluctuations, is of order $2\pi/\sigma$, helps us to choose initial conditions for the iteration process. At small $\nu$, it is numerically difficult to use this method to solve the equations. Therefore, we write ordinary differential equations with these two equations as their nullclines, and follow the their fixed point solution using the program XPPAUT [29,30], starting from the solution at large $\nu$ that we already have computed. Using Eq. 2.24, we calculate the range of stability as a function of $g_{\text{syn}}$ and $\tau_d$.

The different behavioral regimes of the continuous pulse for a square footprint shape are presented in Fig. 4B. The critical delay $\tau_{dc}$ increases with $g_{\text{syn}}$ almost linearly at large $g_{\text{syn}}$. As a result, for a specific $\tau_d$ there is a moderate $g_{\text{syn}}$ value for which the continuous pulse is stable. Lurching pulses are obtained in the region denoted by the gray shading in Fig. 4B. There are two apparent differences between the situation here and the situation for exponential footprint shape. First, lurching pulses exist in an area which is composed of "tongues". Second, a *bistable* regime exists, denoted by the dark-gray shading, in which the two types of pulses can propagate, depending on the initial stimulation [18]. This bistability suggests that the Hopf bifurcation in which the continuous pulse loses stability is subcritical.

## 2.4. Velocity of lurching pulses

How does the velocity of the lurching pulse depend on the synaptic strength? We can calculate this velocity in the case

$$\tau_s \ll \tau_0 \ll \tau_d \tag{2.26}$$

This case corresponds to a large delay and fast EPSCs; note that $\tau_1 = 0$. Simulations of such cases show that neurons fire only during a time period that is small in comparison to the delay, and $L$ is almost unaffected by the delay period as long as the delay is large enough. Therefore, the pulse velocity is $v = L/\tau_d$. For $\tau_d \gg \tau_0$, a neuron that fires during the $n$th lurching period (with length $L$ and time $T_{per}$) is affected only by neurons that have fired during the previous lurching period. The contribution to the potential of that neuron of neuronal EPSCs from neurons in earlier periods has already decayed, mostly because of the large delay and also because of the fact that neuron in earlier lurching periods are more distant from that neuron. The neuron is also not affected by neurons that fire during the same lurching period. We assume that the lurching wave is initiated at very large, negative $x$, and a lurching spatial period starts at $x = 0$. Neurons at a position $0 \le x < L$ will be affected only by neurons located in the interval $-L \le x < 0$. From the definition of the function $f(x)$ (Eq. 2.12), we see that $T(x) = f(x)$ for $0 \le x < L$ and $T(x) = f(x) - T_{per}$ for $-L \le x < 0$. Eq. 2.9 for neurons at $0 \le x < L$ becomes

$$\frac{V_T}{g_{syn}} = \int_{-L}^{0} dx' \, w\left(x - x'\right) G\left[f(x) - f(x') + \hat{T}\right] \tag{2.27}$$

where $\hat{T} = T_{per} - \tau_d$. In all the simulations we have performed, we found that $\hat{T} > 0$. Here we calculate $L$ by considering neurons with $x \lesssim L$ such that the argument of $G$ in Eq. 2.27 is non-negative. Substituting Eq. 2.8 in 2.27 we obtain

$$\frac{V_T}{g_{syn}} = \int_{0}^{L} dx' \, w\left(x - x' + L\right) \frac{\tau_0}{\tau_0 - \tau_s} \times \tag{2.28}$$

$$\left\{ e^{-\left[f(x) - f(x') + \hat{T}\right]/\tau_0} - e^{-\left[f(x) - f(x') + \hat{T}\right]/\tau_s} \right\}$$

We continue by using the condition $\epsilon \equiv \tau_s/\tau_0 \ll 1$. Using the ansatz:

$$\hat{T} = O(\epsilon) \ , \ f(x) = O(\epsilon) \tag{2.29}$$

we define the scaled function and variable

$$\phi(x) = f(x)/\epsilon \ , \ \theta = \hat{T}/\epsilon \tag{2.30}$$

This assumption is supported by the simulations for the exponential footprint shape shown in Fig. 5, where the function $\phi(x)$ is plotted for several values of $\tau_s$ (A), and $\theta$ is plotted as a function of $\tau_s$ (B). In is seen that $\phi(x)$ converges to a constant function as $\epsilon$ decreases, and $\theta$ converges to a limit value. Because $\exp\{\epsilon[\phi(x) - \phi(x') + \theta]/\tau_0\} \simeq 1$, Eq. 2.28 becomes approximately

$$\frac{V_T}{g_{syn}} \simeq \tag{2.31}$$

$$\int_0^L dx' \, w\left(x - x' + L\right) - e^{[\theta - \phi(x)]/\tau_0} \int_0^L dx' \, w\left(x - x' + L\right) e^{\phi(x')/\tau_0}$$

As $x$ approaches $L$, $\phi$ increases. At the point where the firing time $T(x)$ diverges (or at least becomes larger than $\tau_d$), the lurching spatial period ends. In order to find $L$, we let $\phi(x) \to \infty$ as $x \to L$, and hence $\lim_{x \to L} \exp\left[-\phi(x)/\tau_0\right] = 0$. Therefore, taking at the limit $x \to L$, Eq. 2.31 yields, after changing of variables

$$\frac{V_T}{g_{syn}} = \int_L^{2L} dx \, w(x) \tag{2.32}$$

This is an implicit equation for $L$ under the conditions of Eq. 2.26. The RHS of Eq. 2.32 is zero for $L = 0$ and for $L \to \infty$ (because $w(x)$ vanishes at large $L$ values), and is non-negative for finite $L$. Therefore, it reaches a finite maximum at finite $L$. If $g_{syn}$ is smaller than its value at that maximum, no lurching pulse can propagate, and therefore the lurching pulse has a threshold. If $g_{syn}$ is above this threshold value, there are (at least) two solutions to Eq. 2.32. We consider only the solution for which $dL/dg_{syn} > 0$. The second solution is probably unstable; this has yet to be proven. Here we discuss several specific footprint shapes.

**Exponential footprint shape.** By substituting Eq. 2.5 in Eq. 2.32, one obtains

$$L = \sigma \ln 2 - \sigma \ln\left(1 - \sqrt{1 - 8V_T/g_{syn}}\right) \tag{2.33}$$

For large $g_{syn}$, expanding Eq. 2.33 yields

$$L = \sigma \ln\left(\frac{g_{syn}}{2V_T}\right) \tag{2.34}$$

Eq. 2.33 has a solution only if $g_{syn} > 8V_T$, which is the synaptic conductance threshold for enabling the propagation of lurching waves under the conditions of Eq. 2.26. The dependence of the lurching spatial period $L$ (in units of $\sigma$) as a function of $g_{syn}/V_T$ is shown in Fig. 6. The straight line in logarithmic scale shows that except for $g_{syn}$ near the threshold, $L$ increases logarithmically with

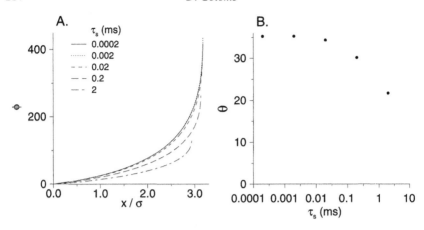

Fig. 5. A. The function $\phi(x)$ versus $x/\sigma$ for several values of $\tau_s$. The values of $\tau_s$ (in ms) are: 0.0002 (solid line), 0.002 (dotted line), 0.02 (dashed line), 0.2 (long-dashed line) and 2 (dotted-dashed line). The curves for $\tau_s = 0.0002$ ms is almost identical to the curve for $\tau_s = 0.002$ ms. B. $\theta$ as a function of $\tau_s$. Results for both (A) and (B) were obtained from simulations with $\tau_d = 1000$ ms, $\tau_0 = 30$ ms, $N = 50,000$, $\rho = 500$. These simulations show that $\phi(x)$ and $\theta$ reaches a limit value as $\tau_s \to 0$. Adapted from [20]

$g_{syn}$. The open circles represent simulation results with $\tau_d = 1000$ ms, $\tau_0 = 30$ ms, and $\tau_s = 0.002$ ms, and they fall exactly on the analytical curve.

Using simulations, we tested the validity of the perturbation calculation when the time constants of the system do not fulfill Eq. 2.26 (not near the threshold). First, we reduced $\tau_d$ to 20 ms. This change has almost no effect for large $g_{syn}$, and mildly increases $L$ for small $g_{syn}$. This velocity increase can be attributed to the excitatory effect on a neuron from neurons in cycles before the immediate previous lurching cycle; it is stronger for small $g_{syn}$ because of the shorter $L$. Second, we increased $\tau_s$ to 2 ms. As a result, $L$ decreases, because the EPSP developed in the post-synaptic cell (Eq. 2.7) is smaller as a result of the interplay between the EPSC and the leaky neuronal integrator. This change of $\tau_s$ has, however, an effect of less then 10% on $L$ in comparison to the analytic result (Eq. 2.33). Third, we simulated a network with $\tau_1 = 0.3$ ms. This change did not change $L$ significantly in comparison to the case $\tau_1 = 0$. These results show that our analytical theory yields a good approximation even beyond the parameter regime for which it is derived.

**Footprint shapes with finite support.** Suppose that the footprint shape has a finite support: $w(x) = 0$ for $x > \sigma$. The RHS of Eq. 2.32 is zero for $L \geq \sigma$, and therefore $L < \sigma$ for every $g_{syn}$. For the square footprint shape, substituting

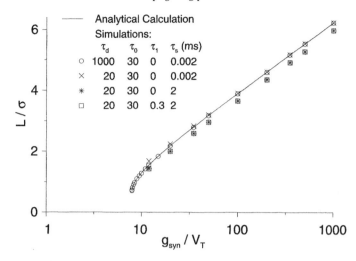

Fig. 6. The normalized length of the lurching period $L/\sigma$ as a function of $g_{\mathrm{syn}}/V_T$. The analytical solution for the case $\tau_1 = 0$, $\tau_s \ll \tau_0 \ll \tau_d$ (Eq. 2.33) is represented by the solid line. Simulations were carried out with $N = 200,000$ and $\rho = 500$. Simulations results with the corresponding parameter set: $\tau_d = 1000$ ms, $\tau_0 = 30$ ms, $\tau_1 = 0$, $\tau_s = 0.002$ ms, denoted by $\circ$, fit the analytical solutions almost exactly. The symbol $\times$ denotes simulations with $\tau_d = 20$ ms, $\tau_0 = 30$ ms, $\tau_1 = 0$, $\tau_s = 0.002$ ms; the symbol $*$ denotes simulations with $\tau_d = 20$ ms, $\tau_0 = 30$ ms, $\tau_1 = 0$, $\tau_s = 2$ ms; and the open square denotes simulations with $\tau_d = 20$ ms, $\tau_0 = 30$ ms, $\tau_1 = 0.3$, $\tau_s = 2$ ms. Adapted from [19]

Eq. 2.6 in Eq. 2.32 yields

$$\frac{V_T}{g_{\mathrm{syn}}} = \begin{cases} \frac{L}{2\sigma} & 0 \le L \le \frac{\sigma}{2} \\ \frac{1}{2}\left(1 - \frac{L}{\sigma}\right) & \frac{\sigma}{2} < L \le \sigma \end{cases} \tag{2.35}$$

The maximum of the RHS of Eq. 2.35 is obtained for $L = \sigma/2$ for $V_T/g_{\mathrm{syn}} = 1/4$. Therefore, a threshold for the propagation of the lurching pulse is $g_{\mathrm{syn}} = 4V_T$, for which $L = \sigma/2$. For larger $g_{\mathrm{syn}}$ values, $L$ is given by

$$L = \sigma\left(1 - \frac{2V_T}{g_{\mathrm{syn}}}\right) \tag{2.36}$$

## 2.5. The nature of lurching pulses

The boundary condition that $\phi(x) \to \infty$ as $x \to L$ is necessary for obtaining Eq. 2.32. The meaning of this condition is that for the spatially-continuous dynamical system defined by Eqs. 2.1, 2.2, the function $T(x)$ for the lurching pulse is continuous in $x$ (although not smooth). The firing time of neurons the edge of each lurching spatial period increases rapidly as $x$ approaches the edge. When this time reaches the value $T_{per}$ (of order $\tau_d$), a new lurching period starts. As seen in Fig. 5, $\phi(x)$ increases considerably only in a narrow spatial regime near $x = L$. The discontinuous lurching period seen in Fig. 2 is a result of the discrete nature of our numerical simulation. On a discrete lattice, The value of $\phi$ (and $f$) must be finite. Our numerical simulations show that for $\tau_d \gg \tau_{dc}$, the value of $f$ for the last neuron in a spatial lurching period is smaller than $T_{per}$, creating an effect which looks like a discontinuity.

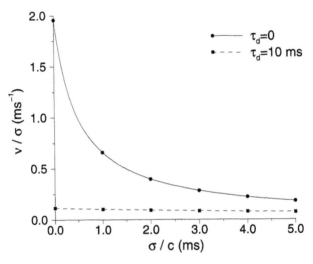

Fig. 7. Effects of finite axonal conduction velocity $c$. The normalized velocity of the continuous pulse $v/\sigma$ is plotted as a function of $\sigma/c$. Analytical results (Eq. 2.38) are denoted by a continuous line for $\tau_d = 0$ and by a dashed line for $\tau_d = 10$ ms. Simulations results are denoted by solid circles ($\tau_d = 0$) and by solid squares ($\tau_d = 10$ ms). Parameters: $\tau_0 = 30$ ms, $\tau_s = 2$ ms, $g_{syn}/V_T = 10$, $N = 5000$, $\rho = 50$. Adapted from [20]

### 2.6. Finite axonal conduction velocity

The effects of finite axonal velocity $c$ on the velocity $v$ is studied by substituting the condition for a continuous pulse into Eq. 2.9 to obtain

$$\int_0^\infty dx' \, w\left(x' + \frac{\tau_d}{\frac{1}{v} - \frac{1}{c}}\right) G\left[x'\left(\frac{1}{v} - \frac{1}{c}\right)\right] = \frac{V_T}{g_{syn}} \tag{2.37}$$

This equation is similar to Eq. 2.14 except that the velocity $v$ in Eq. 2.14 is replaced by the term $v_-$, where

$$\frac{1}{v_-} = \frac{1}{v} - \frac{1}{c} \tag{2.38}$$

The velocity $v_-$ is the pulse velocity in the limit $c \to \infty$. This means that the effects of axonal conduction velocity on the velocity of the continuous waves can be deduced from first studying the properties of a model with $c \to \infty$ and calculating the velocity $v_-$, and then substituting the value of $v_-$ by the value $1/v - 1/c$ (Fig. 7). Similarly, the stability of the continuous pulse is determined by an equation similar to Eq. 2.16, except that the variable $v$ in that equation is replaced by $v_-$. As a result, the value of $\tau_{dc}$ for a specific value of $g_{syn}$ does not depend on $c$.

## 3. Propagating pulses in networks of excitatory and inhibitory neurons

### 3.1. The model

In this section, we consider a one-dimensional network of excitatory (E) and inhibitory (I) neurons (Fig. 8A). A neuron is described by its membrane potential $V_\alpha(x, t)$, $\alpha = E, I$, and its dynamics is governed by the integrate-and-fire scheme in the excitable regime [24]. The dynamical equations and analysis are extensions of the equations and analysis described in the previous section for excitatory population only. The analogue to Eq. 2.1 is

$$\frac{\partial V_\alpha(x, t)}{\partial t} = -\frac{V_\alpha(x, t)}{\tau_{0\alpha}} + I_{syn,E\alpha}(x, t) - I_{syn,I\alpha}(x, t) \tag{3.1}$$

Here, $\tau_{0\alpha}$ is the passive membrane time constant of the neuron, and $I_{syn,E\alpha}$ (resp. $I_{syn,I\alpha}$) is the total synaptic current contributed by the excitatory (resp. inhibitory) population. In the following, we denote by $T_\alpha(x)$ the time at which a neuron from the $\alpha$th population located at $x$ fires. A pre-synaptic spike induces a post-synaptic current that is proportional to the function $\alpha_{\beta\alpha}(t)$, where

$$\alpha_{\beta\alpha}(t) = \begin{cases} \frac{1}{\tau_{s\beta\alpha}} \exp\left(-t/\tau_{s\beta\alpha}\right) & t \geq 0 \\ 0 & \text{otherwise} \end{cases} \tag{3.2}$$

For the excitatory coupling, we consider two types of synaptic current: a fast (f) current and and a slow (s) current, corresponding to the contribution of AMPA and NMDA synaptic receptors in a biological network respectively. The function

**A.**      **B.**

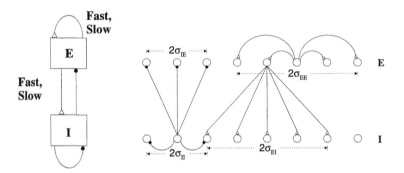

Fig. 8. Schematic diagram of the model architecture. A. Neuronal populations and synaptic types. B. The one-dimensional architecture of the network. Adapted from [22]

$\alpha_{E\alpha}(t)$ and the decay time $\tau_{sE\alpha}$ have a superscript $\gamma = \{f, s\}$ denoting whether the decay of the excitatory current is fast or slow.

The network architecture is shown in Fig. 8B. The contributions to the synaptic current from the excitatory and inhibitory populations are

$$I_{\text{syn},E\alpha}(x, t) = \sum_{\gamma=f,s} g_{E\alpha}^{\gamma} \int_{-\infty}^{\infty} dx' \, w_{E\alpha}(x - x') \alpha_{E\alpha}^{\gamma} \left(t - T_E(x')\right) \tag{3.3}$$

$$I_{\text{syn},I\alpha}(x, t) = g_{I\alpha} \int_{-\infty}^{\infty} dx' \, w_{I\alpha}(x - x') \alpha_{I\alpha} \left(t - T_I(x')\right) \tag{3.4}$$

where $g_{\beta\alpha}$ is the synaptic coupling strength from the $\beta$ population to the $\alpha$ population. The spatial dependence of the synaptic strength on distance ("synaptic footprint shape") is given by

$$w_{\beta\alpha}(x) = \frac{1}{2\sigma_{\beta\alpha}} \exp\left(-|x|/\sigma_{\beta\alpha}\right) \tag{3.5}$$

The spatial variable, $x$, is dimensionless and represents the distance in terms of the excitatory footprint, $\sigma_{EE}$ which is set to 1. As a result, the velocity $v$ has units of ms$^{-1}$.

**Reference parameter set.** In the following, we study the model in its general form. Numerical examples, however, are shown for a particular set of parameters, called the "reference parameter set", with: $\tau_{0E} = \tau_{0I} = 30$ ms, $\tau_{sEE}^{f} = \tau_{sEI}^{f} = 2.5$ ms, $\tau_{sEE}^{s} = \tau_{sEI}^{s} = 50$ ms, $\tau_{sIE} = \tau_{sII} = 8$ ms, $g_{EE}^{f} = 12$, $g_{EE}^{s} = 10$,

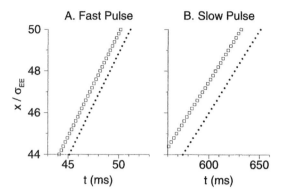

Fig. 9. Fast and slow pulses can propagate for the same set of parameters in the bistable regime. Rastergrams obtained by simulations of the neuronal dynamics are shown. Firing times of excitatory cells are denoted by solid circles, and firing times of inhibitory cells are denoted by open squares. There are $\rho = 50$ neurons from each type within one unit length, and spikes of only one out of every 10 neurons are plotted. The pulses were initiated by initiating a pulse in a group of neurons with $x < 10$ with $\nu$ and $\zeta$ calculated from the theory. Two types of continuous pulses, fast and slow, are shown in A and B for the different initial conditions (note the difference in time scale). Parameters: $\tau_0 E = \tau_0 I = 30$ ms, $\tau_{sEE} = \tau_{sEI} = 2.5$ ms, $\tau_{sIE} = \tau_{sII} = 8$ ms, $g_{EE} = 12$, $\sigma_{EE} = 1$, $\sigma_{IE} = 0.9$, $\sigma_{EI} = 0.8$, $g_{II} = 2$, $\sigma_{II} = 0.5$, $g_{IE} = 5.5$. Adapted from [21]

$\sigma_{EE} = 1$, $g_{IE} = 5.2$, $\sigma_{IE} = 0.5$, $g_{EI}^{f} = 30$, $g_{EI}^{s} = 0$, $\sigma_{EI} = 1$, $g_{II} = 2$, $\sigma_{II} = 0.5$, $V_{TE} = V_{TI} = 1$. These parameters are used unless stated otherwise. Since the slow E-to-I excitation $g_{EI}^{s}$ does not have a strong effect on the network dynamics, we consider it to be zero.

## 3.2. Fast and slow pulses

Figure 9, based on simulations, shows two types of pulses that can propagate in networks of excitatory and inhibitory neurons. One type of pulses is similar to the continuous pulses that propagate in excitatory networks (Fig. 9A). Whereas their velocity, in general, decreases with inhibition, they are called here "fast pulses". A second type of pulses, called "slow pulses", is shown in Fig. 9B. This pulse is characterized by a much smaller velocity and by the fact that inhibitory cells fire before their neighboring excitatory cells. The two simulations in Fig. 9 were carried out with the same parameter set and different initial conditions. Therefore, the two pulses are bistable for this parameter set. In general, however, the bistable parameter regime is narrow and only one type of pulse, or none, can propagate for a specific parameter set. Below, we will investigate the properties of the two pulses theoretically.

### 3.3. Analysis of traveling pulse solutions

#### 3.3.1. Volterra representation
In order to analyze the dynamics, we define the Green's function $G_{\beta\alpha}(t)$ for $t > 0$ as (Eq. 2.7)

$$\frac{dG_{\beta\alpha}}{dt} = -\frac{G_{\beta\alpha}}{\tau_{0\alpha}} + \alpha_{\beta\alpha}(t) \tag{3.6}$$

and $G_{\beta\alpha} = 0$ for $t < 0$. The functions $G_{E\alpha}$ have also a superscript $\gamma$. For $t > 0$, we obtain:

$$G_{\beta\alpha}(t) = \frac{\tau_{0\alpha}}{\tau_{0\alpha} - \tau_{s\beta\alpha}} \left[ \exp(-t/\tau_{0\alpha}) - \exp(-t/\tau_{s\beta\alpha}) \right] \tag{3.7}$$

The integrated form of Eqs. (3.1,3.3,3.4) is given by the two Volterra equations for $\alpha = E, I$:

$$V_{T\alpha} = \sum_{\gamma=f,s} g^{\gamma}_{E\alpha} \int_{-\infty}^{\infty} dx' \, w_{E\alpha}(x') \, G^{\gamma}_{E\alpha} \left[ T_{\alpha}(x) - T_E \left( x - x' \right) \right] \tag{3.8}$$

$$- g_{I\alpha} \int_{-\infty}^{\infty} dx' \, w_{I\alpha}(x') \, G_{I\alpha} \left[ T_{\alpha}(x) - T_I \left( x - x' \right) \right]$$

In addition, the neuronal voltage should be below threshold before spiking:

$$V_{\alpha}(x, t) < V_{T\alpha} \quad \text{for all} \quad t < T_{\alpha}(x), \quad \alpha = E, I \tag{3.9}$$

A necessary, but not sufficient, condition for this (Eq. 2.10) is

$$\left. \frac{dV_{\alpha}[x, t]}{dt} \right|_{t=T_{\alpha}(x)} > 0, \quad \alpha = E, I \tag{3.10}$$

#### 3.3.2. Existence of traveling pulses
We consider a traveling pulse solution with velocity $v$. Without loss of generality, we assume that $v > 0$. The firing time of an inhibitory cell lags after the firing time of an excitatory cell at the same position by $\zeta$:

$$T_E(x) = \frac{x}{v}, \quad T_I(x) = \frac{x}{v} + \zeta \tag{3.11}$$

Negative $\zeta$ means that an I cell fires before a neighboring E cell. Substituting Eq. (3.11) into Eq. (3.8) yields

$$V_{T\alpha} = B^{f}_{E\alpha} + B^{s}_{E\alpha} - B_{I\alpha} \tag{3.12}$$

where

$$B_{\beta\alpha}^{\gamma} = g_{\beta\alpha}^{\gamma} \int_0^{\infty} dx \, w_{\beta\alpha} \left(x + \zeta \, v \, s_{\beta\alpha}\right) G_{\beta\alpha}^{\gamma} \left(\frac{x}{v}\right) \tag{3.13}$$

We define $s_{\beta\alpha} = \left(s_\alpha - s_\beta\right)/2$, where $s_E = 1$ and $s_I = -1$. Substituting the expressions for $G_{\beta\alpha}$ and $w_{\beta\alpha}$ in Eqs. (3.12,3.13), we obtain two algebraic equations for $v$ and $\zeta$, for negative $\zeta$

$$V_{TE} = \frac{\tau_{0E} \, v \, \sigma_{EE}}{2 \left(v \, \tau_{0E} + \sigma_{EE}\right)} \sum_{\gamma} g_{EE}^{\gamma} \frac{1}{\left(v \, \tau_{sEE}^{\gamma} + \sigma_{EE}\right)} - g_{IE} \frac{\tau_{0E} v}{\left(\tau_{0E} - \tau_{sIE}\right)} \times$$

$$\left[\frac{\tau_{0E}^2 \, v}{v^2 \, \tau_{0E}^2 - \sigma_{IE}^2} \exp\left(\frac{\zeta}{\tau_{0E}}\right) - \frac{\tau_{sIE}^2 \, v}{v^2 \, \tau_{sIE}^2 - \sigma_{IE}^2} \exp\left(\frac{\zeta}{\tau_{sIE}}\right) + \tag{3.14}$$

$$\frac{\left(\tau_{0E} - \tau_{sIE}\right) \sigma_{IE}}{2 \left(v \, \tau_{0E} - \sigma_{IE}\right) \left(v \, \tau_{sIE} - \sigma_{IE}\right)} \exp\left(\frac{\zeta v}{\sigma_{IE}}\right)\right]$$

$$V_{TI} = \frac{\tau_{0I} \, v \, \sigma_{EI}}{2 \left(v \, \tau_{0I} + \sigma_{EI}\right)} \exp\left(\frac{\zeta v}{\sigma_{EI}}\right) \sum_{\gamma} g_{EI}^{\gamma} \frac{1}{\left(v \, \tau_{sEI}^{\gamma} + \sigma_{EI}\right)} \tag{3.15}$$

$$-g_{II} \frac{\tau_{0I} \, v \, \sigma_{II}}{2 \left(v \, \tau_{0I} + \sigma_{II}\right) \left(v \, \tau_{sII} + \sigma_{II}\right)}$$

Similarly, for positive $\zeta$, we obtain the following two algebraic equations for $v$ and $\zeta$:

$$V_{TE} = \frac{\tau_{0E} \, v \, \sigma_{EE}}{2 \left(v \, \tau_{0E} + \sigma_{EE}\right)} \sum_{\gamma} g_{EE}^{\gamma} \frac{1}{\left(v \, \tau_{sEE}^{\gamma} + \sigma_{EE}\right)} - \tag{3.16}$$

$$g_{IE} \frac{\tau_{0E} \, v \, \sigma_{IE}}{2 \left(v \, \tau_{0E} + \sigma_{IE}\right) \left(v \, \tau_{sIE} + \sigma_{IE}\right)} \exp\left(\frac{-\zeta v}{\sigma_{IE}}\right)$$

$$V_{TI} = \sum_{\gamma} g_{EI}^{\gamma} \frac{\tau_{0I} \, v}{\left(\tau_{0I} - \tau_{sEI}^{\gamma}\right)} \left[\frac{\tau_{0I}^2 \, v}{v^2 \, \tau_{0I}^2 - \sigma_{EI}^2} \exp\left(\frac{-\zeta}{\tau_{0I}}\right) - \tag{3.17}$$

$$\frac{\left(\tau_{sEI}^{\gamma}\right)^2 v}{v^2 \left(\tau_{sEI}^{\gamma}\right)^2 - \sigma_{EI}^2} \exp\left(\frac{-\zeta}{\tau_{sEI}^{\gamma}}\right) + \frac{\left(\tau_{0I} - \tau_{sEI}^{\gamma}\right) \sigma_{EI}}{2 \left(v \, \tau_{0I} - \sigma_{EI}\right) \left(v \, \tau_{sEI}^{\gamma} - \sigma_{EI}\right)} \times$$

$$\exp\left(\frac{-\zeta v}{\sigma_{EI}}\right)\right] - g_{II} \frac{\tau_{0I} \, v \, \sigma_{II}}{2 \left(v \, \tau_{0I} + \sigma_{II}\right) \left(v \, \tau_{sII} + \sigma_{II}\right)}$$

Propagating pulses exist only if Eqs. (3.14,3.15) have at least one solution $v$ with $\zeta < 0$, or if Eqs. (3.16,3.17) have at least one solution $v$ with $\zeta > 0$.

### 3.3.3. Stability of traveling pulses

Stability of the continuous pulses is calculated by following the growth rate of a small perturbation

$$T_E(x) = x/v + \theta_E(x) \tag{3.18}$$

$$T_I(x) = x/v + \zeta + \theta_I(x) \tag{3.19}$$

Substituting these perturbations in Eq. (3.8), and keeping only the first order terms in $\theta_E, \theta_I$, we obtain two equations for $\alpha = E, I$

$$0 = \sum_{\gamma=f,s} g_{E\alpha}^{\gamma} \times \int_{\zeta v S_{E\alpha}}^{\infty} dx'\, w_{E\alpha}(x')\, G''^{\gamma}_{E\alpha}\left(\frac{x'}{v} - \zeta S_{E\alpha}\right) \times \tag{3.20}$$

$$\left[\theta_\alpha(x) - \theta_E(x - x')\right] - g_{I\alpha} \int_{\zeta v S_{I\alpha}}^{\infty} dx'\, w_{I\alpha}(x')\, G'_{I\alpha}\left(\frac{x'}{v} - \zeta S_{I\alpha}\right) \times$$

$$\left[\theta_\alpha(x) - \theta_I(x - x')\right]$$

where

$$G'(t) = dG(t)/dt \tag{3.21}$$

Assuming that the perturbations evolves as $\theta_E(x) = \theta_{E0}\exp(\lambda x)$ and $\theta_I(x) = \theta_{I0}\exp(\lambda x)$, yields the matrix equation

$$\sum_{\beta=E,I} A_{\beta\alpha}(\lambda)\, \theta_{\beta 0} = 0 \tag{3.22}$$

where

$$A_{EE}(\lambda) = \sum_{\gamma} g_{EE}^{\gamma} \int_{0}^{\infty} dx\, w_{EE}(x)\, G''^{\gamma}_{EE}\left(\frac{x}{v}\right)\left(1 - e^{-\lambda x}\right) \tag{3.23}$$

$$-g_{IE} \int_{\zeta v}^{\infty} dx\, w_{IE}(x)\, G'_{IE}\left(\frac{x}{v} - \zeta\right)$$

$$A_{IE}(\lambda) = g_{IE} \int_{\zeta v}^{\infty} dx\, w_{IE}(x)\, G'_{IE}\left(\frac{x}{v} - \zeta\right) e^{-\lambda x} \tag{3.24}$$

$$A_{EI}(\lambda) = -\sum_{\gamma} g_{EI}^{\gamma} \int_{-\zeta v}^{\infty} dx\, w_{EI}(x)\, G''^{\gamma}_{EI}\left(\frac{x}{v} + \zeta\right) e^{-\lambda x} \tag{3.25}$$

$$A_{II}(\lambda) = \sum_{\gamma} g_{EI}^{\gamma} \int_{-\zeta v}^{\infty} dx\, w_{EI}(x)\, G''^{\gamma}_{EI}\left(\frac{x}{v} + \zeta\right) \tag{3.26}$$

$$-g_{II} \int_0^\infty dx\, w_{II}(x)\, G'_{II}\left(\frac{x}{v}\right)\left(1 - e^{-\lambda x}\right)$$

Eq. (3.22) has non-trivial solutions if

$$\det\left[A(\lambda)\right] = 0 \qquad\qquad (3.27)$$

The value $\lambda = 0$ is always a solution of the characteristic equation (3.27) because of the translation invariance. Apart from this marginal stability, the traveling pulse is stable if all the other solutions of this equation have negative real parts. A pulse can lose stability at a saddle-node bifurcation (SNB), where the $\lambda = 0$ solution to Eq. (3.27) is a double zero, namely

$$d\left\{\det\left[A(\lambda)\right]\right\}/d\lambda|_{\lambda=0} = 0 \qquad\qquad (3.28)$$

Alternatively, a pulse can lose stability at a Hopf bifurcation (HB), where Eq. (3.27) has two imaginary solutions with $\lambda = \pm i\omega$.

### 3.3.4. Voltage profile
For a traveling pulse, the voltage profile of the E and I neurons that have not fired is determined by the voltage profile of the neurons at time $t = 0$

$$V_\alpha(x, t) = V_\alpha(x - vt, 0) \qquad \alpha = E, I \qquad\qquad (3.29)$$

We calculate $V_E(x - vt, 0)$ in the domain $0 \le x \le \infty$ and $V_I(x - vt, 0)$ in the domain $-\zeta v \le x \le \infty$ using Eqs. (3.1,3.3,3.4,3.6) and obtain

$$V_\alpha(x, 0) = \sum_{\gamma = f,s} g_{E\alpha}^\gamma \times \int_0^\infty dx'\, w_{E\alpha}(x + x')\, G_{E\alpha}^\gamma\left(\frac{x'}{v}\right) \qquad\qquad (3.30)$$

$$-g_{I\alpha} \int_{\zeta v}^\infty dx'\, w_{I\alpha}(x + x')\, G_{I\alpha}\left(\frac{x'}{v} - \zeta\right)$$

### 3.4. Theory of propagation of fast and slow pulses

A main goal of this Section is to study the effects of inhibition on pulse propagation. Therefore, we emphasize the effects of the parameter $g_{IE}$ and study how it modifies the system dynamics under various conditions. Effects of other parameters are also studied.

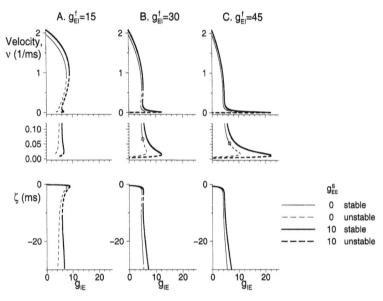

Fig. 10. The velocity $v$ of propagating pulses (upper and middle panels), and the difference $\zeta$ between the firing times of inhibitory and excitatory cells at the same position (lower panels), as functions of inhibitory-to-excitatory synaptic strength $g_{IE}$ for the reference parameter set and three values of the excitatory-to-inhibitory synaptic strength $g_{EI}^f$: A. $g_{EI}^f = 15$. B. $g_{EI}^f = 30$. C. $g_{EI}^f = 45$. Thin lines represent pulses with $g_{EE}^s = 0$, and thick lines represent pulses with $g_{EE}^s = 10$. Solid lines represent stable pulses and dotted lines represent unstable pulses. In the middle panels, which are expansions of the upper panels, Hopf bifurcations are denoted by open circles. Bistability of fast and slow pulses is observed in a small $g_{IE}$ regime for moderate values of $g_{EI}^f$ and $g_{IE}$. Adapted from [22]

### 3.4.1. Effects of I-to-E inhibition and slow E-to-E excitation

The dependence of $v$ (upper and middle panels) and $\zeta$ (lower panels) on $g_{IE}$ is shown in Fig. 10A-C for the reference parameter set, three values of $g_{EI}^f$, and two values of $g_{EE}^s$: 0 (thin lines) and 10 (thick lines). We first describe the situation for $g_{EE}^s = 0$. At low $g_{EI}^f$ values, there is only one stable branch of "fast" pulses, which terminates at a SNB. At intermediate $g_{EI}^f$ values, bistability exists, and at intermediate $g_{IE}$ values both the fast pulse and the slow pulse can propagate. The $g_{IE}$ regime in which slow pulses exists is pretty restricted, because the slow pulse is terminated by a HB. At large $g_{EI}^f$ values, there is a cross-over between fast pulses and a slow pulses as $g_{IE}$ increases. The slow pulse is still destabilized by a HB at a certain $g_{IE}$ value. Whereas our theory cannot determine what happens for $g_{IE}$ larger than its value at the HB, extensive numerical simulations indicate that no pulse can propagate in that regime. This situation is different from the case of excitatory networks with delay, described in the previous section, where the HB leads to the propagation of discontinuous, lurching pulses. Increasing $g_{EE}^s$ to 10 modifies the $\nu$-$g_{IE}$ curve in two aspects. First, the branch in the bifurcation diagram corresponding to the slow pulse extends for wider $g_{IE}$ region. Second, there is no HB. As a result, the slow E-to-E excitation increases the regime where stable slow pulses can propagate. For all the $g_{EE}^s$ and $g_{EI}^f$ values, the time difference $|\zeta|$ increases as $\nu$ decreases, and therefore the time lead of the I cell is larger for the slow pulse than for the fast pulse.

For all the parameter regimes we have examined, E cells fire before or slightly after the I cells during the fast pulse, and the values of $\zeta$ are small positive or negative values, of order 1 ms. In contrast, during the propagation of the slow pulse, E cells fire well after the I cells, and $\zeta$ is negative, on the order of a few tens of ms. The slow pulse can therefore be viewed as a front of I-cells' spikes pushed from behind by the E cells' spikes; because each E cell receives strong inhibition from neurons in front of it, the pulse propagates slowly.

The voltage profile of neurons that have not fired yet at time $t = 0$ (Eq.3.30) is shown in Fig. 11 for the reference parameter set. With this set of parameters, the fast pulse and the slow pulse coexist. During the fast pulse, the membrane potential of a neuron at a position $x$ decays monotonically with distance from the pulse. During the slow pulse, the potential of the I cells also decays monotonically. However, during the slow pulse, the potential of the E cell first decays rapidly and reaches a negative value. Then, it increases to positive values and then decreases again. Note that a mirror image of the same profile, with the abscissa stretched by a factor $1/\nu$, describes the temporal behavior of the pulse at a constant position $x$. This means that during the slow pulse, each excitatory neuron is first excited, but then is affected by strong inhibition and its potential becomes negative. Only when the pulse continues to propagate and the effect of inhibition is diminished, is the neuron again affected by excitation and so can reach threshold and fire.

Plotting the voltage profile as a function of $x$ (Fig. 11) demonstrates a case in which $\nu$ and $\zeta$ are different for the fast pulse and the slow pulse, but the value of $\nu\zeta$ is similar. Since the footprint ranges in the model are of order 1, the value $|\nu\zeta|$ should be of order 1 or less, otherwise the spikes emitted by neurons from one population do not affect neurons in the other population. Hence, lower $\nu$ enables $\zeta$ to have larger values, that can be, in principle, negative or positive.

To further demonstrate that $g_{EE}^s$ increases the regime of slow pulse propagation, we present in Fig. 12 two-dimensional bifurcation diagrams in the $g_{EI}^f$-$g_{IE}$ plane, for $g_{EE}^s = 0$ (A) and $g_{EE}^s = 10$ (B). The fast pulse exists for $g_{IE} \gtrsim 0$ for all $g_{EI}^f$ values. Three lines of SNB are plotted. The lower SNB line (dashed), corresponding to the minimal $g_{IE}$ value above which the fast pulse can propa-

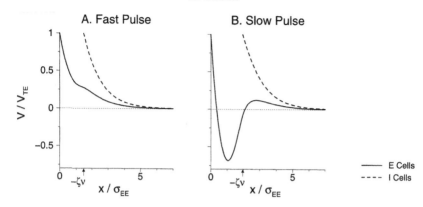

Fig. 11. The potentials $V$ of excitatory neurons (solid line) and inhibitory neurons (dashed line) that have not fired yet at time $t = 0$ are plotted as a function of their position. The reference parameter set is used; in particular, $g_{IE} = 5.2$ and $g^s_{EE} = 10$. A. Potentials during the fast pulse. B. Potentials during the slow pulse. Adapted from [22]

gate, is bounded by two codimension-2 cusp bifurcations [31]. The cusp at low $g^f_{EI}$, denoted by the asterisk, produces the slow pulse branch as a "ripple" on an unstable solution (see Fig. 10A, thick line). The cusp at high $g^f_{EI}$, denoted by the diamond, connects the slow and the fast branches, and eliminates the unstable branch between them. At higher $g^f_{EI}$ values, there is a continuous cross-over between the fast branch and the slow branches as $g^f_{EI}$ increases. For $g^s_{EE} = 0$, but not for $g^s_{EE} = 10$, there is a line of HB representing the maximal $g_{IE}$ above which the slow pulse is unstable. Comparing panels A and B in Fig. 12 shows that in B, the slow-pulse regime, and also the bistable regime (in which the two pulse types can propagate), have larger areas in the 2-parameter space for two reasons. First, the slow branch is terminated by a SNB at higher $g_{IE}$ values. Second, the slow branch is not destabilized by a HB if $g^s_{EE}$ is large enough.

The contribution of the slow E-to-E excitation to the extension of the regime of stable slow pulses is also shown in the $g^s_{EE}$-$g_{IE}$ plane in Fig. 13 for the reference parameter set ($g^f_{EI} = 30$). The HB destabilizes the slow pulse only for $g^s_{EE}$ values smaller than a certain value (which is $g^s_{EE} = 9.24$ here). Note that for a fixed $g^s_{EE}$ value just below that finite value, the slow pulse is stable in two disconnected intervals of $g_{IE}$, between which it is unstable. The robustness of the slow pulse at larger $g^s_{EE}$ values can be explained by the fact that strong, prolonged E-to-E excitation helps excitatory neurons to overcome the inhibition imposed on them by the inhibitory neurons that fire first, such that those excitatory neurons can eventually reach threshold and fire.

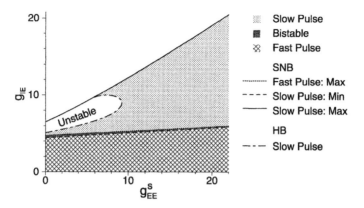

Fig. 12. Regimes of existence and stability of fast and/or slow pulses in the $g_{EI}^f$-$g_{IE}$ plane for the reference parameter set and $g_{EE}^s = 0$ (A) and $g_{EE}^s = 10$ (B). Saddle-node bifurcation curves are denoted by thick lines: dotted line - the maximal $g_{IE}$ value of the fast pulse; dashed line - the minimal $g_{IE}$ value of the slow pulse; solid line - the maximal $g_{IE}$ value of the slow pulse. The Hopf bifurcation curve in (A), is denoted by the dot-dashed line. Such a curve does not appear in (B). For $g_{EI}^s$ value smaller than the thin long-dashed line, only excitatory cells fire, and inhibitory cells are quiescent. Shadings: dark gray - bistable regime; light gray - regime in which only slow pulses can propagate; mesh of diagonal lines - regime in which only fast pulses can propagate; "continuous transition" (or "C.T." in (A)) - regime of continuous transition from a fast-pulse behavior for $g_{IE}$ values near 0 to slow-pulse behavior as $g_{IE}$ increases. In all the other white regimes, no pulse can propagate. The cusps of the SNB lines are denoted by an ∗ (left) and by a ◇ (right). The three arrows below the abscissa in (B) represent the three values of $g_{EI}^f$ in Fig. 10. Slow excitation increases considerably the parameter regime in which slow pulses can propagate. Adapted from [22].

### 3.4.2. *Response to shock initial conditions*

Even if, for a particular set of parameters, a pulse exists and is stable, it does not mean that it can be generated using a particular choice of initial conditions. Since the space of initial conditions has, in principle, infinite dimensions, we cannot determine the volume of the basins of attraction for a particular propagating state in that space. Instead, we chose to use one type of initial condition, the "shock" initial condition. All the neurons in a region $0 < x < 2.5$ are excited at $t = 0$, and we follow which type of pulse is generated, if at all. The shock initial condition is chosen because it replicates the experimental situation, in which propagating discharges are initiated by a brief spatially localized stimulation [1].

The system's response to shocks is described in Fig. 14 for two values of $g_{EE}^s$:

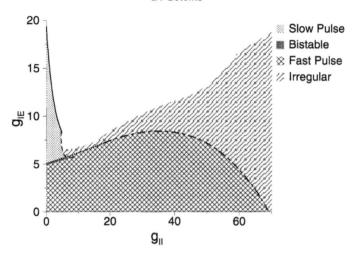

Fig. 13. Regimes of existence and stability of fast and/or slow pulses in the $g^s_{EE}$-$g_{IE}$ plane. The meanings of the lines and the shading are the same as in Fig. 12. In the white regime bounded by the HB line (dot-dashed), the SNB line (solid) and the ordinate, the slow pulse exists, but it is unstable. For $g_{IE}$ values above the solid line, no pulses exist. Adapted from [22]

0 (A and B) and 10 (C and D). We compare the responses for two values of $\nu$. We keep all the parameters at their reference values except for $g_{IE}$, which we tune in order to obtain the desired value of $\nu$. These $g_{IE}$ and $\nu$ values are shown in Fig. 14E, which is the same as Fig. 10B (middle panel). For $g^s_{EE} = 0$ and $\nu = 0.085$ ms$^{-1}$ (Fig. 14A), the slow pulse is the only attracting pulse. A shock stimulus initiates a transient fast pulse which propagates along a considerable distance before it switches to the slow pulse (at about $x = 30$). For $\nu = 0.072$ ms$^{-1}$ (Fig. 14B), the slow pulse is also the only attracting pulse, but a shock stimulus generates a localized activity only which does not propagate. When $g^s_{EE}$ is raised to 10, the same shock stimulus generates a slow pulse for the two values of $\nu$, beyond a small interval of fast propagation (Fig. 14C), or after two periods of "lurching" activity (Fig. 14D).

The effect of the slow excitation can be explained intuitively as following. After a shock stimulus, in order to generate a slow pulse, the firing times of the neurons should reorganize such that the I cells fire before the E cells at the same position. If there is slow excitation, a cell receives inhibition and excitation due to the fast inhibitory and excitatory synapses, and then, for a prolonged amount a time, slow excitation that enables it to overcome the inhibition and fire. Note that when the fast pulse is the only attractor, shock initial conditions generate it with or without $g^s_{EE}$ for all the cases we examined (not shown).

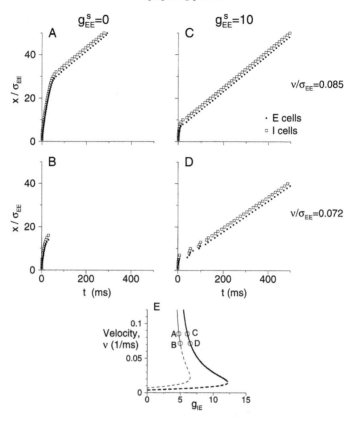

Fig. 14. Network responses to a "shock" stimulus are shown in rastergrams (A-D). Firing times of excitatory cells are denoted by solid circles, and firing times of inhibitory cells are denoted by open squares. There are $\rho = 20$ neurons from each type within each unit length ($\sigma_{EE}$), and spikes of only one out of every 20 neurons are plotted. The number of neurons in each population is $N = 1000$, and the reference parameter set is used. The network is initiated by an abrupt activation of all the excitatory and inhibitory neurons on the "left" ($0 < x < 2.5$). Simulations are carried out for two values of $g^s_{EE}$ and two values of $v$: A. $g^s_{EE} = 0$, $v = 0.085$ ($g_{IE} = 4.8$); B. $g^s_{EE} = 0$, $v = 0.072$ ($g_{IE} = 5$); C. $g^s_{EE} = 10$, $v = 0.085$ ($g_{IE} = 6.08$); D. $g^s_{EE} = 10$ $v = 0.072$ ($g_{IE} = 6.51$). In E, the values of $g_{IE}$ and the velocities $v$ of the slow pulses are shown. The curves, corresponding to slow pulses, are identical to the curve shown in Fig. 10B (middle panel). Thick lines represent pulses with $g^s_{EE} = 10$, and thin lines represent pulses with $g^s_{EE} = 0$. Solid lines represent stable pulses and dashed lines represent unstable pulses. The circles labeled A-D correspond to the value of $g_{IE}$ and $v$ in panels A-D. Without slow excitation, it is difficult to evoke slow pulses, even if they exist and are stable. Adapted from [22]

## 3.4.3. Effects of strength of fast E-to-E excitation

In the cortex, there is a delicate balance between excitation and inhibition, and deviations from this balance can lead to the generation of epileptic-like dis-

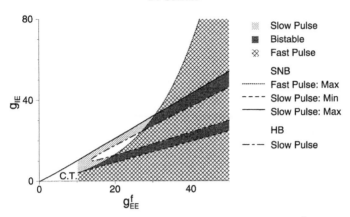

Fig. 15. Regimes of existence and stability of fast and/or slow pulses in the $g^f_{EE}$-$g_{IE}$ plane. The meanings of the lines and the shading are the same as in Fig. 12. The dotted line, below which the fast pulse can propagate, is convex as a function of $g^f_{EE}$ Adapted from [22]

charges [11, 12], corresponding to the fast pulses in our model. The strength of fast (AMPA-mediated) E-to-E excitatory conductance may change as a result of learning. For example, Saar *et al.* found indirect evidence for an increase in $g^f_{EE}$ during olfactory learning in rats on the order of tens of percents [32]. Despite this increase, cortical slices do not become more epileptic after learning. A possible explanation to these facts is the E-to-I excitation and/or the I-to-E inhibition increase as well, and prevent the propagation of the fast pulse. In order to study this hypothesis, we analyze the appearance of propagating pulses in the $g^f_{EE}$-$g_{IE}$ and $g^f_{EE}$-$g^f_{EI}$ planes.

The regimes in which the two pulse types can propagate in the $g^f_{EE}$-$g_{IE}$ plane are shown in Fig. 15. Fast pulses can propagate if $g_{IE}$ is below a certain value, $g_{IE,\max}$, which increases with $g^f_{EE}$ (dotted line). The curve of $g_{IE,\max}$ as a function of $g^f_{EE}$ is convex (*i.e.*, it has a positive curvature) and $d\, g_{IE,\max}\, /\, d\, g^f_{EE} > 0$. Therefore, to compensate for an increase of $g^f_{EE}$ and prevent fast pulse propagation, the enhancement in the level of $g_{IE,\max}$ should itself increase with $g^f_{EE}$.

The minimal and maximal $g_{IE}$ values for which the slow pulses cease to exist both grow almost linearly with $g^f_{EE}$, although with different slopes. Interestingly, for large enough $g^f_{EE}$, there are two HB points on the slow pulse branch. The slow pulse is unstable between them and is stable in two separate $g_{IE}$ intervals.

The regimes in which the two pulse types can propagate in the $g^f_{EE}$-$g^f_{EI}$ plane are shown in Fig. 16A. Except for a small parameter regime at small $g^f_{EE}$ values,

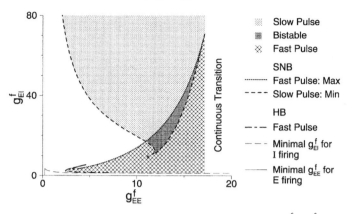

Fig. 16. Regimes of existence and stability of fast and/or slow pulses in the $g^f_{EE}$-$g^f_{EI}$ plane; $g_{IE} =$ 6.5. The meanings of the lines and the shading are the same as in Fig. 12. The asterisk points to a codimension-2 pitchfork bifurcation of two pairs of SNB solutions. In the white regime bounded by the thin long-dashed line, the thin solid line and the abscissa, only excitatory cells fire as a pulse propagates. The dotted line, below which the fast pulse can propagate, is convex as a function of $g^f_{EE}$. Adapted from [22]

the dependence of the maximal $g^f_{EI}$ value for which fast pulses can propagate (dotted line) is convex and has a positive curvature. This behavior is functionally similar to the dependence of $g_{IE,\max}$ on $g^f_{EE}$ for constant $g^f_{EI}$. At low $g^f_{EI}$ values, this regime has a "tail" which points back toward large $g^f_{EE}$ values. Slow pulses can propagate if $g^f_{EI}$ is above a certain value, $g^f_{EI,\min}$, (dashed line) which first decreases and then increases as a function of $g^f_{EE}$. The asterisk points to a codimension-2 pitchfork bifurcation (at $g^f_{EE} = 11.6$, $g_{IE} = 10.6$) of two pairs of SNB solutions.

### 3.4.4. I-to-I conductance and irregular pulses

The I-to-I conductance $g_{II}$ was found to strongly affect the firing properties of networks under steady-state conditions [33, 34]. In order to examine how $g_{II}$ affects pulse propagation, we study the regimes where various types of pulses can propagate in a two-parameter, $g_{II} - g_{IE}$ plane, as shown in Fig. 17. In this figure, solid and dotted lines represent saddle-node bifurcation in which slow and fast pulses are terminated, respectively, as $g_{IE}$ values are increased. In addition, pulses with large enough $g_{II}$ are terminated because the solution violates Eq. (3.9). Specifically,

$$\frac{dV_I\left[-\zeta v, T_I\left(-\zeta v\right)\right]}{dt} > 0 \tag{3.31}$$

*D. Golomb*

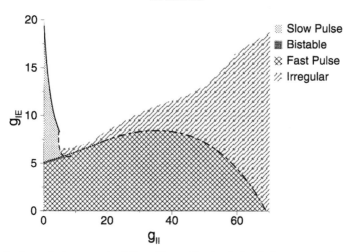

Fig. 17. Regimes of existence and stability of fast and/or slow pulses in the $g_{II}$-$g_{IE}$ plane. Saddle-node bifurcation lines: solid line - termination of the slow pulse; dotted line - termination of the fast pulse. Terminations of solutions because of Eq. (3.9): dashed line - slow pulse; dot-dashed line - fast pulse. Shadings: dark gray - bistable regime; light gray - regime in which only slow pulses can propagate; mesh of diagonal lines - regime in which only fast pulses can propagate; bent diagonal lines - irregular pulses. Adapted from [22]

Such a termination is denoted by a dashed line for the slow pulse and by a dot-dashed line for the fast pulse. The bistable regime and the regimes in which either slow or fast pulses can propagate are shaded as in the other two-parameter figures. The slow pulse can propagate only if $g_{II}$ is small enough, and its regime of existence shrinks rapidly as $g_{II}$ increases. At low $g_{II}$ value, the slow pulse is terminated as $g_{IE}$ increases by a SNB, and at higher values, it is terminated because of Eq. (3.31). Fast pulses are terminated by SNB (as $g_{IE}$ increases) for much larger $g_{II}$ values in comparison with the slow pulses. At even larger $g_{II}$ values, however, these pulses are terminated by the condition of Eq. (3.31).

What happens beyond the curve on which a pulse is terminated by condition (3.31)? Surprisingly, we find in numerical simulations that irregular pulses can propagate. Three examples of such pulses are shown in the rastergrams of Fig. 18. These pulses are characterized by the fact that excitatory cells fire almost as in regular traveling pulses, whereas inhibitory cells segregate into two spatiotemporal clusters. Neurons in the first cluster fire before their excitatory neighbors almost with a constant time delay $|\zeta_1|$. Neurons in the second cluster fire after their inhibitory neighbors from the first clusters, and often (as in Fig. 18A, $g_{II} = 50$) also after their neighboring excitatory neurons. The pulses in Fig. 18A have the characteristics of a fast pulse: inhibitory cells fire either less

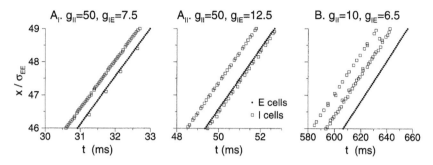

Fig. 18. Examples of neuronal firing times during the propagation of irregular pulses is shown for the reference parameter set and: $A_I$: $g_{II} = 50$, $g_{IE} = 7.5$, $\nu = 1.45$ 1/ms; $A_{II}$: $g_{II} = 50$, $g_{IE} = 12.5$, $\nu = 0.91$ 1/ms; B: $g_{II} = 10$, $g_{IE} = 6.5$, $\nu = 0.06$ 1/ms. Firing times of excitatory cells are denoted by solid circles, and firing times of inhibitory cells are denoted by open squares. There are $\rho = 20$ neurons from each type within each unit length ($\sigma_{EE}$), and spikes of all the neurons are plotted. The number of neurons in each population is $N = 1000$ and the total length of the system is 50 $\sigma_{EE}$. Adapted from [22]

than 1 ms before neighboring excitatory cells or just after them, and $\nu$ is large (1.45 1/ms in (I) and 0.91 1/ms in (II)). In Fig. 18B ($g_{II} = 10$), all the inhibitory neurons fire before the neighboring excitatory neurons, and the segregation into two clusters is less strict. The pulse in Fig. 18B has characteristics of a slow pulse: inhibitory cells fire tens of ms before their neighboring excitatory cells, and $\nu$ is small (0.06 1/ms).

In order to define the border of the appearance of the irregular pulses, we carried out numerical simulations in which we started from a shock initial conditions and found out whether a pulse can propagate. The results are shown in the regime shaded by the bent diagonal lines in Fig. 17. We cannot rule out the possibility that pulses which are excited by different initial conditions propagate also outside of this regime.

We can understand the appearance of irregular pulses using the following argument. The strong mutual inhibition between inhibitory neurons at large $g_{II}$ values prevents the propagation of a regular traveling pulse because when one I cell fires, it reduce the propensity of its neighboring I cell to fire afterward. As a result, neighboring I cells tend to fire with time delays between them.

### 3.4.5. Lurching pulses

Can lurching pulses with more complicated spatiotemporal form be obtained in our model, without synaptic delays? We find such lurching pulse, with spatiotemporal periodicity in the firing pattern, if the E-to-E excitation is made slow, whereas all the other synapses decay fast. A typical rastergram of a lurching

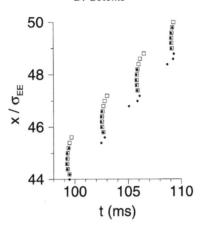

Fig. 19. A lurching pulse is shown in a Rastergram obtained by simulations of the neuronal dynamics; Methods and symbols are as in Fig. 9. The pulses were initiated by a "shock" for $x < 0.5$. Parameters that are different than those in Fig. 9 (B) are: $\tau_{sEE} = 50$ ms, $\tau_{sEI} = 2$ ms, $\tau_{sIE} = \tau_{sII} = 5$ ms, $g_{EE} = 250$, $g_{IE} = 21$, $\sigma_{IE} = 0.7$, $g_{EI} = 20$, $g_{II} = 17$. Adapted from [21]

pulse is shown in Fig. 19; the parameter set is characterized by a slow $\tau_{sEE}$ (50 ms), and large $g_{EE}$ (250). A group of E cells that fire inhibit other excitatory neurons through the neighboring inhibitory neurons, but the E-to-E excitation is prolonged enough to recruit a new group of excitatory cells into the pulse at later time. Surprisingly, during a lurching pulse, the firing time of the spikes is not a monotone function of the spatial position (see Fig. 19). As the parameter $g_{IE}$ increases, the system switches from a continuous to a lurching pulse, although not through instability. Instead, the continuous pulse ceases to exist because the solution violates Eq. (3.9), and a lurching pulse emerges. At even higher $g_{IE}$ values, no pulse can propagate.

### 3.5. Finite axonal conduction velocity

#### 3.5.1. Traveling pulse solution with finite axonal conduction velocity

The response of a post-synaptic neuron to a firing of a pre-synaptic cell is delayed because the conductance velocity of action potential in axons is a finite value, denoted here by $c$. For finite $c$, the Volterra equations for the firing times $T_\alpha(x)$ become

$$V_{T\alpha} = \sum_{\gamma = f, s} g_{E\alpha}^\gamma \int_{-\infty}^{\infty} dx' \, w_{E\alpha}(x') \, G_{E\alpha}^\gamma \left[ T_\alpha(x) - T_E\left(x - x'\right) - \frac{|x'|}{c} \right]$$

$$-g_{I\alpha} \int_{-\infty}^{\infty} dx' \, w_{I\alpha}(x') \, G_{I\alpha} \left[ T_\alpha(x) - T_I \left( x - x' \right) - \frac{|x'|}{c} \right] \qquad (3.32)$$

Assuming a traveling pulse solution, Eq. (3.11), and defining

$$v_- = \left( \frac{1}{v} - \frac{1}{c} \right)^{-1}, \qquad v_+ = \left( \frac{1}{v} + \frac{1}{c} \right)^{-1} \qquad (3.33)$$

one obtains, for $\zeta < 0$:

$$V_{TE} = \sum_{\gamma=f,s} g_{EE}^\gamma \int_0^\infty dx \, w_{EE}(x) \, G_{EE}^\gamma \left( \frac{x}{v_-} \right) - \qquad (3.34)$$

$$g_{IE} \int_{-\zeta v_-}^\infty dx \, w_{IE}(x + \zeta v_-) \, G_{IE} \left( \frac{x}{v_-} \right) -$$

$$g_{IE} \int_0^{-\zeta v_-} dx \, \frac{v_+}{v_-} w_{IE} \left[ \frac{v_+}{v_-} (x + \zeta v_-) \right] G_{IE} \left( \frac{x}{v_-} \right)$$

$$V_{TI} = \sum_{\gamma=f,s} g_{EI}^\gamma \int_0^\infty dx \, w_{EI}(x - \zeta v_-) \, G_{EI}^\gamma \left( \frac{x}{v_-} \right) - \qquad (3.35)$$

$$g_{II} \int_0^\infty dx \, w_{II}(x) \, G_{IE} \left( \frac{x}{v_-} \right)$$

All the terms in Eqs. (3.34,3.35), except for the "backward" term (the third integral in Eq. (3.34), representing the effect of the I cell spike on E cells with smaller $x$), are similar to the corresponding terms in Eqs. (3.12,3.13), but the velocity $v$ there is replaced by $v_-$. Using Eq. (3.5), one can see that in that "backward" term, there is an additional modification: $\sigma_{IE}$ is replaced by $\sigma_{IE} \times v_- / v_+$. Since $v_- > v_+$, this means that the length constant $\sigma_{IE}$ is "stretched" in this term. Similar results are obtained for the EI term in the equations for $\zeta > 0$.

In networks with excitation only, introducing finite axonal conduction velocity $c$ reduces $v$, because the term $v_-$ replaces $v$ in the dynamical equation 2.37. In networks with excitation and inhibition, however, the factor $v_+$ can play a role as well, and in principle, decreasing $c$ may *increase* $v$, as shown below.

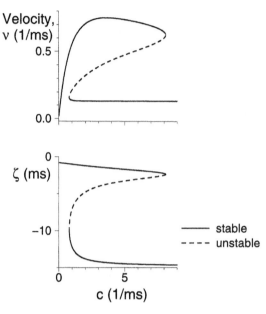

Fig. 20. The velocity $v$ of propagating pulses (upper panel), and the difference $\zeta$ between the firing times of inhibitory and excitatory cells at the same position (lower panel), as functions of the axonal conduction velocity $c$ for the reference parameter and $g_{IE} = 5.4$. The velocity of the fast pulse decreases with $c$ near the SNB. Adapted from [22]

### 3.5.2. Pulse velocity may decrease with c

Conduction velocity of unmyelined axons is of order 1 m/s [35, 36]. Taking it into account is expected to affect mostly the fast pulses, whereas its effect on the slow pulses is expected to be very small. Indeed, numerical solutions of Eqs. (3.34, 3.35) show that in most cases, the velocity of the fast pulses is reduced if $c$ is considered to be finite rather than infinite. There is, however, an exception that is shown in Fig. 20: the fast pulse exists at small $c$ values but ceases to exists at a SNB at a critical $c$ values. For a range of $c$ values smaller than this critical value, the pulse velocity decreases with $c$. For this particular parameter regime, a slow pulse can also propagate if $c$ is not too small, and its velocity is almost independent of $c$.

The decreasing $v$ with increasing $c$ may occur in parameter regimes in which the fast pulse does not exist for $c \to \infty$, but exists if $c$ is smaller than a certain value of a SNB. At a SNB, a stable branch and an unstable branch of solutions should coalesce and eliminate each other. In order to enable this coalescence, the upper, stable branch should "bend" toward the lower, unstable branch, and as a result $\nu$ should decrease at $c$ values lower than the SNB values. This intuitively unexpected result demonstrates the nonlinear dynamics nature of the pulse propagation phenomenon.

## 4. Discussion

### 4.1. Types of propagating pulses

Three main types of traveling pulses were described in this Book Chapter: continuous pulses, lurching pulses and irregular pulses. The continuous pulses belong to two types: fast and slow. The two types of continuous pulses are well-defined only in the bistable regime, where both exist. In other regimes in the parameter space, they are characterized by their similarity to the fast or the slow pulses in the bistable regime. Here we summarize the properties of the various pulse types.

**Fast pulses** can propagate in networks of excitatory neurons. In networks with inhibition, they can be regarded as a continuation of propagating pulse states in the excitatory-only case. They are characterized by E cells firing before or just after neighboring I cells, and by monotonic increase of the neuronal potential before the firing. Fast pulses are robust with respect to initial conditions. If they are stable, a strong-enough initial shock will evoke them. Fast pulses are hardly affected by slow excitation, as was shown in models of networks of excitatory neurons [1]. The parameter regime in which fast pulses can propagate strongly expands, and the velocity strongly increases, as the level of fast E-to-E excitatory conductance $g_{EE}^f$ increases. In particular, the curves of maximal $g_{IE}$ and $g_{EI}^f$ for which fast pulses can propagate are convex as a function of $g_{EE}^f$. At high $g_{II}$ values, fast traveling pulses cannot propagate because of the "repulsive" interaction between inhibitory interneurons. Instead, the network exhibits irregular pulses.

**Slow pulses** are characterized by E cells firing substantially after the I cells, and by a decrease in the potential of the E cells before a subsequent potential increase until the neuron reaches threshold and fires. Slow pulses can propagate even without slow excitation, but the parameter regime in which they are stable strongly expands as the level of slow E-to-E excitatory conductance $g_{EE}^s$ increases. Slow pulse are not robust. Even if they are stable, an initial shock often does not evoke them, and, even if it does, the system dynamics can converge into this state after a long transient with fast or lurching pulse characterization. Slow excitation increases the basin of attraction of slow pulses and the possibility of evoking them with shock initial conditions. The dependence of the parameter regime in which slow pulses can propagate on $g_{EE}^f$ is complicated. There are conditions in which increasing $g_{EE}^f$ supports the propagation of slow pulses, whereas in other conditions it prevents propagation. Enhancing the I-to-I conductance $g_{II}$ decreases, then eliminates, the regime in which slow pulses can propagate.

**Transition between continuous pulses.** As a parameter of the system varies, the transition from a fast-pulse parameter regime to a slow-pulse parameter regime can occur through a bistable regime, in which both types of pulses can propagate. It can also occur continuously, as the velocity of the pulses decreases and the time lead of I cell firing, in comparison with E cell firing, increases. A third possibility is that the fast pulse stops propagating as a parameter is varied, and a slow pulse appears in a distant parameter regime.

**Lurching pulses** are characterized by periodicity in both space and time, and by a time interval between two successive cycles of activity, in which the network is silent. These pulses are generated by an excitatory network if the synaptic delay is large enough. The parameter regime of their existence depends strongly on the footprint synaptic shape: for exponential shape, they appear for lower delay $\tau_d$ than for square shape. To enable the appearance of lurching pulses in networks of excitatory and inhibitory neurons without any delay, the E-to-E excitation should be large and slowly decaying, whereas the E-to-I excitation and the I-to-E inhibition should decay quickly. Under those conditions, the feedback inhibition terminates the activity in each cycle, and the slow inhibition initiates a new cycle of activity. The slow excitation together with fast feedback inhibition mimic a situation of an effective excitation with delay.

**Irregular pulses** appear for large $g_{II}$. They look almost like continuous traveling pulses with respect to the excitatory population, but inhibitory neurons segregate into two groups, which fire with two different delay times with respect to their neighboring excitatory cells. Irregular pulses can be regarded as pulses with "spatiotemporal clustering" of inhibitory cells. These pulses are different than lurching pulses. Lurching pulses are characterized by periods of activity propagation followed by periods of silence and no propagation. During irregular pulses, activity does not stop, there are no silent periods, and the excitatory population is constantly active, cell after cell. The irregular pulses have spatiotemporal periodicity, at least approximately. From this respect, they are similar to lurching pulses.

## 4.2. *Effects of approximations*

The model described here is based on two approximations. First, the subthreshold neuronal dynamics are described by an integrate-and-fire model. Second, each neuron is allowed to fire only one spike. The first approximation does not seem to affect the main results presented here, because the key issue here is that a neuron responds to propagating pulses by integrating the responses of other excitatory and inhibitory neurons and firing if the time-integrated amount of excitation is strong enough. To further support this claim, we replaced the integrate-and-fire scheme by the Morris-Lecar model, a version of a conductance-based model [37], and found regimes of fast and slow pulses with bistability between them (not shown).

The one-spike approximation is exact in the limit of very prolonged refractory period or very strong synaptic depression. In the first case, a neuron cannot fire a second spike before the pulse has completely passed. In the second case, spikes other than the first one do not generate any post-synaptic effect. Far from these limits, however, this approximation can have an effect on the dynamical mechanisms of the slow pulse propagation. In networks with excitatory populations only, the results of this model are qualitatively similar to the results obtained by simulations of conductance-based models (compare [5, 24] and the results described here with [1]). For example, if we assume that only the first spike elicits an EPSP for the parameters of Fig. 8 in [1], the velocity decreases by only 15%. In two-population systems, however, other scenarios can happen. For example, in our model, the potential of excitatory neurons becomes negative (hyperpolarized) before it becomes positive again and the neuron can fire. If the I cells can fire several fast spikes, they can prevent the E cell from firing. The model described in this work can be regarded, therefore, as a paradigm for illuminating a possible mechanism for a slow pulse propagation, which is the advanced firing of I cells. We have carried out preliminary simulations of conductance-based neuronal models, in which cells can fire many spikes, which demonstrated a transition from a fast pulse to a slow pulse as $g_{IE}$ increases. As in the one-spike model, inhibitory cells lead substantially in firing during the slow pulse, but not during the fast pulse, and slow E-to-E excitation was found to be important for propagation of slow, but not fast, pulses. Similar results were obtained in Ref. [38, 39] in simulations of conductance-based models. Further analytical and numerical investigation of models with more spikes should be carried out to see whether there are alternative mechanisms for slow pulse propagation, in addition to the mechanism described here.

### 4.3. Application to biological pulses

#### 4.3.1. Network without inhibition

Our model of excitatory networks relates to experiments on disinhibited cortical slices. Our main theoretical results are consistent with experiments [1]: There is a minimal velocity below which the pulse cannot propagate; above this minimal velocity, the pulse velocity increase with the strength of synaptic excitation. Our theory predicts that transitions from continuous to lurching pulses may occur if the synaptic delay is of order of 10 ms or more. This value is about an order of magnitude larger than the delay between adjacent cortical neurons, which is about 1-2 ms [40,41]. Therefore, we do not expect to observe lurching pulses in disinhibited cortical slices.

Our results on lurching pulse are relevant for thalamic slices, in which spindle oscillations appear in a circuit composed of excitatory thalamocortical (TC) cells and inhibitory reticular (RE) thalamic cells, coupled with reciprocal synaptic connections [17, 18, 42–44]. The excitatory cells possess a post-inhibitory rebound mechanism (see, *e.g.,* [45]). The propagation velocity is around 1 mm/s [46]. Numerical simulations have indicated that these discharges propagate in a non-smooth, periodic, lurching manner [17, 47]. Each recruitment cycle has two stages. At the first stage, a new group of inhibitory RE cells is excited by synapses from TC cells, and this RE group inhibits a new group of silent TC cells. At the second stage, the newly recruited TC cells rebound from hyperpolarization and fire a burst of action potentials. These bursts further recruit more RE cells during the next cycle. Since the RE-to-TC projection is topographic and acts via $GABA_A$ and $GABA_B$ receptors, excitation of one RE cell would result in a delayed barrage of excitatory post-synaptic conductances (EPSPs) in the neighboring RE cells through the disynaptic RE-TC-RE loop. In this idealized view of the isolated thalamic circuit, the RE cell layer acts effectively as a cell population with reciprocal AMPA-mediated excitatory interactions, with an effective delay $\tau_d$ (Eq. 2.3) of order 100 ms caused by the time needed for the TC cell to rebound from inhibition. Since, in this work, we are interested only in the recruitment process, we can model the system by considering only the first "spike" each cell fires. The large effective delay assures that the velocity at moderate coupling is determined mainly by the first spike. Indeed, our theory shows that at such large delay values, propagating pulses are lurching, as was shown before in simulations [17].

### 4.3.2. Network with inhibition

Several results of the present model can be compared with data from experiments on cortical slices. First, our theoretical work shows that slow pulses can propagate even without slow excitation, but slow excitation substantially enhances the parameter regime in which slow pulses can propagate. Experiments showing that blocking the slow NMDA receptors blocks the slow pulse or greatly reduces its intensity [15], are consistent with the prediction. Second, our analysis shows that there can be an abrupt transition from a fast pulse to a slow pulse as inhibition is increased. Preliminary experimental results [48] confirm this prediction. Third, the proposed mechanism for slow pulse propagation demands that inhibitory cells fire before their neighboring excitatory cells. This prediction can be tested by dual intracellular recording from adjacent excitatory and inhibitory neurons. Fourth, according to the theory, during the slow pulse propagation, excitatory cells should be hyperpolarized by inhibitory cells before their potential increases again and they can fire. This theoretical result can be tested experimentally using intracellular recording. Note, however, that the shunting (and not hyperpolarizing) nature of certain types of inhibitory synapses may modulate this behavior. Fifth, if the total fast E-to-E synaptic conductance $g_{EE}^{f}$ is substantially increased, for example during certain types of learning tasks [32], the I-to-E inhibition and the E-to-I excitation should increase considerably in order to prevent the propagation of fast, epileptic-like pulses. This prediction can be tested by doing dual intracellular recording between neurons from the excitatory and inhibitory populations, and comparing the results with or without learning. Sixth, strong and even moderate values of I-to-I inhibition prevent the propagation of the slow pulse. On the other hand, strong I-to-I inhibition seems to be necessary for generating stable states of persistent activity [33,34]. In experiments, the slow pulse is sometimes accompanied by a prolonged state of persistent activity [15]. Theoretical and experimental investigation of such systems should be carried out in order to examine how slow pulses and persistent activity can occur in the same system.

## Acknowledgements

The research described in this Book Chapter has been carried out together with G. Bard Ermentrout. This work is supported by the Israel Science Foundation (grant no. 657/01).

## Appendix A. Stability of the lower branch

Here, following the ideas of Bressloff [25], we prove that the branch of solutions with slow velocity are unstable in networks of excitatory neurons. Recall, the velocity satisfies the implicit equation (Eq. 2.14):

$$\frac{V_T}{g_{\text{syn}}} = \int_0^\infty w(x' + v\,\tau_d)\,G(x'/v)\,dx' \tag{A. 1}$$

We differentiate this implicitly to get $dg_{\text{syn}}/dv$:

$$-\frac{V_T}{g_{\text{syn}}^2}\frac{dg_{\text{syn}}}{dv} = \tag{A. 2}$$

$$\int_0^\infty dx'\left[\tau_d\,w'(x' + v\tau_d)\,G(x'/v) - \frac{x'}{v^2}\,w(x' + v\tau_d)\,G'(x'/v)\right]$$

Using the fact that $G(0) = 0$, an integration by parts of the first term in the integral yields

$$-\frac{V_T}{g_{syn}^2} \frac{dg_{syn}}{dv} = \tag{A. 3}$$

$$\int_0^\infty dx' \left[ -\frac{\tau_d}{v} w(x' + v\tau_d) G'(x'/v) - \frac{x'}{v^2} w(x' + v\tau_d) G'(x'/v) \right]$$

Thus, we have the following result:

$$\frac{V_T}{g_{syn}^2} \frac{dg_{syn}}{dv} = \frac{1}{v^2} \int_0^\infty dx' (x' + v\tau_d) w(x' + v\tau_d) G'(x'/v) \tag{A. 4}$$

This means that on the slow branch (where $dv/dg_{syn} < 0$ and therefore $dg_{syn}/dv < 0$), the integral is negative.

The stability equation (Eq. 2.16) is:

$$0 = H(\lambda) \equiv \int_0^\infty dx' w(x' + \tau_d v) G'(\frac{x'}{v}) \left[ 1 - e^{-\lambda (x' + \tau_d v)} \right] \tag{A. 5}$$

Clearly $H(0) = 0$ (corresponding to translation invariance). Differentiating this with respect to $\lambda$ and evaluating at $\lambda = 0$ yields:

$$\frac{dH}{d\lambda}(0) = \int_0^\infty dx' w(x' + \tau_d v) G'(\frac{x'}{v}) (x' + v\tau_d) \tag{A. 6}$$

and from (A. 4), we see that:

$$\frac{dH}{d\lambda}(0) = \frac{v^2 V_T}{g_{syn}^2} \frac{dg_{syn}}{dv} \tag{A. 7}$$

thus, on the slow branch, $dH/d\lambda(0) < 0$. Thus, for $\lambda$ real and close to zero $H(\lambda) < 0$. Now, clearly,

$$H(\infty) = \int_0^\infty dx' w(x' + v \tau_d) G'(\frac{x'}{v}) \tag{A. 8}$$

Suppose that $w(x)$ is differentiable. Then we can integrate the above by parts obtaining:

$$H(\infty) = -v \int_0^\infty dx' w'(x' + v \tau_d) G(\frac{x'}{v}) \tag{A. 9}$$

If $w(x)$ is monotone decreasing and $G(t) \geq 0$ then we see that $H(\infty) > 0$. Since $H(\lambda) < 0$ for $\lambda$ small and $H(\lambda) > 0$ for sufficiently large $\lambda$, we conclude that there must be at least one positive real root $\lambda_0$ of $H(\lambda)$ and therefore the slow (lower) branch is unstable.

# References

[1] D. Golomb and Y. Amitai, J. Neurophysiol. **78** (1997) 1199.

[2] M.J. Gutnick, B.W. Connors and D.A. Prince, J. Neurophysiol. **48** (1982) 1321.

[3] B.W. Connors, Nature **310** (1984) 685.

[4] D. Golomb, J. Neurophysiol. **79** (1998) 1.

[5] G.B. Ermentrout J. Comput. Neurosci. **5** (1998) 191.

[6] Y. Tsau, L. Guan and J.-Y. Wu, J. Neurophysiol. **80** (1998) 978.

[7] J.-Y. Wu, L. Guan and Y. Tsau, J. Neurosci. **19** (1999) 5005.

[8] R. Miles, R.D. Traub and R.K. Wong, J. Neurophysiol. **60** (1988) 1481.

[9] R.D. Traub, J.G.R Jefferys and R. Miles, J. Physiol. (London) **472** (267) 267.

[10] R. Demir, L.B. Haberly and M.B. Jackson, J. Neurophysiol. **80** (1998) 2727.

[11] Y. Chagnac-Amitai and B.W. Connors, J. Neurophysiol. **61** (1989) 747.

[12] Y. Chagnac-Amitai and B.W. Connors, J. Neurophysiol. **62** (1989) 1149.

[13] N. Laaris, G.C. Carlson and A. Keller, J. Neurosci. **20** (2000) 1529.

[14] J.-Y. Wu and L. Guan, J. Neurophysiol. **86** (2001) 2416.

[15] M.V. Sanchez-Vives and D.A. McCormick, Nature Neurosci. **3** (2000) 1027.

[16] Y. Shu, A. Hasenstaub and D.A. McCormick, Nature **423** (2003) 288.

[17] D. Golomb, X.-J. Wang and J. Rinzel, J. Neurophysiol. **75** (1996) 750.

[18] J. Rinzel, D. Terman, X.-J. Wang and B. Ermentrout, Science **279** (1998) 1351.

[19] D. Golomb and G.B. Ermentrout, Proc. Natl. Acad. Sci. USA **96** (1999) 13480.

[20] D. Golomb and G.B. Ermentrout, Network **11** (2000) 221.

[21] D. Golomb and G.B. Ermentrout, Phys. Rev. Lett. **86** (2001) 4179.

[22] D. Golomb and G.B. Ermentrout, Phys. Rev. E **65** (2002) 061911.

[23] H.C. Tuckwell, *Introduction to Theoretical Neurobiology* (Cambridge University Press, Cambridge, UK, 1988).

[24] P.C. Bressloff, Phys. Rev. Lett. **82** (1999) 2979.

[25] P.C. Bressloff, J. Math. Biol. **40** (2000) 169.

[26] D. Hansel, G. Mato and C. Meunier, Neural Comp. **7** (1995) 307.

[27] D. Hansel, G. Mato, C. Meunier and L. Neltner, Neural Comp. **10** (1998) 467.

[28] W.H. Press, S.A. Teukolsky, W.T. Vetterling and B.P. Flannery, *Numerical Recipes in C* (Cambridge University Press, Cambridge, UK, 1992).

[29] E. Doedel, Cong. Num. **30** (1981) 265.

[30] B. Ermentrout, *Simulating, analyzing, and animating dynamical systems: a guide to XPPAUT for researchers and students (software, environments, tools)* (SIAM, Philadelphia, 2002).

[31] F.C. Hoppensteadt and E.M. Izhikevich, *Weakly Connected Neural Networks* (Springer-Verlag, New-York, 1997).

[32] D. Saar, Y. Grossman and E. Barkai, J. Neurosci. **19** (1999) 8616.

[33] D. Golomb, D. Hansel and G. Mato, in: *Handbook of Biological Physics, Volume 4: Neuro-Informatics and Neural Modelling*, F. Moss and S. Gielen Eds. (Elsevier Science, Amsterdam, 2001), p. 887.

[34] D. Hansel and G. Mato, Phys. Rev. Lett. **86** (2001) 4175.

[35] L.B. Haberly, in: *The Synaptic Organization of the Brain*, 3rd Edition, G.M. Shepherd Ed. (Oxford University Press, Oxford, 1990), p. 317.

[36] Z. Gil and Y. Amitai, J. Neurosci. **16** (1996) 6567.

[37] J. Rinzel and G.B. Ermentrout, in: *Methods in neuronal modeling: From ions to networks*, 2nd Edition, C. Koch and I. Segev Eds. (MIT Press, Cambridge, MA, 1998), p. 251.

[38] M. Bazhenov, I. Timofeev, M. Steriade and T.J. Sejnowski, J. Neurosci **22** (2002) 8691.

[39] A. Compte, M.V. Sanchez-Vives, D.A. McCormick and X.-J. Wang, J. Neurophysiol. **89** (2003) 2707.

[40] A.M. Thomson, J. Deuchars and D.C. West, J. Neurophysiol. **70** (1993) 2345.

[41] H. Markram, J. Lubke, M. Frotscher, A. Roth and B. Sakmann, J. Physiol. (London) **500** (1997) 409.

[42] M. von Krosigk, T. Bal and D.A. McCormick, Science **261** (1993) 361.

[43] T. Bal, M. von Krosigk and D.A. McCormick, J. Physiol. (London) **483** (1995) 641.

[44] T. Bal, M. von Krosigk and D.A. McCormick, J. Physiol. (London) **483** (1995) 665.

[45] M. Steriade, D.A. McCormick and T.J. Sejnowski, Science **262** (1993) 679.

[46] U. Kim, T. Bal and D.A. McCormick, J. Neurophysiol. **74** (1995) 1301.

[47] A. Destexhe, T. Bal, D.A. McCormick and T.J. Sejnowski, J. Neurophysiol. **76** (1996) 2049.

[48] D. Golomb, G.B. Ermentrout and J.-Y. Wu, Soc. Neurosci. Abstr. **26** (2000) 1467.

Course 7

# ACTIVITY-DEPENDENT TRANSMISSION IN NEOCORTICAL SYNAPSES

## M. Tsodyks

*Department of Neurobiology, Weizmann Institute, Rehovot, Israel*

*C.C. Chow, B. Gutkin, D. Hansel, C. Meunier and J. Dalibard, eds.*
*Les Houches, Session LXXX, 2003*
*Methods and Models in Neurophysics*
*Méthodes et Modèles en Neurophysique*

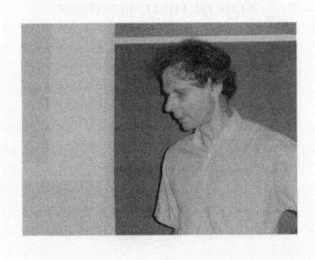

# Contents

**Abstract**

In this lecture series, I describe the recent advances in studying short-term plasticity in synaptic transmission. The material is divided into 3 sections. The first section is dealing with a phenomenological model of synaptic transmission and its underlying biophysical assumptions. In the second section, the model is used to study the implications of synaptic dynamics on the information transmission between ensembles of neocortical neurons. Finally, the last section deals with the effects of short-term synaptic plasticity on dynamics of recurrent networks.

## 1. Introduction

A marked feature of synaptic transmission between neocortical neurons is a pronounced activity dependence of synaptic responses to trains of presynaptic spikes [1]. The precise nature of this dynamic transmission varies between different classes of neurons. In particular, connections between pyramidal neurons typically display pronounced synaptic depression, characterized by fast decrease of synaptic response during the pre-synaptic train. Connections involving interneurons usually exhibit various degrees of synaptic facilitation that depend on the precise class of interneurons involved [2, 3].

In [4] (see also [5, 6]) we studied synaptic depression between neocortical pyramidal neurons using a phenomenological model. It was later generalized to describe facilitating synapses between pyramidal cells and inhibitory interneurons [2]. This approach has two major goals. First, it allows the quantification of the features of the AP activity of the presynaptic neurons and neuronal populations, transmitted by these different types of synapses. Second, it can be used in deriving a modified mean-field dynamics of neocortical networks aimed at understanding the dynamic behavior of large neuronal populations without having to solve an equally large number of equations. Mean-field descriptions were extensively used in order to understand the possible computations of cortical neural networks (see e.g. [6–9]). The novel formulation which uses the generalized phenomenological model of dynamic properties of synaptic connections between different types of neocortical neurons enables one to study the effects of synaptic dynamics and synaptic plasticity on information processing in large neural networks.

249

## 2. Phenomenological model of synaptic depression and facilitation

### 2.1. Synaptic depression

The main idea behind the model is the following: every connection can be characterized by its absolute amount of 'resources', which can be partitioned into three states: effective, inactive and recovered. If all the resources are activated by a presynaptic action potential (AP), this would generate the maximal possible response defined here as the absolute synaptic efficacy ($A$). In reality, each presynaptic AP utilizes a certain fraction ($u$) of resources available in the recovered state which then quickly inactivate with a time constant of a few milliseconds ($\tau_{in}$) and recover with a time constant of about 1 second ($\tau_{rec}$). The model could reflect various possible biophysical mechanisms of synaptic depression, such as receptor de-sensitization [10] or depletion of synaptic vesicles [11], or their combination.

This verbal formulation can be translated into a system of kinetic equations that describes the time course of the fraction of synaptic resources in each of these three states:

$$
\begin{aligned}
\frac{dx}{dt} &= \frac{z}{\tau_{rec}} - ux\delta(t - t_{sp}) \\
\frac{dy}{dt} &= -\frac{y}{\tau_{in}} + ux\delta(t - t_{sp}) \\
\frac{dz}{dt} &= \frac{y}{\tau_{in}} - \frac{z}{\tau_{rec}}
\end{aligned}
\tag{2.1}
$$

where $t_{sp}$ denotes the time of spike arrival and $x$, $y$ and $z$ are the fractions of resources in the recovered, active and inactive states respectively. The post-synaptic current ($PSC$) is proportional to the fraction of resources in the active state, $PSC(t) = Ay(t)$. Every time the presynaptic spike arrives, it triggers an exponential PSC with decay time $\tau_{in}$ and the amplitude $E = Aux$ that is proportional to the amount of recovered resources immediately before the arrival time. The three major parameters of the model are $A$, the absolute synaptic strength which can only be expressed by activating all of the resources, and $u$ and $\tau_{rec}$, kinetic parameters that together determine the time course of the responses to presynaptic train of APs, in particular the rate of synaptic depression. The higher the $u$, the faster synaptic resources are utilized, which effectively leads to more rapid depression. This formulation ignores the stochastic nature of synaptic release (see [12]), and reproduces the average post-synaptic responses generated by any presynaptic spike train $t_{sp}$ for inter-pyramidal synapses [6].

Equations 2.1 can be simplified substantially if one takes into account that realistically $\tau_{in} \ll \tau_{rec}$ (see above), and the rate of the incoming spike train is usually much lower then $1/\tau_{in}$. Taken together with the normalization condition $x + y + z = 1$ this allows to approximate Eqs. 2.1 with a single equation for $x$ only:

$$\frac{dx}{dt} = \frac{1-x}{\tau_{rec}} - ux\delta(t - t_{sp}). \tag{2.2}$$

This equation allows iterative expressions for successive $PSC$ amplitudes $(E_n)$ produced by an arbitrary train of presynaptic APs:

$$\begin{aligned} E_n &= Aux_n \\ x_{n+1} &= x_n(1-u)e^{-\Delta t_n/\tau_{rec}} + 1 - e^{-\Delta t/\tau_{rec}}, \end{aligned} \tag{2.3}$$

where $\Delta t_n$ is the time interval between $n$th and $(n+1)$th AP. Every $PSC$ is given in this approximation by $E_n exp(-t/\tau_{in})$. In a particular case, if the synapse is driven by a regular spike train of frequency $f$ $(\Delta t_n \equiv 1/f)$, response amplitudes reach a plateau value given by

$$E_\infty = E_1 \frac{1 - \exp(-\frac{1}{f\tau_{rec}})}{1 - (1-u)\exp(-\frac{1}{f\tau_{rec}})}. \tag{2.4}$$

An interesting prediction arises from this expression if we consider the limit of high frequencies $(f >> \frac{1}{u\tau_{rec}})$:

$$E_\infty = \frac{A}{f\tau_{rec}}, \tag{2.5}$$

indicating that stationary amplitude of individual PSCs reached during a regular spike train decreases in inverse proportion to the frequency above a certain 'limiting' frequency. This prediction was experimentally confirmed in [6].

## 2.2. Modeling synaptic facilitation

The formulation of Eqs. 2.1 does not account for short-term synaptic facilitation, which is not evident in connections between pyramidal neurons. It is however prominent in synapses between pyramidal neurons and some classes of inhibitory inter-neurons [1]. A standard way of modeling facilitation is by introducing a 'facilitation factor' which is elevated by each spike by a certain amount and decays between spikes, possibly at several rates (see e.g. [13, 14]). To add facilitation into our synaptic model, we therefore assume that the value of $u$ is not fixed but rather increases by a certain amount due to each presynaptic spike. The resulting model therefore includes both facilitating and depressing mechanisms.

Increase in $u$ could reflect e.g. the accumulation of calcium ions caused by spikes arriving in the presynaptic terminal, which is responsible for the release of neurotransmitter [15]). For a simple kinetic scheme, assume that an AP causes a fraction $U$ of calcium channels to open which subsequently close with a time constant of $\tau_{facil}$. The fraction of opened calcium channels determines the current value of $u$. The corresponding kinetic equation therefore reads:

$$\frac{du}{dt} = -\frac{u}{\tau_{facil}} + U(1-u)\delta(t - t_{sp}). \tag{2.6}$$

$U$ determines the increase in the value of $u$ due to each spike and coincides with the value of $u$ reached upon the arrival of the first spike (in other words, at very low frequency of stimulation).

This equation can be transformed into an iterative expression for the value of $u$ reached immediately after the arrival of $n$-th spike in a train, which, together with corresponding expression for $x$, determine the post-synaptic response according to Eqs. 2.1:

$$
\begin{aligned}
E_n &= A u_n x_n \\
u_{n+1} &= u_n(1-U)exp(-\Delta t_n/\tau_{facil}) + U \\
x_{n+1} &= x_n(1-U_{SE})e^{-\Delta t_n/\tau_{rec}} + 1 - e^{-\Delta t/\tau_{rec}}
\end{aligned}
\tag{2.7}
$$

If the presynaptic neuron emits a regular spike train at the frequency $f$, $u$ reaches a steady value of

$$\frac{U}{1-(1-U)exp(-\frac{1}{f\tau_{facil}})}.$$

Thus in this formulation, $u$ is a frequency dependent variable and $U$ is a kinetic parameter characterizing an activity dependent transmission in a given synapse. In the limit of $\tau_{facil} \to 0$ facilitation becomes negligible and the model reduces to the model of depressing synapses described above with $u \equiv U$.

One should keep in mind that facilitating and depressing mechanisms are intricately interconnected since stronger facilitation leads to higher $u$ values which in turn leads to stronger depression. The value of $U$ therefore determines the contribution of facilitation in generating subsequent synaptic responses. Facilitation is marked for small values of $U$ and is not observed for higher $U$. Analysis of experimental data revealed that the main features of synaptic transmission between pyramidal neurons and inhibitory inter-neurons are well captured by this model with $U \sim 0.01 \to 0.05$ and $\tau_{rec}$ is typically several times faster than $\tau_{facil}$ [2] (see also Fig. 1D).

When facilitating synapses are stimulated at progressively higher frequencies, the plateau amplitudes of PSCs exhibit non-monotonic bell-shaped dependency with the frequency. This behavior differs markedly from the frequency dependence of depressing synapses (where $E_\infty$ decreases as the frequency increases, Eq. 2.4) because of simultaneous facilitation of u and growing depression at higher values of u. The peak of the bell-shaped curve is a characteristic feature of a particular facilitating synapse and can be derived from the model equations by finding the frequency where the product of the steady-state values of u and x is at a maximum (see Eqs. 2.7).

Fig. 1A,B shows responses from facilitating and depressing synapses with the same absolute strength to a regular spike train of $20Hz$. Fig. 1C illustrates the buildup of depression in facilitating synapses when they are stimulated at high frequencies. As a result, the stationary level of response exhibits a tuning curve dependence on the frequency, in agreement with experimental results (Fig. 1D).

## 3. Dynamic synaptic transmission on the population level

The phenomenological models presented above can be used to analyze the way activity of large populations of presynaptic neurons is transmitted to post-synaptic targets. To simplify the analysis, we will assume that every neuron's firing obeys Poisson statistics, as indeed supported by experimental observations [16]. Mathematically, the Poisson statistics means that at each moment the probability for a neuron to fire an AP is given by the value of the instantaneous firing rate and is independent of the timing of previous spikes. We also assume that the whole presynaptic population has the same firing rate ($r(t)$) and there is no spike-to-spike correlations between the firing of different neurons. These assumptions allow averaging the equations 2.2, 2.6 over different realizations of Poisson trains (representing different neurons) with a given rate, to obtain the new dynamics for the corresponding mean quantities [17]:

$$\frac{dx}{dt} = \frac{1-x}{\tau_{rec}} - uxr(t)$$

$$\frac{du^-}{dt} = -\frac{u^-}{\tau_{facil}} + U(1-u^-)r(t)$$

$$u = u^-(1-U) + U \tag{3.1}$$

Here $u^-$ denotes the average value of $u$ immediately before the spike. Depressing synapses are described by the first of these equations with the fixed value of $u$. The evolution of post-synaptic current can be obtained from the remaining

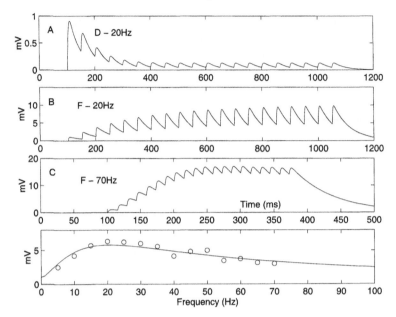

Fig. 1. Phenomenological synaptic model. (A) Simulated post-synaptic potential generated by a regular spike train at a frequency of 20 Hz transmitted through a depressing synapse. (B) Same as in A for facilitating synapse. (C) Same as B but for a presynaptic frequency of 70 Hz. (D) Stationary level of EPSPs vs presynaptic frequency for facilitating synapses. Open circles - experimental results for one of the recorded synaptic connections between pyramidal neuron and inhibitory interneuron (details of experiments were reported in [2]); solid line - model results. The post-synaptic potential is computed using a passive membrane mechanism ($\tau_{mem} \frac{dV}{dt} = -V + R_{in}I_{syn}(t)$) with an input resistance of $R_{in} = 100M\Omega$ for pyramidal target and $1Giga\Omega$ for interneuron. Parameters: (A): $\tau_{mem} = 40$ msec; $\tau_{inact} = 3$ msec; $A = 250$ pA; $\tau_{rec} = 800$ msec; $U = 0.5$; (BCD) $\tau_{mem} = 60$ msec; $\tau_{in} = 1.5$ msec; $A = 1540$ pA; $\tau_{rec} = 130$ msec; $\tau_{facil} = 530$ msec; $U = 0.03$;

equation for $y$ and recalling that $PSC(t) = Ay(t)$:

$$\frac{dy}{dt} = -\frac{y}{\tau_{in}} + uxr(t). \tag{3.2}$$

which can be simplified to $y = r\tau_{in}ux$ if one is interested only in the time scale slower than $\tau_{in}$.

We should mention that while averaging equations 2.1 over different realizations of Poisson spike trains, we assumed that there is no statistical dependence between the variables $x(t)$ and $u(t)$ and the probability of spike emission at time $t$. This is strictly valid only if there is no facilitation since in this case $u$ is a fixed parameter of the model and $x(t)$, which is a function of the spike arrival times prior to the current time, is independent of the probability of a spike at time $t$ due to the Poisson assumption. However, if facilitation is included, both $x(t)$ and $u(t)$ are a function of previous spikes and are thus not statistically independent. This issue was addressed in [17], where it was shown that for realistic values of parameters and rates, the effect of the neglected correlations between variables $x$ and $u$ on the resulting post-synaptic current is indeed weak.

An important advantage of Eqs. 3.1, 3.2 is that they can be solved analytically for an *arbitrary* time course of the firing rates of the presynaptic population. In the case of depressing synapses, the solution takes a particular simple form:

$$y(t) = Ur(t) \int_{-\infty}^{t} dt' exp(-\frac{t - t'}{\tau_{rec}} - U \int_{t'}^{t} dt'' r(t''))$$ (3.3)

This equation can be used to determine which features of the presynaptic AP activity are transmitted by depressing synapses to their targets. Assuming that the presynaptic rate changes gradually ($r'/r << r$), one can write down the expansion over the derivatives of the rate with respect to time ($r, r', r''$, etc). The first two terms of this expansion are

$$y(t) \approx \frac{r}{1 + rU\tau_{rec}} + r' \frac{Ur\tau_{rec}^2}{(1 + rU\tau_{rec})^3} + ...$$ (3.4)

This expression describes the relative contribution of rate and temporal signaling in generating the postsynaptic response. The first term depends on the current rate which is dominant for rates that are small compared to the *limiting frequency* $\lambda \sim 1/(U\tau_{rec})$. As the rate increases this term saturates and thus progressively less rate signaling is possible. The main contribution at higher rates therefore comes from a transient term reflecting the changes in frequencies. In the context of population signaling, this means that only synchronous transitions in the population activity can be signaled to the postsynaptic neuron [6, 18].

The solution of the full set of equations 3.1 for facilitating synapse has the same form as Eq. 3.3 with the single complication due to the fact that $u$ is now itself a functional of the rate:

$$u = U \int_{-\infty}^{t} dt' r(t') exp(-\frac{t - t'}{\tau_{facil}} - U \int_{t'}^{t} dt'' r(t''))$$ (3.5)

which has to be substituted in the Eq. 3.3. One could still analyze the qualitative features of this solution by noting that at very high frequencies, $u \rightarrow 1$ and thus facilitating synapses behave in the same way as depressing ones, transmitting the information about the rate transitions. As the frequency decreases towards the 'peak frequency' (see Fig. 1D)

$$\theta = 1/\tau_{facil} + \sqrt{2/\tau_{facil}^2 + \frac{1+U}{U\tau_{rec}\tau_{facil}}} \approx 1/\sqrt{U\tau_{rec}\tau_{facil}}, \tag{3.6}$$

the presynaptic rate dominates in the postsynaptic response. The reason for this is that at this frequency facilitating and depressing effects compensate each other and the average amplitude of EPSP, which is $\sim xu$, is approximately constant. At even smaller frequencies depressing effects become less relevant since $x$ recovers to almost unity between the subsequent spikes. In this regime the postsynaptic signal mainly reflects the current value of rate amplified by the value of $u$:

$$I_s \sim r(t) \int_{-\infty}^{t} dt' r(t') exp(-(t-t')/\tau_{facil}) \tag{3.7}$$

The integral in this equation is roughly equal to the number of spikes emitted by the presynaptic neuron in the preceding time window of $\tau_{facil}$. In this regime, postsynaptic response is a delayed and amplified transformation of the presynaptic frequency.

As an example, we show in Fig. 2 the post-synaptic current resulting from a series of transitions in the firing rate for both depressing and facilitating synapses. All three regimes of transmission via facilitating synapses are illustrated in Fig. 2B.

## 4. Recurrent networks with synaptic depression

Computations in neocortical circuits are traditionally believed to be carried by selective increase or decrease of average firing rates of single neurons. However, there is a growing number of observations which indicate that information can in fact be represented in the temporal domain. For example, in the auditory cortex neurons have a tendency to respond in a strongly transient manner, often emitting just one spike on average per presentation of the stimulus. Clearly, for these neurons the notion of a firing rate defined as a spike count does not exist at all. One is therefore forced to think about the computational strategies which would utilize the timing of single spikes. The simplest such strategy involves generating a synchronous, near coincident firing of a group of neurons, which would indicate the presence of a certain feature in the sensory stimulus. In this section we will consider the effect of synaptic depression on the activity of recurrent neural networks. In particular, we will show that these networks can generate a 'population

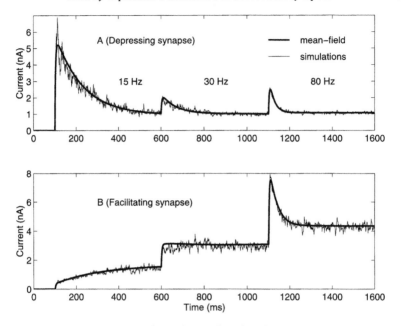

Fig. 2. Post-synaptic current, generated by Poisson spike trains of a
population of 1000 neurons with synchronous transitions from 0 Hz to 15 Hz to 30 Hz and then to 80 Hz, transmitted through facilitating (A) and depressing (B) synapses. Thick line - solution of mean-field equations 3.1,3.2; thin line - simulations of 1000 spike trains with the use of the full model (Eqs. 2.1,2.6). Parameters are the same as in Fig. 1 with $A = 250$ pA;

spike', characterized by a near coincident firing of neurons, either spontaneously or as a response to an external inputs, in an all-or-none fashion.

In the analysis presented below, we will use the dynamic mean-field approximation, which includes the effects of short-term synaptic plasticity [17]. We therefore assume that the network consists of $N$ neuronal 'columns', such that neurons within each column are described by a single rate variable. For simplicity, we consider the network that consists of excitatory neurons only. As shown in [19], this simplification does not alter the main qualitative features of the network activity. We therefore take use of Eqs. 3.1,3.2 and ignore synaptic facilitation:

$$\tau \frac{dh_i}{dt} = -h_i + \frac{J}{N} \sum_j r_j x_j$$

$$\frac{dx_j}{dt} = \frac{1 - x_j}{\tau_{rec}} - u x_j r_j$$

$$r_j = g(e_j + h_j) \tag{4.1}$$

Here $e_i$, $h_i$ denotes external input and network input respectively to neuronal column number $i$ ($i = 1, ..., N$). We assumed that all synaptic parameters ($A$, $u$, $\tau_{rec}$, $\tau_{in}$) are uniform across the network, and denote $Au = J/N$ and $\tau_{in} = \tau$. An important factor that controls the dynamics of the network is the distribution of external inputs $e_i$ across the columns. We will consider the uniform distribution in the interval $e_0 < e_i < e_1$ and use such labeling of the columns that external inputs are increasing with the column index $i$. The gain function $g(h)$ describes the relation between the input and output of the columns. In the following, we take the simple threshold-linear form of this function, $g(h) = [h]_+$. We further assume that all synaptic parameters are the same across the network. This assumption simplifies equations considerably, since the network inputs to all of the neuron quickly converge to the same value ('mean field'), $h_i \rightarrow h$, resulting in the following ($N + 1$)-dimensional non-linear dynamical system:

$$\tau \frac{dh}{dt} = -h + \frac{J}{N} \sum_j g(e_j + h)x_j$$

$$\frac{dx_j}{dt} = \frac{1 - x_j}{\tau_{rec}} - ux_j g(e_j + h) \tag{4.2}$$

The fixed point of this dynamical system can be easily found:

$$H = J/N \sum_i \frac{g(e_i + H)}{1 + u\tau_{rec}g(e_i + H)} \equiv Jg_{st}(H)$$

$$X_i = \frac{1}{1 + u\tau_{rec}g(e_i + H)} \tag{4.3}$$

The first of these equations is a simple algebraic equation for the mean field $H$, that can be solved numerically. Analyzing the stability of these equations is less straightforward. The simple solution is to use the fact that the time constants $\tau$ and $\tau_{rec}$ are vastly different ($\tau \ll \tau_{rec}$), i.e. $x$-variables of the system change much more slower than the mean-field $h$. More precisely, the time constant of $x$-dynamics is given by

$$(\tau_x)_i = \frac{\tau_{rec}}{1 + u\tau_{rec}g(e_i + h)}, \tag{4.4}$$

i.e. it is large compared to $\tau$, provided that the rate of the column is not too big. We can then replace $x$-variables in Eq. 4.2 by their steady state values, reducing it to a one-dimensional system:

$$\tau \frac{dh}{dt} = -h + \frac{J}{N} \sum_j g(e_j + h)X_j \equiv -h + Jg_{dyn}(h; H), \tag{4.5}$$

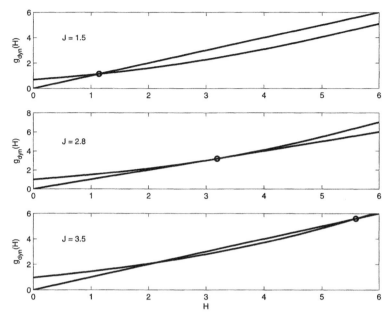

Fig. 3. Graphical solution of the Eq. 4.5 for three different values of $J$. The value of H at the fixed point is marked by the red circle. The stability of the fixed point is determined by the slope of $g_{dyn}(h)$ at the fixed point. Transition from stable to unstable fixed point occurs at $J \approx 2.8$.

that is valid in the vicinity of the fixed point and can be used to analyze its stability. The fixed point is stable if the slope of the $g_{dyn}(h)$ is less then one at the fixed point $h = H$ (i.e. $J g'_{dyn}(H; H) < 1$), and unstable otherwise. This is illustrated in Fig. 3 for 3 different values of synaptic strength J. One concludes from this analysis that the system undergoes a bifurcation from steady solution $h(t) = H$ to the unsteady one at a certain critical value of $J = J_{crit}$. This value satisfies the following approximate equation, that can be derived from the Eq. 4.5, taking into account our choice of the gain function $g(x) = [x]_+$:

$$\frac{J_{crit}}{N} \sum_j{}' X_j = 1, \tag{4.6}$$

where here and in the following $\sum'_i$ stays for the sum over the columns with super-threshold inputs, i.e. with $e_i + H > 0$.

This analysis does not allow one to understand the nature of the bifurcation and the solution of the system above it. Numerical simulations of Eqs. 4.2 are shown on Fig. 5, illustrating a 'population spike' behavior with sharp and narrow

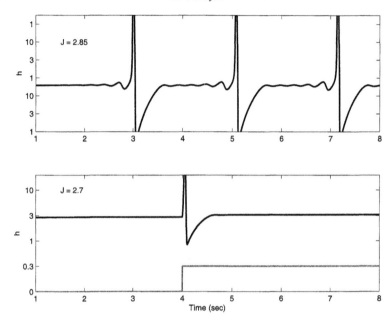

Fig. 4. Numerical solution of Eqs. 4.2. Upper panel: activity of the network for $J > J_{crit}$ with spontaneous appearance of population spikes. Lower panel: activity of the network for $J < J_{crit}$. Red curve shows the external input of the amplitude $A = 0.3$ applied to every neuron in the network.

increases in activity separated by long periods of almost steady state activity. This peculiar behavior is a signature of these networks and the conditions for its existence are analyzed in more details in [20]. Qualitatively, this solution can be understood as following: when the fixed point is unstable, any deviation from it grows exponentially to large values of $h$. When $h$ is large, $x$-variables become fast (see Eq. 4.4), and they quickly decrease, leading to effective weakening of the synapses and therefore fast termination of the population spike. On the same figure, we show the solution of Eqs. 4.2 with step-wise external input ($e_i \rightarrow e_i + A$). While the steady state solution is steady before and after the onset of the stimulus, the system exhibits a transient population spike at the close temporal proximity of the onset. This transient response is closely reminiscent to the response of the neurons in primary cortex to tonic sounds.

To understand the dynamical nature of the bifurcation one has to return to the full $(N + 1)$-dimensional system 4.2 and linearize it around the fixed point solution. The corresponding equations are

$$\tau \frac{dh}{dt} = -h + \frac{J}{N}[\sum_j x_j g(e_j + H) + h \sum_j X_j g'(e_j + H)]$$

$$\frac{dx_j}{dt} = -\frac{x_j}{\tau_{rec}} - u[x_j g(e_j + H) + h X_j g'(e_j + H)] \tag{4.7}$$

where $h$, $x_j$ now denote the deviations of the corresponding variables from their fixed point values $H$, $X_j$. We are looking for time-dependent solutions in the form $exp(\lambda t)$, where $\lambda$ is the eigen-value(s) of the linearized system. Substituting this dependency in Eqs. 4.7, we can deduce a single equation for all the eigen-values of the system:

$$1 + \lambda \tau = \frac{J}{N} \sum_i' X_i \frac{\lambda + 1/\tau_{rec}}{\lambda + 1/\tau_{rec} + ug(e_i + H)} \tag{4.8}$$

The fixed point solution $H$, $X_i$ becomes unstable when one of the eigenvalues acquires the positive real value. Two possible scenarios can therefore happen: (i) saddle-node bifurcation ($\lambda = 0$) and (ii) Hopf bifurcation ($\lambda_{1,2} = \pm i\mu$). Substituting to Eq. 4.8, we get: condition for (i):

$$1 = \frac{J}{N} \sum_i' X_i^2 \tag{4.9}$$

and condition for (ii):

$$1 = \frac{J}{N} \sum_i' X_i \frac{1/X_i + \mu^2 \tau_{rec}^2}{1/X_i^2 + \mu^2 \tau_{rec}^2}$$

$$\tau = \tau_{rec} \frac{J/N'}{\sum_i 1/X_i^2 + \mu^2 \tau_{rec}^2} \frac{1 - X_i}{} \tag{4.10}$$

Using the separation of time scales ($\tau_{rec} \gg \tau$), one can show that Eqs. 4.10 reduce to

$$1 = \frac{J}{N} \sum_i' X_i \tag{4.11}$$

which coincide with the approximate expression from Eq. 4.6. Since $X_i < 1$, this condition happens before the corresponding condition for saddle-node bifurcation (Eq. 4.9), i.e. we conclude that the fixed point loses its stability via the Hopf bifurcation when condition of Eqs. 4.10 is satisfied. The imaginary part of the eigenvalue at the bifurcation point can be estimated as

$$\mu^2 = \frac{1}{\tau \tau_{rec}} \frac{J}{N} \sum_i^{'} (1 - X_i), \tag{4.12}$$

i.e. $\mu \approx 1/\sqrt{(\tau \tau_{rec})}$. Numerical solution of Eqs. 4.2 show that immediately above the bifurcation, the system exhibits population spikes of high amplitude with very low frequency and oscillations in between (see Fig. 4). This means that the Hopf bifurcation at $J = J_{crit}$ is of sub-critical type. We don't know how to prove that this is indeed the case for the $(N + 1)$-dimensional system of Eqs. 4.2. An interesting question is what is the minimal number of dimensions for which the behavior of the system is qualitatively similar to many-dimensional system. We first consider the case of $N = 1$ (two-dimensional system):

$$\tau \frac{dh}{dt} = -h + Jg(e + h)x$$

$$\frac{dx}{dt} = \frac{1 - x}{\tau_{rec}} - uxg(e + h) \tag{4.13}$$

In order to exhibit a Hopf bifurcation, we have to choose the gain function that is smooth everywhere , e.g.

$$g(h) \quad = \quad h + \sigma + \frac{1}{3\sigma^2}, \quad h > 0$$

$$= \quad \sigma(1 + \tanh(\frac{h}{\sigma})) + \frac{1}{3\sigma^2}e^{h^3}, \quad h < 0 \tag{4.14}$$

where $\sigma$ is a parameter that controls the smoothness of the gain function at $h = 0$. It can be shown that the system indeed exhibits a super-critical Hopf bifurcation, at which the fixed point becomes unstable. The stable limit cycle that replaces the fixed point, is shown on Fig. 5. It consists of a slow part, during which the system rises along the null-cline of the $h$-variable, until reaching the upper knee point of this null-cline. At this point, the fast phase of the limit cycle begins, that consist of a large increase and decrease of $h$. If the initial point of the system is chosen very close to the fixed point, it oscillates for several times until approaching the limit cycle (Fig. 5, inset). However, after the system reached the limit cycle, it cannot return to the vicinity of the fixed point, because this would require the trajectory in the phase plane to cross itself. Thus, the system will not exhibit the behavior that we observed in the many-dimensional system, characterized by population spikes with low frequency and inter-spike oscillations. We therefore conclude that the minimal system that can exhibit this behavior is the system

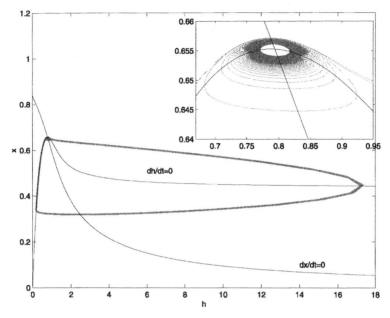

Fig. 5. Numerical solution of Eqs 4.13 for one self-connected neuron. Red - limit cycle solution on the phase plane (h,x); two black curves are the null-clines of the system along which the corresponding time derivatives are zero. Inset: zoom on of the graph near the fixed point of the system.

with $N = 2$, i.e. the system with 3 degrees of freedom. Indeed, in this case, the trajectory of the system can approach the vicinity of the fixed point at every cycle via the 3rd dimension without having to cross itself. This type of bifurcation is called the Shilnikov bifurcation [21]. The proof that three-dimensional system indeed exhibits Shilnikov bifurcation will be given elsewhere (Tsodyks & Hansel, in preparation). Since the behavior of three-dimensional system is qualitatively similar to the behavior with large $N$, we conjecture that also many-dimensional dynamical system of Eqs. 4.2 exhibits Shilnikov bifurcation.

## 5. Conclusion

As we demonstrated at these lectures, synaptic dynamics has a profound effect on the activity of neuronal circuits and on the way they transmit information to their post-synaptic targets. In terms of the neural code, synaptic dynamics determines what features of presynaptic population activity are transmitted to the post-synaptic targets. In particular, depressing synapses at high rates of depression work as high-pass filters, since they cut the transmission of the time-averaged firing rates but emphasize the fast common fluctuations in the rates of pre-synaptic neurons, i.e. neural synchrony. The exact form of transmission depends on synaptic parameters that characterize synaptic depression. Facilitating synapses have different transmission properties, that depend on the level of pre-synaptic activity, by gradually changing from transmitting the integrals of the rates, to rates, and finally to derivatives of the rates at very high level of activity. It therefore transpires, that population activity, when transmitted to its post-synaptic targets, undergoes complex temporal transformations that may differ for different pathways. In other words, the same population may send different 'messages' along the different pathways, depending on the properties of the synaptic connections that characterize these pathways. We also saw that synaptic dynamics have a strong effect on the synchronization pattern obtained in recurrent cortical circuits. In particular, synaptic depression between excitatory (pyramidal) neurons tend to generate fast synchronized response during which most of the neurons in the network fire a spike with high temporal precision. It would be of interest to consider the effects of synaptic facilitation of various types of connections on the temporal activity of recurrent circuits.

## Acknowledgements

Most of the results described in this contribution are based on the joing work with Henry Markram and Alex Loebel. I am also grateful to David Hansel for discussions and collaborations. This work is supported by Israeli Science Foundation, Anne P. Lederer Research Institute and Pearl M. Levine trust

## References

[1] Thomson, A. M. and Deuchars, J. "Temporal and spatial properties of local circuits in neocortex", Trends in Neuroscience, 17 (1994) 119-126.

[2] Markram H., Wang Y. and Tsodyks M. "Differential signaling via the same axon from neocortical layer 5 pyramidal neurons", PNAS, 95 (1998) 5323-5328.

[3] Gupta A., Wang Y. and Markram H. "Organizing principles for a diversity of GABAergic interneurons and synapses in the neocortex", Science, 287 (2000) 273-278.

[4] Tsodyks M. and Markram H. "Plasticity of neocortical synapses enables transitions between rate and temporal coding", Lecture Notes in Comp. Sci., 1112 (1996) 445-450.

[5] Abbott L., Varela J., Sen K. and Nelson S. "Synaptic depression and cortical gain control", Science, 275 (1997) 220-224.

[6] Tsodyks M. and Markram H. "The neural code between neocortical pyramidal neurons depends on neurotransmitter release probability", PNAS, 94 (1997) 719-723.

[7] Wilson H. and Cowan J. "Excitatory and inhibitory interactions in localized populations of model neurons", Biophys. Journal, 12 (1972) 1-24.

[8] Tsodyks M. and Amit D. "Quantitative study of attractor neural network retrieving at low spike rates: I Substrate-spikes, rates and neuronal gain", Network, 2 (1991) 259-274.

[9] Ginsburg I. and Sompolinsky H. "Theory of correlations in stochastic neural networks", Phys. Rev. E, 50 (1994) 3171-3191.

[10] Destexhe A., Mainene Z. F. and Sejnowski T. J. "Synthesis of models for excitable membranes, synaptic transmission and neuromodulation using a common kinetic formalism", Journ. of Comp. Neuroscii, 1 (1994) 195-230.

[11] McNaughton B. L., "Neuronal Mechanisms for spatial computation and information storage", in "Neural Connections, ental Computations", eds. L. Nadel and L. A. Cooper and P. Culicover and R. M. Harnish", The MIT Press, Cambridge, Massachusetts 1989, 285-350.

[12] Fuhrmann G., Segev I, Markram H. and Tsodyks M., "Coding of temporal information by activity-dependent synapses", J. of Neurophysiology, 87 (2002) 140-148.

[13] A. Mallart and A. R. Martin, "Two components of facilitation at the neuromuscular junction of the frog.", Journal of Physiology 1933 (1967) 677-694.

[14] Zengel J. E. and Magleby K.L., "Augmentation and facilitation of transmitter release. A quantitative description at the frog neuromuscular junction", J. Gen. Physiol., 80 (1982) 583-611.

[15] Bertram R., Sherman A. and Stanley E. F. ""Single-domain/bound calcium hypothesis of transmitter release and facilitation", J. Neurophysiol., 75 (1996) 1919-1931.

[16] Softky W. R. and Koch C., "The higly irregular firing of cortical cells isinconsistent with temporal integration of random EPSPs", J. Neurosci. 13 (1993) 334-350.

[17] Tsodyks M., Pawelzik K. and Markram H., "Neural networks with dynamic synapses", Neural Computations, 10 (1998) 821-835.

[18] Senn W., Segev I. and Tsodyks M., "Reading neuronal synchrony with depressing synapses", Neural Computation, 10 (1998) 815-819.

[19] Tsodyks M., Uziel A. and Markram H. "Synchrony generation in recurrent network with frequency-dependent synapses", J. Neurosci 20 (2000) 1-5.

[20] Loebel A. and Tsodyks M. "Computation by Ensemble Synchronization in Recurrent Networks with Synaptic Depression", J. of Comp. Neuroscience, 13 (2002) 111-124.

[21] Kuznetsov Y. A. "Elements of applied theory", Springer-Verlag New York, 1998.

Course 8

# THEORY OF LARGE RECURRENT NETWORKS: FROM SPIKES TO BEHAVIOR

## H. Sompolinsky[1,2] and O.L. White[3]

[1] *Racah Institute of Physics and Center for Neural Computation, Hebrew University, Jerusalem, 91904, Israel (permanent address)*
[2] *Department of Molecular and Cellular Biology, Harvard University, Cambridge, MA 02138, USA*
[3] *Department of Physics, Harvard University, Cambridge, MA 02138, USA*

*C.C. Chow, B. Gutkin, D. Hansel, C. Meunier and J. Dalibard, eds.*
*Les Houches, Session LXXX, 2003*
*Methods and Models in Neurophysics*
*Méthodes et Modèles en Neurophysique*

# Contents

## 1. Introduction

In these notes we review the role of network dynamics in brain function. Some of the classical models of brain function assume a feedforward flow of signals from low to high brain areas. Here, on the other hand, we will focus on the role of *recurrent* dynamics within local circuits in cortex and other brain structures.

The computational characterization of a system involves describing the kind of input-output transformation that it performs. Thus quite generally, a neuronal network can be viewed as transforming incoming spike trains to outgoing spike trains. The space of possible spike trains, however, is enormous and the dynamical description of the transformation between them is in general quite complex. We may expect that the nature of *relevant* inputs and outputs is not fixed but depends on the class of computations under consideration. In these notes we focus on a class of computations in which both the relevant inputs and the relevant outputs are firing rates. A neuron may code a variable by the number of spikes it emits within a given window of time. This coding may sometimes be binary. A neuron may signal a yes/no decision depending upon whether its firing rate is higher or lower than some threshold. These are examples of what is known as 'rate coding'.

The rate coding approach is controversial (see *e.g.*, Shadlen & Newsome, 1998; Prut et al., 1998; Mainen & Sejnowski, 1995). Some hold that it is misguided, missing the point, which is the importance of the fine temporal pattern of spikes. Nevertheless, the rate approach is consistent with the available data in several important brain areas (*e.g.*, visual and motor cortex) at least in the restricted conventional laboratory settings. Furthermore, our understanding of both the dynamics and the function of rate based network models far surpasses our understanding of spike based models. It is therefore an extremely useful starting point for building a more comprehensive understanding of the relation between dynamics and computation in the brain.

At the microscopic level the dynamics of neuronal circuits is cast in terms of conductances, currents, membrane potentials and spikes. Sections 2 and 3 will be devoted to the circumstances under which this microscopic dynamics can be reduced to relatively simple rate based dynamics. The simplest rate models are linear networks. Despite their simplicity they have received much attention recently, particularly in modeling the problem of persistence and temporal integration. This is the topic of Section 4. There we also discuss an alternative recently developed model for persistence and integration, which is based on regenerative calcium dynamics in single neurons. The minimal nonlinearity present in rate based models is threshold rectification of the firing rates. In Section 5 we show that this nonlinearity is sufficient to endow networks with novel functions not present in linear models. In particular, we discuss the *ring model*, which is the simplest recurrent network model for feature selectivity. Finally, binary networks have been used extensively in computational models. Among them is the well known Hopfield model of associative memory, discussed in Section 6. In these notes we focus on highly connected networks, which are governed by conventional mean-field like equations. Another important class of recurrent networks, in which the activity is dominated by fluctuations, is that of *balanced networks*. This class of networks is described in Carl van Vreeswijk's contribution to this volume. These notes are not intended as a comprehensive survey of the vast literature on this topic (the theory of recurrent networks is treated in detail in Hertz et al., 1991; Amit, 1989; and in Chapter 7 of Dayan & Abbott, 2001).

## 2. From spikes to rates I: rates in asynchronous states

### 2.1. Introduction: rate models

Rate equations for neuronal networks often take the form of a set of first order ODEs similar to the following,

$$\tau_0 \frac{dr_i}{dt} = -r_i + f_i \left( \sum_{j=1}^{N} J_{ij} r_j + I_i^0 \right), \quad i = 1, \ldots, N \tag{2.1}$$

where $r_i$ is the rate of the $i^{\text{th}}$ neuron, $\tau_0$ the time constant, $f$ an input-output function, $I^0$ the external input, and $J_{ij}$ the coupling matrix (Wilson & Cowan, 1972; Amari, 1977; Hopfield, 1982). Some convenient choices for the form of $f$ are:

$$\textit{sigmoidal 1}: \quad f(x) = \frac{1}{1 + \exp[\beta(\theta - x)]} \tag{2.2}$$

where $\theta$ represents a soft threshold and $\beta$ the neuronal gain. With this function, rate $r$ is normalized so that it ranges between 0 and 1, where 1 stands for the neuron's maximum firing rate. Sometimes it useful to shift the definition of $r$ so it takes also negative values. For instance, a common $f$ is

$$\textit{sigmoidal 2}: \quad f(x) = \tanh[\beta(x - \theta)] \tag{2.3}$$

which corresponds to shifting $r$ so that it ranges between $-1$ and $1$. In the limit of large $\beta$ we obtain the famous binary neuron that assumes two values (0 and 1; or $-1$ and 1). Another important choice of $f$ is the threshold linear function

$$\textit{threshold linear}: \quad f(x) = [x - \theta]_+ \tag{2.4}$$

where $[x]_+ = x$ if $x$ is positive and is 0 otherwise. As discussed later, this is a useful choice for cortical networks. Finally, if we assume that all the neurons in the network are above the firing threshold in the particular setting with which we are concerned, then we can ignore the thresholding operation in equation (2.4) and arrive at a linear network

$$\textit{linear}: \quad f(x) = x - \theta. \tag{2.5}$$

This approach raises several questions:

(1) When can a general rate description of the network dynamics be justified? How can we derive rate equations from the underlying microscopic spiking dynamics?

(2) Under which of these conditions do the resultant rate equations have a relatively simple form of the type given above?

(1) What is the biophysical interpretation of the parameters appearing in the rate equations? For instance, to what real time constant does $\tau_0$ in equation (2.1) correspond? What is the meaning of the coupling matrix $J_{ij}$ or the function $f$?

Until recently, models of this sort were viewed primarily as abstract and simplified *neural* network models, inspired or motivated by real *neuronal* networks. Usually, no claims were made for a precise relationship between the simplified rate equations and the biophysical reality. Today we understand the relationship between these two things better and under certain conditions can derive a precise mapping between the parameters appearing in the rate equations and those that play a role in the conductance based equations of spiking neurons (Amit & Tsodyks, 1991; Abbott & Kepler, 1990; Ermentrout, 1994; Shriki et al., 2003). We discuss these developments in what follows.

## 2.2. *Synchronous and asynchronous states*

When can we justify describing the network state merely by the neuronal firing rates? As we will see, when a highly connected large network is in an asynchronous state, this state can be captured in terms of rates. In the case of a synchronous state, however, a rate description may miss important aspects of the system's behavior. Intuitively this is because in the asynchronous case, there is no interesting temporal information *per se*. There is no dynamical coherence in the system and therefore the spike times of different neurons are essentially independent. Although the above intuition is right, it is not easy to formalize, partly because the definition of asynchrony is non-trivial.

## Measures of synchrony

Defining synchronous or asynchronous states is difficult for the simple reason that in any network with a significant degree of connectivity, the spiking of one neuron influences that of others via by monosynaptic or polysynaptic pathways, leading to some degree of correlation. Thus a connected network always shows *some* synchrony. The difference between synchronous and asynchronous states lies in the *degree* of synchrony. We therefore need to define a quantitative measure of synchrony (Ginsburg & Sompolinsky, 1994; Hansel & Sompolinsky, 1996).

Traditionally coherence between the activities of neurons has been quantified experimentally by their *cross-correlograms*. Given two neurons A and B, the cross-correlogram measures the average number of spikes per unit time emitted by neuron B at time $t + \tau$ where $t$ is the time of a spike emitted by cell A. In many cortical neurons, the cross-correlogram will show a narrow peak centered at $\tau = 0$. The area under this peak measures the number of excess spikes of B per spike in A due to the correlations between the two neurons. An area of $\sim 0.1$ (*e.g.* (5Hz)*(10msec)), is often quoted for cortical pairs (see *e.g.* Abeles, 1991).

There are two major difficulties in using cross-correlograms as a measure of synchrony in a network. First, there is enormous inhomogeneity in the values of the cross-correlograms even within the same brain area. This indicates the need for some sort of a global measure of synchrony. Secondly, it is unclear how to normalize a cross-correlogram. For instance, does the above quoted value of 0.1 excess spikes represent strong or weak synchrony? Naturally we would like to derive an appropriate correlation coefficient for spike trains. But the usual procedure of normalizing by standard deviations is inapplicable in this case since the associated variances are delta functions.

To define a global measure of synchrony we begin by choosing a local dynamical variable, $V_i(t)$, for instance the membrane potential, the synaptic currents or filtered spiking activity, which we then spatially average to construct an 'order parameter' of the system,

$$V(t) \equiv \frac{1}{N} \sum_{i=1}^{N} V_i(t). \tag{2.6}$$

Experimentally, these global measures are related to the on-going activity measured in optical imaging as well as the Local Field Potential (LFP) measured by extra cellular recordings. In some cases the choice of the local dynamical variable may matter. For instance, consider the state in which the neurons are grouped into two subpopulations which are exactly out of phase. Obviously spatial averages of the time dependent potential or activity over the entire population will miss this pattern of synchrony. Exact cancellations of this kind, however, are relatively rare.

The time-delayed autocorrelation of $V(t)$ is

$$C_V(\tau) = \langle \delta V(t) \delta V(t + \tau) \rangle_t \qquad (2.7)$$

where the angular brackets denote time average and $\delta V(t) \equiv V(t) - \langle V \rangle$ gives the deviation of the mean potential from its time average. $C_V$ is maximal at $\tau = 0$ where it is simply the variance of $V$. Its dependence on $\tau$ quantifies the time extent of temporal correlations. We can construct a dimensionless measure of synchrony from $C_V$ by

$$\chi \equiv \frac{C_V(0)}{\frac{1}{N} \sum_{i=1}^{N} \langle [\delta V_i(t)]^2 \rangle_t} \qquad (2.8)$$

where we have divided the variance of $V$ by the mean variance of single neurons $V_i$. In general,

$$C_V(0) = N^{-2} \sum_{ij} \langle \delta V_i \delta V_j \rangle. \qquad (2.9)$$

If the neurons are completely independent, then only the $i = j$ terms contribute to $C_V(0)$ and hence $\chi = 1/N$. In the other extreme, where all the neurons behave identically, $\chi = 1$, so in general

$$\frac{1}{N} \leq \chi \leq 1. \qquad (2.10)$$

In an asynchronous network $\chi$ will be much smaller than 1, while in a synchronous network, we expect that $\chi$ will be of order 1. We use the way that $\chi$ depends upon the size of the averaged population $N$ to give a more formal definition of asynchrony. An asynchronous state is defined as one in which:

$$\lim_{N \to \infty} \chi = 0 \qquad (2.11)$$

Typically, in the spirit of the Central Limit Theorem, $\chi$ is expected to fall off as $1/N$ or faster. Note that equation (2.11) hides the question of how to take the infinite $N$ limit. Although this is an important formal issue, we would hope that the definition of an asynchronous state does not depend heavily on such details.

## 2.3. Synchronous and asynchronous states in highly connected networks

Although the concepts of synchrony and asynchrony are not restricted to a specific pattern of connectivity, the conditions for asynchrony do depend on both connectivity and synaptic dynamics. Here we focus on the conditions for asynchrony in highly connected networks, similar to those present in cortex and many subcortical structures. Experiments show that cortical neurons are often only weakly synchronized. As explained below, this is not surprising given our understanding of synchrony in large highly connected networks. We begin by considering a coupled pair of neurons.

### Phase relationship between a pair of neurons

Carl van Vreeswijk (1996) has analyzed the phase diagram of a class of pulse coupled neurons. The phase relationship between a pair of neurons with either excitatory or inhibitory coupling is shown in Figure 2.1. In the case of *excitatory coupling*, illustrated in Figure 2.1a, the in-phase state is always unstable. For sufficiently large synaptic time constants ($\alpha < \alpha_c$), the state in which the two neurons are exactly out of phase is stable and dictates the long term dynamics of the system. However, the system undergoes a supercritical pitchfork bifurcation as the parameter $\alpha = 1/$(synaptic time constant) increases through a critical value $\alpha_c$. The antiphase solution becomes unstable and two stable phase locked states at intermediate values of the phase emerge. Note that for small values of the synaptic time constant, these are very near to completely in phase.

In the case of *inhibitory coupling*, illustrated in Figure 2.1b, the in-phase solution is always stable. For sufficiently large synaptic time constants ($\alpha < \alpha_c$), the state in which the two neurons are exactly out of phase is unstable and hence in steady state the neuron pair will always fire in phase. However, the system undergoes a subcritical pitchfork bifurcation as the parameter $\alpha = 1/$(synaptic time constant) increases through a critical value $\alpha_c$. The antiphase solution becomes stable and there are unstable phase locked states at intermediate phase values separating its basin of attraction from that of the in-phase state. Now the steady state of the two neurons will depend upon initial conditions. Note that as synaptic time constant decreases, the basin of attraction of the in-phase solution becomes smaller and smaller.

One can gain insight into the behavior of a large, highly connected network by analogy to the system of two coupled neurons. In the case of excitatory coupling, a system of two coupled neurons is never completely synchronized and thus we expect that large networks will also never be synchronized entirely. Nevertheless we expect the degree of network synchrony to increase with decreasing synaptic time constant. Simulations on excitatory networks with all-to-all connectivity with various time constants bare out these expectations.

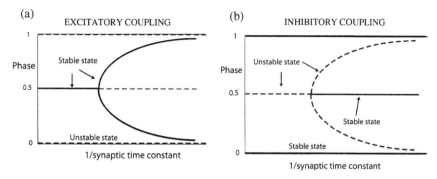

Fig. 2.1. Phase relationship of coupled neurons.

*Asynchronous state in an excitatory network*

Figures 2.2a and 2.2b illustrate the stable asynchronous state observed in a simulation of an excitatory network with a sufficiently large time constants $\tau_S$, here 3msec. In Figure 2.2a the lower trace illustrates the network firing rate per neuron, without any coupling. The higher trace shows firing rate per neuron when all-to-all coupling has been added. Since the coupling is not negligible, the average rate increases. However, the effective delay between the firing of connected neurons is large and thus they do not fire synchronously, as seen in the small variation in the network firing rate. Figure 2.2b depicts the membrane potentials of two cells as a function of time. Their phase relations drift, also indicating network asynchrony.

*Synchronous state in excitatory network*

Figure 2.2c illustrates the stable synchronous state observed in a simulation of an excitatory network with a sufficiently small time constants $\tau_S$, here 1msec. The fast excitatory synapses between neurons lead to strong network synchrony as seen in the normalized network rate as a function of time.

## Inhibitory networks

In the case of inhibitory coupling, since only the synchronous state is stable for a pair of coupled neurons with a long synaptic time constant, we expect synchronization in a large and highly connected network with slow inhibitory coupling. For faster inhibitory couplings, however, one expects both synchronized network states and states in which at least some of the neurons are out of phase with one another. We present some simulations on all-to-all inhibitory networks.

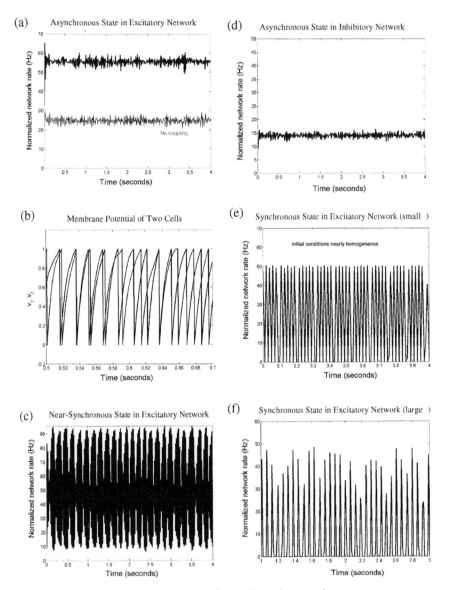

Fig. 2.2. Synchrony and asynchrony in networks.

*Asynchronous state in inhibitory network*
Figure 2.2d illustrates the stable asynchronous state observed in a simulation of an inhibitory network with a sufficiently small time constant $\tau_S$, here 1msec. As a result of the inhibition, the network firing rate decreases from the uncoupled rate (compare to Figure 2.2a). However there is little synchrony as seen in the small variation of this rate.

*Synchronous states in inhibitory network*
Figure 2.2e illustrates that networks with inhibitory coupling and sufficiently small synaptic time constants display bi-stability, with both stable synchronous and asynchronous states. In Figure 2.2e we plot the network firing rate versus time for the stable synchronous state, again for $\tau_S = 3$ msec but with a different set of initial conditions.

Figure 2.2f illustrates the uniquely stable synchronous state observed in a simulation of an inhibitory network with a sufficiently large time constants $\tau_S$, here 30msec. The average network rate oscillates with time, indicating synchrony.

## 2.4. Rate equations for asynchronous networks

In what follows, we show that asynchronous states in highly connected networks can be characterized using rate equations. Our analysis follows Shriki et al., (2003).

*Homogeneous Network*
Consider the simplest example of a fully connected homogeneous network of identical spiking neurons. Each neuron $i$ is driven by an external static current $I^0$ as well as by a synaptic current. Describe its membrane potential by a Hodgkin-Huxley-type equation,

$$C_m \frac{dV_i}{dt} = -g_L(V_i(t) - E_L) + I_i^{spike} + I^0 + I_i^{syn}. \tag{2.12}$$

Here $I^{syn}$ is a synaptic current and for each $i$ is given by

$$I^{syn}(t) = Js(t) \tag{2.13}$$

with

$$s(t) = \frac{1}{\tau_S} \sum_{j=1}^{N} \sum_{\{t_j\}} K(t - t_j) \tag{2.14}$$

and where, for reasons which will become clear later on, the strength of the synapse is denoted by $J/\tau_S$ with $\tau_S$ the synaptic time constant defined below in equation (2.15). $J$ has units of [current][time] = [total charge of the synaptic current]. In what follows, we assume homogenous all-to-all connectivity of neurons and write the synaptic current $I^{syn}$ as a product of a constant synaptic strength $J$ times a time-dependent synaptic rate $s(t)$ given by equation (2.14). The function $K(t)$ is a filter describing the time evolution of a single synaptic current and is normalized so that its maximum is 1. The synaptic time constant $\tau_S$ is defined as

$$\tau_S \equiv \int_0^\infty K(t)dt. \tag{2.15}$$

The simplest example of the synaptic rate is

$$\tau_s \frac{ds}{dt} = -s + \sum_{j=1}^N r_j(t) \tag{2.16}$$

with

$$r_j(t) = \sum_{\{t_j\}} \delta(t - t_j). \tag{2.17}$$

This corresponds to a simple low pass filter

$$K(t) = \exp(-t/\tau_s), \quad t > 0. \tag{2.18}$$

For a general synaptic rate function $s(t)$ and a network in an asynchronous state with sufficiently long synaptic integration time, the distribution of synaptic input amplitudes to a given neuron within one synaptic integration time is the same as the distribution over time of the amplitudes of one synaptic input. Therefore in an asynchronous state, if the firing rate large is enough that the total number of pre-synaptic spikes within one synaptic integration time is large, then we can replace $s(t)$ by its time average,

$$s(t) = \langle s(t) \rangle_t = Nr \tag{2.19}$$

where $r$ denotes the time averaged firing rate of each of the neurons. To see how to calculate the time average, note that

$$s(t) = \left\langle \frac{1}{\tau_S} \sum_{j=1}^N \sum_{t_j} K(t - t_j) \right\rangle_t \tag{2.20}$$

In a large time interval of length $T$, each input neuron will spike on average $rT$ times where $r$ is the neuronal firing rate in an homogeneous system. For a given input spike time $t_j$, the average value of $K(t - t_j)$ over time $T$ is $\tau_S/T$. Since there are $N$ input synapses the quantity in brackets in equation (2.20) is $(1/\tau_S)(rT)(\tau_S/T)(N)$, yielding equation (2.19).

We thus conclude that each neuron receives a static current which is the sum of the external current and the current from the internal feedback. Since $I^{syn}(t) = Js(t)$ by equation (2.13), the feedback contribution is proportional to the firing rate $r$ of an average neuron in the network. This quantity is yet unknown. However, we can write down the self-consistent equation for $r$, assuming that all neurons are equivalent. This is straightforward, since the firing rate of a single neuron with a constant current is given simply by its f-I curve. Therefore the self-consistency equation is,

$$r = f(I^0 + NJr) \tag{2.21}$$

where $f(I)$ is the f-I curve of an individual neuron in isolation.

*Inhomogeneous Network*
We can extend the above reasoning to an inhomogeneous network with a high degree of connectivity so that each neuron receives many synaptic inputs from the network. Denote the matrix of synaptic strengths (*i.e.*, the peak synaptic currents of single spikes) from a neuron $j$ to a neuron $i$ by $J_{ij}/\tau_{ij}$. Then if the network settles into an asynchronous state with static rates $\{r_i\}$, the self-consistent equations for these rates, are

$$r_i = f_i\left(\sum_{j=1}^{N} J_{ij}r_j + I_i^0\right), \quad i = 1, \dots, N \tag{2.22}$$

where $f_i(I)$ is the f-I curve of the $i^{\text{th}}$ neuron in isolation, and $I_i^{app}$ is the external current (*i.e.* coming from outside the network).

The above equations (2.21) and (2.22) represent an enormous simplification for the analysis of networks. All we have to know for the description of an asynchronous state is the f-I curve of the single neurons in isolation.

## Examples of f-I curves:

*The Hodgkin-Huxley model (HH)*
A schematic of the f-I curve for a point neuron with the original HH dynamics is shown in Figure 2.3. Note that this f-I curve has a pronounced discontinuity at the threshold current. The HH cell has a limited dynamic range (Tuckwell, Chapter 3, 1988). Following current injection, the firing rates of most cortical neurons range from a few spikes per seconds to several hundred Hz. Often the observed dependence of the rates on current amplitude is roughly linear. This

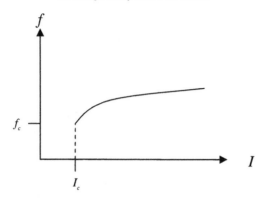

Fig. 2.3. Hodgkin-Huxley neuron f-I curve.

means that the original Hodgkin-Huxley model is not a good model for the response properties of a cortical cell.

### The Integrate and Fire Neuron

The equation for the f-I curve of the Integrate and Fire neuron is

$$f(I) = \begin{cases} \dfrac{1}{-\tau \log(1 - I_C/I)}, & I > I_C \\ 0, & I < I_C \end{cases} \tag{2.23}$$

where $I_C$ is the threshold current and $\tau$ the characteristic time scale. This function is plotted in Figure 2.4.

*Low frequencies:* The frequency of firing starts from zero continuously but the rise is very steep, with a slope which diverges at the threshold.

*High frequencies:* This f-I curve is linear at high frequencies. For $I/I_C \gg 1$ one can Taylor expand the log term using $\log(1 - x) \sim -x$ for $x \ll 1$ to derive the result

$$f \simeq \frac{1}{\tau} \frac{I}{I_C}, \quad I \gg I_C \tag{2.24}$$

The slope of the f-I curve for large $I/I_C$ is thus $1/(\tau - I_C) = 1/(C - \theta)$, which means that in the high frequency limit the leak current is negligible compared to the capacitance current.

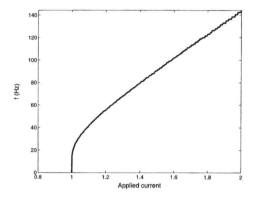

Fig. 2.4. Integrate and fire neuron f-I curve.

### *Hodgkin-Huxley-type model for cortical cells*

Various modifications of the HH model for cortical cells exist in the literature, most of which are for compartmental neuron models. Here we restrict our attention to a point neuron model. If one retains the HH currents but changes appropriate parameters for them one can obtain a continuous f-I curve starting from zero frequency with an initial rise which is not linear but of a square root form ($f = A(I - I_C)^{1/2}$). Similar behavior is obtained in more realistic models which incorporate additional active currents (*e.g.*, Wang & Buzsaki, 1996). If some of these currents contain sufficiently strong and moderately slow negative feedback then the resultant f-I curve is close to linear. An example is the model of Shriki et al., which obeys the following dynamics,

$$C_m \frac{dV}{dt} = -g_L(V(t) - E_L) - I^{Na} - I^K - I^A + I^0. \tag{2.25}$$

The sodium and delayed potassium rectifier current are defined in the standard way with parameters such that a saddle-node bifurcation occurs at firing onset. The slow A-current serves to linearize the f-I relationship as shown in Figure 2.5 where f-I curves are plotted using three different parameter values: $g_A = 0$ (dashed line); $g_A = 20$mS/cm$^2$ and $\tau_A = 0$msec (dash dotted line); $g_A = 20$mS/cm$^2$ and $\tau_A = 20$msec (solid line) where $g_A$ and $\tau_A$ are the A-current conductance and time constant, respectively. As indicated by these three curves, linearization at small firing rates is due to the long A-current time constant. The inset shows a voltage trace of a model neuron with injection of constant current of amplitude $I = 1.6\mu$A/cm$^2$ for $g_A = 20$mS/cm$^2$ and $\tau_A = 20$msec. For further details of the model and the parameters used, see the paper of Shriki et al. (2003).

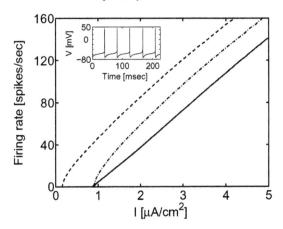

Fig. 2.5. f-I curve for Shriki's model. (Taken from Shriki et al., figure 1.)

## 3. From spikes to rates II: dynamics and conductances

### 3.1. Introduction

In Section 2, we derived a set of self consistent equations for the steady-state firing rates in asynchronous networks of the form,

$$r_i^{s.s.} = f_i\left(\sum_{j=1}^{N} J_{ij} r_j^{s.s.} + I_i^0\right), \quad i = 1, \ldots, N \tag{3.1}$$

where, $J_{ij}$ is the amplitude of the synaptic connection between a presynaptic neuron $j$ and a postsynaptic neuron $i$, $f_i(I)$ is the f-I curve of the $i^{\text{th}}$ neuron in isolation, and $I_i^0$ is the external current coming from outside the network. The derivation of these equations relied on spatial averaging; we average over the many spikes that bombard a cell in a highly connected network. We have shown that in the asynchronous state this averaging is equivalent to temporal averaging of the synaptic inputs. This approach is deficient in multiple ways: it is limited to the description of (1) asynchronous states, (2) the effects of time varying inputs, and (3) even in the asynchronous case with step inputs, we can only describe the steady state dynamics itself and not the approach to this state or (4) dynamic perturbations to it. (5) In particular, we cannot even discuss the stability of the asynchronous solution that we find from the above equations because generally the stability depends upon the underlying dynamics. There could be, for instance, a stable synchronous state or multiple asynchronous states.

Unfortunately extending the above approach to the dynamics is difficult because it requires that we know how single neurons respond to a time-dependent current. In general the spiking patterns that time dependent inputs elicit can be rather complex, and are not well captured by rate variables.

## 3.2. Dynamics of firing rates—linear response theory

We can, however, extend the rate theory of asynchronous networks to the regime where dynamic modulation is relatively weak (see Shriki et al., 2003). Specifically, we assume that the external inputs have a small amplitude time dependent component and consider the response of neuronal firing rates to these temporal modulations. We use the familiar tools of linear response theory.

To apply linear response theory to the rate variables we will assume that there is a small amount of external local noise so that the instantaneous rates are smooth variables defined by averaging the instantaneous rate functions (defined as a sum of delta functions, equation (2.17)) over this noise at a fixed time. We write a general expression for the instantaneous firing rate of a neuron as a functional of its synaptic current plus the external current,

$$r_i(t) = F\{I_i^{syn}(t') + I_i^0(t')\} \tag{3.2}$$

where

$$I_i^{syn}(t) = \sum_j J_{ij} s_j(t) \tag{3.3}$$

with

$$s_i(t) = \frac{1}{\tau_S} \sum_{t_j} K(t - t_j) \tag{3.4}$$

the synaptic activation rates. As in Section 2, $K(t)$ is a filter describing the time evolution of a single synaptic current, normalized so that its integral over time defines the time constant $\tau_s$. Note that in general the functional $F$ will depend upon the full time-history of currents $I^{syn}$ and $I^0$.

We now use linear response theory to examine behavior under application of a small amplitude, time varying current. Write $I^0$ explicitly as a sum of a constant component and a small time-varying component,

$$I_i^0(t) = I_i^0 + \delta I_i^0(t). \tag{3.5}$$

Assuming smoothness, we expand the expression given by equation (3.2) in the time dependent amplitudes,

$$r_i(t) = r_i^0 + \int_0^\infty dt' \chi_i(t - t')(\delta I_i^{syn}(t') + \delta I_i^0(t')) + O((\delta I_i^0)^2) \tag{3.6}$$

where $r_i^0$ is the steady state value of the rates (given by the self-consistency equations (3.1)) and

$$\delta I_i^{syn}(t) = \sum_j J_{ij}\delta s_j(t) \tag{3.7}$$

are the time dependent perturbations of the synaptic currents. It is convenient to work in Fourier domain where to first order,

$$r_i(\omega) = \chi_i(\omega)\left[\sum_j J_{ij}s_j(\omega) + I_i^0(\omega)\right]. \tag{3.8}$$

Since by equation (3.4)

$$s_i(\omega) = K_i(\omega)r_i(\omega) \tag{3.9}$$

we can write

$$s_i(\omega) = \tilde{\chi}_i(\omega)\left[\sum_j J_{ij}s_j(\omega) + I_i^0(\omega)\right] \tag{3.10}$$

where

$$\tilde{\chi}_i(\omega) = K_i(\omega)\chi_i(\omega). \tag{3.11}$$

In matrix form, equation (3.10) can be written as

$$\mathbf{s}(\omega) = \tilde{\chi}[\mathbf{J} \cdot \mathbf{s}(\omega) + \mathbf{I}^0(\omega)] \tag{3.12}$$

where $\tilde{\chi}(\omega)$ is the matrix given by

$$\tilde{\chi}_{ij}(\omega) = \tilde{\chi}_i(\omega)\delta_{ij}. \tag{3.13}$$

Equation (3.12) can be solved for $\vec{s}(\omega)$ to give

$$\mathbf{s}(\omega) = [\tilde{\chi}^{-1}(\omega) - \mathbf{J}]^{-1} \cdot \mathbf{I}^0(\omega) \tag{3.14}$$

which is the Fourier transform of the synaptic rate vector to first order in $\delta I^0$.

**Single neuron response**

Equations (3.12) and (3.14) are useful provided we have some knowledge of the single neuron dynamic susceptibility, $\chi_i(\omega)$. As we did in Section 2 for the case of static input, we relate the network rate equations to single neuron rate properties, in this case the time dependent modulations of the single neuron rates.

As an example we examine the results from a study of the rate response of a conductance based neuron, the Shriki model. This model is a simple HH model that produces an almost linear f-I curve similar to that of many cortical cells as discussed in Section 2 and shown in Figure 2.5. The behavior can be summarized in a linear-threshold f-I curve,

$$f(I) = \beta[I - I_C]_+. \tag{3.15}$$

Shriki et al. study numerically the response of the model neuron to a time dependent input current under the conditions: (1) the neuron is 'above threshold', in the linear regime, and (2) there is a moderate amount of noise, smoothing the singular response of a noiseless neuron in the neighborhood of its resonance frequency. Furthermore, they use a simple low pass filter model for the synaptic dynamics given by

$$K(\omega) = (I + i\omega\tau_s)^{-1}. \tag{3.16}$$

Note that the study of Shriki et al. incorporated time dependent synaptic conductances. For the sake of simplicity we ignore this difference for now.

Using numerical simulations of the single neuron model described above in a range of parameters Shriki et al. find a linear filter

$$\chi_i(\omega) = \frac{(1 + ia\omega/\omega_0^i)}{1 + i\omega/(\omega_0^i Q) - \omega^2/(\omega_0^i)^2}, \tag{3.17}$$

which is the complex response function of a damped harmonic oscillator. The parameter $\omega_0^i$ is the resonance frequency of the filter. It increases linearly with the DC firing rate of the neuron:

$$\omega_0^i/2\pi = r_i^0 + \rho(Hz) \tag{3.18}$$

where $\rho$ is a constant with units of firing rate. $Q$ has the meaning of a Q-factor in a damped oscillator: it equals the resonance frequency divided by the half-width half-max frequency. $Q$ is noise dependent and controls broadening due to noise. The parameter $a$ denotes the sensitivity of the rates to the *rate of change* of the input current. We can see this by noting that the factor $i\omega$ arises from a derivative in the time domain and so firing rate is sensitive to the instantaneous current's time

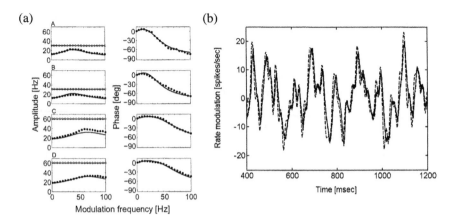

Fig. 3.1. Dynamic linear response in an all-to-all excitatory network. (Taken from Shriki et al., figures 9 and 11.)

derivative. For the conductance-based network of single-compartment neurons in their simulations, Shriki et al. find $\rho = 13.1$Hz, $Q \approx 0.85$ and $a \approx 1.9$.

Sample results from simulations and analytic evaluation of the linear response of an all-to-all excitatory network to an oscillatory input are shown in Figure 3.1. Figure 3.1a shows mean amplitude and phase of the single neuron response as a function of frequency. The left panels show mean output rate (hollow circles) and response amplitude (filled circles). The right panels show the phase of the response. The solid curves are analytic predictions as discussed above. The external input was designed to produce a mean output rate of $\sim$30 spikes/sec in A and B and $\sim$60 spikes/sec in C and D. A and C show the response to inputs with a small noise coefficient of variation, $\Delta = 0.18$ while B and D show responses for high noise, $\Delta = 0.3$.

Figure 3.1b shows a single neuron response to a broadband signal. The solid curve shows the rate of the conductance based neuron and the dotted curve shows the prediction of the rate model. The input consisted of a superposition of 10 cosine functions with random frequencies, amplitudes and phases. Only fluctuations around the mean firing rate ($\sim$30 spikes/sec) are shown.

## Rate equations at low frequencies

To gain some insight into the salient dynamic properties of the rates, consider perturbations limited to low frequency. We expand the frequency dependent terms in equation (3.17) and retain only linear terms in frequency. The resulting expression can be written in the time domain as

$$\tau_i \frac{d\delta s_i}{dt} = -\delta s_i + \sum_j J_{ij} \delta s_j(t) + \delta I_i^0(t) \tag{3.19}$$

where $\tau_i$ is an effective time constant of the neuron given by

$$\tau_i = \tau_s^i - \tau_f^i. \tag{3.20}$$

Here $\tau_s^i$ is the synaptic time constant and $\tau_f^i$ is the time scale set by the cell's resonance

$$\tau_f^i = \frac{a - 1/Q}{\omega_0^i}. \tag{3.21}$$

With our parameters $\tau_f^i$ is positive so firing reduces the cell's effective time constant.

*Non-linearity*
It is important to note that although the above calculations are performed within linear response theory, the final result is nonlinear. This is because the resonance frequencies depend strongly on the mean firing rates, that is upon the DC firing rates of the cells.

### 3.3. *Synaptic inputs as conductances*

Until now we have assumed that the inputs can be described as temporal sums of synaptic current pulses. In real neurons, however, synaptic inputs are conductance pulses: they are transient increases in the conductance of the post synaptic cell following the arrival of an action potential at the presynaptic terminal. We will assume again a temporal linear summation of spikes with low pass filtering.

The current induced in neuron $i$ by an action potential in the presynaptic neuron $j$ is given by

$$I_i^{syn}(t) = g_j^{syn}(t)(V_s^j - V_i(t)) \tag{3.22}$$

where $V_s^j$ is the reversal potential of the synapse – it is a constant which characterizes the synapse. $V_i(t)$ is the membrane potential at time $t$ of the postsynaptic neuron. There are two major classes of synapses, differing in their reversal potential

$$\begin{aligned}
\text{Excitatory Synapse}: \quad & V_E > \theta \\
\text{Inhibitory Synapse}: \quad & V_I > \theta
\end{aligned} \tag{3.23}$$

where $\theta$ is the threshold membrane potential of the postsynaptic neuron. The total synaptic current is given by

$$I^{syn}(t) = \sum_j I_j^{syn}(t) \tag{3.24}$$

where there is a term in the sum from the contribution of each individual synapse.

## Synaptic conductances in asynchronous states

Consider an all-to-all network with only excitatory synapses in an asynchronous state:

$$C_m \frac{dV_i}{dt} = -g_L(V_i(t) - E_L) + I_i^{spike} + I_i^{app} - g_s^E(t)(V_i(t) - E_L) \tag{3.25}$$

Here we assume that all excitatory synapses have the same reversal potential $V_E$. The total synaptic conductance is given by the sum of individual synaptic conductances

$$g_s^E(t) = \sum_{j=1}^{N_E} g_{ij}^E(t). \tag{3.26}$$

In the asynchronous state, neither the membrane potential nor the total synaptic current impinging upon a cell is constant. However, in the limit of a very large all-to-all network, the *total* synaptic conductance is constant,

$$g_s^E(t) \approx \langle g_s^E(t) \rangle_t = \sum_{j=1}^{N_E} g_{ij}^0 \tau_{ij}^E r_j^E \tag{3.27}$$

where we define

$$\int_0^\infty g_{ij}^E(t) dt = g_{ij}^0 \tau_{ij}. \tag{3.28}$$

We can take the average in equation (3.27), as we did with the synaptic rate in Lecture 1, because we assume that the system is in an asynchronous state. The total current impinging on the neuron in the asynchronous state (apart from active currents) is:

$$\begin{aligned} I^{tot} &= -g_L(V - E_L) - g_s^E(V - V_E) + I^0 \\ &= -\hat{g}_L(V - E_L) - \hat{I}^0 \end{aligned} \tag{3.29}$$

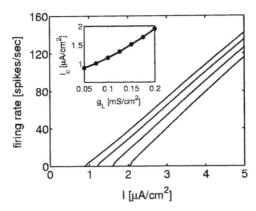

Fig. 3.2. f-I curve dependence on leak conductance. (Taken from Shriki et al., figure 2.)

where

$$\hat{I}^0 = I^0 + g_s^E(V_E - E_L)$$
$$\hat{g}_L = g_L + g_s^E. \qquad\qquad (3.30)$$

So in summary, asynchronous synaptic inputs generate: (1) a stationary additive contribution to the applied current and (2) an increase in the passive conductance of the postsynaptic cell. We need to know how a cell's firing is affected by changes in its applied current and the passive conductance.

In many cortical cells, increasing the leak conductance does not change the form of the f-I curve substantially but does increase its threshold current, essentially shifting the entire f-I curve to higher currents. This behavior is captured in the Shriki et al. neuron model as shown in Figure 3.2. In addition, the change in the threshold current is approximately linear with the change in the leak conductance over a broad range, as shown in the inset.

This behavior can be summarized in the following formula,

$$f(I, g_L) = \beta[I - I_c(g_L)]_+$$
$$= \beta[I - (I_c^0 + V_c g_L)]_+ \qquad\qquad (3.31)$$

where $V_c$ is the threshold gain potential. It measures the rate of increase in the current threshold with the increase of the leak conductance. Therefore the firing rate of a neuron in this network is

$$r = \beta[\hat{I}^0 - (I_c^0 + V_c \hat{g}_L)]_+ \qquad\qquad (3.32)$$

where $\hat{I}^0$ and $\hat{g}_L$ are given by equation (3.30). Substituting these expressions into equation (3.32) and using equation (3.27) yields

$$r_i = \beta \left[ I + \sum_{j=1}^{N_E} J_{ij}^E r_j^E - \theta \right]_+ \tag{3.33}$$

where

$$J_{ij}^E = g_{ij}^o \tau_{ij}^E (V_E - E_L - V_c) \tag{3.34}$$

and $\theta = I_c^0 + V_c g_L$ is the threshold of the original f-I curve.

**Example: Fully connected excitatory networks.**

Shriki et al. applied this theory to the simple case of a fully connected excitatory network of neurons. The network consists of point neurons of the type described in Section 2 connected via excitatory synaptic conductances. The synaptic conductance dynamics is a simple low pass filter of the presynaptic firing activity (equation (3.16)) with synaptic time constant $\tau_s = 5 msec$. The results are depicted in Figure 3.3, which shows the firing rate versus excitatory synaptic strength in a large network of excitatory neurons with all-to-all connectivity. The dashed line shows the analytical result from equations (3.33) and (3.34) for a biologically relevant choice of parameter values. The solid line shows the analytical result using a modified f-I curve with a quadratic correction. The circles are the results from simulations on the conductance based model with a network of size $N = 1000$. Note that it is reasonable that at high firing rates the above approximation breaks down and non-linearities must be taken into account. The inset shows analytic results for firing rate versus external input for strong excitatory feedback. Note that there is a range of inputs for which the system is bi-stable; in this range it can be either quiescent or in a stable sustained active state.

### 3.4. Rate dynamics for slow synapses

We consider again the example of synaptic dynamics described by a simple low pass filtering that we discussed in Section 2,

$$\tau_s \frac{ds_i}{dt} = -s_i + r_i(t). \tag{3.35}$$

To close the equations, however, we have to calculate the neuronal firing rates. For slow synapses, we make the crucial assumption that the synaptic time constants are much slower than any other time constant in the problem (see, for example, Ermentrout, 1994). That is, on the rapid time scales of spiking (milliseconds or tens of milliseconds) we take the synaptic currents as roughly constant

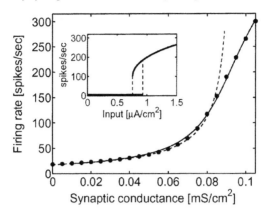

Fig. 3.3. Firing rate vs. synaptic strength in an all-to-all excitatory network. (Taken from Shriki et al., figure 3.)

in time and hence at any given time the neurons will fire as if the instantaneous synaptic currents are constant, yielding

$$r_i(t) = f\left( I^0 + \sum_j J_{ij} s_j(t) \right). \tag{3.36}$$

Combining equations (3.35) and (3.36) yields first order dynamics similar to the familiar rate equations. Note that this approximation is limited to slow synapse dynamics but *not* to asynchronous networks.

### 3.5. Summary

**Rate equations via spatial or temporal summation**

In the two previous sections we discuss two averaging scenarios for reducing the network dynamics to relatively simple rate equations. The first, which we have focused on, is spatial averaging. The other is temporal averaging, mentioned in the Section 3.4. Both have their weaknesses and advantages, which we summarize below.

*Spatial summation*
- Limited to large highly connected networks.
- Limited to asynchronous or nearly asynchronous states.
- Steady state equations are rate fixed point equations.

- The time dependent rate equations are more complicated, even in the supra-threshold linear response regime.
- Dynamics of rates reflects the combined effect of spiking and synaptic dynamics.

*Temporal summation*
- Limited to slow synapses and to the dynamics on slow time scales.
- Time constant is simply the synaptic time constant.
- Arbitrary size networks.
- Network can be synchronized.

**Biophysical interpretation of rate equations**

We began our discussion of rate models (Section 2.1) by presenting the familiar equations of the form,

$$\tau_0 \frac{ds_i}{dt} = -s_i + f_i \left( \sum_{j=1}^{N} J_{ij} s_j + I_i^0 \right), \quad i = 1, \dots, N \tag{3.37}$$

We asked about the biophysical meaning of the various terms in these equations. Using the theory of highly connected asynchronous networks we can identify $f_i$ as the f-I curve of the $i^{\text{th}}$ neuron. The relation between the connection parameters $J_{ij}$ and the synaptic conductances is the following

$$J_{ij} = g_{ij}^0 \tau_{ij} (V_s^j - E_L - V_C) \tag{3.38}$$

where $g_{ij}^0$ is the peak synaptic conductance, $\tau_{ij}$ is the synapse time constant (see equation (2.15)), $V_s^j - E_L$ is the synaptic reversal potential relative to the leak potential and $V_C$ is the current threshold potential defined by the change of the current threshold with the change in the leak conductance.

Finally, we saw that this mapping is valid for the fixed point solution of the rate equations. The real rate dynamics is more complicated in general. Even if one considers slow dynamics, neglecting higher order time derivatives, the effective time constant $\tau_0$ is neither a membrane time constant nor a synaptic time constant but a combination of the synaptic time constant and contributions from the single neuron spiking dynamics (see equations (3.20) and (3.21)).

## 4. Persistent activity and neural integration in the brain

### 4.1. Introduction

In this section we discuss recent modeling efforts to explain the origin of persistent activity and temporal integration in neuronal circuits. Persistent activity is a term denoting neuronal firing that persists for seconds or more, long after the sensory stimulus that triggered the activity has ceased (for a review, see Wang, 2001). Persistent activity is selective to the stimulus variables, hence it is viewed as a neural correlate of short-term or working memory. Models of long term memory will be the topic of Section 6. As we discuss there, activation of memorized patterns in the retrieval of long term memory can also lead to persistent activity. This is the case if the memorized patterns are attractors of the network dynamics. However, patterns embedded in long term memory are usually thought of as discrete states. A distinguishing feature of persistent activity in working memory tasks is that the tasks often form a continuum of states where the persistent rates assume graded activity levels. From a dynamical perspective, persistent graded activity is far more challenging to explain than discrete persistent states. Whereas existence of discrete multiple attractors is a generic feature of nonlinear dynamical systems, continuous manifolds of attractors are structurally unstable. In general, their existence requires a substantial degree of fine tuning of system parameters.

Some systems that display graded persistent activity also exhibit the property of neural integrators; in the presence of an appropriate input their firing rate varies in proportion to the integral of this input. Thus their instantaneous firing rate can be thought of as representing the computation of position variables by integration of velocity signals. In the absence of inputs the firing rate persists at a steady value which codes the memorized value of the position variable. Persistence and integration are essential in spatial navigation and motor control.

A well-studied example is the oculomotor integrator in the goldfish (Seung, 1996; Seung et al., 2000; Aksay et al., 2001). Neurons in the integrator nuclei receive inputs about head velocity from the vestibular system and output the desired eye position. In the absence of movement, the integrator neurons exhibit persistent firing rates which vary linearly with the stable eye position, as shown in Figure 4.1. In the dark, the animal maintains stable gaze for intervals of several seconds, followed by rapid eye movements. Recordings of eye positions and firing activity are shown in Figure 4.2. The firing rates of these cells can be viewed as integrals over the bursts that occur in on and off burst cells which trigger the rapid eye movements.

What mechanism might underlie this behavior? As already indicated, it must provide both for a continuous manifold of stable states (representing *e.g.* the eye

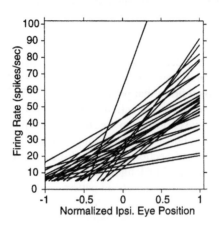

Fig. 4.1. Rate vs. eye position in a goldfish integrator nuclei. (Adapted from Aksay et al., figure 5.)

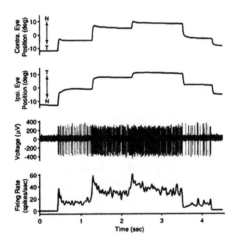

Fig. 4.2. Eye position and firing activity in a goldfish integrator nuclei. (Adapted from Aksay et al., figure 1.)

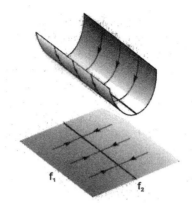

Fig. 4.3. Dynamic trajectories in a line attractor.

position or head direction) and for the possibility of an input signal which directs the system from one state to another on that manifold, such that the change of position represents an integral over velocity. A schematic picture of such a one-dimensional attractor manifold, called a line attractor, is shown in Figure 4.3.

Persistence can arise from a long time constant associated with the single neuron dynamics. Over times smaller than this time constant, the system would be effectively stable in one dynamical direction. However, the membrane potential of the goldfish integrator cells responds to injected current at the soma with normal time constants, and does not exhibit bistability at least under these conditions. An alternative mechanism is recurrent network dynamics which has been studied by Robinson and collaborators (Cannon et al. 1983; Arnold & Robinson, 1997), Seung and colleagues (Seung, 1996; Seung et al., 2000) and others (*e.g.*, Koulakov et al., 2002). The simplest realization of the network mechanism for a line attractor is in networks in which the rate equations are close to linear. We begin by reviewing some basic properties of the dynamics of linear networks. We then describe the recurrent network model for line attractors and integration. Finally we discuss a recently developed alternative model which is based on single neuron properties (Lowenstein & Sompolinsky, 2003).

## 4.2. Dynamics of linear recurrent networks

### Fixed points and their stability

*Existence of fixed points*
In the rate model,

$$\tau_0 \frac{dr_i}{dt} = -r_i + W_{ij}r_j + h_i; \quad r_i(t=0) \equiv r_i^0 \tag{4.1}$$

where $h_i$ is the external input which, for the moment, we assume time-independent and $r_i^0$ is the initial condition. Furthermore note that in writing this linear equation, we assume that we are in a regime in which most neurons are above threshold. In vector notation, this equation can be written as

$$\tau_0 \frac{d\mathbf{r}}{dt} = -(\mathbf{I} - \mathbf{W}) \cdot \mathbf{r} + \mathbf{h} \tag{4.2}$$

where $\mathbf{r}$ and $\mathbf{h}$ are vectors, $\mathbf{W}$ is the matrix of connections and $\mathbf{I}$ the identity matrix. We first look for time-independent solutions, which satisfy the equation

$$0 = -\mathbf{r} + \mathbf{W} \cdot \mathbf{r} + \mathbf{h} \tag{4.3}$$

yielding the fixed point equation,

$$\mathbf{r}^* = (\mathbf{I} - \mathbf{W})^{-1} \cdot \mathbf{h}. \tag{4.4}$$

If $\mathbf{I} - \mathbf{W}$ is a regular matrix (*i.e.* has no zero eigenvalues), a unique solution always exists.

*Stability of fixed points*
A fixed point is stable provided that $Re\,\lambda_l < 1$ for $l = 1, \ldots, N$ where $\{\lambda_l\}$ are the eigenvalues of matrix $\mathbf{W}$. To see why this is so, first write $\mathbf{r}(t) = \mathbf{r}^* + \boldsymbol{\delta}\mathbf{r}(t)$ and substitute this expression into equation (4.1), yielding

$$\tau_0 \frac{d\boldsymbol{\delta}\mathbf{r}(t)}{dt} = -[\mathbf{I} - \mathbf{W}]\delta\mathbf{r}^l(t); \qquad \boldsymbol{\delta}\mathbf{r}(t = 0) = \boldsymbol{\delta}\mathbf{r}^0. \tag{4.5}$$

Denote the eigenvalues and eigenvectors of the matrix $\mathbf{W}$ as $\{\lambda_l\}$ and $\{\mathbf{u}_l\}$ respectively. Decompose the vector $\boldsymbol{\delta}\mathbf{r}(t)$ in terms of the eigenvectors $\mathbf{u}_l$

$$\boldsymbol{\delta}\mathbf{r}(t) = \sum_{l=1}^{N} \delta r^l(t)\mathbf{u}_l. \tag{4.6}$$

Substituting in equation (4.5), and collecting terms proportional to each of the eigenvectors yields

$$\sum_{l=1}^{N} \left[ \tau_0 \frac{d\delta r^l}{dt} + (1 - \lambda_l)\delta r^l \right] \vec{u}_l = 0. \tag{4.7}$$

Since the eigenvectors $\mathbf{u}_l$ are linearly independent this sum is zero if and only if each of the coefficients vanishes, hence,

$$\delta r^l(t) = \delta r^l(0) \exp(-t/\tau_l) \tag{4.8}$$

where

$$\tau_l = \frac{\tau_0}{1 - \lambda_l}. \tag{4.9}$$

Therefore, we can conclude that a fixed point is stable provided that $Re\,\lambda_l < 1$ for $l = 1, \ldots, N$.

### Energy function

We will define a scalar function of the system's state $E(r)$ which is minimized by the dynamics. More specifically we would like the dynamics to follow a path that maximizes the rate of decrease of $E$, given by

$$\frac{dE}{dt} = \sum_{i=1}^{N} \frac{\partial E}{\partial r_i} \frac{dr_i}{dt} = \nabla E^T \frac{d\vec{r}}{dt}. \tag{4.10}$$

In order to guarantee that $E$ is always non-increasing, we need to ensure that

$$\nabla E^T \frac{d\mathbf{r}}{dt} \leq 0 \tag{4.11}$$

that is, $d\mathbf{r}/dt$ must have a negative projection on the gradient of $E$. A stringent condition, that of *gradient descent*, is that $d\mathbf{r}/dt$ is proportional to the negative gradient of $E$,

$$\frac{d\mathbf{r}}{dt} = -\eta \nabla E. \tag{4.12}$$

Using the gradient descent condition and equation (4.2), we require

$$[\mathbf{I} - \mathbf{W}] \cdot \mathbf{r} - \mathbf{h} = \eta \nabla E. \tag{4.13}$$

Differentiating both sides with respect to $\mathbf{r}$ gives

$$[I - W]_{ij} = \eta \frac{\partial^2 E}{\partial r_i \partial r_j}. \tag{4.14}$$

Since the right hand-side of equation (4.14) is a symmetric matrix the left hand side must be as well. Thus in order for an energy function satisfying gradient descent to exist,

$$\mathbf{W} = \mathbf{W}^T. \tag{4.15}$$

Equation (4.15) is also a sufficient condition. To see this, note that if $\mathbf{W}$ is symmetric then we can construct the following energy function,

$$
\begin{aligned}
E(r) &= \frac{1}{2} \sum_{i,j=1}^{N} r_i [\delta_{ij} - W_{ij}] r_j - \sum_{i=1}^{N} h_i r_i \\
&= \frac{1}{2} \mathbf{r}^{\mathsf{T}} \mathbf{r} - \mathbf{r}^{\mathsf{T}} \mathbf{W} \mathbf{r} - \mathbf{h}^{\mathsf{T}} \mathbf{r}.
\end{aligned}
\tag{4.16}
$$

By direct differentiation, it is easy to see that

$$
\tau_0 \frac{d\mathbf{r}}{dt} = -\nabla E.
\tag{4.17}
$$

**General solution**

Now consider the more general case, in which the input $h = h(t)$ is a function of time.

Decompose $\mathbf{r}(t)$ and $\mathbf{h}(t)$ as

$$
\mathbf{r}(t) = \sum_{l=1}^{N} r^l(t) \mathbf{u}^l; \qquad \mathbf{h}(t) = \sum_{l=1}^{N} h^l(t) \mathbf{u}^l
\tag{4.18}
$$

Proceeding as above, using the linear independence of the eigenvectors $\mathbf{u}_l$ yields $N$ decoupled scalar differential equations,

$$
\tau_0 \frac{dr^l}{dt} = -r^l + \lambda_l r^l + h^l(t).
\tag{4.19}
$$

The general solution of this equation is of the form

$$
r^l(t) = r_l^0 \exp(-t/\tau_l) + \frac{1}{\tau_0} \int_0^t dt' \exp(-(t-t')/\tau_l) h^l(t').
\tag{4.20}
$$

In the case that the input $h$ is time-independent,

$$
r^l(t) = r_0^l \exp(-t/\tau_l) + G_l h^l [1 - \exp(-t/\tau_l)]
\tag{4.21}
$$

with

$$
\tau_l = \frac{\tau_0}{1 - \lambda_l}, \qquad G_l = \frac{1}{1 - \lambda_l}
\tag{4.22}
$$

which we can see is the same result found above by remarking that equation (4.4) implies that

$$
r^{*l} = G_l h^l.
\tag{4.23}
$$

Note that the spectrum of eigenvalues $\{\lambda_l\}$ yields a corresponding spectrum of relaxation times $\{\tau_l\}$ corresponding to each of the spatial modes. Furthermore, in general $\mathbf{W}$ may have complex eigenvalues, which always come in complex conjugate pairs. When this is the case, it is the real part of the eigenvalues that determines the speed of the corresponding relaxation.

## Noise filtering

### Derivation in the frequency domain

We apply the general theory to the case in which the external input is white noise in time. For this as for other applications it is easiest to work in the Fourier representation,

$$
\begin{aligned}
\mathbf{r}(\omega) &= \int_{-\infty}^{+\infty} dt\, \mathbf{r}(t) \exp(i\omega t); \quad -\infty < \omega < \infty \\
\mathbf{r}(t) &= \frac{1}{2\pi} \int_{-\infty}^{+\infty} dt\, \mathbf{r}(t) \exp(-i\omega t); \quad \mathbf{r}^*(\omega) = \mathbf{r}(-\omega)
\end{aligned}
\tag{4.24}
$$

and similarly for the input $\mathbf{h}(t)$. Recalling that $\frac{d\mathbf{r}}{dt} \to -i\omega \mathbf{r}(\omega)$ in the Fourier domain,

$$
[(-i\omega\tau_0 + 1)\mathbf{I} - \mathbf{W}] \cdot \mathbf{r}(\omega) = \mathbf{h}(\omega)
\tag{4.25}
$$

and thus,

$$
\mathbf{r}(\omega) = [(-i\omega\tau_0 + 1)\mathbf{I} - \mathbf{W}]^{-1}\mathbf{h}(\omega).
\tag{4.26}
$$

We are interested in the steady state solution so we ignore the initial condition. Diagonalizing $\mathbf{W}$ yields $N$ equations for the $N$ spatial modes of the system,

$$
r^l(\omega) = \frac{h^l(\omega)}{-i\omega\tau_0 + 1 - \lambda_l}.
\tag{4.27}
$$

In the case in which the input is white noise in time,

$$
\begin{aligned}
\langle h^l(t)h^{l'}(t')\rangle &= C_{ll'}\tau_0\delta(t - t') \\
\langle h^l(\omega)h^{l'}(\omega')\rangle &= C_{ll'}2\pi\tau_0\delta(\omega + \omega').
\end{aligned}
\tag{4.28}
$$

Therefore,

$$
\langle r^l(\omega)r^{l'}(\omega')\rangle = \frac{C_{ll'}}{(-i\omega\tau_0 + 1 - \lambda_l)(i\omega\tau_0 + 1 - \lambda_{l'})}2\pi\tau_0\delta(\omega + \omega').
\tag{4.29}
$$

We can transform back to the time domain to obtain

$$
\langle r^l(t)r^l(t + \Delta)\rangle = \frac{C_{ll}}{2(1 - \lambda_l)}\exp(-|\Delta|/\tau_0).
\tag{4.30}
$$

It is also straightforward to obtain time correlations in the case that $l \neq l'$, making sure to close the contour of integration so that the exponential term decays to zero.

*Special case: white noise in space* and *time*
In this case, the correlation of inputs is given by

$$\langle h_i(t)h_j(t') \rangle = 2T\delta_{ij}\tau\delta(t - t') \tag{4.31}$$

and the matrix $\mathbf{W}$ is symmetric, so that the eigenvectors are orthogonal. Hence the projections onto them is also white: $C_{ll'} = 2T\delta_{ll'}$ and for equal time correlations we can consider $\mathbf{r}$ as Gaussian with zero mean and covariance

$$\langle \mathbf{r}(t)\mathbf{r}^{\mathbf{T}}(t) \rangle = T[\mathbf{I} - \mathbf{W}]^{-1}. \tag{4.32}$$

*Derivation in the time domain*
Consider the steady state solution in the presence of white noise and recall that we are not interested in the decay of the initial condition. For completeness, we also consider an additional constant input vector $\mathbf{h}_0$:

$$r^l(t) = \frac{1}{\tau_0} \int_{-\infty}^{t} dt' \exp(-(t - t')/\tau_l)h^l(t') + \frac{h_o^l}{1 - \lambda_l} \tag{4.33}$$

Therefore $\mathbf{r}$ itself is Gaussian with mean $[\mathbf{I} - \mathbf{W}]^{-1} \cdot \mathbf{h}_0$ and variance derived from

$$\langle [\delta r^l(t)]^2 \rangle = \frac{1}{\tau_0^2} \int_{-\infty}^{t} dt' \int_{-\infty}^{t} dt'' \exp\left[ -\frac{2t - t' - t''}{\tau_l} \right] \langle h^l(t')h^l(t'') \rangle. \tag{4.34}$$

Substituting the white noise correlation function one obtains

$$\langle [\delta r^l(t)]^2 \rangle = \frac{2T\tau_0}{\tau_0^2} \int_{-\infty}^{t} dt' \exp[-2(t - t')/\tau_l] = \frac{T\tau_l}{\tau_0} = \frac{T}{1 - \lambda_l}. \tag{4.35}$$

Substituting back into the vector $\mathbf{r}$ yields

$$\langle \delta \mathbf{r}(t)\delta \mathbf{r}^{\mathbf{T}}(t) \rangle = T[\mathbf{I} - \mathbf{W}]^{-1}. \tag{4.36}$$

## Relation to statistical mechanics

*Gibbs distribution*

Under white-noise inputs, the steady state probability distribution of rates is given by

$$P(\mathbf{r}) \propto \exp\left[-\frac{1}{2T}(\mathbf{r}^{\mathbf{T}} - \mathbf{h}_0^{\mathbf{T}} \cdot [\mathbf{I} - \mathbf{W}]^{-1})[\mathbf{I} - \mathbf{W}](\mathbf{r} - [\mathbf{I} - \mathbf{W}]^{-1} \cdot \mathbf{h}_0)\right].$$

(4.37)

The quadratic term can be written up to a constant as

$$\frac{1}{2}\mathbf{r}^{\mathbf{T}}\mathbf{r} - \frac{1}{2}\mathbf{r}^{\mathbf{T}}\mathbf{W}\mathbf{r} - \mathbf{h}_0^{\mathbf{T}}\mathbf{r}$$

(4.38)

which is exactly energy function discussed above. Thus we can rewrite our finding as a Gibbs distribution,

$$P(\mathbf{r}) = \frac{1}{Z} \exp[-\beta E(\mathbf{r})]$$

(4.39)

where

$$Z = \int d^N \mathbf{r} \exp[-\beta E(\mathbf{r})].$$

(4.40)

Note that equation (4.40) has the same form as a partition function in statistical mechanics with $\beta = 1/T$ where here $T$ gives the amplitude of the input correlations.

*Free Energy, average energy, and Entropy*

The partition function defined by equation (4.40) allows us to formally define other quantities familiar from statistical mechanics. The free-energy $F$ is defined through

$$Z \equiv \exp(-\beta F).$$

(4.41)

The average energy can then be defined as

$$\langle E \rangle \equiv -\beta \frac{dF}{d\beta} = \frac{1}{Z} \int d^N \mathbf{r} E(\mathbf{r}) \exp[-\beta E(\mathbf{r})].$$

(4.42)

And the entropy can be defined to satisfy $F = \langle E \rangle - TS$ so that

$$S = -\int d^N \mathbf{r} Z^{-1} \exp[-\beta E(\mathbf{r})] \log[Z^{-1} \exp[-\beta E(\mathbf{r})]]$$

$$= \log Z + \beta \langle E \rangle.$$

(4.43)

## 4.3. *Recurrent network model of persistence and integration*

We now turn to a linear recurrent network model that exhibits a continuum of stable states in the absence of input and performs temporal integration under application of a time dependent input. For concreteness we focus on the example of the oculomotor integrator. We begin by formalizing the line attractor associated with eye position.

### Neural representation of memorized eye position

The firing rate of the $i^{\text{th}}$ neuron is taken to be

$$x_i = k_i(E - E_i), \quad i = 1, \ldots, N \tag{4.44}$$

where $E$ denotes azimuthal eye position, and $k_i$ and $E_i$ are the $i^{\text{th}}$ neuron's gain and threshold respectively (Figure 4.1). Note that this representation of firing rate only accounts for the suprathreshold regime. In the state space defined by the neuronal firing rates $x_1, \ldots, x_N$ each $E$ uniquely specifies a point which lies along a line in $\mathbf{R}^N$. The equation of this line is

$$\mathbf{x} = E\mathbf{k} + \mathbf{x}^0 \tag{4.45}$$

where $x_i^0 = k_i E_i$. Since each of the states along this line is stable (while external input is constant), this line is composed of (marginally) stable fixed points.

Such a line attractor is generated by the following dynamical mechanism,

$$\tau_0 \frac{d\mathbf{r}}{dt} = -\mathbf{r} + \mathbf{W} \cdot \mathbf{r} + \mathbf{h}; \qquad \mathbf{r}(t = 0) \equiv \mathbf{r}^0. \tag{4.46}$$

Assume for simplicity that $\mathbf{W}$ is symmetric. In order to generate a continuum of fixed points we assume that the dynamics along exactly one of the eigenvectors, say $\mathbf{u}_1$, is marginal

$$\mathbf{W} \cdot \mathbf{u}_1 = \mathbf{u}_1. \tag{4.47}$$

We decompose vectors $\mathbf{r}$ and $\mathbf{h}$,

$$\begin{aligned} \mathbf{r} &= r^1\mathbf{u}_1 + \mathbf{r}^+ \\ \mathbf{h} &= r^1\mathbf{u}_1 + \mathbf{h}^+ \end{aligned} \tag{4.48}$$

where $\mathbf{r}^+$ and $\mathbf{h}^+$ are the projections of vectors $\mathbf{r}$ and $\mathbf{h}$ respectively onto the subspace perpendicular to $\mathbf{u}_1$. Therefore, decomposing the steady state equation (of the form of equation (4.3)) yields

$$\begin{aligned} 0 &= h^1 = \mathbf{h} \cdot \mathbf{u}_1 \\ \mathbf{r}^+ &= [\mathbf{I} - \mathbf{W}]^{-1} \cdot \mathbf{h}^+. \end{aligned} \tag{4.49}$$

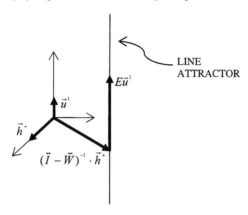

Fig. 4.4. Continuum of solutions forming a line attractor.

Thus there is no static fixed point solution if $h^1$ is non zero. If $h^1$ is zero then there is a continuum of solutions parameterized by $r_0^1$,

$$\mathbf{r}^* = [\mathbf{I} - \mathbf{W}]^{-1} \cdot \mathbf{h}^+ + r_0^1 \mathbf{u}_1 \tag{4.50}$$

where $r_0^1$ is the initial condition in the direction $\mathbf{u}_1$. This is illustrated schematically in Figure 4.4. Note that $\mathbf{h}^+$ is perpendicular to $\mathbf{u}_1$ and that the transformation $[\mathbf{I} - \mathbf{W}]^{-1}\mathbf{h}^+$ also lies in the plane perpendicular to $\mathbf{u}_1$. Eye position $E$ parameterizes position along the line attractor. In the case of no input, *i.e.* $h^1 = 0$, eye position is just given by the initial condition in the $\mathbf{u}_1$ direction, $r_0^1$.

**Integration of velocity signals**

Suppose that at time $t = 0$, eye position is given by $E = r_1^1$. Furthermore, suppose that $h^1$ is turned on at $t = 0$ and consider the time dependent equations,

$$\begin{aligned} \tau_0 \frac{dr^1}{dt} &= -(1 - \lambda_1)r^1 + h^1(t) = h^1(t), \\ \tau_0 \frac{dr^+}{dt} &= 0 \end{aligned} \tag{4.51}$$

with initial conditions

$$r^1(0) = E(0); \qquad \mathbf{r}^+(0) = [\mathbf{I} - \mathbf{W}]^{-1} \cdot \mathbf{h}^+. \tag{4.52}$$

Therefore all time evolution occurs along the (marginal) direction $\mathbf{u}_1$,

$$\mathbf{r}(t) = r^1(t)\mathbf{u}_1 + \mathbf{r}^+(0) \tag{4.53}$$

with

$$r^1(t) = E(0) + \frac{1}{\tau_0} \int_0^t dt' h^1(t') = E(t) \tag{4.54}$$

where $E(t)$ is eye position at time $t$. So according to this model, the projection of the rate vector onto the direction of the marginal eigenvector gives the readout of eye position at time $t$,

$$E(t) = \mathbf{u}_1 \cdot \mathbf{r}(t). \tag{4.55}$$

## Discussion

While the network model of a line attractor is simple mathematically, it is not clear how it might be implemented by the brain.

*Sensitivity*
This system requires a great deal of fine tuning since it requires that $\lambda_1$ be very near one. In the case that

$$\lambda_1 < 1 \tag{4.56}$$

the system is a leaky integrator, which will move to a stable fixed point, with a finite memory of initial position of length of order the time scale $\tau_1$. On the other hand, if

$$\lambda_1 > 1 \tag{4.57}$$

then the system is unstable and eye position will always move to some physically limiting value.

Furthermore, note that the existence of the line attractor depends crucially upon the linearity of the above system. Non-linear terms will break the line attractor into discrete states. Only if these states are dense, would the effect be similar to that of the line attractor itself.

*Learning the line attractor*
Do there exist learning mechanisms for the matrix of connections $W$ that will yield a system with the needed properties? For instance, might some Hebbian mechanism work or is some other learning process needed?

## 4.4. *Single neuron model for integration and persistence*

The network hypothesis for persistence and integration in the goldfish oculomotor integrator is not supported by recent experimental observations. First, so far recurrent collateral axons have not been found in these nuclei. Secondly, lesions of a substantial part of the nucleus have only a mild effect on the integrator performance. For these and other reasons, it is important to study alternative mechanisms for these functions. Recently Loewenstein and Sompolinsky (2003) have proposed a single neuron model for persistence and integration.

The model assumes that the intracellular nonlinear calcium dynamics generate local bistability in individual dendritic compartments of the integrator neuron. Intracellular calcium dynamics generate rich spatio-temporal patterns, including oscillations and propagating waves. Of particular interest to us are calcium wavefronts, which have been observed in several cell types including neurons. These phenomena result from regenerative calcium dynamics that involves the release of calcium from internal stores into the cytoplasm. Indeed, models show that the diffusion of calcium coupled with the local nonlinear calcium dynamics generate a family of stationary front solutions, where the level of calcium changes gradually from one side to the other side of the front, with the location of the front being arbitrary within the bulk of the dendrite. It is further assumed that synaptic inputs onto the dendrite interact with the regenerative calcium dynamics and produce a moving front of calcium with a velocity proportional to integral of the synaptic input. Finally, calcium-dependent currents translate the changing front location into changes in the firing rate of the cell. In what follows, we describe the model in greater detail.

**Generation of calcium fronts**

A simple model of the local nonlinear calcium dynamics is

$$\frac{dc}{dt} = f(c) \tag{4.58}$$

where $c = [\text{Ca}^{2+}]$ and the reaction function $f(c)$ is N-shaped as illustrated schematically in figure 4.5a. In their paper, Loewenstein and Sompolinsky make explicit calculations using a third order polynomial function,

$$f(c) = -K(c - c_1)(c - c_2)(c - c_3) \tag{4.59}$$

where $K$ is a positive constant and $c_1 < c_2 < c_3$. This reaction equation yields two stable states, one at concentration $c_1$ and one at concentration $c_3$. For later use, we define $c_m$ as the midpoint of $c_1$ and $c_3$

$$c_m = \frac{c_1 + c_3}{2}. \tag{4.60}$$

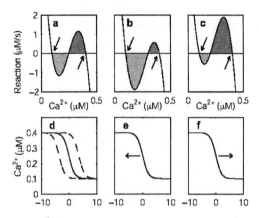

Fig. 4.5. Generation of a [Ca$^{2+}$] front. (Taken from Loewenstein & Sompolinsky, figure 1.)

In a long dendrite, calcium diffuses between compartments, yielding a space-dependent dynamics

$$\frac{\partial c(x,t)}{\partial t} = f(c(x,t)) + D\frac{\partial^2 c}{\partial x^2} \tag{4.61}$$

where $c(x,t)$ denotes the concentration of Ca$^{2+}$ at position $x$ along the dendrite at time $t$ and $D$ is the cytoplasmic calcium diffusion constant. If the calcium concentration at one end of the dendrite, say the left end, is held high, and the concentration at the other end held low, then a front will develop between these two ends. To solve equation (4.61) explicitly using the cubic reaction function defined by equation (4.59), we specify boundary conditions

$$c(-\infty, t) = c_1;$$
$$c(-\infty, t) = c_3 \tag{4.62}$$

where $+1$ denotes the high Ca$^{2+}$ concentration state and $-1$ the low concentration state. In the case that $c_m = c_2$, equation (4.61) has the *time-independent* solution

$$c(x) = \frac{c_1 + c_3}{2} + \frac{c_3 - c_1}{2}\tanh\left(\frac{x - x_0}{\lambda}\right) \tag{4.63}$$

where the width of the front $\lambda$ is given by

$$\lambda = \frac{2\sqrt{2D}}{(c_3 - c_1)\sqrt{K}} \tag{4.64}$$

and front position $x_0$ can be anywhere along the dendrite. In the case of a general N-shaped reaction function $f$, equation (4.61) has a stationary solution whenever the negative area under the curve (between $c_1$ and $c_2$) equals the positive area under the curve (between $c_2$ and $c_3$), as is the case in the schematic curve shown in Figure 4.5a. It is easy to check that this is equivalent to $c_m = c_2$ for the cubic equation. For a general reaction function, the front location in the stationary state will be arbitrary (as shown in Figure 4.5d) and thus this special case of $Ca^{2+}$ dynamics yields a line of attractors parameterized by the front position $x_0$.

The idea behind this single neuron model is to use front position to encode the continuous variable in question, say eye position. Front position plays a role analogous to that which $r^1$ filled in the case of the recurrent networks discussed in Section 4.3. According to Loewenstein and Sompolinsky's model, when there is no external input the front remains stationary at some location $x_0$ along the dendrite. External input drives the front to move in one direction or another, affecting the proportion of the dendrite with a high $Ca^{2+}$ concentration. This in turn dictates the persistent firing rate of the neuron.

To see how this works, we again consider the specific case of the cubic reaction function given by equation (4.59) which is associated with a general, *time-dependent* solution to equation (4.61),

$$c(x, t) = \frac{c_1 + c_3}{2} + \frac{c_1 - c_3}{2} \tanh\left(\frac{x - L(t)}{\lambda}\right) \tag{4.65}$$

where $\lambda$ is given by equation (4.64) and the location of the front center is

$$L(t) = L(0) + vt \tag{4.66}$$

where $v$ is the front velocity

$$v = \sqrt{sDK}(c_2 - (c_1 + c_3)/2). \tag{4.67}$$

Note that when $c_2 = c_m$, this reduces to the stationary solution given by equation (4.63). Similarly for a more general reaction function, when the area under the negative part of the curve exceeds that under the positive part (see Figure 4.5b), the front will propagate left (see Figure 4.5e) and when the area under the positive part of the curve exceeds that under the negative part (see Figure 4.5c), the front will propagate right (see Figure 4.5f).

In their model, Loewenstein and Sompolinsky propose making the effective value of $c_2$ input dependent. They consider the modified dynamics

$$\frac{\partial c(x, t)}{\partial t} = f(c(x, t)) + D\frac{\partial^2 c}{\partial x^2} + g(c)I(t) \tag{4.68}$$

where they take the function $g$ to be

$$g(c) = K\frac{c_3 - c_1}{2}(c - c_1)(c - c_3) \tag{4.69}$$

and $I(t)$ is a time-dependent but spatially homogeneous input in dimensionless units between $-1$ and $+1$. The function $g$ models the $Ca^{2+}$ concentration dependence on the response of receptors and stores to incoming input. Note that $g(c = c_1) = g(c = c_3) = 0$, so that $Ca^{2+}$ concentration is always bounded between $c_1$ and $c_3$. This is consistent with experimental findings that both low and high $Ca^{2+}$ concentration block ionic channels. Furthermore, they assume that the reaction function $f$ is such that negative and positive areas under the curve are equal. When input is zero, this equation reduces to equation (4.61) and has the stationary front solution given by equation (4.63). Non-zero input, however, is equivalent to changing the value of $c_2$. In fact, the solution to equation (4.68) is the same as equation (4.65) with the time-dependent front position now given by

$$L(t) = S\int_0^t dt' I(t') + L(0). \tag{4.70}$$

Therefore, in the absence of input the front is stationary but when the dendrite receives synaptic input, the front moves with a velocity proportional to the instantaneous input and as a result its location is proportional to a time integral of the input. So in the context of the goldfish oculomotor integrator discussed above, front location plays the role that $r^1$ played in the network model, representing eye position and input pulses $I(t)$ play the role of the input $h(t)$ in the above formulation of the network model, representing bursty saccadic commands.

### Readout mechanism

Loewenstein and Sompolinsky postulate a read-out mechanism to translate the position of the $Ca^{2+}$ front into neuronal firing rate. They incorporate calcium-activated cationic channels, open only in regions of high $Ca^{2+}$ concentration. Therefore the total amount of depolarizing current is a function of the location of the front along the dendrite. When the front moves to the right, increasing the total $Ca^{2+}$ concentration, firing rate increases and likewise, when it moves left, firing rate decreases. In particular, *in-vitro* studies of the goldfish oculomotor integrator neurons indicate that their f-I curves are roughly linear over a large range of input currents. Incorporating these f-I curves, Loewenstein and Sompolinsky find that except very close to dendritic tips, neuronal firing rate varies nearly linearly with front location.

The way in which this model cell integrates inputs is shown schematically in Figure 4.6. Here the cell receives three input pulses of the form indicated

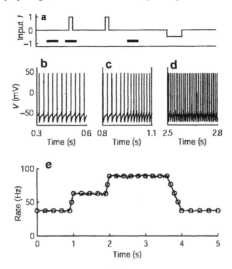

Fig. 4.6. Integration of a time dependent spatially homogeneous stimulus. (Adapted from Loewenstein & Sompolinsky, figures 2 and 4.)

in Figure 4.6a. The resulting membrane potential at the times indicated by the three black bars are shown in Figures 4.6b-d. In the absence of input, the cell fires at a constant rate (Figure 4.6b). At the onset of the first input, the firing rate increases (Figure 4.6c) and persists at its new level even after the input is terminated (Figure 4.6d). Figure 4.6e shows how the firing rate changes with time under the input protocol depicted in 4.6a.

## Extension to real dendrites

In the above discussion, we have considered an idealized individual homogeneous dendrite. Granularity (or inhomogeneities) in the locations of synaptic input sites and intracellular calcium-secreting organelles introduce an input threshold $I_T$ below which the front is stationary, as seen in Figure 4.7. There the dotted line indicates the front velocity in a homogenous dendrite and the solid line shows front velocity found in a simulation of a granular dendrite. While this means that the system will not be sensitive to very small inputs, it also confers robustness to input noise, and to small changes in the reaction function.

The fact that real neurons have many dendrites also increases the robustness of this single neuron model. Heterogeneities in each individual dendrite lead to errors in that dendrite's integration. However, the total calcium concentration is

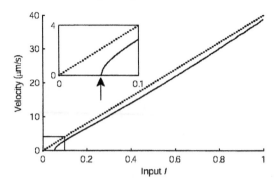

Fig. 4.7. Front velocity as a function of constant external input in a homogeneous dendrite. (Taken from Loewenstein & Sompolinsky, figure 5.)

given by a sum over that in each dendrite, which averages out individual errors and leads to a firing rate of an individual neuron that is relatively stable.

The biological realism of the present model has yet to be determined, however. First, the stationary calcium fronts assumed in this model have yet to be observed in neuronal dendrites. Second, it is not clear whether the model is sufficiently robust to perturbation to withstand biologically realistic levels of inhomogeneities.

## 5. Feature selectivity in recurrent networks—the ring model

### 5.1. Introduction

Neuronal systems show two types of coding scheme for representing sensory or motor variables. The first can be termed monotonic coding, such as the coding of eye position in the oculomotor integrator of the goldfish (see Section 4) or in the coding of contrast in the visual system. Another example is the representation of limb velocity in the motor system. In all these cases neuronal firing rates vary monotonically with the coded variables. The second coding scheme is feature selectivity, in which each neuron has a preferred spatial location at which it fires with highest rate. Examples are the representation of space in place cells, coding of head direction in head direction systems, and orientational selectivity in V1. In each of these cases each neuron responds to a restricted range of stimulus values forming a bell shape tuning curve. In this section we discuss the possible role of recurrent networks in shaping feature selectivity.

The question of the relevant contribution of recurrent dynamics has been discussed extensively in the context of orientation tuning in V1 (Sompolinsky &

Shapley, 1997; Ferster and Miller, 2002). Various mechanisms have been postulated to account for orientation tuning: (1) the Hubel-Wiesel mechanism, (2) feedforward cortical inhibition, (3) recurrent excitation, and (4) recurrent inhibition. Similar questions arise with regard to the emergence of sharp feature selectivity in other brain areas. The ring network is a very useful framework in which to address these questions and in particular to study the possible role of recurrent dynamics in shaping selectivity to an angular variable, such as visual orientation, head direction, or arm movement velocity vector (Ben Yishai et al., 1995; Hansel and Sompolinsky, 1998; Zang, 1996; Lukashin et al., 1996). Here we will focus on the parameter regime in which recurrent excitation plays a decisive role. The regime in which recurrent inhibition dominates is best described by a *balanced ring network* which is outside the scope of these notes. (See, for instance, van Vreeswijk's contribution.)

### 5.2. The ring network

Here we describe the simplest version of the ring network. It consists of N neurons, arranged around a ring and equally spaced from 0 to $2\pi$, so that the neuron $i$ is labeled by an angle $\theta_i = 2\pi i / N$, for $i = 0, \ldots, N - 1$. Note that in the case of orientation the underlying period is $\pi$. Nevertheless, for the sake of generality and notational simplicity, we will use a periodicity of $2\pi$.

In the simplest ring architecture, each neuron is connected to all the others with symmetric synaptic weight $J$ that depends only on the angular distance $\theta$ between them:

$$J_{ij} = N^{-1} J(\theta_i - \theta_j) \tag{5.1}$$

where $J(\theta)$ is a $2\pi$-periodic function of its angular variable. In Fourier components the general form of $J(\theta)$ is

$$J(\theta) = \sum_{n=0}^{\infty} J_n \cos n\theta. \tag{5.2}$$

We examine a minimal dynamics, in which the activities of neurons in the network evolve according to the equation

$$\dot{r}_i = -r_i + \left[ N^{-1} \sum_{j=1}^{N} J(\theta_i - \theta_j) r_j + I^0(\theta_i - \theta_0) \right]_+ \tag{5.3}$$

where $I^0$ is an external input to the network relative to threshold, conveying information about the angular coordinate of the external event. We use the simplest thresholding non-linearity. In general,

$$I^0(\theta) = \sum_{n=0}^{\infty} I_n^0 \cos n\theta. \tag{5.4}$$

**The simplest ring**

The minimal non-trivial model is given by:

$$J(\theta) = J_0 + J_1 \cos \theta \tag{5.5}$$

and

$$I^0(\theta) = I_0^0 + I_1^0 \cos \theta. \tag{5.6}$$

**Order parameters**

Let $a(\theta, t)$ denote the activity profile on the ring in the continuum limit, which is achieved when $N$ diverges, while the angular width of $J$ and $I^0$ remains fixed (as is the case in the simplest ring). In this limit, the dynamics of the activity profile reduces to,

$$\frac{\partial a(\theta, t)}{\partial t} = -a(\theta, t) + [I(\theta, t)]_+ \tag{5.7}$$

where

$$I(\theta, t) = \int_{-\pi}^{\pi} \frac{d\theta'}{2\pi} J(\theta - \theta') a(\theta', t) + I^0(\theta - \theta^0). \tag{5.8}$$

Substituting

$$\cos(\theta - \theta') = \operatorname{Re} \exp(i(\theta - \theta')) \tag{5.9}$$

the integral becomes

$$\int_{\pi}^{\pi} \frac{d\theta'}{2\pi} J(\theta - \theta') a(\theta', t)$$
$$= J_0 \int_{-\pi}^{\pi} \frac{d\theta'}{2\pi} a(\theta', t) + J_1 \operatorname{Re} \exp(i\theta) \int_{-\pi}^{\pi} \frac{d\theta'}{2\pi} \exp(-i\theta') a(\theta', t). \tag{5.10}$$

This motivates definition of the following order parameters

$$a_0(t) = \int_{-\pi}^{\pi} \frac{d\theta'}{2\pi} a(\theta', t)$$

$$a_1(t) \exp(-i\psi(t)) = \int_{-\pi}^{\pi} \frac{d\theta'}{2\pi} \exp(-i\theta') a(\theta', t); \quad a_1(t) \geq 0. \tag{5.11}$$

Note that $a_0(t)$ gives the mean activity at time $t$ while $a_1(t)$ and $\psi(t)$ give the amplitude and phase, respectively, of the first Fourier component of the activity profile.

We can now write an equation for the dynamics of $a(\theta, t)$ in terms of the order parameters. By equation (5.8)

$$I(\theta, t) = I_0(t) + \text{Re}\{I_1(t) \exp(i\theta)\} \tag{5.12}$$

where

$$I_0(t) = J_0 a_0(t) + I_0^0 \tag{5.13}$$

and

$$I_1(t) = J_1 a_1(t) \exp(-i\psi(t)) + I_1^0 \exp(-i\theta^0). \tag{5.14}$$

In summary we have,

$$\frac{\partial a(\theta, t)}{\partial t} = -a(\theta, t) + [J_0 a_0(t) + I_0^0 + J_1 a_1(t) \cos(\theta - \psi(t))$$
$$+ I_1^0 \cos(\theta - \theta^0)]_+. \tag{5.15}$$

The network dynamics is exhausted by the two equations determining the order parameter dynamics. These are derived by Fourier transforming the above equations. Below we consider the stationary limit.

### 5.3. Stationary solution of the ring model

The stationary profile of the above equations is simple in that the stationary profile of the network activity peaks at the peak of the external input, *i.e.*,

$$\psi = \theta^0. \tag{5.16}$$

(Note that it is easy to see that this is at least a possible solution by assuming that it holds and then solving the resultant equations. Furthermore, it can be shown that this is the only solution.) We then write the steady state solution in the form

$$a(\theta) = A(\theta - \theta^0). \tag{5.17}$$

Substituting equation (5.17) into equation (5.15) and setting the time derivative equal to zero yields

$$A(\theta) = [J_0 A_0 + I_0^0 + (J_1 A_1 + I_1^0) \cos(\theta)]_+. \tag{5.18}$$

**Amplification of signals in the linear regime**

In this simplest model, if we assume that everything is above threshold, equation (5.18) becomes linear. By Fourier transforming one obtains,

$$A_0 = \frac{I_0^0}{1 - J_0} \qquad A_1 = \frac{I_1^0}{2 - J_1}. \tag{5.19}$$

These results demonstrate the potential signal amplification provided by the recurrent network even in the linear regime. One measure of feature selectivity is the ratio between the spatially modulated and the unmodulated components of the neural activity. In our case at the level of inputs, this Selectivity Index (SI) is:

$$SI_{input} = \frac{I_1^0}{I_0^0}. \tag{5.20}$$

On the other hand the SI of the recurrent network is:

$$SI_{network} = \frac{1 - J_0}{2 - J_1} SI_{input}. \tag{5.21}$$

So if $J_1$ is close to 2, the selectivity of the network can be substantially bigger than the selectivity of the input.

The results of equation (5.19) yield two critical values of interaction at $J_0 = 1$ and $J_1 = 2$. When $J_0 > 1$ amplitude instability occurs. In the context of this model, this means that the activity profile diverges. This would correspond to saturation in real neurons. The implication of the second instability is different. When $J_1 > 2$ spatial symmetry breaking occurs, as explained below.

*5.4. Symmetry breaking in the ring model*

Symmetry breaking refers to the phenomenon in which a pattern dynamically develops in a system in the absence of a spatially modulated external input; that is it develops spontaneously. To study this phenomenon in our system let us assume that $I_1^0 = 0$. The naïve solution is a homogeneous activity pattern with $A_1 = 0$. However, for $J_1 > 2$ this pattern becomes unstable. What happens as we cross the line into this region of instability? We must consider a *nonlinear* solution with $A_1 > 0$. Because we are in the nonlinear regime, the activity profile is a cosine truncated at some distance from its peak, denoted as $\theta_c$. With this form we can integrate equation (5.18) and derive the following equations for the order parameters

$$A_1 = \frac{1}{2\pi} \int_{-\theta_c}^{\theta_c} d\theta \cos \theta A(\theta)$$

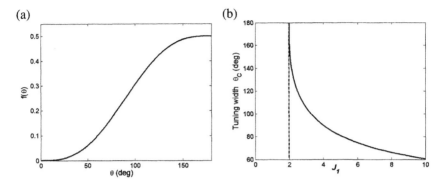

Fig. 5.1. $f(\theta)$ and tuning width vs. $J_1$ for ring model with spontaneously broken symmetry.

$$= J_1 A_1 \frac{1}{2\pi} \int_{-\theta_c}^{\theta_c} d\theta \cos\theta (\cos\theta - \cos(\theta_c)). \tag{5.22}$$

Excluding the unstable solution $A_1 = 0$, this equation reduces to

$$1 = J_1 f(\theta_c) \tag{5.23}$$

where

$$f(\theta_c) = \frac{1}{2\pi} \left[ \theta_c - \frac{1}{2} \sin(\theta_c) \right]. \tag{5.24}$$

The function $f$ is depicted in Figure 5.1a. Clearly there is a solution to equation (5.23) only for $J_1$ larger than 2. For $J_1 > 2$ there is a solution for $\theta_c$, which decreases monotonically with the value of $J_1$, as shown in Figure 5.1b.

### The ring attractor

In deriving the tuned solution above, we used equation (5.18). But this equation was derived from the more general equation (5.15) by substituting $\psi = \theta^0$. In the present case, the external input is uniform, so $\theta^0$ is ill defined. So what is $\psi$? The answer is that the location of the peak $\psi$ is arbitrary. Thus when $J_1 > 2$ the system possesses a manifold of states called *the ring attractor*. In this parameter regime the system is *in a marginal phase* since the stationary states along the ring are stable against all perturbations except those that cause a translation of the activity profile along the ring. This marginal stability is inherent in any system with a continuous attractor manifold, similar to the case of the line attractor discussed in Section 4.

**Contrast invariance**

An important property of the ring attractor is the invariance of activity profile
shape to change in the intensity of the stimulus (which in the case of vision can
be taken as the stimulus contrast). This feature is important since, at least in the
case of orientation, contrast invariance of tuning is observed experimentally and
it is not easy to obtain by simple feedforward scenarios. The difficulty comes
from the 'iceberg effect', caused by the presence of the neuronal firing threshold.
When the stimulus intensity changes, $I_0^0$ changes by a smaller amount than $I_1^0$.
This occurs because the former is the stimulus DC value relative to threshold and
presumably remains constant when contrast changes.

To see that the ring attractor yields contrast invariance, first notice that the
equation for profile width $\theta_c$ (equation (5.23)) does not involve $I_0^0$. It is completely determined by the recurrent interactions and hence the intensity of the
external input does not affect the shape of the activity profile. It only scales its
overall amplitude.

Next, we consider the effect of a tuned stimulus on the network state. Assume
a non-uniform but weakly tuned input,

$$0 < I_1^0 \ll I_0^0. \tag{5.25}$$

In this case, the location of the peak of the stationary solution is not arbitrary but
pinned to the stimulus angle $\theta^0$. On the other hand, the shape of the solution is
not substantially different from the one calculated with $I_1^0 = 0$. Thus as long as
the tuning of the input does not dominate the response of the network, the shape
of the network profile (and hence also the tuning curves of the neurons) will be
approximately invariant to the stimulus contrast.

In Fig. 5.2 we show the activity profile of the ring network when input with
angle $\theta_0 = \pi$ is presented. The input has the shape of $I_0$ given by equation (5.6)
with $SI_{input} = 0.18$ and two values of intensity ('contrast'): (1) $I_0^0 = 1$; and 2)
$I_0^0 = 1.4$. The threshold of firing is 1. The interaction has the cosine profile of
the simplest ring, given by equation (5.5) with $J_0 = -0.4$ and $J_1 = 4$. In the
absence of recurrent interactions the network's stationary profile is narrow for the
low contrast because of the threshold effect but becomes very broad at the higher
contrast, as indicated by the dashed curves. In the ring attractor regime, the activity profile remains with roughly the same shape in both cases although the amplitude of the response increases almost five fold, as indicated by the solid curves.

**Phase diagram of the ring network**

So far we have focused on the role of $J_1$. A full study of the role of $J_0$ in the stability of the marginal phase yields the phase diagram of the simplest ring shown
in Figure 5.3.

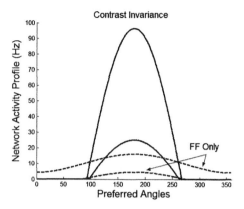

Fig. 5.2. Activity profile of the ring network for different contrasts.

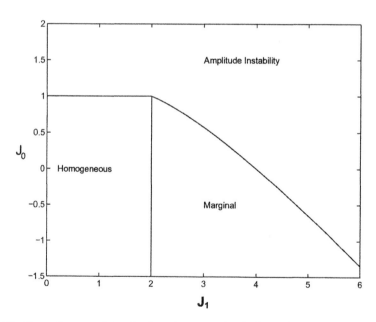

Fig. 5.3. Phase diagram of the simplest ring network. (Adapted from Hansel and Sompolinsky, figure 7.)

**Symmetry breaking and spontaneous activity in the brain**

There has been a considerable amount of theoretical work on pattern formation in cortical dynamics (see *e.g.* Ermentrout & Cowan, 1979). Such spatial or spatio-temporal activity patterns emerge spontaneously due to some sort of dynamical instabilities (see Bressloff's contribution to this volume). Naively it would seem that a central feature of the ring attractor is that even in the spontaneous state of the brain, the cortical circuit should generate spontaneously patterns of activity resembling those activated by tuned stimuli. To address this issue properly, however, we must elaborate on the role of external input and noise. Above we showed that the shape of the activity profile in the ring attractor is independent of stimulus contrast. This holds only as long as the contrast is suprathreshold, *i.e.* when $I_0^0$ is positive. In the above model, if $I_0^0$ is less than zero, corresponding to a subthreshold stimulus, the only stable state is the quiescent one, in which $a(\theta) = 0$ for all $\theta$. To incorporate spontaneous activity in the absence of external stimuli, we need to add noise to the model. This has been studied recently by Goldberg et al. (2003).

**The ring model with noise**

Let us consider adding a white noise to the input of each neuron,

$$\dot{r}_i = -r_i + \left[ \frac{1}{N} \sum_{j=1}^{N} J(\theta_i - \theta_j) r_j + I^0(\theta_i - \theta_0) + \eta(\theta_i, t) \right]_+ \qquad (5.26)$$

where

$$\langle \eta(\theta_i, t)\eta(\theta_j, t + \tau) \rangle = \sigma^2 \delta_{ij} \delta(\tau). \qquad (5.27)$$

As before, if the external inputs and the interactions change slowly with the distance along the ring, one can write the continuum limit of these equations by averaging out the local noise (since each section on the ring contains many neurons and their noise is uncorrelated). This yields the following deterministic equations for the activity profile,

$$\frac{\partial a(\theta, t)}{\partial t} = -a(\theta, t) + \sigma f(I(\theta, t)/\sigma) \qquad (5.28)$$

where, as in the noiseless case, $I$ is the total input to the column (given in equation (5.6)) and $f$ is the effective input-output transfer function resulting from smoothing the linear-threshold nonlinearity,

$$f(x) = \int_{-\infty}^{x} \phi(u)du; \qquad \phi(x) = \int_{-\infty}^{x} \frac{dz}{\sqrt{2\pi}} e^{-z^2/2}. \qquad (5.29)$$

Let us now consider the simplest ring architecture with constant input (apart from the noise). Then the equation for the order parameter is

$$\frac{d}{dt}a_1(t) = -a_1(t) + \sigma \int_{-\pi}^{\pi} \frac{d\theta}{2\pi} f([J_1 a_1(t) \cos\theta + I]/\sigma) \cos\theta. \qquad (5.30)$$

$I$ is a constant term replacing a DC input relative to threshold and we have assumed for simplicity that $J_0 = 0$.

Solving for the stationary solution of this equation and its stability, yields the phase diagram shown in Figure 5.4a. The homogeneous ($H$) regime in the phase diagram corresponds to the case in which the steady-state solution $a_1 = 0$ is stable. The marginal ($M$) regime corresponds to a stable solution where $a_1 > 0$. In the unstable ($I$) regime $a_1$ has no finite stable solution. When the DC input is positive and large compared to the noise amplitude, the transition between the ($H$) and ($M$) regions occurs at $J_1 = 2$ as in the noiseless suprathreshold case discussed above. When the DC input is small compared to the noise amplitude, the system is close to threshold and the rounding of the $f$-$I$ curve due to the noise becomes pronounced, resulting in an increase in the critical value of $J_1$. This reflects the decrease of the neuronal gain, as shown in Fig. 5.4d. In general the critical $J_1$ is given by the point where the solution $a_1 = 0$ loses stability, which is at

$$I/\sigma_n = \phi^{-1}(2/J_1) \qquad (5.31)$$

where $\phi^{-1}(\cdot)$ is the inverse gain function (i.e, the derivative of the input-output function f). Examples of snapshots of the local field $I(\theta, t)$ in the two regimes (with parameters corresponding to the symbols in the phase diagram) are shown in Figures 5.4b and c.

The transition between the ($M$) and ($I$) regimes is given by the line $J_1 = 4$ and represents the line where the solution $a_1 > 0$ diverges. This instability line corresponds to the amplitude instability of Figure 5.3 at $J_0 = 1$. Inspection of this phase diagram reveals that when the system is subthreshold, *i.e.* $I < 0$, there is no value of $J_1$ such that a stable ring attractor appears. Thus, in this model, the spontaneous state is characterized by small fluctuations of activity around a single homogeneous state, even in a parameter regime where the activated state exhibits a ring attractor.

## The ring model with conductance based dynamics

The above analysis of the ring network was based on a simplified rate dynamics. Using the results of Section 3 we can compare the predictions of the rate dynamics with the behavior of a conductance based network with a Hudgkin-Huxley type single neuron dynamics and ring architecture. Indeed such a study

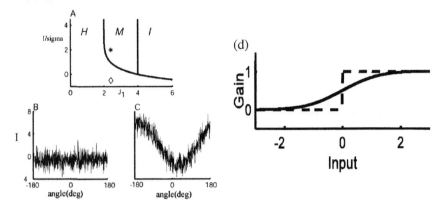

Fig. 5.4. Network states of the ring model with noise. (Adapted from Goldberg et al.)

is carried out by Shriki et al. (2003). In that study, the recurrent excitatory and inhibitory connections have exponential dependence on the distance along the ring. The magnitude of the peak synaptic conductances of these interactions are chosen according to the equations that map between the parameters of the simplified rate dynamics and the conductance-based dynamics, equations (3.33) and (3.34). The single neuron model is that described in section 2. Indeed Shriki et al. find very good agreement between the analytical solution of the rate fixed point equations (with exponentially decaying connections) and the simulations of the conductance based dynamics.

An example of symmetry breaking in the conductance based network is shown in the raster plots in Figure 5.5a. The upper figure shows a raster plot for a simulation of the conductance-based network at a synaptic conductance predicted to lead to a homogeneous state by the analytical solution to the corresponding rate model. The response is indeed homogeneous. The lower figure shows a raster plot at a conductance predicted to yield a bimodal state. Again, the response is in good agreement with predication. Note that the noise in the system leads to slow, random wondering of the activity pattern. The resultant tuning curves in the marginal phase regime in the presence of weakly tuned input are shown in Figure 5.5b together with the predictions of the rate model. Note that no free fitting parameters have been employed. The dashed line shows the tuning curves of the LGN input oriented at $0°$ and the solid line shows the analytical solution of the rate model. The circles are from numerical solutions of networks composed of the Shriki model neuron. The remarkable agreement between the two models shows that the reduction of the asynchronous state of conductance based spiking

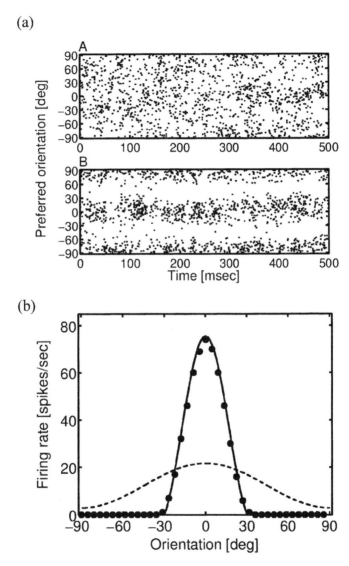

Fig. 5.5. Comparison of rate equations and conductance-based simulations of the ring network. (Taken from Shriki et al., figures 5 and 6.)

networks to simple rate equations is powerful even for relatively complicated spatial organization of synaptic connectivity.

## Conclusion

It should be noted that in the present model, the ring attractor exists as long as $I_0^0$ is positive, or equivalently the contrast is above threshold. If the external inputs to the network are subthreshold the system will remain in a quiescent state which is the unique stable state for this system. Thus, to model cases of persistent activity with ring symmetry, additional nonlinear elements need to be considered (Wang, 2001). An important issue is the robustness of the ring attractor. The presence of this attractor manifold depends on the isotropy of the system, an unrealistic feature for biological systems. A small degree of anisotropy will not matter much for orientation selectivity where an oriented stimulus is present. However, other model systems such as networks for working memory or head directions are highly sensitive to these perturbations. A possible homeostatic mechanism for compensating for the effects of inhomogeneities has been recently proposed (Renart et al., 2003).

## 6. Models of associative memory

### 6.1. Introduction

Associative memory has been one of the most active topics in neural modeling. Recurrent network models of associative memory have been extensively studied, following the seminal work of Hopfield (Hopfield, 1984; see also Amit, 1989; Herz et al., 1991; Rolls & Treves, 1998). In this section we will review some aspects of Hopfield's model and discuss one variant thereof that is closer to biological reality.

Although the Hopfield model has been extended to analog neurons we will limit the discussion here to the original binary version. This will serve as an example of a recurrent network operating in the strong nonlinear regime where each neuron takes only two values. We first recall some basic properties of classes of binary networks.

### 6.2. Binary networks

#### Network dynamics: single neuron updates

In this and the next section we will use $\pm 1$ to represent the two states of each neuron.

At each time step, a single neuron is chosen as a candidate for state updating

$$S_i(t+1) = \text{sgn}(h_i(t)). \tag{6.1}$$

The local fields $h_i$ are the total synaptic inputs to the $i^{\text{th}}$ neuron at time $t$ relative to threshold,

$$h_i(t) = \sum_{j=1}^{N} J_{ij} S_j(t) + h_i^0(t). \tag{6.2}$$

$J$ is the synaptic matrix and $h_i^0$ the external inputs or biases relative to thresholds. Update is sequential. It could occur either in a random order or with fixed order. Additionally, if $h_i(t) = 0$ then $S_i(t+1)$ is chosen to be $\pm 1$ at random with equal probability.

**Energy function**

If the $h_i^0$'s are time independent, then under certain conditions the network dynamics can be described as moving downhill with respect to an energy surface. $E(S)$ is an energy function of the network if and only if

$$\Delta E(t+1) = E(t+1) - E(t) \le 0 \tag{6.3}$$

and equality holds only if $\Delta S_i(t+1) = 0$ for the updating neuron or $h_i(t) = 0$. These properties imply that the system's attractors are fixed points and that these fixed points are local minima of $E$. A state $S$ is a local minimum if its energy cannot be further decreased by a change of the states of any one of the neurons.

A well known theorem states that if the connection matrix is symmetric and there is no self coupling (*i.e.* (1) $\mathbf{J} = \mathbf{J}^{\text{T}}$ and (2) $J_{ii} = 0$ for all $i$) then the following is an energy function of the binary network

$$E(S) = -\frac{1}{2} \sum_{i,j}^{N} J_{ij} S_i S_j - \sum_{i=1}^{N} h_i^0 S_i. \tag{6.4}$$

Here $S$ denotes one of the $2^N$ states of the network.

[*Proof:*
Suppose at time $t+1$ the state of the $i^{\text{th}}$ neuron changes from $S_i$ to $-S_i$. This induces a corresponding change in $E$

$$\Delta E(t+1) = -(S_i(t+1) - S_i(t)) \left[ \sum_{j \ne i}^{N} \frac{1}{2}(J_{ij} + J_{ji}) S_j(t) + h_i^0 \right]$$

$$= 2S_i(t) h_i(t). \tag{6.5}$$

But according to the update rule, the state of $S_i$ changes only if either (a) $h_i(t) = 0$ or (b) $S_i(t)h_i(t) < 0$, from which it follows that

$$\Delta E(t+1) = -2|h_i(t)|. \tag{6.6}$$

Therefore, by definition $E$ is an energy function of the network.]

## Stochastic dynamics in binary networks

Now we consider the case in which the dynamics is not entirely deterministic. When the local field on a spin $S_i$ is small in magnitude, the spin has non-zero probability both to flip and to remain in its original direction. We express this by introducing a temperature $T > 0$ and setting $\beta = 1/T$. Now the probability distribution for $S_i(t+1)$ can be expressed in terms of the field at spin $i$ at time $t$ as

$$\Pr(S_i, t+1) = \frac{\exp(\beta S_i h_i(t))}{\exp(\beta h_i(t)) + \exp(-\beta h_i(t))}$$

$$= \frac{1}{2}(1 + S_i \tanh(\beta h_i(t))). \tag{6.7}$$

Note that in the case of zero temperature ($\beta = \infty$), this expression reduces to the deterministic update rule.

*Detailed balance*

$$\frac{\Pr(S_i \rightarrow -S_i)}{\Pr(-S_i \rightarrow S_i)} = \frac{\Pr(-S_i)}{\Pr(S_i)}$$

$$= \exp(-2\beta h_i S_i)$$

$$= \exp(-\beta \Delta E_i). \tag{6.8}$$

*Equilibrium distribution*

In the infinite time limit, the probability distribution tends to its equilibrium form, given by

$$\Pr_{eq}(S) = \frac{1}{Z} \exp(-\beta E(S)) \tag{6.9}$$

where $Z$ is the partition function,

$$Z = \sum_{\text{states } \tilde{S}} \exp(-\beta E(\tilde{S})). \tag{6.10}$$

Equation (6.9) is just a Gibbs distribution. At zero temperature ($\beta = \infty$), this is a delta function at the minimal energy state, as derived in the above analysis of the deterministic case. As temperature increases, it becomes an increasingly wide distribution, still peaked at the minimal energy state. As $T \to \infty$, the equilibrium probability distribution approaches a uniform distribution.

*Mean-field dynamics of averages*

Define $m_i(t) = \langle S_i(t) \rangle_T$ where the average is taken over the probability distribution at a given temperature, given by equation (6.7). The dynamics of $m_i(t)$ are given by

$$\tau \frac{dm_i}{dt} = -m_i + \langle \tanh(\beta h_i(t)) \rangle. \tag{6.11}$$

In highly connected networks when each neuron receives a large number of inputs, $K \gg 1$, and each of the existing connections is scaled as $O(1/K)$ then equation (6.11) can be to

$$\tau \frac{dm_i}{dt} = -m_i + \tanh(\beta h_i^{MF}(t)) \tag{6.12}$$

where

$$h_i^{MF}(t) = \sum_j J_{ij} m_j(t). \tag{6.13}$$

The reason for that is that at any given time the local field is a sum of many weakly correlated random variables and hence fluctuations from the mean are small. The resultant deterministic equations (6.12) and (6.13) are similar to the rate dynamics discussed in previous sections but now these equations describe the dynamics of averages of variables obeying an underlying stochastic dynamics.

## 6.3. The Hopfield model

The basic idea behind the Hopfield model is to build a system where a given set of states, the memorized patterns, are the energy minima of the network dynamics. Each given initial condition will be associated with a different memory pattern, towards which it will tend in steady state. The convergence of the network state from an initial state to the nearby memory state represents the retrieval of the full memory on the basis of partial information about it.

### The Hebb synapse

In this framework, a given memory is just one particular state out of all possible network states and can be denoted as an $N$-component binary vector $\xi$ where $\xi_i$ denotes the state of the $i^{th}$ neuron in that memory. Suppose that we wish to store the following $P$ memory patterns in our system

$$\Sigma = \begin{pmatrix} \xi_1^1 & \cdots & \xi_1^P \\ \vdots & & \vdots \\ \xi_N^1 & \cdots & \xi_N^P \end{pmatrix} \tag{6.14}$$

where each column contains one memory pattern. The aim is to define recurrent connections between the neurons in our network such that each of the above P memory patterns is at a global energy minimum of our system. In the Hopfield model, the matrix of connection is chosen as

$$J_{ij} = \frac{1}{N} \sum_{\mu=1}^{P} \xi_i^\mu \xi_j^\mu, \quad i \neq j. \tag{6.15}$$

Although far from the original associative rule proposed by Hebb, this rule is 'Hebbian' in the sense that the change incurred by a single memory state at any synapse depends only on the activity of the pre- and post-synaptic cells.

### Analysis of the Hopfield Model

We would like to know whether the matrix of connections defined by the Hebb rule given by equation (6.15) guarantees that the specified memory patterns are global minima of the energy function. The answer depends upon the *particular* choice of memory patterns. As one may guess, the overlap or similarity between different memories is an important factor in system performance. We therefore need to ask how well the model does for a *typical* selection of memory patterns, where typicality is defined relative to a statistical ensemble of neurons. The Hopfield model performs best when the memories are as far apart from each other (and their reflections) as possible. This is achieved naturally by choosing them from a uniform random distribution so that each element $S_i^\mu = \pm 1$ with equal probability. We can write the distribution for memories formally,

$$\Pr(\vec{\xi}^1, \dots, \vec{\xi}^P) = \prod_{\mu=1}^{P} \prod_{i=1}^{N} \left( \frac{1}{2} \delta_{\xi_i^\mu, 1} + \frac{1}{2} \delta_{\xi_i^\mu, -1} \right). \tag{6.16}$$

Given a probabilistic description of inputs, we must evaluate performance in terms of probabilities. We do not demand perfect performance at all instances but only good performance with high probability. This probabilistic notion of performance became a very important concept in modern learning theory. As we will see, the Hopfield model performs well with the above uniform distribution of input patterns. It does very badly, however, in the case of other input distributions.

## Hopfield model far below saturation

Let us first concentrate on the limit of $N \to \infty$ for finite $P$. It is useful to write the energy of the system (substituting equation (6.15) into equation (6.4) with no external input) as

$$E(S) = -\frac{N}{2} \sum_{\mu=1}^{P} M_\mu^2 + \frac{1}{2} P \tag{6.17}$$

where

$$M_\mu = \frac{1}{N} \sum_{i=1}^{N} \xi_i^\mu S_i \tag{6.18}$$

These quantities measure the overlap between the network states and the memory patterns and are the system's natural order parameters. For a state which is one of the memory states, say $\xi^1$, we have $M_1 = 1$ while the remaining $M_l$ are of order $1/N^{1/2}$, because the different memories are statistically independent. Hence,

$$E(\xi^1) = -\frac{N}{2} + O(1). \tag{6.19}$$

It can be shown that $\sum_{\mu=1}^{P} M_\mu^2 \le 1$ and thus for large $N$ and finite $P$ the memory states are global minima of the system. What happens as $P$ increases? Several things may occur. For example, the memory states may become local minima of E rather than global minima. This may not be catastrophic as they still may have a sizable basin of attraction. Alternatively, the memory states may be unstable altogether, in which case the retrieval process will (almost) never end up in these states. We address this question below using 'signal to noise' analysis, which is useful in this and many other contexts.

## Hopfield model near saturation – signal to noise analysis

In estimating the capacity of the system, we will limit ourselves to the question of stability of memory states. Recall from our analysis of the $T = 0$ case that since $S_i(t + 1) = \mathrm{sgn}(h_i(t))$, if there exists a ground state, $S_i = \mathrm{sgn}(h_i)$ for all $i$ after enough time. Suppose that we start in a memory state. In order for it to be stable, it must be the case that for all $i$, $h_i S_i > 0$.

Assume we are in memory state 1 so that $S = \xi^1$ and let $N$ be large. Therefore, by equations (6.2) and (6.15),

$$S_i^1 h_i^1 = 1 + \frac{1}{N} \sum_{\substack{\mu \neq 1}}^{P} \sum_{\substack{j=1 \\ j \neq i}}^{N} S_i^1 S_i^\mu S_j^\mu S_j^1 \equiv 1 + z_i. \tag{6.20}$$

Here $z_i$ is made up of many uncorrelated random terms so it has effectively Gaussian statistics with

$$\langle z_i \rangle = 0$$

$$\langle (z_i)^2 \rangle = \frac{(P-1)(N-1)}{N^2} \simeq \alpha \qquad (6.21)$$

where

$$\alpha \equiv \frac{P}{N}. \qquad (6.22)$$

The average of $z_i$ is zero because memories are made up of random, uncorrelated, binary bits. To calculate the second moment, note that

$$\langle (z_i)^2 \rangle = \frac{1}{N^2} \sum_{\mu,\nu \neq 1}^{P} \sum_{j,j' \neq i}^{N} \langle (S_i^1)^2 S_i^\mu S_i^\nu S_j^\mu S_j^1 S_{j'}^1 S_{j'}^\nu \rangle$$

$$= \frac{1}{N^2} \sum_{\mu \neq 1}^{P} \sum_{j \neq i}^{N} \langle (S_i^1)^2 (S_i^\mu)^2 (S_j^\mu)^2 (S_j^1)^2 \rangle$$

$$= \frac{(P-1)(N-1)}{N^2} \simeq \alpha \qquad (6.23)$$

The second line follows from the first because averages over sums such that $\mu \neq \nu$ and $i \neq j$ give zero, again because memories are made up of random uncorrelated binary bits.

We conclude, therefore, that for sufficiently small $\alpha$, the local field has the right sign for an arbitrarily large number of $i$'s and the memory states are stable. On the other hand when $\alpha$ is order 1, the standard deviation of the noise generated by the random overlaps between memories is of the same order as the signal, hence with high probability it will destabilize the memory states. To evaluate the value of $\alpha$ for which errors begin to appear with significant probability, we perform the following analysis.

The probability of error is given by $P = 1 - \mathrm{Pr}\,(all\ bits\ retrieved)$. Since there are $N$ independent bits that need to be retrieved,

$$P_\varepsilon = 1 - \prod_{i=1}^{N} [\mathrm{Pr}(h_i^1 S_i^1 > 0)]$$

$$= 1 - \left[ \int_{-\infty}^{1} \frac{dz}{\sqrt{2\pi\alpha}} \exp\left(-\frac{z^2}{2\alpha}\right) \right]^N. \qquad (6.24)$$

Define

$$H(x) \equiv \int_x^\infty \frac{du}{\sqrt{2\pi}} \exp(-u^2/2). \tag{6.25}$$

Note that $H(-\infty) = 1$, $H(0) = 1/2$, and $H(+\infty) = 1$. Thus we can write,

$$\begin{aligned} P_\varepsilon &= 1 - [1 - H(1/\sqrt{\alpha})]^N \\ &= 1 - \exp[N \log(1 - H(1/\sqrt{\alpha}))]. \end{aligned} \tag{6.26}$$

When $\alpha$ is small we can use the asymptotic result (when $\alpha$ is of order 1, the error is obviously large as can be seen from equation (6.26)),

$$\lim_{x \to \infty} H(x) = \frac{\exp(-x^2/2)}{\sqrt{2\pi} x}. \tag{6.27}$$

Retaining only the dominant exponential dependence on $\alpha$ yields

$$\begin{aligned} P_\varepsilon &\simeq 1 - \exp[-NH(1/\sqrt{\alpha})] \\ &\simeq 1 - \exp[-N \exp(-1/2\alpha)]. \end{aligned} \tag{6.28}$$

Setting a performance criterion

$$P_\varepsilon \simeq N \exp(-1/2\alpha) < c \tag{6.29}$$

$c < 1$, places a bound on $\alpha$,

$$\alpha < \frac{1}{2 \log(N/c)}. \tag{6.30}$$

This bound on $\alpha$ insures that with high probability each memory pattern is near a ground state of the system.

We can also impose the stronger requirement that the probability of having an erroneous memory is small,

$$\begin{aligned} P_\varepsilon &\simeq 1 - \exp[-NPH(1/\sqrt{\alpha})] \\ &\simeq 1 - \exp[-NP \exp(-1/2\alpha)] \end{aligned} \tag{6.31}$$

which requires that

$$\alpha < \frac{1}{4 \log(N/c)}. \tag{6.32}$$

The weakness of this analysis is that it does not tell us what happens to system states if the number of memories is larger than the above limits. If we start from a given memory state, does the system converge to a state far from that memory or

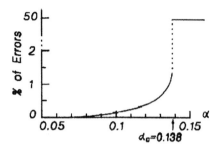

Fig. 6.1. Average percentage of errors as a function of $\alpha$. (Taken from Amit et al., 1985, figure 1.)

to a nearby state? If indeed there are stable states near the memories, how much error do they incur? Unfortunately, in order to address these questions, one has to use rather complex statistical mechanical analysis. The result of this analysis is shown in Figure 6.1. As long as $\alpha$ is smaller than a critical value, $\alpha_c \approx 0.14$ there are local minima which differ from the memory states by only a few bits. Beyond this value, the system will converge to a stable state far away from the memory state, and the system stops functioning as associative memory. In the next section we will discuss these issues in a model which is more interesting from a biological perspective and at the same time simpler to analyze.

### 6.4. Networks with sparse memories

Although the capacity of the Hopfield model is not unreasonable (compared to that of other models) it is restricted to uncorrelated memories. When the memories have significant overlap between them, it fails completely even when the number of memories is small ($P$ of order one). This is a significant limitation since real-life memories have non-trivial statistical structure. A related problem is the fact that in that model a memory state corresponds to half of the neurons being active. In the brain, the activity patterns putatively associated with memory states involve only a small fraction of the local neuronal population. Furthermore there are computational advantages to using a sparse representation of memories, where each memory corresponds to a small number of active bits. An interesting variant of the Hopfield model introduced by Tsodyks and Feigelman (1988) addresses the storage of sparse memories.

**Tsodyks and Feigelman model**

In the Tsodyks and Feigelman model, sparseness of memories is achieved by choosing uncorrelated random binary vectors but with biased statistics,

$$\Pr(\vec{\xi}^1, \dots, \vec{\xi}^P) = \prod_{\mu=1}^{P} \prod_{i=1}^{N} (f \delta_{\xi_i^\mu, 1} + (1 - f) \delta_{\xi_i^\mu, 0}) \tag{6.33}$$

where here we switch to 0,1 notation for the binary states, and we work in the sparse limit $f \ll 1$. To minimize the overlap between the memories induced by their sparseness, use a modified connectivity matrix,

$$J_{ij} = \frac{1}{Nf(1-f)} \sum_{\mu=1}^{P} (\xi_i^\mu - f)(\xi_j^\mu - f), \quad i \neq j. \tag{6.34}$$

This choice of weights guarantees that the overlapping terms in the local fields will average to zero, as we will show below. Finally, the neural dynamics is a sequential update

$$S_i(t+1) = \Theta(h_i(t) - \theta) \tag{6.35}$$

where $\Theta(x)$ is the Heaviside step function, and $\theta$ a threshold. As before,

$$h_i(t) = \sum_j J_{ij} S_j(t). \tag{6.36}$$

Signal-to-noise analysis similar to that described above shows that if the number of patterns per neuron is larger than $O(1/\log(N))$ then memory states are unstable with high probability. This is already an improvement over the behavior of the Hopfield model when memories are overlapping as they are here. Next, we investigate the behavior of the system when $P/N$ is finite.

**Mean field equations near the capacity limit**

When $P/N$ is finite, memory states are no longer stable. To study the existence of nearby stable states, we consider large $N$ and make the following *ansatz*. Let us assume that of the 'active' neurons in a memory state, say 1, a fraction $f_+$ are active in the nearby stable state. Likewise a fraction $f_-$ of neurons quiescent in the memory state are active in the nearby stable state. The overlap between this stable state and the memory state is

$$m \equiv \frac{1}{Nf(1-f)} \sum_{i=1}^{N} (\xi_i^1 - f) S_i = f_+ - f_-. \tag{6.37}$$

Obviously in the case of perfect overlap, $f_+ = 1$ and $f_- = 0$. Note that the mean fractional error is given by

$$\varepsilon = f(1 - f_+) + (1 - f)f_- \approx f\left(1 - f_+ + \frac{f_-}{f}\right). \tag{6.38}$$

The mean activity of this state is given by

$$q \equiv \frac{1}{N} \sum_{i=1}^{N} S_i = ff_+ + (1 - f)f_-. \tag{6.39}$$

To evaluate these quantities, we note that the local fields generated in this state on sites with $\xi_i^1 = 1$ are

$$h_i^{(1)} = (1 - f)m + z_i \tag{6.40}$$

where the first term is the signal and the second a Gaussian noise with variance

$$\langle z_i^2 \rangle = \frac{(P - 1)(N - 1)}{N^2 f^2 (1 - f)^2} \langle (\xi_i^\mu - f)^2 \rangle \langle (\xi_j^\mu - f)^2 \rangle \langle S_j^2 \rangle. \tag{6.41}$$

Note that here we have neglected the correlations between the stationary state $S_j$ and the patterns $\xi_i^\mu$, $\mu \neq 1$. This is justified in the sparse limit but not in general. Using the fact that $\langle S_j^2 \rangle = \langle S_j \rangle = q$ the averages of equation (6.41) yield

$$\Delta \equiv \langle z_i^2 \rangle / \alpha$$
$$= ff_+ + (1 - f)f_- \tag{6.42}$$

Likewise, the field on the sites with $\xi_i^1 = 0$ is

$$h_i^{(0)} = -fm + z_i \tag{6.43}$$

where the variance of the noise is also given by equation (6.41).

To derive equations for $f_\pm$ we use the fact that neurons are active if their field is above the threshold. Hence,

$$f_+ = H((\theta - m)/\sqrt{\alpha\Delta});$$
$$f_- = H(\theta/\sqrt{\alpha\Delta}) \tag{6.44}$$

where we have substituted $\theta + fm \approx \theta$. These equations, together with equation (6.37) constitute self-consistent equations for the unknown $f_\pm$.

**Dynamic mean field equations**

Extending the above argument to the dynamics, one find that the time dependent $f_\pm$ obey the following equations,

$$
\frac{df_+}{dt} = -f_+ + H((\theta - m)/\sqrt{\alpha\Delta})
$$
$$
\frac{df_-}{dt} = -f_- + H(\theta/\sqrt{\alpha\Delta})
$$

(6.45)

with

$$
m(t) = f_+(t) - f_-(t)
$$
$$
\Delta(t) = f\left(f_+(t) + \frac{f_-(t)}{f}\right).
$$

(6.46)

The stationary state is the stable fixed point solution to these equations. The dynamical equations are similar to the 'statistical neurodynamic' theory of Amari and Maginu (1988). It should be pointed out that outside the sparse limit of small $f$, neglecting the correlations in the expression for the noise is not justified.

**Solution of the stationary state**

We now analyze the above equations in the limit of small $f$. Inspection of the equations reveals that in the regime of interest, where $1 - f_+$ and $f_- \ll 1$, the following scaling holds,

$$
\alpha = \alpha_0/f; \qquad \Delta = \Delta_0 f
$$

(6.47)

in terms of which the self-consistent equations are

$$
1 - f_+ \approx \exp(-(\theta - m)^2/2\alpha_0\Delta_0)
$$
$$
f_- \approx \exp(-\theta^2/2\alpha_0\Delta_0)
$$

(6.48)

together with

$$
m = f_+ - f_-; \qquad \Delta_0 = f_+ + \frac{f_-}{f}.
$$

(6.49)

When $\alpha_0$ goes to zero and $0 < \theta < 1$, $f_+$ approaches unity and $f_-$ vanishes. In this limit, the noise amplitude (per memory) is given by $\Delta_0 = 1$. When $\alpha_0$ increases from zero, depending on the value $\theta$ the noise amplitude $\Delta_0$ may initially decrease. However upon further increase in $\alpha_0$, the noise begins to increase, due to an increase in $f_-$, eventually causing the state to move far from the memory. For $f_-$ to overcome the decrease in $f_+$, $f_-/f$ must be larger than $1 - f_+$. This means that we have to search for a regime in which the $H$ terms are power law in $f$. This suggests writing,

$$\alpha_0 = \frac{x\theta^2}{2|\log f|} \tag{6.50}$$

which yields

$$f_- \approx \exp(\log f/x) \approx f^{1/x} \tag{6.51}$$

(where we have substituted $\Delta_0 \approx 1$). This solution is consistent as long as $x < 1$. When $x$ increases above 1, the contribution of $f_-$ to the noise diverges, destroying the solution. Hence we conclude that the critical value of $\alpha$ is

$$\alpha_C = \frac{\theta^2}{2f|\log f|}. \tag{6.52}$$

Below this value, the stationary state and the memory state have large overlap. The largest error occurs at $\alpha_C$ where

$$
\begin{aligned}
1 - f_+ &\approx \exp((\theta - 1)^2\theta^{-2}\log f) = f^{(\theta^{-1}-1)^2} \\
f_- &\approx \exp(\log f) = f.
\end{aligned} \tag{6.53}
$$

Our analysis implies that in the sparse limit, the error near the capacity limit is dominated by the fraction of neurons that are active at sites with $\xi_i^1 = 0$.

The above theoretical results also highlight the role of the threshold. On one hand, the closer $\theta$ is to 1 the larger the system's capacity. On the other hand, increased $\theta$ increases the error in retrieval (see equation (6.53)).

We illustrate the above results by numerical integration of the dynamic mean field equations (6.45) and (6.46) starting from a memory state. In Figure 6.2a we show the order parameters as function of the scaled number of examples, $x = 2Pf|\log f|/N\theta^2$, for $f = 0.02$ and $\theta = 0.5$. There is a sharp first order transition at $x \approx 0.97$, close to the sparse limit prediction of $x = 1$. Note the large increase in the noise parameter $\Delta_0$ before the transition, signaling the increase in $f_-$.

The basin of attraction of the retrieval state can be studied by solving the dynamic mean field equations from an initial condition in the vicinity of the memory state. An example is shown in Fig. 6.2b where the evolution of $m$ with time is shown for two values of $x$ starting from the initial condition $f_+ = 0.9$, $f_- = 0.1$ for $f = 0.02$. Note that already at $x \approx 0.85$ the basin of attraction of the retrieval state is too narrow to accommodate these initial conditions.

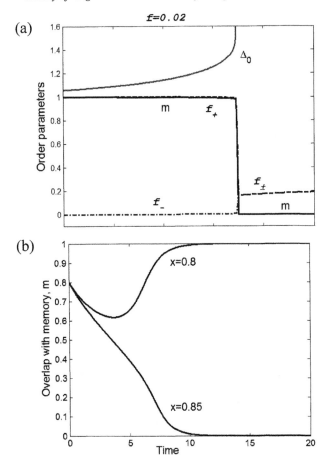

Fig. 6.2. (a) Order parameters as a function of number of examples. (b) Overlap with memory vs. time for two values for number of examples, x.

## Discussion

Though the sparse model has some attractive features, it shares some basic deficiencies with the Hopfield model. First, information is coded in binary states, a feature which is not seen usually in the cortex. Second, when the system is loaded over its capacity all memories degrade, an undesired feature for a memory system. Lastly, the sparse model does not incorporate temporal relations between memories. These issues are addressed by other models of associative memory (see *e.g.*, Hertz et al., 1991; Amit, 1989).

## 7. Concluding remarks

In these notes we have reviewed some of the models of recurrent networks that have been used in modeling brain function. Our approach was based on a simplified rate description of the network dynamics. Rate based network models have found a broad spectrum of applications from questions related to experimentally observed features of neuronal sensory representations in concrete brain areas to more abstract topics in perception, motor control, learning, and memory. These far ranging applications indicate that the framework provided by rate models serves as a flexible and fruitful description at an intermediate level of complexity, highlighting the roles of connectivity and nonlinearity in the emergent computation. As our understanding of realistic network dynamics advances we may be able to build and study improved models without sacrificing too much underlying simplicity. Nevertheless, it is clear that some domains of brain processing, for instance auditory processing, require models explicitly addressing the spiking dynamics of the associated neuronal circuits. And it may turn out that even in areas such as vision, somatosensation and motor control, the temporal patterns of neuronal activity play a more dominant role than is currently apparent.

Another simplification in our approach here is that we describe single neurons as point neurons, ignoring their morphology and the associated linear and nonlinear dynamics. Indeed most current models of neural information processing ignore the role of dendritic spatio-temporal dynamics. In the last part of Section 4, however, we do describe a concrete model of integration in which the dendritic morphology and nonlinear calcium dynamics of a single neuron are essential for the computation. Other models for the functional role of dendritic processing have also been proposed. The metaphor of neuronal networks as consisting of a large number of simple processors obeying simple dynamical rules has dominated the field of neural modeling since the work of McCulloch and Pitts. Hopefully theoretical and experimental studies in the near future will elucidate the extent to which this metaphor needs revision. In any event, some of the theories described in this article will continue to serve as useful guides in the search for principles that underlie information processing in the brain.

## Acknowledgements

Research of HS is partially supported by grants from the Israeli Science Foundation (Center of Excellence Grant-8006/00) and the U.S.-Israel Binational Science Foundation. OW acknowledges support from the National Science Foundation via DMR 02209243 and the National Institute of Health via R01 EY014737-01.

HS would like to thank David Hansel for many fruitful collaborations and helpful discussions. Sections 2, 3 and 5 are based upon work done in collaboration with David Hansel, Oren Shriki, Rani BenYishai, and Ruthy Lev Bar-Or. The analysis of the role of noise in Section 5 is based on work with Josh Goldberg and Uri Rokni. The second portion of Section 4 is based upon work with Yonatan Loewenstein. The authors also thank the École de Physique des Houches for the opportunity that lead to this contribution.

## References

Abbott, L.F., & Kepler, T. (1990). Model neurons: From Hodgkin-Huxley to Hopfield. In L. Garrido (Ed.), *Statistical mechanics of neural networks* (pp. 5-18). Berlin, Springer-Verlag.

Abeles, M. (1991). *Corticonics: Neural circuits of the cerebral cortex.* Cambridge University Press, Cambridge.

Aksay, E., Baker, R., Seung, H.S., & Tank, D.W. (2001). Anatomy and discharge properties of premotor neurons in the goldfish medulla that have eye-position signals during fixations. *J. Neurophysiol.*, **84**, 1035-1049.

Amari, S. (1977). Dynamics of pattern formation in lateral-inhibition type neural fields, *Biol. Cybern.*, **27**, 77-87.

Amari, S.I., & Maginu, K. (1988). Statistical neurodynamics of associative memory. *Neural Networks*, **1**, 63-73.

Amit, D.J., Gutfreund, H., & Sompolinsky, H. (1985). Storing infinite number of patterns in a spin glass model for neural networks. *Phys. Rev. Lett.*, **55**, 1530-3.

Amit, D.J. (1989). *Modeling Brain Function.* Cambridge University Press, Cambridge.

Amit, D.J., & Tsodyks, M.V. (1991). Quantitative study of attractor neural networks retrieving at low spike rates I: Substrate—spikes, rates and neuronal gain. *Network*, **2**, 259-273.

Arnold, D.B., & Robinson, D.A. (1997). The oculomotor integrator: testing a neural network model. *Exp. Brain Res.*, **113**, 57-74.

Ben-Yishai, R., Lev Bar-Or, R., & Sompolinsky, H. (1995). Theory of orientation tuning in visual cortex. *Proc. Natl. Acad. Sci. USA*, **92**, 3844-3848.

Cannon, S.C., Robinson, D.A., & Shamma, S. (1983). A proposed neural network for the integrator of the oculomotor system. *Biol. Cybern.*, **49**, 127-136.

Churchland, P.S., & Sejnowski, T.J. (1992). *The computational brain.* MIT Press, Cambridge.

Dayan, P., & Abbot, L.F. (2001). *Theoretical neuroscience: Computational and mathematical modeling of neural systems.* MIT Press, Cambridge.

Ermentrout, G.B. & Cowan, J.D. (1979). A mathematical theory of visual hallucination patterns. *Biol. Cybern.*, **34**, 137-150.

Ermentrout, B. (1994). Reduction of conductance-based models with slow synapses to neural nets. *Neural Comp.*, **6**, 679-695.

Ferster, K.D. & Miller, D. (2002). Neural mechanisms of orientation selectivity in the visual cortex. *Am. Rev. Neurosci.*, **23**, 441-471.

Ginsburg, I., & Sompolinsky, H. (1994). Theory of correlations in stochastic neuronal networks. *Phys. Rev. E*, **50**, 3172-3191.

Goldberg, J., Rokni, U., & Sompolinsky, H. (2004). Patterns of ongoing activity and the functional architecture of the primary visual cortex. *Neuron*, **42**, 489-500.

Hansel, D., & Sompolinsky, H. (1996). Chaos and Synchrony in a Model of a Hypercolumn in Visual Cortex. *J. Comp. Neurosci.*, **3**, 7-34.

Hansel, D., & Sompolinsky, H. (1998). Modeling feature selectivity in local cortical circuits. In C. Koch & I. Segev (Eds.), *Methods in neuronal modeling: From synapses to networks* (2$^{nd}$ ed., pp. 499-567). Cambridge, MA: MIT Press.

Hertz, J., Krogh, A., & Palmer, R.G. (1991). *Introduction of the Theory of Neural Computation.* Addison-Wesley, Redwood City.

Hopfield, J.J. (1982). Neural networks and physical systems with emergent collective properties, *Proc. Natl. Acad. Sci. USA*, **79**, 2554-2558.

Hopfield, J.J. (1984). Neurons with graded response have collective computational properties like those of two-state neurons. *Proc. Natl. Acad. Sci.*, **81**, 3088-3092.

Koulakov, A.A., Raghavachari, S., Kepecs, A., & Lisman, J.E. (2002). Model for a robust neural integrator. *Nat. Neurosci.*, **5**, 775-782.

Loewenstein, Y. & Sompolinsky, H. (2003). Temporal integration by calcium dynamics in a model neuron. *Nat. Neurosci.*, **6**, 961-967.

Lukashin, A.V., Amirikian, B.R., Moxahev, V.L., Wilcox, G.L., & Georgopoulos, A.P. (1996). Modeling motor cortical operations by an attractor network of stochastic neurons, *Biol. Cybern.*, **74**, 255-261.

Mainen, Z.F. & Sejnowski, T.J. (1995). Reliability of spike timing in neocortical neurons. *Science*, **268**, 1503-1506.

Prut, Y., Vaadia, E., Bergman, H., Haalman, I., Slovin, H., & Abeles, M. (1998). Spatiotemporal Structure of Cortical Activity: Properties and Behavioral Relevance. *J. Neurophysiol.*, **79**, 2857-2874.

Renart, A., Sing, P., Wang, X.-J. (2003). Robust spatial working memory through homeostatic synaptic scaling in heterogeneous cortical networks. *Neuron*, **38**, 473-485.

Rolls, E.T. & Treves, A. (1998). *Neural Networks and Brain Function.* Oxford University Press, Oxford.

Seung, H.S. (1996). How the brain keeps the eyes still. *Proc. Natl. Acad. Sci. USA*, **93**, 13339-13344.

Seung, H.S., Lee, D.D., Reis, B.Y., & Tank, D.W. (2000). Stability of the memory of eye position in a recurrent network of conductance-based model neurons. *Neuron*, **26**, 259-271.

Shadlen, M.N. & Newsome, W.T. (1998). The variable discharge of cortical neurons: implications for connectivity, computation and information coding. *J. Neurosci.*, **18**, 3870-3896.

Shriki, O., Hansel, D., & Sompolinsky, H. (2003). Rate models for conductance-based cortical neuronal networks. *Neural Comp.*, **15**, 1809-1841.

Sompolinsky, H. & Shapley, R. (1997) New perspectives on the mechanisms for orientation selectivity. *Current Opinion in Neurobiology*, **7**, 514-522.

Tsodyks, M.V. & Feigelman, M.V. (1988). The enhanced storage capacity in neural networks with low activity level. *Europhys. Let.*, **6**, 101-105.

Tuckwell, H.C. (1988). *Introduction to Theoretical Neurobiology.* Cambridge University Press, Cambridge.

van Vreeswijk, C. (1996). Partial synchronization in populations of pulse-coupled oscillators. *Phys. Rev. E*, **54**, 5522-5537.

Wang, X.-J. & Buzsáki, G. (1996). Gamma oscillations by synaptic inhibition in a hippocampal interneuronal network. *J. Neurosci.* **16**, 6402-6413.

Wang, X.-J. (2001). Synaptic reverberation underlying mnemonic persistent activity. *Trends Neurosci.*, **24**, 455-63.

Wilson, H.R. & Cowan, J. (1972). Excitatory and inhibitory interactions in localized populations of model neurons. *Biophys. J.*, **12**, 1-24.

Zhang, K. (1996). Representation of spatial orientation by the intrinsic dynamics of the head-direction cell ensemble: a theory. *J. Neurosci.*, **16**, 2112-2126.

Course 9

# IRREGULAR ACTIVITY IN LARGE NETWORKS OF NEURONS

C. van Vreeswijk[1] and H. Sompolinsky[2,3]

[1] *CNRS UMR 8119, Université Paris 5 René Descartes, 45 rue des Saints Pères,*
*75270 Paris Cedex 06,*
[2] *Racah Institute of Physics and Center for Neural Computation, Hebrew University, Jerusalem,*
*91904, Israel (permanent address)*
[3] *Department of Molecular and Cellular Biology, Harvard University, Cambridge, MA 02138, USA*

*C.C. Chow, B. Gutkin, D. Hansel, C. Meunier and J. Dalibard, eds.*
*Les Houches, Session LXXX, 2003*
*Methods and Models in Neurophysics*
*Méthodes et Modèles en Neurophysique*
© *2005 Elsevier B.V. All rights reserved*

341

# Contents

# 1. Introduction

The firing patterns of neurons *in vivo* almost always have a high temporal irregularity. The inter-spike interval (ISI) distribution of many neurons is, after a short refractory period, very close to that of a Poisson process [1–5]. Intracellular recordings show that the voltage of these neurons also fluctuates irregularly [3, 6]. The origins of these irregularities and their functional implications are only poorly understood [7–9]. *In vitro* experiments show that when a constant current is applied, neurons fire highly regularly. Thus the irregularity of the firing has to be due to irregular fluctuations in the in put [11, 12]. To some extent these fluctuations may be due to stochastic nature of the synaptic release, or due to fluctuations in the stimuli intensity. But since neurons in the vertebrate central nervous system receive inputs from thousands of synaptic boutons one would expect the summation of these inputs at the soma to average out most of these fluctuations. This should be true in particular when the local network is firing vigorously, so that the neuron receives many synaptic inputs within a single integration time-constant [4, 11]. This can be captured by the following simple toy model. Let $S(t)$ be the sum of $K$ binary variables, $\sigma_i$ which take the value 1 with probability $m$ and are 0 otherwise. In the sum $S$ each contributes a amount $J/K$,

$$S(t) = \frac{J}{K} \sum_{i=1}^{K} \sigma_i(t). \tag{1.1}$$

Using the central limit theorem one sees that the average value of $S$ is $< S > = Jm$, while the variance, var($S$) of $S$ is var($S$) $= J^2(m - m^2)/K$ if the variables $\sigma_i$ are uncorrelated. Thus, if $K$ is large the fluctuations will be small. One way to avoid this is to have a substantial correlation between the neurons, so that the fluctuations do not average out. In our example, if $< \sigma_i(t)\sigma_j(t) > = (1 + C)m^2$ instead of $m^2$ the variance in the input will be var($S$) $= J^2Cm^2 + J^2[m - (1 + C)m^2]/K$. Cortical neurons often have positive pair-wise correlations on the time scale of a single integration time constant (about 10 msec) [2, 13–16].

Since neurons are strongly non-linear, it may seem natural for highly connected networks of neurons to evolve to a state that exhibits chaotic behavior, but in most large model networks with high connectivity show highly ordered

345

behavior, if there is not strongly stochastic component in the external input into the cells [17–23]. There are however a few examples of numerical studies of networks of fairly complex neurons in which deterministic dynamics evolves to a state in which the cells fire irregularly [24–26]. These are different from the network we describe here in that they require strong synchrony to induce synchronous chaos into the network.

However, if one considers larger populations, the correlations are often much weaker, possibly due to weak negative correlations between many other pairs, which do not have measurable effect on the individual pair-wise correlations. Another way to get fluctuations that do become negligible small in the large $K$ limit, which is explored in this course, is that the individual synapses are not that small. This potentially results in a enormous average, with reasonably sized fluctuations. But if one has both excitatory and inhibitory inputs, one can have a net input that has a of order 1, even though the total excitation and the total inhibition are large, because these two will have opposite sign, and can nearly cancel each other if the variables are chosen in the right way [27–33]. To extend to our toy model: We have $K$ uncorrelated excitatory units $\sigma_i^E$, which are active with probability $m_E$, and $K$ uncorrelated inhibitory units $\sigma_i^I$, which are active with probability $m_I$, and these contribute to our sum $S$ as

$$S(t) = \frac{J_E}{\sqrt{K}} \sum_i \sigma_i^E(t) - \frac{J_I}{\sqrt{K}} \sum_j \sigma_j^I(t). \tag{1.2}$$

Under these assumptions $S$ is on average $<S> = \sqrt{K}(J_E m_E - J_I m_I)$, while the variance is given by $\mathrm{var}(S) = J_E^2(m_E - m_E^2) + J_I^2(m_I - m_I^2)$. If the inputs are balanced, $m_I = J_E m_E / J_I + O(1/\sqrt{K})$, both the mean, $<S>$, and the variance, $\mathrm{var}(S)$, will be of order 1, even in the large $K$ limit.

This leads naturally to the question how the rates can be organized in such a way that the inputs are balanced. In a network with both excitatory and inhibitory neurons one would need to have a balance of the excitatory and inhibitory inputs both for the excitatory and inhibitory neurons. This suggests that one may have to fine-tune the connection strengths to obtain a balance in both populations at the same time. On the other hand, in a network with two populations of neurons, one has two rates that can be varied, so that maybe no fine-tuning is necessary.

Several studies have investigated numerically the balanced state in networks of integrate and fire neurons with a stochastic feedforward input [34–36]. In these studies the variability of the neuronal output is at least partially due the the fluctuations in the external input into the network. One would like to know whether the variability of the network response is merely an amplification of the external noise, or whether it is intrinsic to the network, and even present in a network with constant external drive.

In this course we will study how strongly coupled network of neurons can evolve to an asynchronous chaotic state, characterized by strongly irregular activity and very weak global correlations. Since irregular activity is ubiquitous in the central nervous system, one of the underlying assumptions is that this chaotic state does not depend on special properties of the neuron. Therefore it should be possible to study this state with networks of very simple model neurons. Such models may also have the advantage to be amenable to analytical treatment. As it turns out, the basic behavior can already be observed in models with binary neurons. This is the case provided that the connectivity is fairly sparse. Each cell should receive input only from a small percentage of the other cells. A simple way in which this can be done is by assuming sparse random connection matrices.

Statistical mechanics has developed many tools to study networks of units that are connected with random matrices [37–39] and we will use this course to review some of these. The simplest model, a randomly connected network with an excitatory and an inhibitory population is studied in section 2. Most of the basic techniques can already be brought to bare on this model. In section 3 we investigate and extension of the simple model that can be used as a long term memory model. The advantage of this model is that it can be studied in detail analytically. A disadvantage is that the memory model is not very robust, and some of its features are unrealistic. As will already be made clear in section 2, the quenched disorder in the connectivity induces a strong heterogeneity in the activity of the neurons. In section 4 we study a model for a hypercolumn in the primary visual cortex. In this model the activity of the neurons shows a similar heterogeneity, and we will see how this allows us to study heterogeneity in conjunction with invariance with respect of some stimulus features, contrast invariance in this case. Al these analytical results hold for binary neurons, and after one understands the mechanisms that lead the asynchronous chaotic state, one can be fairly confident that these results also hold for models with more neuron-like dynamics. However, to be sure, one needs to investigate whether this is the case. In section 5 this is done for networks of integrate-and-fire (IF) neurons through numerical simulations of large networks of such neurons. An issue that comes up often in such extensions, is that the that usually one can only make qualitative predictions about more realistic models based on the analysis of the simpler system. One of the goal in section 5 is to demonstrate by example, how this can be done fruitfully.

## 2. A simple binary model

We begin the investigation of the balanced state by considering a generic model of neuronal network. Since irregular activity is a feature of all cortical and subcortical areas, one would expect that the mechanisms that lead to such activity does not depend on the details of the connectivity or non-generic properties of the neurons. Let us therefore consider a model with the simplest possible connectivity architecture and the simplest possible model neurons.

## 2.1. The model

Before we describe the model it is useful introduce notation conventions that make the equations simple to follow. If a variable refers to an excitatory (inhibitory) population it is given a sub- or superscript $E$ ($I$). If we want to refer to either population, we will use variable $A, B, \ldots$. The neurons of both populations are numbered. We will use $i, j, k, \ldots$ to denote the different cells.

The model, introduced in [40,41], consists of $N_E$ excitatory and $N_I$ inhibitory neurons. The neurons are binary variables. The cells are updated asynchronously, in random order with one update per cell in population $A$ per time constant $\tau_A$. If at the time $t$ of the update of neuron $i$ of population $A$ its input, $u_i^A(t)$, exceeds the neurons threshold, $\theta_i^A$, the state $\sigma_i^A$ is set to 1, otherwise it is set to 0

$$\sigma_i^A = \Theta(u_i^A(t) - \theta_i^A), \tag{2.1}$$

where $\Theta$ is the Heaviside function. The total input into the cell consists of three components, an excitatory external input, excitatory feedback and inhibitory feedback. It can be written as

$$u_i^A(t) = u_A^0 + \sum_B \sum_j J_{ij}^{AB} \sigma_B^j(t), \tag{2.2}$$

Here $J_{ij}^{AB}$ is the matrix that describes the connections from neuron $j$ in population $B$ to neuron $i$ in population $A$. The connections are random with a probability $K/N_B$ that there is a connection, so that a neuron receives input from on average $K$ excitatory and $K$ inhibitory cells. If there is a connection, $J_{ij}^{AB} = J_{AB}/\sqrt{K}$, otherwise $J_{ij}^{AB} = 0$. $J_{AB}$ is positive for $B = E$ and negative for $B = I$, to ensure that these refer to excitatory and inhibitory populations respectively.

Because the neuron receives feedback input from on average $K$ cells per population, and the individual input contribute of order $1/\sqrt{K}$, the total input from each population scales as $\sqrt{K}$. We want the feedforward input into the cells to have the same dependence on the average number of inputs, so we will write

$$u_A^0 = \sqrt{K} E_A m_0. \tag{2.3}$$

where $E_A$ is of order unity and $0 \le m_0 < 1$. As noted in the introduction, we will assume that the external input is not fluctuating, and identical for all cells.

Since for binary units the absolute scale of the input is irrelevant, we can rescale all inputs and the threshold by the same amount, without affecting the activity of the cells. Thus the dependence of the activity on the connection matrices should be only through the ratios of the coupling strengths for of the synapses projecting to each population. We will use the variables $E \equiv E_E/J_E E$, $I \equiv E_I/J_{IE}$, $J_E \equiv -J_{EI}/J_{EE}$ and $J_I \equiv -J_{II}/J_{IE}$ in this dependence. (The minus sign is used in the latter two to ensure that these variables are all positive.)

Even neurons from the same class have different properties, no two pyramidal cells are alike. So besides the fact that different neurons receive different inputs, their intrinsic properties are also different. This heterogeneity may affect the behavior of the network, and we want to take this into account. This is done by assuming that the thresholds of the neurons vary. The thresholds $\theta_i^A$ are randomly drawn from a distribution $\rho_A$.

## 2.2. Population averaged activity

In describing the behavior of the system we encounter the problem that the network has random connectivity. Therefore the detailed behavior of the system will depend on the exact choice of the connection matrices. However, if one looks at the population average variables in large networks, these quantities will be almost the same for almost all realizations of the network. As one considers larger and larger networks, this difference becomes smaller and smaller, the system is self-averaging. Thus if one calculates these properties for one realization of the network, one has determined them for (almost) all networks. This is still not very useful, because the analytical expressions one obtains will involve the network architecture in a complicated manner, but since the results for all realizations is the same, one can take these expressions, and *average over the realizations* of the network. Usually this leads to much simpler expressions.

To study large networks of neurons, each with many connection, it is usually easiest to take the limit where $N_A$ and $K$ go to infinity, and add finite size corrections. A subtlety here is how to take these limits. In standard mean field theory one takes $N_A$ and $K$ to infinity, keeping $K/N_A$ fixed. A difficulty with this limit is that, because there are loops in the network graph, correlations can build up, and these need to be taken into account. If the connectivity is sparse, $K/N_A$ is small, one can also approximate the system by an infinitely diluted network, where one first takes the limit $N_A \to \infty$ and the limit $K \to \infty$ afterward. The result of this is that, in the graph, the loops become of infinite length (and can be ignored) after the limit $N_A \to \infty$ is taken, so that the correlations are negligible [39, 42]. This simplifies the analysis considerably.

A typical computational unit in the cortex, for example a hypercolumn in the primary visual cortex (V1), has approximately $10^5$ neurons, each receiving input from $10^3$ - $10^4$ other cells, so that the probability of a connection is rather small, between 0.1 and 0.01. Therefore we will study the model in the sparse limit.

We denote by $m_A$ the population averaged activity of neurons in population $A$.

$$m_A(t) = \frac{1}{N_A} \sum_i \sigma_i^A(t). \tag{2.4}$$

If and arbitrarily chosen neuron of population $A$ is updated at time $t$, the probability that its state is set to $\sigma_i^A = 1$, $P_A$ depends on the population averaged rates, $m_E(t)$ and $m_I(t)$. $P_A$ is the probability that the total input is above the neurons threshold

$$P_A = \sum_{n_E,n_I} p(n_E|m_E)p(n_I|m_I)\langle \Theta \left( \sqrt{K}E_Am_0 + \sum_B \frac{J_{AB}}{\sqrt{K}}n_A - \theta_i^A \right) \rangle_{\theta_i^A}. \tag{2.5}$$

Here $\langle \cdots \rangle_{\theta_i^A}$ denotes averaging over the threshold distribution, and $p(n|m)$ is the probability of receiving $n$ active inputs from a population, given that a fraction $m$ of the cells in this population is active. In the large $N_A$ limit the probability that $k$ synapses from one of the population projecting to the cell is $K^k e^{-K}/k!$. In a sparse network the input are uncorrelated, so that the probability that $n$ of these are active is given by the binomial distribution. Therefore

$$p(n|m) = \sum_{k=n}^{\infty} K^k \frac{e^{-K}}{k!} \binom{k}{n} m^n (1-m)^{k-n}$$

$$= (mK)^n \frac{e^{-mK}}{n!} \tag{2.6}$$

is a Poisson distribution with mean $mK$. In the large $K$ limit this $p(n|m)$ can be replaced by a Gaussian with mean and variance $mK$. Inserting this into Eqn. (2.5) one obtains that, for large $K$, $P_A(m_E, m_I)$ is given by

$$P_A(m_E, m_I) = \int d\theta \, \rho_A(\theta) \int Dx \, \Theta(u_A - \theta + \sqrt{\alpha_A}x) \tag{2.7}$$

where $\rho_A(\theta)$ is the distribution of thresholds, $Dx$ is the Gaussian measure, $Dx = e^{-x^2/2}dx/\sqrt{2\pi}$. The mean, $u_A$, and variance, $\alpha_A$, of the inputs satisfy

$$u_A = \sqrt{K} \left( E_Am_0 + \sum_B J_{AB}m_B \right) \tag{2.8}$$

and

$$\alpha_A = \sum_B (J_{AB})^2 m_B \qquad (2.9)$$

respectively.

Between times $t$ and $t + dt$ a fraction $dt/\tau_A$ of the neurons of population $A$ is updated. Before the update the probability that they were in the active state was $m_A(t)$, after the update this probability is $P_A(m_E(t), m_I(t))$. Hence the average rates satisfy

$$\tau_A \frac{d}{dt} m_A = -m_A + \int d\theta \rho_A(\theta) \int Dx \Theta(u_A - \theta + \sqrt{\alpha}x), \qquad (2.10)$$

where $u_A$ and $\alpha_A$ are given by Eqns. (2.8) and (2.9). This is the familiar Glauber dynamics [43, 44].

If the input is constant, the network may evolve to a solution with constant rate. These can be simplified using the complimentary error-function, $H(x) = \int_x^\infty dy\, e^{-y^2/2}/\sqrt{2\pi}$, with which the equilibrium rates can be written as

$$m_A = \int d\theta\, \rho_A(\theta) H\left(\frac{\theta - u_A}{\sqrt{\alpha_A}}\right). \qquad (2.11)$$

Because we use synapses that scale as $1/\sqrt{K}$ the variances $\alpha_A$ stay finite in the large $K$ limit, the fluctuations do not go to zero as $K$ is increased. But this also has the effect of making the mean input potentially of order $\sqrt{K}$. However, if the population rates are such that the inhibitory feedback to leading order cancels the feedforward input and excitatory feedback, both the mean and variance of the input can be finite in the large $K$ limit.

In the large $K$ limit there exist two kind of solutions for the equilibrium rates. Unbalanced solution in which $m_A = 0$ and $u_A = -\infty$, or $m_A = 1$ and $u_A = +\infty$, for at least one of the populations, and a balanced solution in which $0 < m_A < 1$ and $u_A$ is finite.

To see what a balanced state implies we write $m_A$ as a power series in $1/\sqrt{K}$

$$m_A = m_A^{(0)} + \frac{m_A^{(1)}}{\sqrt{K}} + \frac{m_A^{(2)}}{K} + \dots \qquad (2.12)$$

The average input can now be written as

$$u_A = u_A^{(-1)}\sqrt{K} + u_A^{(0)} + \frac{u_A^{(1)}}{\sqrt{K}} + \dots \qquad (2.13)$$

where

$$u_A^{(n)} = \sum_B J_{AB} m_B^{(n+1)} \tag{2.14}$$

for $n = 0, 1, 2, \ldots$, while

$$u_A^{(-1)} = E_A m_0 + \sum_B J_{AB} m_B^{(0)}. \tag{2.15}$$

In the balanced state $u_A$ stays finite in the large $K$ limit, $U_A^{-1)} = 0$, which implies that, the balance conditions

$$E m_0 + m_E^{(0)} - J_E m_I^{(0)} = 0 \tag{2.16}$$

$$I m_0 + m_E^{(0)} - J_I m_I^{(0)} = 0, \tag{2.17}$$

are satisfied. Solving these equations one finds

$$m_E^{(0)} = \frac{J_I E - J_E I}{J_E - J_I} m_0 \equiv A_E m_0 \tag{2.18}$$

$$m_I^{(0)} = \frac{E - I}{J_E - J_I} m_0 \equiv A_I m_0. \tag{2.19}$$

Since the activities cannot be negative, $A_E$ and $A_I$ must be positive. This puts constraints synaptic strengths. If one also requires that there is no unbalanced equilibrium state (no self-consistent state with $m_E = 0$ and $u_E^{(-1)}$ negative, $m_E = 1$ and $u_E^{(-1)}$ positive, etc.), one finds that the synaptic strengths have to satisfy

$$\frac{E}{I} > \frac{J_E}{J_I} > 1, \quad \text{and} \quad J_E > 1. \tag{2.20}$$

Under these constraints one can also show that the balanced state is stable, provided that the inhibitory time constant $\tau_I$ is sufficiently small. Generally $\tau_I$ can be of the same order as the excitatory time constant, $\tau_E$. For details about the stability analysis we refer to [41].

Equations (2.18) and (2.19) determine the leading order of the population rates, $m_A$, and therefore the rates in the large $K$ limit. From these rates one obtains the variance in the input to leading order, $\alpha_A^{(0)}$, but not the leading term in mean input, $u_A^{(0)}$, since this depends on the next leading order term in the rates, $m_A^{(1)}$. Instead, $u_A$, in the large $K$ limit, can be determined using Eqn. (2.11) and is the solution of

$$A_A m_0 = \int d\theta \, \rho_A(\theta) H\left(\frac{\theta - u_A}{\sqrt{\alpha_A}}\right), \tag{2.21}$$

where $\alpha_A = \sum_B (J_{AB})^2 A_B m_0$.

Surprisingly, considering that transfer function of the neurons is extremely non-linear, the population rates $m_A$ scale linearly with the rate of the external input, $m_0$. This is because the network finds an operating point at which the inputs balance and the linear input/output relation reflects the linear summation of the inputs.

In a realistic neural network, each neuron will receive input from 1000 to 10000 other cells. While this is relatively large, the effects of finite number of synaptic inputs, $K$ cannot, in general, be neglected. This is the case because, in the large $K$ expansion, the large term is not $K$, but $m_A K$. In the cortex neurons fire typically at a rate of a few to at most a few tens of spikes per second, while in vitro one can easily drive these cells to a sustained firing rate of several hundred Hertz. Thus biologically relevant is the regime where the cells fire much below their maximum rate, and $m_A$ is small. Therefore one cannot neglect the finite $K$ corrections. These corrections can be incorporated, to high accuracy, by observing that the finite $K$ effects on the mean input is much more appreciable than its effect on the distribution around its mean. The latter is still close to a Gaussian with a variance $\alpha_A = \sum_B (J_{AB})^2 m_B$, so that for finite $K$ we can use that the rates, $m_A$, are the solutions of

$$m_A = \int d\theta \rho_A(\theta) H\left(\frac{\theta - u_A}{\sqrt{\alpha_A}}\right), \qquad (2.22)$$

where $u_A = \sqrt{K}(E_A m_0 + \sum_b J_{AB} m_B)$.

Figure 1 shows the $K \to \infty$ rates $m_A$ as a function of $m_0$, also shown are the finite $K$ solutions with $K = 1000$. In the large $K$ limit the network response is linear with the input. The finite $K$ corrections cause the output to have a small threshold. Near the threshold the response is quite non-linear, reflecting the non-linearity of the I-f curve of the (binary) neurons, while sufficiently above the threshold the response is indeed quite linear, due to the linear summation of the synaptic inputs.

### 2.3. Heterogeneity of the firing rate.

Above we have determined the average probability $m_A$ that a cell in population $A$ is active. However a probability $m_A$ that a cell can be active can be achieved in many ways. At the extremes we can have: 1) a fraction $m_A$ of the cells is always active, $\sigma_i^A(t) = 1$, while the others are always inactive, $\sigma_i^A(t) = 0$, or 2) each cell has an equal probability $m_A$ to be active at any given time. To characterize the state in which the network is, we need to determine how the activity is distributed.

To understand this disorder, lets us first consider a homogeneous network, where $\theta_i^A = \theta_A$. In such a network the variability in the average firing rate

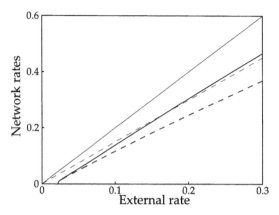

Fig. 1. The average excitatory (solid) and inhibitory (dashed) rates as a function of the external input $m_0$. The thin straight lines represent the response in the large $K$ limit, while the bold curves show the response in the case where $K = 1000$. Other parameters are, $E = 1$, $I = 0.8$, $J_E = 2$, $J_I = 1.8$, $\theta_E = 1$ and $\theta_I = 0.7$. (Figure adapted from [41])

is due to a variability in the time averaged input. The latter has two sources. Firstly, because the synaptic connections are random, the standard deviation of the number of inputs is $\sqrt{K}$. Because each of these inputs contributes an amount of order $1/\sqrt{K}$ to the input, this contributes an amount of order 1 to the disorder in the time averaged input. The second contribution comes from the variability of the rates itself. If the standard deviation in the rates of population $A$ is $\delta m_A$, the sum of $K$ input rates will have a variance $\sqrt{K}\delta m_A$, which also contributes an amount of order 1 to the variability of the average input. Thus, the disorder in the input causes a disorder in the firing rates, but is also caused by it. To determine the disorder we have relate these two self-consistently.

The mean field theory that we are using here can be used to do this. The input into neuron $i$ of population $A$ can be written as

$$u_i^A(t) = u_A + \Delta u_i^A + \delta_i^A(t), \tag{2.23}$$

where the fluctuating part of the input $\delta u_i^A$ has a temporal average $< \delta_i^A(t) >_t = 0$. Both quenched disorder $\Delta u_i^A$ and the temporal fluctuations $\delta u_i^A$ are Gaussian variables. If we write the variance of the quenched disorder as $\beta_A$, the equal time correlation of the fluctuation will satisfy $< [\delta_i^A(t)]^2 >_i = \alpha_A - \beta_A$.

Thus we can write for the input

$$u_i^A(t) = u_A + \sqrt{\beta_A} x_i^A + \sqrt{\alpha_A - \beta_A} y_i^A(t), \tag{2.24}$$

where the random variables $x_i^A$ and $y_i^A(t)$ are drawn from zero mean Gaussian distributions with $< x_i^A x_i^B >=< y_i^A(t) y_j^B(t) >= \delta_{i,j} \delta_{A,B}$, while $< y_i^A(t) y_i^A(t') >= 0$ for $|t - t'| \to \infty$.

We will use $m_i^A$ to denote the temporal average of the activity of neuron $i$ of population $A$, $m_i^A =< \sigma_i^A(t) >_t$. This local rate is determined by the neurons threshold, $\theta_i^A$ and the quenched variable $x_i^A$, and satisfies

$$
\begin{aligned}
m_i^A &= \int Dy \, \Theta \left( u_A + \sqrt{\beta_A} x_i^A - \theta_i^A + \sqrt{\alpha_A - \beta_A} y \right) \\
&= H \left( \frac{\theta_i^A - u_A - \sqrt{\beta_A} x_i^A}{\sqrt{\alpha_A - \beta_A}} \right).
\end{aligned}
\tag{2.25}
$$

The temporal average of the input $u_i^A$, $< u_i^A >_t = u_i^A + \sqrt{\beta_A} x_i^A$ can be written in terms of the local firing rates, $m_i^A$, as

$$
< u_i^A >_t = E_A m_0 \sqrt{K} + \sum_B \sum_j J_{ij}^{AB} m_j^B.
\tag{2.26}
$$

Since $\beta_A$ is given by $\beta_A = N_A^{-1} \sum_i (< u_i^A >_t)^2 - u_A^2$, we can determine the amplitude of the quenched disorder from the distribution of the local rates, $m_i^A$. The average of $(< u_i^A >_t)^2$ is given by

$$
(< u_i^A >_t)^2 = \left( E_A m_0 \sqrt{K} + \sum_B \sum_j J_{ij}^{AB} m_j^B \right)^2,
\tag{2.27}
$$

averaged over the synaptic matrices $J_{ij}^{AB}$. On average $(< u_i^A >_t)^2$ is given by is given by

$$
< (< u_i^A >_t)^2 >_J = K E_A^2 m_0^2 + 2\sqrt{K} E_A m_0 < \sum_B \sum_j J_{ij}^{AB} m_j^B >_J +
$$

$$
+ < \sum_{B,C} \sum_{j,k} J_{ij}^{AB} J_{ik}^{AC} m_j^B m_k^C >_J
\tag{2.28}
$$

We will perform the average in two steps, first we fix the number of inputs, $k_B$ from neurons from population $B$ and average over all matrices with this number of inputs, and then we average over $k_B$. With $k_A$ fixed, the probability of having an input from any of the cells is equally likely, so that in the second term we can in the right hand side of Eqn. (2.28) we can replace $m_j^B$ by $m_B$. Likewise each pair of neurons is equally likely to project to the cell, so $m_j^B m_k^C$ can be replaced by $m_B m_C$ in the third term, except if $B = C$ and $j = k$. Thus $< (< u_i^A >_t)^2 >_J$ can be written as

$$
< (< u_i^A >_t)^2 >_J = K E_A^2 m_0^2 + 2 E_A m_0 \sum_B < k_B >_k J_{AB} m_B +
$$

$$
+ \sum_{B,C} \frac{< k_B (k_C - \delta_{B,C}) >_k}{K} J_{AB} J_{AC} m_B m_C +
$$

$$
+ \sum_B \frac{k_B}{K} J_{AB}^2 \frac{1}{N_B} \sum_j (m_j^B)^2, \tag{2.29}
$$

where we have used $< \cdots >_k$ to denote the average over $k_A$. At this point it is convenient to introduce a new order parameter, $q_A$, defined are the mean square of the local rate

$$
q_A = \frac{1}{N_A} \sum_i (m_i^A)^2. \tag{2.30}
$$

With this order parameter and the statistics of $k_B$, $< k_B >_k = K$ and $< k_B (k_B - 1) >_k = K^2$ we obtain averaged input as

$$
< (< u_i^A >_t)^2 >_J = K \left( E_A m_0 + \sum_B J_{AB} m_B \right)^2 + \sum_B J_{AB}^2 q_B. \tag{2.31}
$$

Therefore $\beta_A$ is given by

$$
\beta_A = \sum_B (J_{AB})^2 q_B. \tag{2.32}
$$

At the same time we have for $q_A$, using Eqn. (2.25)

$$
q_A = \int Dx \int d\theta \, \rho_A(\theta) H^2 \left( \frac{\theta - u_A - \sqrt{\beta_A} x}{\sqrt{\alpha_A - \beta_A}} \right). \tag{2.33}
$$

Thus $\beta_E$ and $\beta_I$ can be written as functions of $q_E$ and $q_I$, while at the same time $q_E$ and $q_I$ can be written as functions of $\beta_E$ and $\beta_I$. In general self-consistent solutions of these equations satisfy $m_A^2 \leq q_A \leq m_A$. The limit $q_A \to m_A$ indicates a frozen state solution in which a fraction $m_A$ of the neurons is always active and $q_A = m_A^2$ indicates a state in which all cells have an identical rate.

Note that Eqns. (2.32) and (2.33) have always one solution in which $q_A = m_A$. This is because for $\beta_A = \alpha_A$

$$
H([\theta - u_A - \sqrt{\beta_A} x] / \sqrt{\alpha_A - \beta_A}) = \Theta(u_A + \sqrt{\alpha_A} x - \theta). \tag{2.34}
$$

However, one can show that this state is unstable and that there is always one other solution, in which the activity fluctuates, and $q_A < m_A$.

Once $m_A$ and $q_A$ are determined, and thereby $u_A$, $\alpha_A$ and $\beta_A$, one can determine the mean input into the cells, the amplitude of quenched fluctuations and the size of the temporal fluctuations. Since these fluctuations are all Gaussian, this is sufficient to fully determine the distribution of the firing rates. Details of how to do this can be found in [41]. Here we will briefly describe the main findings, concentrating on the biologically interesting regime, where $m_A$ is small.

In a network with identical inhibitory and identical excitatory neurons, $\theta_i^A = \theta_A$ one can show that if $m_A$ becomes very small, $q_A$ approaches $q_A = m_A^2 + O(m_A^3 |\log(m_A)|)$. The distribution of firing rates becomes very narrow. For a network in which the thresholds are uniformly distributed, between $\theta = \theta_A$ and $\theta = \theta_A + \Delta$ the order parameter $q_A$ satisfies $q_A \propto m_A^{3/2}$, for small $m_A$. This means that, for low rates, the rate distribution becomes very skewed, and extremely broad relative to the mean rate.

This is shown in Fig. 2A where the distribution of the activity rates for the excitatory population is depicted for two different values of the average rate, $m_E = 0.01$ and $m_E = 0.1$, in a network of neurons with uniformly distributed thresholds. For both mean rates the distribution is skewed, but for the lower rate the distribution is both more skewed and broader relative to the mean. For comparison, Fig. 2B shows an experimentally determined firing rate distribution from a monkey right prefrontal cortex (Abeles *et al.* [45] unpublished data). During the measurement the monkey was attending to a complex stimulus with visual as well as a auditory component, while executing a reaching movement. In agreement with the distribution predicted by the simple model, this distribution is unimodal and broad and has a substantial skewness.

## 2.4. Temporal correlations

To complete the description of the structure of the balanced state, one needs to determine the temporal correlations of input into the cell, and thereby its output statistics. The quenched disorder could be determined from the average $< u_i^A(t) u_i^A(t') >$ in the limit $|t - t'| \to \infty$. To determine the temporal structure of the input we need to know $< u_i^A(t) u_i^A(t') >$ also for intermediate values of $t - t'$. Using arguments similar to those above, one obtains that

$$\beta_A(\tau) \equiv < u_i^A(t) u_i^A(t + \tau) > -u_A^2 = \sum_B (J_{AB})^2 q_B(\tau), \tag{2.35}$$

where $q_A(\tau) = < \sigma_i^A(t) \sigma_i^A(t + \tau) >$. Obviously $q_A(-\tau) = q_A(\tau)$, and because $\sigma_i^A$ is a binary variable, $q_A(0) = m_A$. Keeping in mind that the neuron was last

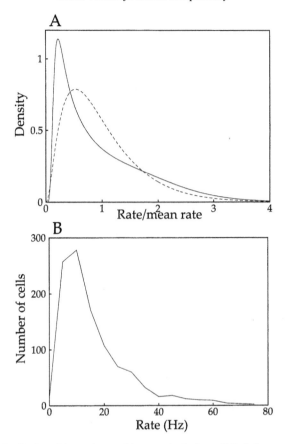

Fig. 2. A) The distribution of the excitatory firing rates in the large $K$ limit for two different values of the mean rate, $m_E = 0.01$ (solid line) and $m_E = 0.1$ (dashed line), in a network with uniformly distributed thresholds. Parameters are as in figure 1, except that $\Delta = 0.2$. B) Firing rate distribution of neurons in the prefrontal cortex of a monkey attending to a complex signal and performing a reaching movement. Rates were averaged over duration of event showing significant response. (Figure adapted from [41])

updated a time $t$ before the reference-point $\tau = 0$ with probability $\tau_A^{-1} e^{-t/\tau_A}$, one can show [41], that for $\tau > 0$ $q_A(\tau)$ satisfies

$$\tau_A \frac{d}{dt} q_A(\tau) = -q_A(\tau) +$$
$$\int_0^\infty \frac{dt}{\tau_A} e^{-t/\tau_A} \int Dx \int d\theta \, \rho_A(\theta) H^2 \left( \frac{\theta - u_A - \sqrt{\beta_A(\tau + t)}x}{\sqrt{\alpha_A - \beta_A(\tau + t)}} \right). \quad (2.36)$$

As it should $q_A(\tau)$ decays to $q_A$ as $\tau \to \infty$. A numerical solution of $q_E(\tau)$ is shown in Fig. 3. As can be seen the correlations fall of slower than predicted by Poisson statistics. This increase in the short-time correlations reflects the finite update rate of the cells that project into the cell.

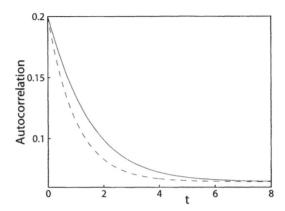

Fig. 3. Population averaged autocorrelation $q_E(\tau)$ in the large $K$ limit; The dashed lime shows the autocorrelation for a population with the same mean rate-distribution, but Poisson inputs. Parameters as in Fig. 1, and $m_0 = 0.1$, $\tau_E = 1$ and $\tau_I = 0.9$. (Figure adapted from [41])

Once the correlations $q_A(\tau)$ are determined, a full statistical description of the input into the cells is obtained. To get an idea of the total input into the neurons, Fig. 4 shows a sample of the input statistics for an excitatory neuron in a network with $K = 1000$. Both the total excitatory input (external plus excitatory feedback) and the total inhibitory input, as well as the net input are shown. To obtain the inputs it was assumed that its fluctuations can still be described by a Gaussian variable, with a correlation identical to that of the large $K$ limit, while the mean input was adjusted to agree with the finite $K$ approximation of $m_E$ and $m_I$.

The upper panel in Fig. 4 shows the total excitatory feedback (upper trace), and the total inhibitory feedback (lower trace), as well as the total net input (middle trace) The total excitatory input is far above the threshold (dashed line) and has fluctuations that are small compared to the mean. But the inhibitory input is negative by about the same amount, resulting in a net input that is somewhat below the threshold, but close enough so that the fluctuations can bring the cell to threshold at irregular intervals. The lower panel indicates the times at which the cell is updated from the passive to the active state, which corresponds roughly to a spike. Note that not every time that the input exceeds the threshold the neuron fires, this is because at such time the cell may not be updated.

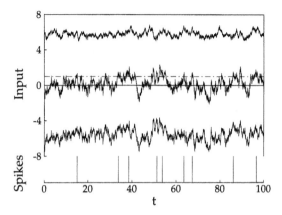

Fig. 4. Sample of total input into a cell in a network with $K = 1000$. Total excitatory (upper trace), total inhibitory (lower trace) and net input are shown in upper panel. Lower panel indicates times at which the cell is updated into the active state. Parameters as in Fig. 1, and $m_0 = 0.1$, $\tau_E = 1$ and $\tau_I = 0.9$. (Figure adapted from [41])

## 2.5. Dependence on initial conditions

The strong fluctuation in the total input into the cells in the absence of fluctuations in the external input, strongly suggest that the network is in a chaotic state. Strictly speaking this cannot be true because our model is not deterministic since the neurons are updated stochastically. At every update step a randomly chosen cells is updated. However one can construct a model with a fully deterministic dynamics in which the network evolves to a state in which the neurons receive inputs with exactly the same structure [41]. In that model the neurons are updated periodically, but different neurons have different update rates, with the update periods, $\tau$, randomly drawn from a distribution $R_A(\tau) = \tau e^{-\tau/\tau_A}/\tau_A^2$. To the extend that that model is chaotic, our model can also be said to be chaotic.

To show that the system is chaotic, one has to show that small perturbations in the state of the network grow at least exponentially. We consider two copies of the network, in one copy the states of the neurons are given by $\sigma_{i,1}^A$, in the other by $\sigma_{i,2}^A$. The network architectures are identical, they have the same $J_{ij}^{AB}$ and the networks receive identical external input $m_0(t) = m_0$. To negate the effects of the stochasticity in the update rule, we have to assume that the update schedule for the two copies is the same. The network is assumed to have reached its equilibrium, so that

$$\frac{1}{N_A} \sum_i \sigma_{i,p}^A = m_A, \tag{2.37}$$

with $p = 1, 2$. For both populations we define the distance, $d_A$, between the two networks as the fraction of neurons state is different in the two copies.

$$d_A(t) = \frac{1}{N_A} \sum_i \left( \sigma_{i,1}^A - \sigma_{i,2}^A \right)^2. \tag{2.38}$$

The network is in a chaotic state if, on average, a small perturbation of the input grows at least exponentially. To assess whether this is the case we determine the evolution of the average distance, $D_A = < d_A >$, conditioned on the constraint that at time $t = 0$, each copy of the network individually is at equilibrium (macroscopic variables such as $m_A$ have their stationary values), and the initial distances are given by $d_A(0)$.

To evaluate the dynamics of $D_A$ if is useful to write this as

$$\begin{aligned} D_A(t) &= \frac{1}{N_A} \sum_i \langle \left( \sigma_{i,1}^A(t) - \sigma_{i,2}^A(t) \right)^2 \rangle \\ &= \frac{1}{N_A} \sum_i \langle \sigma_{i,1}^A(t) \rangle + \langle \sigma_{i,2}^A(t) \rangle - 2\langle \sigma_{i,1}^A(t)\sigma_{i,2}(t) \rangle \\ &= 2 \left( m_A - Q_A(t) \right), \end{aligned} \tag{2.39}$$

where $Q_A \equiv N_A^{-1} \sum_i < \sigma_{i,1}^A \sigma_{i,2}^A >$ is the overlap between the two copies. The overlap $Q_A$ is similar to $q_A$, except rather than looking at the overlap of the neuron in the same copy at different times, one looks at the overlap of the neuron in different copies at the same time. Because of this similarity, the dynamics of $Q_A$ is similar to that of $q_A$, and is given by

$$\tau_A \frac{dQ_A}{dt} = -Q_A + \int Dz H^2 \left( \frac{\theta_A - u_A - \sqrt{\gamma_A(t)}z}{\sqrt{\alpha_A - \gamma_A(t)}} \right), \tag{2.40}$$

where $\gamma_A$ is given by $\gamma_A(t) = \sum_B (J_{AB})^2 Q_B(t)$.

The overlap $Q_A$ has two stationary solutions, $Q_A = q_A$ and $Q_A = m_A$. The solution $Q_A = m_A$ corresponds to two fully locked trajectories, where the two copies behave exactly the same. This solution is unstable, as shown below, and if the two copies start with $D_A(0) \neq 0$, the network will eventually reach $Q_A(t) = q_A$. The distance between the two copies at the same time will be the same as the distance in the same copy an infinite time apart.

Starting with nearly identical states, $D_A(0) \approx 0$, the evolution of $D_A$ for small $t$ can be determined by inserting $Q_A(t) = m_A - D_A(t)$ in Eqn. (2.40) and expanding in $D_A$. One finds that to leading order

$$\tau_A \frac{d}{dt} D_A = \frac{2}{\pi} \frac{e^{-u_A^2/2\alpha_A}}{\sqrt{\alpha_A}} \sqrt{\alpha_A - \gamma_A}. \tag{2.41}$$

Since $\alpha_A - \gamma_A = \sum_B (J_{AB})^2 D_A/2$, the distance grows faster than exponential, and even for an infinitesimal initial perturbation $D_A(0)$ the distance reaches a final value in finite time.

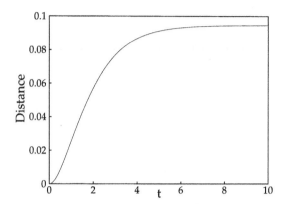

Fig. 5. Evolution of the distance $D_E$ starting from a small initial distance $D_E(0)$. Parameters as in Fig. 1 and $\tau_E = 1$ and $\tau_I = 0.9$. (Figure adapted from [41])

Figure 5 shows the evolution of $D_E$ starting from an initial condition $D_E(0) = D_I(0) \approx 0$. The distance very rapidly reaches the equilibrium rate. In contrast, for a system with a finite Lyapunov exponent, $\lambda_L$, the time to reach the saturation value for the distance should be proportional to $- \log(D_E(0))/\lambda_L$, and hence go to infinity as $D_E(0) \to 0$. It should be noted however, that this is an artifact of binary neurons. In similar systems with units that are represented as soft spins, that their input output relation is $\sigma = H(g(\theta - u))$ rather than $\sigma = \Theta(u - \theta)$, one obtains that the perturbations grow exponentially with and Lyapunov exponent proportional to $g$ [46]. Therefore the fact that the Lyapunov exponent for binary neurons is infinite, does not necessarily imply that the Lyapunov exponent is infinite for networks with more realistic model neurons.

### 2.6. Non-linearity of response of individual cells

Remarkably, in the large $K$ limit, the population averaged activities $m_E$ and $m_I$ vary linearly with the external activity rate $m_0$. As we have explained above, this reflects the linearity of synaptic integration, and negates the non-linearity of the individual cells response. It should be kept in mind, that the linearity of the population averaged response does *not* imply that the individual neurons respond linearly. Indeed that the cells cannot all behave linearly with the input can be seen in Fig. 2A. This figure clearly shows that the distribution of the firing rates changes as the average rate is changed. But if the activity of all neurons would scale linearly with the input, the distribution of the rate relative to the average rate would not change.

To investigate how to what extend the response of the individual neurons is non-linear, we will study the response of a network to two different external inputs, with rates $m_0$ and $\tilde{m}_0$ respectively. Using the methods outlined above, the variables mean rates $m_A$, mean input $u_A$ and variances $\alpha_A$ and $\beta_A$ for the first input, and means $\tilde{m}_A$, $\tilde{u}_A$ and variances $\tilde{\alpha}_A$ and $\tilde{\beta}_A$ for the second input can be determined.

Since the quenched and temporal fluctuations are the result of many small contributions, they are distributed in a Gaussian manner. To determine the relationship between the local firing rates $m_i^A$ and $\tilde{m}_i^A$ we need to determine how the quenched fluctuations in $<u_i^A>_t$ and $<\tilde{u}_i^A>_t$ are related. But since these are both due to many small contributions, their joint distribution must be a correlated Gaussian. Thus we can write

$$u_i^A(t) = u_A + \sqrt{\alpha_A - \beta_A}\, y_i^A(t) + \sqrt{\beta_A - B_A}\, x_i^A + \sqrt{B_A}\, z_i^A, \qquad (2.42)$$

and

$$\tilde{u}_i^A(\tilde{t}) = \tilde{u}_A + \sqrt{\tilde{\alpha}_A - \tilde{\beta}_A}\, \tilde{y}_i^A(\tilde{t}) + \sqrt{\tilde{\beta}_A - B_A}\, \tilde{x}_i^A + \sqrt{B_A}\, z_i^A. \qquad (2.43)$$

Here the variables $y_i^A(t)$, $\tilde{y}_i^A(\tilde{t})$, $x_i^A$, $\tilde{x}_i^A$, and $z_i^A$ are uncorrelated unitary Gaussians. To describe the relationship between the activity of the neuron for the inputs $m_0$ and $\tilde{m}_0$ we need to find the value of $B_A$. First observe that $B_A = <u_i^A \tilde{u}_i^A >_i - u_A \tilde{u}_A$. Furthermore $m_i^A$ and $\tilde{m}_j^B$ are uncorrelated for $\{i, A\} \neq \{j, B\}$. Now we introduce a new order-parameter, $Q_A$, defined by

$$Q_A = \frac{1}{N_A} \sum_i m_i^A \tilde{m}_i^A. \qquad (2.44)$$

Using this order-parameter one obtains for $B_A$

$$B_A = \sum_B (J_{AB})^2 Q_B. \qquad (2.45)$$

In a homogeneous network, $\theta_i^A = \theta_A$, this order parameter satisfies

$$\begin{aligned}
Q_A =\ & \int Dx \int Dy \int Dz \\
& \times \Theta(u_A + \sqrt{\alpha_A - \beta_A}\, y + \sqrt{\beta_A - B_A}\, x + \sqrt{B_A}\, z - \theta_A) \\
& \times \int D\tilde{x} \int D\tilde{y}\, \Theta(\tilde{u}_A + \sqrt{\tilde{\alpha}_A - \tilde{\beta}_A}\, \tilde{y} + \sqrt{\tilde{\beta}_A - B_A}\, \tilde{x}
\end{aligned}$$

*C.van Vreeswijk and H. Sompolinsky*

$$+ \sqrt{B_A} z - \theta_A)$$

$$= \int Dz \, H \left( \frac{\theta_A - u_A - \sqrt{B_A} z}{\sqrt{\alpha_A - B_A}} \right) H \left( \frac{\theta_A - \tilde{u}_A - \sqrt{B_A} z}{\sqrt{\tilde{\alpha}_A - B_A}} \right). \quad (2.46)$$

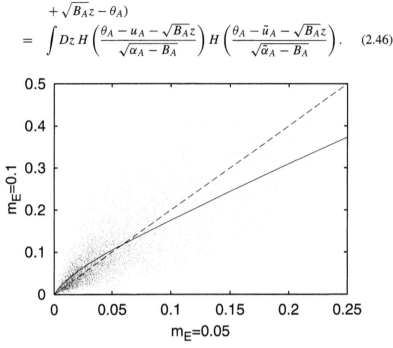

Fig. 6. Scatter-plot of the local rates of the excitatory cells in a large $K$ network. The local activity for $m_0 = 0.05$ is plotted against the neurons rate for $m_0 = 0.1$. The dashed line shows the line $y = 2x$, while the solid line shows $m_i^E$ at $m_0 = 0.1$ averaged over neurons with the same $m_i^E$ for input $m_0 = 0.05$.

These two equations determine $Q_A$ and $B_A$, so that the input statistics is now fully determined. This allows one to determine the joint probability distribution $\rho(m_i^A, \tilde{m}_i^A)$, using some straightforward, but tedious algebra. This does not lead to any useful insight which cannot be obtained much simpler. Figure 6 shows a scatter-plot of $m_i^E$ versus $\tilde{m}_i^E$ for $10^4$ excitatory neurons in a network in the large $K$ limit in which the input activities are $m_0 = 0.05$ and $\tilde{m}_0 = 0.1$ respectively. The dashed curve shows the line $y = 2x$, on which the neurons would lie if they had had a linear input-output relationship. The figure shows that the rates for the neurons in the population are scattered around this line. It clearly shows that whose rate are high, relative to the population, at the low input rate, are also the cells that fire most vigorously for the high input rate, while low rate neurons stay at low rate. Nevertheless, there is a considerable amount of scatter in the ratio of the firing rates.

There is also a trend visible in which the low rate neurons tend to fall above the line $y = 2x$, while the neurons with high rate tend to fall below this line.

This can be confirmed by calculation the average value of $\tilde{m}_i^A$, for neurons with the same $m_i^A$. For these neurons the difference between the threshold and the temporally averaged input, $\theta_i^A - u_A - \sqrt{\beta_A} x_i^A$ satisfies

$$\theta_i^A - u_A - \sqrt{\beta_A} x_i^A = \sqrt{\alpha_A - \beta_A} H^{inv}(m_i^A). \tag{2.47}$$

Since $< \Delta u_i^A \Delta \tilde{u}_i^A > = B_A$, we can write for the temporal average of the input into this cell with external input $\tilde{m}_0$,

$$< \tilde{u}_i^A > = u_A + B_A x_i^A / \sqrt{\beta_A} + \sqrt{\tilde{\beta}_A - B_A^2 / \beta_A} z_i^A,$$

where $z_i^A$ is drawn from a univariate Gaussian, and uncorrelated with $x_i$. Taking this into consideration, we can write for the average rate with the second input, conditioned on the rate first the first input, $< \tilde{u}_i^A >_{m_i^A}$,

$$
\begin{aligned}
< \tilde{u}_i^A &>_{m_i^A} \\
&= \int d\theta \, \rho_A(\theta) \int Dx \, \delta[\theta - u_A - \sqrt{\beta_A} x - \sqrt{\alpha_A - \beta_A} H^{inv}(m_i^A)] \\
&\quad \times \int Dz \, H\left( \frac{\theta - \tilde{u}_A - B_A x / \sqrt{\beta_A} - \sqrt{\tilde{\beta}_A - B_A^2 / \beta_A} z}{\sqrt{\tilde{\alpha}_A - \tilde{\beta}_A}} \right) \\
&= \int d\theta \, \frac{\rho_A(\theta)}{\sqrt{2\pi\beta_A}} \exp\left( -\frac{[\theta - u_A - \sqrt{\alpha_A - \beta_A} H^{inv}(m_i^A)]^2}{2\beta_A} \right) \\
&\quad \times H\left( \frac{[1 - b_A]\theta - \tilde{u}_A + b_A u_A - b_A \sqrt{\alpha_A - \beta_A} H^{inv}(m_i^A)}{\sqrt{\tilde{\alpha}_A - b_A^2 \beta_A}} \right), \tag{2.48}
\end{aligned}
$$

where we have used $b_A$ to denote the ratio between $B_A$ and $\beta_A$, $b_A = B_A / \beta_A$. This relationship between the conditional average of $\tilde{m}_i^A$ and $m_i^A$ is shown as the solid line in Fig. 6.

We have shown how to calculate the cross-correlation of the quenched input between two different external inputs $m_0$. This is also sufficient to investigate the statistics of the activity for more levels of the external input. Consider an array of inputs, $m_0(i)$, with $i = 1, 2, \ldots, n$ than the statistics of the quenched disorder is fully described by the matrix $q_A(i, j)$ with $i, j = 1, \ldots, n$, where $q_A(i, j)$ is the correlation in the local rate at external input $m_0(i)$ and external input $m_0(j)$. For each pair of these the preceding paragraphs describe how to calculate this, so we can determine the whole matrix. Once we have these matrices, the statistics of the quenched disorder is fully determined.

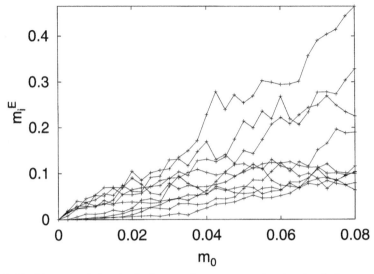

Fig. 7. The firing rates of 10 randomly chosen excitatory neurons plotted against the input rate $m_0$.

To see how a typical cell behaves as the external input is changed, we can determine $u_A$, $\alpha_A$, and $\beta_A$ for all values of $m_0$, and calculate the correlation matrix for the quenched disorder. For a typical cell we draw the quenched disorder for each level $m_0$ from a Gaussian distribution with this correlation matrix, and calculate the activity with this disorder. In Fig. 7 we show the activity $m_i^E$ for 10 excitatory neurons plotted against an input that goes $m_0 = 0$ to $m_0 = 0.08$ in 32 steps. As expected, the general trend is that the firing rates of the neurons increase with $m_0$. For each value of $m_0$ here is also a large heterogeneity on the activity of the cells. For each of the cells, the rate of the increase of the activity varies considerably with $m_0$, with the firing range even decreasing over some range of external inputs.

## 3. A memory model

So far we have only considered the the balanced state in the simplest possible network model. In the next two sections we will study networks that may have functional relevance, and which still behave like balanced networks. The difference between these model networks and the network we have described so far is that the structure of the network connectivity and the external inputs is richer than in the network we have studied up to now. It the next section we will consider a model for a V1 hypercolumn. Here we will describe a rather simple model for long term memory. Modeling of associative memory has a long history [47–50] and here we will concentrate on the persistent activity as it is observed in delayed response tasks [51–53]. In a typical delayed activity task the subject, usually a monkey, is shown a stimulus from a familiar set, has to wait during a delay period, after which he receives a signal to initiate an action which depends on the stimulus that was presented. Persistent activity can be found in many cortical areas, and typically neurons in these areas show the following behavior: they are spontaneously active, during the presentations of the stimuli they increase their activity for a small fraction of the stimuli but not for others, and during the delay period their firing rates stay elevated for these stimuli. The firing rate during the delay period is 3 to 6 times the background firing rate [54]. During all of these episodes the activity of the neurons is irregular and the total input into the cells fluctuate. This suggests that these areas are in a balanced state. We will therefore look at an extension of the model from the previous section that may account for delay activity.

### 3.1. A modified Willshaw model

The model we will describe here is a modification of the Willshaw model [55, 56]. In the Willshaw model there a $p$ patterns that are stored. Each pattern $\mu$ consisting of a 'word' in which a fraction $f$ of the elements $s_i^\mu$ is 1, while all the others are 0. All cells are excitatory and there is a synapse with strength $J$ between cell $i$ and $j$ if for at least 1 pattern $s_i^\mu$ and $s_j^\mu$ are both 1, otherwise there is no connection. For this model Willshaw *et al.* have shown that if $p$ is not to large, and the coupling strength $J$ is chosen correctly, there is multi-stability, with a stable state where all cells are in the rest state $\sigma_i = 0$, but also stable fixed points for which $\sigma_i = s_i^\mu$.

In the balanced network model of a long term memory module we will study here we assume that there are $p$ patterns, where pattern $\mu$ consists of a word, $\{s_i^\mu\}$, in which the $N_E$ elements $s_i^\mu$ are chosen randomly, and set to 1 with probability $f$, and 0 otherwise. The model of the memory network is the same as the simple network model described in section 2 except that the excitatory to excitatory synapses code the memory patterns. With probability $K/N_E$ there is a connection from excitatory neuron $j$ to excitatory neuron $i$ and if there is a connection, *and* both $s_i^\mu$ and $s_j^\mu$ are equal to 1 for at least pattern $\mu$, $\sum_\mu s_i^\mu s_j^\mu > 0$, the synapse is strong, with $J_{ij}^{EE} = a J_{EE}/\sqrt{K}$, while if there is a connection and $\sum_\mu s_i^\mu s_j^\mu = 0$ the synapse has the normal strength, $J_{ij}^{EE} = J_{EE}/\sqrt{K}$, while if there is no connection, $J_{ij}^{EE} = 0$.

The system is supposed to act as a device for delay activity in the following manner: There is always a constant background external input, $u_A^0 = \sqrt{K} E_A m_0$, which is the same for all neurons in population $A$. During a delay activity task, in which a pattern $\mu$ has to be remembered, the neurons for which $s_i^\mu = 1$ transiently receive and extra input, and after the transient the external input drops back to $u_A^0$, but the activity of the excitatory neurons for which $s_i^\mu = 1$ stays elevated relative to the rest of the network.

We will not study the dynamics of this network during, and immediately after the transient input, but we will look at the rates after the network has reached equilibrium. A necessary requirement for the system to be able to work as a delay activity device is that there is multi-stability, with a state in which all cells are active at the background rate, as well as states in which the excitatory cells in one of the patterns is elevated.

### 3.2. Mean field equations

To study this, we look at the average firing rate, $m_E^{(+)}$, of the excitatory neurons in pattern $\mu$, which can be determined using

$$m_E^{(+)} = \frac{\sum_i s_i^\mu m_i^E}{\sum_i s_i^\mu}, \tag{3.1}$$

and the average rate of the excitatory neurons in the background, $m_E^{(-)}$, given by

$$m_E^{(-)} = \frac{\sum_i (1 - s_i^\mu) m_i^E}{\sum_i 1 - s_i^\mu}, \tag{3.2}$$

as well the average inhibitory rate, $m_I$. The inputs into the two excitatory and into the inhibitory population all have a Gaussian distribution, and we need to determine the means, $u_E^{(+)}$, $u_E^{(-)}$ and $u_I$, as well as their respective variances, $\alpha_E^{(+)}$, $\alpha_E^{(-)}$, and $\alpha_I$. To calculate these we need the probability, $P(p)$, that a synapse between excitatory neurons in the background, or between a foreground and a background cell, is strengthened. This is equal to the probability that $s_i^\nu$ and $s_j^\nu$ are both non-zero for any of the patterns $\nu \neq \mu$. The probability that they are not both 1 in a particular pattern is $(1 - f^2)$, so that, since the patterns are uncorrelated, $P(p) = 1 - (1 - f^2)^{p-1}$. So neurons in the background receive input from $(1 - f)K$ excitatory neurons in the background and from $fK$ excitatory cells in the foreground. The strength of their connections is on average $(P(p)a + 1 - P(p))J^{EE}/\sqrt{K}$. These neurons also receive input from, on average, $K$ inhibitory neurons, with strength $J_{II}/\sqrt{K}$ and an external input $\sqrt{K} E_E m_0$. Putting all of these together, we obtain for $u_E^{(-)}$

$$u_E^{(-)} = \sqrt{K}\Big\{E - E m_0 + [P(p)a + 1 - P(p)]$$

$$\times J_{EE}[(1-f)m_E^{(-)} + fm_E^{(+)}] + J_{II}m_I \Big\}. \tag{3.3}$$

The neurons in the foreground receive, on average, the same input from background excitatory cells, the inhibitory cells and the external input, as the background ones. But they receive a larger input from the $fK$ foreground neurons because their synapses will all be enhanced to $aJ_{EE}/\sqrt{K}$, so that

$$u_E^{(+)} = u_E^{(-)} + \sqrt{K} J_{EE}[a-1][1-P(p)]fm_E^{(+)}. \tag{3.4}$$

The excitatory to inhibitory connections, and inhibitory to inhibitory connections are not changed by the patterns, so that $u_I$ will simply be given by

$$u_I = \sqrt{K}\left\{E_I m_0 + J_{IE}[(1-f)m_E^{(-)} + fm_E^{(+)}] + J_{II}M_I\right\}. \tag{3.5}$$

In section 2.2 we saw that $\alpha_A$ satisfies $\alpha_A = \sum_B < (J_{ij}^{AB})^2 m_j^B >_j$. Using this one obtains

$$\alpha_E^{(-)} = (J_{EE})^2[P(p)a^2 + 1 - P(p)][(1-f)m_E^{(+)} + fm_E^{(+)}] + (J_{EI})^2 m_I, \tag{3.6}$$

$$\alpha_E^{(+)} = \alpha_E^{(-)} + (J_{EE})^2[a^2-1][1-P(p)]fm_E^{(+)} \tag{3.7}$$

and

$$\alpha_I = (J_{IE})^2[(1-f)m_E^{(+)} + fm_E^{(+)}] + (J_{II})^2 m_I. \tag{3.8}$$

The variance of the input is of order 1 for the foreground, the background and the inhibitory populations, while the mean inputs all have a contribution of order $\sqrt{K}$. To have a solution where input in all three population is balanced, this leading order term has to be zero in all three population averages. But the difference between $u_E^{(+)}$ and $u_E^{(-)}$ is of order $\sqrt{K}$ unless $m_E^{(+)}$, or $1 - P(p)$, or $f$ is of order $1/\sqrt{K}$. If $m_E^{(+)}$ is of order $1/\sqrt{K}$, so is $m_E^{(-)}$ and the network is quiescent. While $1 - P(p)$ may be of order $1/\sqrt{K}$, is a sufficiently large number, $p$, of patterns are stored, in this regime the foreground and background can not be both balanced if a smaller number of patterns is stored, which is not desirable.

The model becomes interesting if we assume that $f = F/\sqrt{K}$. If this is the case, $P(p)$ starts to deviate significantly from 0 only if of order $K$ pattern are stored and, with $p = \alpha K$, $P(p)$ is given by $P(p) = 1 - e^{-\alpha F^2}$. The average inputs are given by

$$u_E^{(-)} = \sqrt{K}\left\{E_E m_0 + [(a-1)P(p) + 1]J_{EE}m_E^{(-)} + J_{EI}m_I\right\} +$$
$$+ [(a-1)P(p) + 1]J_{EE}F(m_E^{(+)} - m_E^{(-)}), \tag{3.9}$$

$$u_I = \sqrt{K} \left\{ E_E m_0 + J_{IE} m_E^{(-)} + J_{II} m_I \right\} + J_{IE} F(m_E^{(+)} - m_E^{(-)}), \qquad (3.10)$$

while $u_E^{(+)} = u_E^{(-)} + [a - 1][1 - P(p)]J_{EE} F m_E^{(+)}$. In the balanced state the terms of order $\sqrt{K}$ have to cancel, so to leading order the rates $m_E^{(-)}$ and $m_I$ are given by

$$\begin{aligned} E_E m_0 + \tilde{J}_{EE} m_E^{(-)} + J_{EI} m_I &= 0 \\ E_I m_0 + J_{IE} m_E^{(-)} + J_{II} m_I &= 0, \end{aligned} \qquad (3.11)$$

where $\tilde{J}_{EE} = [(a - 1)P(p) + 1]J_{EE}$ is the average excitatory to excitatory connection strength. There exists a unique fixed point with a balanced solution provided that

$$\frac{E_E}{E_I} > \frac{J_{EI}}{J_{II}} > \frac{\tilde{J}_{EE}}{J_{IE}} \quad \text{and} \quad J_{IE} < -\tilde{J}_{EE}. \qquad (3.12)$$

With these balanced rates the variance of the different inputs is also determined and $\alpha_E^{(+)} = \alpha_E^{(-)} = \alpha_E$, where $\alpha_E$ is given by

$$\alpha_E = [(a^2 - 1)P(p) + 1](J_{EE})^2 m_E^{(-)} + (J_{EI})^2 m_I. \qquad (3.13)$$

The variance of the inhibitory population satisfies

$$\alpha_I = (J_{IE})^2 m_E^{(-)} + (J_{II})^2 m_I. \qquad (3.14)$$

The mean rate and the variance of the input also determine the average input for the background and inhibitory population. In particular $u_E^{(-)}$ is given by

$$u_E^{(-)} = \theta_E - \sqrt{\alpha_E} H^{inv}(m_E^{(-)}), \qquad (3.15)$$

where $H^{inv}$ is the inverse of $H$.

### 3.3. Multiple solutions

The average input into the foreground neurons depends on $m_E^{(+)}$ and is given by $u_E^{(+)} = u_E^{(-)} + \Delta u$, with

$$\Delta u = [a - 1][1 - P(p)]J_{EE} F m_E^{(+)}. \qquad (3.16)$$

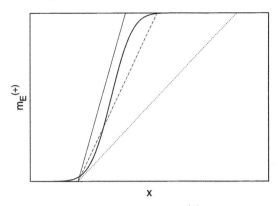

Fig. 8. Graphical determination of the foreground rate $m_E^{(+)}$. The thick curve plots $m_E^{(+)} = H(H^{inv}(m_E^{(-)}) - x)$ while the thin lines show $m_E^{(+)} = Cx$, where for large $C$ (solid line) there is one low rate solution, with $m_E^{(+)} \approx m_E^{(-)}$. For small $C$ (dashed line) there is only a high rate solution, while for intermediate $C$ (dotted line) there are three solutions, a stable low rate solution, a stable high rate solution and an intermediate solution that is unstable.

The mean rate of the foreground neurons is given by

$$
\begin{aligned}
m_E^{(+)} &= H\left(\frac{\theta_E - u_E^{(-)} - \Delta u}{\sqrt{\alpha_E}}\right) \\
&= H\left(H^{inv}(m_E^{(-)}) - \frac{\Delta u}{\sqrt{\alpha_E}}\right).
\end{aligned} \tag{3.17}
$$

Equations (3.16) and (3.17) together determine the average foreground rate, $m_E^{(+)}$. Figure 8 shows this solution graphically. In this graph $y = m_E^{(+)}$ is plotted against $x = \Delta u/\sqrt{\alpha_E}$. The thick curve corresponds of Eqn. (3.17), while the thin lines show Eqn. (3.16) in three possible cases. Equation (3.16) can be written as $y = Cx$ with

$$
C = \frac{\sqrt{\alpha_E}}{[a - 1][1 - P(p)]J_{EE}F}. \tag{3.18}
$$

If $C$ is large (solid line) there is only one intersection, at a low rate. For intermediate values of $C$ there are three solutions. One can show that the low rate solution and the high rate solution are the stable solutions, while the intermediate one is unstable. If $C$ is to small, the low rate solution disappears, and the foreground neurons are always in a high activity mode.

From Fig. 8 we learn the following: $m_E^{(+)}$ is always larger than $m_E^{(-)}$, but that was to be expected, because the foreground neurons participate in pattern $\mu$, while the background neurons do not, and both have the same probability of participating in other patterns, so that foreground neurons participate in, on average 1 more pattern, and therefore have more strengthened excitatory synapses.

If $a$ is to close to 1, $C$ will always be large, and even if very few patterns are stored, $P(p) \approx 0$, the foreground neurons will alway fire at a low rate. On the other hand, if $a$ is very large, the foreground neurons are always extremely active if only a small number of patterns are stored, and only for $P(p)$ sufficiently large, only if these is a sufficiently large number of stored patterns, can the foreground have both high and low average activity. This this is because for small $p$ the excitatory feedback in the foreground population if so much higher than in the background, that there is not enough feedback inhibition to keep the foreground neurons at a low rate. For larger $p$ the excitatory feedback to the background is strengthened, which strengthens the inhibitory feedback, and this allows foreground neurons to be in either a low or a high activity state. For intermediate values of $a$ there is low as well as a high rate solution for small $P(p)$.

But both for intermediate values of $a$ and for large $a$ there is a maximum number of patterns that can be stored. If $P(p)$ gets too close to 1, $C$ becomes too large and only the foreground neurons always evolve to the low rate state. This is how the capacity of the network is limited. If there are multiple stable solution for $m_E^{(+)}$ as long as $P(p) < P_{max}$, the maximum number of pattern that can be stored is $p = \alpha_c K$ with $\alpha_c = -F^{-2} \log(1 - P_{max})$.

One feature of this model of a neuronal network with delay activity that is extremely unrealistic is that, if there are multiple solutions for $m_E^{(+)}$, the stable high rate always has $m_E^{(+)} > 1/2$. >From Fig. 8 it is clear that the highest rate solution has to occur beyond the inflection point and, whatever the value of $m_E^{(-)}$, at the inflection point $m_E^{(+)} = 1/2$. Experimental data shows that, while the activity is enhanced during the delay period in the activated pattern, the elevated rates are nowhere near the maximum firing rate of the neurons [57]. It is highly likely that this unrealistic feature of the model presented here is due to the fact that binary are to non-linear near the threshold, and for neurons with more realistic transfer functions one may overcome this problem. But discussing such models is beyond the scope of this course. For a more detailed discussion of the modeling of delayed activity we refer to the course by N. Brunel in this volume.

## 4. A model of visual cortex hypercolumn

Most neurons in the primary visual cortex (V1) respond preferentially to stimuli the consist of bars or gratings at a specific orientation [58], with neurons with similar orientation preference grouped together [59, 60]. It is well known [8, 61–

63] that the tuning curves of these neurons are independent of contrast. There has been much interest in the mechanism that leads to contrast invariance, with many competing models trying to account for it [64–75]. Less well known is that there is a very broad distribution of the level of tuning of the cells, with some cells only having a response for 30 degrees of input orientation, while others respond for all orientations [77]. Taking this into account, one realizes that the data for the contrast invariance of the tuning curves is less clear-cut than it seems at first. It is notable that all contrast invariance measurements that have been reported, test the contrast invariance of the tuning of well tuned cells. No measurements of the contrast dependence of the tuning curves of badly tuned cells have been published. What would one expect of the contrast dependence of the tuning curves for these cells. Obviously this will depend on the origin of tuning width heterogeneities.

Since this has never been considered in experimental work, or in any theoretical modeling, it may be interesting if one can construct a simple model in which this can be investigated. In section 2 we saw that heterogeneity occurs naturally in the balanced state in a network with synaptic strengths that scale as $1/\sqrt{K}$. We also saw that the population averaged activity rates scale linearly with the input rate $m_0$. We will here consider a model for a visual column hypercolumn that uses these features.

### 4.1. The model

The model is an extension of the model described in section 2. With each neuron is associated a preferred orientation, $\phi_i^A$. These orientations are uniformly distributed between 0 and $\pi$. As before the connections are still random, but the probability of a synaptic connection depend on the preferred orientation of the pre- and post-synaptic cells. The probability $P_{AB}^{ij}$ of a connection from cell $j$ in population $B$ to cell $i$ of population $A$ is given by

$$P_{AB}^{ij} = \frac{K}{N_B}\left[1 + 2p\cos\left(2[\phi_i^A - \phi_j^B]\right)\right]. \tag{4.1}$$

If there is a connection its strength is $J_{AB}/\sqrt{K}$. Note that, as before, the average cell still receives input from $K$ cells from each population. The difference with the earlier model is that the cell is more likely to receive input from cells with similar preferred orientation. The level with which the probability is modulated is encoded by the parameter $p$, which can vary between 0 (no modulation) and 1/2 (maximum modulation). The external input into the cells, $u_{A,0}$, depends on the cells preferred orientation and the orientation of the stimulus, $\Phi_0$, and satisfies

$$u_{A,0}(\phi_i^A, \Phi_0) = u_{A,0}(\phi_i^A - \Phi_0) = \sqrt{K}E_A m_0\left[1 + \mu\cos\left(2[\Phi_0 - \phi_i^A]\right)\right].$$

$$(4.2)$$

The external rate, $m_0$, depends on the contrast and $\mu$ described the amount of modulation of the input. It is convenient to describe the total input, $u_i^A$, into the cell as a function of the difference, $\phi$, between the stimulus orientation and the preferred orientation of the cell, $\phi = \Phi_0 - \phi_i^A$. It is given by

$$u_i^A(\phi, t) = u_A^0(\phi) + \sum_B \sum_j J_{ij}^{AB} \sigma_j^B(t).$$

$$(4.3)$$

Finally, to keep the model as simple as possible, we will not consider here the effect of heterogeneities of the properties of the individual cell and assume that all neuron in the each population have the same threshold, $\theta_i^A = \theta_A$.

### 4.2. Population averaged response

In the large limit $N_A$, we want to consider the average activity, $m_A(\phi)$ of all cells with approximately the same preferred orientation (relative to the input orientation), and relate this to the statistics of the total input into cells with the same preferred orientation. As before, this input will have a Gaussian distribution, and is suffices to determine the mean, $u_A(\phi, t)$ and the variance, $\alpha_A(\phi, t)$, of this input as a function of $\phi$.

The average input, $u_A(\phi, t)$, can be determined by taking $u_i^A(\phi, t)$ and average over the connection matrices, $J_{ij}^{AB}$. On average the neuron will receive input from $K$ excitatory and $K$ inhibitory cells. Each of these cells will have a probability $p = \pi^{-1}[1 + 2p\cos(2[\phi_i^A - \phi_j^B])]d\phi$ that its preferred input orientation is between $\phi_j^B$ and $\phi_j^B + d\phi$. For neuron $j$ of population $B$ the probability that it is active is $m_j^B$. Averaging over the connections projecting to this cell, we have $< m_j^B >_J = m_B(\Phi_0 - \phi_j^B)$, which can be written as $< m_j^B >_J = m_B(\phi + \phi_i^A - \phi_j^B)$. Thus one obtains for $u_A$

$$
\begin{aligned}
u_A(\phi, t) &= u_A^0(\phi, t) \\
&\quad + \sqrt{K} \sum_B J_{AB} \frac{1}{\pi} \int_0^\pi d\phi' \left[1 + 2p\cos(2\phi')\right] m_B(\phi - \phi').
\end{aligned}
$$

$$(4.4)$$

Writing $m_A(\phi, t)$ in its Fourier components, $m_A(\phi, t) = \sum_n m_A^{(n)} \cos(2n\phi)$, one sees that $u_A(\phi, t)$ can be written as

$$u_A(\phi, t) = u_A^{(0)}(t) + u_A^{(1)}(t) \cos(2\phi),$$

$$(4.5)$$

with

$$u_A^{(0)}(t) = \sqrt{K}\left(E_A m_0 + \sum_B J_{AB} m_B^{(0)}(t)\right) \tag{4.6}$$

and

$$u_A^{(1)}(t) = \sqrt{K}\left(\mu E_A m_0 + p \sum_B J_{AB} m_B^{(1)}(t)\right). \tag{4.7}$$

Note that in writing the Fourier expansion of $m_A$ we have assumed that this was symmetric in $\phi$. That this is indeed the case is easily checked.

In section 2 we saw that the variance in the input was given by

$$\alpha_A = \langle \sum_B \sum_j (J_{ij}^{AB})^2 \sigma_j^B \rangle \tag{4.8}$$

The same is true here, so that $\alpha_A$ can be written as

$$\begin{aligned}
\alpha_A(\phi, t) &= \sum_B (J_{AB})^2 \frac{1}{\pi} \int d\phi' \left[1 + 2p \cos(2\phi')\right] m_B(\phi - \phi', t) \\
&= \sum_B (J_{AB})^2 \left[m_B^{(0)}(t) + p m_B^{(1)}(t) \cos(2\phi)\right].
\end{aligned} \tag{4.9}$$

Equations (4.5) and (4.9) can be used to determine the probability that a neurons is active after an update and this determines the evolution equations for the mean rate. These are given by

$$\tau_A \frac{d}{dt} m_A(\phi, t) = -m_A(\phi, t) + H\left(\frac{\theta_A - u_A(\phi, t)}{\sqrt{\alpha_A(\phi, t)}}\right). \tag{4.10}$$

The mean $u_A$ and variance $\alpha_A$ depend only on the Fourier moments $m_A^{(0)}$ and $m_A^{(1)}$. So that the network equilibrium rates are fully determined by the equilibrium activity of these two order parameters.

$$m_A^{(0)} = \int \frac{d\phi}{\pi} H\left(\frac{\theta_A - u_A(\phi)}{\sqrt{\alpha_A(\phi)}}\right) \tag{4.11}$$

and

$$m_A^{(1)} = 2 \int \frac{d\phi}{\pi} H\left(\frac{\theta_A - u_A(\phi)}{\sqrt{\alpha_A(\phi)}}\right) \cos(2\phi). \tag{4.12}$$

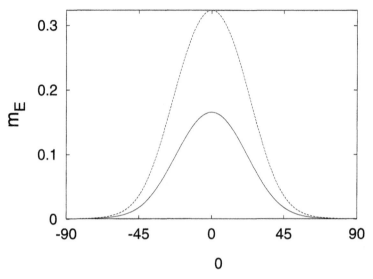

Fig. 9. Mean firing rate $m_E$ if the excitatory population as function of the preferred orientation $\phi$ in the large $K$ limit for three levels of contrast, $m_0 = 0.05$ (bottom curve), $m_0 = 0.1$ (middle curve) and $m_0 = 0.2$ (top curve).

The network is in the balanced state if the leading order in $K$ of the mean input cancels. For this to be the case $m_A^{(0)}$ and $m_A^{(1)}$ have to satisfy

$$m_A^{(0)} = A_A m_0 + \frac{\tilde{m}^{(0)}}{\sqrt{K}} \qquad (4.13)$$

and

$$m_A^{(1)} = \frac{\mu}{p} A_A m_0 + \frac{\tilde{m}^{(1)}}{\sqrt{K}}, \qquad (4.14)$$

where as before, $A_A = -\sum_B (J^{-1})_{AB} E_B$ [76]. The requirement that the equilibrium rates have to be positive lead to the same restrictions on the connection parameters we had before, but we have the extra requirement that $\mu \le p$, since if $m_A(\phi) \ge 0$, $m_A^{(0)} \ge m_A^{(1)}$.

This balance condition determines $m_A^{(0)}$ and $m_A^{(1)}$, and also $\alpha_A^{(0)}$ and $\alpha_A^{(1)}$. These can be inserted in Eqns. (4.11) and (4.12) to determine $u_A^{(0)}$ and $u_A^{(1)}$. Once these are given, the equilibrium firing rates $m_A(\phi)$ at any orientation can be computed.

Figure 9 shows the population activity of the excitatory population as a function of $\phi$ for two levels of contrast. Up to a scaling factor, the population response is the same for both contrast. This should not surprise us, since due to the balance conditions, Eqns. (4.13) and (4.14), the first two Fourier components both scale linearly with the input and therefore $m_A^{(1)}/m_A^{(0)}$ is constant. It is true that $m_A^{(n)}/m_A^{(0)}$ is not exactly constant for $n$ larger than 1, but as long as the mean rate does not become extremely small, or the maximum rate gets to close to 1, the higher Fourier moment scale fairly linearly with the mean.

It is also the case that even though the first two Fourier components of the population activity scale linearly with the input, there can be a significant sharpening tuning, because both scale their input component by a different amount. The zero-th Fourier moment of the population is $A_A$ as large as the zero-th Fourier moment of the input, while for their first Fourier moment this ratio is $A_A/p$. Somewhat counter-intuitively, the weaker the modulation of the synaptic connectivity, $p$, the large the amplification of the activity modulation. Upon reflection, this is what one would expect, because the weaker the modulation in the connection strength, the stronger the activity has to be modulated, to enable the feedback to balance the input modulations.

## 4.3. Quenched disorder

Because the balance conditions impose a linearity between the different modes of the input and output, the population tuning is close to contrast invariant. So we have used the linearity to good effect. But in section 2 we also saw that, even neurons with the same single cell properties, the activity can be quite heterogeneous.

We will now determine the heterogeneity in the ring model. First the fluctuations in input can be decomposed in a quenched and temporally fluctuating part,

$$u_i^A(\phi, t) = u_A(\phi) + \Delta u_i^A(\phi) + \delta u_i^A(\phi, t) \tag{4.15}$$

where the average of $\delta u_i^A(\phi, t)$ over time is zero. Both the quenched disorder $\Delta u_i^A(\phi)$ and the temporal fluctuations $\delta u_i^A(\phi, t)$ are Gaussian variables, so all we need to determine average amplitude of either fluctuation. In fact, because

$$
\begin{aligned}
\alpha_A(\phi) &= < [\Delta u_i^A(\phi) + \delta u_i^A(\phi, t)]^2 >_{i,t} \\
&= < [\Delta u_i^A(\phi)]^2 >_i + < [\delta u_i^A(\phi, t)]^2 >_{i,t} \tag{4.16}
\end{aligned}
$$

it suffices to determine the variance of $\Delta u_i^A(\phi)$.

The quenched disorder $\Delta u_i^A$ will be correlated for different orientations, and if we want to be able to say something about the tuning curves of individual cells, we need to take this into account. We denote by $\beta_A(\phi, \Delta)$ the correlations of the quenched fluctuations at angles $\phi + \Delta$ and $\phi - \Delta$

$$\beta_A(\phi, \Delta) = <\Delta u_i^A(\phi + \Delta)\Delta u_i^A(\phi - \Delta)>_i . \tag{4.17}$$

With these correlations, the local rates, $m_i^A$ can be written as

$$m_i^A(\phi) = H\left(\frac{\theta_A - u_A(\phi) - \Delta u_i^A(\phi)}{\sqrt{\alpha_A(\phi) - \beta_A(\phi, 0)}}\right). \tag{4.18}$$

To calculate the correlations $\beta_A$ we introduce the order parameter $q_A$, defined by

$$q_A(\phi, \Delta) = \frac{1}{N_A} \sum_i m_i^A(\phi + \Delta)m_i^A(\phi - \Delta). \tag{4.19}$$

Following closely the steps in section 2, one obtains for $\beta_A$

$$\beta_A(\phi, \Delta) = \sum_B \sum_j (J_{ij}^{AB})^2 m_j^B(\phi + \Delta)m_j^B(\phi - \Delta). \tag{4.20}$$

which can be written as

$$\beta_A(\phi, \Delta) = \sum_B \sum_j (J_{AB})^2 [q_B^{(0)}(\Delta) + pq_B^{(1)}(\Delta)\cos(2\phi)]. \tag{4.21}$$

Here $q_A^{(n)}(\Delta)$ is the $n$th Fourier component of $q_A$ with respect to $\phi$

$$q_A(\phi, \Delta) = \sum_n q_A^{(n)}(\Delta)\cos(2n\phi). \tag{4.22}$$

Using the definition of $\beta_A$ it is straightforward to show that

$$
\begin{aligned}
q_A(\phi, \Delta) &= \int Dz H\left(\frac{\theta_E - u_A(\phi_1) - \sqrt{\beta_A(\phi, \Delta)}z}{\sqrt{\alpha_A(\phi_1) - \beta_A(\phi, \Delta)}}\right) \times \\
&\quad \times H\left(\frac{\theta_E - u_A(\phi_2) - \sqrt{\beta_A(\phi, \Delta)}z}{\sqrt{\alpha_A(\phi_2) - \beta_A(\phi, \Delta)}}\right),
\end{aligned}
\tag{4.23}
$$

where we have used $\phi_1$ and $\phi_2$ to denote $\phi + \Delta$ and $\phi - \Delta$ respectively. Equation (4.23) can be used to express $q_A^{(0)}(\Delta)$ and $q_A^{(1)}(\Delta)$ as functions of $\beta_A(\phi, \Delta)$. Together with with Eqn. (4.21) this determines $q_A^{(0)}(\Delta)$ and $q_A^{(1)}(\Delta)$.

Note that these self-consistency equations are uncoupled for different $\Delta$. This means that, for example, if one is only interested in distribution of firing rates at each orientation, and not in how these are correlated between orientations, one only has to evaluate the self-consistent solutions for $\Delta = 0$, without having to consider other values.

Having determined $\beta_A(\phi, \Delta)$ we can return to the quenched fluctuations $\Delta u_i^A$. As mentioned before, these are drawn from a Gaussian distribution with correlations between the different orientations. We can write the quenched disorder in using its Fourier components as

$$\Delta u_i^A(\phi) = \sum_n \Delta u_{A,i}^{(n)} \cos(2n\phi) + \Delta v_{A,i}^{(n)} \sin(2n\phi). \qquad (4.24)$$

Inserting this in Eqn. (4.17) one finds for $\beta_A$

$$
\begin{aligned}
\beta_A(\phi, \Delta) &= \sum_{n,m} < \Delta u_{A,i}^{(n)} \Delta u_{A,i}^{(m)} >_i \cos[2n(\phi + \Delta)] \cos[2n(\phi - \Delta)] + \\
&\quad < \Delta u_{A,i}^{(n)} \Delta v_{A,i}^{(m)} >_i \cos[2n(\phi + \Delta)] \sin[2n(\phi - \Delta)] + \\
&\quad < \Delta v_{A,i}^{(n)} \Delta u_{A,i}^{(m)} >_i \sin[2n(\phi + \Delta)] \cos[2n(\phi - \Delta)] + \\
&\quad < \Delta v_{A,i}^{(n)} \Delta v_{A,i}^{(m)} >_i \sin[2n(\phi + \Delta)] \sin[2n(\phi - \Delta)].
\end{aligned}
$$
$$(4.25)$$

After some manipulation, and using $\beta_A(\phi, -\Delta) = \beta_A(\phi, \Delta)$, one obtains from this

$$< \Delta u_{A,i}^{(n)} \Delta u_{A,i}^{(m)} >_i = < \Delta v_{A,i}^{(n)} \Delta v_{A,i}^{(m)} >_i = \beta_A^{(|n-m|,n+m)} \qquad (4.26)$$

and

$$< \Delta u_{A,i}^{(n)} \Delta v_{A,i}^{(m)} >_i = 0. \qquad (4.27)$$

Here we have used $\beta_A^{(|n-m|,n+m)}$ to denote $n, m$ the Fourier component of $\beta_A$

$$\beta_A(\phi, \Delta) = \sum_{n,m} \beta_A^{(n,m)} \cos(2n\phi) \cos(2m\Delta). \qquad (4.28)$$

Since one the Fourier components of $\beta_A$ with $n = 0$ and $n = 1$ are non- zero, the correlation matrices $< \Delta u_{A,i}^{(n)} \Delta u_{A,i}^{(m)} >_i$ and $< \Delta v_{A,i}^{(n)} \Delta v_{A,i}^{(m)} >_i$ are tri-diagonal matrices.

Note that the fluctuations $\Delta u_{A,i}^{(n)}$ and $\Delta v_{A,i}^{(n)}$ are uncorrelated, and both have the same correlation matrices, this means that the position of the maximum of the quenched fluctuations is independent of $\phi$. This is easiest to see by writing $\Delta_i^A(\phi)$ as

$$\Delta_i^A(\phi) = \sum_n \Delta w_{A,i}^{(n)} \cos[2n(\phi - \phi_{A,i}^{(n)})]. \qquad (4.29)$$

The random angle $\phi_{A,i}^{(n)}$ in this representation satisfies

$$\phi_{A,i}^{(n)} = \frac{1}{2n} \tan\left(\frac{\Delta v_{A,i}^{(n)}}{\Delta u_{A,i}^{(n)}}\right), \tag{4.30}$$

and is uniformly distributed between $-\pi/(4n)$ and $\pi/(4n)$. This shows that the position of the quenched disorder is uncorrelated with the preferred orientation.

One result of this is that the maximum response of a cell need not be at $\phi = 0$, the orientation at which the external input is maximum. Strictly speaking, $\phi_i^A$ is not the preferred orientation of the cell, as this is normally defined, but rather, it is the preferred orientation of its external input.

The quenched disorder of the input is fully described by its correlation structure, but because the neurons are so non-linear, the disorder in the neurons' activity cannot be captured by a few simple numbers. To get some idea of the variability of the response, one can look at some examples of the neurons' tuning curves, which can be generated by drawing $\Delta u_i^A(\phi)$ randomly from a Gaussian with the correct correlations, and calculating the cells' activity from this. The Fourier expansion of quenched disorder has infinitely many moments, but their amplitude decreases rapidly as the order of the moments become higher. Thus one can approximate the disorder by only taking the lowest few moments into account.

Figure 10 show the activity profile from three excitatory neurons. To generate these only the lowest 5 Fourier moments of $\Delta u_i^A$ were taken into account. Adding more moments did not noticeably change the results. The figure clearly shows that these three samples have different response properties. Their response amplitudes are different, they have different tuning widths, and the preferred orientation of their response are different, even though their preferred inputs are the same.

### 4.4. Contrast invariance

So far we have shown the the the population response is almost contrast invariant, while we have also shown that there is a large heterogeneity in the tuning curves of the individual cells. Experimental data shows a large heterogeneity in the response, and contrast invariance for individual cells, not the population. In a heterogeneous network contrast invariance of the whole population does not imply that the individual cells have contrast invariant tuning curves. One could, for example have that well tuned cells become even better tuned as the contrast is increased, and badly tuned cells worse, and still have contrast invariance of the average tuning. Thus we want to study the tuning of individual cells.

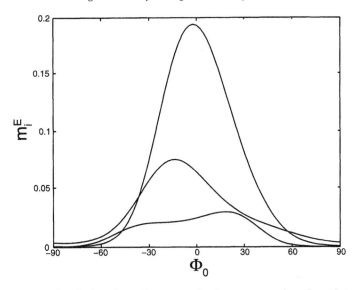

Fig. 10. Three samples of orientation tuning curves of excitatory neurons from the excitatory population. The activity of the cells is plotted against the difference between the stimulus orientation and the cells preferred input orientation.

Looking back to section 2.6 it becomes clear how we can approach this problem. We can consider the network activity for two different contrasts, and look at correlation in the quenched disorder with these contrasts.

Assume that the two contrasts are given by the external rate, which is denoted by $m_0$ for one contrast and by $\tilde{m}_0$ for the other. Let the mean rate, total variance and quenched variance be given by $m_A$, $\alpha_A$ and $\beta_A$ in the former case and by $\tilde{m}_A$, $\tilde{\alpha}_A$ and $\tilde{\beta}_A$ in the latter, while the total input into cell $i$ of population $A$ satisfies

$$u_i^A(\phi, t) = u_A(\phi) + \Delta u_i^A(\phi) + \delta u_i^A(\phi, t) \tag{4.31}$$

for contrast $m_0$, and

$$u_i^A(\phi, t) = \tilde{u}_A(\phi) + \Delta \tilde{u}_i^A(\phi) + \delta \tilde{u}_i^A(\phi, t) \tag{4.32}$$

for contrast $\tilde{m}_0$. We need to determine correlation, $B_A(\phi, \Delta)$, between the quenched disorder with the first contrast level at orientation $\phi + \Delta$ and the second contrast at orientation $\phi - \Delta$

$$B_A(\phi, \Delta) = < \Delta u_i^A(\phi + \Delta) \Delta \tilde{u}_i^A(\phi - \Delta) >_i . \tag{4.33}$$

This can be done with the use of the order parameter $Q_A$ defined as

$$Q_A(\phi, \Delta) = < m_i^A(\phi + \Delta) \tilde{m}_i^A(\phi - \Delta) >_i, \tag{4.34}$$

where $m_i^A$ and $\tilde{m}_i^A$ are the local rates in the first and second contrast respectively.

The correlation $B_A$ depends on the zeroth and first Fourier component of $Q_A(\phi, \Delta)$

$$B_A(\phi, \Delta) = \sum_B (J_{AB}^2 [Q_B^{(0)} + p Q_B^{(1)} \cos(2\phi)], \tag{4.35}$$

while $Q_A$ and $B_A$ are also related through

$$Q_A(\phi, \Delta) = \int Dz \, H \left( \frac{\theta_A - u_A(\phi_1) - \sqrt{B_A(\phi, \Delta)} z}{\sqrt{\alpha(\phi_1) - B_A(\phi, \Delta)}} \right) \times$$

$$H \left( \frac{\theta_A - \tilde{u}_A(\phi_2) - \sqrt{B_A(\phi, \Delta)} z}{\sqrt{\tilde{\alpha}(\phi_2) - B_A(\phi, \Delta)}} \right). \tag{4.36}$$

Here $\phi_1 = \phi + \Delta$ and $\phi_2 = \phi - \Delta$. These equations can be solved self-consistently in the same manner as was done for $q_A$ and $\beta_A$ in section 4.3.

Finally, $B_A$ can be used to calculate the correlations between $\Delta u_{A,i}^{(n)}$, $\Delta v_{A,i}^{(n)}$, $\Delta \tilde{u}_{A,i}^{(n)}$, and $\Delta \tilde{v}_{A,i}^{(n)}$.

These correlations fully describe the statistics of the quenched disorder in the total input. This allows one to draw random samples with the correct statistics and from these calculate the tuning curves for both contrast is for the same cell.

To compare the response at both contrasts, we will compare the tuning strength, defined as ratio between the first and zeroth Fourier moment of the activity. A scatter plot of the tuning strength at high contrast (vertical axis) against the tuning strength at low contrast (horizontal axis) is shown in Fig. 11. This figure show a scatter plot for 10000 excitatory neurons. There is a clear trend in which neurons which are well tuned for low contrast (for which $m^{(1)}/m^{(0)}$ is high), have even better tuning at high contrast, and appear above the line $x = y$ in the figure, while badly tuned cells become even more badly tuned. For the very well tuned cells the absolute change in tuning strength is rather small, and is not necessarily contradicted by experimental data, if the measurement error is taken into account. For the badly tuned cells, the absolute change in tuning strength is much larger, and it should be possible to resolve a change of this magnitude experimentally. Thus once experimentalists start to look at contrast invariance in badly tuned cells, this balanced network model of the V1 hypercolumn predicts that there should be badly tuned cells that do not have contrast independent tuning curves. On average these cells should become more badly tuned as the contrast is increased.

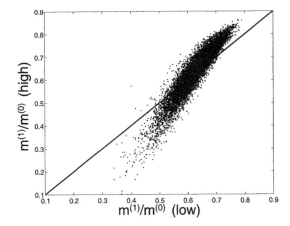

Fig. 11. Scatter plot of the tuning strength $m^{(1)}/m^{(0)}$ at high contrast (vertical axis, $m_0 = 0.1$) against the tuning strength at low contrast (horizontal axis, $m_0 = 0.05$). Results for 10000 sample neurons from the excitatory population are shown.

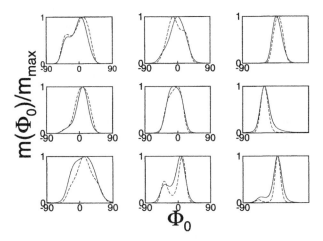

Fig. 12. examples of normalized tuning curves for different neurons. The response of the neuron, normalized by its maximum is plotted against the stimulus orientation. For each cell the response for high contrast, $m_0 = 0.1$, (solid line) and low contrast, $m_0 = 0.05$ (dashed line) are shown.

The activity for 9 randomly selected neurons is shown in Fig. 12. This figure shows the activity of the cells plotted against the stimulus orientation. The activity is normalized by the cells maximum activity. The response of the cells at high (solid line) and low contrast (dashed line) are shown. These examples confirm the trend shown in Fig. 11 according to which badly tuned cells broaden their response as contrast is increased (see lower left example) and well tuned cells get sharper tuning (middle right example). Small changes in preferred orientation also occur (upper right example), and secondary bumps in the tuning curves may appear (lower middle example) or disappear (top right example) if the contrast is increased.

## 5. Adding realism: integrate-and-fire network

The work that has been presented so far has all been analytical. To be able to carry out this analysis some severe simplifications had to be made. In particular we have assumed infinitely diluted coupling matrices, and binary neurons. In this section we will investigate to what extend the behavior of the network is changed if these simplifications are relaxed. Unfortunately this can only be done numerically. We want to be able to simulate large networks of neurons, so the model of the neuron we will use has to be simple. We will use the simplest model of a spiking cell available, the integrate-and-fire (IF) neurons [78]. Model networks with IF neurons can be simulated efficiently [79,80], so that sufficiently large networks can be studied numerically in a realistic amount of time. A further advantage of the IF neuron is that it is sufficiently simple that, while the behavior of the network cannot be fully determined analytically, we can still analyze some aspects of the network activity in approximation.

To further speed up the simulations, we observe that many aspects of the balanced network can already be studied in a network with only inhibitory neurons and a strong excitatory external input [81]. The balanced state is the state in which the neurons in the network fire with such rate that the inhibitory feedback nearly offsets the external drive. We will first study such a network.

### 5.1. Current based synapses

We consider network of $N$ IF neurons which are randomly connected. The connection strength, $J_{ij}$ between the $i$th post- and $j$th pre-synaptic cell is $J/\sqrt{K}$ is probability $K/N$, and zero otherwise. The state of the neuron is characterized by it voltage $V_i$ which satisfies

$$C_M \frac{dV_i}{dt} = -g_L V_i + I_0 + I_i(t). \tag{5.1}$$

Here $C_M$ is the membrane capacitance, $g_L$ the leak conductance, $I_0$ the external input, and $I_i$ the inhibitory feedback current. The membrane time constant,

$\tau_m \equiv C_M/g_L$ is 10 *msec*. The neurons integrate their input until the voltage reaches the threshold, $\theta$, after which they are immediately reset to the potential, $V_{reset}$, which we assume to be $V_{reset} = 0$. For the input currents we will consider two models. We will first consider a model in which they are described by currents and independent of the post synaptic voltage. In the second model we will take the dependence on the post-synaptic voltage into account, and the synaptic activity as changes of the conductance of a sodium or potassium channel.

In the model we consider first the external current is given by $I_0 = I_c \sqrt{K} \tau_m v_0$ and the inhibitory feedback satisfies

$$I_i(t) = -I_c \sum_j J_{ij} E_j(t), \tag{5.2}$$

where $I_c = \theta g_L$. The activity of the synapses projecting from neuron $j$ are modeled as the sum of the contributions of the preceding pre-synaptic spikes

$$E_j(t) = \sum_{t_j^{(k)} < t} \varepsilon(t - t_j^{(k)}), \tag{5.3}$$

where $t_j^{(k)}$ is the time of the $k$th spike on neuron $j$ and single spike response, $\varepsilon$, is described by a difference of exponents

$$\varepsilon(t) = \frac{1}{\tau_1 - \tau_2} \left( e^{-t/\tau_1} - e^{-t/\tau_2} \right), \tag{5.4}$$

for $t > 0$. The prefactor is chosen such that the integral of $\varepsilon$ is 1, and hence the average of $E_j$ is equal to neuron $j$'s firing rate. We will use for the synaptic time constants $\tau_1 = 3$ *msec*, and $\tau_2 = 1$ *msec*. The synaptic weight $J$ is expressed in units of time, and throughout this section we will use $J = 36$ *msec*.

If we assume that the network connectivity is sufficiently sparse, so that the inputs from the different pre-synaptic neurons can be considered as uncorrelated, the cell receives $K$ uncorrelated fluctuating inputs of strength $1/\sqrt{K}$. The mean net input, $I_{net}$, into a cell is

$$I_{net} = I_c \sqrt{K} (\tau_m v_0 - J v_I), \tag{5.5}$$

where $v_I$ is the average firing rate. In the large $K$ limit the net input will be much large than the threshold current $I_c$ if $v_I$ is less than $\tau_m v_0/J$. This would lead to a raped increase of the firing rate for all cells so that $v_I$ would increase rapidly, and the neurons' input decrease, until the activity reaches the point where the difference between the net input into the cell and the threshold reaches 1. For large but finite $K$ the mean firing rate will deviate from $\tau_m v_0/J$ by an amount of order $1/\sqrt{K}$.

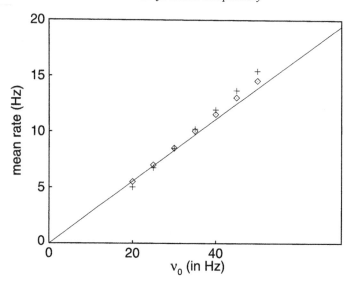

Fig. 13. Mean firing rate $\nu_i$ as function of the external rate $\nu_0$ in an inhibitory network in which the synapses are modeled as currents. The solid line shows the theoretical prediction in the infinite $K$ limit. Pluses show results from simulations of a network with $K = 100$ and $N = 1000$, diamonds for a network with $K = 1000$ and $N = 10000$

These predictions were tested with simulations of networks with $K = 100$ and $N = 1000$ and $K = 1000$ and $N = 1000$. In both networks the average firing rate was determined for different external input rates $\nu_0$. The results are plotted if Fig. 13. In both the network rate $\nu_I$ varies linearly with $\nu_0$. The deviation from the theoretical infinite $K$ is smaller in the large network than in the small network. Indeed the deviation is decreased by about a factor of $\sqrt{10}$ as one would expect if these deviations reflect $1/\sqrt{K}$ corrections.

The temporal correlations induced in by the network are more complex than those in the binary network, due to the interplay between the temporal correlations induce by the finite synaptic time constants, and the reset of the potential. For large $K$ the evolution of the voltage can be approximated by

$$C_M \frac{dV_i}{dt} = -g_L V_i + I_c \sqrt{K}(\tau_m \nu_0 - J\nu_I) + I_c J \sqrt{\nu_I/\tau_m} z_i(t) \qquad (5.6)$$

where the $z_i$s are (for sparsely coupled networks) independent, temporally correlated Gaussians. The temporal correlations in the input both determine the temporal correlations in the spike statistics and are determined by it. They have to satisfy complicated self-consistency equations. Qualitatively the picture is as follows: Immediately after a spike, the cells voltage is reset to 0 and it takes a time of order $\tau_m$ before the cell can fire again. This leads to negative correlation in the cell's spike correlation up to a time of order $\tau_m$. For the fluctuation in the input these spike correlations are filtered through the synaptic response $\varepsilon$, so that the $z_i$s have a positive correlation up to time of order $\tau_1$ and then negative up to a time of order $\tau_m$, after which the correlation is close to zero.

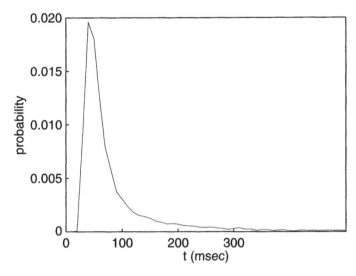

Fig. 14. Inter-spike interval histogram for a neuron in the network with $K = 1000$, and $N = 10000$ and $\nu_0 = 40$ Hz.

Figure 14 shows the inter-spike interval distribution for a neuron in a network with $K = 1000$, $N = 10000$ and $\nu_0 = 40$ Hz. Up to a time of order $\tau_m$ (10 *msec*) there are no intervals, then the spike probability increases rapidly, after which the probability decays exponentially, as would expects if there are no long range correlations in the input.

In networks of finite size, there is a finite probability that neurons have common inputs. If $K/N$ is small, the average number of common inputs for two cells is $K^2/N$. These will lead to small correlations in their input that are of order $J^2K/N$. Figure 15 show the correlation of the network activity in a network with $K = 100$ and $\nu_0 = 40$ Hz, both for $N = 1000$ (solid line) and $N = 5000$ (dashed line). The number of times that two spikes with a delay between $t$ and $t + \Delta t$ divided by $N^2\nu_I^2T\Delta t$ is plotted against $t$. Here $T$ is the time of the simulation over which was measured. If there are no correlation the expected value of this quantity is one. For a network with $N = 1000$ there are positive equal time correlations, followed by a negative correlation for about 40 *msec*, another small bump of positive correlations, after which the correlation have died out. On sees the same for $N = 5000$, except that the correlations are much smaller.

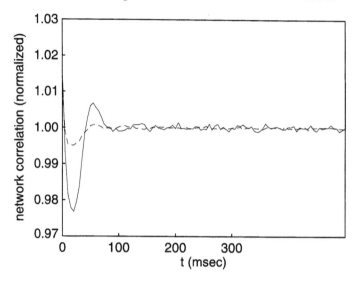

Fig. 15. Network correlations for networks with $K = 100$ and $\nu_0 = 40$ Hz. The solid line shows the result for $N = 1000$, the dashed line for $N = 5000$.

## 5.2. Conductance based synapses

Synaptic currents depend on the voltage of the post synaptic cell, a feature that we have neglected so far. We can incorporate this in the model by modifying the external input and feedback currents. We will now use that the external input is given by

$$I_0 = C_m \sqrt{K} \nu_0 (V_E - V_i) \tag{5.7}$$

while the inhibitory feedback satisfies

$$I_i(t) = g_L \sum_j J_{ij} E_j (V_I - V_i). \tag{5.8}$$

The synaptic activation variables are unchanged and still governed by Eqn. (5.3). In the simulations we choose for the reversal potentials $V_E = 5\theta$ and $V_i = 0$. In this model the net current is on average

$$I_{net} = g_L \sqrt{K} \left[ \tau_m \nu_0 (V_E - V_i) + J \nu_I (V_I - V_i) \right]. \tag{5.9}$$

The neuron will only fire at a low rate, if the net input is of order 1 when the voltage is at the threshold, $V_i = \theta$. Therefore, the average network firing rate is, in the large $K$ limit given by

$$\nu_I = \frac{\tau_m}{J} \frac{V_E - \theta}{\theta - V_I} \nu_0. \tag{5.10}$$

Simulations show that this relation holds in network simulations, although the $1/\sqrt{K}$ corrections are larger in this network than in the previous model.

For large $K$ the evolution of the voltage can be written as

$$\tau_m \frac{dV_i}{dt} = -[1 + \sqrt{K}(\tau_m \nu_0 + J\nu_I) + J\sqrt{\nu_I/\tau_m}z_i(t)]V_i + \sqrt{K}[\tau_m \nu_0 V_E + J\nu_I V_I] + J\sqrt{\nu_I/\tau_m}V_I z_i(t). \tag{5.11}$$

Here the random variables $z_i$ are again independently drawn from a Gaussian with temporal correlations, for which the temporal correlation has to be drawn self-consistently. A big difference with the previous model is that the effective membrane time constant, $\tau_{eff}$, which is given by

$$\tau_{eff} = \frac{\tau_m}{1 + \sqrt{K}(\tau_m \nu_0 + J\nu_I)}, \tag{5.12}$$

is in the large $K$ limit much shorter than the synaptic time constant. Thus already at a time $\tau_{eff}$ after a spike the cell is ready to fire again. But even if the neurons that would project to the cell fire in a Poissonian manner, the synaptic input would be positively correlated over a time $\tau_1$, which means that the output of the target cell would be correlated positively over this interval. This implies that *its* synaptic currents would have an even longer correlation time. One can see how a self-consistently temporal correlation can have a very long time constant.

In simulations one does indeed observe that activity of the neurons is correlated over a very long time, this correlation time increases with increasing $K$. As the number of synapses is increased, the activity of the neurons becomes more and more bursty. Even though the neurons have a very strong tendency to burst in this state, their are no strong correlations between the neurons. Figure 16 shows the voltage trace for a neuron. The upper trace shows the voltage of a cell in a network in which the synaptic currents are independent of the post-synaptic voltage, while the lower trace is for a neuron in a network with voltage dependent synapses. The cell in the first model spikes irregularly, but has no tendency to burst. The cell in the lower panel has very pronounced bursts.

It is likely that the burstiness in the model with synapses described with conductances is an artifact of the integrate and fire neuron. Conductance based model

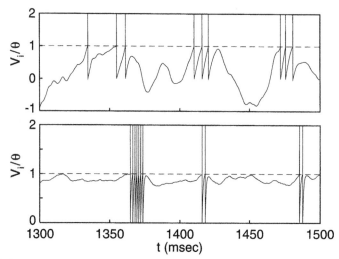

Fig. 16. Voltage traces for models with $K = 1000$ and $N = 10000$. Top trace, in model with voltage independent current, and $v_0 = 40$ Hz. Bottom trace for voltage dependent synapses and $v_0 = 10$ Hz.

neurons cannot fire spikes with an arbitrary short interval. To take this into account, one can add an absolute refractory period during which the cell is held at the reset potential, and to implement the relative refractory period we will assume that after a spike the neurons threshold, $\theta_i$ is increased by an amount $\Delta\theta$, after which the threshold decays according to

$$\tau_R \frac{d\theta_i}{dt} = -\theta_i + \theta, \tag{5.13}$$

Figure 17 shows the logarithm of the inter-spike interval distribution in a model without refractoriness (solid line), and in a model with a 2 *msec* absolute refractoriness, $\Delta\theta = \theta/10$ and $\tau_R = 10$ *msec* (dashed line). Even though $K$ is only 100, and therefore the cell without refractoriness is much less bursty that the cell in Fig. 16, it is clearly more irregular than a Poisson. With refractoriness in inter spike interval distribution is very close to that of a Poisson neuron with refractory period.

## 5.3. Networks with two populations

One can perform simulations with both excitatory and inhibitory neurons in a randomly connected network with constant external input. Based on the fact that networks of IF neurons with only an inhibitory population closely track that of a network with binary cells, one would expect that a network of sufficiently sparsely coupled excitatory and inhibitory IF neurons in which the synaptic current that are independent of the post-synaptic voltage behave similarly to binary

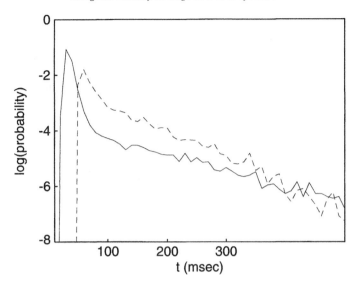

Fig. 17. Logarithm of the ISI histograms in networks with $K = 100$, $N = 1000$ and $\nu_0 = 10$ Hz. Solid line: Network without refractoriness. Dashed line: Network with 2 *msec* absolute refractoriness, $\Delta\theta = \theta/10$ and $\tau_R = 10$ *msec*.

networks. If the time constants of the inhibitory synapses and the membrane time constant of the inhibitory cells are fast enough (typically slightly faster than the excitatory ones), the network evolves to a balanced state. In this case firing rates of the inhibitory and excitatory populations vary linearly with the external rate, and the neurons fire irregularly, with a very small amount of correlation between the cells. Extensive simulations have been done for such networks with fluctuating external inputs, where the input into the cells is modeled as Poisson spike trains with synapses similar to those of the excitatory cells. Brunel and Amit [35, 36] have studied such a system with infinitely fast synapses with random delays. This scenario was analyzed in more detail in [82]. In [83] an IF system with synapses with finite time constants was simulated. Amit and Brunel have also investigated an IF model of associative memory similar to that in section 3 in the case where few patterns are present [36]. This study shows that if the parameters are chosen carefully, models with IF neurons can low background rates and in activated patterns rates that are reasonable. In [84] a network of integrate and fire neurons with constant inputs, but infinitely fast synapses was studied, and the irregularity of the neuronal spiking investigated. It was found that, depending on coupling strength, the neurons can fire more irregular than, less irregular than, or as irregular as Poisson neurons. There is even study that used conductance based neurons [85]. In this study a network of Hodgkin-Huxley neurons [86–90] with random synaptic connections was investigated. Unfortunately the computational complexity of this model has, so far, restricted these simulations to networks of only 250 neurons per population in which $K$ was kept very small, $K = 12$, so that the finite $K$ corrections are considerable. As a result, the network response is not to linear, but the sensitivity to initial conditions was confirmed for this model.

## 5.4. A model of a hypercolumn

Finally we will consider a model of a visual cortex hypercolumn [91]. The model has $N_E$ excitatory and $N_I$ inhibitory neurons. The inhibitory cells are standard IF neurons, and the state of inhibitory neuron $i$ is characterized by its voltage, $V_i^I$, which satisfies

$$C_m \frac{d}{dt} V_i^I = -g_L^I V_i^I + I_i^I(t) \tag{5.14}$$

where $C_M$ is the capacitance, $g_L^I$ the leak conductance of the inhibitory cells, and $I_i^I$ is the total synaptic current into the cell, which is assumed to be independent of the cell's voltage. Immediately upon reaching the threshold voltage, $\theta_I$, which is the same for all cells, the voltage is reset to $V_{reset} = 0$.

The excitatory neurons are IF neurons to which a spike adaptation [92] is added to take into account the adaptation that is observed in excitatory neurons in the cortex. The voltage, $V_i^E$, of excitatory neuron $i$ is given by

$$C_M \frac{d}{dt} V_i^E = -g_L^E V_i^E + I_i^E(t) - A_i. \tag{5.15}$$

Here $g_L^E$ is the leak conductance of the excitatory cells, $I_i^E$ is the total synaptic input into excitatory cell $i$ and $A_i$ is the adaptation current, which satisfies between spikes

$$\tau_A \frac{d}{dt} A_i = -A_i \tag{5.16}$$

When the voltage reaches the threshold, $\theta_E$, it is reset to $V_i = 0$, while the spike adaptation current is increased by an amount $g_A/\tau_A$.

Each neuron has a preferred orientation $\phi_i^A$, which is drawn from a uniform distribution between 0 and $\pi$. The probability of a synaptic connection from neuron $j$ in population $B$ to neuron $i$ in population $A$, $P_{ij}^{AB}$, satisfies

$$P_{ij}^{AB} = \frac{K}{N_E} \left( 1 + 2p \cos[2(\phi_i^A - \phi_j^B)] \right), \tag{5.17}$$

where $p$ is the synaptic modulation. With this choice of the coupling probability a cell receives input from on average $K$ excitatory and $K N_I / N_E$ inhibitory cells. If these is a synaptic connection its strength is $J_{AB}/\sqrt{K}$.

The total input, $I_i^A$, into neuron $i$ of population $A$ is given by

$$I_i^A(t) = I_{i,0}^A + \sum_B \sum_j J_{ij}^{AB} E_j^B(t), \tag{5.18}$$

where $I_{i,0}^A$ is the input from the LGN, which is assumed to be constant and satisfies $I_{i,0}^A = E_A \sqrt{K} v_0 (1 + \mu \cos[2(\Phi_0 - \phi_i^A)])$, for a stimulus with orientation $\Phi_0$. The overall strength of the input, $v_0$, is an increasing function of the contrast, while the input modulation $\mu$ is contrast independent. The activation variable $E_i^A$ of the synapses projecting from cell $i$ in population $A$ is given by

$$E_i^A(t) = \frac{1}{\tau_{\text{syn}}^A} \sum_{t_i^A(k) < t} e^{-(t - t_i^A(k))/\tau_{\text{syn}}^A}, \tag{5.19}$$

where $t_i^A(k)$ is the time of the $k$ th spike of the neuron and $\tau_{\text{syn}}^A$ is the synaptic time constant. Note that for simplicity we have here assumed an instantaneous rise for the synapse. We have also assumed that the synaptic input is independent of the voltage.

Throughout this section we will show result for simulations in which the values in Table 1 were used.

Table 1

Parameters of the model of the visual cortex use for the figures this section.

| | | | |
|---|---|---|---|
| $N_E$ | 10000 | $N_I$ | 2000 |
| $\tau_E$ | 10 msec | $\tau_I$ | 10 msec |
| $\tau_A$ | 100 msec | $g_A$ | $0.1\theta_E C_M$ |
| $\tau_{\text{syn}}^E$ | 2 msec | $\tau_{\text{syn}}^I$ | 1 msec |
| $E_E$ | $3\theta_E C_M$ | $E_I$ | $\theta_I C_M$ |
| $J_{EE}$ | $3.5\theta_E C_M$ | $J_{IE}$ | $3.5\theta_I C_M$ |
| $J_{EI}$ | $7\theta_E C_M$ | $J_{II}$ | $5\theta_I C_M$ |
| $p$ | 0.143 | $\mu$ | 0.2 |

In the large $K$ limit the net input into a cell, averaged over all cells with the same difference $\phi$ between their preferred orientation and the orientation of the stimulus, $I_A$, can be written as $I_A(\phi) = I_A^{(0)} + I_A^{(1)} \cos(2\phi)$, where $I_A^{(0)}$ and $I_A^{(1)}$ are given by

$$I_A^{(0)} = \sqrt{K} \left[ E_A v_0 + \frac{N_A}{N_E} \sum_B J_{AB} v_B^{(0)} \right]$$

$$I_A^{(1)} = \sqrt{K}\left[\mu E_A v_0 + p\frac{N_A}{N_E}\sum_B J_{AB}v_B^{(1)}\right]. \tag{5.20}$$

Here $v_A(\phi)$ is the average firing rate of neurons in population $A$ with a relative preferred orientation $\phi$ and $v_A^{(n)}$ is the $n$th Fourier moment of $v_A$, $v_A(\phi) = \sum_n v_A^{(n)}\cos(2n\phi)$.

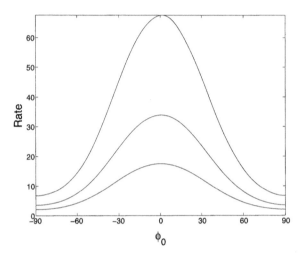

Fig. 18. Tuning of the population averaged response at three contrast levels, corresponding to $v_0 = 5$ Hz (bottom curve), $v_0 = 10$ Hz (middle curve) and $v_0 = 20$ Hz (top curve).

This is similar to what we found in section 4.2, except that we have an extra term $N_I/N_E$ on the right-hand side in the inhibitory component. As in that section, the ratios between the zeroth and first Fourier component of the population response is constant in the large $K$ limit. That this leads to an approximate contrast invariance of the population averaged tuning is a network with $K = 1000$ is shown in Fig. 18. This figure shows average firing rate in the excitatory neurons in the network as a function of the relative preferred orientation $\phi$. The average was obtained by only keeping the first 8 Fourier components of the firing rate as function of $\phi$. The average firing rate is shown for three contrast levels, corresponding to $v_0 = 5$ Hz, $v_0 = 10$ Hz and $v_0 = 20$ Hz respectively.

Equation 5.20 also implies that the zeroth and first Fourier moments of of the firing rate vary linearly with the input rate, This is also true in a network with $K = 3000$ as can be seen in Fig. 19. Here $v_E^{(0)}$ and *twice* $v_E^{(1)}$ are shown as function of $v_0$. The relationship is close to linear for both. Also, for the parameters used here, $v_E^{(1)}$ should be 0.35 times $v_E^{(0)}$, according to Eqn. 5.20.

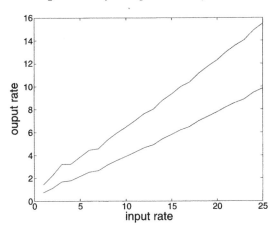

Fig. 19. Linearity of the average response and the modulation against the external input. The top curve shows $\nu_E^{(0)}$ as a function of $\nu_1$ while the bottom curve traces $2\nu_E^{(1)}$ against $\nu_1$.

This also fits reasonably well.

Figure 20 shows the voltage trace and the trace of the total input for three neurons. A cell with a preferred orientation orthogonal to the stimulus (top two panels), a cell with preferred orientation at 45 degrees from the stimulus (middle two panels), and a cell whose orientation coincides with the stimulus orientation (bottom two panels). In all three cells the input has large fluctuations, leading to irregular firing. The average input in the cell that is stimulated at the preferred orientation is higher that for the other two, but for all three the distance between the average input and the threshold input is of the same order as the amplitude of the fluctuations. All three cells are in the balanced state.

To characterize the irregularity of the neurons, we look at how the coefficient of variation, $C_V$, is distributed in the excitatory population. The $C_V$ is defined as $C_V = (\sqrt{< (\Delta t_i)^2 > - < \Delta t_i >^2})/ < \Delta t_i >$ [11], where $\Delta t_i$ is the $i$th interspike interval. The coefficient of variation is 1 for a Poisson process, less than 1 for a more regularly spiking cell, and larger than 1 for cells that are more bursty. Figure 21 shows the distribution of the coefficient of variation for the excitatory neurons. At one fixed stimulus orientation we determined the coefficient of variation for all neurons that fired at least 5 spikes during a 2 *sec* simulation. The $C_V$s are distributed around 1, with most cells having a $C_V$ between 0.7 and 1.3. This is slightly higher than what is measured experimentally.

Figure 22 show 9 examples of neurons of the excitatory population at three different contrast levels. Clearly there is a high level of heterogeneity, with some cells having broad tuning curves, and other having narrower tuning curves. For

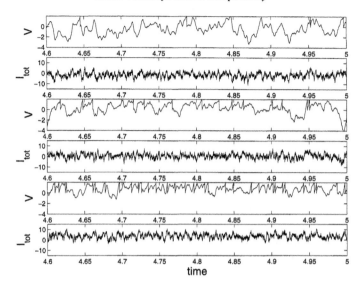

Fig. 20. Voltage and input current traces for three neurons in a network with $K = 3000$ and $\nu_0 = 10$ Hz. The top two panels are for a neuron with an orientation orthogonal to the stimulus, the middle two for a neuron 45 degrees away from the stimulus, and the bottom one for a cell at the preferred orientation.

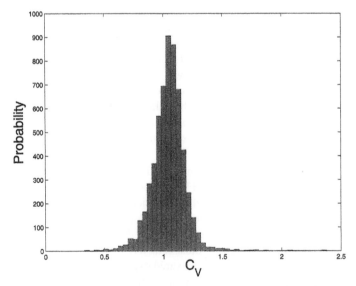

Fig. 21. The distribution of the inter-spike interval coefficient of variation of the excitatory population in a network with $K = 3000$ and $\nu_0 = 10$ Hz.

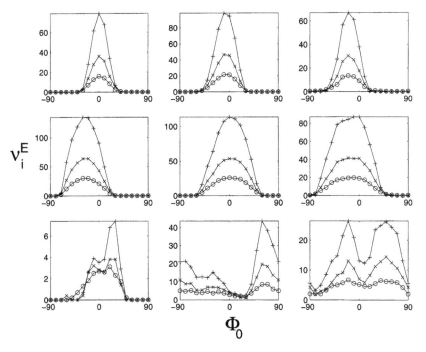

Fig. 22. Sample tuning curves of excitatory neurons at low ($\nu_0 = 5$ Hz.), intermediate ($\nu_0 = 10$ Hz.) and high ($\nu_0 = 20$ Hz.) contrast.

all cells, the responses are plotted against the difference between the stimulus and preferred orientation. The maximum response of the cells does often not occur at this orientation, but can be shifted by up to 30 degrees. Nevertheless well tuned cells (top row) and cells with an intermediate tuning (middle row) have contrast independent normalized tuning curves. Only for very badly tuned cells (bottom) row do the tuning curves change appreciably. But even for these, it is not so much the tuning width that changes as the shape of the tuning curve.

This is also shown in Fig. 23, which show a scatterplot of the circular variance for $\nu_0 = 40$ Hz. against the circular variance for $\nu_0 = 10$ Hz. The circular variance is defined as $Circ - Var = 1 - \nu_E^{(1)}/\nu_E^{(0)}$, and is smaller for narrower tuning curves. $Circ - Var = 0$ for an infinitely narrow tuning curve, and $Circ - Var = 1$ for an untuned cell. Figure 23 shows that for nearly all cells the circular variance stays the same or increases as the contrast is increased. But the circular variance does not change that much with the contrast. If we compare Fig. 23 with Fig. 11 which plotted the tuning strength (which is defined as 1 minus the

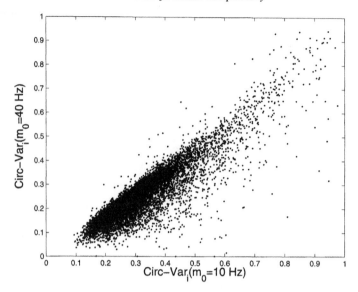

Fig. 23. Scatterplot of the circular variance at high contrast ($\nu_0 = 40$ Hz.) against the circular variance at low contrast ($\nu_0 = 10$ Hz. in the excitatory population of a network with $K = 3000$.

circular variance) in a binary network model for $V1$ we see that in the binary cells the circular variance is more invariant than in the model with IF neurons presented here. It is likely that this is due to the fact that the transfer functions of binary neurons are much more non-linear than those of IF neurons.

### 5.5. Beyond the Poisson assumption

In section 5.1 we approximated the behavior of a network in inhibitory IF neurons with synaptic currents that are independent of the post-synaptic voltage. In this approximation we made the *Ansatz* that the statistics of the inputs into the cells is Poissonian. In a set of recent papers the group of John Hertz has improved on this [93–96]. In their calculations they make sure that the input into the neurons has the same statistics as the output of the cells.

If the statistics of the neurons is not Poisson, the input $I_i^A(t)$ of cell $i$ in population $A$ is given by [97, 98]

$$I_i^A(t) = \sqrt{K}\left(E_A \nu_0 + \sum_B J_{AB} \nu_B\right) + \sqrt{\beta_A} x_i^A + \sum_B J_{AB} z_i^{AB}(t). \qquad (5.21)$$

where $\beta_A = \sum_B (J_{AB})^2 < (r_j^B)^2 >_j$, and $x_i$ is random variable drawn from a unitary Gaussian. The variable $z_i^{AB}$ is a Gaussian that satisfies

$$< z_i^{AB}(t) z_i^{AB}(t') >= C_B(t - t'),$$

where $C_B(t)$ is the average correlation of the activity of synapses projecting from neurons in population $B$.

The self-consistency requirement is that the correlations in the synaptic activation is consistent with the neuronal spike statistics. The spike-train of neuron $i$ of population $A$ can be written as $S_i^A(t) = \sum_k \delta(t - t_i^A(k))$, where $t_i^A(k)$ is the $k$th time the neuron spikes. For this spike-train the synaptic current due to this cell is

$$E_i^A(t) = \int_0^\infty dt' \, S(t - t') \epsilon_A(t'). \qquad (5.22)$$

The temporal correlation of the synaptic activity satisfies

$$
\begin{aligned}
C_A(t) &= \frac{1}{N_A} \sum_i < E_i^A(t + t') E_i^A(t') >_{t'} - \left( < E_i^A(t') >_{t'} \right)^2 \\
&= \int dt' \int dt'' \, \tilde{C}_A(t - t' + t'') \epsilon_A(t') \epsilon_A(t''), \qquad (5.23)
\end{aligned}
$$

where we have used $\tilde{C}$ to denote the average spike correlation for neurons in population $A$

$$\tilde{C}_A(t) = \frac{1}{N_A} \sum_i < S_i^A(t + t') S_i^A(t') >_{t'} - (\nu_i^A)^2. \qquad (5.24)$$

This relates the statistics of the synaptic activation to that of the spikes.

As before, the rates $\nu_E$ and $\nu_I$ are to leading order determined by the requirement that the population averaged input should be of order 1. The remaining average input $u_A$, the mean square rates $q_A = < (r_i^A)^2 >_i$ and the correlations in the currents, $C_A(t)$ have to be solved self-consistently.

This can be done numerically in the following manner: We start with a trial solution in which $u_A(0)$ is given, $q_A(0) = \nu_A^2$ and $\tilde{C}_A(0, t) = \nu_A \delta(t)$. From this we first calculate correlation in the synaptic activity, $C_A(0, t)$, using Eqn. (5.23). Then we drawn $x_i^A$ from a uniform Gaussian, and draw $z_i^{AB}(t)$ for $0 < t < T$ from a Gaussian with temporal correlation $C_B(0, t)$. This determines $I_i^A(t)$. This current is injected in to a IF neuron and we record the spike times $t_i^A(k)$. >From these spike times we can estimate $\nu_i^A$ and to correlation, $\tilde{C}_i^A(t)$, for this neuron. We repeat this many times for both $A = E$ and $A = I$ to obtain an estimate $\nu_A(1)$ for the average rate, an estimate $q_A(1)$ for the mean square rate, and and estimate $\tilde{C}_A(1, t)$ for the spike correlation.

If the initial choice for $u_A(0)$, $q_A(0)$ and $\tilde{C}_A(0,t)$ was correct, we would have gotten that $v_A(1) = v_A$, $q_A(1) = q_A(0)$, and $\tilde{C}_A(1,t) = \tilde{C}_A(0,t)$, but in general this will not be the case. We can repeat this procedure taking $u_A(1) = u_A(0) + \gamma[r_A - r_A(1)]$, $q_A(1)$ and $\tilde{C}(1,t)$ to obtain a new estimate, use this estimate for the next iteration, etc. If the constant $\gamma$ is not to large the system will eventually converge to a fixed point $u_A(\infty)$, $q_A(\infty)$, $\tilde{C}_A(\infty,t)$, which is the self-consistent solution we were looking for.

The group of Hertz did essentially the same thing, except that they assumed that the external input comes from Poisson neuron, and the synapses were instantaneous. The Poissonian nature of the external input adds a extra Gaussian noise term with correlation $C_0(t) = E_A v_0 \delta(t)$. Because the synapses are instantaneous Eqn. (5.23) reduces to $C_A(t) = \tilde{C}_A(t)$. For the self-consistent solution the dependence of the Fano factor on the different variables was investigated. The Fano factor, $F_i^A$, the ratio between the variance and the mean of the spike count, is given by

$$F_i^A = \frac{1}{v_i} \int_{-\infty}^{\infty} dt \, \tilde{C}_i^A(t), \tag{5.25}$$

and measures the variability of the spike train. $F_i^A$ is 1 for a Poisson process, less than 1 for a process that is less irregular, and larger than 1 for a more irregular process. Numerically determined self-consistent solutions indicate that the Fano factor increases is the firing rate is increased. This is consistent experimentally measured variability in the spike count in the primary visual cortex where the variance, $var(N)$, in the spike count depends on the mean $E(N)$ as $var(N) \propto E(N)^\gamma$ with $\gamma$ between 1.2 and 1.4 [99, 100].

## 6. Discussion

In this course we have described a model for irregular activity in the central nervous system that relies on strong feedback between the cells. We have used methods from statistical mechanics of disordered systems to study this mechanism and analyze the activity in this state for model networks with simple binary neurons, under the assumption that the network is sparse. We have shown that many of the features of this state also show up in networks of IF neurons, and that the sparseness is not an extremely strong constraint. Indeed even with neurons connection to 10 percent of the other cells, the network behaves very similarly to an infinitely diluted network.

To keep the number of variable in check we have made several simplifications that can be relaxed easily. For example, we have throughout that on average each cell in both populations receive the same number $K$ of excitatory and inhibitory inputs. Models closer to biology will assume that on average neurons receive input from a larger number of excitatory cells, and that excitatory neurons receive mode inputs that inhibitory once. This can easily be implemented in the model by assuming that the probability of a connection from a neuron in population $B$ to a neuron in population $A$ is $K_{AB}/N_B$, where $K_{AB} = C_{AB}K$. Similarly in sections 4 and 5.4 we have assumed that the modulation in the connectivity $p$ was the same for all couplings, and that the modulation $\mu$ is the same for the excitatory and inhibitory connections. It is straightforward to replace these by $p_{AB}$ and $\mu_A$, which can be different for different populations.

To highlight that the irregular activity is an emergent property of the network, we have assumes that the feedforward input is constant. Quenched and temporal disorder in the external input is biologically plausible and can be incorporated without problem if they have Gaussian statistics. Let us assume that for the external input, $m_{i,0}^A$, we have

$$m_{i,0}^A(t) = m_0 + \sqrt{q_A(\infty)}x_{i,0}^A + \sqrt{q_A(0) - q_A(\infty)}z_{i,0}^A(t), \tag{6.1}$$

where the $x_{i,0}^A$ and $z_{i,0}^A(t)$ are Gaussian variables which satisfy $< (x_{i,0}^A)^2 >= 1$, $< (z_{i,0}^A(t))^2 >= 1$ and $< z_{i,0}^A(t)z_{i,0}^A(t') > \to 0$ for $|t-t'| \to \infty$. These contribute to the total and quenched variances in the input, $\alpha$ and $\beta$ so that these will be given by

$$\alpha_A = (E_A)^2 q_A(0) + \sum_B (J_{AB})^2 m_B \tag{6.2}$$

and

$$\beta_A = (E_A)^2 q_A(\infty) + \sum_B (J_{AB})^2 q_B. \tag{6.3}$$

Using this further development of the theory is straightforward.

There are several other ways in which this work can be extended. Most importantly, we have no good theoretical understanding why the sparse limit fits so well even for relatively highly connected networks. Theoretically, the sparse limit is exact, is $K$ scales at most as $\log(N)$ [42]. It is not clear why networks with $K/N \approx 0.1$ fit this theory so well.

A problem that needs to be solved if we want to study models with more realistic neurons, is how to deal with synapses that scale as $1/\sqrt{K}$ in a model in which the synapses are described as conductances. In point neurons this leads to an effective membrane time constant that is of the order of $1/\sqrt{K}$ and the large $K$ limit is not sensible. A possible source of this problem is that one first approximates the neuron by an object without spatial extend, in effect assuming that the neurons is an object that is smaller than its electrotonic length. After that one takes the large $K$ limit. But if one describes synapses by conductances, the electrotonic length decreases as $K$ is increased, so that one can no longer treat the neuron as a point object. If one wants to simulate large networks of neurons, one needs to have a simple description of the individual units, without having to keep track of the neurons voltage at different locations in the cell. A better understanding of the large $K$ limit in such cells is needed.

## Acknowledgements

The results of sections 4 and 5 have only been presented previously in abstract form. Section 3 appears here for the first time. Reseach of HS is partially supported by grants from the Israeli Science Foundation (Center of Excellence Grant-8006/00) and the U.S.–Israel Binational Science Foundation.

## References

[1] Burns B.D. and Webb A.C., "The spontaneous activity of neurons in the cat's visual cortex", Proc. R. Soc. Lond. B 194 (1976) 211-223.

[2] Abeles M., "Corticonics: Neural Circuits of the Cerebral Cortex", (Cambridge Univ. Press, Cambridge, 1991).

[3] Douglas R.J., Miles K.A.C. and Whitteridge D., "An intracellular analysis of visual response of neurons in the cat visual cortex", J. Physiol. 440 (1991) 659-696.

[4] Softky W.R. and Koch C., "The highly irregular firing of cortical cells is inconsistent with temporal integration of random EPSPs", J. Neurosci. 13 (1993) 334-350.

[5] Bair W., Koch C., Newsome W., and Britten K., "Power spectrum analysis of bursting cells in area MT in the behaving monkey", J. Neurosci. 14 (1994) 2870-2892.

[6] Borg-Graham L.J., Monier C. and Fregnac Y., "Visual input invokes transient and strong shunting inhibition in visual cortical neurons", Nature 393 (1998) 369-373.

[7] Douglas R.J. and Martin K.A.C., "Opening the Grey Box", Trends in Neurosci. 14 (1991) 286-293.

[8] Ferster D. and Jagadeesh B., "EPSP-IPSP interactions in cat visual cortex studied with *in vivo* whole-cell patch recording", J. Neurosci. 12 (1992) 1262-1274.

[9] Anderson J.S., Lampl I., Gillespie D.C., and Ferster D., "The contribution of noise to contrast invariance of orientation tuning in cat visual cortex", Science 290 (2000) 1969-1972.

[10] Hansel D. and van Vreeswijk C.,"How noise contributes to contrast invariance of orientation tuning in cat visual cortex", J. Neurosci. 22 (2002) 5118-5128.

[11] Holt G.R., Softky W.R., Koch C., and Douglas R.J., "A comparison of discharge variability *in vitro* and *in vivo* in cat visual cortex", J. Neurophysiol. 75 (1996) 1806-1814.

[12] Mainen Z.J. and Sejnowski T., "Reliability of spike timing in neocortical neurons", Science 268 (1995) 1503-1506.

[13] Perkel D.H., Gerstein G.l. and Moore G.P., "Neuronal spike trains and stochastic point-processes. I. the single spike train", Biophys. J. 7 (1967) 391-418.

[14] Perkel D.H., Gerstein G.l. and Moore G.P., "Neuronal spike trains and stochastic point-processes. II. simultaneous spike trains", Biophys. J. 7 (1967) 419-440.

[15] Gray C.M. and Singer W., "Oscillatory responses in cat visual cortex exhibits inter-columnar synchronization which reflects global stimulus properties", Nature 338 (1989) 334-337.

[16] Vaadia E., Haalman I. Abeles M., Prut Y., Slovin H., and Aertsen A., "Dynamics of neural interaction in monkey cortex in relation to behavioral events", Nature 373 (1995) 515-518.

[17] Wilson H.R. and Cowan J.D.,"Excitatory and inhibitory interactions in a localized population of model neurons", Biophys. J. 12(1972) 1-21.

[18] Grannan E., Kleinfeld D. and Sompolinsky H., "Stimulus-dependent synchronization of neuronal assemblies", Neural Comput. 4 (1992) 550-569.

[19] Abbott L.F. and van Vreeswijk C.,"Asynchronous states in networks of pulse-coupled neurons", Phys. Rev. E48 (1993) 1483-1490.

[20] Gerstner W. and van Hemmen J.L., "Associative memory in a network of 'spiking' neurons", Network 3 (1993) 139-164.

[21] Hansel D., Mato G. and Meunier C., "Synchronization in excitatory neural networks", Neural Comput. 3 (1995) 307-338.

[22] van Vreeswijk C., "Partial Synchrony in populations of pulse-coupled oscillators", Phys. Rev. E54 (1996) 5522-5537.

[23] van Vreeswijk C., "Stability of the asynchronous state in networks of non-linear oscillators", Phys. Rev. Lett. 84 (2000) 5110-5113.

[24] Bush P.C. and Douglas R.J., "Synchronization of bursting action potential discharges in a model network of neocortical neurons", Neural Comput. 3 (1991) 19-30.

[25] Hansel D. and Sompolinsky H., "Synchronization and computation in a chaotic neural network", Phys. Rev. Lett. 68 (1992) 718-721.

[26] Hansel D. and Sompolinsky H., "Chaos and synchrony in a model of a hypercolumn in visual cortex", J. Comput. Neursci. 3 (1996) 7-34.

[27] Gerstein G.L. and Mandelbrod B., "Random walk models for the spike activity of a single cell", Biophys. J. 4 (1964) 41-68.

[28] Bell A., Mainen Z.F., Tsodyks M., and Sejnowski T., Soc. Neurosci. Abstr. 20 (1994) 1527.

[29] Shadlen M.N. and Newsome W.T., "Noise, neural codes and cortical organization". Curr. Opin. Neurobiol. 4 (1994) 569-579.

[30] Shadlen M.N. and Newsome W.T., "Is there a signal in the noise?", Curr. Opin. Neurobiol. 5 (1995) 248-250.

[31] Softky W.R., "Simple codes versus efficient codes', Curr. Opin. Neurobiol. 5 (1995) 239-247.

[32] Gutkin B.S. and Ermentrout G.B., "Dynamics of membrane excitability determine inter-spike interval variability: A link between spike generating mechanisms and cortical spike train statistics", Neural Comput. 10 (1998) 1047-1065.

[33] Troyer T.W. and Miller K.D., "Physiological gain leads to high ISI variability in a simple model of a cortical regular firing cell", Neural Comput. 9 (1997) 733-745.

[34] Tsodyks M. and Sejnowski T., "Rapid state switching in balanced cortical network models", Network 6 (1995) 111-124.

[35] Amit D.J. and Brunel N., "Model of global spontaneous activity and local structured activity during delay periods in the cerebral cortex", Cereb. Cortex 7 (1996) 237-252.

[36] Amit D.J. and Brunel N., "Dynamics of a recurrent network of spiking neurons before and following learning", Network: Comput. in Neural Sys. 8 (1997) 373-404.

[37] Amit D.J., "Modeling brain function", (Cambridge University Press, 1989, Cambridge).

[38] Hertz J., Krogh A. and Palmer R.G., "Introduction to the theory of neural computation", (Addison Wesley, 1991, Redwood City).

[39] Mézard M., Parisi G. and Virasoro M.A., "Spin Glass Theory and Beyond", (World Scientific, Singapore, 1987).

[40] van Vreeswijk C. and Sompolinsky H., "Chaos in neuronal networks with excitatory and inhibitory activity", Science 274 (1996) 1724-1726.

[41] van Vreeswijk C. and Sompolinsky H., "Chaotic Balanced State in a Model of Cortical Circuits", Neural Comput. 10 (1998) 1321-1371.

[42] Derrida B., Gardner E. and Zippelius A., "An exactly soluble asymmetric neural network model", Europhys. Lett. 4 (1987) 167-173.

[43] Glauber R.J., "Time-dependent statistics of the Ising model", J. Math. Phys. 4 (1963) 294-307.

[44] Ginzburg I. and Sompolinsky H., "Theory of correlations in stochastic neural networks", Phys. Rev E 50 (1994) 3171-3190.

[45] Abeles M., Bergman H. and Vaadia E., unpublished data.

[46] Hackenbracht and Schuster H.J., "Critical behavior of the spherical in a random magnetic field", Phys. Rev. B 27 (1983) 6961.

[47] Caianiello E.R., "Outline of a theory of thought-processes and thinking machines", J. Theo. Biol. 1 (1961) 204-235.

[48] Amari, S.-I., "Learning pattern sequences by self-organizing nets of threshold elements", IEEE trans. Comput. C 21 (1972) 1197-1206.

[49] Little W.A., "The existence of persistent states in the brain", Math. Biosci. 19 (1974) 101-119.

[50] Hopfield J.J., "Neural networks and physical systems with emergent collective computational abilities", Proc. Natl. Acad. Sci. USA 79 (1982) 2554-2558.

[51] Fuster J.M. and Alexander G.E., "Firing changes in cells of the nucleus medialis dorsalis associated with delayed response behavior", Brain Res. 61 (1973) 79-91.

[52] Kubota K. and Niki H., "Prefrontal cortical unit activity and delayed alternation performance in monkeys", J. Neurophysiol. 34 (1971) 337-347.

[53] Bliss T.V.P. and Lomo T., "Long-lasting potentiation of synaptic transmission in the dentate area of anaesthetized rabbit following stimulation of the perforant path", J. Physiol. (London) 232 (1973) 331-356.

[54] Nakamura K. and Kubota K., "Mnemonic firing of neurons in the monkey temporal pole during a visual recognition memory task", J. Neurophysiol. 75 (1995) 162-178.

[55] Willshaw D., Bunemann O.P. and Longuet-Higgins H., "Non-holographic associative memory", Nature 222 (1969) 960-962.

[56] Golomb D., Rubin N. and Sompolinsky H., "Willshaw model: associative memory with sparse coding and low firing rate", Phys. Rev. A 41 (1990) 1943-1854.

[57] Miyashita Y., "Neuronal correlate of visual associative long-term memory in the primary temporal cortex", Nature 335 (1988)817-820.

[58] Hubel D.H. and Wiesel T.N., "Receptive fields, binocular interaction and functional architecture in the cat's visual cortex", J. Neurosci. 3 (1962) 1116-1133.

[59] Hubel D.H. and Wiesel T.N., "Sequence regularity and geometry of orientation columns in the monkey striate cortex", J. Comp. Neurol. 158 (1974) 267-297.

[60] Bonhoeffer T. and Grinvald A., "Orientation columns in cat are organized in pinwheel like patterns", Nature 364 (1991) 166-169.

[61] Sclar G. and Freeman R.D., "Orientation selectivity in the cat's striate cortex in invariant with stimulus contrast", Exp. Brain Res. 46 (1982) 457-461.

[62] Skottun B.C., Bradley A., Sclar G., Ohzawa I, and Freeman R.D., "The effects of contrast on visual orientation and spatial frequency discrimination: a comparison of single cells and behavior", J. Neurophysiol. 57 (1987) 773-786.

[63] Ferster D., Chung S. and Wheat H., "Orientation selectivity of thalamic inputs to simple cells of the cat visual system", Nature 380 (1996) 249-252.

[64] Nelson S., Toth L.J., Seth B., and Sur M., "Orientation selectivity of cortical during extra-cellular blockade of inhibition", Science 265 (1994) 774-777.

[65] Somers D.C., Nelson S. and Sur M., "An emergent model of orientation selectivity in cat visual cortical simple cells", J. Neurosci. 15 (1995) 5448-5465.

[66] Ben-Yishai R., Lev Bar-Or R. and Sompolinsky H., "Theory of orientation tuning in visual cortex", Proc. Natl. Acad. Sci. USA 92 (1995) 3844-3848.

[67] Sompolinsky H. and Shapley R., "New perspectives on the mechanisms for orientation selectivity", Curr. Opin. Neurobiol. 7 (1997) 514-522.

[68] Ben Yishai R., Hansel D. and Sompolinsky H., "Traveling waves and the processing of weakly tuned inputs in a cortical network module", J. Comput. Neurosci. 4 (1997) 57-77.

[69] Troyer T.W., Krukowski A., Probe N.J., and Miller K.D., "Contrast-invariant orientation tuning in cat visual cortex: feedforward tuning and correlation-based intra cortical connectivity", J. Neurosci. 18 (1998) 5908-5927.

[70] Hansel D. and Sompolinsky H., "Modeling feature selectivity in local cortical circuits", In: C. Koch and I. Segev "Methods of Neural Modeling (2nd edition)", pp 499-567. (MIT press, Cambridge, 1998).

[71] McLaughlin D.R., Shapley R., Shelley M., and Wielaard D.J., "A neural network model of macaque primary visual cortex (V1): orientation selectivity and dynamics in the input layer $4C\alpha$", Proc. Natl. Acad. Sci. USA 97 (2000) 8087-8092.

[72] Ferster D. and Miller K.D., "Neural mechanisms of orientation selectivity in the visual cortex", Ann. Rev. Neursci. 23 (2000) 441-471.

[73] Wielaard D.J., Shelley M., McLaughlin D.W. and Shapley R., "How simple cells are made in a non-linear network model of the visual cortex", J. Neurosci. 21 (2001) 5203-5211.

[74] Shelley M. and McLaughlin D.J., "Course-grained reduction and analysis of a network model of cortical response: I drifting grating stimuli", J. Comput. Neursci. 12 (2002) 97-122.

[75] Kang K., Shelley M. and Sompolinsky H., "Mexican hats and pinwheels in visual cortex", Proc. Natl. Acad. Sci. USA 100 (2003) 2848-2853.

[76] Wolt F., van Vreeswijk C. and Sompolinsky H., "Chaotic activity induces contrast invariant orientation tuning", Soc. Neurosci. Abstr. 12.7 (2001).

[77] Ringach D.L., Shapley R.M. and Hawken M.J., "Orientation selectivity in macaque V1: diversity and laminar dependence", J. Neurosci. 22 (2002) 5639-5651.

[78] Lapicque L., "Recherches quantitatives sur l' excitabilité électrique des nerfs traitée comme une polarisation", J. Physiol.Pathol. Gen. 9 (1907) 620-635.

[79] Hansel D., Mato G., Meunier C. and Neltner L., "Numerical simulations of integrate and fire neural networks", Neural Comput. 10 (1998) 467-483.

[80] Shelley M. and Tao L., "Efficient and accurate time-integration schemes for integrate-and-fire neural networks", J. Comput. Neurosci. 11 (2001) 111-119.

[81] van Vreeswijk C., Kim J.W. and Sompolinsky H., "Stochastic activity in balanced networks of integrate and fire neurons", Technical report, 1998.

[82] Brunel N., "Persistent activity and the single cell f-I curve in cortical network models", Network 11 (2000) 261-280.

[83] Hertz J. and Solinas S., "Stability of the asynchronous firing state in networks with synaptic adaptation", Neurocomput. 38-40 (2001) 915-902.

[84] Hertz J., Richmond B. and Nilson K., "Anomalous response of variability in a balanced network model", Neurocomput. 52-54 (2002) 787-792.

[85] Lauritzen T.Z., "Biologically realistic modelling of cortical network dynamics", PhD thesis (Niels Bohr Inst.) 1998.

[86] Hodgkin A.L., Huxley A.F. and Katz B., "Measurements of current-voltage relations in the membrane of the giant axon in loligo", J. Physiol. 116 (1952) 424-448.

[87] Hodgkin A.L. and Huxley A.F., "Currents carried by sodium and potassium ions through the membrane of the giant axon of loligo", J. Physiol. 116 (1952) 449-472.

[88] Hodgkin A.L. and Huxley A.F., "The components of membrane conductance in the giant axon of loligo", J. Physiol. 116 (1952) 473-496.

[89] Hodgkin A.L. and Huxley A.F., "The dual effect of membrane potential on sodium conductance in the giant axon of the squid", J. Physiol. 116 (1952) 497-506.

[90] Hodgkin A.L. and Huxley A.F., "A quantitative description of membrane current and it application to conduction and excitation in nerve", J. Physiol. 117 (1952) 500-544.

[91] van Vreeswijk C. and Sompolinsky H., "Heterogeneity and contrast invariance in a model of primary visual cortex", preprint (2004).

[92] van Vreeswijk C. and Hansel D., "Patterns of synchrony in neural networks with spike adaptation", Neural Comput. 13 (2001) 959-992.

[93] Lerchner A., Ursta C., Hertz J., Ahmadi M., and Ruffiott P., "Response variability in balanced cortical netrworks", preprint (2004).

[94] Hertz J., Lerchner A. and Ahmadi M., "Mean field methods for cortical network dynamics", preprint (2004).

[95] Lerchner A., Ahmadi M. and Hertz J., "High conductance states in a mean field cortical network model", preprint (2004).

[96] Lerchner A., Sterner G., Hertz J., and Ahmadi M., "Mean field theory for a balanced hypercolumn modl of orientation selectivity in primary visual cortex", preprint (2004).

[97] Fulvi Mari C., "Random networks of spiking neurons: instability in the *xenopur* tadpole motoneuron pattern", Phys. Rev. Lett. 85 (2000) 210-213.

[98] Kree R. and Zippelius A., "Continous time dynamics of assymetrically diluted neural networks", Phys Rev A 36 (1987) 4421-4427.

[99] Vogels R., Spileers W. and Orban G.A., "The response variability of striate cortical neurons in behaving monkey", Exp. Brain Res. 77 (1989) 414-419.

[100] Gershon E., Wiener M.C., Latham P.E., and Richmond B.J., "Coding strategies in monkey V1 and inferior temporal cortex", J. Neurophysiol. 79 (1998) 1135-1144.

Course 10

# NETWORK MODELS OF MEMORY

## Nicolas Brunel

*CNRS UMR 8119, Université Paris 5 René Descartes, 45 rue des Saints Pères,*
*75270 Paris Cedex 06, nicolas.brunel@univ-paris5.fr*

*C.C. Chow, B. Gutkin, D. Hansel, C. Meunier and J. Dalibard, eds.*
*Les Houches, Session LXXX, 2003*
*Methods and Models in Neurophysics*
*Méthodes et Modèles en Neurophysique*
© *2005 Elsevier B.V. All rights reserved*

# Contents

## 1. Introduction

The main hypothesis that has guided modellers interested in the neuronal and network mechanisms underlying memory since Hebb [67] is the separation between two forms of memory: a short term memory component, maintained by the persistent activation of an ensemble of neurons (the 'reverberatory activity' of a 'cell assembly'); and a long-term memory component, maintained by persistent changes in synaptic efficacies. The idea that neuronal activity can persist beyond the stimulation that caused it, due to excitatory feedback loops, dates back from Lorente de No [86]. Building on these ideas, theorists have investigated neuronal network models with many *attractors* (stable states of the network). In these models, each attractor is a network state in which a subset of neurons exhibits sustained activity, after selective inputs to this subset of neurons have disappeared. Each attractor state can be thought as an internal representation of a stimulus. The attractor property means that this internal representation is maintained stably in *short-term (or working) memory*, after the disappearance of the stimulus that caused it. In addition, when large basins of attraction are present around each attractor, the network can perform as an *associative memory*: it can retrieve the stored information from a partial cue.

Hebb further conjectured that specific synaptic plasticity mechanisms could be responsible for the formation, or enhance the stability of persistent activity patterns ('cell assemblies'). In this view, plasticity (learning) mechanisms allow the network to store information in long-term memory. In the attractor picture, synaptic plasticity modifies the 'attractor landscape' (number and/or location of attractors and/or size of basins of attraction). Formation of an attractor is identified with storage of an item in long-term memory. Conversely, the disappearance of an attractor from the 'attractor landscape' of the network represents forgetting of the corresponding memory.

Both phenomena of persistent neuronal activity and synaptic plasticity remained purely theoretical hypotheses until the seventies. In 1971, Fuster, Alexander [59], Kubota and Niki [79] discovered persistent activity in the awake monkey during delayed response tasks. In 1973, Bliss and Lomo [19] discovered long-term potentiation in hippocampal slices. Since these pioneering studies, a huge amount of data has been accumulating in both fields. On the theoretical side, the ideas of Hebb were first implemented in network models of binary neurons [4, 36, 43, 73, 85]. In particular, Hopfield's associative memory model in 1982 led to detailed analytical investigations and consequently a deep understanding of the properties of such models [5, 11]. In the last decade, modeling efforts have led to models which are much closer to neurobiology than the early models. These efforts have allowed theorists to perform detailed comparisons between models and experiment [7, 8, 137, 146].

This chapter is an attempt to give an overview of 'Hebbian' neural network models of memory. Section 2 presents an overview of electrophysiological data on persistent activity during short-term memory tasks. Section 3 presents scenarios for multistability in neuronal systems. Section 4 is devoted to networks models of binary neurons with discrete attractors: the Hopfield model and related models. Section 5 describes how learning allows such systems to form and/or destroy attractors. Section 6 describes models of networks of spiking neurons with discrete attractors. Section 7 is devoted to various phenomena of plasticity of persistent activity patterns due to synaptic plasticity mechanisms. Finally, models with continuous attractors are briefly considered in section 8.

## 2. Persistent neuronal activity during delayed response experiments

Experiments performed with monkeys have focused on two different classes of working memory: (i) 'Discrete' working memory, i.e. memory for discrete items (identity of colors, objects, faces, abstract patterns, etc.); and (ii) 'Continuous' working memory, i.e. memory for continuous variables (space, frequency of stimulation, etc). In case (i), appropriate models are networks with discrete attractors. In case (ii), networks with continuous attractors are needed.

Persistent activity is observed in several areas of the cortex, as shown in figure 1: areas in the temporal lobe, prefrontal cortex, and parietal cortex. Furthermore, persistent activity has also been documented in the thalamus and in the basal ganglia.

### 2.1. Discrete working memory: persistent activity in the delayed match to sample experiments

The delay match to sample experiment is the simplest experiment that probes short-term memory (see figure 2). In this protocol, a given trial starts by the presentation of an image (the sample) on a screen, typically chosen at random from a set of images experienced repeatedly by an animal. This image presentation, sometimes called cue period, lasts hundreds of ms. The image then disappears from the screen, and a delay period of several seconds follows. Finally, the test period begins. In some experiments, the same image that was used as a sample is

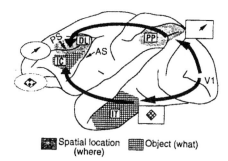

Spatial location (where)  Object (what)

Fig. 1. Areas in cortex where cells with persistent activity are observed. These areas include inferior temporal (IT) cortex and the inferior convexity (IC) of prefrontal cortex (PFC) during object working memory tasks; posterior parietal (PP) cortex and dorsolateral (DL) PFC during spatial working memory tasks. Thick arrows indicate the main streams of visual information processing from area V1, the ventral (V1 → IT → IC) and dorsal (V1 → PP → DL) streams. Adapted from ref. [143].

DMS                                        correct

                                          error

Fig. 2. Delay match to sample (DMS) protocol. After a pre-stimulus interval, a stimulus (sample, or cue) appears on the screen for several hundred ms. The stimulus disappears, and a delay period of several seconds follows. Finally, two stimuli appear on the screen: the sample together with a distractor. The monkey has to touch the stimulus that corresponds to the sample to get a reward. Adapted from ref. [101].

shown together with a distractor, at random locations on the screen, and the monkey has to choose the image that matches the sample. In other experiments, only one image is shown: the image is the sample with 50% probability, otherwise it is a distractor. In this case, the monkey has to press a lever or not, depending on whether the image was the same as the sample or different.

During the experiment, neuronal activity is recorded through one or several electrodes. One of the most striking characteristics of neuronal activity during these experiments is the phenomenon of selective persistent activity, illustrated in Figure 3. Typically, recorded cells have 'spontaneous' (also called background or baseline) activity of several Hz before the trial begins. During the cue presentations, some of the recorded cells exhibit significant visual responses to some of the shown stimuli. In IT cortex and other areas of the temporal lobe, these visual responses are typically very selective and convey a large amount of information about the identity of the stimulus. Then, in the delay period, a subset of cells that exhibit visual response keep on firing at a significantly higher than baseline rate

Table 1

Characteristics of spontaneous and persistent activity in various studies. $n$: number of cells with significant delay period activity. Indicated firing rates are average over the $n$ recorded cells. IT: inferior temporal cortex (area TE); TP: temporopolar cortex; PRh: perirhinal cortex (area 36); EC: entorhinal cortex; PFC: prefrontal cortex.

| Ref | Area | Type of stimulus | Duration of delay | $n$ | Baseline activity | Delay activity |
|-----|------|------------------|-------------------|-----|-------------------|----------------|
| [100] | TP | photographs | 0.5-5s | 20 | 4Hz | 15Hz |
| [100] | PRh | photographs | 0.5-5s | 19 | 8Hz | 22Hz |
| [100] | EC | photographs | 0.5-5s | 5 | 2Hz | 9Hz |
| [100] | IT | photographs | 0.5-5s | 22 | 6Hz | 17Hz |
| [92] | PFC | images | 1s | 82 | 12Hz | 16Hz |
| [101] | IT/PRh | Fourier descriptors | 5s | 15 | 2Hz | 6Hz |
| [144] | IT | fractals/Fourier | 5-8s | 23 | 5Hz | 12Hz |

throughout the whole delay period. Some cells that do not have visual responses also have persistent activity. Persistent activity can be highly selective, as shown in figure 3, but many cells have non-selective persistent activity.

Table 1 summarizes several studies where quantitative estimates can be found on the magnitude of spontaneous and persistent activity. This table shows that the basic phenomenon of persistent activity has been seen in a number of areas, from IT cortex, perirhinal, entorhinal, to prefrontal cortex. The typical firing rates are 1-10Hz in baseline activity and 5-25Hz in delay activity, with a typical ratio of 3 between the two. There are systematic differences in this ratio between studies, presumably due to different quantitative criteria for selecting cells. Finally, the tested delay periods vary from 0.5s to 25s, with no reported change in the magnitude of persistent activity as a function of duration.

Several studies have tested whether delay period activity is invariant with respect to simple manipulations of the stimuli. It is shown in ref. [95] that delay period activity is invariant with respect to manipulations of the images such as rescaling, translation and transformation from color to black-and-white. It is shown in ref. [10] that persistent activity is invariant with respect to degradation by random RGB noise, until it abruptly disappears. It is unknown whether this invariance of persistent activity is due to processing before the signal reaches the observed area, or if it is due to attractor dynamics in the observed area.

## 2.2. *Plasticity of persistent activity patterns induced by learning*

Miyashita [94] subjected a monkey to a DMS task in which about 100 sample stimuli were ordered in a sequence and always shown in the same order. After a lengthy training, recordings revealed that the patterns of delay activity following

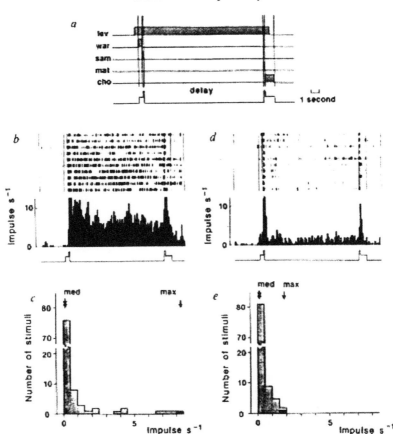

Fig. 3. Persistent activity in a DMS experiment. *a* experimental protocol (lev: lever press; war: warning signal; sam: sample presentation; mat: match presentation; cho: choice period. *b-c* data using familiar pictures. *b* rasters and trial-averaged firing rate of a cell in trials where the picture that elicits the highest delay activity is shown; *c* histogram of firing rates of the cell following presentation of 97 distinct familiar pictures; *d-e* same as *b-c* but with new pictures. From ref. [94].

stimuli that were neighbors in the sequence were correlated, up to a distance of about 5 in the sequence (see figure 4).

Sakai and Miyashita [120] devised a pair-associate task in which visual patterns where organized in fixed pairs: the pair-associate task (see figure 5). In this task, one of the patterns is shown as a cue; then, following the delay period, the pair associate of the cue is shown together with a distractor. Sakai and Miyashita found that (i) as a result of the pairing, the visual responses to pair associate stimuli became correlated; (ii) in the delay period following presentation of a given

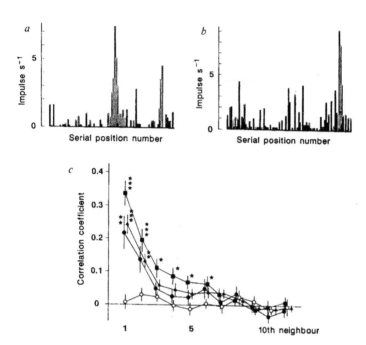

Fig. 4. Characteristics of delay activity when the DMS task is performed with a fixed sequence of 97 sample stimuli. *a*: Firing rate of a neuron during the delay period as a function of serial position of sample stimulus (grey: familiar stimuli; black: new stimuli). Note that for familiar stimuli, the neuron tends to respond for stimuli which are neighbors in the sequence. *b*: Same as *a* for a different neuron. *c*: correlation coefficient between delay activity following two sample stimuli, as a function of distance in the sequence. Black circles: averaged over 17 neurons with significant delay activity that were tested with both familiar and new stimuli, familiar stimuli. White circles: average over 17 neurons, new stimuli. Black triangles: average over 57 neurons, familiar stimuli. Black squares: averaged over neurons that have a significant ($P < 0.05$) nearest neighbor correlation coefficient, familiar stimuli. From ref. [94].

cue A, some neurons selective for the pair associate A' increased their firing rates, in apparent anticipation of the presentation of A. This type of activity has later been called 'prospective activity', since it is correlated with a stimulus which is to appear in the future, while the standard persistent activity observed in DMS task has been called 'retrospective activity', since it maintains the memory of a stimulus shown in the past. Refs [102, 103] further characterize the properties of neurons in IT and perirhinal cortices during such tasks. In the experiment of Naya et al [101], the monkey has to perform both DMS and PA tasks. The choice

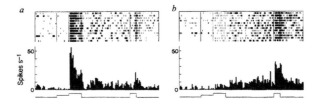

Fig. 5. Delay activity in the the pair-associate (PA) task. *a*: rastergram and trial average firing rate of a cell in trials in which its 'best' stimulus A is shown as a cue. The cell has strong visual response and persistent activity following presentation of A. *b* rasters and trial-averaged firing rate for the same cell, in trials in which the pair-associate A' is shown as a cue. The firing rate of the cell increases during the delay period that precedes the 'best' stimulus. From ref. [120].

of the task was indicated in the middle of the delay period by a color switch. In this task 'prospective activity' was triggered by the color switch which signalled the monkey he had to perform the pair-associate task.

In the tasks of refs. [101–103, 120], the monkey needs many training sessions in order to learn the associations between stimuli. Hence, it is difficult to observe directly the changes in prospective and/or retrospective activity induced by learning. In order to circumvent these difficulties, Erickson et al [54] devised a task that the monkeys can learn easily. They observed that in the first day in which new stimuli are shown to the monkey (about 500 trials), retrospective activity is already present, but not prospective activity. During the second day with the same set of stimuli, retrospective activity grows stronger, but more importantly, prospective activity appears. Moreover, during the second day, visual responses during associated stimuli becomes correlated. The main lesson of this experiment is that retrospective activity appears before prospective activity.

The recordings of all the above mentioned studies were done in the temporal lobe (IT cortex and perirhinal cortex). Similar phenomena have been observed more recently in prefrontal cortex [16, 109].

### 2.3. Spatial short-term memory: persistent activity in the oculomotor delayed response task

The classical experiment used to probe spatial short-term memory is the oculomotor delayed response (ODR) task [56–58]. In this task (see figure 6), the monkey has to maintain fixation on a central spot on the screen. A light is flashed at one location, out of 8 possible locations, on the screen (cue). This light disappears, and the monkey has to maintain fixation on the central spot throughout the whole delay period. Disappearance of the fixation point marks the beginning of the response period, during which the monkey has to perform a saccade to the memorized position of the cue. A striking feature of neuronal activity during this

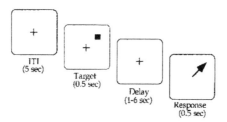

Fig. 6. The oculomotor delayed response task. The monkey has to maintain fixation on a central spot on the screen during the ITI (inter-trial interval), target and delay periods. After the ITI, a target is shown at one location for 500ms. After the delay period, the fixation point disappears, and the monkey has to perform a saccade to the memorized location. From ref. [56].

task is again selective persistent activity following a cue at a particular location, as shown in figure 7. This persistent activity is maximal at a particular cue location and decreases away from that preferred cue location. The set of locations for which a neuron is active above baseline during the delay period has been termed the 'memory field' of the neuron. Such 'memory fields' (tuned activity during the delay period) have been demonstrated in PFC and in posterior parietal cortex [38], with no significant difference between the two areas.

The group of Patricia Goldman-Rakic have further characterized the properties of memory fields using iontophoretic application of specific agonists and/or antagonists during the ODR task. In particular, it was shown that D1 antagonists can potentiate the 'memory fields' of prefrontal neurons [140]; blockade of GABA tends to destroy spatial tuning in the delay period, though it actually creates tuning in a small minority of cells [111]; finally, 5-HT2A antagonists attenuate memory fields, while 5-HT2A stimulation increases spatial tuning [141].

Persistent activity is typically characterized by firing rate of single cells. More recently, several studies have characterized activity at the cellular and network levels beyond average firing rates. In area LIP, spatially tuned elevated power in the gamma band (25-90 Hz) was found in both LFP and spiking activity during the memory period [106], while such oscillatory activity seems to be lacking in dorsolateral PFC [40]. It was further shown in ref [40] that most recorded neurons behave approximately as a Poisson process, with high variability of interspike intervals in baseline as well as in the delay period.

## 2.4. Parametric short-term memory

Persistent activity has also been observed in relationship to working memory of a parametric variable: the frequency of a vibration applied to the fingertips [118].

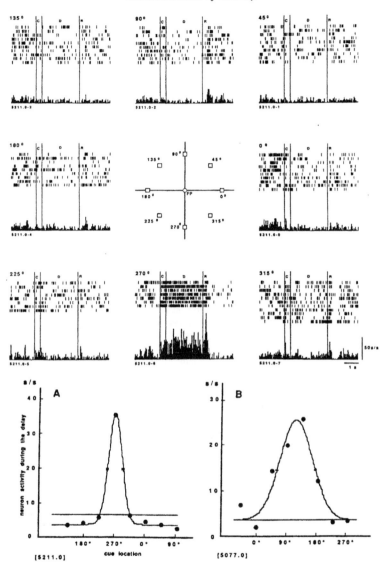

Fig. 7. Memory fields of neurons in prefrontal cortex during the delay period of a delayed oculomotor task. The eight top panels show rasterplots and trial-averaged firing rate of a neuron when a cue is flashed at 8 different locations, indicated in the center. The neuron responds preferentially when the cue is flashed at 270 degrees. A and B: 'memory fields' of two neurons (average firing rate in the delay period vs cue location). The neuron in A is the same as the one shown in the top panels. Horizontal line: baseline firing rate. Full curve: Gaussian fit of the data points. From ref. [56].

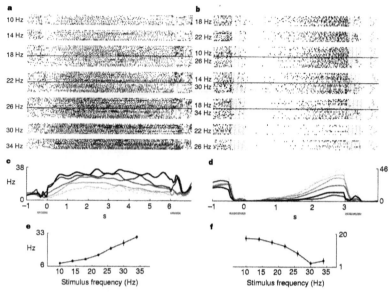

Fig. 8. Persistent activity in the Romo et al experiment [118]. **a, b** Rastergrams of two PFC cells, arranged according to frequency of the base stimulus (indicated to the left) Frequency of the comparison stimulus is shown in the center. **c, d** Trial-averaged firing rate of the two cells for several stimulation frequencies showing the monotonic response as a function of stimulation frequency. Grey level indicates the base frequency, with the lightest indicating 10Hz, while the darkest indicates 34Hz. **e, f** Tuning curve of the two cells in the delay period (average firing rate vs stimulation frequency).

In the experiment of ref. [118], the monkeys have to perform a vibrotactile discrimination task. A first (base) stimulus is applied to the finger of the animal. After a delay period, a second (comparison) stimulus is applied, and the monkey needs to indicate whether the second frequency was largest or smallest than the first. Figure 8 shows two representative cells from the experiment of ref. [118]. The firing rate of some cells in the PFC are a monotonously increasing or decreasing function of stimulation frequency. Persistent activity in many cells was found not to be stationary, but rather to increase or decrease with time.

## 2.5. Persistent activity in slices

Persistent neuronal activity has also been recently documented in cortical slice preparations [40, 41, 52, 122, 125]. Egorov et al [52] showed that individual neurons from layer V of entorhinal cortex can have several stable firing rates, and can switch from one to the other in an input specific way. This behavior was linked to cholinergic muscarinic receptor activation, and to activity-dependent changes of a calcium sensitive cationic current. McCormick and others [40, 122, 125] observed persistent network activity in slices of prefrontal and occipital cortex. The slices show an alternation of 'up' (all cells firing between 2 and 50Hz) and 'down' (all cells silent or firing at low rates <5Hz) states, both lasting several seconds. Transitions between these states were provoked by electrical stimulation of the slice. Finally, Cossart et al [41] have shown that spatially localized states can appear spontaneously and last several seconds in visual cortex slices. The advantage of these preparations is that they give better access to the mechanisms (synaptic or cellular) of persistent neuronal activation. They demonstrate the existence of mechanisms for maintenance of activity (at least for several seconds) at the single cell [52] or local network [40, 41, 122, 125] level. However, it is still unclear whether the mechanisms at work in slices are the same as those responsible for the in vivo persistent activity.

## 3. Scenarios for multistability in neural systems

What mechanisms can explain the observed multistability? Most if not all proposed scenarios explain multistability using a combination of non-linearity and excitatory feedback (see ref [137] for a review). What distinguishes these scenarios is the physical implementation of the positive feedback mechanism. The elemental feedback loop can be implemented all the way from ionic currents in a single cell to long range connections between brain structures.

### 3.1. Single cell

In the simplest single cell scenario, a single cell is bistable due to active currents providing a positive feedback to the cell. A large class of models exhibit bistability in some range close to the firing threshold. The prototypical model in this class is the original Hodgkin-Huxley model [71] (see also ref [49]). One important limitation of these models is the fact that bistability is between a quiescent ('down') state and a repetitively firing ('up') state. In the cortex of awake animals, bistability is observed between a state of low and irregular activity and a state of higher, but still irregular, activity. Another scenario is a two compartment model, in which a dendritic compartment is endowed with an active current (a calcium current for example) leading to bistability of the compartment. The dendritic compartment then imposes its state upon the somatic compartment where spike generation occurs [20, 21]. In this scenario, bistability between two active states can be achieved. Finally, one can achieve a continuous attractor at the single neuron level using regenerative calcium dynamics in the dendritic tree (see Sompolinsky, this volume).

## 3.2. Local network

Networks of excitatory neurons can become bistable by virtue of their excitatory synaptic connections: once a sufficiently large fraction of neurons is activated, they can maintain their activity through large excitatory synaptic currents. These local excitatory connections are plentiful in cortex (typically several thousands per neuron [2, 22]). Unlike the single cell models in which multistability is usually obtained through fine tuning, multistability with a large number of attractors can easily be obtained in networks with plastic synapses. This makes this scenario attractive for a network functioning both as a working memory (due to attractor neural network dynamics) and long-term memory (due to synaptic plasticity). This is the scenario that has been most studied by theorists. Sections 4-8 are entirely devoted to it.

Note that bistability, or multistability, can be obtained not only through the sheer strength of excitatory synapses, but also by other known properties of synapses: voltage dependence of NMDA currents [83]; and short-term facilitation [137]. These mechanisms offer the advantage that a network state can be maintained persistently active even in the absence of long-term modifications in the synaptic matrix due to a stimulus presentation. Hence, these mechanisms could explain how a novel item (that has never been experienced before) can be maintained in short term memory, in the absence of fast one-shot learning through synaptic plasticity mechanisms.

## 3.3. Networks involving several areas

The combined observation of persistent activity in several cortical areas and of long-range excitatory connections in both directions between these areas suggests a scenario in which persistent activity is stabilized by the presence of these long-range connections. For example, working memory of the identity of an object might be maintained by way of direct or indirect connections between IT and PFC, while working memory of the spatial location of an object might be maintained by connections between PPC and PFC (see fig 1).

One piece of experimental data however argues against this scenario. In an experiment in which distractors were presented in between the sample and match pictures, Miller et al [92] found that persistent activity does not survive the first distractor in IT, while it persists until the match presentation in PFC. This difference suggests that these two areas perform relatively independent functions (memory of the last shown stimulus in IT independently of behavioral relevance vs memory of behaviorally relevant stimuli in PFC). For these distinct functions to hold, the feedback loop allowing maintenance of activity must be localized at a lower level in both areas, perhaps at the local network level of section 3.2.

Finally, persistent activity could be a result of positive feedback loops between cortex and sub-cortical structures. One natural candidate is the cortico-basalo-thalamo-cortical (CBTC) loop, where positive feedback is due to a double inhibitory connection. This scenario is consistent with the observed pattern of persistent activity in basal ganglia [69, 70] and thalamus [60]. However, this pattern does not constitute a proof that the loop is required, since activity in basal ganglia and thalamus could simply be a reflection of persistent activity generated in cortex. Finally, a simpler feedback loop would be the thalamo-cortical loop.

## 4. Networks of binary neurons with discrete attractors

### 4.1. The Hopfield model

The Hopfield model (sometimes called Amari-Hopfield model) represents a formalization of Hebb ideas in a simple setting of a fully connected network of binary neurons [4, 36, 73, 85]. The simplicity of the model is such that the multiplicity of stable states in this model can be analyzed in great detail using tools of statistical mechanics. This model is also discussed at length in the chapter by Sompolinsky and White and in several textbooks [5, 68, 117].

The network is composed of $N$ neurons whose state as time $t$ is described by a binary variable $S_i(t)$ $(i = 1, \ldots, N) = 1$ (active neuron) or $-1$ (inactive neuron). The network is fully connected with a matrix of synaptic weights $\{J_{ij}\}$, where $J_{ij}$ is the synaptic strength from neuron $j$ to neuron $i$. $p$ patterns (or 'memories'), i.e. random network configurations $\xi_i^\mu = \pm 1$ with equal probability, $\mu = 1, \ldots, p$, $i = 1, \ldots, N$, are stored in the synaptic matrix using a 'Hebbian' prescription:

$$J_{ij} = \frac{1}{N} \sum_\mu \xi_i^\mu \xi_j^\mu \tag{4.1}$$

for $i \neq j$, $J_{ij} = 0$ if $i = j$. Updating of neuron states can be synchronous, or asynchronous. Neuron $i$ is updated according to its 'synaptic input' (or 'local field')

$$h_i = \sum_j J_{ij} S_j. \tag{4.2}$$

In the absence of noise, updating is done according to:

$$S_i(t + 1) = \text{sign}(h_i(t)) \tag{4.3}$$

In the presence of noise, updating is stochastic, with

$$\text{Prob}\,[S_i(t+1) = 1] = \frac{\exp(\beta h_i(t))}{\exp(\beta h_i(t)) + \exp(-\beta h_i(t))} \tag{4.4}$$

where $\beta = 1/T$ is an inverse temperature.

It is interesting in particular to investigate the network dynamics when the initial configuration of the network $(S_i(0), i = 1, \ldots, N)$ is correlated with one of the stored patterns, e.g. $\mu = 1$. To characterize the quality of retrieval, it is useful to introduce the overlaps of network state at time $t$ with patterns $\mu$,

$$m^\mu(t) = \frac{1}{N} \sum_i S_i \xi_i^\mu \tag{4.5}$$

Retrieval of a given pattern, means that the network reaches an attractor in which there is a macroscopic overlap with that pattern only, i.e.

$$m^1(t) = O(1) \tag{4.6}$$

$$m^\mu(t) = O\left(\frac{1}{\sqrt{N}}\right) \quad \mu \neq 1 \tag{4.7}$$

In particular, perfect retrieval means that the network state in which $S_i = \xi_i^1$ for all $i = 1, \ldots, N$ is stable.

What makes the Hopfield model amenable to detailed analysis is that, due to the symmetry of synaptic matrix and the absence of a self-coupling term, one can write down an energy (or Lyapunov) function

$$E = -\frac{1}{2} \sum_{i \neq j} J_{ij} S_i S_j \tag{4.8}$$

This function decreases monotonically as the network follows the asynchronous dynamics, Eqs. (4.2,4.3). When a given neuron, say $k$, is updated, its new state becomes $S_k(t+1) = \text{sign}(h_k(t))$. To see how the energy changes from $t$ to $t+1$, we rewrite the energy as

$$E(t) = -S_k(t)h_k(t) - \frac{1}{2} \sum_{i \neq j \neq k} J_{ij} S_i S_j \tag{4.9}$$

Hence,

$$E(t+1) - E(t) = -(S_k(t+1) - S_k(t))h_k(t)$$

If neuron $k$ does not flip, then there is no energy change. If it flips, then

$$E(t+1) - E(t) = -2S_k(t+1)h_k(t) < 0$$

Hence, $E$ decreases monotonically to a local minima in the space of network states.

### 4.1.1. Signal-to-noise ratio (SNR) analysis

One of the basic questions regarding this model is: are the stored patterns fixed points of the network dynamics? A first hint can be provided by a simple signal-to-noise analysis. We suppose the network is in one of its stored patterns, $S_i = \xi_i^1$. The local field of neuron $i$ is

$$h_i = \sum_{j \neq i} J_{ij} \xi_j^1 \tag{4.10}$$

$$= \frac{1}{N} \sum_{j \neq i} \sum_{\mu=1}^{p} \xi_i^\mu \xi_j^\mu \xi_j^1 \tag{4.11}$$

$$= \xi_i^1 + \frac{1}{N} \sum_{j} \sum_{\mu=2}^{p} \xi_i^\mu \xi_j^\mu \xi_j^1 \tag{4.12}$$

In Eq. (4.12), the first term in the r.h.s., $\xi_i^1$, represents the 'signal', while the second term, $\xi_i^1 \delta_i$ where

$$\delta_i = \frac{1}{N} \xi_i^1 \sum_{j \neq i} \sum_{\mu=2}^{p} \xi_i^\mu \xi_j^\mu \xi_j^1 \tag{4.13}$$

represents the 'noise' due to the other patterns stored in the synaptic matrix. For the network state corresponding to pattern 1 to be stable, we need $\delta_i > -1$ for $i = 1, \ldots, N$. For large $N$, $\delta_i$ has a Gaussian distribution with mean 0 and SD $\sqrt{\alpha}$, where $\alpha = p/N$ is the number of stored patterns per neuron. Hence, the signal to noise ratio is equal to $1/\sqrt{\alpha}$. As the number of stored patterns $\alpha$ grows, the noise due to other stored patterns increases.

The probability that the noise is smaller than the signal in a given neuron is

$$\rho = \int_{-\sqrt{1/\alpha}}^{\infty} \frac{dx}{\sqrt{2\pi}} \exp\left(-\frac{x^2}{2}\right)$$

For small $\alpha$,

$$\rho = 1 - \exp\left(-\frac{1}{2\alpha}\right)\left(\sqrt{\alpha} + O\left(\alpha^{3/2}\right)\right)$$

Neglecting correlations between local fields, the probability that the noise is smaller than the signal in all neurons is

$$\rho^N \quad \sim \quad \left(1 - \sqrt{\alpha}\exp\left(-\frac{1}{2\alpha}\right)\right)^N \tag{4.14}$$

$$\sim \quad \exp\left(-N\sqrt{\alpha}\exp\left(-\frac{1}{2\alpha}\right)\right) \tag{4.15}$$

For

$$\alpha < \frac{1}{2\ln N} \quad \rightarrow \quad \rho^N \sim 1 \tag{4.16}$$

For

$$\alpha > \frac{1}{2\ln N} \quad \rightarrow \quad \rho^N \sim 0 \tag{4.17}$$

In terms of the signal-to-noise ratio (SNR), the condition for perfect storage is SNR $< 1/\sqrt{\ln N}$. Hence, $\alpha = 1/2\ln N$ represents the maximal capacity, defined as the maximal number of patterns that can be stored as exact fixed points of the dynamics [138]. Beyond this number of patterns, the patterns are no longer fixed points. However, as we will see later, network states which are highly correlated with the stored attractors are still fixed points of the dynamics until the SNR is of order 1, i.e. $\alpha$ becomes of order 1. Note that when the storage level $\alpha$ is small ($\alpha \ll 1$), the memorized patterns have huge basins of attraction: any network configuration which has a macroscopic overlap with one of the stored pattern eventually flows to a configuration close to the pattern.

### 4.1.2. Statistical mechanics of the Hopfield model
The full statistical mechanics analysis of the Hopfield model was performed by Amit, Gutfreund and Sompolinsky [11, 12]. The following is a sketch of the main steps of the calculation, together with the final results. For details, see refs. [5, 12, 68, 117].

A system with energy

$$E = -\frac{1}{2}\sum_{i\neq j} J_{ij} S_i S_j \tag{4.18}$$

can be described by the partition function at inverse temperature $\beta$

$$Z = \mathrm{Tr}_{S_i}\exp(-\beta E).$$

The probability of the system to be in a state $\{S_i\}$ at temperature $1/\beta$ is

$$\frac{\exp(-\beta E)}{Z},$$

and the free energy of the system is

$$F = -\frac{1}{\beta} \ln Z$$

To get the 'typical' free energy of the system, we need to compute the average of the free energy over the disorder, here the realizations of the random patterns $\{\xi_i^\mu\}$ [91]. Hence, what we need to compute is the free energy per neuron, averaged over the patterns

$$f = \lim_{N \to \infty} -\frac{1}{\beta N} \langle\langle \ln Z \rangle\rangle$$

To perform the average of $\ln Z$ we need to use the 'replica method' [91]: calculate the average of $Z^n$ for integers $n = 1, 2, ...,$ and then perform an analytic continuation to $n = 0$.

$$f = \lim_{n \to 0} \lim_{N \to \infty} -\frac{1}{\beta n N} \left( \langle\langle Z^n \rangle\rangle - 1 \right)$$

$Z^n$ is the partition function of $n$ replicas of the system with the same quenched disorder (i.e. the same set of random patterns $\{\xi_i^\mu\}$). The replica method has been applied successfully in a wide variety of situations in disordered systems, but it still lacks a rigorous mathematical foundation [91].

What we need to compute is

$$\langle\langle Z^n \rangle\rangle = \langle\langle \mathrm{Tr}_{S_i^a} \exp \left( \frac{\beta}{2N} \sum_{i \neq j, \mu, a} \xi_i^\mu \xi_j^\mu S_i^a S_j^a \right) \rangle\rangle$$

where $a = 1, \ldots, n$ is a replica index. As usual in statistical mechanics, it is convenient to introduce *order parameters* that characterize the macroscopic order of the system:

- Overlaps with a finite number $s$ of patterns in replica $a$ $m_a^\nu$,

$$m_a^\nu = \frac{1}{N} \sum_i \xi_i^\mu S_i^a$$

Such patterns have been termed 'condensed' patterns in ref [11]. These parameters are the analog of the magnetization of the Ising model.

- Overlap between network states in two different replicas $q_{ab}$

$$q_{ab} = \frac{1}{N} \sum_i S_i^a S_j^b$$

This parameter is similar to the Edward-Anderson order parameter in spin glasses [91].

- Parameters to describe noise due to 'uncondensed' patterns $r_{ab}$

$$r_{ab} = \frac{1}{\alpha} \sum_{\mu=s+1}^{p} m_a^\mu m_b^\mu$$

To perform explicitly the calculation, the replica symmetry ansatz is introduced. It consists in setting $m_a^\nu = m^\nu$ for all $a$, and $q_{ab} = q$, $r_{ab} = r$, for all $a$, $b$. A discussion of the validity of this ansatz is beyond the scope of this chapter. It has been shown, using both numerical simulations and computations using a one-step symmetry breaking ansatz [44] that for this system the replica symmetric ansatz gives a very good approximation of the behavior of the system. This replica-symmetric ansatz leads to a free energy per neuron

$$
\begin{aligned}
f \quad = \quad &\min_{m^\nu, q, r} \left( \frac{\alpha}{2} + \frac{1}{2} \sum_\nu (m^\nu)^2 \right. \\
&+ \frac{\alpha}{2\beta} \left( \ln(1 - \beta + \beta q) - \frac{\beta q}{1 - \beta + \beta q} \right) \\
&+ \frac{\alpha \beta r (1 - q)}{2} \\
&\left. - \frac{1}{\beta} \int \frac{dz}{\sqrt{2\pi}} \exp\left( -\frac{z^2}{2} \right) \langle\langle \ln 2\cosh\beta \left( \sqrt{\alpha r} z + \sum_\nu m^\nu \xi^\nu \right) \rangle\rangle \right)
\end{aligned}
$$

(4.19)

where the average $\langle\langle . \rangle\rangle$ in the above equation is an average over the 'condensed' patterns $\xi^\nu$, $\nu = 1, \ldots, s$. To get the minima of the free energy, we should solve the saddle-point equations

$$m^\nu = \langle\langle \xi^\nu \tanh\beta \left( \sqrt{\alpha r} z + \sum_{\nu'} m^{\nu'} \xi^{\nu'} \right) \rangle\rangle \tag{4.20}$$

$$q = \langle\langle \tanh^2\beta \left( \sqrt{\alpha r} z + \sum_{\nu'} m^{\nu'} \xi^{\nu'} \right) \rangle\rangle \tag{4.21}$$

$$r = \frac{q}{(1 - \beta + \beta q)^2} \tag{4.22}$$

In the limit $T \to 0$, $q \to 1$ but $C = \beta(1 - q)$ is finite. The saddle-point equations become:

$$m^{\nu} = \left\langle\left\langle \xi^{\nu} \mathrm{erf} \left( \frac{1}{\sqrt{2\alpha r}} \sum_{\nu'} m^{\nu'} \xi^{\nu'} \right) \right\rangle\right\rangle \qquad (4.23)$$

$$C = \sqrt{\frac{2}{\pi \alpha r}} \left\langle\left\langle \exp \left( -\frac{1}{2\alpha r} \left( \sum_{\nu'} m^{\nu'} \xi^{\nu'} \right)^2 \right) \right\rangle\right\rangle \qquad (4.24)$$

$$r = \frac{1}{(1 - C)^2} \qquad (4.25)$$

These equations can be solved numerically. Three qualitatively different types of solutions can be found:

• The **spin-glass state**, in which all overlaps are zero. This solution exists for all $\alpha$.

• The **retrieval states**, in which there is a single non-vanishing overlap $m$. This solution exists up to $\alpha < \alpha_c = 0.138$, where $\alpha_c$ is the critical capacity of the system.

• The **mixture states**, in which there are finite overlaps with an odd number $s \geq 3$ of patterns. For example, the 3-mixtures are stable up to $\alpha = 0.03$.

The full phase diagram is shown in Fig. 9.

### 4.1.3. Critiques of the Hopfield model

The solution of the Hopfield model represented a triumph of statistical physics. However, the model is so far from the neurobiological reality in almost any of its components that it soon attracted a deluge of criticisms. Here is a summary of some of the complaints that have been made:

• Architecture: There is no separation between excitatory and inhibitory neurons, unlike in cortex; The network is fully connected, while cortical networks are diluted.

• Memories use standard coding: half neurons are active in a given memory. Electrophysiological data rather suggests sparse coding, with a small fraction of neurons being active in a given memory.

• Neurons: A very unrealistic feature is the symmetry between active and inactive states (-1,1). Furthermore, real neurons are not binary, but are best described by analog (firing rate) variables, or even better, by spiking units.

• Synapses are assumed to have a very large number of stable states (of order $N$). In real synapses, it is hard to imagine how such a large number of states (scaling with network size) could be achieved in a robust way. All synapses in the model can switch between excitatory and inhibitory, and all are plastic. In

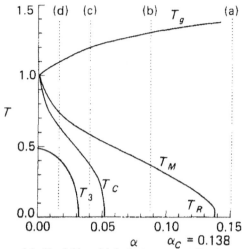

Fig. 9. Phase diagram of the Hopfield model, in the temperature $T$ - storage level $\alpha$ plane. Above the line marked $T_g$ only the paramagnetic state ($q = m = 0$) exists. Below this line, a second-order transition occurs, and the spin-glass state with $q > 0$ becomes stable. Below the line marked $T_M$, the retrieval states appear (through a first-order transition). Below $T_c$, these states become the absolute minima of the free energy. Finally, below $T_3$, the 3-mixture states become stable. At low $T$/low $\alpha$ mixture states combining large number of patterns are also stable. From ref. [11].

cortex, synapses can be either excitatory or inhibitory, and synaptic plasticity mechanisms have been well characterized only at excitatory-excitatory synapses. On the functional point of view, the model exhibits blackout catastrophe: if patterns are learned sequentially, the network operates well when the number of patterns does not exceed its capacity. However, beyond that number, a catastrophe occurs: none of the patterns can be retrieved anymore. The system becomes useless as a memory device.

Stimulated by these criticisms, theorists worked hard in the decade following the proposal of the Hopfield model to show that the essential features of the model are robust to many modifications. In the next sections, I review some of the main findings that have emerged from these studies.

## 4.2. Robustness to perturbations of the synaptic matrix

Sompolinsky showed that the Hopfield model is robust to many perturbations of the synaptic matrix. In particular, it is very robust to quenched symmetric noise in synaptic efficacies [126], $J_{ij} = J_{ij}^{Hebb} + \eta_{ij}$, where $\eta_{ij}$ is drawn randomly and independently with mean 0 and standard deviation $\Delta$ for $i < j$, as shown in

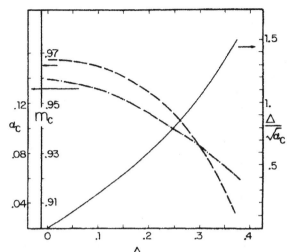

Fig. 10. Critical capacity $\alpha_c$ as a function of standard deviation $\Delta$ of quenched symmetric noise in synaptic matrix (dot-dashed line). Also shown are $m_c$, the overlap with retrieved pattern at $\alpha_c$ (dashed line), and $\Delta/\sqrt{\alpha_c}$ (solid line). From ref. [126].

figure 10.

The model is also remarkably robust to clipping [126], $J_{ij} = \text{sign}\left(J_{ij}^{Hebb}\right)$: the critical capacity $\alpha_c$ decreases mildly, from 0.14 to 0.10. Note that this form of clipping does not solve the problem of analog vs discrete synapses mentioned in the previous section. In fact, every synapse still needs an analog variable to specify its value. Networks with truly binary synapses have very different properties, as will be shown below.

The next question is robustness with respect to dilution of the synaptic matrix. Neighboring neurons in cortex are connected monosynaptically with about 10% probability [72, 88, 89]. There are two ways of diluting the synaptic matrix: symmetrically or asymmetrically. If the diluted synaptic matrix is symmetric, there is still an energy function, and the standard statistical mechanics techniques hold. This allowed Sompolinsky [126] to compute the capacity of a network with a synaptic matrix of the form

$$J_{ij} = \frac{c_{ij}}{c} J_{ij}^{Hebb}$$

where $c_{ij} = 1$ with probability $c$, $c_{ij} = 0$ otherwise. The main conclusion of this study is that storage capacity *per available synapse* increases as the degree of dilution increases. Capacity as a function of dilution is shown in figure 11. In

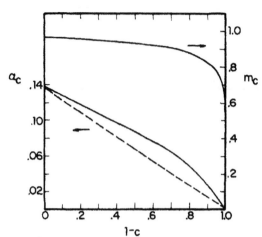

Fig. 11. Capacity as a function of random symmetric dilution of the synaptic matrix. The dashed line represents to the line $\alpha_c(c) = c.\alpha_c(0)$. Hence, the storage capacity *per available synapse* increases as the degree of dilution increases. The quality of retrieval at critical capacity $m_c$ decreases as $c$ decreases. From ref. [126].

cortex, the synaptic matrix is however far from symmetric. Asymmetry hinders the application of statistical mechanics techniques, since an energy function no longer exists. However, Derrida and collaborators [50] were able to solve exactly the dynamics of an extremely diluted network, with $c \ll \log N/N$. They have shown that in this limit the capacity per synapse is $\alpha_c = 2/\pi \sim 0.64$, a more than 4-fold increase in capacity compared to the Hopfield model. Hence, random symmetric or asymmetric dilution generally enhance performance in terms of capacity per available synapse.

Taken together, these results demonstrate a remarkable robustness of the Hopfield model to perturbations of its synaptic matrix.

### 4.3. Models with 0,1 neurons and low coding levels

Another important issue is what happens to the performance of the model when the spurious $-1/1$ symmetry in neurons and patterns is removed. This issue was considered independently by Tsodyks and Feigel'man, and by Buhmann and collaborators [35, 131].

The model they studied contains discrete 0,1 neurons $S_i$ ($i = 1, \ldots, N$). The synaptic matrix stores $p$ patterns (random network configurations) $\xi_i^\mu = 1$ with probability $f$, $\xi_i^\mu = 0$ with probability $1 - f$ ($\mu = 1, \ldots, p$, $i = 1, \ldots, N$, where $f$ is the coding level, using a Hebbian rule:

$$J_{ij} = \frac{1}{f(1-f)N} \sum_{\mu} (\xi_i^{\mu} - f)(\xi_j^{\mu} - f) - \frac{\gamma}{fN} \tag{4.26}$$

for $i \neq j$, where $\gamma$ represents a uniform inhibitory component, and $J_{ij} = 0$ if $i = j$. In the absence of noise, updating is done according to:

$$S_i(t+1) = \theta(h_i(t) - U) \tag{4.27}$$

where $U$ is a neuronal threshold.

A simple signal-to-noise ratio (SNR) analysis along the lines of section 4.1.1 shows that the average synaptic input in foreground neurons (neurons whose activity is 1 in the retrieved pattern) is equal to $1 - f - \gamma$, while the average synaptic input in background neurons (neurons whose activity is 0 in the retrieved pattern) is $-f - \gamma$. The standard deviation of the synaptic input in both types of neurons is $\sqrt{f\alpha}$. In the Hopfield model, the capacity is reached when the SNR becomes of order one. Assuming that this is also true in the model with 0,1 neurons predicts that the capacity increases as the coding level $f$ decreases as $\alpha \sim 1/f$. However, a more careful analysis gives $\alpha \sim 1/(2f|\ln(f)|)$, provided the threshold is tuned appropriately [131]. Hence, the capacity increases with $f$ and diverges when $f$ goes to zero. This is another example where a modification of the original Hopfield model towards more neurobiological realism improves its performance. The information content per synapse

$$I = \alpha \left[ -f \log_2 f - (1-f) \log_2(1-f) \right]$$

increases as $f$ decreases and reaches $1/(2\ln 2) = 0.72$ bits per synapse when $f \to 0$ [131]. Remarkably, it turns out that this is the optimal capacity in the space of all possible synaptic matrices in the limit $f \to 0$ [61], as mentioned in section 4.6.

This model has additional interesting properties. In addition to the 'memory states' (states with macroscopic overlap with one of the stored patterns), the silent state in which no neurons are active is stable. This state is useful from a computational point of view to indicate that a shown stimulus is not familiar. Another interesting property is that strong inhibition in this model destabilizes mixture states, since strong inhibition tends to favor states in which the activity is low.

### 4.4. Summary—models with binary neurons and analog synapses

The major conclusion from the studies considered until now is that a network with binary neurons with a large number $C$ of modifiable analog synapses per

neuron, that stores in its synaptic matrix random patterns with low coding level
$f$ has a capacity that scales as

$$p_c \sim \frac{C}{f|\ln f|}$$

If we consider a local network in cortex with $C \sim 10,000$, and a coding level of
order 1%, this estimate leads to a storage capacity of about $10^6$ patterns, or about
1 bit per synapse.

### 4.5. A model with binary synapses: the Willshaw model

All the models considered until now have analog synapses, i.e. the number of
states of at least one variable describing the synapse is assumed to grow unbounded as network size $N$ (or number of patterns $p$) grows to infinity. What
happens when synapses are assumed to have only a finite number of states? In
particular, what happens in the extreme case of binary synapses? A model with
0,1 neurons, sparsely coded patterns stored with a binary (0,1) synaptic matrix,
was introduced by Willshaw, Bunemann and Longuet-Higgins in 1969, and studied further by many authors [62, 90, 98, 99, 104, 142]. The Willshaw model is
characterized by the following synaptic matrix

$$J_{ij} = \frac{1}{fN} \begin{cases} 1 & \text{if } \xi_i^\mu \xi_j^\mu = 1 \text{ in at least one pattern } \mu. \\ 0 & \text{otherwise.} \end{cases} \tag{4.28}$$

where the normalization factor $1/fN$ is chosen for the sake of simplicity of the
analysis (see below).

A relatively simple analysis is possible when patterns activate exactly $fN$
neurons:

$$\sum_i \xi_i^\mu = fN$$

The SNR analysis in this case proceeds as follows. Suppose the network state is
in one of the learned patterns, say $\mu = 1$. The local field in foreground neurons is
$h_i = 1$ since all synapses connecting foreground neurons have efficacy $1/(fN)$.
The average local field in background neurons is equal to the probability that
a synapse $J_{ij}$ that connects a foreground and a background neuron is positive,
which is equal to

$$\langle h_i \rangle_0 = C = 1 - (1 - f^2)^{p-1}$$

while the standard deviation around this number goes to zero in the limit $fN \to \infty$.

For the network to function properly as an associative memory, the neuronal
threshold $U$ should be in between the field in the foreground neurons and the

field in the background neurons. The requirement that $1 > U > C$ leads to a first estimate of critical capacity $p \sim 1/|\ln(1 - f^2)|$. For small $f$, $C \sim 1 - \exp(-pf^2)$, and the capacity can be expected to scale as $p \sim 1/f^2$.

If the threshold can be fine tuned to be just below $h_i = 1$, then we can compute the capacity by computing the probability $\rho$ that all the background neurons have their field strictly smaller than 1. Neglecting the effect of correlations between the local fields, this leads to

$$\rho = \left(1 - C^{fN}\right)^{(1-f)N} \tag{4.29}$$

$$= \exp\left(-(1 - f)N \exp\left(fN \ln C\right)\right) \tag{4.30}$$

The probability that a pattern is stable is equal to one if

$$C < \exp\left(-\frac{\ln N}{fN}\right)$$

or

$$p < \frac{1}{f^2} \left| \ln\left(1 - \exp\left(-\frac{\ln N}{fN}\right)\right)\right| \tag{4.31}$$

Inspection of equation (4.31) shows that there are three different regimes:

1. For $f \gg \frac{\ln N}{N}$, we obtain $p \sim \ln(fN)/f^2$. The factor $\ln(fN)$ is what one gains by fine tuning the threshold, compared to a threshold which would be at a finite distance from 1. The information capacity per synapse in this regime is $I \sim \ln(fN)|\ln(f)|/(fN)$ and hence goes to zero in the large $N$ limit. For example, $f \sim 1/\sqrt{N}$ leads to $p \sim N\ln(N)$, $I \sim (\ln N)^2/\sqrt{N}$.

2. For $f \sim \frac{\ln N}{N}$ we obtain $p \sim 1/f^2$, and $I \sim 1$. This is the optimal scaling in terms of information capacity. Remarkably, the capacity in terms of bits per synapse becomes comparable to networks with analog synapses.

3. Finally, for $f \ll \frac{\ln N}{N}$, one gets $p \sim (1/f^2)\exp\left(-\frac{\ln N}{fN}\right)$ and $I \sim |\ln f|\exp\left(-\frac{\ln N}{fN}\right)/(fN)$. The information capacity again goes to zero in the large $N$ limit.

Hence, the capacity of this model is in general drastically reduced compared to a network with analog synapses. Only in the special case $f \sim \ln N/N$ is the network operating with a capacity which is of the same order of magnitude as the network with analog synapses. Sparse coding has again a beneficial effect on capacity, both in terms of number of patterns and information per synapse. However, there is a lower limit $\ln N/N$ on coding level below which performance drops significantly.

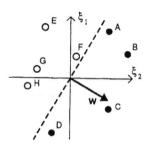

Fig. 12. The perceptron problem: finding a synaptic vector that correctly classifies the $p$ input patterns. Patterns are represented as circles (marked A-H) in a 2-dimensional input space (here continuous patterns are represented). To each point is associated either a $+1$ output (black circles) or a $-1$ output (empty circles). Finding a weight vector that correctly classifies all the patterns is equivalent to finding a hyperplane separating white from black circles. From ref. [68].

### 4.6. Optimal capacity of networks of binary neurons

The networks we have considered until now have been characterized by particular forms of the synaptic matrix (typically a hebbian form). The question was, given a particular form of the synaptic matrix, what is the maximal number of patterns that can be stored as attractors of the network? We now turn to a different question: what is the maximal number $p$ of patterns, for which a synaptic matrix $J_{ij}$ can be found such as the patterns are fixed points of the dynamics?

This problem can be decomposed in $N$ independent single neuron problems. For each neuron $i$, we consider the synaptic vector $J_{ij}$ composed of all the weights from its presynaptic neighbors $j = 1, \ldots, N$. An individual neuron together with its plastic synaptic weights is a *perceptron* with $N$ inputs and 1 output [93, 119]. Each of the $p$ patterns is an association between a random point in the $N$ dimensional input space, and a random $\pm 1$ output. In the $N$ dimensional space, the problem faced by the perceptron is therefore to find a hyperplane (defined by the synaptic vector $J$) separating the points with 1 output from the points with -1 output, as illustrated in figure 12.

Such a hyperplane should satisfy simultaneously the $p$ conditions

$$\text{sign} \sum_j J_{ij} \xi_j^\mu = \xi_i^\mu \tag{4.32}$$

for $\mu = 1, \ldots, p$, in order for all the patterns to be stable as far as neuron $i$ is concerned. If we repeat this operation for all neurons, storage of the $p$ patterns will be secured.

The capacity problem for a perceptron with patterns in 'general position' was solved by Cover in 1965 [42]. The answer is that a synaptic vector can be found with probability 1 in the large $N$ limit when $p < 2N$. Hence, $\alpha = p/N = 2$ is the optimal capacity of patterns with standard coding. The proof of this result can be found e.g. in ref. [68].

The approach of Cover does not easily generalize to sparsely coded patterns (which are no longer in general position) and to the introduction of a requirement of basins of attraction with finite size. Using a different approach, Gardner [61] was able to compute the optimal capacity for networks with arbitrary coding level, and arbitrary size of basins of attraction.

The Gardner calculation consists in computing the fraction of the volume in the space of all weights of those weights that satisfy the $p$ constraints, Eq. (4.32)

$$ V = \frac{\mathrm{Tr}_J \delta(\sum_j J_{ij}^2 - N) \prod_\mu \Theta\left(\frac{\xi_i^\mu}{\sqrt{N}} \sum_j J_{ij} \xi_j^\mu\right)}{\mathrm{Tr}_J \delta(\sum_j J_{ij}^2 - N)} $$

where the normalization term $\delta(\sum_j J_{ij}^2 - N)$ is introduced since the problem is invariant through rescaling of all synapses by the same factor. To get the typical value of the volume, one needs again to perform an average over the random patterns. The typical value is obtained through the computation of $\langle\langle \ln V \rangle\rangle$ using the replica technique,

$$ \langle\langle V^n \rangle\rangle = \frac{\mathrm{Tr}_{J^a} \prod_a \delta(\sum_j (J_{ij}^a)^2 - N) \prod_{\mu,a} \Theta\left(\frac{\xi_i^\mu}{\sqrt{N}} \sum_j J_{ij}^a \xi_j^\mu\right)}{\mathrm{Tr}_{J^a} \prod_a \delta(\sum_j (J_{ij}^a)^2 - N)} $$

The calculation proceeds along the following lines:
- Introduce integral representations for the Heaviside and delta functions;
- Average over random patterns;
- Introduce order parameters, and in particular the 'Edwards-Anderson' order parameter

$$ q^{ab} = \frac{1}{N} \sum_j J_{ij}^a J_{ij}^b $$

which is now an overlap between two synaptic matrices that are solution to the storage problem. It measures the extent of available volume in weight space. Optimal capacity is reached when $q$ goes to 1, since at this value of $q$ the volume shrinks to a single point in the space of weights.

The results are summarized in figure 13 and compared with specific synaptic matrices in table 2. In particular, the optimal capacity increases as sparseness increases, consistent with the results for Hebbian synaptic matrices, while the information content per synapse decreases from 2 bits to 0.72 bits from standard

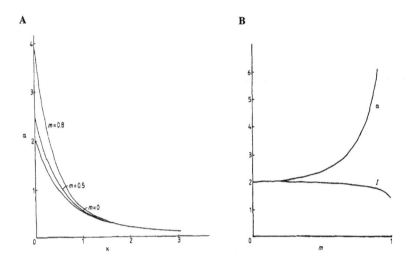

Fig. 13. **A**: Optimal capacity $\alpha$ as a function of basin of attraction size $\kappa$, for several values of the 'magnetization' $m$. $m$ is related to coding level $f$ by $m = 1 - 2f$. Hence, $m = 0$ corresponds to standard coding, while $m \to 1$ corresponds to the sparse coding limit $f \to 0$. **B**: Capacity as a function of pattern 'magnetization' $m$. Also shown is the information content per synapse $I$. From ref. [61].

Table 2

Bounds on capacity compared with capacity of specific models. For standard coding, models with a Hebbian matrix are far from being optimal, while they become optimal in the sparse coding limit. The Willshaw model with binary synapses performs poorly in standard coding, while its performance becomes of the same order of magnitude as the models with analog synapses in sparse coding. Note that in the case of the Willshaw model, and more generally for models with excitatory synapses, the capacity depends on whether patterns activate exactly or in average $fN$ neurons [25, 98].

| Patterns per neuron | Optimal synaptic matrix | Hebbian analog matrix | Willshaw (binary) matrix |
|---|---|---|---|
| $f = 0.5$ | 2 | 0.14 | 0 |
| $f \to 0$ | $\dfrac{1}{2f\lvert\ln(f)\rvert}$ | $\dfrac{1}{2f\lvert\ln(f)\rvert}$ | $\sim \dfrac{1}{f\lvert\ln(f)\rvert}$ |
| Information per synapse | | | |
| $f = 0.5$ | 2 | 0.14 | 0 |
| $f \to 0$ | $\dfrac{1}{2\ln 2} = 0.72$ | $\dfrac{1}{2\ln 2} = 0.72$ | 0.24 |

coding ($f = 0.5$) to sparse coding ($f \to 0$). Interestingly, Hebbian matrices become optimal in the sparse coding limit. The advantage of the Gardner approach is that it can easily be generalized to constraints on the couplings (bounded, discrete, etc). More details on this approach can be found in ref. [53].

### 4.7. Other generalizations of the Hopfield model

To conclude this chapter on associative memory models with binary neurons, it is worth mentioning that several generalizations of the Hopfield model have been introduced that have dealt with storage of more complex structures of data. In particular, simple generalizations are able to store and retrieve sequences of patterns; other models store hierarchical structures of data. For lack of space, we refer the interested reader to ref. [5].

## 5. Learning

We have considered until now the properties of networks with fixed synapses. We now turn to the dynamical process by which the synaptic matrix evolves – *learning*. What defines learning is an algorithm, a differential equation or a stochastic process according to which the synaptic efficacy evolves, often as a function of the activities of pre and post synaptic neurons, sometimes as a function of an additional supervisory signal. The presence or absence of this additional supervisory signal separates learning processes in two classes: supervised or unsupervised learning. In supervised learning (section 5.1) processes or algorithms, the supervisor tells the synapse whether a pattern has already been learned or not; synaptic modification occurs only in case the pattern has not been learned yet. In unsupervised learning processes (section 5.2), there is no such supervisor. For unsupervised Hebbian rules, the synaptic efficacy typically evolves according to the activities of the pre and post synaptic neuron and to the current synaptic efficacy only. Here, we will focus on processes that are stable on the long term, i.e. those that do not lead to the black-out catastrophe. There are two ways to avoid the black-out catastrophe. The first possibility is for a system to stop learning before the capacity is reached. The second possibility is to learn continuously new patterns at the expense of forgetting the older ones. Models with such property are sometimes called palimpsest models.

At the end of this section, I will reconsider the issue of synapses with a finite number of stable states (section 5.4). The motivations to consider discrete rather than analog synapses are severalfold: (i) An analog synapse means in practice achieving at the level of a single synapse a dynamical system with a continuous attractor. This means fine tuning, and poor resistance to noise. Systems with discrete attractors typically do not require fine tuning, and are known to be much more robust, especially at long time scales. (ii) There are known biochemical models for bistability at the single synaptic site. One candidate for a switch at the synaptic level is provided by CamKII autophosphorylation [82, 84, 145]. There exist currently no biochemical models for an analog synapse. (iii) There are experimental indications suggesting that individual synapses might be binary [107]. Learning dynamics with such synapses is best described by a stochastic process whose transition probabilities depend on pre and postsynaptic activities.

## 5.1. Supervised learning: the perceptron learning algorithm

Learning a synaptic matrix that stores patterns as fixed points of its dynamics in a recurrent network is equivalent to learning a set of $N$ perceptrons, one for each neuron, as already mentioned in section 4.6. Each perceptron is a feed-forward network with $N$ input neurons and one output neuron.

The problem we consider here is the perceptron learning problem: we wish to find a synaptic vector $J$ that associates to $p$ input patterns its correct outputs. The perceptron learning algorithm guarantees that such a synaptic vector will be found, provided it exists, starting from any initial synaptic weight [93, 119]. The algorithm is applied independently to all neurons $i = 1, \ldots, N$. Consider neuron $i$ as a perceptron with synaptic weight $(J_{ij})$, $j = 1, \ldots, N$. The algorithm goes as follows:

1. Pick up a pattern $\mu$.

2. Compute the 'synaptic current' $h_i = \sum_j J_{ij} \xi_j^\mu$.

3. If $\xi_i^\mu h_i > 0$ go to 1. Else, update $J_{ij}$ according to

$$J_{ij} = J_{ij} + \xi_i^\mu \xi_j^\mu \tag{5.1}$$

   Go to 1.

This simple algorithm is guaranteed to converge with a finite number of iterations, provided a solution exists.

Note that the resulting synaptic matrix is of the form

$$J_{ij} = \sum_\mu a_i^\mu \xi_i^\mu \xi_j^\mu \tag{5.2}$$

where the $a_i^\mu$ depends on the presynaptic neuron $i$ and on $\mu$. Hence, allowing different weights between patterns and neurons can lead to a gain of a factor 14 in capacity (from 0.14 to 2).

The proof of the algorithm goes as follows. Suppose that a solution $\vec{J}^{\star}$ exists such that $\vec{J}^{\star}.\vec{X}^{\mu} > \delta$ for all $\mu$, where

$$X_j^{\mu} = \frac{\xi_i^{\mu} \xi_j^{\mu}}{\sqrt{N}}$$

Both $\vec{J}^{\star}$ and $\vec{X}^{\mu}$ are normalized to 1 for convenience.

Define a function of the couplings $\vec{J}$ as

$$F(\vec{J}) = \frac{\vec{J}^{\star}.\vec{J}}{|\vec{J}|}. \tag{5.3}$$

$F(\vec{J})$ represents the cosine of the angle between the two vectors $\vec{J}^{\star}$ and $\vec{J}$. Hence, $F(\vec{J}) < 1$.

Let us consider first how the numerator of Eq. (5.3) changes with a single update:

$$\vec{J}^{\star}.\vec{J}^{t} = \vec{J}^{\star}.(\vec{J}^{t-1} + \vec{X}^{\mu}) \tag{5.4}$$

$$\geq \vec{J}^{\star}.\vec{J}^{t-1} + \delta \tag{5.5}$$

$$\geq \vec{J}^{\star}.\vec{J}^{0} + t\delta \tag{5.6}$$

The denominator of Eq. (5.3) changes as:

$$\left|\vec{J}^{t}\right|^2 = \left|\vec{J}^{t-1} + \vec{X}^{\mu}\right|^2 \tag{5.7}$$

$$\leq \left|\vec{J}^{t-1}\right|^2 + 1 \tag{5.8}$$

$$\leq \left|\vec{J}^{0}\right|^2 + t \tag{5.9}$$

As the number of updates becomes large, the numerator grows faster than the denominator ($t$ vs $\sqrt{t}$), and, for large $t$

$$F(\vec{J}) > \sqrt{t}\delta$$

Since $F(\vec{J})$ is smaller than 1 by construction, the algorithm must reach a solution (in which no more updates are done) in a finite number of updates,

$$t < \frac{1}{\delta^2}$$

The advantage of such supervised learning rules is the ability to reach optimal capacity. The drawback is that one needs a supervisor. In neocortex, it is not clear what could be the neurobiological equivalent of a supervisor. A candidate for a supervisor has been described in at least neural structure: the climbing fibers in the cerebellum [74].

## 5.2. Unsupervised learning—palimpsest models

We now turn to unsupervised learning, i.e. learning processes without error signals. The scenario we consider here is the following: the network is exposed to an infinite stream of random patterns $\xi_i^t$ in discrete time $t \in ]-\infty, \infty[$. Hence, time corresponds in fact to number of presentations of discrete patterns. At each time step, each synapse changes according to its current state and pre and post synaptic activities. Here, we consider the case of analog synapses, and deterministic changes of the synaptic efficacy. In the next section, we consider binary (or discrete) synapses and stochastic changes. The learning process can then be described as a Markov process. The questions we ask are the following:

• Does a pattern shown at time $t$ become an attractor of the system? If not, how many presentations of the same pattern do we need until it is learned?

• How long a given learned pattern stays an attractor of the system, if it is no longer presented? In other terms, how many presentations of other patterns will erase the memory of this pattern?

## 5.3. Palimpsest models with analog synapses

As already mentioned, the Hopfield model needs modifications in order to perform adequately as an associative memory in the face of a continuous stream of incoming stimuli. In the eighties, several authors proposed simple modifications of the learning process that allows the network to become a palimpsest.

Parisi [105] introduced an upper and lower bound to each synaptic efficacy,

$$J_{ij}(t+1) = F\left( J_{ij}(t) + \frac{1}{N}\xi_i^{t+1}\xi_j^{t+1} \right)$$

where $F(x) = x$ in $[-A, A]$, $F(x) = A\ (-A)$ for $x > A\ (< -A)$ This simple modification allows the network to perform as a palimpsest: the last shown patterns are attractors of the system, while older patterns gradually fade away. This is because each time a synaptic efficacy hits one of the bounds, memory of the previous individual modifications is partially erased.

Mézard, Nadal and Toulouse [87] proposed a learning rule of the form

$$J_{ij}(t+1) = J_{ij}(t)\exp\left( -\frac{\epsilon^2}{2N} \right) + \frac{\epsilon}{N}\xi_i^{t+1}\xi_j^{t+1}$$

This learning rule leads to a matrix of the type

$$J_{ij}(t) = \sum_{\mu=-\infty}^{t} \frac{\epsilon}{N}\exp\left( -\frac{\epsilon^2(t-\mu)}{2N} \right)\xi_i^\mu\xi_j^\mu$$

The exponential decay produces a network that recalls the most recent patterns only, while it forgets the old ones.

Such a network can be studied using the usual statistical mechanics techniques. Optimal performance is obtained for $\epsilon \sim 4$, for which $\alpha_c \sim 0.05$. Hence, there is a reduction of critical capacity by factor 3 compared to the Hopfield model. This is a price to pay for avoiding the black-out catastrophe.

## 5.4. Unsupervised learning in networks with discrete synapses

We now turn to networks with discrete synapses (i.e. characterized by a discrete and finite number of states), focusing mainly on the simplest scenario of binary synapses [9, 15, 30, 130]. The scenario is again an infinite string of binary patterns $(\xi_i^t)$ presented to the network. Each synapse has two states, an 'up' (high efficacy) state $J_1$ and a 'down' (low efficacy) state $J_0$. At each time step $t$ (presentation number), the synaptic element $J_{ij}$ undergoes transitions between the up and down states, with probabilities conditioned by the state of the pre and post synaptic neurons

$$ M = \begin{pmatrix} 1 - D_{ij}(\xi_i^t, \xi_j^t) & P_{ij}(\xi_i^t, \xi_j^t) \\ D_{ij}(\xi_i^t, \xi_j^t) & 1 - P_{ij}(\xi_i^t, \xi_j^t) \end{pmatrix} $$

where $P_{ij}$ represents the transition probability from the down to the up state (similar to experimentally observed long term potentiation (LTP)). Typically, LTP occurs when there is coincidence of pre and postsynaptic activity, or

$$ P_{ij}(\xi_i^t, \xi_j^t) = q_+ \xi_i^t \xi_j^t, $$

where $0 < q_+ < 1$ is the 'LTP' transition probability. $D_{ij}$ represents the transition from the up to the down state (similar to experimentally observed long term depression (LTD)). Typically, LTD occurs when there is either pre-synaptic activity in the absence of post-synaptic activity (homosynaptic LTD) or post-synaptic activity in the absence of pre-synaptic activity (heterosynaptic LTD), i.e.

$$ D_{ij}(\xi_i^t, \xi_j^t) = q_- \left[ (1 - \xi_i^t)\xi_j^t + (1 - \xi_j^t)\xi_i^t \right] $$

where $0 < q_- < 1$ is the 'LTD' transition probability.

The Willshaw model is a particular case of this learning process. Starting from all synapses at zero at time $t = 0$, this model has LTP transition probability $q_+ = 1$, and LTD transition probability $q_- = 0$. This particular choice leads to the blackout catastrophe. The reason is simple: after some number of presentations, all synapses become potentiated! The network becomes useless as a memory device. In fact, to get efficient learning, LTD must balance LTP. When LTD is present, one generically gets a 'palimpsest' network: more recent patterns are retrievable, while older patterns are progressively erased from memory.

Tsodyks [130], and Amit and Fusi [15] showed that this type of learning leads to a drastic drop in capacity, compared with palimpsest models with analog synapses, when patterns have standard coding and/or transition probabilities are finite. To show this, consider again the case of $(+1, -1)$ neurons, $(+1, -1)$ synapses.

The transition matrix becomes

$$M = \begin{pmatrix} 1 - q\left[\frac{1-\xi_i'\xi_j'}{2}\right] & q\left[\frac{1+\xi_i'\xi_j'}{2}\right] \\ q\left[\frac{1-\xi_i'\xi_j'}{2}\right] & 1 - q\left[\frac{1+\xi_i'\xi_j'}{2}\right] \end{pmatrix}$$

The probability of a synapse to be in an up (down) state is then

$$P_\pm(t) = \sum_{u \le t} q\left[\frac{1 \pm \xi_i^u \xi_j^u}{2}\right] (1-q)^{t-u},$$

and the average synaptic value is

$$\langle J_{ij} \rangle = \sum_{u \le t} q \xi_i^u \xi_j^u (1-q)^{t-u}$$

A signal-to-noise (SNR) analysis for a pattern presented a number $t$ of presentations in the past leads to a mean signal

$$S = q(1-q)^t$$

and a noise

$$R = \frac{1 - q^2(1-q)^{2t}}{N}$$

For patterns to be fixed points of the dynamics we should have (see section 4.1.1)

$$\frac{S^2}{N^2} > 2 \ln N$$

This condition leads to

$$t < \frac{\ln N + 2 \ln q - \ln \ln N}{|\ln(1-q)|}$$

For a finite transition probability $q$, the network can only retrieve patterns if followed by no more than $\ln N$ presentations of other stimuli! Hence, the binary nature of synapses leads to a drastic decrease in capacity.

There are way outs, however. A solution to the problem is to make some parameter depend on $N$, as $N$ goes to infinity [9]

• Choose a small transition probability $q$: $q$ can be taken as small as $\sqrt{\ln N}/N$, for which the capacity of the system becomes

$$p_c \sim \sqrt{\frac{N}{\ln N}}$$

This is still not optimal.

• Choose a small coding level $f$, a finite LTP probability $q_+$, but a small LTD probability $q_- = O(f)$. This choice makes the number of LTD transitions of the same order as the number of LTP transitions. For $f \sim \ln N/N$, the capacity becomes

$$p_c \sim \frac{N^2}{(\ln N)^2}$$

This is the optimal scaling when $f \sim \ln N/N$, since it leads to a finite information content per synapse. Network performance becomes comparable to the Willshaw model, but with the advantage of the palimpsest property.

• Choose a large number of synaptic states $n$: when $n$ is of order $\sqrt{N/\ln N}$, we recover the behavior of the Hopfield model,

$$p_c \sim \frac{N}{\ln N}.$$

Synapses become effectively analog.

Interestingly, a network with such a stochastic learning process can be shown to have two modes of behavior [30]: For sufficiently high transition probabilities, it learns in one shot individual items, while for low transition probabilities, it learns slowly class prototypes, when individual patterns are structured in classes.

## 5.5. Summary

Palimpsest networks with analog synapses have a capacity which is of the same order as the Hopfield network. On the other hand, networks with discrete synapses (and hence more robust to noise) experience a severe drop in capacity, unless coding is sparse, and LTD balances LTP. In this situation, the capacity becomes on the same order as the Willshaw network.

## 6. Networks of spiking neurons with discrete attractors

Investigations of networks of binary neurons have led to a deep understanding of the properties of systems with a large number of attractors. However, these studies do not answer the questions raised by the experiments described in section

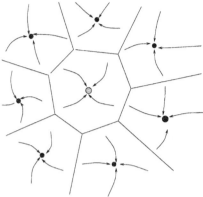

Fig. 14. Discrete attractor landscape, with one background state (red), and many 'memory' states (green).

2, and these models cannot be compared directly with the experimental data. To answer such questions, it is necessary to analyze networks of spiking neurons. Before turning to such analysis, it is useful to recall some basic experimental observations from extracellular recordings in the awake monkey:

• In the baseline period, recorded neurons fire at low rates, in an irregular fashion. The coefficient of variation (CV), that measures the irregularity of the spike trains (standard deviation of inter-spike interval divided by the mean) is close to one.

• In the delay period, some neurons fire at higher rates. However, these 'persistent' rates are not far from background activity (ratio typically about 3). Persistent rates are much lower than saturation rates, as they are measured in vitro. Firing is also highly irregular.

• Only a small fraction of recorded neurons have increased firing rates in the delay period.

This phenomenology is consistent with a model that has

• One 'background' network state, with all neurons firing at low rates;

• 'Memory' network states, with a small fraction of neurons active at higher rates.

Such an attractor landscape is represented in Figure 14. Note that the attractor landscape is similar to what is seen in the models with 0,1 neurons [35, 131], except that: neurons are not silent in the background state; and the firing rate of neurons is low even when they are in the foreground. Using networks of spiking neurons,

• one can relate background/persistent rate to the underlying biophysical single

neuron and synaptic parameters;
- one can investigate stability of asynchronous state;
- one can investigate statistics of spike trains (CV, cross-correlations between neurons, etc) in different types of states.

Here, we have chosen to jump directly from binary to spiking neurons. An intermediate step between binary and spiking neurons are networks of analog 'firing rate' neurons (see e.g. refs [14, 127, 129]). Such networks have properties which are qualitatively similar to the networks of binary neurons.

Many models of recurrent networks of spiking neurons appear in the literature. One way to classify these models is through how the number of connections per neuron $C$ and the connection strength $J$ scale. In most models, there is a large number of connections per neuron, and the synaptic strength scales as $1/C$. In these models, the synaptic input to each neuron becomes deterministic in the large $C$ limit, and the fluctuations vanish in that limit. Hence, such models are not appropriate to describe network states with irregular activity, unless external noise is added by hand. Two alternative strategies are possible. The first is to take $C$ and $J$ large but finite. This is the approach e.g. of ref. [7], where synaptic inputs are described by both their mean and their variance, which is finite in this scenario. Fluctuations of finite amplitude around the mean synaptic input allows the network to reach states with irregular activity, as we will see below. Another approach is to take again the large $C$ limit, but with synaptic strengths of order $1/\sqrt{C}$ [134, 135]. Using such a scaling, both mean and variance of synaptic inputs can be made finite in the large $C$ limit, provided excitation and inhibition are balanced (see van Vreeswijk, this volume). In this section, we will mostly consider the case in which $C$ and $J$ are finite ($C \sim 10,000$, $J \sim 0.01$ times the threshold), except in section 6.5, where the large $C$, $J \sim 1/\sqrt{J}$ limit is discussed.

### 6.1. A cortical network model

Our discussion on models of networks of spiking neurons will be based on the model proposed by Amit and Brunel [7]. This model represents a good compromise between two opposite requirements: It is realistic enough so that, on the one hand it incorporates some important anatomical and physiological features of real cortical networks, and on the other hand its output can be compared directly with experimental data; It is simple enough to be studied analytically.

The model is a large randomly connected network of $N_E$ excitatory pyramidal cells (80%) and $N_I$ inhibitory interneurons (20%). Each neuron receives a large number of connections ($\sim 10,000$), $C_E$ excitatory connections from other pyramidal cells of the circuit, $C_I$ inhibitory connections from interneurons, and $C_E$ excitatory connections from outside the circuit. The network is randomly connected, with a small connection probability (of order 10%). Connection strengths

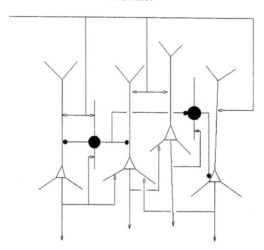

Fig. 15. The cortical circuit. A large number of pyramidal cells (open triangles) and interneurons (black circles) are connected randomly. Both pyramidal cells and interneurons receive excitatory connections from outside the network.

are such that about 100 simultaneous inputs are needed to reach threshold. These basic features make the network similar to cortical networks as characterized by slice studies [72, 88, 89]. The architecture of the network is represented schematically in Figure 15. Neurons are modelled as leaky integrate-and-fire (LIF) neurons [76, 81, 132]. Each neuron $i = 1, \ldots, N$ is described by its membrane potential $V_i$ that evolves according to

$$\tau \dot{V}_i(t) = -V_i(t) + I_{i,syn}(t)$$

where $\tau$ is the membrane time constant and $I_{i,syn}(t)$ is the synaptic input of the neuron. The neuron fires a spike whenever the voltage reaches a threshold $V_t$, and is reset after a refractory period $\tau_{rp}$ to a voltage $V_r$. The synaptic input is the sum of contributions of individual synapses, modelled as a sum of delta functions at a latency $\delta$ after each pre-synaptic spike.

## 6.2. Dynamics of networks of spiking neurons—single population analysis

We start by a short presentation of the analytical methods that are used to study sparsely connected networks of spiking neurons in parameter regimes where single neurons fire irregularly. This is done in three steps: First, the formalism is described in the case of a single population; Second, we briefly describe excitatory-inhibitory networks (section 6.3); Finally, networks with selective subpopulations are considered (section 6.4).

We assume that the synaptic inputs to a neuron in a network can be described by a (possibly time-dependent) Poisson process, and consider a population in which all synaptic weights are equal to $J$. Then the synaptic inputs of any neuron in the network is a sum of delta functions with weight $J$, at random times according to a Poisson process with rate $C(v(t - \delta) + v_{ext})$, where $C$ is the number of recurrent synapses, $v(t)$ is the instantaneous firing rate of the network, and $Cv_{ext}$ is the external rate. $J$ is the amplitude of the post-synaptic potential (PSP) following a single spike.

We consider the probability density function (p.d.f.) $P(V, t)$ that a neuron is at voltage $V$ at time $t$. With the synaptic input described above, this p.d.f. evolves according to

$$\tau \frac{\partial P(V, t)}{\partial t} = Cv(t)\tau \left[ P(V - J) - P(V) \right] + \frac{\partial}{\partial V} \left[ V P(V, t) \right] \tag{6.1}$$

To convert equation 6.1 to a more convenient second order partial differential equation, one uses the diffusion approximation. When the PSP amplitude $J$ is small compared to threshold (the relevant situation in cortex), we can expand the right-hand side of equation 6.1 up to second order in $J$. This leads to

$$\tau \frac{\partial P(V, t)}{\partial t} = \frac{\sigma^2(t)}{2} \frac{\partial^2 P(V, t)}{\partial V^2} + \frac{\partial}{\partial V} \left[ (V - \mu(t)) P(V, t) \right] \tag{6.2}$$

where

$$\mu(t) = C(v(t - \delta) + v_{ext})\tau J, \tag{6.3}$$
$$\sigma^2(t) = C(v(t - \delta) + v_{ext})\tau J^2 \tag{6.4}$$

This is the Fokker-Planck equation describing the p.d.f $P(V, t)$ of a neuron, receiving a deterministic input $\mu(t)$ plus white noise with variance $\sigma^2(t)$. If the connection probability of the network is small, the noise terms $\sigma(t)$ become uncorrelated from neuron to neuron. Hence, $P(V, t)$ becomes the p.d.f. of the voltage of the neurons in the network.

Equation 6.2 can be written as a continuity equation

$$\frac{\partial P(V, t)}{\partial t} = -\frac{\partial S(V, t)}{\partial V} \tag{6.5}$$

where the probability flux $S(V, t)$ is

$$S(V, t) = -\frac{\sigma^2(t)}{2\tau} \frac{\partial P(V, t)}{\partial V} - \frac{(V - \mu(t))}{\tau} P(V, t) \tag{6.6}$$

The boundary conditions of the problem are the following:

- At threshold $V_t$, the probability flux is equal to the instantaneous firing rate, leading to

$$\frac{\partial P}{\partial V}(V_t, t) = -\frac{2\nu(t)\tau}{\sigma^2(t)} \tag{6.7}$$

- At reset $V_r$, the difference between probability fluxes between $V_r^+$ and $V_r^-$ is equal to the reinjected instantaneous firing rate at time $t - \tau_{rp}$ where $\tau_{rp}$ is the refractory period

$$\frac{\partial P}{\partial V}(V_r^+, t) - \frac{\partial P}{\partial V}(V_r^-, t) = -\frac{2\nu(t - \tau_{rp})\tau}{\sigma^2(t)} \tag{6.8}$$

- Finally, there are natural boundary conditions at $V \to -\infty$,

$$\lim_{V \to -\infty} P(V, t) = 0 \qquad \lim_{V \to -\infty} V P(V, t) = 0. \tag{6.9}$$

The p.d.f $P(V, t)$ must be normalized to 1. Taking into account the probability for a neuron to be refractory at time $t$, $p_r(t)$, this leads to

$$\int_{-\infty}^{V_t} P(V, t)dV + p_r(t) = 1 \tag{6.10}$$

where the probability to be refractory is

$$p_r(t) = \int_{t-t_{rp}}^{t} \nu(u)du$$

The population firing rate $\nu(t)$ can be obtained by solving Eqs. (6.2-6.10), as a function of the system parameters $C$, $J$ and $\nu_{ext}$, and the single cell parameters $\tau$, $V_t$, $V_r$ and $\tau_{rp}$. In particular, the stationary states of the system are obtained by setting the time derivatives to zero in Equation (6.2) and the associated boundary conditions. The stationary membrane potential distribution is

$$P_0(V) = 2\frac{\nu_0\tau}{\sigma_0(\nu_0)}\exp\left(-\frac{(V-\mu_0(\nu_0))^2}{\sigma_0(\nu_0)^2}\right)\int_{\frac{V-\mu_0(\nu_0)}{\sigma_0(\nu_0)}}^{\frac{V_t-\mu_0(\nu_0)}{\sigma_0(\nu_0)}}\Theta(u-V_r)e^{u^2}du,$$

$$p_{r,0} = \nu_0\tau_{rp} \tag{6.11}$$

with

$$\mu_0(\nu_0) = \mu_{ext} + CJ\tau\nu_0, \tag{6.12}$$
$$\sigma_0(\nu_0)^2 = \sigma_{ext}^2 + CJ^2\tau\nu_0 \tag{6.13}$$

The stationary firing rate is the solution of

$$\nu_0 = \Phi(\mu_0(\nu_0), \sigma_0(\nu_0))$$

$$= \left(\tau_{rp} + \tau\sqrt{\pi}\int_{\frac{V_r-\mu_0(\nu_0)}{\sigma_0(\nu_0)}}^{\frac{V_t-\mu_0(\nu_0)}{\sigma_0(\nu_0)}}due^{u^2}(1 + \text{erf}(u))\right)^{-1} \tag{6.14}$$

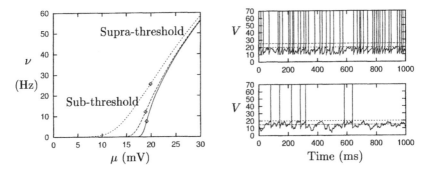

Fig. 16. Left: Single neuron transfer function, $\Phi(\mu, \sigma)$ vs $\mu$, for several values of $\sigma = 1$ mV (full line), 2mV (long-dashed), 5mV (short-dashed). The threshold is $V_t = 20$ mV. Diamonds indicate the inflexion point of the transfer function (change of sign of curvature). Right: Representative spike trains in the sub-threshold (bottom) and supra-threshold (top) ranges, for $\sigma_0 = 5$ mV.

where $\Phi$ is the single neuron transfer function (or f-I curve) in the presence of noise (see e.g. [7, 14, 27, 116]). It is depicted in figure 16. Eq. (6.14) describes the firing rate of a neuron in a network with instantaneous synapses. If synapses have a finite width, the firing rate can no longer be computed exactly. However, various asymptotic expansions have been proposed in cases in which the synaptic times are either much smaller or much larger than the membrane time constant, and interpolations between these two limits have been shown to provide a good approximation of the firing rates for a wide range of synaptic time constants [32, 33, 55, 97].

Finally, the coefficient of variation (CV) of inter-spike intervals (SD divided by mean ISI) can also be computed (see e.g. [27, 132])

$$
CV^2 = 2\pi \nu_0^2 \int_{\frac{V_r - \mu_0(\nu_0)}{\sigma_0(\nu_0)}}^{\frac{V_t - \mu_0(\nu_0)}{\sigma_0(\nu_0)}} e^{x^2} dx \int_{-\infty}^{x} e^{y^2} (1 + \mathrm{erf} y)^2 dy
$$

The stability of the stationary (asynchronous) state described by Eqs. (6.11-6.14) can be checked using linear stability analysis. See refs. [1, 128] and (Mato, this volume) for analysis in fully connected networks, refs. [27, 31] for analysis in sparsely connected networks.

### 6.3. Two population networks

It is easy to generalize the approach developed in the previous sections to two population networks - one excitatory and one inhibitory. There are now four types of synaptic connections. The strength (PSP amplitude) of a synapse from a

$b = E, I$ neuron to an $a = E, I$ neuron is $J_{ab}$. Membrane time constants are $\tau_E$ and $\tau_I$. The self-consistent equations for stationary states are

$$\nu_E = \Phi_E(\mu_E, \sigma_E) \tag{6.15}$$
$$\mu_E = C_E J_{EE} \tau_E [\nu_E + \nu_{ext}] - C_I J_{EI} \tau_E \nu_I$$
$$\sigma_E^2 = C_E J_{EE}^2 \tau_E [\nu_E + \nu_{ext}] + C_I J_{EI}^2 \tau_E \nu_I$$
$$\nu_I = \Phi_I(\mu_I, \sigma_I)$$
$$\mu_I = C_E J_{IE} \tau_I [\nu_E + \nu_{ext}] - C_I J_{II} \tau_I \nu_I$$
$$\sigma_I^2 = C_E J_{IE}^2 \tau_I [\nu_E + \nu_{ext}] + C_I J_{II}^2 \nu_I \tau_I \tag{6.16}$$

The analysis of two population networks [27] shows that asynchronous state at low rates ($\sim 5$ Hz) and high CVs ($\sim 1$) are stable if:
- Recurrent inhibition dominates over recurrent excitation;
- External inputs are above threshold;
- The synaptic delay is small compared to $\tau$.

In the large $C$ limit, if synaptic couplings scale as $J \sim 1/\sqrt{C}$, Equations (6.15-6.16) become equivalent to the equations of the 'balanced' network (see refs [134, 135], and van Vreeswijk, this volume). Equations (6.15-6.16) generalize the equations of a 'balanced' network to a network of LIF neurons with finite connectivity and finite synaptic strengths.

### 6.4. Storing binary non-overlapping patterns in the cortical network model

The first analytical study to consider associative memory properties of networks of spiking neurons was performed by Amit and Brunel [7]. The model was analyzed further in [28]. The architecture of the network is shown in Figure 17. The starting point of the architecture is the two population network of section 6.3. Each pattern shown to the network is assumed to activate a sub-population which represents a small fraction $f$ of the whole excitatory population. For the sake of simplicity of the analytical study, the patterns are assumed to activate non-overlapping populations of neurons. As a result of a Hebbian learning process, connections between cells which belong to in the same sub-population have a strength $J_{EE}g_+$, where $g_+ \geq 1$ represents a synaptic potentiation parameter, while connections between cells that belong to different sub-populations have a strength $J_{EE}g_-$, where $g_- \leq 1$ represents a synaptic depression parameter. $g_-$ can be chosen such that the average (overall) excitatory-to-excitatory synaptic strength in the network remains constant as $g_+$ is varied. This has the advantage that the background firing rate remains constant as $g_+$ is varied.

Depending on the potentiation parameter $g_+$, three types of states may exist in such networks:

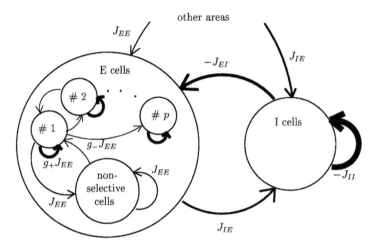

Fig. 17. Non-overlapping patterns in the two-population network. See text and refs. [7,28] for details.

- A background states in which neurons in all sub-populations fire at similar rates. The equations for the firing rates in such a state are similar to Eqs. (6.15-6.16). This state is functionnally similar to the 'null' state in the networks of 0,1 neurons. This state exists and is stable for $g_+$ smaller than a critical value $g_s$.
- Memory states, in which one sub-population (the last one that received 'sensory inputs') has an average firing rate $\nu_{act}$ which is higher than the other selective sub-populations (whose rate is $\nu_+$) and non-selective neurons (whose rate is $\nu_0$). The rates in these different states are the solutions of the self-consistent equations

$$\nu_{act} = \Phi_E(\mu_{act}, \sigma_{act}), \tag{6.17}$$

$$\nu_+ = \Phi_E(\mu_+, \sigma_+), \tag{6.18}$$

$$\nu_0 = \Phi_E(\mu_0, \sigma_0), \tag{6.19}$$

$$\nu_I = \Phi_I(\mu_I, \sigma_I), \tag{6.20}$$

$$\mu_{act} = C_E J_{EE} \tau_E \left[ f g_+ \nu_{act} + (p-1) f g_- \nu_+ + (1-pf) g_- \nu_0 + \right.$$
$$\left. + \nu_{ext} \right] - C_I J_{EI} \tau_E \nu_I \tag{6.21}$$

$$\sigma_{act}^2 = C_E J_{EE}^2 \tau_E \left[ f g_+^2 \nu_{act} + (p-1) f g_-^2 \nu_+ + (1-pf) g_-^2 \nu_0 + \right.$$
$$\left. + \nu_{ext} \right] + C_I J_{EI}^2 \tau_E \nu_I \tag{6.22}$$

$$\mu_+ = C_E J_{EE} \tau_E \left[ f g_- \nu_{act} + f [g_+ + (p-2) g_-] \nu_+ + (1-pf) g_- \nu_0 \right.$$
$$\left. + \nu_{ext} \right] - C_I J_{EI} \tau_E \nu_I \tag{6.23}$$

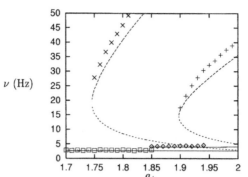

$\nu$ (Hz)

Fig. 18. Bifurcation diagram: Firing rates in background (solid lines) and memory states (long-dashed lines) as a function of the synaptic potentiation parameter $g_+$, for two parameter sets, one for which background rates are 2.5 Hz (bifurcation at $g_+ = 1.75$), the other for which background rates are 4 Hz (bifurcation at $g_+ = 1.9$). Lines show the analytical results, Eqs. (6.17-6.28). Short-dashed line is an unstable solution, separating basins of attraction of memory and background states. Symbols show the results of simulations (squares and diamonds, background rates; crosses and pluses, persistent rates). From ref. [28].

$$\sigma_+^2 = C_E J_{EE}^2 \tau_E \left[ f g_-^2 \nu_{act} + f[g_+^2 + (p-2)g_-^2]\nu_+ + (1-pf)g_-^2 \nu_0 \right.$$
$$\left. + \nu_{ext} \right] + C_I J_{EI}^2 \tau_E \nu_I \qquad (6.24)$$

$$\mu_0 = C_E J_{EE} \tau_E \left[ f \nu_{act} + f(p-1)\nu_+ + (1-pf)\nu_0 + \nu_{ext} \right]$$
$$- C_I J_{EI} \tau_E \nu_I \qquad (6.25)$$

$$\sigma_0^2 = C_E J_{EE}^2 \tau_E \left[ f \nu_{act} + f(p-1)\nu_+ + (1-pf)\nu_0 + \nu_{ext} \right]$$
$$+ C_I J_{EI}^2 \tau_E \nu_I \qquad (6.26)$$

$$\mu_I = C_E J_{IE} \tau_I \left[ f \nu_{act} + (p-1)f \nu_+ + (1-pf)\nu_0 + \nu_{ext} \right]$$
$$- C_I J_{II} \tau_I \nu_I \qquad (6.27)$$

$$\sigma_I^2 = C_E J_{IE}^2 \tau_I \left[ f \nu_{act} + (p-1)f \nu_+ + (1-pf)\nu_0 + \nu_{ext} \right]$$
$$+ C_I J_{II}^2 \nu_I \tau_I \qquad (6.28)$$

These correspond to the retrieval states in the networks of binary neurons. These states exist beyond a critical value of the potentiation parameter, $g_+ > g_1$.

• Multi-item memory states, in which more than one population is active at higher than background rates. These states are similar to the mixture states of networks of binary neurons. A $k$-memory state, where $k$ is the number of co-activated patterns, exist beyond a critical value of the potentiation parameter $g_+ > g_k$, where $g_1 < g_2 < g_3 \ldots$. This type of state is not discussed further in this section (see Discussion).

Bifurcation diagrams of such networks obtained from both analytical and sim-

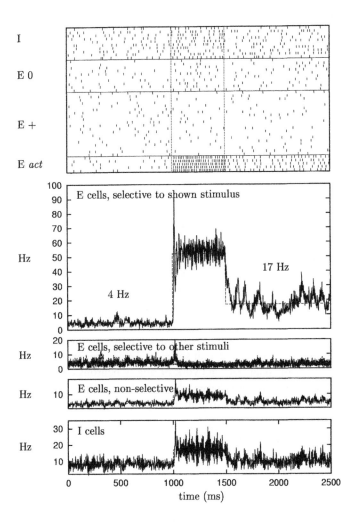

Fig. 19. Simulation of the initial part of a delayed-match-to sample task (inter-trial interval of 1s, sample presentation of 500ms, delay of 1s) in a network of 20,000 integrate-and-fire neurons with 10% connection probability. Top panel: rasterplots of neurons from various subpopulations (E, excitatory neurons; act, neurons activated by sample stimulus; +, neurons selective for other stimuli; 0, non-selective neurons; I, inhibitory neurons). Bottom panels, average firing rates in distinct subpopulations. Dashed line: analytical results. From ref. [28].

ulation results are shown in figure 18. As the potentiation parameter increases beyond a critical value $g_1$ (equal to 1.75 and 1.9 in the two examples shown in figure 18), memory states become stable. The spontaneous activity destabilizes beyond a second critical value $g_s$ (about 2 in figure 18). For a potentiation parameter between $g_1$ and $g_s$, both types of states are stable. A simulation of a DMS task in this region is shown in figure 19. The figure shows that transient inputs in one of the sub-populations make the network switch from the background state to the corresponding memory state. For these parameters, the spontaneous and persistent activity have the same magnitude as reported in neurophysiological experiments.

### 6.5. *Spontaneous and persistent activity in the large C, sparse coding limit*

Equations (6.17-6.28) can be solved numerically to obtain background and persistent firing rates in various subpopulations of the network (see Figure 18). However, it is difficult to analyze further such a system of four coupled non-linear equations. To gain additional insight in the emergence of memory states in such networks, it is useful to consider again the 'balanced network' limit, in which the number of connections $C$ goes to infinity, while coupling strengths $J$ scale as $1/\sqrt{C}$.

Here, we consider the case where only one memory is embedded in the network, with a sparse coding level $f$ that scales as $1/\sqrt{C}$ (see van Vreeswijk, this volume for analysis of 'balanced' network of binary neurons close to capacity). Excitatory to excitatory synapses have two states, $J_{EE}g_+$ and $J_{EE}g_-$. The value of $g_-$ is chosen to ensure that the overall average synaptic efficacy is unaffected by the potentiation parameter $g_+$.

The network is composed of three populations: the foreground neurons (characterized by $\nu_+, \mu_+, \sigma_+$), background neurons ($\nu_0, \mu_0, \sigma_0$), and inhibitory neurons ($\nu_I, \mu_I, \sigma_I$).

The analysis of Equations (6.17-6.28) yields:

- The firing rates of background and inhibitory neurons $\nu_0$ and $\nu_I$ are given by the balanced condition (as in the unstructured network); hence, these neurons are unaffected by persistent activity in the foreground population.
- The firing rate in the foreground population $\nu_+$ is a solution of

$$\nu_+ = \Phi \left( \mu_0 + C_E f J_{EE}(g_+ - 1)\tau[\nu_+ - \nu_0], \sigma \right) \tag{6.29}$$

Hence, firing rates in stable network states are given by intersections between the f-I curve,

$$\nu_+ = \Phi(I, \sigma)$$

and a straight line intersecting the f-I curve at $\mu_0, \nu_0$ with slope

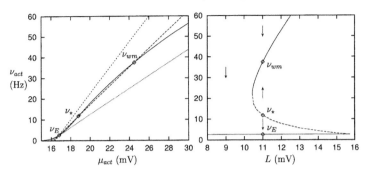

Fig. 20. Left: solutions of equation (6.29) as intersections of f-I curve (solid line) and a straight line intersecting f-I curve at the background firing rate $\nu_E$ (here 3 Hz) with a slope which is inversely proportional to $g_+$ (dashed/dotted lines). For large enough $g_+$, there are three intersections: the one at higher rate corresponds to persistent activity; the intermediate one is unstable. Right: firing rates in stable solutions (solid lines) and unstable solution (dashed line), as a function of $L = C_E f J_{EE}(g_+ - 1)$.

$1/(C_E f J_{EE}(g_+ - 1))$

$$\nu_+ = \nu_0 + \frac{I - \mu_0}{C_E f J_{EE}(g_+ - 1)\tau}.$$

Such intersections are shown in figure 20 for several values of the potentiation parameter $J_+$. This analysis shows that there are two scenarios for the emergence of persistent states. If background activity, $\nu_0$, is such that the curvature of the f-I curve at $\nu_0$ is positive (i.e. $\nu_0$ is below the inflexion point of the f-I curve), then persistent states appear at $g_+ = g_1$ through a saddle-node bifurcation. Increasing $g_+$ further leads to an increase of the persistent firing rate, and at $g_+ = g_s$ there is a transcritical bifurcation: the background state $\nu_0$ meets the unstable state and they exchange stability. On the other hand, if background activity is such that the curvature of the f-I curve at $\nu_0$ is negative (i.e. $\nu_0$ is below the inflexion point of the f-I curve), persistent states emerge continuously from the background state (through a transcritical bifurcation) at $g_+ = g_1 = g_s$.

This shows that the minimal persistent firing rate that can be achieved in such a network is intimately linked with the single neuron f-I curve and the magnitude of background activity. The persistent firing rate in particular cannot be lower than the inflexion point of the f-I curve. Note however that this lower bound depends critically on the assumption that synaptic currents depend linearly on pre-synaptic firing rates. In the presence of non-linearities, that can be due to saturation and/or short-term depression properties of synapses, persistent activity can be achieved at lower rates. In this case, the firing rate in the persistent state is controlled by saturation properties of synapses [136].

## 6.6. Capacity

What is the capacity of such networks? Very few studies have considered the storage of a large number of patterns in networks of spiking neurons (but see refs [17, 47]). Here, we only give orders of magnitudes of the capacity of such networks. In the large $C$ limit, when the coding level $f \sim 1/\sqrt{C}$, and synapses are binary, using the Willshaw or its stochastic LTP/LTD version [30], one expects the capacity to scale as $p \sim 1/f^2 \sim C$. Note that in a network of spiking neurons with background activity, $f$ cannot be decreased below $1/\sqrt{C}$ (unlike in binary networks, where it can decrease down to $\ln N/C$), otherwise the selective signal becomes drowned in noise coming from the Poissonian background inputs.

Hence, in the case of binary synapses, it seems that the information stored in the network can never be of order 1 bit per synapse in the large $C$ limit. However, if the number of synaptic states increases, one expects to reach optimal scaling (finite information per synapse) when the number of synaptic states $n \sim C^{1/4}/\sqrt{\ln C}$. For realistic numbers of synapses $C \sim 10,000$, $C^{1/4}/\sqrt{\ln C} \sim 3$. Hence, networks with relatively simple synapses might be able to operate close to the theoretical bounds for information storage.

## 6.7. Stability of persistent states vs synchronized oscillations

Are persistent states stable with respect to synchronized oscillations? This question has not been addressed analytically in the network described until now. Simulations show that in such a network persistent activity states are stable. However, synapses in such simulations are oversimplified delayed delta function, with equal delays in excitatory and inhibitory synapses. Synaptic time constants are among the parameters that are most critical to determine the stability of asynchronous states in networks of spiking neurons (see Mato, this volume). This fact motivated the study of the dynamics of memory networks with more realistic time courses. In particular, Wang studied a network in which synapses have realistic time constants (AMPA $\sim$ 2ms, GABA $\sim$ 5ms, NMDA $\sim$ 50ms), where the bistability between a silent state and a state in which all neurons fire repetitively is due to unstructured recurrent excitation. He observed that the dynamics is strongly influenced by the ratio of NMDA (slow excitation) to AMPA (fast excitation). At low NMDA to AMPA ratios, an oscillatory instability develops due to the fact that AMPA excitation is faster than GABA inhibition. Increasing the NMDA to AMPA ratio, persistent activity is stabilized by the long time constant of NMDA receptors [136, 137]. Similar results were obtained in simulations of networks with selective memory states [34, 39]. These studies also found that there is a regime at intermediate values of the NMDA/AMPA ratio in which persistent activity is stable but oscillatory. However, synapses with long time constants, even though they potentially stabilize asynchronous activity, are not a necessary requirement for stability, as shown analytically by Hansel and Mato [65, 66] in an unstructured two population network of quadratic integrate-and-fire neurons (see Mato, this volume), and numerically by Gutkin et al [64] in a network of conductance-based neurons with spatially decaying connectivity. Strong synchronization of neurons in a persistent activity state has been proposed as a mechanism for memory erasure by Laing and Chow [80] and Gutkin et al [64].

## 7. Plasticity of persistent activity

Models based on the Hebbian cell assembly hypothesis predict that the patterns of persistent activity should be plastic over time, and that they should be influenced by the presentation protocols in delayed response experiments. The experiments reviewed in section 2.2 have provided clear evidence that persistent activity is indeed plastic, and that this plasticity reflects the learning of associations between arbitrary patterns. In this section, I describe how the theoretical models account for these findings.

### 7.1. Learning correlations between patterns separated by delay periods

How can correlations between stimuli separated by delay periods be formed during the course of a pair-associate task? A possible scenario is shown in Figure 21. In this scenario, the network architecture is as shown in figure 17, but with initially no structure in the synaptic matrix. Hence, the synaptic efficacies between neurons are in average equal, regardless of the selectivity properties of pre and post synaptic neurons. In a pair associate task, the cue stimulus elicits a strong visual response of neurons selective for that stimulus. The assumptions we make on synaptic dynamics are as follows. (i) synaptic changes are triggered by a 'Hebbian' variable that is proportional to a 'covariance' term, i.e. the product of instantaneous firing rates (minus baseline) of pre and post synaptic neurons. (ii) for LTP to occur, such a variable must be above a finite threshold, to ensure stability of long-term memory of synaptic changes. If no such threshold exists, transitions occurs when the network is in the background state, in the absence of any stimulus presentations, leading to a rapid erasure of any synaptic structure. The presence of a finite threshold allows synaptic transitions to occur only during presentations of external stimuli that elicit strong visual response. The consequence of such a learning dynamics is that synapses between neurons which are selective for the same stimulus (here A) will grow, at a rate proportional to the probability of LTP during a single presentation. This corresponds to moving towards the right in the bifurcation diagram of Fig. 18.

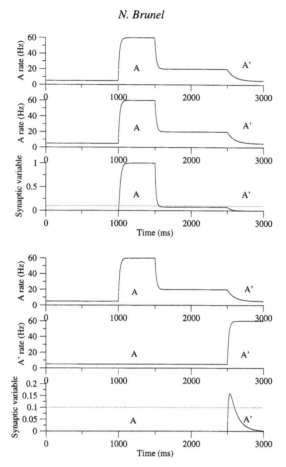

Fig. 21. Synaptic structuring in the pair-associate task. Top 3 panels: average firing rate of two cells with strong visual response to a stimulus A and persistent activity following presentation of A. The third panel from the top shows a 'Hebbian' synaptic variable driving synaptic potentiations, that occur only when this variable is above a threshold, indicated by a dotted line. Transitions occur only during the presentation of the cue. Bottom 3 panels: average firing rate of two cells, one of which is selective to A, the other to its pair associate A'. Due to persistent activity in cell A, both cells have elevated firing rates in a short interval at the beginning of presentation of pair-associate A', leading to synaptic potentiations between these cells, as shown in the bottom panel.

Initially, during the delay period, these neurons come back to their background firing rate, since the synaptic efficacies between them are not yet at the level at which they can sustain persistent activity (the 'potentiation parameter' $g_+$ has not yet reached its critical value $g_1$). However, after a large enough number of presentations of a given stimulus, and $g_+$ has increased beyond the critical value

$g_1$, persistent activity becomes stable. This allows neurons selective for A to maintain rates significantly higher than background, up to the presentation of the next stimulus. Hence, there is now a temporal window (at the beginning of the presentation of the next stimulus) during which the two neuronal populations (corresponding to A and A') are at significantly higher rates than background, leading potentially to LTP transitions in synapses connecting neurons in these two populations. Enhanced synapses between neuronal populations selective for distinct populations then provoke changes in the patterns of delay activity, as shown in the next subsections.

## 7.2. Network subjected to DMS task with a fixed sequence of samples

Training of such a network using the protocol of the Miyashita experiment [94] leads to the patterns of persistent activity shown in figure 22 [8, 24, 26, 29, 144], provided delay activity following a given sample picture can survive until the next sample presentation. This has been observed experimentally by Yakovlev et al [144]. As the strength of the connections between nearest-neighbor populations (the 'sequence learning' parameter) grows stronger, activation of one neuronal population during the delay period leads to activation of more and more neighbors in the sequence (see patterns of activation shown in the insets of figure 22B), with decreasing firing rates as the distance from the cue population increases. These patterns are similar to the patterns observed in the Miyashita experiment (see figure 4a,b).

Correlations between delay activity patterns have been initially investigated in a fully connected network of binary neurons using a synaptic matrix originally proposed by Griniasty et al [63]

$$J_{ij} = \frac{1}{N} \sum_{\mu} \left( \xi_i^{\mu} \xi_j^{\mu} + a \xi_i^{\mu+1} \xi_j^{\mu} + a \xi_i^{\mu} \xi_j^{\mu+1} \right)$$

where $a$ is the 'sequence learning' parameter of figure 22. This network was studied analytically by Cugliandolo and Tsodyks [45, 46]. At low loading, one finds that:

- for $a \in [0, 0.5]$, uncorrelated patterns are the attractors of the system.
- for $a \in [0.5, 1]$, attractors have finite overlaps with 9 patterns, the overlaps being

$$\frac{1}{128}(0, \ldots, 0, 1, 3, 13, 51, 77, 51, 13, 3, 1, 0, \ldots, 0)$$

The standard correlations between delay activity patterns are, as a function of distance between patterns,

$$(1, 0.66.0.33, 0.12, 0.04, 0.01, 0, \ldots)$$

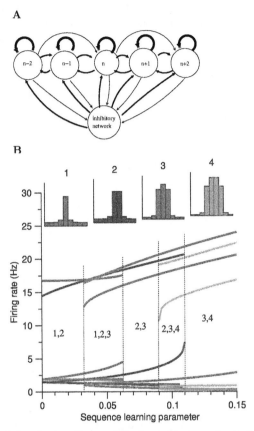

Fig. 22. **A.** Network functional architecture following training according to the protocol of ref [94]. The strength of the connections between nearest neighbor populations is parameterized by the 'sequence learning parameter' $0 < a < 1$. $a$ represents the fraction of such synapses which are in the potentiated state. For $a = 1$, these synapses have the same strength as the intra-population synapses. **B.** Bifurcation diagrams showing patterns of persistent activity as a function of sequence learning parameter $a$. Persistent activity attractor states are shown in the insets (firing rate vs population serial position number), labelled by the number of populations which are strongly active in such states (1: red, 2: blue, 3: green, 4: brown, see insets), as a function of sequence learning index. The bifurcations where specific states appear or disappear are marked with dotted vertical lines. E.g. for $a \sim 0.03$ states with 3 strongly active populations appear. Similar transitions where more populations get activated occur at higher values of $a$. From ref. [29].

Note that this Hopfield-type network has a very different behavior than the more realistic network with separate excitation and inhibition. In the Hopfield-type network, the correlations vanish at distance 6 exactly, while in the network of figure 22, the distance at which correlations vanish varies with the sequence learning parameter (that increases the number of populations/patterns that are activated in an attractor) and inhibition (that decreases this number).

## 7.3. Network subjected to pair-associate tasks

A network trained using a pair-associate task has a synaptic structure as shown in figure 23A [26, 29, 96]. The strength of the connections between pair associate populations is parameterized by the 'pair learning' parameter $0 < a < 1$, which plays a similar role as the 'sequence learning' parameter. It is the fraction of such synapses which are in the potentiated state. This type of synaptic structure leads to two kinds of selective attractor states (shown in figure 23B). The first type of persistent activity is an 'individual' attractor state: it evolves continuously from the $a = 0$ memory state. One of the objects of the pair has strong activity, while the other has intermediate activity (between background and persistent rate). The second type of persistent activity pattern is a network state in which both pair associates have equal elevated persistent activity – a pair attractor. The behavior of the network in a pair associate task depends on the value of $a$:

(i) For $a < 0.06$: after presentation of the cue, the network goes to the memory state corresponding to the cue. The neurons corresponding to the pair-associate are active at levels slightly above spontaneous activity. This corresponds to a weak 'prospective' activity (activity correlated with the stimulus that will be shown after the delat period) that is essentially constant during the delay, since the attractor is reached very rapidly after the end of the visual presentation. Upon presentation of the pair associate, the network can either switch to the memory state corresponding to the pair associate, or to the pair, since both types of states exist and are stable for this level of pair learning.

(ii) For $a > 0.06$: after presentation of the cue, the network goes to the pair memory state and stays there after presentation of the pair associate. Individual attractors (memory states) disappear.

## 7.4. Transitions between states in the absence of external inputs in networks of spiking neurons

Simulations of networks of spiking neurons have been shown to agree with the analytical results (see figures 19, 18). An important difference between a system of a finite number of spiking neurons and mean-field analysis, however, is the presence of 'random' fluctuations of the global activity of each population, due to the finite size of the system. In other words, each 'stable' state of the system is stable only in the limit of an infinite system. In a finite system, transitions between states can take place:

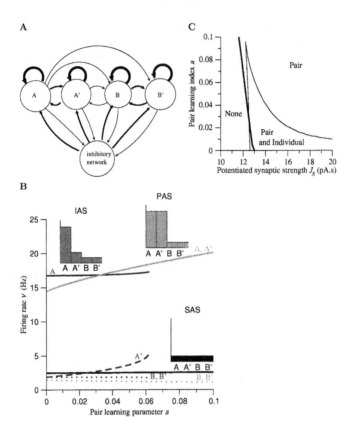

Fig. 23. **A**. Architecture of a network after learning of pair-associates. The strength of the connections between neurons in pair associate populations (e.g. between A and A') are parameterized by the 'pair learning' parameter $a$, representing the fraction of potentiated synapses among this type of synapses. **B**. Firing rates in memory states, as a function of pair learning index. All the states whose rates are plotted in this graph are stable states. Black curve: spontaneous activity (activity in subpopulations A, A', B, B' shown schematically in inset SAS). Individual attractor state (IAS, see inset) between $a = 0$ and $a \approx 0.06$ : red line (neurons selective for cue A), red dashed (neurons selective for pair associate A'), and red dotted (other neurons). Pair attractor state (PAS, see inset) state: green line (neurons selective for both cue and pair-associate of cue), green dotted line (other neurons). **C**. Diagram showing the regions where the different attractors live, in the space of synaptic variables $(J_S, a)$. The thick black curve shows the boundary of the region of existence of the pair states. Thin lines show the boundary of the region of existence of the individual states. From ref. [29].

• The network can jump from the background 'spontaneous' state to a memory state, especially if the basin of attraction of the spontaneous state is small. This is equivalent to a 'spontaneous activation' of a memory in the absence of external cue [6];

• A memory state can decay back to the background state, indicating a loss of short-term memory [6, 77];

• Last, transitions can occur between selective memory states. In particular, the probability of such transitions can be enhanced by learning of associations between stimuli, that leads to strengthened synapses between the corresponding selective populations. For example, in the pair-associate task, transitions are likely to occur between a state corresponding to an individual memory, and a state corresponding to the pair to which that individual memory belongs. Since these transitions occur at random times in individual trials, the trial averaged firing rate of neurons selective for the pair associate increases gradually during the delay period, giving rise to pronounced 'prospective' activity [96]. This ramping up of prospective activity during the delay period is a hallmark of neurophysiological recordings during pair-associate tasks [16, 110, 120].

## 8. Models with continuous attractors

The models that we have considered until now have discrete attractors. These models might not be appropriate for maintenance of working memory of a continuous variable such as the spatial location of a stimulus. Note that one could store a continuous variable in working memory with reasonable accuracy using a network with discrete attractors representing a discretized, coarse grained spatial variable. The attractor landscape would then look like the one of the network learning of a fixed sequence of stimuli (Fig. 22): attractors which are correlated together as a function of distance. This scenario could account for most experimental data on 'continuous' working memory since in these experiments there is actually only a rather small number of possible stimuli (e.g. eight in the ODR task of Funahashi et al [56–58]). In this section, we rather consider the scenario of networks with continuous attractors.

Most theoretical studies of continuous working memory belong to the class of 'ring models'. Ring models denote networks in which neurons are selective for an angular variable $\theta$. Neurons are labelled corresponding to their 'best stimulus' $\theta$. A combination of short-range excitation and long-range inhibition ('Mexican-hat' type profile, see figure 24) leads to attractor profiles which have the shape of a localized bump, that can peak at any stimulus location, hence a continuum of attractor states. This class of networks is reviewed extensively in other chapters (Bressloff, Sompolinsky and White). Here, we just mention that such models have been proposed by several authors in the context of working memory of a spatial variable [37, 39, 64, 80].

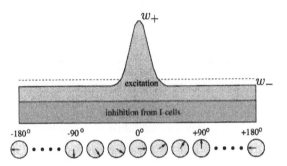

Fig. 24. Architecture of a network sustaining a continuum of attractors. Shown is the strength of excitatory connections from a cell whose preferred location is at $\theta = 0$ as a function of preferred location of post-synaptic cell (red). Excitatory connections are structured in the same way as in the discrete attractor network: structure is introduced keeping the average synaptic strength (dashed line) constant. Inhibition is uniform. From ref. [39].

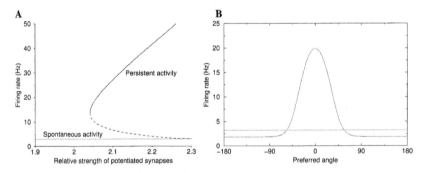

Fig. 25. A: bifurcation diagram for the spatial working memory network, obtained from mean-field analysis. Red: persistent activity of neurons whose preferred location is at the cue location vs synaptic potentiation parameter. Green: background activity. Dashed: unstable state. Bottom: 'memory fields' of cells in spatial working memory network: Firing rate in the delay period vs preferred location (red). Background rate is indicated in green.

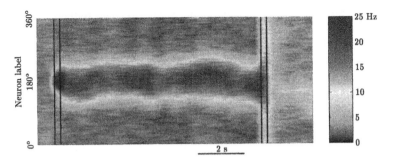

Fig. 26. Simulation of the oculomotor delayed response task in a fully connected network of integrate-and-fire neurons receiving noisy external inputs. Color-coded firing rates of neurons vs time. A cue is presented at 180 degrees in the interval marked by the two vertical lines on the left. A bump state peaked around 180 degrees has small random drifts during the delay period due to noise. Finally, the bump is switched off by strong external inputs to all neurons in the 'response' interval, marked by the two vertical lines on the right. From ref. [39].

The tuned state in such networks appear when the strength of recurrent excitation exceeds a critical value, as shown in Figure 25, much as in the discrete case (compare figure 18 and 25). As in the discrete case, the bifurcation can be supercritical or subcritical (as in figure 25).

One important difference between continuous and discrete attractors is their behavior with respect to noise. Discrete attractors can be very resistant to noise if 'energy barriers' between different attractors are large, as in the Hopfield model. On the other hand, there are no energy barriers separating nearby states in a continuous attractor. Any state on such continuous attractor is therefore only marginally stable with respect to perturbations. In the presence of noise, this leads to a drift of the memorized position with respect to noise, as shown in figure 26. This drift of the memorized position during the delay period leads in turn to an error in the memorized position whose variance grows linearly with time. Interestingly, psychophysical data on both humans and monkeys indicate that the variance of the error in the position of a saccade to a memorized location grows linearly with time, up to delays of about 4s [108, 139].

Other types of one dimensional attractor models have been proposed. Models in which the attractor has the topology of an open segment rather than a circle have been proposed in the context of the oculomotor integrator by Seung and collaborators [3, 123, 124]. Two-dimensional attractors have been hypothesized to subserve memory of 2D environments in the rat hippocampus [18, 121].

All these models suffer from a fine tuning problem which is absent in discrete attractor models: in the presence of any type of heterogeneity, the attractor landscape breaks down to a few discrete attractors. Several investigators have considered this problem recently. Camperi and Wang [37], and Koulakov et al [78] showed that bistable elements can yield robustness to a continuous attractor network that would otherwise be very sensitive to inhomogeneities.

Renart et al showed that homeostasis mechanisms [133] can also give stability to a continuous attractor network [114]. The idea is that the 'energy landscape' can be modified by homeostasis. If initially, there is a rugged energy landscape as a function of the underlying memorized variable, with a discrete number of minima, the network spends long intervals close to these minima. This leads to elevated firing rates of neurons whose preferred stimulus is close to these minima. These elevated firing rates cause a decrease in the synaptic efficacies of these neurons through homeostatic mechanisms [133]. The decrease of the synaptic efficacies leads in turn to a progressive flattening of the energy landscape. Effectively, the network becomes able to perform as an analog memory device even though heterogeneities are present.

## 9. Conclusions

In this review, I have attempted to give an account of the properties of a variety of 'Hebbian' models, from network models of binary neurons to network models of spiking neurons. Models with binary neurons have been particularly helpful in obtaining a deep understanding of the properties of attractor networks. Models with spiking neurons have been shown to share these basic properties, and in addition to reproduce a wealth of experimental data in the awake monkey. These features include: magnitude of background and persistent activity (even though the range of coexistence between spontaneous and persistent activity at realistic rates is rather small in present models); appearance of prospective activity in tasks where associations between stimuli are learned by the animal; correlations between patterns of delay activity. For each phenomenon, the models give insights about possible underlying mechanisms: balance between strong excitation and recurrent inhibition to generate irregular background activity; strong recurrent excitation, possibly induced by synaptic plasticity mechanisms, in sparse sub-populations to generate persistent activity; combination of persistent activity and synaptic plasticity mechanisms to generate prospective activity. These successes are encouraging, but rapid experimental progress constantly challenges these theoretical models and raises new questions. I conclude with a short selection of such open questions (see also recent reviews in refs [23,51,114,137,146])

• What would be an experimental test that persistent activity is maintained by synaptic circuitry rather than by intrinsic neuronal mechanisms? A possible idea would be to perturb a single neuron during persistent activity, in such a way as to decrease its firing rate to baseline or below. Then, one would observe whether the neuron's firing rate comes back to persistent rate (indicating the presence of strong synaptic inputs due to collective activity) or not. This idea was tested in the oculomotor integrator of the goldfish using intracellular recordings [3]. In the awake monkey, it is likely that such an experimental test will not be possible before intracellular recordings become available.

• Most models consider the issue of short-term memory of a single item or spatial location. Primates are able to maintain working memory of several items simultaneously. How can this be achieved?

– A proposal is that this is achieved by the mixture/multi-item memory states. As items are presented to a network, the network goes to attractors in which more and more selective sub-populations are active. One then expects an upper bound on the number of populations that can be simultaneously activated, due to recurrent inhibition. This upper bound would correspond to short-term (working) memory capacity. Such a scenario has been implemented in networks of LIF neurons by ref. [6].

– Another proposal is that the items are stored as a sequence of selective population discharges. This could be achieved in a network with both slow and fast oscillations (see e.g. [75]).

Unfortunately, there is currently no experimental data on neuronal activity during delayed response tasks in which more than one item needs to be maintained in working memory.

• Recent analysis of neuronal data during the delay period of a delayed oculomotor task has shown that the irregularity of the discharge of neurons in persistent activity is similar to the irregularity in background activity (in both cases the CV is close to 1) [40]. Most models ignore completely the issue of irregularity. In the few models in which background irregular activity is present (e.g. [7, 28, 34, 39]), the CV in persistent activity is significantly below 1. An alternative model in which a sub-population of both pyramidal cells and interneurons increase their firing rate in a memory states was proposed by Renart [112, 113, 115], but it suffers from a fine tuning problem: the range of bistability is typically extremely small. Hence, there seems to be no satisfactory solution to date to this problem.

• Learning rules that have been used in associative memory networks are typically firing rate dependent rules. In recent years, spike-timing dependent learning rules have been introduced to account for the increasingly available experimental data (see Gerstner and Abbott chapter in this issue). What patterns of persistent activity, or what kind of multistability scenarios, can one expect with such learning rules? This question remains largely open (for preliminary studies on how discrete attractors can be generated with such rules, see refs [13, 48]).

## Acknowledgements

I am indebted to Daniel Amit, Stefano Fusi and Claude Meunier for their numerous insightful comments on a previous version of this manuscript.

## References

[1] L. F. Abbott and C. van Vreeswijk. Asynchronous states in a network of pulse-coupled oscillators. *Phys. Rev. E*, 48:1483–1490, 1993.

[2] M. Abeles. *Corticonics.* New York: Cambridge University Press, 1991.

[3] E. Aksay, G. Gamkrelidze, H. S. Seung, R. Baker, and D. W. Tank. In vivo intracellular recording and perturbation of persistent activity in a neural integrator. *Nat. Neurosci.*, 4:184–193, 2001.

[4] S.-I. Amari. Learning pattern sequences by self-organizing nets of threshold elements. *IEEE Trans. Comput. C*, 21:1197–1206, 1972.

[5] D. J. Amit. *Modeling brain function.* Cambridge University Press, 1989.

[6] D. J. Amit, A. Bernacchia, and V. Yakovlev. Multiple-object working memory - a model for behavioral performance. *Cerebral Cortex*, 13:435–443, 2003.

[7] D. J. Amit and N. Brunel. Model of global spontaneous activity and local structured activity during delay periods in the cerebral cortex. *Cerebral Cortex*, 7:237–252, 1997.

[8] D. J. Amit, N. Brunel, and M. V. Tsodyks. Correlations of cortical hebbian reverberations: experiment *vs* theory. *J. Neurosci.*, 14:6435–6445, 1994.

[9] D. J. Amit and S. Fusi. Dynamic learning in neural networks with material sysnapses. *Neural Computation*, 6:957–982, 1994.

[10] D. J. Amit, S. Fusi, and V. Yakovlev. A paradigmatic working memory (attractor) cell in IT cortex. *Neural Comp.*, 9:1071–1092, 1997.

[11] D. J. Amit, H. Gutfreund, and H. Sompolinsky. Storing infinite numbers of patterns in a spin-glass model of neural networks. *Phys. Rev. Lett.*, 55:1530–1531, 1985.

[12] D. J. Amit, H. Gutfreund, and H. Sompolinsky. Statistical mechanics of neural networks near saturation. *Ann. Phys.*, 173:30–67, 1987.

[13] D. J. Amit and G. L. Mongillo. Spike driven synaptic plasticity generating working memory states. *Neural Comput*, 15:565–96, 2003.

[14] D. J. Amit and M. V. Tsodyks. Quantitative study of attractor neural network retrieving at low spike rates II: Low-rate retrieval in symmetric networks. *Network*, 2:275, 1991.

[15] D. J. Amit and M. V. Tsodyks. Effective neurons and attractor neural networks in cortical environment. *Network*, 3:121–, 1992.

[16] W. F. Asaad, G. Rainer, and E. K. Miller. Neural activity in the primate prefrontal cortex during associative learning. *Neuron*, 21:1399–1407, 1998.

[17] Y. Aviel, M. Abeles, and D. Horn. Memory capacity of balanced networks. submitted manuscript, 2004.

[18] F. P. Battaglia and A. Treves. Attractor neural networks storing multiple space representations: A model for hippocampal place fields. *Phys. Rev. E*, 58:7738–7753, 1998.

[19] T. V. P. Bliss and T. Lomo. Long-lasting potentiation of synaptic transmission in the dentate area of the anaesthetized rabbit following stimulation of the perforant path. *J. Physiol. London*, 232:331–356, 1973.

[20] V. Booth and J. Rinzel. A minimal, compartmental model for a dendritic origin of bistability of motoneuron firing patterns. *J Comput Neurosci*, 2:299–312, 1995.

[21] V. Booth, J. Rinzel, and O. Kiehn. Compartmental model of vertebrate motoneurons for Ca2+-dependent spiking and plateau potentials under pharmacological treatment. *J Neurophysiol*, 78:3371–85, 1997.

[22] V. Braitenberg and A. Schütz. *Anatomy of the cortex*. Springer-Verlag, 1991.

[23] C. D. Brody, R. Romo, and A. Kepecs. Basic mechanisms for graded persistent activity: discrete attractors, continuous attractors, and dynamic representations. *Curr Opin Neurobiol*, 13:204–11, 2003.

[24] N. Brunel. Dynamics of an attractor neural network converting temporal into spatial correlations. *Network*, 5:449–, 1994.

[25] N. Brunel. Storage capacity of neural networks: effect of the fluctuations of the number of active neurons per memory. *J. Phys. A: Math. Gen.*, 27:4783–4789, 1994.

[26] N. Brunel. Hebbian learning of context in recurrent neural networks. *Neural Computation*, 8:1677–1710, 1996.

[27] N. Brunel. Dynamics of sparsely connected networks of excitatory and inhibitory spiking neurons. *J. Comput. Neurosci.*, 8:183–208, 2000.

[28] N. Brunel. Persistent activity and the single cell f-I curve in a cortical network model. *Network*, 11:261–280, 2000.

[29] N. Brunel. Dynamics and plasticity of stimulus-selective persistent activity in cortical network models,. *Cerebral Cortex*, 13:1151–1161, 2003.

[30] N. Brunel, F. Carusi, and S. Fusi. Slow stochastic Hebbian learning of classes in recurrent neural networks. *Network*, 9:123–152, 1998.

[31] N. Brunel and V. Hakim. Fast global oscillations in networks of integrate-and-fire neurons with low firing rates. *Neural Computation*, 11:1621–1671, 1999.

[32] N. Brunel and P. Latham. Firing rate of noisy quadratic integrate-and-fire neurons. *Neural Computation*, 15:2281,2306, 2003.

[33] N. Brunel and S. Sergi. Firing frequency of integrate-and-fire neurons with finite synaptic time constants. *J. Theor. Biol.*, 195:87–95, 1998.

[34] N. Brunel and X. J. Wang. Effects of neuromodulation in a cortical network model of object working memory dominated by recurrent inhibition. *J. Comput. Neurosci.*, 11:63–85, 2001.

[35] J. Buhmann, R. Divko, and K. Schulten. Associative memory with high information content. *Phys. Rev. A*, 39:2689–2692, 1989.

[36] E. R. Caianiello. Outline of a theory of thought-processes and thinking machines. *J Theor Biol*, 1:204–35, 1961.

[37] M. Camperi and X.-J. Wang. A model of visuospatial short-term memory in prefrontal cortex: recurrent network and cellular bistability. *J. Comput. Neurosci.*, 5:383–405, 1998.

[38] M. V. Chafee and P. S. Goldman-Rakic. Matching patterns of activity in primate prefrontal area 8a and parietal area 7ip neurons during a spatial working memory task. *J. Neurophysiol.*, 79:2919–2940, 1998.

[39] A. Compte, N. Brunel, P. S. Goldman-Rakic, and X.-J. Wang. Synaptic mechanisms and network dynamics underlying spatial working memory in a cortical network model. *Cerebral Cortex*, 10:910–923, 2000.

[40] A. Compte, C. Constantinidis, J. Tegnér, S. Raghavachari, M. Chafee, P. S. Goldman-Rakic, and X.-J. Wang. Temporally irregular mnemonic persistent activity in prefrontal neurons of monkeys during a delayed response task. *J. Neurophysiol.*, 90:3441–3454, 2003.

[41] R. Cossart, D. Aronov, and R. Yuste. Attractor dynamics of network up states in the neocortex. *Nature*, 423:283–288, 2003.

[42] T. M. Cover. Geometrical and statistical properties of systems of linear inequalities with applications in pattern recognition. *IEEE Trans. Electron. Comput.*, 14:326, 1965.

[43] B. G. Cragg and H. N. V. Temperley. The organization of neurones: a cooperative analogy. *Electroencephalogr. Clin. Neurophysiol.*, 6:85–92, 1954.

[44] A. Crisanti, D. J. Amit, and H. Gutfreund. Saturation level of the Hopfield model for neural network. *Europhys. Lett.*, 2:337–341, 1986.

[45] L. F. Cugliandolo. Correlated attractors from uncorrelated stimuli. *Neural Comp.*, 6:220–, 1994.

[46] L. F. Cugliandolo and M. V. Tsodyks. Capacity of networks with correlated attractors. *J. Phys. A*, 27:741–756, 1994.

[47] E. Curti, G. Mongillo, G. La Camera, and D. Amit. Mean-field and capacity in realistic networks of spiking neurons storing sparsely coded random memories. to appear in Neural Computation, 2004.

[48] P. del Giudice, S. Fusi, and M. Mattia. Modeling the formation of working memory with networks of integrate-and-fire neurons connected with plastic synapses. *J. Physiol. Paris*, 97:659–681, 2003.

[49] B. Delord, A. J. Klaassen, Y. Burnod, R. Costalat, and E. Guigon. Bistable behaviour in a neocortical neurone model. *Neuroreport*, 8:1019–23, 1997.

[50] B. Derrida, E. Gardner, and A. Zippelius. An exactly solvable asymmetric neural network model. *Europhys. Lett.*, 4:167–173, 1987.

[51] D. Durstewitz, J. K. Seamans, and T. J. Sejnowski. Neurocomputational models of working memory. *Nat. Neurosci.*, Suppl:1184–1191, 2000.

[52] A. V. Egorov, B. N. Hamam, E. Fransen, M. E. Hasselmo, and A. A. Alonso. Graded persistent activity in entorhinal cortex neurons. *Nature*, 420:173–178, 2002.

[53] A. Engel and C. van den Broeck. *Statistical mechanics of learning*. Cambridge University Press, 2001.

[54] C. A. Erickson and R. Desimone. Responses of macaque perirhinal neurons during and after visual stimulus association learning. *J. Neurosci.*, 19:10404–10416, 1999.

[55] N. Fourcaud and N. Brunel. Dynamics of firing probability of noisy integrate-and-fire neurons. *Neural Computation*, 14:2057–2110, 2002.

[56] S. Funahashi, C. J. Bruce, and P. S. Goldman-Rakic. Mnemonic coding of visual space in the monkey's dorsolateral prefrontal cortex. *J. Neurophysiol.*, 61:331–349, 1989.

[57] S. Funahashi, C. J. Bruce, and P. S. Goldman-Rakic. Visuospatial coding in primate prefrontal neurons revealed by oculomotor paradigms. *J. Neurophysiol.*, 63:814–831, 1990.

[58] S. Funahashi, C. J. Bruce, and P. S. Goldman-Rakic. Neuronal activity related to saccadic eye movements in the monkey's dorsolateral prefrontal cortex. *J. Neurophysiol.*, 65:1464–1483, 1991.

[59] J. M. Fuster and G. Alexander. Neuron activity related to short-term memory. *Science*, 173:652–654, 1971.

[60] J. M. Fuster and G. E. Alexander. Firing changes in cells of the nucleus medialis dorsalis associated with delayed response behavior. *Brain Res*, 61:79–91, 1973.

[61] E. J. Gardner. The phase space of interactions in neural network models. *J. Phys. A: Math. Gen.*, 21:257–270, 1988.

[62] D. Golomb, N. Rubin, and H. Sompolinsky. Willshaw model: Associative memory with sparse coding and low firing rates. *Phys. Rev. A*, 41:1843–1854, 1990.

[63] M. Griniasty, M. V. Tsodyks, and D. J. Amit. Conversion of temporal correlations between stimuli to spatial correlations between attractors. *Neural Computation*, 5:1–17, 1993.

[64] B. S. Gutkin, C. R. Laing, C. L. Colby, C. C. Chow, and G. B. Ermentrout. Turning on and off with excitation: the role of spike time asynchrony and synchrony in sustained neural activity. *J. Comput. Neurosci.*, 11:121–134, 2001.

[65] D. Hansel and G. Mato. Existence and stability of persistent states in large neuronal networks. *Phys. Rev. Lett.*, 10:4175–4178, 2001.

[66] D. Hansel and G. Mato. Asynchronous states and the emergence of synchrony in large networks of interacting excitatory and inhibitory neurons. *Neural Comp.*, 15:1–56, 2003.

[67] D. O. Hebb. *Organization of behavior*. New York: Wiley, 1949.

[68] J. Hertz, A. Krogh, and R. G. Palmer. *Introduction to the Theory of Neural Computation*. Addison-Wesley, Redwood City, 1991.

[69] O. Hikosaka, M. Sakamoto, and S. Usui. Functional properties of monkey caudate neurons. III. Activities related to expectation of target and reward. *J Neurophysiol*, 61:814–32, 1989.

[70] O. Hikosaka and R. H. Wurtz. Visual and oculomotor functions of monkey substantia nigra pars reticulata. III. Memory-contingent visual and saccade responses. *J Neurophysiol*, 49:1268–84, 1983.

[71] A. L. Hodgkin and A. F. Huxley. A quantitative description of membrane current and its application to conduction and excitation in nerve. *J. Physiol.*, 117:500–544, 1952.

[72] C. Holmgren, T. Harkany, B. Svennenfors, and Y. Zilberter. Pyramidal cell communication within local networks in layer 2/3 of rat neocortex. *J Physiol*, 551:139–153, 2003.

[73] J. J. Hopfield. Neural networks and physical systems with emergent collective computational abilities. *Proc. Natl. Acad. Sci. U.S.A.*, 79:2554–2558, 1982.

[74] M. Ito. *The cerebellum and neural control*. Raven Press, New York, 1984.

[75] O. Jensen and J. E. Lisman. Theta/gamma networks with slow NMDA channels learn sequences and encode episodic memory : Role of NMDA channels in recall. *Learning and Memory*, 3:264–278, 1996.

[76] B. W. Knight. Dynamics of encoding in a population of neurons. *J. Gen. Physiol.*, 59:734–766, 1972.

[77] A. A. Koulakov. Properties of synaptic transmission and the global stability of delayed activity states. *Network*, 12:47–74, 2001.

[78] A. A. Koulakov, S. Raghavachari, A. Kepecs, and J. E. Lisman. Model for a robust neural integrator. *Nat. Neurosci.*, 5:775–782, 2002.

[79] K. Kubota and H. Niki. Prefrontal cortical unit activity and delayed alternation performance in monkeys. *J. Neurophysiol.*, 34:337–347, 1971.

[80] C. R. Laing and C. C. Chow. Stationary bumps in networks of spiking neurons. *Neural Computation*, 13:1473–1494, 2001.

[81] L. Lapicque. Recherches quantitatives sur l'excitabilité électrique des nerfs traitée comme une polarisation. *J. Physiol. Pathol. Gen.*, 9:620–635, 1907.

[82] J. E. Lisman. A mechanism for memory storage insensitive to molecular turnover: a bistable autophosphorylating kinase. *P. N. A. S. USA*, 82:3055–3057, 1985.

[83] J. E. Lisman, J.-M. Fellous, and X.-J. Wang. A role for NMDA-receptor channels in working memory. *Nat. Neurosci.*, 1:273–275, 1998.

[84] J. E. Lisman and A. M. Zhabotinsky. A model of synaptic memory: a CaMKII/PP1 switch that potentiates transmission by organizing an AMPA receptor anchoring assembly. *Neuron*, 31:191–201, 2001.

[85] W. A. Little. The existence of persistent states in the brain. *Math. Biosci.*, 19:101–119, 1974.

[86] R. Lorente de Nó. Vestibulo-ocular reflex arc. *Arch. Neurol. Psych.*, 30:245–291, 1933.

[87] M. Mézard, J.-P. Nadal, and G. Toulouse. Solvable models of working memories. *J. Physique*, 47:1457–, 1986.

[88] H. Markram, J. Lubke, M. Frotscher, A. Roth, and B. Sakmann. Physiology and anatomy of synaptic connections between thick tufted pyramidal neurones in the developing rat neocortex. *J. Physiol. (London)*, 500:409–440, 1997.

[89] A. Mason, A. Nicoll, and K. Stratford. Synaptic transmission between individual pyramidal neurons of the rat visual cortex *in vitro*. *J. Neurosci.*, 11:72–84, 1991.

[90] C. Meunier, H.-F. Yanai, and S. Amari. Sparsely coded associative memories: capacity and dynamical properties. *Network*, 2:469–487, 1991.

[91] M. Mézard, G. Parisi, and M. A. Virasoro. *Spin Glass Theory and beyond*. World Scientific: Singapore, 1987.

[92] E. K. Miller, C. A. Erickson, and R. Desimone. Neural mechanisms of visual working memory in prefrontal cortex of the macaque. *J. Neurosci.*, 16:5154–5167, 1996.

[93] M. Minsky and S. Papert. *Perceptrons: An Introduction to Computational Geometry*. MIT Press, Cambridge, Ma, 1969.

[94] Y. Miyashita. Neuronal correlate of visual associative long-term memory in the primate temporal cortex. *Nature*, 335:817–820, 1988.

[95] Y. Miyashita and H. S. Chang. Neuronal correlate of pictorial short-term memory in the primate temporal cortex. *Nature*, 331:68–70, 1988.

[96] G. Mongillo, D. J. Amit, and N. Brunel. Retrospective and prospective persistent activity induced by Hebbian learning in a recurrent cortical network. *Eur J Neurosci*, 18:2011–24, 2003.

[97] R. Moreno and N. Parga. Role of synaptic filtering on the firing response of simple model neurons. *Phys. Rev. Lett.*, 92:028102, 2004.

[98] J.-P. Nadal. Associative memory: on the (puzzling) sparse coding limit. *J. Phys. A: Math. Gen.*, 24:1093, 1991.

[99] J.-P. Nadal and G. Toulouse. Information storage in sparsely-coded memory nets. *Network*, 1:61–74, 1990.

[100] K. Nakamura and K. Kubota. Mnemonic firing of neurons in the monkey temporal pole during a visual recognition memory task. *J. Neurophysiol.*, 74:162–178, 1995.

[101] Y. Naya, K. Sakai, and Y. Miyashita. Activity of primate inferotemporal neurons related to a sought target in pair-association task. *Proc. Natl. Acad. Sci. USA*, 93:2664–2669, 1996.

[102] Y. Naya, M. Yoshida, and Y. Miyashita. Backward spreading of memory-retrieval signal in the primate temporal cortex. *Science*, 291:661–4, 2001.

[103] Y. Naya, M. Yoshida, and Y. Miyashita. Forward processing of long-term associative memory in monkey inferotemporal cortex. *J Neurosci*, 23:2861–2871, 2003.

[104] G. Palm. On associative memory. *Biol. Cybern.*, 36:19–31, 1980.

[105] G. Parisi. A memory which forgets. *J. Phys. A: Math. Gen.*, 19:L617, 1986.

[106] B. Pesaran, J. S. Pezaris, M. Sahani, P. P. Mitra, and R. A. Andersen. Temporal structure in neuronal activity during working memory in macaque parietal cortex. *Nat. Neurosci.*, 5:805–811, 2002.

[107] C. C. Petersen, R. C. Malenka, R. A. Nicoll, and J. J. Hopfield. All-or-none potentiation at CA3-CA1 synapses. *Proc.Natl.Acad.Sci.USA*, 95:4732–4737, 1998.

[108] C. J. Ploner, B. Gaymard, S. Rivaud, Y. Agid, and C. Pierrot-Deseilligny. Temporal limits of spatial working memory in humans. *Eur. J. Neurosci.*, 10:794–797, 1998.

[109] G. Rainer, W. F. Assad, and E. K. Miller. Memory fields of neurons in the primate prefrontal cortex. *Proc Natl Acad Sci (USA)*, 95:15008–15013, 1998.

[110] G. Rainer, S. C. Rao, and E. K. Miller. Prospective coding for objects in primate prefrontal cortex. *J. Neurosci.*, 19:5493–5505, 1999.

[111] S. G. Rao, G. V. Williams, and P. S. Goldman-Rakic. Destruction and creation of spatial tuning by disinhibition: $GABA_A$ blockade of prefrontal cortical neurons engaged by working memory. *J. Neurosci.*, 20:485–494, 2000.

[112] A. Renart. *Multi-modular memory systems*. PhD thesis, Universidad Autónoma de Madrid, 2000.

[113] A. Renart, N. Brunel, and X-J. Wang. *Mean-field theory of recurrent cortical networks: from irregularly spiking neurons to working memory*, chapter 15, pages 431–490. CRC Press, Boca Raton, 2003.

[114] A. Renart, R. Moreno, and X. J. Wang. Robust spatial working memory through homeostatic synaptic scaling in heterogenous cortical networks. *Neuron*, 38:473–485, 2003.

[115] A. Renart, P. Song, and X. J. Wang. Bistability in balanced recurrent networks. unpublished ms, 2004.

[116] L. M. Ricciardi. *Diffusion processes and Related topics on biology*. Springer-Verlag, Berlin, 1977.

[117] E. T. Rolls and A. Treves. *Neural Networks and Brain Function*. Oxford University Press, 1998.

[118] R. Romo, C. D. Brody, A. Hernández, and L. Lemus. Neuronal correlates of parametric working memory in the prefrontal cortex. *Nature*, 399:470–474, 1999.

[119] F. Rosenblatt. *Principles of neurodynamics*. Spartan Books, New York, 1962.

[120] K. Sakai and Y. Miyashita. Neural organization for the long-term memory of paired associates. *Nature*, 354:152–155, 1991.

[121] A. Samsonovich and B. L. McNaughton. Path integration and cognitive mapping in a continuous attractor neural network model. *Journal of Neuroscience*, 17:5900–5920, 1997.

[122] M. V. Sanchez-Vives and D. A. McCormick. Cellular and network mechanisms of rhythmic recurrent activity in neocortex. *Nat Neurosci*, 3:1027–34, 2000.

[123] H. S. Seung. How the brain keeps the eyes still. *Proc Natl Acad Sci USA*, 93:13339–13344, 1996.

[124] H. S. Seung, D. D. Lee, B. Y. Reis, and D. W. Tank. Stability of the memory of eye position in a recurrent network of conductance-based model neurons. *Neuron*, 26:259–271, 2000.

[125] Y. Shu, A. Hasenstaub, and D. A. McCormick. Turning on and off recurrent balanced cortical activity. *Nature*, 423:288–93, 2003.

[126] H. Sompolinsky. Neural networks with nonlinear synapses and a static noise. *Phys. Rev. A*, 34:2571–2574, 1986.

[127] A. Treves. Graded-response neurons and information encodings in auto-associative memories. *Phys. Rev. A*, 42:2418–, 1990.

[128] A. Treves. Mean-field analysis of neuronal spike dynamics. *Network*, 4:259–284, 1993.

[129] A. Treves and E. T. Rolls. Computational constraints suggest the need for two distinct input systems to the hippocampal CA3 network. *Hippocampus*, 2:625–, 1992.

[130] M. Tsodyks. Associative memory in neural networks with binary synapses. *Mod. Phys. Lett. B*, 4:713–, 1990.

[131] M. Tsodyks and M. V. Feigel'man. The enhanced storage capacity in neural networks with low activity level. *Europhys. Lett.*, 46:101–, 1989.

[132] H. C. Tuckwell. *Introduction to theoretical neurobiology.* Cambridge: Cambridge University Press, 1988.

[133] G. G. Turrigiano and S. B. Nelson.   Hebb and homeostasis in neuronal plasticity. *Curr. Opin. Neurobiol.*, 10:358–364, 2000.

[134] C. van Vreeswijk and H. Sompolinsky. Chaos in neuronal networks with balanced excitatory and inhibitory activity. *Science*, 274:1724–1726, 1996.

[135] C. van Vreeswijk and H. Sompolinsky. Chaotic balanced state in a model of cortical circuits. *Neural Computation*, 10:1321–1371, 1998.

[136] X.-J. Wang. Synaptic basis of cortical persistent activity: the importance of NMDA receptors to working memory. *J. Neurosci.*, 19:9587–9603, 1999.

[137] X.-J. Wang. Synaptic reverberation underlying mnemonic persistent activity. *Trends Neurosci.*, 24:455–463, 2001.

[138] G. Weisbuch and F. Fogelman-Soulié. Scaling laws for the attractors of hopfield networks. *Journal de Physique-lettres*, 46:623–630, 1985.

[139] J. M. White, D. L. Sparks, and T. R. Stanford. Saccades to remembered target locations: and analysis of systematic and variable errors. *Vision Res.*, 34:79–92, 1994.

[140] G. V. Williams and P. S. Goldman-Rakic. Modulation of memory fields by dopamine D1 receptors in prefrontal cortex. *Nature*, 376:572–575, 1995.

[141] G. V. Williams, S. G. Rao, and P. S. Goldman-Rakic. The physiological role of 5-HT2A receptors in working memory. *J Neurosci*, 22:2843–54, 2002.

[142] D. Willshaw, O. P. Buneman, and H. Longuet-Higgins. Non-holographic associative memory. *Nature*, 222:960–962, 1969.

[143] F. A. W. Wilson, S. P. Ó Scalaidhe, and P. S. Goldman-Rakic. Dissociation of object and spatial processing domains in primate prefrontal cortex. *Science*, 260:1955–1958, 1993.

[144] V. Yakovlev, S. Fusi, E. Berman, and E. Zohary. Inter-trial neuronal activity in inferior temporal cortex: a putative vehicle to generate long-term visual associations. *Nature Neurosci.*, 1:310–317, 1998.

[145] A. M. Zhabotinsky. Bistability in the $Ca^{2+}$/calmodulin-dependent protein kinase-phosphatase system. *Biophys. J.*, 79:2211–2221, 2000.

[146] Special issue: Persistent neural activity: Experiments and theory. *Cerebral Cortex*, 13:1123–1269, 2003.

Course 11

# PATTERN FORMATION IN VISUAL CORTEX

## Paul C. Bressloff

*Department of Mathematics, University of Utah, 155 S 1400 E, Salt Lake City, Utah 84112*

This work is supported by NSF grant DMS-0209824

*C.C. Chow, B. Gutkin, D. Hansel, C. Meunier and J. Dalibard, eds.*
*Les Houches, Session LXXX, 2003*
*Methods and Models in Neurophysics*
*Méthodes et Modèles en Neurophysique*

# Contents

## 1. Introduction

When studying the large-scale functional and anatomical structure of cortex two distinct questions naturally arise: (I) how did the structure develop? and (II) what forms of spontaneous and stimulus-driven neural dynamics are generated by such a cortical structure? It turns out that in both cases the Turing mechanism for spontaneous pattern formation plays an important role. Turing originally considered the problem of how animal coat patterns develop, suggesting that chemical markers in the skin comprise a system of diffusion-coupled chemical reactions among substances called morphogens [98]. He showed that in a two-component reaction-diffusion system, a state of uniform chemical concentration can undergo a diffusion-driven instability leading to the formation of a spatially inhomogeneous state. Ever since the pioneering work of Turing on morphogenesis [98], there has been a great deal of interest in spontaneous pattern formation in physical and biological systems [28, 71]. In the neural context, Wilson and Cowan [105] proposed a non-local version of Turing's diffusion–driven mechanism, based on competition between short-range excitation and longer-range inhibition. Here interactions are mediated, not by molecular diffusion, but by long-range axonal connections. Since then this neural version of the Turing instability has been applied to a number of problems concerning the dynamics and development of cortex. Examples in visual neuroscience include the Marr–Poggio model of stereopsis [67], developmental models of retinotopic, ocular dominance and iso–orientation maps [48, 88, 89, 100, 101], the ring model of orientation tuning [5, 86] and cortical models of geometric visual hallucinations [14, 34]. In most cases there exists some underlying symmetry in the model that plays a crucial role in the selection and stability of the resulting patterns.

In these lectures we review various theoretical approaches to studying spontaneous pattern formation in non–local neural models. Throughout we emphasize the important role that symmetries play. For concreteness, we focus our discussion on activity–based patterns generated in primary visual cortex (V1), exploring the links between spontaneous cortical activity and the underlying functional architecture of V1. Such a link has recently been indicated experimentally using single-unit recordings and real-time optical imaging, where it was found that very similar spatial patterns of ongoing population activity occur both when a neuron fires spontaneously and when it is driven by its optimal stimulus [97]. As origi-

481

Fig. 1. Periodic activity patterns in visual cortex consisting of alternating regions of low and high activity.

nally proposed by Ermentrout and Cowan [34], spontaneous activity patterns in visual cortex could also provide an explanation for the occurrence of certain basic types of geometric visual hallucinations. In Fig. 1 we show some examples of cortical activity patterns generated by the Ermentrout–Cowan model, consisting of alternating regions of low and high cortical activity that cover the cortical plane in regular stripes, squares, rhomboids or hexagons. The corresponding images that would be generated by mapping such activity patterns back into visual field coordinates are shown in Fig. 2. Such images bear a striking resemblance to some of the so–called form constants or basic hallucinations classified by the Chicago neurologist Kluver [59].

One of the basic features of the cortical patterns shown in Fig. 1 is that they form a regular tiling of the cortical plane, that is, they are doubly–periodic with respect to some regular planar lattice (square, rhomboid or hexagonal). This is a common property of pattern forming instabilities in systems with Euclidean symmetry that are operating in the weakly nonlinear regime [28]. In the neural context, Euclidean symmetry reflects the invariance of synaptic interactions with respect to rotations, translations and reflections in the cortical plane. The emerg-

Fig. 2. Corresponding hallucinatory images generated using the inverse retinocortical map

ing patterns spontaneously break Euclidean symmetry down to the discrete symmetry group of the lattice, and this allows techniques from bifurcation theory to be used to analyze the selection and stability of the patterns. The global position and orientation of the patterns are still arbitrary, however, reflecting the hidden Euclidean symmetry.

The Ermentrout–Cowan theory of visual hallucinations is over–simplified in the sense that V1 is represented as if it were effectively a cortical version of the retina, in which cells only signal the level of local light illumination. However, most V1 cells are selective to a variety of features of a local stimulus such as its orientation, whether it originates from the left or right eye (ocular dominance), color etc. Moreover, these feature preferences are distributed in an approximately periodic fashion across cortex forming a set of overlapping feature maps. Such feature maps are also correlated with the distribution of long–range patchy horizontal connections within V1. A common computational representation of the action of V1 is that it decomposes a visual image into a set of local edges as illustrated in Fig. 3. Interestingly, there are also a number of basic hallucinatory images that appear to be ordered versions of these edge-like structures, one of which is shown in Fig. 3b. One interpretation of the contoured hallucination is

Fig. 3. (a) Under normal conditions the visual cortex analyzes an image by breaking it up into various local features such as edges. (b) The spontaneous formation of a cortical activity pattern that is commensurate with the underlying orientation feature map could lead to an ordered lattice of oriented edges during an hallucination.

that it corresponds to an activity pattern in cortex that is somehow commensurate with the underlying orientation feature map. Such commensurability suggests that the position of the activity pattern is no longer arbitrary, in contrast to the Euclidean symmetric theory of Ermentrout and Cowan. An extended theory of hallucinations that takes into account the presence of periodic feature maps underlying the functional architecture of cortex has been developed in a series of recent papers [14–17, 19, 22].

The lectures are organized as follows. In section 2 we review the functional architecture of V1, emphasizing the correlations between the long–range recurrent circuitry and the various feature maps. In section 3 we describe a number of large–scale continuum models of V1 that take into account such correlations. In section 4 we introduce some of the basic methods for analyzing neural pattern formation including perturbation methods, the Fredholm alternative, amplitude equations, symmetric bifurcation theory, and mean field theory. We illustrate these techniques by considering the ring model of orientation tuning in a cortical hypercolumn. In sections 5 and 6 we extend our analysis to continuum models of V1 and show how the symmetry properties of the long–range horizontal connections are reflected by the cortical activity patterns. We conclude in section 7 by discussing an example of pattern formation in a model of cortical development. Here the pattern is in the distribution of feedforward synaptic weights rather than in neural activity.

## 2. The functional architecture of V1

The primary visual cortex (V1) is the first cortical area to receive visual information from the retina (see Fig. 4). The output from the retina is conveyed by

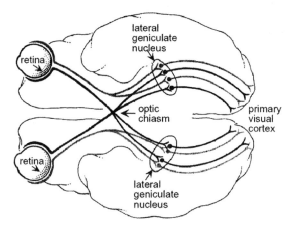

Fig. 4. Visual pathways from the retina through the lateral geniculate nucleus (LGN) of the thalamus to the primary visual cortex (V1)

ganglion cells whose axons form the optic nerve. The optic nerve conducts the output spike trains of the retinal ganglion cells to the lateral geniculate nucleus (LGN) of the thalamus, which acts as a relay station between retina and primary visual cortex (V1). Prior to arriving at the LGN, some ganglion cell axons cross the midline at the optic chiasm. This allows the left and right sides of the visual fields from both eyes to be represented on the right and left sides of the brain, respectively. Note that signals from the left and right eyes are segregated in the LGN and in input layers of V1. This means that the corresponding LGN and cortical neurons are monocular, in the sense that they only respond to stimuli presented to one of the eyes but not the other (ocular dominance).

### 2.1. Retinotopic map

One of the striking features of the visual system is that the visual world is mapped onto the cortical surface in a topographic manner. This means that neighboring points in a visual image evoke activity in neighboring regions of visual cortex. Moreover, one finds that the central region of the visual field has a larger representation in V1 than the periphery, partly due to a non–uniform distribution of

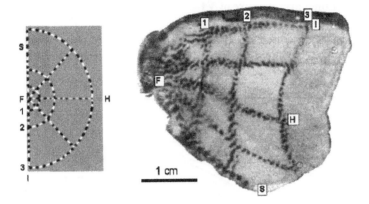

Fig. 5. An autoradiograph from the primary visual cortex in the left side of a macaque monkey brain. The pattern is a radioactive trace of the activity evoked by the image shown to the left. Adapted from [93].

retinal ganglion cells. The retinotopic map is defined as the coordinate transformation from points in the visual world to locations on the cortical surface. In order to describe this map, we first need to specify visual and cortical coordinate systems. Since objects located a fixed distance from one eye lie on a sphere, we can introduce spherical coordinates with the "north pole" of the sphere located at the fixation point, the image point that focuses onto the fovea or center of the retina. In this system of coordinates, the latitude angle is called the eccentricity $\epsilon$ and the longitudinal angle measured from the horizontal meridian is called the azimuth $\varphi$. In most experiments the image is on a flat screen such that, if we ignore the curvature of the sphere, the pair $(\epsilon, \varphi)$ approximately coincides with polar coordinates on the screen. One can also represent points on the screen using Cartesian coordinates $(X, Y)$. In primary visual cortex the visual world is split in half with the region $-90^o \leq \varphi \leq 90^o$ represented on the left side of the brain, and the reflection of this region represented on the right side brain. Note that the eccentricity $\epsilon$ and Cartesian coordinates $(X, Y)$ are all based on measuring distance on the screen. However, it is customary to divide these distances by the distance from the eye to the screen so that they are specified in terms of angles. The structure of the retinotopic map in monkey is shown in Fig. 5, which was produced by imaging a radioactive tracer that was taken up by active neurons while the monkey viewed a visual image consisting of concentric circles and radial lines. The fovea is represented by the point F on the left hand side of the cortex, and eccentricity increases to the right. Note that concentric circles are approximately mapped to vertical lines and radial lines to horizontal lines.

Motivated by Fig. 5, we assume that eccentricity is mapped on to the horizontal coordinate $x$ of the cortical sheet, and $\varphi$ is mapped on to its $y$ coordinate An approximate equation for the retinotopic map can then be obtained through specification of a quantity known as the cortical magnification factor $M(\epsilon)$. This determines the distance across a flattened sheet of cortex separating the activity evoked by two nearby image points. First suppose that the two image points in question have eccentricities $\epsilon$ and $\epsilon + \Delta\epsilon$ but the same azimuthal coordinate $\varphi$. The corresponding distance on cortex is $\Delta x = M(\epsilon)\Delta\epsilon$ so that

$$\frac{dx}{d\epsilon} = M(\epsilon) \tag{2.1}$$

Using experimental data such as shown in Fig. 5 suggests that

$$M(\epsilon) = \frac{\lambda}{\epsilon_0 + \epsilon} \tag{2.2}$$

with $\lambda \approx 12\ mm$ and $\epsilon_0 \approx 1^o$ in macaque monkey. It follows that

$$x = \lambda \ln(1 + \epsilon/\epsilon_0) \tag{2.3}$$

assuming $x = 0$ when $\epsilon = 0$. Similarly, for two image points with the same eccentricity but different azimuthal coordinates we find that

$$\frac{dy}{d\epsilon} = -\frac{\epsilon\pi}{180^o}M(\epsilon) \tag{2.4}$$

and hence

$$y = -\frac{\lambda\epsilon a\pi}{(\epsilon_0 + \epsilon)180^0} \tag{2.5}$$

The minus sign appears because the visual field is inverted in cortex. For eccentricities greater than $1^o$,

$$x \approx \lambda \ln(\epsilon/\epsilon_0), \quad y \approx -\frac{\lambda\pi\varphi}{180^o} \tag{2.6}$$

and the retinotopic map can be approximated by a complex logarithm [80]. That is, introducing the complex representations $Z = (\epsilon/\epsilon_0)e^{-i\pi\varphi/180^o}$ and $z = x + iy$ then $z = \lambda \log Z$.

## 2.2. Feature maps

Superimposed upon the retinotopic map are additional maps reflecting the fact that neurons respond preferentially to stimuli with particular features [90]. Neurons in the retina, LGN and primary visual cortex respond to light stimuli in restricted regions of the visual field called their classical receptive fields (RFs). Patterns of illumination outside the RF of a given neuron cannot generate a response directly, although they can significantly modulate responses to stimuli within the RF via long–range cortical interactions (see below). The RF is divided into distinct ON and OFF regions. In an ON (OFF) region illumination that is higher (lower) than the background light intensity enhances firing. The spatial arrangement of these regions determines the selectivity of the neuron to different stimuli. For example, one finds that the RFs of most V1 cells are elongated so that the cells respond preferentially to stimuli with certain preferred orientations (see Fig. 6). The RFs of retinal ganglion neurons and LGN neurons, on the other hand, are circularly symmetric and hence these neurons do not exhibit any stimulus orientation preference.

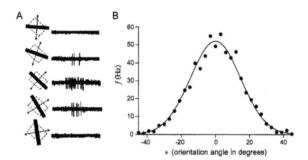

Fig. 6. (A) Recordings from a neuron in the primary visual cortex of a monkey in response to a bar of light moved across the receptive field of the cell (shown by the dashed rectangle) at different angles. (B) Average firing rate of a cat V1 neuron plotted as a function of the orientation of the bar stimulus. The peak of the orientation tuning curve corresponds to the orientation preference of the cell

In recent years much information has accumulated about the spatial distribution of orientation selective cells in V1 [43]. In Fig. 7 is given a typical arrangement of such cells, obtained *via* microelectrodes implanted in cat V1. The first panel shows how orientation preferences rotate smoothly over the surface of V1, so that approximately every $300\mu m$ the same preference reappears, i.e. the distribution is $\pi$–periodic in the orientation preference angle. The second panel shows

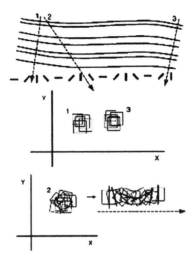

Fig. 7. Orientation tuned cells in layers of cat V1 which is shown in cross-section. Note the constancy of orientation preference at each cortical location [electrode tracks 1 and 3], and the rotation of orientation preference as cortical location changes [electrode track 2]. Redrawn from [43].

the receptive fields of the cells, and how they change with V1 location. The third panel shows more clearly the rotation of such fields with translation across V1. One also finds that cells with similar feature preferences tend to arrange themselves in vertical columns so that to a first approximation the layered structure of cortex can be ignored. For example, electrode track 1 in Fig. 7 is a vertical penetration of cortex that passes through a single column of cells with the same orientation preference and ocular dominance. The situation regarding orientation columns in macaque V1 is more complicated [63, 64]. For example, input layer 4 has an additional sublaminar structure that reflects amongst other things the division of the LGN afferents into parvocellular (P) and magnocellular (M) pathways (see Fig. 8). One finds that many cells in layer $4C\beta$ are not orientation selective: orientation preference emerges in a graded fashion as one moves to mid and upper layer $4C\alpha$. The $M$ pathway is thought to contribute primarily to motion perception, whereas the $P$ pathway contributes primarily to form and color perception. However, there is some mixing of the two pathways.

A more complete picture of the two–dimensional distribution of both orientation preference and ocular dominance in layers 2/3 has been obtained using optical imaging techniques [7–9]. The basic experimental procedure involves

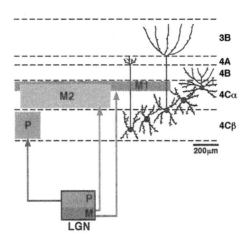

Fig. 8. The parvocellular (P) and magnocellular (M) pathways of macaque V1. The *P* pathway innervates layer $4C\beta$, consisting of cells with smaller receptive fields and slower responses. These send axons to layer $4A$ which feeds into the CO blob regions of superficial layers 2/3. The *M* pathway innervates layer $4C\alpha$, consisting of cells that have larger receptive fields and faster responses. Upper $4C\alpha$ cells connect to layer $4B$ which itself connects to the blob regions of layers 2/3. Mid-layer $4C$ has a mixture of M and P neurons and connects to the interblob regions of layers 2/3.

shining light directly on to the surface of the cortex. The degree of light absorption within each patch of cortex depends on the local level of activity. Thus, when an oriented image is presented across a large part of the visual field, the regions of cortex that are particularly sensitive to that stimulus will be differentiated. The topography revealed by these methods has a number of characteristic features [74], see Fig. 9: (i) Orientation preference changes continuously as a function of cortical location, except at singularities or *pinwheels*. (ii) There exist *linear zones*, approximately $750 \times 750 \ \mu m^2$ in area (in Macaque), bounded by pinwheels, within which iso–orientation regions form parallel slabs. (iii) Linear zones tend to cross the borders of ocular dominance stripes at right angles; pinwheels tend to align with the centers of ocular dominance stripes. (iv) Approximately half the pinwheels have a strong association with the cytochrome oxidase (CO) *blobs*, which are regions of cortical cells that are more metabolically active and hence richer in their levels of CO [50]; cells in the CO blobs tend to be color selective and have a preference for low spatial frequencies. These experimental findings suggest that there is an underlying periodicity in the mi-

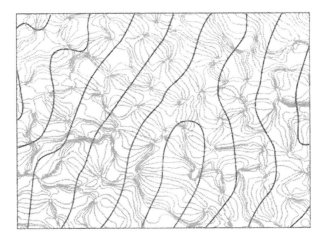

Fig. 9. Iso–orientation (light) and ocular dominance (dark) contours in a region of Macaque V1. Redrawn from [8]. A cortical hypercolumn consisting of two orientation singularities or pinwheels per ocular dominance column is indicated by the rectangular region.

crostructure of V1 with a period of approximately $1mm$ (in cats and primates). The fundamental domain of this periodic tiling of the cortical plane is the hyper-column [53,54,60], which contains two sets of orientation preferences $\theta \in [0, \pi)$ per eye, organized around a pair of singularities (see Fig. 9). The distribution of spatial frequency preference across cortex is less clear than that of orientation preference. However early anatomical studies in macaque established that there is a cortical map for spatial frequency in which low spatial frequency preference regions tend to coincide with CO blobs [31,92]. Intriguingly, recent optical imaging data concerning spatial frequency maps in cat suggest that those orientation singularities that do not coincide with CO blobs correspond to regions of high spatial frequency [10,55,56].

## 2.3. Long–range horizontal and feedback connections

Given the existence of a regularly repeating set of feature preference maps, how does such a periodic structure manifest itself anatomically? Two cortical cir-cuits have been fairly well characterized: There is a local circuit operating at sub–millimeter dimensions in which cells make connections with most of their neighbors in a roughly isotropic fashion. It has been suggested that such circuitry provides a substrate for the recurrent amplification and sharpening of the tuned response of cells to local visual stimuli. The other circuit operates between hyper-columns, connecting cells separated by several millimetres of cortical tissue. The axons of these connections make terminal arbors only every 0.7 mm or so along their tracks [42, 78], such that local populations of cells are reciprocally connected in a patchy fashion to other cell populations. Optical imaging combined

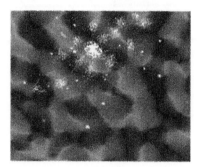

Fig. 10. Lateral Connections made by a cells in Tree Shrew (Left panel) and owl monkey (Right panel) V1. A radioactive tracer is used to show the locations of all terminating axons from cells in a central injection site, superimposed on an orientation map obtained by optical imaging. (Patches with the same coarse-grained orientation preference are shown in the same color – this is purely for visualization purposes). The patchy distribution of the lateral connections is clearly seen, linking regions of like orientation preference along a particular visuotopic axis. The local axonal field, on the other hand, is isotropic and connects all neurons within a neighborhood ($\approx$ 0.7 mm) of the injection site. Redrawn from [12] and [85].

with labelling techniques has generated considerable information concerning the pattern of these connections in superficial layers of V1 [12, 65, 107], see Fig. 10. In particular, one finds that the patchy horizontal connections tend to link cells with similar feature preferences. Moreover, in tree shrew and cat there is a pronounced anisotropy in the distribution of patchy connections, with differing iso–orientation patches preferentially connecting to neighboring patches in such a way as to form continuous contours following the topography of the retino–cortical map [12]. That is, the major axis of the horizontal connections tends to run parallel to the visuotopic axis of the connected cells' common orientation preference. There is also a clear anisotropy in the patchy connections of owl [85] and macaque [1] monkeys. However, in these cases most of the anisotropy can be accounted for by the fact that V1 is expanded in the direction orthogonal to ocular dominance columns [1]. It is possible that when this expansion is factored out, there remains a weak anisotropy correlated with orientation selectivity but this remains to be confirmed experimentally. (Interestingly, the recently observed patchy feedback connections from extrastriate areas in macaque tend to be more strongly anisotropic [1]). Stimulation of a hypercolumn via lateral connections modulates rather than initiates spiking activity [49, 94], suggesting that the long-range interactions provide local cortical processes with contextual information about the global nature of stimuli. As a consequence the horizontal connections have been invoked to explain a wide variety of context-dependent visual processing phenomena [38, 44].

Finally, there are extensive feedforward and feedback connections linking V1 to extrastriate areas such as V2, V3 and MT [24, 79]. As one proceeds to higher cortical areas the receptive field size of neurons increases. Hence, feedback from these areas provides information to V1 neurons from a much larger region of visual space than expected from the classical feedforward receptive fields. It is also important to note that the feedforward and feedback connections are layer specific and reciprocal. Ongoing studies of feedback connections from points in extrastriate areas back to area V1 [1, 2], show that the feedback connectional fields are also distributed in highly regular geometric patterns, having a topographic spread of up to 13mm that is significantly larger than the spread of intrinsic lateral connections. It is likely that the patchiness again signifies that feedback correlates cells with similar feature preferences [83].

## 3. Large–scale models of V1

### 3.1. Planar model of V1

Wilson and Cowan [104, 105] introduced a population model of cortical tissue based on the idea that the identity of individual presynaptic neurons is not important, only the distribution of their level of activity. This leads to a statistical description of cortical activity in which the proportion of active excitatory and inhibitory neurons are chosen as model variables. A justification for a statistical approach can be given in terms of the spatial clustering of neurons with similar response properties within cortical columns. Neurons within a column share many inputs and are tightly interconnected, so that some form of local spatial or population averaging appears reasonable. However, in order to construct a closed set of equations for population activity, it is necessary that the total input into a population is slowly varying relative to the time-scale of action potential generation [41]. Hence, an implicit assumption of most population models of cortex is that neurons within a population have spike trains that are temporally incoherent at the millisecond timescale. Another consequence of averaging over cortical columns is that the cortex is reduced to a two–dimensional neural medium. However, it is important to note that there can be functional differences between neurons in distinct layers within the same column as illustrated in Fig. 8), so that certain care has to be taken in ignoring the third dimension [46,63,64]. Given that optical imaging studies measure the response properties of superficial layers of cortex, we shall consider a model based on the structure of these layers.

Consider a two–dimensional, unbounded cortical sheet. Let $a_E(\mathbf{r}, t)$ be the activity of excitatory neurons in a given volume element of a slab of neural tissue located at $\mathbf{r} \in \mathbf{R}^2$, and $a_I(\mathbf{r}, t)$ be the corresponding activity of inhibitory neurons. The neural fields $a_E, a_I$ are taken to evolve according to the following voltage–based population equations (see [36] for a discussion of different versions of population models):

$$
\begin{aligned}
\tau_E \frac{\partial a_E(\mathbf{r}, t)}{\partial t} &= -a_E(\mathbf{r}, t) + \int_{\mathbf{R}^2} w_{EE}(\mathbf{r}|\mathbf{r}') f_E[a_E(\mathbf{r}', t)] d^2\mathbf{r}' \\
&\quad + \int_{\mathbf{R}^2} w_{EI}(\mathbf{r}|\mathbf{r}') f_I[a_I(\mathbf{r}', t)] d^2\mathbf{r}' + h(\mathbf{r}, t) \\
\tau_I \frac{\partial a_I(\mathbf{r}, t)}{\partial t} &= -a_I(\mathbf{r}, t) + \int_{\mathbf{R}^2} w_{IE}(\mathbf{r}|\mathbf{r}') f_E[a_E(\mathbf{r}', t)] d^2\mathbf{r}' \\
&\quad + \int_{\mathbf{R}^2} w_{II}(\mathbf{r}|\mathbf{r}') f_I[a_I(\mathbf{r}', t)] d^2\mathbf{r}'
\end{aligned}
\tag{3.1}
$$

where $w_{ij}(\mathbf{r}|\mathbf{r}')$ is the weight per unit volume of all synapses to the $i$th population at $\mathbf{r}$ from neurons of the $j$th population at $\mathbf{r}'$, $h$ is the feedforward (excitatory) input from the LGN or other cortical layers, and $\tau_{E,I}$ are synaptic time constants. The nonlinearities $f_E$ and $f_I$ are taken to be smooth output functions of the form

$$
f_j(a) = \frac{1}{1 + e^{-\eta_j(a - \kappa_j)}}, \quad j = E, I
\tag{3.2}
$$

where $\eta_j$ determines the slope or sensitivity of the input–output characteristics of the population and $\kappa_j$ is a threshold. Note that $w_{jE} \geq 0$ whereas $w_{jI} \leq 0$.

From a mathematical viewpoint, it is often convenient to reduce the above two-population model to an effective one-population model. In particular, suppose that $\tau_I \ll \tau_E$, $w_{II} = 0$ and $f_I$ is a linear function. We can then eliminate $a_I$ in terms of $a_E$ such that

$$
\tau \frac{\partial a(\mathbf{r}, t)}{\partial t} = -a(\mathbf{r}, t) + \int_{\mathbf{R}^2} w(\mathbf{r}|\mathbf{r}') f[a(\mathbf{r}', t)] d\mathbf{r}' + h(\mathbf{r}, t)
\tag{3.3}
$$

where we have dropped the index $E$, and set

$$
w(\mathbf{r}|\mathbf{r}') = w_{EE}(\mathbf{r}|\mathbf{r}') + \int w_{EI}(\mathbf{r}|\mathbf{r}'') w_{IE}(\mathbf{r}|\mathbf{r}'') d^2\mathbf{r}''
\tag{3.4}
$$

However, even the two–population model is an over simplification, since it lumps together a number of different types of interneuron, for example, interneurons with horizontally distributed axonal fields of the basket cell subtype and inhibitory interneurons that have predominantly vertically aligned axonal fields.

It is likely that these different classes of interneuron serve different dynamical and computational roles [63].

It remains to determine the form of the feedforward input distribution $h$ in response to a visual stimulus and to specify the form of the synaptic weight distributions $w_{ij}$. We will show how both depend on a set of regularly repeating feature maps that underly the functional architecture of V1, see section 2.

### 3.2. Receptive fields and the feedforward input h

A visual stimulus is typically described in terms of a function $s(X, Y, t)$ that is proportional to the difference between the luminance at point $(X, Y)$ in the visual field at time $t$ and the average or background level of luminance (since the visual system adapts to the background illumination). Often $s$ is divided by the background luminance level, making it a dimensionless quantity called the contrast. Suppose that the neurons at cortical position $\mathbf{r}$ have a set of RF properties $\mathcal{F}(\mathbf{r})$. Assuming a linear relationship between the feedforward input $h$ to the neuron and the stimulus $s$, we take

$$h(\mathbf{r}, t) = h(\mathcal{F}(\mathbf{r}), \bar{X}(\mathbf{r}), \bar{Y}(\mathbf{r}), t) \tag{3.5}$$

with

$$h(\mathcal{F}, \bar{X}, \bar{Y}, t) = \int_0^\infty \int_{\mathbf{R}^2} D(X - \bar{X}, Y - \bar{Y}, \tau | \mathcal{F}) s(X, Y, t - \tau) dX dY d\tau \tag{3.6}$$

where $D$ is the space–time RF profile of the neuron and $(\bar{X}(\mathbf{r}), \bar{Y}(\mathbf{r}))$ is the RF center in visual coordinates. (Neurons that carry out a linear RF summation are termed simple cells, whereas neurons with nonlinear RF properties are called complex cells [103]). It is clear that the distribution $h(\mathbf{r})$ depends both on the RF properties of single cells and on how these properties are distributed across cortex. The latter is specified by the retino–cortical map $(\bar{X}(\mathbf{r}), \bar{Y}(\mathbf{r}))$ and the associated feature maps $\mathcal{F}(\mathbf{r})$. Note, however, that some features at the single cell level are distributed randomly across cortex, so that at the population level $h(\mathbf{r})$ should be averaged with respect to these particular properties. One such example for simple cells is spatial phase [68, 82]. In the following we will identify $\mathcal{F}(\mathbf{r})$ with the set of feature maps $(\theta(\mathbf{r}), p(\mathbf{r}), \chi(\mathbf{r}))$ where $\theta(\mathbf{r})$ and $p(\mathbf{r})$ represent the orientation and spatial frequency preference maps and $\chi(\mathbf{r})$ is the (binary–valued) ocular dominance map. [Depending on the cortical layer and the species, other possible feature maps that could be included in $\mathcal{F}$ are those associated with direction of motion, binocular disparity and color].

*Recurrent versus feedforward processing*   The RF profile $D$ is typically measured experimentally using reverse correlation methods [30]. Consider a neuron that is driven by a time dependent stimulus $s$ and suppose that every time a spike occurs, a recording is made of the time course of the stimulus in a time window of about 100 msec immediately before the spike. Averaging the results for several spikes yields the typical time course of the stimulus just before a spike, known as the spike–triggered average. The RF profile $D$ is then identified as the optimal linear filter obtained by measuring the spike-triggered average for a white-noise stimulus that is uncorrelated in both space and time. It is important to note that since the reverse correlation method is based on the output response of a neuron, it does not determine the relative contributions of feedforward and recurrent inputs to the generation of the RF. Only the bare RF generated by feedforward processes should be used in equation (3.6).

The degree to which recurrent process contribute to the RF properties of V1 neurons is a matter of ongoing debate [40, 87]. The classical model of Hubel and Wiesel [52] proposes that the orientation preference and selectivity of a cortical neuron in input layer 4 arises primarily from the geometric alignment of the receptive fields of thalamic neurons in the lateral geniculate nucleus (LGN) projecting to it. (Orientation selectivity is then carried to other cortical layers through vertical projections). This has been confirmed by a number of recent experiments [39, 76]. However, there is also growing experimental evidence suggesting the importance of recurrent cortical interactions in orientation tuning [87]. For example, the blockage of extracellular inhibition in cortex leads to considerably broader tuning [72, 84]. Moreover, intracellular measurements indicate that direct inputs from the LGN to neurons in layer 4 of the visual cortex provide only a fraction of the total excitatory inputs relevant to orientation selectivity [33]. There is also evidence that orientation tuning takes about 50-60 msec to reach its peak, and that the dynamics of tuning has a rather complex time course [77] suggesting some cortical involvement. Another major argument against a purely feedforward model is that it cannot account for the experimental observation that the width of orientation tuning curves are approximately contrast invariant. That is, increasing the strength of the LGN excitatory input would raise more neurons above threshold and thus broaden the tuning curve - the so-called iceberg effect. One way to obtain contrast invariant tuning curves is through strong recurrent interactions. (An alternative mechanism is through some form of *push-pull* cortical inhibition [40, 96]). We shall assume in the following that the feature preferences of a neuron are determined primarily by feedforward inputs, whereas variations in the degree of feature selectivity arise from recurrent processes [87].

*RF of a V1 simple cell*   In experimental recordings from visual neurons, a commonly used stimulus is a sinusoidal grating with orientation $\Theta$, spatial frequency

$P$, and spatial phase $\Psi$. The time variation of the stimulus can be introduced in two distinct ways: for a counterphase grating

$$s(X, Y, t) = A \cos(PX \cos \Theta + PY \sin \Theta - \Psi) \cos(\omega t) \tag{3.7}$$

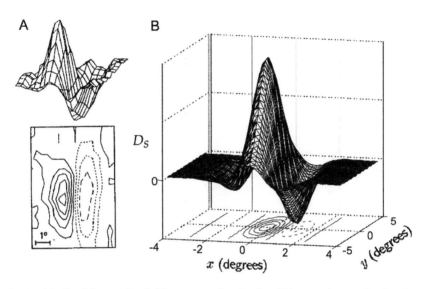

Fig. 11. (a). Spatial receptive field structure of a simple cell in cat primary visual cortex. (b). Gabor function with $\sigma_x = 1^o$, $\sigma_y = 2^o$, $1/p = 0.56^o$ and $\psi = 1 - \pi/2$. Adapted from [57].

whereas for a drifting grating

$$s(X, Y, t) = A \cos(PX \cos \Theta + PY \sin \Theta - \Psi - \omega t) \tag{3.8}$$

We only consider spatial RF properties here by assuming that $D$ is separable, that is, it can be decomposed into the product of space and time contributions

$$D(X, Y, \tau|\mathcal{F}) = D_S(X, Y|\mathcal{F})D_T(\tau) \tag{3.9}$$

Fig. 11 shows the spatial structure of a separable RF for a simple cell in the primary visual cortex of the cat. It can be seen that there are elongated ON and OFF regions corresponding to domains where $D_S$ is positive and negative, respectively. As a result such a neuron responds most vigorously to light-dark edges positioned along the border between the ON and OFF regions, oriented parallel to this border and to the elongated direction of the RFs. Also shown in Fig. 11 is a mathematically generated approximation of the RF based on a Gabor function, which is a product of a Gaussian function and a sinusoidal function [29, 57]:

$$G(X, Y) = \frac{1}{2\pi\sigma_X\sigma_Y} \exp\left(-\frac{X^2}{2\sigma_X^2} - \frac{Y^2}{2\sigma_Y^2}\right) \cos(pX - \psi) \qquad (3.10)$$

Here $\sigma_X$ and $\sigma_Y$ determine the extent of the RF in the $X$ and $Y$ directions; $p$ is the preferred spatial frequency (which determines the spacing of light and dark bars that produce the maximum response) and $\psi$ is the preferred spatial phase (which determines where the ON-OFF boundaries fall within the RF). For this

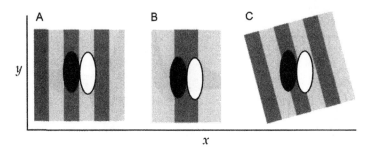

Fig. 12. Grating stimuli superimposed on spatial receptive fields similar to the one shown in Fig. 11. The RF is idealized as two oval regions, with dark representing the OFF region and light the ON region. The grating in (A) is the optimal stimulus whereas those in (B) and (C) are nonoptimal due to a mismatch in spatial frequencies and orientations respectively.

particular RF, the orientation of the light-dark edges that produces the maximal response is parallel to the $Y$-axis, so that the preferred orientation is $0^o$. This is illustrated geometrically in Fig. 12. An arbitrary preferred orientation $\theta$ can be generated by rotating the coordinates according to $X \to X\cos\theta + Y\sin\theta$ and $Y \to Y\cos\theta - X\sin\theta$. The number of subregions within the RF is controlled by the products $p\sigma_X$, $p\sigma_Y$. Increasing these products leads to more subregions and consequently greater spatial frequency selectivity.

*Orientation and spatial frequency dependence of the input h*  We now consider in more detail the orientation and spatial frequency dependence of the stationary input $h$ into a V1 simple cell. The linear RF is taken to be a Gabor function with a fixed orientation bias $\theta$, spatial frequency preference $p$ and zero spatial phase $\psi = 0$. As a further simplification, the receptive field center $(\bar{X}, \bar{Y})$ of the neuron is located at the origin, and $\sigma_x = \sigma_y = \sigma$. Suppose that the stimulus consists of

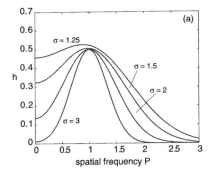 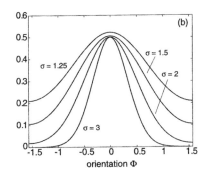

Fig. 13. Linear response $h$ of a Gabor filter to a cosine grating with $\sigma_x = \sigma_y = \sigma$, $p = 1$ and $C = 1$. Figure (a) shows $h$ as a function of stimulus frequency $P$ at the preferred orientation, $\Theta = \theta$ The units of spatial frequency are taken to be cycles/deg. Figure (b) shows $h$ as a function of stimulus orientation $\Theta$ at the preferred spatial frequency $P = p$.

a sinusoidal grating with a fixed spatial frequency $P$, orientation $\Theta$, zero spatial phase $\Psi = 0$ and zero temporal frequency $\omega$, see equation (3.7). The associated stationary input can then be calculated explicitly using equation (3.6):

$$h(\theta, p) = Ce^{-\sigma^2[p^2+P^2]/2}\cosh(\sigma^2 pP\cos[\theta - \Theta]) \tag{3.11}$$

For a given stimulus spatial frequency $P$, it is clear that the maximum response occurs at the stimulus orientation $\Theta$. At the optimal stimulus orientation, the maximum response occurs when $p \approx P$ assuming that the spatial frequency is sufficiently large so that $e^{-\sigma^2 pP} \approx 0$ and hence $h \approx (C/2)e^{-\sigma^2(p-P)^2/2}$. In Fig. 13 we plot the linear response $h$, equation (3.11), as a function of (a) stimulus spatial frequency at the preferred orientation $\Theta = \theta$ and (b) stimulus orientation at the preferred spatial frequency $P = 1$. The spatial frequency of the Gabor filter is $p = 1$, and $\psi = \Psi = 0$. Results are shown for various values of the space constant $\sigma$. It can be seen that for relatively low values of $\sigma$, the RF acts like a lowpass spatial frequency filter with a shallow maximum at $p = P$. When $\sigma$ is increased, however, the profile is sharpened and the RF acts more like a bandpass filter. Similarly, the orientation selectivity increases with increasing $\sigma$.

Using the identity

$$e^{x\cos\theta} = I_0(x) + 2\sum_{n\geq 1} I_n(x)\cos(n\theta) \tag{3.12}$$

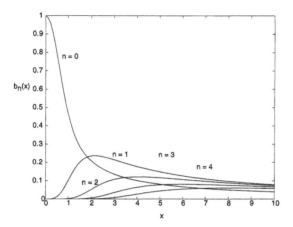

Fig. 14. Plot of function $b_n(x)$ for $n = 0, 1, 2, 3$.

where $I_n(x)$ is the modified Bessel function of integer order $n$ and $I_n(-x) = (-1)^n I_n(x)$, equation (3.11) can be expanded as

$$h(\theta, p) = \sum_{n \in \mathbf{Z}} h_n(p) e^{2in[\theta - \Theta]} \tag{3.13}$$

with

$$h_n = C e^{-\sigma^2 [p^2 + P^2]/2} I_{2n}(\sigma^2 p P) \tag{3.14}$$

We can then consider the relative strength of the different harmonic components $h_n$ of the input at the preferred spatial frequency $P = p$, for which $h_n = C b_n(\sigma P)$ with $b_n(x) = e^{-x^2} I_{2n}(x^2)$. The functions $b_n(x)$ for $n = 0, 1, 2, 3$ are plotted in Fig. 14. It can be seen that in the case of a weakly biased input, small $\sigma P$, the major contributions come from the zeroth and first-order harmonics. This is precisely one of the simplifying assumptions used in the analysis of the ring model of orientation tuning [5, 47], see section 4.5. The expansion of the external input in terms of Bessel functions also provides information regarding the spatial frequency content of these harmonic components. For example, it can be seen that at higher spatial frequencies, contributions from higher harmonic components become important since $\sigma P$ increases in size. However, this effect would be removed if the size $\sigma$ of the RF decreases with spatial frequency according to $\sigma = \sigma_0/p$ for fixed $\sigma_0$.

### 3.3. Orientation pinwheels as organizing centers for feature map $\mathcal{F}(\mathbf{r})$

The optical imaging studies described in section 2 show that the orientation, ocular dominance and spatial frequency maps have a number of regularly repeating structures, which motivates the partitioning of cortex into hypercolumns. Each hypercolumn contains two sets of orientation preferences $\theta \in [0, \pi)$ per eye organized around a pair of orientation pinwheels. Motivated by recent studies of cat V1 [10, 56], we further assume that high frequency preferences are located at one orientation pinwheel and low spatial frequencies at the other, and around the pinwheels iso–orientation and iso–frequency preference contours are approximately orthogonal, see Fig. 15. Although the partitioning of the cortex into

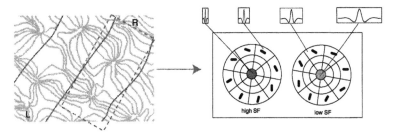

Fig. 15. Schematic diagram of a hypercolumn (for a given eye preference) consisting of orientation and spatial frequency preferences organized around a pair of pinwheels.

hypercolumns has a degree of disorder, we assume that to a first approximation each hypercolumn is a fundamental domain $\mathcal{U}_\mathcal{L}$ of a regular planar lattice $\mathcal{L}$. The lattice $\mathcal{L}$ is generated by two linearly independent vectors $\boldsymbol{\ell}_1$ and $\boldsymbol{\ell}_2$:

$$\mathcal{L} = \{(m_1\boldsymbol{\ell}_1 + m_2\boldsymbol{\ell}_2)d : m_1, m_2 \in \mathbf{Z}\} \tag{3.15}$$

The lattice spacing $d$ is taken to be the average width of a hypercolumn, that is, $d \approx 1mm$. Let $\psi$ be the angle between the two basis vectors $\boldsymbol{\ell}_1$ and $\boldsymbol{\ell}_2$. We can then distinguish three types of lattice according to the value of $\psi$: square lattice ($\psi = \pi/2$), rhombic lattice ($0 < \psi < \pi/2$, $\psi \neq \pi/3$) and hexagonal ($\psi = \pi/3$). After rotation, the generators of the planar lattices are given in Table 1. Also shown are the generators of the *dual lattice* $\widehat{\mathcal{L}}$ satisfying $\widehat{\boldsymbol{\ell}}_i.\boldsymbol{\ell}_j = \delta_{i,j}$ for $i, j = 1, 2$. Sites on the lattice can be identified with the set of orientation pinwheels having the same ocularity and spatial frequency preference.

Given the partitioning of cortex into a lattice of hypercolumns, the set of feature maps $\mathcal{F}(\mathbf{r})$ is periodic with respect to the lattice $\mathcal{L}$, that is, $\mathcal{F}(\mathbf{r} + \boldsymbol{\ell}) = \mathcal{F}(\mathbf{r})$ for all $\boldsymbol{\ell} \in \mathcal{L}$. Thus, specification of $\mathcal{F}(\mathbf{r})$ reduces to determining the distribution of orientation and spatial frequency preferences within a single hypercolumn.

Table 1

Generators for the planar lattices and their dual lattices.

| Lattice | $\ell_1$ | $\ell_2$ | $\hat{\ell}_1$ | $\hat{\ell}_2$ |
|---------|----------|----------|----------------|----------------|
| Square | $(1, 0)$ | $(0, 1)$ | $(1, 0)$ | $(0, 1)$ |
| Hexagonal | $(1, 0)$ | $\frac{1}{2}(1, \sqrt{3})$ | $(1, \frac{-1}{\sqrt{3}})$ | $(0, \frac{2}{\sqrt{3}})$ |
| Rhombic | $(1, 0)$ | $(\cos \eta, \sin \eta)$ | $(1, -\cot \eta)$ | $(0, \csc \eta)$ |

Further simplification occurs if one restricts attention to a region surrounding a single pinwheel. Suppose that the two-dimensional cortical coordinate $\mathbf{r}$ is represented in polar coordinates $\mathbf{r} = (r, \phi)$ where the origin is the closest pinwheel center. The relationship between $(r, \phi)$ and the feature preferences $(p, \theta)$ can then be approximated by the coordinate transformations

$$p = p(r) \equiv p_\pm e^{\mp r/\xi}, \quad \theta = \pm\frac{\phi}{2} \tag{3.16}$$

where $(+)$ applies to a high spatial frequency pinwheel and $(-)$ applies to a low spatial frequency pinwheel. We are assuming that the two pinwheels have opposite indices and that radial separation varies as log spatial frequency; the latter is consistent with experimental data [31, 56]. A mathematical representation of the feature maps across the whole hypercolumn is more complicated. It should also be noted that the detailed form of the spatial frequency map in macaque is not currently known. It is possible, for example, that the distribution of spatial frequency preferences varies across cortical layers due to different combinations of the M and P pathways, see Fig. 8.

The effects of the pinwheel architecture of V1 on orientation tuning has recently been investigated using a detailed computational model of a local patch of layer 4Cα in macaque, consisting of a large-scale network of conductance-based integrate-and-fire neurons [68]. The cells are organized around a set of four pinwheel centers such that the angular position of a cell relative to the nearest pinwheel determines its orientation preference and hence the input it receives from the lateral geniculate nucleus (LGN). The spatial frequency dependence of the inputs is not considered. Local recurrent connections are assumed to depend only on cortical distance, with inhibition dominant at short range. When operated in a strongly inhibitory regime it is found that cells close to pinwheel centers are more orientation selective than those away from the centers; this has also been established analytically using nonlinear rate models [82]. However, experimental evidence suggests that single cell orientation selectivity is either approximately uniform within a hypercolumn or is actually broader near pinwheel centers [56, 66]. A recent analysis using linear rate equations has established that it is possible to obtain variations in selectivity that are more consistent with experimental data, provided that the spatially narrow inhibition is fast [58].

### 3.4. Long–range connections and the weight distribution $w$

Motivated by the functional architecture of V1, we decompose the weight distribution $w_{ij}$, $i, j = E, I$, according to

$$w_{ij}(\mathbf{r}|\mathbf{r}') = W_{ij}(|\mathbf{r} - \mathbf{r}'|) + \beta_i \delta_{j,E} W^\Delta(\mathcal{F}(\mathbf{r}), \mathcal{F}(\mathbf{r}')), \quad i, j = E, I \quad (3.17)$$

where $W_{ij}$ represents isotropic and homogeneous local connections that depend on the Euclidean distance $|\mathbf{r} - \mathbf{r}'| = \sqrt{(x - x')^2 + (y - y')^2}$, $W^\Delta$ represents the dependence of excitatory horizontal connections on the feature preferences of the presynaptic and postsynaptic neuron populations and $\beta_i$ is a positive coupling parameter. Experimentally it is found that the horizontal connections modulate rather than drive a neuron's response to a visual stimulus [38, 49], suggesting that $\beta_i$ is small. The local connections span a single hypercolumn, whereas the patchy horizontal connections link cells with similar feature preferences in distinct hypercolumns.

In the absence of long–range connections ($\beta_i = 0$), the resulting weight distribution is invariant under the action of the Euclidean group $\mathbf{E}(2)$ of rigid motions in the plane, that is,

$$\gamma \cdot w_{ij}(\mathbf{r}|\mathbf{r}') = w_{ij}(\gamma^{-1} \cdot \mathbf{r}|\gamma^{-1} \cdot \mathbf{r}') = w_{ij}(\mathbf{r}|\mathbf{r}')$$

for all $\gamma \in \mathbf{E}(2)$. The Euclidean group is composed of the (semi-direct) product of $\mathbf{O}(2)$, the group of planar rotations $\mathbf{r} \to R_\varphi \mathbf{r}$ and reflections $(x, y) \to (x, -y)$, with $\mathbf{R}^2$, the group of planar translations $\mathbf{r} \to \mathbf{r} + \mathbf{s}$. Here

$$R_\varphi = \begin{pmatrix} \cos\varphi & -\sin\varphi \\ \sin\varphi & \cos\varphi \end{pmatrix}, \quad \varphi \in [0, 2\pi) \quad (3.18)$$

Most large–scale models of cortex assume Euclidean symmetric weights [36, 105]. However, it is clear that the long–range connections break Euclidean symmetry due to correlations with the feature map $\mathcal{F}(\mathbf{r})$. A certain degree of symmetry still remains under the approximation that the feature map is perdiodic with respect to the planar lattice $\mathcal{L}$ (see section 3.3). The resulting weight distribution for $\beta_i \neq 0$ is then doubly periodic with respect to $\mathcal{L}$:

$$w_{ij}(\mathbf{r} + \boldsymbol{\ell}|\mathbf{r}' + \boldsymbol{\ell}) = w_{ij}(\mathbf{r}|\mathbf{r}') \quad (3.19)$$

for all $\boldsymbol{\ell} \in \mathcal{L}$. Additional symmetries may also exist depending on the particular form of $W^\Delta$.

There are number of distinct ways in which $W^\Delta$ may depend on the underlying feature maps $\mathcal{F}$. The first reflects the "patchiness" of the horizontal connections that link cells with similar feature preferences. This may be implemented by taking

$$W^\Delta(\mathcal{F}(\mathbf{r}), \mathcal{F}(\mathbf{r}')) = \sum_{\boldsymbol{\ell} \in \mathcal{L}} J_{\boldsymbol{\ell}} \Delta(\mathbf{r} - \mathbf{r}' - \boldsymbol{\ell}) \tag{3.20}$$

where $\Delta(\mathbf{r})$ is some localized unimodal function that is maximal when $\mathbf{r} = 0$, thus ensuring that presynaptic and postsynaptic cells with similar feature preferences are connected. The width of $\Delta$ determines the size of the patches and $J_{\boldsymbol{\ell}}, \boldsymbol{\ell} \neq 0$ is a monotonically decreasing function of $\boldsymbol{\ell}$. In this particular example, the patchy horizontal connections break continuous rotation symmetry down to the discrete rotation symmetry of the lattice. On the other hand, continuous translation symmetry (homogeneity) still holds, since $W^\Delta$ only depends on the relative separation $\mathbf{r} - \mathbf{r}'$ in cortex. However, it is likely that continuous translation symmetry may also be broken due to a periodic modulation in the distribution of the patchy connections. For example, suppose that the patchy connections depended on separation in feature space $W^\Delta(\mathcal{F}, \mathcal{F}') \sim \delta(\mathcal{F} - \mathcal{F}')$. Under the further assumption that $\mathcal{F}$ is invertible within a fundamental domain, this would introduce an additional Jacobian factor in converting from distance in feature space to distance in cortical space:

$$W^\Delta(\mathcal{F}(\mathbf{r}), \mathcal{F}(\mathbf{r}')) = \frac{1}{|\mathcal{F}'(\mathbf{r})|} \sum_{\boldsymbol{\ell} \in \mathcal{L}} J_{\boldsymbol{\ell}} \delta(\mathbf{r} - \mathbf{r}' - \boldsymbol{\ell}) \tag{3.21}$$

Another striking source of periodic inhomogeneity arises from anisotropy in the patchy connectional field as seen in Fig. 10. In some animals the direction of anisotropy is correlated with the orientation preference map and thus rotates periodically across cortex [12].

Anisotropy can be incorporated into the model by modifying the weight distribution $W^\Delta$ along the following lines [14, 16, 19, 27]:

$$W^\Delta(\mathcal{F}(\mathbf{r}), \mathcal{F}(\mathbf{r}')) = \sum_{\boldsymbol{\ell} \in \mathcal{L}} J_{\boldsymbol{\ell}} \mathcal{A}_{\mathcal{F}(\mathbf{r})}(\boldsymbol{\ell}) \Delta(\mathbf{r} - \mathbf{r}' - \boldsymbol{\ell}) \tag{3.22}$$

with

$$\mathcal{A}_{\mathcal{F}}(\boldsymbol{\ell}) = \frac{\mathcal{B}(p)}{4\eta(p)} \left( H[\eta(p) - |\arg \boldsymbol{\ell} - \theta|] + H[\eta(p) - |\arg \boldsymbol{\ell} - \theta - \pi|] \right) \tag{3.23}$$

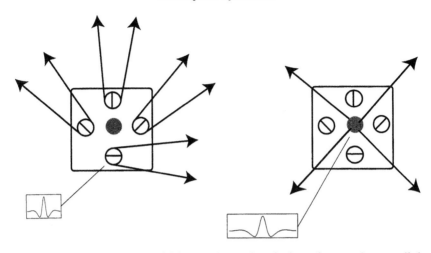

Fig. 16. Cells at intermediate spatial frequencies send out horizontal connections to cells in other hypercolumns in a direction parallel to their common preferred orientation, whereas cells at low (high) spatial frequency pinwheels connect to other low (high) pinwheels in an isotropic fashion.

where $\mathcal{F} = (\theta, p)$ and $H$ is a Heaviside function. The second terms takes account of the fact that $\theta \in [0, \pi)$ whereas $\arg \boldsymbol{\ell} \in [0, 2\pi)$. The parameter $\eta$ determines the degree of anisotropy, that is the angular spread of the horizontal connections around the axis joining cells with similar orientation preferences and $\mathcal{B}$ is an additional normalization factor. Both $\eta$ and $\mathcal{B}$ are taken to be spatial frequency dependent as we now explain. First, note that at the population level there is zero mean selectivity for orientation at the pinwheels. This implies that the horizontal weight distribution has to be isotropic at the pinwheels, that is, at low and high spatial frequencies. In order to incorporate any anisotropy away from the pinwheels, we conclude that the spread parameter has to be $p$–dependent, $\eta = \eta(p)$ with $\eta(p_+) = \eta(p_-) = \pi/2$. This is illustrated in Fig. 16. The additional factor $\mathcal{B}(p)$ allows for a possible correlation between the strength of the lateral connections and the spatial frequency of cells linked by these connections. There is recent data indicating that some cells located outside cytochrome oxidase (CO) blobs (regions of cells that are more metabolically active) have very little in the way of lateral connections [106], thus leading to an effective reduction in connectivity at the population level. Since the CO blobs have a strong association with the orientation singularities corresponding to low spatial frequencies [55,61], the coupling may be larger around the low frequency pinwheels.

### 3.5. Coupled hypercolumn model of V1

One of the difficulties with the above formulation is that it is necessary to specify the feature map $\mathcal{F}(\mathbf{r})$. One way to avoid this problem is to collapse each hypercolumn into a single point (through some form of spatial coarse-graining) and to treat V1 as a continuum of hypercolumns [14–16]. Thus cortical position $\mathbf{r}$ is replaced by the pair $\{\mathbf{r}, \mathcal{F}\}$ with $\mathbf{r} \in \mathbf{R}^2$ now labeling the hypercolumn at (coarse-grained) position $\mathbf{r}$ and $\mathcal{F}$ labeling the feature preferences of neurons within the hypercolumn. Let $a_i(\mathbf{r}, \mathcal{F}, t)$ denote the activity of the excitatory ($i = E$) or inhibitory ($i = I$) population at $(\mathbf{r}, \mathcal{F})$, and suppose that $a_i$ evolves according to

$$\tau_i \frac{\partial a_i(\mathbf{r}, \mathcal{F}, t)}{\partial t} = -a_i(\mathbf{r}, \mathcal{F}, t) + h_i(\mathbf{r}, \mathcal{F}, t) \tag{3.24}$$

$$+ \sum_j \int_{\mathbf{R}^2} \int w_{ij}(\mathbf{r}, \mathcal{F}|\mathbf{r}', \mathcal{F}') f(a_j(\mathbf{r}', \mathcal{F}', t)) D\mathcal{F}'$$

with $D\mathcal{F}'$ an appropriately defined measure on feature space. We decompose $w_{ij}$ into local and long-range parts by assuming that the local connections mediate interactions within a hypercolumn whereas the patchy horizontal connections mediate interactions between hypercolumns:

$$w_{ij}(\mathbf{r}, \mathcal{F}|\mathbf{r}', \mathcal{F}') = \delta(\mathbf{r} - \mathbf{r}')w_{ij}(\mathcal{F}|\mathcal{F}') \tag{3.25}$$

$$+ \beta_i \delta_{j,E} J(|\mathbf{r} - \mathbf{r}'|) \mathcal{A}_{\mathcal{F}}(\mathbf{r} - \mathbf{r}') w^\Delta(\mathcal{F}|\mathcal{F}')$$

where $w_{ij}(\mathcal{F}|\mathcal{F}')$ and $w^\Delta(\mathcal{F}|\mathcal{F}')$ represent the dependence of the local and long-range interactions on the feature preferences of the pre– and post–synaptic cells, and $J(\mathbf{r})$ with $J(0) = 0$ is a positive function that determines the variation in the strength of the long-range interactions with cortical distance. We have also included the anisotropy factor $\mathcal{A}_{\mathcal{F}}$ of equation (3.23). The advantage of collapsing each hypercolumn to a single point in the cortical plane is that a simpler representation of the internal structure of a hypercolumn can be developed that captures the essential tuning properties of the cells as well as incorporating the modulatory effects of long–range connections.

*Ring model and* $\mathbf{O}(2)$ *symmetry* For the sake of illustration, suppose that we identify $\mathcal{F}$ in equations (3.24) and (3.25) with the orientation preference $\theta \in [0, \pi)$ of cells within a hypercolumn. The weight distribution (3.25) is taken to have the form

$$w_{ij}(\mathbf{r}, \theta|\mathbf{r}', \theta') = \delta(\mathbf{r} - \mathbf{r}')w_{ij}(\theta - \theta') \tag{3.26}$$

$$+ \beta_i \delta_{j,E} J(|\mathbf{r} - \mathbf{r}'|) \mathcal{P}(\arg(\mathbf{r} - \mathbf{r}') - \theta) w^\Delta(\theta - \theta')$$

with

$$P(\psi) = \frac{1}{4\eta}[H(\eta - |\psi|) + H(\eta - |\psi - \pi|)] \tag{3.27}$$

We have neglected the spatial frequency dependence of the anisotropy factor (3.23). The functions $w_{ij}(\theta)$ and $w^{\Delta}(\theta)$ are assumed to be even, $\pi$-periodic functions of $\theta$, with corresponding Fourier expansions

$$w_{ij}(\theta) = W_{ij}(0) + 2\sum_{n \geq 1} W_{ij}(n)\cos 2n\theta$$

$$w^{\Delta}(\theta) = W^{\Delta}(0) + 2\sum_{n \geq 1} W^{\Delta}(n)\cos 2n\theta \tag{3.28}$$

The distribution $w^{\Delta}(\theta)$ is taken to be a positive, narrowly tuned distribution with $w^{\Delta}(\theta) = 0$ for all $|\theta| > \theta_c$ and $\theta_c \ll \pi/2$; the long-range connections thus link cells with similar orientation preferences. Equation (3.24) then describes a continuum of coupled ring networks, each of which corresponds to a version of the so–called ring model of orientation tuning [5, 86].

If there is no orientation–dependent anisotropy then the weight distribution (3.26) is invariant with respect to the symmetry group $\mathbf{E}(2) \times \mathbf{O}(2)$ where $\mathbf{O}(2)$ is the group of rotations and reflections on the ring $S^1$ and $\mathbf{E}(2)$ is the Euclidean group acting on $\mathbf{R}^2$. The associated group action is

$$\begin{aligned} \zeta \cdot (\mathbf{r}, \theta) &= (\zeta\mathbf{r}, \theta), & \zeta \in \mathbf{E}(2) \\ \xi \cdot (\mathbf{r}, \theta) &= (\mathbf{r}, \theta + \xi) \\ \kappa \cdot (\mathbf{r}, \theta) &= (\mathbf{r}, -\theta) \end{aligned} \tag{3.29}$$

Invariance of the weight distribution can be expressed as

$$\gamma \cdot w_{ij}(\mathbf{r}, \theta | \mathbf{r}', \theta) = w_{ij}(\gamma^{-1} \cdot (\mathbf{r}, \theta)|\gamma^{-1} \cdot (\mathbf{r}', \theta')) = w_{ij}(\mathbf{r}, \theta | \mathbf{r}', \theta')$$

for all $\gamma \in \Gamma$ where $\Gamma = \mathbf{E}(2) \times \mathbf{O}(2)$. Anisotropy reduces the symmetry group $\Gamma$ to $\mathbf{E}(2)$ with the following *shift–twist* action on $\mathbf{R}^2 \times S^1$ [14, 15]:

$$\begin{aligned} \mathbf{s} \cdot (\mathbf{r}, \theta) &= (\mathbf{r} + \mathbf{s}, \theta) \\ \xi \cdot (\mathbf{r}, \theta) &= (R_{\xi}\mathbf{r}, \theta + \xi) \\ \kappa \cdot (\mathbf{r}, \theta) &= (R_{\kappa}\mathbf{r}, -\theta) \end{aligned} \tag{3.30}$$

where $R_{\xi}$ denotes the planar rotation through an angle $\xi$ and $R_{\kappa}$ denotes the reflection $(x_1, x_2) \mapsto (x_1, -x_2)$. It can be seen that the rotation operation comprises a translation or *shift* of the orientation preference label $\theta$ to $\theta + \xi$, together with a rotation or *twist* of the position vector $\mathbf{r}$ by the angle $\xi$.

It is instructive to establish explicitly the invariance of anisotropic long–range connections under shift–twist symmetry. Let us define

$$w_{hoz}(\mathbf{r}, \theta | \mathbf{r}', \theta') = J(|\mathbf{r} - \mathbf{r}'|)\mathcal{P}(\arg(\mathbf{r} - \mathbf{r}') - \theta)w^\Delta(\theta - \theta') \qquad (3.31)$$

Translation invariance of $w_{hoz}$ follows immediately from the spatial homogeneity of the interactions, which implies that

$$w_{hoz}(\mathbf{r} - \mathbf{s}, \theta | \mathbf{r}' - \mathbf{s}, \theta') = w_{hoz}(\mathbf{r}, \theta | \mathbf{r}', \theta').$$

Invariance with respect to a rotation by $\xi$ follows from

$$\begin{aligned}
w_{hoz}&(R_{-\xi}\mathbf{r}, \theta - \xi | R_{-\xi}\mathbf{r}', \theta' - \xi) \\
&= J(|R_{-\xi}(\mathbf{r} - \mathbf{r}')|)\mathcal{P}(\arg[R_{-\xi}(\mathbf{r} - \mathbf{r}')] - \theta + \xi)w^\Delta(\theta - \xi - \theta' + \xi) \\
&= J(|\mathbf{r} - \mathbf{r}'|)\mathcal{P}(\arg(\mathbf{r} - \mathbf{r}') - \theta)w^\Delta(\theta - \theta') \\
&= w_{hoz}(\mathbf{r}, \theta | \mathbf{r}', \theta').
\end{aligned}$$

We have used the conditions $|R_\xi \mathbf{r}| = |\mathbf{r}|$ and $\arg(R_{-\xi}\mathbf{r}) = \arg(\mathbf{r}) - \xi$. Finally, invariance under a reflection $\kappa$ about the $x$-axis holds since

$$\begin{aligned}
w_{hoz}&(\kappa\mathbf{r}, -\theta | \kappa\mathbf{r}', -\theta') \\
&= J(|\kappa(\mathbf{r} - \mathbf{r}')|)\mathcal{P}(\arg[\kappa(\mathbf{r} - \mathbf{r}')] + \theta)w^\Delta(-\theta + \theta') \\
&= J(|\mathbf{r} - \mathbf{r}'|)\mathcal{P}(-\arg(\mathbf{r} - \mathbf{r}') + \theta)w^\Delta(\theta - \theta') \\
&= w_{hoz}(\mathbf{r}, \theta | \mathbf{r}', \theta').
\end{aligned}$$

We have used the conditions $\arg(\kappa\mathbf{r}) = -\arg(\mathbf{r})$, $w^\Delta(-\theta) = w^\Delta(\theta)$, and $\mathcal{P}(-\psi) = \mathcal{P}(\psi)$. The fact that the weight distribution is invariant with respect to this shift–twist action has important consequences for the global dynamics of V1 in the presence of anisotropic horizontal connections.

*Spherical model and* $\mathbf{O}(3)$ *symmetry*   The above ring model effectively reduces a hypercolumn to an annular region of orientation selective cells surrounding a single pinwheel and having a fixed spatial frequency. Recently this feature–based model has been extended to take into account the distribution of spatial frequency preferences as illustrated in Fig. 15 [18, 19]. Neurons within a given hypercolumn are now labeled by the pair $(\theta, p)$ where $p \in [p_-, p_+]$ denotes the spatial frequency preference and $\theta \in [0, \pi)$ the orientation preference of a cell. (For simplicity, we neglect ocular dominance here: this could be incorporated by introducing an additional discrete label for left/right eye dominance). Typically, the bandwidth of a hypercolumn is between three and four octaves, that is, $p_+ \approx 2^n p_-$ with $n = 3.5$. This is consistent with the observations of Hubel and Wiesel [54], who found a two octave scatter of receptive field sizes at each cortical location. Rather than attempting to accurately model the distribution of orientation and spatial frequency preferences within a hypercolumn, we consider a reduced model in which only a subset of neurons are explicitly represented. That is, we consider the union $D_- \cup D_+$ of two disc regions enclosing the low ($-$) and high ($+$) spatial frequency pinwheels respectively. As a further abstraction, we sew together the two disc boundaries $\partial D_+$ and $\partial D_-$. The resulting topological space is then a sphere. Identifying the north and south poles of the sphere with the low and high spatial frequency pinwheels leads to a spherical model of a cortical hypercolumn [18, 19], see Fig. 17. It is important to emphasize that the spherical (or ring) topology is a feature–based representation of a single hypercolumn, and does not correspond to the actual distribution of cells in the cortical plane.

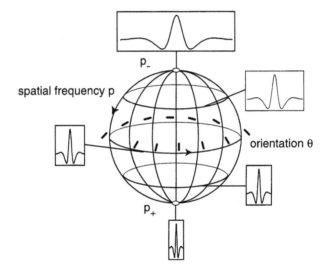

Fig. 17. Spherical network topology. Orientation and spatial frequency labels are denoted by $(\phi, p)$ with $0 \leq \theta < \pi$ and $p_- \leq p \leq p_+$.

Taking $(\theta, \phi)$ to be the angular coordinates on the sphere with $\theta \in [0, \pi)$, $\phi \in [0, \pi)$, we set

$$\phi \equiv \mathcal{Q}(p) = \pi \frac{\log(p/p_-)}{\log(p_+/p_-)} \tag{3.32}$$

That is, $\phi$ varies linearly with $\log p$. This is consistent with experimental data that suggests a linear variation of $\log p$ with cortical separation [56]. We then identify $\mathcal{F}$ in equations (3.24) and (3.25) with the pair of angular coordinates

$(\theta, \phi)$. Given the spherical topology, it is natural to construct weight distributions $W_{ij}$, $W^{\triangle}$ that are invariant with respect to coordinate rotations and reflections of the sphere, that is, the symmetry group $\mathbf{O}(3)$. This spherical symmetry, which generalizes the $\mathbf{O}(2)$ circular symmetry of the ring model, implies that the weight distributions only depend on the relative distance of cells on the sphere as determined by their angular separation along geodesics or great circles. Given two points on the sphere $(\theta, \phi)$ and $(\theta', \phi')$ their angular separation $\alpha$ is determined from the equation

$$\cos \alpha = \cos \phi \cos \phi' + \sin \phi \sin \phi' \cos(2[\theta - \theta']) \tag{3.33}$$

The simplest non–trivial form for the local weight distributions is thus

$$w_{ij}(\theta, \phi|\theta', \phi') = W_{ij}(0) + W_{ij}(1)(\cos \phi \cos \phi' + \sin \phi \sin \phi' \cos(2[\theta - \theta'])) \tag{3.34}$$

which is equivalent to an expansion in terms of zeroth and first order spherical harmonics. More generally, one can expand in terms of higher order spherical harmonics:

$$w_{ij}(\theta, \varphi|\theta', \varphi') = \sum_{n=0}^{\infty} W_{ij}(n) \sum_{m=-n}^{n} Y_n^{m*}(\theta', \varphi') Y_n^m(\theta, \varphi) \tag{3.35}$$

with $W_n$ real and

$$Y_n^m(\theta, \varphi) = (-1)^m \sqrt{\frac{2n+1}{4\pi} \frac{(n-m)!}{(n+m)!}} P_n^m(\cos \phi) e^{2im\theta} \tag{3.36}$$

for $n \geq 0$ and $-n \leq m \leq n$, where $P_n^m(\cos \phi)$ is an associated Legendre function. The lowest order spherical harmonics are

$$Y_0^0(\theta, \varphi) = \frac{1}{\sqrt{4\pi}}, \quad Y_1^0(\theta, \varphi) = \sqrt{\frac{3}{4\pi}} \cos \phi,$$

$$Y_1^{\pm}(\theta, \phi) = \mp \sqrt{\frac{3}{8\pi}} \sin \phi e^{\pm 2i\theta}. \tag{3.37}$$

Note that around the equator $\phi = \phi' = \pi/2$ equation (3.35) reduces to the corresponding harmonic expansion of the ring model. On the other hand, close to the pinwheel centers interactions are approximately $\theta$–independent. This is a consequence of the fact that there is a large deal of spatial scatter in orientation preference around a pinwheel so that some form of averaging occurs at the population level. Given that the long range horizontal connections tend to link neurons with similar feature preferences, one can also construct an $\mathbf{O}(3)$–invariant long range distribution $\Delta$ of the form

$$w^{\Delta}(\theta, \phi | \theta', \phi') = H[\cos\theta \cos\theta' + \sin\theta \sin\theta' \cos(2[\phi - \phi']) - \cos\alpha_0] \tag{3.38}$$

where $H$ is the Heaviside function and the angle $\alpha_0$ determines the patch size.

It follows that in the absence of anisotropy, the full weight distribution (3.25) for the spherical model is invariant under the action of the group $\mathbf{E}(2) \times \mathbf{O}(3)$. Inclusion of spatial-frequency and orientation dependent anisotropy then breaks the symmetry group down to $\mathbf{E}(2)$ with the same group action on the coordinates $(\mathbf{r}, \theta)$ as the ring model and a trivial group action with respect to the spatial frequency label $\phi$:

$$\begin{array}{rcl} \mathbf{s} \cdot (\mathbf{r}, \theta, \phi) & = & (\mathbf{r} + \mathbf{s}, \theta, \phi) \\ \xi \cdot (\mathbf{r}, \theta, \phi) & = & (R_\xi \mathbf{r}, \theta + \xi, \phi) \\ \kappa \cdot (\mathbf{r}, \theta, \phi) & = & (R_\kappa \mathbf{r}, -\theta, \phi) \end{array} \tag{3.39}$$

## 4. Pattern formation in a single hypercolumn

In order to introduce some of the analytical methods used in the study of cortical pattern formation, we begin by considering the example of orientation tuning in a single isolated hypercolumn.

### 4.1. Orientation tuning in a one–population ring model

Suppose that we represent a hypercolumn by the following one–population ring model (see section 3.5):

$$\frac{\partial a(\theta, t)}{\partial t} = -a(\theta, t) + \int_0^\pi \frac{d\theta'}{\pi} w(\theta | \theta') f(a(\theta', t)) + h(\theta, t) \tag{4.1}$$

where $a(\theta, t)$ denotes the activity at time $t$ of a local population of cells with orientation preference $\theta \in [0, \pi)$, $w(\theta | \theta')$ is the strength of synaptic weights between cells with orientation preference $\theta'$ and $\theta$, and $h(\theta, t)$ is the feedforward input expressed as a function of $\theta$ (see section 3.2). We take an $\mathbf{O}(2)$ symmetric weight distribution by setting $w(\theta | \theta') = w(\theta - \theta')$ with $w(\theta)$ an even $\pi$-periodic function. It follows that $w(\theta)$ has the Fourier series expansion

$$w(\theta) = W_0 + 2\sum_{n \geq 0} W_n \cos(2n\theta) \tag{4.2}$$

with $W_n$ real.

In the case of a constant input $h(\theta, t) = \bar{h}$ there exists at least one equilibrium solution of equation (4.1), which satisfies the algebraic equation

$$\bar{a} = W_0 f(\bar{a}) + \bar{h} \tag{4.3}$$

with $W_0 = \int_0^\pi W(\theta)d\theta/\pi$. If $\bar{h}$ is sufficiently small relative to the threshold $\kappa$ of the neurons then the equilibrium is unique and stable. Under the change of coordinates $a \to a - \bar{h}$, it can be seen that the effect of $\bar{h}$ is to shift the threshold by the amount $-\bar{h}$. Thus there are two ways to increase the excitability of the network and thus destabilize the fixed point: either by increasing the external input $\bar{h}$ or reducing the threshold $\kappa$. The latter can occur through the action of drugs on certain brain stem nuclei, which provides a mechanism for generating geometric visual hallucinations [14–16, 34].

The stability of the fixed point can be determined by setting $a(\theta, t) = \bar{a} + a(\theta)e^{\lambda t}$ and linearizing about $\bar{a}$. This leads to the eigenvalue equation

$$\lambda a(\theta) = -a(\theta) + \mu \int_0^\pi w(\theta - \theta')a(\theta') \frac{d\theta'}{\pi} \tag{4.4}$$

where $\mu = f'(\bar{a})$. The linear operator on the right–hand side of equation (4.4) has a discrete spectrum (since it is a compact operator) with eigenvalues

$$\lambda_n = -1 + \mu W_n, \ n \in \mathbf{Z} \tag{4.5}$$

and corresponding eigenfunctions

$$a(\theta) = z_n e^{2in\theta} + z_n^* e^{-2in\theta} \tag{4.6}$$

where $z_n$ is a complex amplitude with complex conjugate $z_n^*$. It follows that for sufficiently small $\mu$, corresponding to a low activity state, $\lambda_n < 0$ for all $n$ and the fixed point is stable. However, as $\mu$ is increased beyond a critical value $\mu_c$ the fixed point becomes unstable due to excitation of the eigenfunctions associated with the largest Fourier component of $w(\theta)$, see equation (4.2). We refer to such eigenfunctions as *excited modes*.

Two examples of discrete Fourier spectra are shown in Fig. 18a. In the first case $W_1 = \max_m\{W_m\}$ so that $\mu_c = 1/W_1$ and the excited modes are of the form

$$a(\theta) = Z e^{2i\theta} + Z^* e^{-2i\theta} = |Z| \cos(2[\theta - \theta_0]) \tag{4.7}$$

with complex amplitude $Z = |Z|e^{-2i\theta_0}$. Since these modes have a single maximum around the ring, the network supports an activity profile consisting of a tuning curve centered about the point $\theta_0$. The location of this peak is arbitrary and depends only on random initial conditions, reflecting the $\mathbf{O}(2)$ symmetry of the weight distribution $w$. Such a symmetry is said to be spontaneously broken by the action of the pattern forming instability. Since the dominant component is $W_1$, the distribution $w(\theta)$ is excitatory (inhibitory) for neurons with sufficiently similar (dissimilar) orientation preferences. (This is analogous to the Wilson-Cowan "Mexican Hat" function [105], see section 6.1). The inclusion of an additional small amplitude input $\Delta h(\theta) \sim \cos[2(\theta - \Theta)]$ explicitly breaks $\mathbf{O}(2)$ symmetry, and locks the peak of the tuning curve to the stimulus orientation, that is, $\theta_0 = \Theta$. As one moves further away from the point of instability, the amplitude of the tuning curve increases and sharpening occurs due to the nonlinear effects of the firing rate function (3.2). This is illustrated in Fig. 18b, where the input and output (normalized) firing rate of the excitatory population of a single hypercolumn are shown. Thus the local intracortical connections within a hypercolumn serve both to amplify and sharpen a weakly oriented input signal from the LGN [6,86]. On the other hand, if the local level of inhibition is reduced such that $W_n$ is a monotonically decreasing function of $|n|$ (see Fig. 18a), then the homogeneous fixed point undergoes a bulk instability at $\mu_c = 1/W_0$, resulting in a broadening of the tuning curve. This is consistent with experimental data demonstrating a loss of stable orientation tuning in cats with blocking of intracortical inhibition [75].

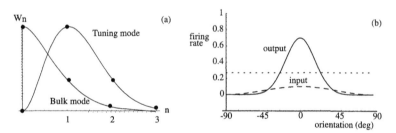

Fig. 18. (a) Spectrum $W_n$ of local weight distribution with a maximum at $n = 1$ (tuning mode) and a maximum at $n = 0$ (bulk mode). (b) Sharp orientation tuning curve in a single hypercolumn. Local recurrent excitation and inhibition amplifies a weakly modulated input from the LGN. Dotted line is the base–line output without orientation tuning.

## 4.2. *Derivation of amplitude equation using the* Fredholm *alternative*

So far we used linear theory to show to how a hypercolumn can undergo a pattern forming instability through the spontaneous breaking of $\mathbf{O}(2)$ symmetry, leading to the growth of an orientation tuning curve. However, as the activity profile increases in amplitude the linear approximation breaks down and nonlinear theory is necessary in order to investigate whether or not a stable pattern ultimately forms. Sufficiently close to the bifurcation point $\mu = \mu_c$ where the homogeneous state becomes unstable, we can treat $\mu - \mu_c = \varepsilon\Delta\mu$ as a small parameter and carry out a perturbation expansion in powers of $\varepsilon$. This generates a dynamical equation for the amplitude of the pattern that can be used to investigate stability as well as the effects of a weakly biased external input [13, 36].

First, Taylor expand the nonlinear firing-rate function in equation (4.1) about the fixed point $\bar{a}$:

$$f(a) - f(\bar{a}) = \mu(a - \bar{a}) + g_2(a - \bar{a})^2 + g_3(a - \bar{a})^3 + \ldots \tag{4.8}$$

where $g_2 = f''(\bar{a})/2$, $g_3 = f'''(\bar{a})/6$. Next substitute into equation (4.1) the perturbation expansion

$$a = \bar{a} + \varepsilon^{1/2}a_1 + \varepsilon a_2 + \varepsilon^{3/2}a_3 + \ldots \tag{4.9}$$

The dominant temporal behavior just beyond bifurcation is the slow growth of the excited mode at a rate $e^{\varepsilon\Delta\mu t}$. This motivates the introduction of a slow time-scale $\tau = \epsilon t$ (not to be confused with a synaptic time constant). Finally, assuming that the input is only weakly orientation–dependent, we perform the rescaling $h(\theta, t) = \bar{h} + \varepsilon^{3/2}\Delta h(\theta, t)$ (so that the input couples to the cubic amplitude equation). Collecting terms with equal powers of $\varepsilon$ then leads to a hierarchy of equations of the form

$$\bar{a} = W_0 f(\bar{a}) + \bar{h} \tag{4.10}$$

$$\hat{L}a_1 = 0 \tag{4.11}$$

$$\hat{L}a_2 = v_2$$
$$\equiv -g_2 w * a_1^2 \tag{4.12}$$

$$\hat{L}a_3 = v_3 \tag{4.13}$$
$$\equiv -g_3 w * a_1^3 - 2g_2 w * a_1 a_2 + \left[\frac{\partial a_1}{\partial \tau} - \Delta\mu\, w * a_1\right] - \Delta h$$

where

$$\hat{L} = -a + \mu_c w * a \tag{4.14}$$

and $*$ denotes the convolution operation

$$w * a(\theta) = \int_0^\pi w(\theta - \theta')a(\theta')\frac{d\theta'}{\pi} \tag{4.15}$$

The $\mathcal{O}(1)$ equation determines the fixed point $\bar{a}$. The $\mathcal{O}(\epsilon^{1/2})$ equation has solutions of the form

$$a_1 = z(\tau)\, e^{2i\theta} + z^*(\tau)\, e^{-2i\theta} \tag{4.16}$$

A dynamical equation for the complex amplitude $z(\tau)$ can be obtained by deriving solvability conditions for the higher–order equations, a method known as the Fredholm alternative. These equations have the general form

$$\hat{L} a_n = v_n(a_0, a_1, \ldots a_{n-1}) \tag{4.17}$$

$n \geq 2$, so that they can be solved sequentially. We proceed by taking the inner product of equations (4.12) and (4.13) with the linear eigenmode (4.16). We define the inner product of two periodic functions $U$, $V$ according to

$$\langle U | V \rangle = \int_0^\pi U^*(\theta) V(\theta)\, \frac{d\theta}{\pi} \tag{4.18}$$

Performing an integration by parts it is simple to show that the linear operator $\widehat{L}$ is self-adjoint with respect to this inner product, so that

$$\langle \tilde{a} | \hat{L} a_n \rangle = \langle \hat{L}\tilde{a} | a_n \rangle = 0 \tag{4.19}$$

for $\tilde{a} = e^{\pm 2i\theta}$. Since $\hat{L} a_n = v_n$, we obtain the set of solvability conditions

$$\langle \tilde{a} | v_n \rangle = 0, \quad n \geq 2 \tag{4.20}$$

The $\mathcal{O}(\epsilon)$ solvability condition $\langle \tilde{a} | v_2 \rangle = 0$ is automatically satisfied, whereas the $\mathcal{O}(\epsilon^{3/2})$ solvability condition $\langle \tilde{a} | v_3 \rangle = 0$ can be written

$$\langle \tilde{a} | \frac{\partial a_1}{\partial \tau} - \Delta\mu\, w * a_1 \rangle = g_3 \langle \tilde{a} | w * a_1^3 \rangle + 2g_2 \langle a_1 | w * a_1 a_2 \rangle + \langle \tilde{a} | \Delta h \rangle \tag{4.21}$$

Taking $\tilde{a} = e^{2i\theta}$ then generates a cubic amplitude for $z$. First, we have

$$\langle e^{2i\theta} | \frac{\partial a_1}{\partial \tau} - \Delta\mu\, w * a_1 \rangle = \frac{dz}{d\tau} - \mu_c^{-1} \Delta\mu\, z \tag{4.22}$$

and

$$\langle e^{2i\theta} | w * a_1^3 \rangle = \mu_c^{-1} \int_0^\pi \frac{d\theta}{\pi} \left( z e^{2i\theta} + z^* e^{-2i\theta} \right)^3 e^{-2i\theta} = 3\mu_c^{-1} z|z|^2 \tag{4.23}$$

We have used the result that

$$\langle e^{2i\theta} | w * b \rangle = \langle w * e^{2i\theta} | b \rangle = W_1 \langle e^{2i\theta} | b \rangle$$

for any $b$ and $W_1 = \mu_c^{-1}$. The next step is to determine $a_2$. From equation (4.12) we have

$$
\begin{aligned}
a_2(\theta) &- \mu_c \int_0^\pi \frac{d\theta'}{\pi} w(\theta - \theta') a_2(\theta') = g_2 \int_0^\pi \frac{d\theta'}{\pi} w(\theta - \theta') a_1(\theta')^2 \\
&= g_2 \left[ z^2 W_2 e^{4i\theta} + z^{*2} W_2 e^{-4i\theta} + 2|z|^2 W_0 \right]
\end{aligned}
\tag{4.24}
$$

Let

$$
a_2(\theta) = A_+ e^{4i\theta} + A_- e^{-4i\theta} + A_0 + \zeta a_1(\theta)
\tag{4.25}
$$

The constant $\zeta$ remains undetermined at this order of perturbation but does not appear in the amplitude equation for $z(\tau)$. Substituting equation (4.25) into (4.24) yields

$$
A_+ = \frac{g_2 z^2 W_2}{1 - \mu_c W_2}, \quad A_- = \frac{g_2 z^{*2} W_2}{1 - \mu_c W_2}, \quad A_0 = \frac{2 g_2 |z|^2 W_0}{1 - \mu_c W_0}
\tag{4.26}
$$

We then find that

$$
\begin{aligned}
\langle e^{2i\theta} | w * a_1 a_2 \rangle &= \mu_c^{-1} \int_0^\pi \frac{d\theta}{\pi} \left( z e^{2i\theta} + z^* e^{-2i\theta} \right) \\
&\quad \times \left( A_+ e^{4i\theta} + A_- e^{-4i\theta} + A_0 + \kappa a_1(\theta) \right) e^{-2i\theta} \\
&= \mu_c^{-1} [z^* A_+ + z A_0] \\
&= z |z|^2 g_2 \mu_c^{-1} \left[ \frac{W_2}{1 - \mu_c W_2} + \frac{2 W_0}{1 - \mu_c W_0} \right]
\end{aligned}
\tag{4.27}
$$

Finally, combining equations (4.22), (4.23) and (4.27), we obtain the Stuart-Landau equation

$$
\frac{dz}{d\tau} = z(\Delta\mu - \Lambda|z|^2) + \overline{\Delta h}
\tag{4.28}
$$

where $\overline{\Delta h} = \langle a_1 | \Delta h \rangle$ and

$$
\Lambda = -3g_3 - 2g_2^2 \left[ \frac{W_2}{1 - \mu_c W_2} + \frac{2 W_0}{1 - \mu_c W_0} \right]
\tag{4.29}
$$

(We have also absorbed a factor of $\mu_c$ into $\tau$ and $\overline{\Delta h}$).

In the case of a uniform external input ($\Delta h = 0$), the phase of $z$ is arbitrary (reflecting a marginal state) whereas the amplitude is given by $|z| = \sqrt{|\mu - \mu_c|/\Lambda}$. It is clear that a stable marginal state will bifurcate from the homogeneous state if and only if $\Lambda < 0$. In the case of a sigmoidal firing rate functions $f$, one finds that the bifurcation is indeed supercritical. Now suppose that there is a weakly biased input of the form $\Delta h(\theta, t) = C \cos(2[\theta - \omega t])$ for some slow frequency $\omega$. Then $\overline{\Delta h} = C e^{-2i\omega t}$. Writing $z = v e^{-2i(\phi + \omega t)}$ (with the phase $\phi$ defined in a rotating frame) we obtain the pair of equations

$$
\begin{aligned}
\dot{v} &= v(\Delta \mu - \Lambda v^2) + C \cos 2\phi \\
\dot{\phi} &= -\omega - \frac{C}{2v} \sin(2\phi)
\end{aligned} \tag{4.30}
$$

Thus, provided that $\omega$ is sufficiently small, equation (4.30) will have a stable fixed point solution in which the peak $\theta$ of the pattern is entrained to the signal.

### 4.3. Bifurcation theory and O(2) symmetry

It is possible to derive higher order contributions to the amplitude equation (4.28) by analyzing higher order solvability conditions. This would be necessary, for example, in order to investigate the possibility of secondary bifurcations. However, considerable information about the structure of the amplitude equation can be deduced from the underlying symmetries of the weight distribution $w$. Since symmetric bifurcation theory proves an invaluable tool in more complicated cortical models, we introduce some of the basic ideas here.

Suppose that the weight distribution $w$ is invariant with respect to the action of the group O(2) on $S^1$:

$$
\gamma \cdot w(\theta|\theta') = w(\gamma^{-1} \cdot \theta|\gamma^{-1} \cdot \theta') = w(\theta|\theta').
$$

with $\gamma \in \{\xi, \kappa\}$ such that $\xi \cdot \theta = \theta + \xi$ (rotations) and $\kappa \cdot \theta = -\theta$ (reflections). Consider the corresponding action of $\gamma$ on equation (4.1) for uniform input $h = \bar{h}$:

$$
\begin{aligned}
\frac{\partial a(\gamma^{-1}\theta, t)}{\partial t} &= -a(\gamma^{-1}\theta, t) + \int_0^\pi w(\gamma^{-1}\theta|\theta') f[a(\theta', t)] \frac{d\theta'}{\pi} \\
&= -a(\gamma^{-1}\theta, t) + \int_0^\pi w(\theta|\gamma\theta') f[a(\theta', t)] \frac{d\theta'}{\pi} \\
&= -a(\gamma^{-1}\theta, t) + \int_0^\pi w(\theta|\theta'') f[a(\gamma^{-1}\theta'', t)] \frac{d\theta''}{\pi}
\end{aligned}
$$

since $d[\gamma^{-1}\theta] = \pm d\theta$ and $w$ is O(2) invariant. If we rewrite equation (4.1) as an operator equation, namely,

$$
F[a] \equiv \frac{da}{dt} - G[a] = 0, \tag{4.31}
$$

then it follows that $\gamma F[a] = F[\gamma a]$. Thus F commutes with $\gamma \in O(2)$ and F is said to be *equivariant* with respect to the symmetry group $O(2)$ [45].

The equivariance of the operator F with respect to the action of $O(2)$ has major implications for the nature of solutions bifurcating from a uniform resting state $\bar{a}$. Sufficiently close to the bifurcation point these states are characterized by (finite) linear combinations of eigenfunctions of the linear operator $\hat{L} = D_0 G$ obtained by linearizing equation (4.1) about $\bar{a}$. (Since the linear operator commutes with the $O(2)$ group action, the eigenfunctions consist of irreducible representations of the group, that is, the Fourier modes $a_n(\theta) = z_n e^{2in\theta} + z_n^* e^{-2in\theta}$). The original infinite–dimensional equation (4.1) can then be projected on to this finite–dimensional space leading to a system of ODEs that constitute the amplitude equation (the so–called center manifold reduction). The major point to note is that the resulting amplitude equation for these modes is also equivariant with respect to $O(2)$ but with a different group action. For example, suppose that there is a single bifurcating mode given by $n = 1$. Under the action of $O(2)$,

$$a_1(\theta + \xi) = ze^{2i\xi}e^{2i\theta} + z^*e^{-2i\xi}e^{-2i\theta}, \quad a_1(-\theta) = ze^{-2i\theta} + z^*e^{2i\theta} \quad (4.32)$$

It follows that the action of $O(2)$ on $(z, z^*)$ is

$$\xi \cdot (z, z^*) = (ze^{2i\xi}, z^*e^{-2i\xi}), \quad \kappa \cdot (z, z^*) = (z^*, z) \quad (4.33)$$

Equivariance of the amplitude equation with respect to these transformations implies that quadratic and quartic terms are excluded from equation (4.28) and the quintic term is of the form $z|z|^4$. Once the basic form of the amplitude equation has been obtained, it still remains to determine the different types of patterns that are selected and their stability. In the case of the cubic amplitude equation (4.28) for a single eigenmode this is relatively straightforward. On the other hand, if more than one eigenmode is excited above the bifurcation point (due to additional degeneracies), then finding solutions is more involved. Again group theoretic methods can be used to identify the types of solution that are expected to occur (see section 5.5).

As another illustration of the use of symmetry arguments, we can directly write down the form of the cubic amplitude equation for a one–population version of the $O(3)$ symmetric spherical model described at the end of section 3.5:

$$\frac{\partial a(\theta, \phi, t)}{\partial t} = -a(\theta, \phi, t)$$
$$+ \int_0^\pi \int_0^\pi w(\theta, \phi | \theta', \phi') f(a(\theta', \phi', t)) \sin \phi' \frac{d\phi' d\theta'}{2\pi}$$
$$(4.34)$$

with

$$w(\theta, \varphi | \theta', \varphi') = \sum_{n=0}^{\infty} W_n \sum_{m=-n}^{n} Y_n^{m*}(\theta', \varphi') Y_n^m(\theta, \varphi) \tag{4.35}$$

Suppose that a homogeneous equilibrium undergoes a stationary bifurcation due to the excitation of the first order spherical harmonics, which will occur if $W_1 = \max_n \{W_n\}$. The excited modes can be written as the linear combination

$$a(\theta, \phi, \tau) = c_0(\tau) \cos \phi + c_1(\tau) \sin \phi \cos 2\theta + c_{-1}(\tau) \sin \phi \sin 2\theta \tag{4.36}$$

for real coefficients $c_0, c_{\pm 1}$. Since the associated amplitude equation is equivariant with respect to $\mathbf{O}(3)$ it follows that at cubic order it takes the form

$$\frac{dc_m}{dt} = c_m \left( \Delta\mu - \Lambda \sum_{p=0,\pm 1} c_p^2 \right) \tag{4.37}$$

The mode–dependent coefficient $\Lambda$ can be calculated explicitly using the same perturbation method as used for the ring model [19].

### 4.4. Two–population model

The above analysis can be extended to the case of a two–population (or a multi–population) model [17,35,91]. The major difference is that it is now possible for oscillatory patterns to occur. Consider for concreteness a two–population ring model of the form

$$\frac{\partial a_i(\theta, t)}{\partial t} = -a_i(\theta, t) + \sum_{j=E,I} \int w_{ij}(\theta - \theta') f(a_j(\theta', t)) \frac{d\theta'}{\pi} + h_i \tag{4.38}$$

For simplicity, the excitatory and inhibitory populations are assumed to have the same firing rate function $f$. Let $(\bar{a}_E, \bar{a}_I)$ denote an equilibrium solution of equation (4.38) in the case of uniform inputs $\bar{h}_i$, which satisfies the algebraic equation

$$\bar{a}_i = \sum_{j=E,I} W_{ij}(0) f(\bar{a}_j) + \bar{h}_i \tag{4.39}$$

with $W_{ij}(0) = \int_0^{\pi} w_{ij}(\theta) d\theta / \pi$. The local stability of the equilibrium $(\bar{a}_E, \bar{a}_I)$ is found by setting $a_i(\theta, t) = \bar{a}_i + a_i(\theta) e^{\lambda t}$ and linearizing:

$$(1 + \lambda) a_i(\theta) = \mu \sum_{j=E,I} \int_0^{\pi} w_{ij}(\theta - \theta') a_j(\theta') \frac{d\theta'}{\pi} \tag{4.40}$$

For convenience we have performed a rescaling of the weights according to $f'(\bar{a}_i)w_{ij} \rightarrow \mu w_{ij}$ where $\mu$ measures the degree of network excitability. Equation (4.40) has solutions of the form

$$a_i(\theta) = A_i \left[ ze^{2in\theta} + z^* e^{-2in\theta} \right] \tag{4.41}$$

where $\lambda$ satisfies the eigenvalue equation

$$(1 + \lambda)\mathbf{A} = \mu \mathbf{W}(n)\mathbf{A} \tag{4.42}$$

with $\mathbf{A} = (A_E, A_I)^T$, and $W_{ij}(n)$ is the $n$th Fourier coefficient in the expansion of the $\pi$-periodic weights kernels $w_{ij}(\theta)$, see equation (3.28). It follows that

$$\lambda_n^{\pm} = -1 + \mu W_n^{\pm} \tag{4.43}$$

for integer $n$, where

$$W_n^{\pm} = \frac{1}{2} [W_{EE}(n) + W_{II}(n) \pm \Sigma(n)] \tag{4.44}$$

are the eigenvalues of the weight matrix with

$$\Sigma(n) = \sqrt{[W_{EE}(n) - W_{II}(n)]^2 + 4W_{EI}(n)W_{IE}(n)} \tag{4.45}$$

The corresponding eigenvectors (up to an arbitrary normalization) are

$$\mathbf{A}_n^{\pm} = \begin{pmatrix} -W_{EI}(n) \\ \frac{1}{2}[W_{EE}(n) - W_{II}(n) \mp \Sigma(n)] \end{pmatrix} \tag{4.46}$$

Equation (4.43) implies that, for sufficiently small $\mu$ (low network excitability), $\lambda_n^{\pm} < 0$ for all $n$ and the homogeneous resting state is stable. However, as $\mu$ increases an instability can occur leading to the spontaneous formation of an orientation tuning curve.

For the sake of illustration, suppose that the Fourier coefficients are given by the Gaussians

$$W_{ij}(n) = \alpha_{ij} e^{-n^2 \xi_{ij}^2 / 2}, \tag{4.47}$$

with $\xi_{ij}$ determining the range of the axonal fields of the excitatory and inhibitory populations. We consider two particular cases.

*Case A*  If $W_{EE}(n) = W_{IE}(n)$ and $W_{II}(n) = W_{EI}(n)$ for all $n$, then $W_n^- = 0$ and

$$W_n^+ = \alpha_{EE} e^{-n^2 \xi_{EE}^2/2} - |\alpha_{II}| e^{-n^2 \xi_{II}^2/2} \tag{4.48}$$

Suppose that $\xi_{II} > \xi_{EE}$ and $0 < |\alpha_{II}| < \alpha_{EE}$, which corresponds to a mixture of short range excitation and longer range inhibition, such that $W_n^+$ has a unique (positive) maximum at $n = 1$, see Fig. 18a. The homogeneous state then desta-bilizes at the critical point $\mu = \mu_c \equiv 1/W_1^+$ due to excitation of the eigenmodes $\mathbf{a}(\theta) = \mathbf{A}a(\theta)$ with $\mathbf{A} = (1, 1)^T$ and

$$a(\theta) = z e^{2i\phi} + \bar{z} e^{-2i\phi} = |z| \cos(2[\theta - \theta_0]) \tag{4.49}$$

with $z = |z| e^{-2i\theta_0}$.

*Case B*  If $\Sigma(n)$ is pure imaginary, $\Sigma(n) = i\Omega(n)$, then

$$W_n^{\pm} = \alpha_{EE} e^{-n^2 \xi_{EE}^2/2} - \alpha_{II} e^{-n^2 \xi_{II}^2/2} \pm i\Omega(n) \tag{4.50}$$

Assume, as in case $A$, that the difference of Gaussians has a maximum at $n = 1$. Then an instability will occur at the critical point $\mu_c = 1/\Re(W_1^+)$ due to excitation of the oscillatory eigenmodes

$$\mathbf{a}(\theta, t) = \left[ z_L e^{i(\Omega_0 t - 2\theta)} + z_R e^{i(\Omega_0 t + 2\theta)} \right] \mathbf{A} + c.c. \tag{4.51}$$

where $\mathbf{A} = \mathbf{A}_1^+$, $z_L$ and $z_R$ represent the complex amplitudes for anti-clockwise ($L$) and clockwise ($R$) rotating waves around the ring, and $\Omega_0 = \mu_c \Omega(1)$.

### 4.5. Amplitude equation for oscillatory patterns

We shall use the Fredholm alternative to derive an amplitude equation for the pair $(z_L, z_R)$. The corresponding equation for stationary patterns is then obtained by setting $\Omega_0 = 0$ and $z_R = z_L^*$. Suppose that each isolated hypercolumn is $\varepsilon$-close to the point of marginal stability of the homogeneous fixed point $(\bar{a}_E, \bar{a}_I)$ due to the excitation of the stationary modes (4.51). That is, $\mu = \mu_c + \varepsilon \Delta \mu$ where $\mu_c = 1/\Re(W_1^+)$, see equation (4.43). Perform a Taylor expansion of equation (4.38) about the equilibrium $\bar{a}_i$ with $b_i = a_i - \bar{a}_i$:

$$\frac{\partial b_i}{\partial t} = -b_i + \sum_{j=E,I} w_{ij} * \left[ \mu b_j + \gamma_j b_j^2 + \gamma_j' b_j^3 + \ldots \right] + \Delta h_i \tag{4.52}$$

where $\Delta h_i = h_i - \bar{h}_i$, $\mu = f'(\bar{a}_E)$, $\gamma_j = f''(\bar{a}_j)/2$, $\gamma'_j = f'''(\bar{a}_j)/6$. Substitute into equation (4.52) the perturbation expansion

$$a_j = \bar{a}_j + \varepsilon^{1/2} a_j^{(1)} + \varepsilon a_j^{(2)} + \varepsilon^{3/2} a_j^{(3)} + \dots \tag{4.53}$$

In order to distinguish between the oscillatory part of the solution and the slow rate of growth in the amplitude close to the bifurcation point we introduce both a fast and a slow time scale by setting $a_j = a_j(\theta, t, \tau)$ with $\tau = \mathcal{O}(\varepsilon)$ and

$$\frac{\partial}{\partial t} \to \frac{\partial}{\partial t} + \varepsilon \frac{\partial}{\partial \tau} \tag{4.54}$$

Finally, perform the rescaling $\Delta h_i \to \varepsilon^{3/2} \Delta h_i$. Collecting terms with equal powers of $\varepsilon$ then leads to a hierarchy of equations of the form (up to $\mathcal{O}(\varepsilon^{3/2})$)

$$[\hat{L}_t \mathbf{a}^{(1)}]_i = 0 \tag{4.55}$$

$$[\hat{L}_t \mathbf{a}^{(2)}]_i = v_i^{(2)}$$

$$\equiv \sum_{j=E,I} \gamma_j w_{ij} * [a_j^{(1)}]^2 \tag{4.56}$$

$$[\hat{L}_t \mathbf{a}^{(3)}]_i = v_i^{(3)}$$

$$\equiv -\frac{\partial a_i^{(1)}}{\partial \tau} + \sum_{j=E,I} w_{ij} * \left[ \Delta \mu a_j^{(1)} + \gamma'_j [a_j^{(1)}]^3 + 2\gamma_j a_j^{(1)} a_j^{(2)} \right]$$

$$+ \Delta h_i \tag{4.57}$$

with the linear operator $\hat{L}_t$ defined according to

$$\hat{L}_t \mathbf{a}_i = \frac{\partial a_i}{\partial t} + a_i - \mu_c \sum_{j=E,I} w_{ij} * a_j \tag{4.58}$$

The first equation in the hierarchy, equation (4.55), has solutions of the form (4.51). We obtain an amplitude equation for the pair $(z_L, z_R)$ by deriving solvability conditions for the higher order equations using a Fredholm alternative. Introduce the generalized inner product

$$\langle \mathbf{u} | \mathbf{v} \rangle = \lim_{T \to \infty} \frac{1}{T} \int_{-T/2}^{T/2} \int_0^\pi [\bar{u}_E(\theta, t) v_E(\theta, t) + \bar{u}_I(\theta, t) v_I(\theta, t)] \frac{d\theta}{\pi} dt \tag{4.59}$$

and the dual vectors $\widetilde{\mathbf{a}}_L = \widetilde{\mathbf{A}}^* e^{i(\Omega_0 t - 2\theta)}$, $\widetilde{\mathbf{a}}_R = \widetilde{\mathbf{A}}^* e^{i(\Omega_0 t + 2\theta)}$ with

$$\widetilde{\mathbf{A}} = \begin{pmatrix} W_{IE}(1) \\ -\frac{1}{2}[W_{EE}(1) - W_{II}(1) - \Sigma(1)] \end{pmatrix} \tag{4.60}$$

such that

$$\langle \tilde{\mathbf{a}}_{L,R} | \widehat{L}_t \mathbf{a}^{(n)} \rangle = \langle \widehat{L}_t^\dagger \tilde{\mathbf{a}}_{L,R} | \mathbf{a}^{(n)} \rangle = 0 \tag{4.61}$$

We thus obtain the pair of solvability conditions

$$\langle \tilde{\mathbf{a}}_L | \mathbf{v}^{(n)} \rangle = \langle \tilde{\mathbf{a}}_R | \mathbf{v}^{(n)} \rangle = 0 \tag{4.62}$$

for each $n \geq 2$. As in the the stationary case, the $n = 2$ solvability conditions are identically satisfied. The $n = 3$ solvability conditions then generate cubic amplitude equations for $z_L, z_R$ of the form [17]

$$\frac{dz_L(\tau)}{d\tau} = (1 + i\Omega_0)z_L(\tau)(\Delta\mu - \Lambda|z_L(\tau)|^2 - 2\Lambda|z_R(\tau)|^2) + \overline{\Delta h_-} \tag{4.63}$$

and

$$\frac{dz_R(\tau)}{d\tau} = (1 + i\Omega_0)z_R(\tau)(\Delta\mu - \Lambda|z_R(\tau)|^2 - 2\Lambda|z_L(\tau)|^2) + \overline{\Delta h_+} \tag{4.64}$$

where

$$\overline{\Delta h_\pm} = \lim_{T \to \infty} \frac{\mu_c}{T} \int_{-T/2}^{T/2} \int_0^\pi e^{-i(\Omega_0 t \pm 2\theta)} \sum_{j=E,I} \tilde{A}_j \Delta h_j(\theta, t) \frac{d\theta}{\pi} dt \tag{4.65}$$

and

$$\Lambda = -\frac{3}{\tilde{\mathbf{A}}^T \mathbf{A}} \sum_{j=E,I} \tilde{A}_j \gamma_j' A_j^3 \tag{4.66}$$

Note that the amplitudes only couple to time-dependent inputs from the LGN. As in the stationary case, group theoretic methods can be used to determine generic aspects of the bifurcating solutions. As with other Hopf bifurcation problems [32, 45], the amplitude equations have an extra phase-shift symmetry in time that was not in the original problem. This takes the form $\psi : (z_L, z_R) \to (e^{i\psi} z_L, e^{i\psi} z_R)$ with $\psi \in \mathbf{S}^1$. This is distinct from rotations $\theta \to \theta + \xi$, which corresponds to the action $\psi : (z_L, z_R) \to (e^{2i\xi} z_L, e^{-2i\xi} z_R)$. Thus the full symmetry group is now $\mathbf{O}(2) \times \mathbf{S}^1$.

### 4.6. Mean-field theory

One of the limitations of the above analysis is that it is restricted to a regime close to the bifurcation point. Therefore, it is difficult to establish analytically that the resulting tuning curves are contrast invariant, although one finds this to be approximately true numerically. Therefore, we consider here an alternative, activity-based version of the ring model in which the output function $f$ is taken to be a linear-threshold function. It is then possible to derive exact solutions for contrast-invariant orientation tuning curves provided that restrictions are imposed on the form of the weight distribution $W$ [5, 47]. We will restrict our discussion to stationary tuning curves in a one population model. Extensions to oscillatory patterns in a two population model have been developed in some detail elsewhere [6].

Consider the activity-based ring model

$$\frac{\partial a(\theta, t)}{\partial t} = -a(\theta, t) + [I(\theta, t) - \kappa]_+ \tag{4.67}$$

where $[\ ]_+$ denotes half-wave rectification, $\kappa$ is a threshold and

$$I(\theta, t) = h(\theta, t) + \int_{-\pi/2}^{\pi/2} w(\theta - \theta')a(\theta', t)\frac{d\theta'}{\pi} \tag{4.68}$$

Following Hansel and Sompolinsky [47], we assume for analytical convenience that

$$w(\theta) = W_0 + 2W_1 \cos(2\theta) \tag{4.69}$$

If $W_0 - 2W_1 < 0$ and $W_0 + 2W_1 > 0$ then this represents a Mexican hat type of weight distribution. Similarly, we take

$$h(\theta) = H_0 + 2H_1 \cos(2[\theta - \Theta]) \tag{4.70}$$

where $\Theta$ is stimulus orientation. (Conditions under which higher-order harmonic contributions to the LGN input are negligible were derived in section 3.2). Under these simplifying various assumptions, stationary solutions of the ring model (4.67) can be written in the form

$$a(\theta) = [A_0 + 2A_1 \cos(2[\theta - \Theta])]_+ \tag{4.71}$$

where

$$A_0 = H_0 + W_0 R_0 - \kappa \tag{4.72}$$

$$A_1 = H_1 + W_1 R_1 \tag{4.73}$$

and $R_0$, $R_1$ are the order parameters

$$R_0 = \int_{-\pi/2}^{\pi/2} a(\theta) \frac{d\theta}{\pi} \tag{4.74}$$

$$R_1 = \int_{-\pi/2}^{\pi/2} a(\theta) \cos(2[\theta - \Theta]) \frac{d\theta}{\pi} \tag{4.75}$$

First, consider the linear regime in which we have a broad activity profile $a(\theta) > 0$ for all all $-\pi/2 \le \theta \le \pi/2$. Then $R_0 = A_0$ and $R_1 = A_1$ so that

$$A_0 = \frac{H_0 - \kappa}{1 - W_0}, \quad A_1 = \frac{H_1}{1 - W_1} \tag{4.76}$$

In order that this solution is self-consistent, we require $A_0 > 2A_1$, which leads to the condition

$$g := \frac{H_0 - \kappa}{2H_1} > \frac{[1 - W_0]}{1 - W_1} \tag{4.77}$$

If this condition is not satisfied then it is necessary to take into account the effects of the rectifying nonlinearity. Suppose, therefore, that $a(\theta)$ is only non-zero over a proper subinterval of the ring so that the response is narrowly tuned with respect to orientation. That is, there exists $0 < \theta_c < \pi/2$ such that $a = 0$ for all $\theta_c \le |\theta - \Theta| \le \pi/2$. It follows that the width $\theta_c$ of the activity profile satisfies the equation

$$A_0 + 2A_1 \cos(2\theta_c) = 0 \tag{4.78}$$

Taking moments of the fixed point equation (4.71) with respect to the zeroth and first order harmonics,

$$R_0 = A_0 \int_{-\theta_c}^{\theta_c} \frac{d\theta}{\pi} + 2A_1 \int_{-\theta_c}^{\theta_c} \cos(2\theta) \frac{d\theta}{\pi} \tag{4.79}$$

and

$$R_1 = A_0 \int_{-\theta_c}^{\theta_c} \cos(2\theta) \frac{d\theta}{\pi} + 2A_1 \int_{-\theta_c}^{\theta_c} \cos^2(2\theta) \frac{d\theta}{\pi} \tag{4.80}$$

Performing the integration over $\theta$ and using equation (4.78) then gives

$$R_0 = 2A_1 c_0(\theta_c) \tag{4.81}$$

and

$$R_1 = A_1 c_1(\theta_c) \tag{4.82}$$

where

$$c_0(\theta_c) = \frac{1}{\pi} [\sin 2\theta_c - 2\theta_c \cos 2\theta_c] \tag{4.83}$$

and

$$c_1(\theta_c) = \frac{2}{\pi} \left[ \theta_c - \frac{1}{4} \sin 4\theta_c \right] \tag{4.84}$$

Finally, substituting equations (4.72) and (4.73) into equations (4.81) and (4.82) shows that

$$A_0 = \frac{H_0 - \kappa}{1 + \dfrac{W_0 c_0(\theta_c)}{\cos(2\theta_c)}} \tag{4.85}$$

and

$$A_1 = \frac{H_1}{1 - W_1 c_1(\theta_c)} \tag{4.86}$$

where the width $\theta_c$ is determined from the equation

$$-\cos 2\theta_c \equiv \frac{A_0}{2A_1} = \frac{H_0 - \kappa}{2H_1} [1 - W_1 c_1(\theta_c)] + W_0 c_0(\theta_c)$$

This can be rearranged into the form

$$G(\theta_c) := \frac{\cos 2\theta_c + W_0 c_0(\theta_c)}{1 - W_1 c_1(\theta_c)} = -g \tag{4.87}$$

with $g$ defined by equation (4.77). Given the critical angle $\theta_c$ and amplitude $A_1$, the resulting narrowly tuned state takes the form

$$a(\theta) = [2A_1(\cos(2[\theta - \Theta]) - \cos 2\theta_c]_+ \tag{4.88}$$

Following Hansel and Sompolinsky [47], it is useful to distinguish between two phases within the nonlinear regime:

*Weak cortical modulation* $(W_0 < 1, W_1 < 1)$    If the recurrent cortical modulation is sufficiently weak then the width $\theta_c$ of the activity profile is sensitive to the size of $\Gamma$, and hence to the size of the stimulus spatial frequency $P$ and the contrast $C$ (see dashed curve in Fig. 19b). Setting $\theta_c = \pi/2$ in equation (4.87) and using $c_0(\pi/2) = 1, c_1(\pi/2) = 1$, it follows that when $W_0, W_1 < 1$ a narrowly tuned state $(\theta_c < \pi/2)$ only exists if

$$g < \frac{[1 - W_0]}{1 - W_1} \tag{4.89}$$

otherwise the network operates in the linear regime.

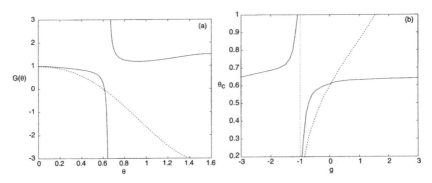

Fig. 19. (a) Plot of function $G(\theta)$ defined by equation (4.87) as a function of width $\theta$. (b) Plot of width $\theta_c$ as a function of $g$, which is obtained by solving the equation $G(\theta_c) = -g$. Solid curves: strong cortical modulation with $W_0 = -2$, $W_1 = 3$. Dashed curves: weak cortical modulation with $W_0 = -2$, $W_1 = 0.1$.

*Strong cortical modulation ($W_0 < W_c$, $W_1 > 1$)*   Inspection of equation (4.86) shows that a narrowly tuned state persists even when $H_1 = 0$ provided that

$$1 = W_1 c_1(\theta_c) \qquad (4.90)$$

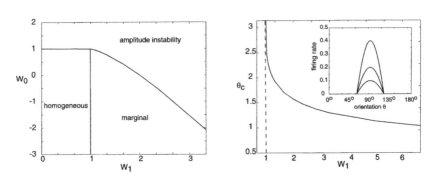

Fig. 20. (a) Phase diagram for ring model in the case of a homogeneous input $H_1 = 0$. (b) Variation of critical angle $\theta_c$ as a function of cortical modulation $W_1$ in the case of a homogeneous input $H_1 = 0$. Inset shows contrast invariant tuning curves.

Since $c_1(\theta_c) < 1$ for $0 \le \theta_c \le \pi/2$, it follows that $W_1 > 1$ is a necessary condition for a narrowly tuned activity profile to occur when $H_1 = 0$. The location of the peak of the orientation tuning curve is now arbitrary since the LGN input is independent of orientation. In other words, there is a continuum of activity profiles that form a manifold of marginally stable fixed points, and the system is said to be in a *marginal phase*. In such a phase, a narrowly tuned state *spontaneously breaks* the underlying $O(2)$ symmetry of the network, which is possible because the spatial modulation of the cortical interactions is sufficiently strong. In the marginal phase, the critical angle $\theta_c$ is determined by equation (4.90) and is thus independent of the contrast. Equations (4.85) and (4.78) imply that

$$A_1 = -\frac{H_0 - \kappa}{\cos(2\theta_c) + W_0 c_0(\theta_c)} \qquad (4.91)$$

In order that the amplitude be finite, we require $W_0 < W_c$ where

$$W_c = -\frac{\cos(2\theta_c)}{c_0(\theta_c)} \qquad (4.92)$$

Performing a stability analysis shows that as $W_0$ approaches $w_c$ the system undergoes an amplitude instability. The variation of the critical angle $\theta_c$ as a function of $W_1$ is plotted in Fig. 20b. When $H_1 \neq 0$ one still finds the width $\theta_c$ to be approximately independent of $g$ over a significant range of values (see solid curve in Fig. 19b). The phase diagram for the stability of the various states in the presence of a homogeneous input $H_1 = 0$ is shown in Fig. 20a.

Finally, note that the above mean field approach can be extended to the spherical model (4.34), under the constraint that the weight distribution and LGN input only involve zeroth and first order spherical harmonics [21].

## 5. Pattern formation in a coupled hypercolumn model of V1

In this section we extend the perturbation methods of the previous section to derive an amplitude equation for a coupled hypercolumn model of V1 in which each hypercolumn is modeled as a ring of orientation selective cells. (For a corresponding analysis of a coupled spherical model see [17, 19]). Consider the one–population coupled ring model

$$
\begin{aligned}
\frac{\partial a(\mathbf{r}, \theta, t)}{\partial t} &= -a(\mathbf{r}, \theta, t) + \int_0^\pi w(\theta - \theta') f(a(\mathbf{r}, \theta', t)) \frac{d\theta'}{\pi} + \bar{h} \qquad (5.1) \\
&+ \beta \int_{\mathbf{R}^2} \int_0^\pi w_{hoz}(\mathbf{r}, \theta | \mathbf{r}', \theta') f(a(\mathbf{r}', \theta', t)) \frac{d\theta'}{\pi} d^2\mathbf{r}'
\end{aligned}
$$

with $w_{hoz}$ given by equation (3.31). Since we are interested in spontaneous pattern formation, we assume that the feedforward input is uniform. Following along similar lines to section 4.3, it is simple to establish that in the case of isotropic weights ($\eta = \pi/2$) the coupled hypercolumn model (5.1) is equivariant with respect to the $\mathbf{E}(2) \times \mathbf{O}(2)$ group action (3.29) and for anisotropic weights it is equivariant with respect to the Euclidean shift–twist group action (3.30). We shall focus on the anisotropic case.

### 5.1. Linear stability analysis

Linearizing equation (5.1) about a uniform equilibrium $\bar{a}$ leads to the eigenvalue equation

$$\lambda a = \widehat{L} a \tag{5.2}$$

with $\widehat{L}$ the linear operator

$$\widehat{L} a = -a + \mu(w * a + \beta w_{hoz} \circ a) \tag{5.3}$$

The convolution operation $*$ and $\circ$ are defined according to

$$w * a(\mathbf{r}, \theta) = \int_0^\pi w(\theta - \theta') a(\mathbf{r}, \theta') \frac{d\theta'}{\pi} \tag{5.4}$$

and

$$w_{hoz} \circ a(\mathbf{r}, \theta) = \int_{\mathbf{R}^2} J(\mathbf{r} - \mathbf{r}', \theta) w^\Delta * a(\mathbf{r}', \theta) d^2\mathbf{r}' \tag{5.5}$$

with $J(\mathbf{r}, \theta) = J(|\mathbf{r}|)\mathcal{P}(\arg(\mathbf{r}) - \theta)$ and $\mathcal{P}$ given by equation (3.27).

Translation symmetry implies that in the case of an infinite domain, the eigenfunctions of equation (5.3) can be expressed in the form

$$a(\mathbf{r}, \theta) = u(\theta - \varphi)e^{i\mathbf{k}\cdot\mathbf{r}} + c.c. \tag{5.6}$$

with $\mathbf{k} = q(\cos\varphi, \sin\varphi)$ and

$$\lambda u(\theta) = -u(\theta) + \mu \left[ w * u(\theta) + \beta \widehat{J}(\mathbf{k}, \theta + \varphi) w^\Delta * u(\theta) \right]. \tag{5.7}$$

Here $\widehat{J}(\mathbf{k}, \theta)$ is the Fourier transform of $J(\mathbf{r}, \theta)$,

$$\widehat{J}(\mathbf{k}, \theta) = \int_{\mathbf{R}^2} e^{-i\mathbf{k}\cdot\mathbf{r}} J(\mathbf{r}, \theta) d^2\mathbf{r}. \tag{5.8}$$

Invariance of the full weight distribution under the Euclidean group action (3.30) restricts the structure of the solutions of the eigenvalue equation (5.7):

(i) The Fourier transform $\widehat{J}(\mathbf{k}, \theta + \varphi)$ is independent of the direction $\varphi = \arg(\mathbf{k})$. This is easy to establish as follows:

$$\begin{aligned}
\widehat{J}(\mathbf{k}, \theta + \varphi) &= \int_{\mathbf{R}^2} e^{-i\mathbf{k}\cdot\mathbf{r}} J(\mathbf{r}, \theta + \varphi) d^2\mathbf{r} \\
&= \int_0^\infty \int_{-\pi}^\pi e^{-iqr\cos(\psi-\varphi)} J(r)\mathcal{P}(\psi - \theta - \varphi) d\psi r\, dr \\
&= \int_0^\infty \int_{-\pi}^\pi e^{-iqr\cos(\psi)} J(r)\mathcal{P}(\psi - \theta) d\psi r\, dr \\
&= \widehat{J}(q, \theta)
\end{aligned} \tag{5.9}$$

Therefore, $\lambda$ and $u(\theta)$ only depend on the magnitude $q = |\mathbf{k}|$ of the wave vector $\mathbf{k}$ and there is an infinite degeneracy due to rotational invariance. Note, however, that the eigenfunction (5.6) depends on $u(\theta - \varphi)$, which reflects the shift–twist action of the rotation group.

(ii) For each $\mathbf{k}$ the associated subspace of eigenfunctions

$$V_{\mathbf{k}} = \{u(\theta - \varphi)e^{i\mathbf{k}\cdot\mathbf{r}} + c.c\} \tag{5.10}$$

decomposes into two invariant subspaces

$$V_{\mathbf{k}} = V_{\mathbf{k}}^+ \oplus V_{\mathbf{k}}^-, \tag{5.11}$$

corresponding to even and odd functions respectively

$$V_{\mathbf{k}}^+ = \{v \in V_{\mathbf{k}} : u(-\theta) = u(\theta)\}, \quad V_{\mathbf{k}}^- = \{v \in V_{\mathbf{k}} : u(-\theta) = -u(\theta)\}. \tag{5.12}$$

This is a consequence of reflection invariance, as we now indicate. That is, let $\kappa_{\mathbf{k}}$ denote reflections about the wavevector $\mathbf{k}$ so that $\kappa_{\mathbf{k}}\mathbf{k} = \mathbf{k}$. Then $\kappa_{\mathbf{k}}a(\mathbf{r}, \theta) = a(\kappa_{\mathbf{k}}\mathbf{r}, 2\varphi - \theta) = u(\varphi - \theta)e^{i\mathbf{k}\cdot\mathbf{r}} + c.c.$ Since $\kappa_{\mathbf{k}}$ is a reflection, any space that it acts on decomposes into two subspaces – one on which it acts as the identity $I$ and one on which it acts as $-I$. The even and odd functions correspond to scalar and pseudoscalar representations of the Euclidean group studied in a more general context by Bosch *et al* [11].

A further reduction of equation (5.7) can be achieved by expanding the $\pi$-periodic function $u(\theta)$ as a Fourier series with respect to $\theta$

$$u(\theta) = \sum_{n\in\mathbf{Z}} U_n e^{2in\theta} \tag{5.13}$$

This then leads to the matrix eigenvalue equation

$$\lambda U_n = [\mathbf{L}(q)\mathbf{U}]_n \equiv (-1 + \mu W_n)U_n + \beta \sum_{m\in\mathbf{Z}} \widehat{J}_{n-m}(q)\mathcal{P}_{n-m}W_m^\Delta U_m, \tag{5.14}$$

where a factor of $\mu$ has been absorbed into $\beta$ and

$$\widehat{J}_n(q) = \int_0^\infty \int_{-\pi}^\pi e^{-iqr\cos(\psi)} e^{-2in\psi} J(r) d\psi \, r dr \tag{5.15}$$

We have used equation (5.9) together with the Fourier series expansions

$$W(\theta) = \sum_{n\in\mathbf{Z}} e^{2in\theta} W_n, \quad W^\Delta(\theta) = \sum_{n\in\mathbf{Z}} e^{2in\theta} W_n^\Delta,$$

$$\mathcal{P}(\psi) = \sum_{n\in\mathbf{Z}} e^{2in\psi} \mathcal{P}_n \tag{5.16}$$

In the following we will take $W^\Delta(\theta) = \delta(\theta)$ so that $W_n^\Delta = 1$ for all $n$. Equation (3.27) implies that

$$\mathcal{P}_n = \frac{\sin 4n\eta}{4n\eta} \tag{5.17}$$

We now exploit the experimental observation that the long–range horizontal connections appear to be weak relative to the local connections. Equation (5.14) can then be solved by expanding as a power series in $\beta$ and using Rayleigh–Schrödinger perturbation theory. In the limiting case of zero horizontal interactions we recover the eigenvalues of the ring model, see section 4.1. In particular, suppose that $W_1 = \max\{W_n, n \in \mathbf{Z}\} > 0$. The homogeneous fixed point is then stable for sufficiently small $\mu$, but becomes marginally stable at the critical point $\mu_c = 1/W_1$ due to the vanishing of the eigenvalue $\lambda_1$. In this case both even and odd modes $\cos(2\phi)$ and $\sin(2\phi)$ are marginally stable. Each hypercolumn spontaneously forms an orientation tuning curve of the form $a(\mathbf{r}, \theta) = A_0 \cos(2[\theta - \theta_0(\mathbf{r})])$ such that the preferred orientation $\theta_0(\mathbf{r})$ is arbitrary at each point $\mathbf{r}$.

If we now switch on the lateral connections, then there is a $q$–dependent *splitting* of the degenerate eigenvalue $\lambda_1$ that also separates out odd and even solutions. Denoting the characteristic size of such a splitting by $\delta\lambda = \mathcal{O}(\beta)$, we impose the condition that $\delta\lambda \ll \mu\Delta W$, where $\Delta W = \min\{W_1 - W_m, m \neq 1\}$. This ensures that the perturbation does not excite states associated with other eigenvalues of the unperturbed problem. We can then restrict ourselves to calculating perturbative corrections to the degenerate eigenvalue $\lambda_1$ and its associated eigenfunctions. Therefore, introduce the power series expansions

$$\lambda = -1 + \mu W_1 + \beta\lambda^{(1)} + \beta^2\lambda^{(2)} + \cdots \tag{5.18}$$

and

$$U_n = z_{\pm 1}\delta_{n,\pm 1} + \beta U_n^{(1)} + \beta^2 U_n^{(2)} + \cdots \tag{5.19}$$

where $\delta_{n,m}$ is the Kronecker delta function. Substitute these expansions into the matrix eigenvalue equation (5.14) and systematically solve the resulting hierarchy of equations to successive orders in $\beta$ using (degenerate) perturbation theory. This analysis leads to the following result valid to $\mathcal{O}(\beta)$ [14]: $\lambda = \lambda_\pm$ for even (+) and odd (−) solutions where

$$\lambda_\pm = -1 + \mu W_1 + \beta \left[ \widehat{J}_0(q) \pm \mathcal{P}_2 \widehat{J}_2(q) \right] \tag{5.20}$$

with corresponding eigenfunctions

$$u_+(\phi) = \cos(2\phi) + \beta \sum_{m \geq 0, m \neq 1} u_m^+(q) \cos(2m\phi) \tag{5.21}$$

$$u_-(\phi) = \sin(2\phi) + \beta \sum_{m > 1} u_m^-(q) \sin(2m\phi) \tag{5.22}$$

with

$$u_0^+(q) = \frac{\mathcal{P}_1 \widehat{J}_1(q)}{W_1 - W_0},$$

$$u_m^\pm(q) = \frac{\mathcal{P}_{m-1} \widehat{J}_{m-1}(q) \pm \mathcal{P}_{m+1} \widehat{J}_{m+1}(q)}{W_1 - W_m}, \quad m > 1 \tag{5.23}$$

### 5.2. Marginal stability

Before using equation (5.20) to determine how the horizontal interactions modify the condition for marginal stability, we need to specify the form of the weight distribution $J(r)$. From experimental data based on tracer injections it appears that the patchy lateral connections extend several mm on either side of a hypercolumn and the field of axon terminals within a patch tends to diminish in size the further away it is from the injection site [42, 65, 78, 107]. The total extent of the connections depends on the particular species under study. In the continuum hypercolumn model we assume that

$$J(r) = e^{-(r-r_0)^2/2\xi^2} \Theta(r - r_0) \tag{5.24}$$

where $\xi$ determines the range and $r_0$ the minimum distance of the (non-local) horizontal connections. Recall that there is growing experimental evidence to suggest that horizontal connections tend to have an inhibitory effect in the presence of high contrast visual stimuli but an excitatory effect at low contrasts [44]. It is possible that during the experience of hallucinations there are sufficient levels of activity within V1 for the inhibitory effects of the lateral connections to predominate. Many subjects who have taken LSD and similar hallucinogens report seeing bright white light at the centre of the visual field which then explodes into a hallucinatory image in about 3 sec, corresponding to a propagation velocity in V1 of about 2.5 cm per sec. suggestive of slowly moving epileptiform activity. In light of this, we assume that $\beta < 0$ during the experience of a visual hallucination. An important point to note in the following is that it is possible to generate a pattern forming instability using a purely inhibitory weight distribution with a gap at the center. Thus it is not necessary to take $J(r)$ to be the standard Mexican hat function consisting of short–range excitation and longer range inhibition (see also section 6).

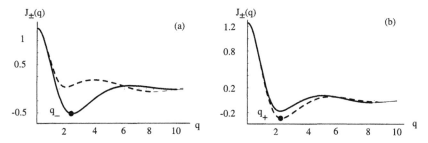

Fig. 21. (a) Plot of functions $J_-(q)$ (solid line) and $J_+(q)$ (dashed line) in the case $\mathcal{P}_2 = 1$ (strong anisotropy) and $J(r)$ defined by (5.24) for $\xi = 1$ and $r_0 = 1$. The critical wavenumber for spontaneous pattern formation is $q_-$. The marginally stable eigenmodes are odd functions of $\theta$. (b) Same as (a) except that $\mathcal{P}_2 = \sin 4\eta/4\eta$ with lateral spread of width $\eta = \pi/3$. The marginally stable eigenmodes are now even functions of $\theta$.

In Fig. 21 we plot $J_\pm(q) = \widehat{J}_0(q) \pm \mathcal{P}_2 \widehat{J}_2(q)$ as a function of $q$ for the given weight distribution (5.24) and two values of $\mathcal{P}_2$. *(a) Strong anisotropy:* If $\eta < \pi/4$ then $J_\pm(q)$ has a unique minimum at $q = q_\pm \neq 0$ and $J_-(q_-) < J_+(q_+)$. This is shown in the limit $\eta \to 0$ for which $\mathcal{P}_2 = 1$. If $\beta < 0$ then the homogeneous state becomes marginally stable at the modified critical point $\mu'_c = \mu_c[1 - \beta J_-(q_-)]$. The corresponding marginally stable modes are of the form

$$a(\mathbf{r}, \theta) = \sum_{j=1}^{N} z_j e^{i\mathbf{k}_j \cdot \mathbf{r}} \sin(2[\theta - \varphi_j]) + c.c. \tag{5.25}$$

where $\mathbf{k}_j = q_-(\cos \varphi_j, \sin \varphi_j)$ and $z_j$ is a complex amplitude. These modes will be recognized as linear combinations of plane waves modulated by *odd* (phase-shifted) $\pi$-periodic functions $\sin[2(\theta - \varphi_j)]$. The infinite degeneracy arising from rotation invariance means that all modes lying on the circle $|\mathbf{k}| = q_-$ become marginally stable at the critical point. However, this can be reduced to a finite set of modes by restricting solutions to be doubly periodic functions as will be explained in section 5.3. *(b) Weak anisotropy.* If $\eta > \pi/4$ then $J_+(q_+) < J_-(q_-)$ as illustrated in Fig. 21b for $\eta = \pi/3$. It follows that the homogeneous state now becomes marginally stable at the critical point $\mu'_c = \mu_c[1 - \beta J_+(q_+)]$ due to excitation of *even* modes given by

$$a(\mathbf{r}, \theta) = \sum_{j=1}^{N} z_j e^{i\mathbf{k}_j \cdot \mathbf{r}} \cos(2[\theta - \varphi_j]) + c.c. \tag{5.26}$$

where $\mathbf{k}_j = q_+(\cos(\varphi_j), \sin(\varphi_j))$.

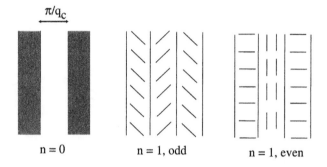

$$\pi/q_c$$

n = 0          n = 1, odd          n = 1, even

Fig. 22. Three classes of rolls found in cortical pattern formation

In the above analysis we assumed that each isolated hypercolumn spontaneously forms an orientation tuning curve; the long–range horizontal connections than induce correlations between the tuning curves across the cortex. Now suppose that each hypercolumn undergoes a bulk instability for which $W_0 = \max_n\{W_n\}$. Repeating the above linear stability analysis, we find that there are now only even eigenmodes. These are $\theta$-independent (to leading order), and take the form

$$a(\mathbf{r}) = \sum_{j=1}^{N}[z_j e^{i\mathbf{k}_j \cdot \mathbf{r}} + c.c.] \tag{5.27}$$

The corresponding eigenvalue equation is

$$\lambda = -1 + \mu W_0 + \beta \widehat{J}_0(q) + \mathcal{O}(\beta^2) \tag{5.28}$$

Thus $|\mathbf{k}_j| = q_0$ where $q_0$ is the minimum of $\widehat{J}_0(q)$. It follows that there are three classes of eigenmode that can bifurcate from the homogeneous fixed point. These are represented, respectively, by linear combinations of one of the three classes of roll pattern shown in Fig. 22. The $n = 0$ roll corresponds to modes of the form (5.27), and consists of alternating regions of high and low cortical activity in which individual hypercolumns do not amplify any particular orientation: the resulting patterns are said to be *non-contoured*. The $n = 1$ rolls correspond to the odd and even oriented modes of equations (5.25) and (5.26). These are constructed using a winner-take-all rule in which only the orientation with maximal response is shown at each point in the cortex (after some coarse-graining). The resulting patterns are said to be *contoured*. To lowest order in $\beta$ one finds that the preferred orientation alternates between parallel and orthogonal directions relative to the stripe boundary in the case of even modes, whereas the preferred orientation alternates between $\pi/4$ and $-\pi/4$ relative to the stripe boundary in the case of odd modes. (One would still obtain stripes of alternating orientations $\theta_0$ and $\theta_0 + \pi/2$ in the case of isotropic horizontal connections. However, the direction of preferred orientation $\theta_0$ relative to the stripe boundaries would now be arbitrary so that the distinction between even and odd modes would disappear). The particular class of mode that is selected depends on the detailed structure of the local and horizontal weights. The $n = 0$ type will be selected when the local inhibition within a hypercolumn is sufficiently weak, whereas the $n = 1$ type will occur when there is strong local inhibition, with the degree of anisotropy in the horizontal connections determining whether the patterns are even or odd.

## 5.3. Doubly-periodic planforms

Rotation symmetry implies that in the case of non-zero critical wavenumber $q_c$, the space of marginally stable eigenfunctions is infinite–dimensional, consisting of all solutions of the form $u(\theta - \varphi)e^{i\mathbf{k}_\varphi \cdot \mathbf{r}}$ where $u(\theta)$ is either an even or odd function of $\theta$, $\mathbf{k}_\varphi = q_c(\cos\varphi, \sin\varphi)$ and $0 \leq \varphi < 2\pi$. However, translation symmetry allows us to restrict the space of solutions of the original equation (5.1) to that of doubly–periodic functions. That is, we impose the condition

$$a(\mathbf{r} + \boldsymbol{\ell}, \theta) = a(\mathbf{r}, \theta)$$

for every $\boldsymbol{\ell} \in \mathcal{L}$ where $\mathcal{L}$ is some regular planar lattice, see equation (3.15). In contrast to the planar model introduced in section 3.1, this lattice does not have a direct physical interpretation in terms of periodic feature maps, rather it is introduced as a mathematical convenience. The restriction to doubly periodic solutions is standard in many treatments of spontaneous pattern formation, but as yet it has no formal justification. However, there is a wealth of evidence from experiments on convecting fluids and chemical reaction-diffusion systems [102] indicating that such systems tend to generate doubly–periodic patterns in the plane when the homogeneous state is destabilized. Restriction to double periodicity means that the original Euclidean symmetry group is now restricted to the symmetry group of the lattice, $\Gamma = \mathbf{D}_n \dot{+} \mathbf{T}^2$, where $\mathbf{D}_n$ is the *holohedry* of the lattice, the subgroup of $\mathbf{O}(2)$ that preserves the lattice, and $\mathbf{T}^2$ is the two torus of planar translations modulo the lattice. Thus, the holohedry of the rhombic lattice is $\mathbf{D}_2$, the holohedry of the square lattice is $\mathbf{D}_4$ and the holohedry of the hexagonal lattice is $\mathbf{D}_6$, see Fig. 23. There are only a finite number of shift-twists

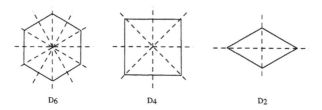

D6                              D4                              D2

Fig. 23. Holohedries of the plane

and reflections to consider for each lattice (modulo an arbitrary rotation of the whole plane). Consequently, a finite set of specific eigenfunctions can be identified as candidate planforms, in the sense that they approximate time–independent solutions of equation (5.1) sufficiently close to the critical point where the homogeneous state loses stability.

Imposing double periodicity on the marginally stable eigenfunctions restricts the lattice spacing $d$ such that the critical wavevector $\mathbf{k}$ lies on the dual lattice. There are infinitely many choices for the lattice size that satisfies this constraint—we select the one for which $q_c$ is the shortest length of a dual wave vector. Linear combinations of eigenfunctions that generate doubly–periodic solutions corresponding to dual wave vectors of shortest length are then given by

$$a(\mathbf{r}, \theta) = \sum_{j=1}^{N} z_j u(\theta - \varphi_j) e^{i\mathbf{k}_j \cdot \mathbf{r}} + c.c \tag{5.29}$$

where the $z_j$ are complex amplitudes. Here $N = 2$ for the square lattice with $\mathbf{k}_1 = \mathbf{k}_c$ and $\mathbf{k}_2 = R_{\pi/2}\mathbf{k}_c$, where $R_\xi$ denotes rotation through an angle $\xi$. Similarly, $N = 3$ for the hexagonal lattice with $\mathbf{k}_1 = \mathbf{k}_c$, $\mathbf{k}_2 = R_{2\pi/3}\mathbf{k}_c$ and $\mathbf{k}_3 = R_{4\pi/3}\mathbf{k}_c = -\mathbf{k}_1 - \mathbf{k}_2$. It follows that the space of marginally stable eigenfunctions can be identified with the $N$–dimensional complex vector space spanned by the vectors $(z_1, \ldots, z_N) \in \mathbf{C}^N$ with $N = 2$ for square or rhombic lattices and $N = 3$ for hexagonal lattices. It can be shown that these form irreducible representations of the group $\Gamma = \mathbf{D}_n \dot{+} \mathbf{T}^2$ whose action on $\mathbf{C}^N$ is induced by the corresponding

Table 2

(Left) $\mathbf{D}_2 \dot{+} \mathbf{T}^2$ action on rhombic lattice; (Center) $\mathbf{D}_4 \dot{+} \mathbf{T}^2$ action on square lattice; (Right) $\mathbf{D}_6 \dot{+} \mathbf{T}^2$ action on hexagonal lattice. For $u(\phi)$ even, $\epsilon = +1$; for $u(\phi)$ odd, $\epsilon = -1$. In each case the generators of $\mathbf{D}_n$ are a reflection and a rotation. For the square and hexagonal lattices, the reflection is $\kappa$, the reflection across the $x$ axis where $\mathbf{r} = (x, y)$. For the rhombic lattice, the reflection is $\kappa_\eta$. The counterclockwise rotation $\xi$, through angles $\frac{\pi}{2}$, $\frac{\pi}{3}$, and $\pi$, is the rotation generator for the three lattices.

| $\mathbf{D}_2$ | Action | $\mathbf{D}_4$ | Action | $\mathbf{D}_6$ | Action |
|---|---|---|---|---|---|
| 1 | $(z_1, z_2)$ | 1 | $(z_1, z_2)$ | 1 | $(z_1, z_2, z_3)$ |
| $\xi$ | $(z_1^*, z_2^*)$ | $\xi$ | $(z_2^*, z_1)$ | $\xi$ | $(z_2^*, z_3^*, z_1^*)$ |
| $\kappa_\eta$ | $\epsilon(z_2, z_1)$ | $\xi^2$ | $(z_1^*, z_2^*)$ | $\xi^2$ | $(z_3, z_1, z_2)$ |
| $\kappa_\eta \xi$ | $\epsilon(z_2^*, z_1^*)$ | $\xi^3$ | $(z_2, z_1^*)$ | $\xi^3$ | $(z_1^*, z_2^*, z_3^*)$ |
| | | $\kappa$ | $\epsilon(z_1, z_2^*)$ | $\xi^4$ | $(z_2, z_3, z_1)$ |
| | | $\kappa\xi$ | $\epsilon(z_2^*, z_1^*)$ | $\xi^5$ | $(z_3^*, z_1^*, z_2^*)$ |
| | | $\kappa\xi^2$ | $\epsilon(z_1^*, z_2)$ | $\kappa$ | $\epsilon(z_1, z_3, z_2)$ |
| | | $\kappa\xi^3$ | $\epsilon(z_2, z_1)$ | $\kappa\xi$ | $\epsilon(z_2^*, z_1^*, z_3^*)$ |
| | | | | $\kappa\xi^2$ | $\epsilon(z_3, z_2, z_1)$ |
| | | | | $\kappa\xi^3$ | $\epsilon(z_1^*, z_3^*, z_2^*)$ |
| | | | | $\kappa\xi^4$ | $\epsilon(z_2, z_1, z_3)$ |
| | | | | $\kappa\xi^5$ | $\epsilon(z_3^*, z_2^*, z_1^*)$ |
| $[\theta_1, \theta_2]$ | | $(e^{-2\pi i \theta_1} z_1, e^{-2\pi i \theta_2} z_2)$ | | $(e^{-2\pi i \theta_1} z_1, e^{-2\pi i \theta_2} z_2, e^{2\pi i (\theta_1 + \theta_2)} z_3)$ | |

shift–twist action (3.30) of $\Gamma$ on $a(\mathbf{r}, \theta)$. For example, on a hexagonal lattice, a translation $a(\mathbf{r}, \theta) \to a(\mathbf{r} - \mathbf{s}, \theta)$ induces the action

$$\gamma \cdot (z_1, z_2, z_3) = (z_1 e^{-i\xi_1}, z_2 e^{-i\xi_2}, z_3 e^{i(\xi_1 + \xi_2)}) \tag{5.30}$$

with $\xi_j = \mathbf{k}_j \cdot \mathbf{s}$, a rotation $a(\mathbf{r}, \theta) \to a(R_{-2\pi/3}\mathbf{r}, \theta - 2\pi/3)$ induces the action

$$\gamma \cdot (z_1, z_2, z_3) = (z_3, z_1, z_2), \tag{5.31}$$

and a reflection across the $x$–axis (assuming $\mathbf{k}_c = q_c(1, 0)$) induces the action

$$\gamma \cdot (z_1, z_2, z_3) = (z_1, z_3, z_2) \tag{5.32}$$

The full shift–twist action of $\mathbf{D}_n \dot{+} \mathbf{T}^2$ on $\mathbf{C}^N$ for the various lattices has been calculated elsewhere [14, 15] and is given in Table 2.

The next important observation is that using weakly nonlinear analysis and perturbation methods, it is possible to reduce the infinite–dimensional system (5.1) to a finite set of coupled ODEs constituting an amplitude equation for $\mathbf{z}$,

$$\frac{dz_j}{dt} = F_j(\mathbf{z}), \quad j = 1, \ldots, N, \tag{5.33}$$

This is carried out explicitly to cubic order in section 5.4. Calculating higher order terms is considerably more involved. Nevertheless, the general form of the amplitude equation can be obtained by exploiting the fact that it is equivariant with respect to the induced shift–twist action of the group $\Gamma$ on $\mathbf{C}^N$. One can then use techniques from symmetric bifurcation theory to determine the equilibrium solutions that are likely to bifurcate from the homogeneous fixed point $\mathbf{z} = 0$ [45], as discussed in section 5.5.

### 5.4. Amplitude equation

We now derive an amplitude equation for the odd and even eigenmodes (5.25) or (5.26) in the case of anisotropic horizontal connections. (The non–contoured patterns can be treated as a special case of even modes). Suppose that each hypercolumn is $\varepsilon$-close to the point of marginal stability of the homogeneous fixed point $\bar{a}$. That is, $\mu = \mu_c' + \varepsilon \Delta \mu$ where $\mu_c' = \mu_c + \mathcal{O}(\beta)$. Following along similar lines to section 4.2, we perform a Taylor expansion of equation (5.1) about the equilibrium $\bar{a}$ :

$$\begin{aligned}
\frac{\partial a}{\partial t} &= -a + w * \left[ f(\bar{a}) + \mu(a - \bar{a}) + g_2(a - \bar{a})^2 + g_3(a - \bar{a})^3 + \ldots \right] \\
&\quad + \beta w^\Delta \circ \left[ f(\bar{a}) + \mu(a - \bar{a}) + g_2(a - \bar{a})^2 + g_3(a - \bar{a})^3 + \ldots \right]
\end{aligned} \tag{5.34}$$

Substitute into equation (5.34) the perturbation expansion (4.9) and introduce the slow time-scale $\tau = \varepsilon t$. Collecting terms with equal powers of $\varepsilon$ then leads to a hierarchy of equations of the form (up to $\mathcal{O}(\varepsilon^{3/2})$)

$$\widehat{L}a_1 = 0 \tag{5.35}$$

$$\begin{aligned}
\widehat{L}a_2 &= v_2 \\
&\equiv -g_2[w * a_1^2 + \beta w^\Delta \circ a_1^2]
\end{aligned} \tag{5.36}$$

$$\begin{aligned}
\widehat{L}a_3 &= v_3 \\
&\equiv \frac{\partial a_1}{\partial \tau} - w * \left[ \Delta \mu a_1 + g_3 a_1^3 + 2g_2 a_1 a_2 \right] \\
&\quad - \beta w^\Delta \circ \left[ \Delta \mu a_1 + g_3 a_1^3 + 2g_2 a_1 a_2 \right]
\end{aligned} \tag{5.37}$$

with the linear operator $\widehat{L}$ defined according to equation (5.3) for $\mu = \mu'_c$.

The first equation in the hierarchy, equation (5.35), has solutions of the form

$$a_1(\mathbf{r}, \theta) = \sum_{j=1}^{N} u(\theta - \varphi_j) \left[ z_j(\tau) e^{i\mathbf{k}_j \cdot \mathbf{r}} + z_j^*(\tau) e^{-i\mathbf{k}_j \cdot \mathbf{r}} \right] \tag{5.38}$$

with $\varphi_j = \arg(\mathbf{k}_j)$ and $u(\theta)$ either an odd or even function. We obtain a dynamical equation for the complex amplitude $z_j(\tau)$ in terms of solvability conditions for the higher order equations. Introduce the inner product for two arbitrary periodic functions $U(\mathbf{r}, \theta)$, $V(\mathbf{r}, \theta)$ according to

$$\langle U | V \rangle = \int_{\Lambda_{\mathcal{L}}} \int_0^\pi U(\mathbf{r}, \theta) V(\mathbf{r}, \theta) \frac{d\theta}{\pi} \tag{5.39}$$

where $\Lambda_{\mathcal{L}}$ is a fundamental domain of the lattice $\mathcal{L}$. The solvability conditions then take the form

$$\langle \widetilde{a}_l | v_n \rangle = 0, \quad l = 1, \ldots, N \tag{5.40}$$

where

$$\widetilde{a}_l(\mathbf{r}, \theta) = u(\theta - \varphi_l) e^{i\mathbf{k}_l \cdot \mathbf{r}} \tag{5.41}$$

We now find a significant difference between the odd and even cases. In the former case, the solvability conditions $\langle \widetilde{a}_l | v_2 \rangle = 0$ are automatically satisfied due the fact that $u(\theta)$ is an odd function of $\theta$. On the other hand, when $u(\theta)$ is an even function $\langle \widetilde{a}_m | v_2 \rangle = g_2 \beta \langle \widetilde{a}_m | w^\Delta \circ a_1^2 \rangle \neq 0$. The simplest way to handle this case is to assume that $g_2 = \mathcal{O}(\varepsilon^{1/2})$ so that this terms contributes to the cubic solvability condition instead. The latter becomes

$$\langle \widetilde{a}_l | \frac{\partial a_1}{\partial \tau} \rangle = \langle \widetilde{a}_l | w * [\Delta \mu a_1 + g_3 a_1^3] \rangle + \beta \langle \widetilde{a}_l | w^\Delta \circ [\Delta \mu a_1 + g_3 a_1^3 + g_2 a_1^2] \rangle \tag{5.42}$$

We shall also consider the same solvability for the odd case, since it does not alter the basic form of the resulting amplitude equation.

Further simplification can be achieved by noting that

$$\langle \widetilde{a}_l | w * b + \beta w^\Delta \circ b \rangle = \langle w * \widetilde{a}_l + \beta w^\Delta \circ \widetilde{a}_l | b \rangle = \mu_c'^{-1} \langle \widetilde{a}_l | b \rangle$$

for arbitrary $b$. Thus,

$$\mu_c'^{-1} \langle \widetilde{a}_l | \frac{\partial a_1}{\partial \tau} \rangle = \langle \widetilde{a}_l | \Delta \mu a_1 + g_3 a_1^3 + g_2 a_1^2 \rangle \tag{5.43}$$

The various inner products can now be evaluated using the identity

$$\int_{\Lambda_{\mathcal{L}}} e^{i(\mathbf{k}_1+\dots+\mathbf{k}_p)\cdot\mathbf{r}} d^2\mathbf{r} = \delta_{\mathbf{k}_1+\dots+\mathbf{k}_p,0} \tag{5.44}$$

We thus obtain the cubic amplitude equation (after appropriate rescalings)

$$\frac{dz_l}{d\tau} = \Delta\mu z_l + \gamma_2 \sum_{i,j=1}^{N} z_i^* z_j^* \delta_{\mathbf{k}_i+\mathbf{k}_j+\mathbf{k}_l,0}$$

$$+3z_l \left[ \gamma_3(0)|z_l|^2 + 2\sum_{j\neq l} \gamma_3(\varphi_j - \varphi_l)|z_j|^2 \right] \tag{5.45}$$

with $\gamma_2$ and $\gamma_3$ given by

$$\gamma_2 = \int_0^{\pi} u(\theta)u(\theta - 2\pi/3)u(\theta + 2\pi/3)\frac{d\theta}{\pi} \tag{5.46}$$

and

$$\gamma_3(\varphi) = \int_0^{\pi} u(\theta - \varphi)^2 u(\theta)^2 \frac{d\theta}{\pi} \tag{5.47}$$

Note that for odd eigenmodes $\gamma_2 \equiv 0$ whereas for even eigenmodes $\gamma_2 \neq 0$ so that there is a quadratic term in the even mode amplitude equation in the case of a hexagonal lattice.

### 5.5. Bifurcation theory and shift–twist symmetry

As in the simpler case of the ring model, the basic structure of the amplitude equation (5.45) including higher order terms can be determined from its equivariance under the shift–twist action of the symmetry group $\Gamma = \mathbf{D}_n \dot{+} \mathbf{T}^2$. This also allows us to systematically explore the different classes of equilibrium solutions $\mathbf{z} = (z_1, \dots, z_N)$ of the amplitude equation (5.45) and their associated bifurcations. In order to understand how this is carried out, it is first necessary to review some basic ideas from symmetric bifurcation theory [45]. In the following we consider a general system of ODE'S

$$\dot{z} = F(\mathbf{z}) \tag{5.48}$$

where $\mathbf{z} \in V$ with $V = \mathbf{R}^n$ or $\mathbf{C}^n$ and $F$ is assumed to be equivariant with respect to some symmetry group $\Gamma$ acting on the vector space $V$. We also assume that $F(0) = 0$ so that the origin is an equilibrium that is invariant under the action of the full symmetry group $\Gamma$.

*Isotropy subgroups* The symmetries of any particular equilibrium solution **z** form a subgroup called the *isotropy* subgroup of **z** defined by

$$\Sigma_{\mathbf{z}} = \{\sigma \in \Gamma : \sigma\mathbf{z} = \mathbf{z}\} \tag{5.49}$$

More generally, we say that $\Sigma$ is an isotropy subgroup of $\Gamma$ if $\Sigma = \Sigma_{\mathbf{z}}$ for some $\mathbf{z} \in V$. Isotropy subgroups are defined up to some conjugacy. A group $\Sigma$ is conjugate to a group $\widehat{\Sigma}$ if there exists $\sigma \in \Gamma$ such that $\widehat{\Sigma} = \sigma^{-1}\Sigma\sigma$. The *fixed-point subspace* of an isotropy subgroup $\Sigma$, denoted by $\mathrm{Fix}(\Sigma)$, is the set of points $\mathbf{z} \in V$ that are invariant under the action of $\Sigma$,

$$\mathrm{Fix}(\Sigma) = \{\mathbf{z} \in V : \sigma\mathbf{z} = \mathbf{z} \,\forall\, \sigma \in \Sigma\} \tag{5.50}$$

Finally, the *group orbit* through a point **z** is

$$\Gamma\mathbf{z} = \{\sigma\mathbf{z} : \sigma \in \Gamma\} \tag{5.51}$$

If **z** is an equilibrium solution of equation (5.48) then so are all other points of the group orbit (by equivariance). One can now adopt a strategy that restricts the search for solutions of equation (5.48) to those that are fixed points of a particular isotropy subgroup. In general, if a dynamical system is equivariant under some symmetry group $\Gamma$ and has a solution that is a fixed point of the full symmetry group then we expect a loss of stability to occur upon variation of one or more system parameters. Typically such a loss of stability will be associated with the occurrence of new solution branches with isotropy subgroups $\Sigma$ smaller than $\Gamma$. One says that the solution has spontaneously broken symmetry from $\Gamma$ to $\Sigma$. Instead of a unique solution with the full set of symmetries $\Gamma$ a set of symmetrically related solutions (orbits under $\Gamma$ modulo $\Sigma$) each with symmetry group (conjugate to) $\Sigma$ is observed.

*Equivariant branching lemma* Suppose that the system of equations (5.48) has a fixed point of the full symmetry group $\Gamma$. The *equivariant branching lemma* [45] basically states that generically there exists a (unique) equilibrium solution bifurcating from the fixed point for each of the axial subgroups of $\Gamma$ under the given group action—a subgroup $\Sigma \subset \Gamma$ is *axial* if $\dim\mathrm{Fix}(\Sigma) = 1$. The heuristic idea underlying this lemma is as follows. Let $\Sigma$ be an axial subgroup and $\mathbf{z} \in \mathrm{Fix}(\Sigma)$. Equivariance of $F$ then implies that

$$\sigma F(\mathbf{z}) = F(\sigma\mathbf{z}) = F(\mathbf{z}) \tag{5.52}$$

for all $\sigma \in \Sigma$. Thus $F(\mathbf{z}) \in \mathrm{Fix}(\Sigma)$ and the system of coupled ODE's (5.48) can be reduced to a single equation in the fixed point space of $\Sigma$. Such an equation is expected to support a codimension one bifurcation. Thus one can systematically identify the various expected primary bifurcation branches by constructing the associated axial subgroups and finding their fixed points.

*Example*    For the sake of illustration, consider the full symmetry group $\mathbf{D}_3$ of an equilateral triangle acting on the plane. The action is generated by the matrices (in an appropriately chosen ortonormal basis)

$$R = \begin{pmatrix} 1/2 & -\sqrt{3}/2 \\ \sqrt{3}/2 & 1/2 \end{pmatrix}, \quad S = \begin{pmatrix} 1 & 0 \\ 0 & -1 \end{pmatrix} \tag{5.53}$$

Here $R$ is a rotation by $\pi/3$ and $S$ is a reflection about the $x$-axis. Clearly, $R$ fixes only the origin, while $S$ fixes any point $(x, 0)$. We deduce that the isotropy subgroups are as follows: (i) the full symmetry group $\mathbf{D}_3$ with single fixed point $(0, 0)$; (ii) the two–element group $\mathbf{Z}_2(S)$ generated by $S$, which fixes the $x$-axis, and the groups that are conjugate to $\mathbf{Z}_2(S)$ by the rotations $R$ and $R^2$; (iii) the identity matrix forms a trivial group in which every point is a fixed point. The isotropy subgroups form the hierarchy

$$\{I\} \subset \mathbf{Z}_2(S) \subset \mathbf{D}_3$$

It follows that up to conjugacy the only axial subgroup is $\mathbf{Z}_2(S)$. Thus we expect the fixed point $(0, 0)$ to undergo a symmetry breaking bifurcation to an equilibrium that has reflection symmetry. Such an equilibrium will be given by one of the three points $\{(x, 0), R(x, 0), R^2(x, 0)\}$ on the group orbit generated by discrete rotations. Which of these states is selected will depend on initial conditions, that is, the broken rotation symmetry is hidden. Note that a similar analysis can be carried out for the symmetry group $\mathbf{D}_4$ of the square. Now, however, there are two distinct types of reflection axes: those joining the middle of opposite edges and those joining opposite vertices. Since these two types of reflections are not conjugate to each other, there are now two distinct axial subgroups.

Let us now return to the amplitude equation (5.45). Since it is equivariant with respect to the shift-twist action of the group $\mathbf{D}_n \dotplus \mathbf{T}^2$, it follows from the equivariant branching lemma that the primary patterns (planforms) bifurcating from the homogeneous state are expected to be fixed points of the corresponding axial subgroups. The calculation of these subgroups is considerably more involved than the above example [14, 15]. Here we simply list the resulting even and odd planforms in Tables 3 and 4.

### 5.6. Selection and stability of patterns

We now discuss solutions of the cubic amplitude equation (5.45) for each of the basic lattices, supplementing our analysis with additional information that can be gained using group theoretic arguments.

Table 3

Even planforms with $u(-\theta) = u(\theta)$. The hexagon solutions (0) and ($\pi$) have the same isotropy subgroup, but they are not conjugate solutions.

| Lattice | Name | Planform Eigenfunction |
|---|---|---|
| square | even square | $u(\theta)\cos x + u\left(\theta - \frac{\pi}{2}\right)\cos y$ |
| | even roll | $u(\theta)\cos x$ |
| rhombic | even rhombic | $u(\theta)\cos(\mathbf{k}_1 \cdot \boldsymbol{\ell}) + u(\theta - \eta)\cos(\mathbf{k}_2 \cdot \boldsymbol{\ell})$ |
| | even roll | $u(\theta)\cos(\mathbf{k}_1 \cdot \boldsymbol{\ell})$ |
| hexagonal | even hexagon (0) | $u(\theta)\cos(\mathbf{k}_1 \cdot \boldsymbol{\ell}) + u\left(\theta + \frac{\pi}{3}\right)\cos(\mathbf{k}_2 \cdot \boldsymbol{\ell}) + u\left(\theta - \frac{\pi}{3}\right)\cos(\mathbf{k}_3 \cdot \boldsymbol{\ell})$ |
| | even hexagon ($\pi$) | $u(\theta)\cos(\mathbf{k}_1 \cdot \boldsymbol{\ell}) + u\left(\theta + \frac{\pi}{3}\right)\cos(\mathbf{k}_2 \cdot \boldsymbol{\ell}) - u\left(\theta - \frac{\pi}{3}\right)\cos(\mathbf{k}_3 \cdot \boldsymbol{\ell})$ |
| | even roll | $u(\theta)\cos(\mathbf{k}_1 \cdot \boldsymbol{\ell})$ |

Table 4

Odd planforms with $u(-\theta) = -u(\theta)$.

| Lattice | Name | Planform Eigenfunction |
|---|---|---|
| square | odd square | $u(\theta)\cos x - u\left(\theta - \frac{\pi}{2}\right)\cos y$ |
| | odd roll | $u(\theta)\cos x$ |
| rhombic | odd rhombic | $u(\theta)\cos(\mathbf{k}_1 \cdot \boldsymbol{\ell}) + u(\theta - \eta)\cos(\mathbf{k}_2 \cdot \boldsymbol{\ell})$ |
| | odd roll | $u(\theta)\cos(\mathbf{k}_1 \cdot \boldsymbol{\ell})$ |
| hexagonal | odd hexagon | $u(\theta)\cos(\mathbf{k}_1 \cdot \boldsymbol{\ell}) + u\left(\theta + \frac{\pi}{3}\right)\cos(\mathbf{k}_2 \cdot \boldsymbol{\ell})u\left(\theta - \frac{\pi}{3}\right)\cos(\mathbf{k}_3 \cdot \boldsymbol{\ell})$ |
| | triangle | $u(\theta)\sin(\mathbf{k}_1 \cdot \boldsymbol{\ell}) + u\left(\theta + \frac{\pi}{3}\right)\sin(\mathbf{k}_2 \cdot \boldsymbol{\ell}) + u\left(\theta - \frac{\pi}{3}\right)\sin(\mathbf{k}_3 \cdot \boldsymbol{\ell})$ |
| | patchwork quilt | $u\left(\theta + \frac{\pi}{3}\right)\cos(\mathbf{k}_2 \cdot \boldsymbol{\ell}) - u\left(\theta - \frac{\pi}{3}\right)\cos(\mathbf{k}_3 \cdot \boldsymbol{\ell})$ |
| | odd roll | $u(\theta)\cos(\mathbf{k}_1 \cdot \boldsymbol{\ell})$ |

*Square or rhombic lattice* First, consider planforms corresponding to a bimodal structure of the square or rhombic type ($N = 2$). Take $\mathbf{k}_1 = k_c(1, 0)$ and $\mathbf{k}_2 = k_c(\cos(\varphi), \sin(\varphi))$, with $\varphi = \pi/2$ for the square lattice and $0 < \varphi < \pi/2$, $\varphi \neq \pi/3$ for a rhombic lattice. The amplitudes evolve according to a pair of equations of the form

$$\frac{dz_1}{d\tau} = z_1\left[1 - \gamma_3(0)|z_1|^2 - 2\gamma_3(\varphi)|z_2|^2\right] \tag{5.54}$$

$$\frac{dc_2}{d\hat{\tau}} = z_2\left[1 - \gamma_3(0)|z_2|^2 - 2\gamma_3(\varphi)|z_1|^2\right] \tag{5.55}$$

Since $\gamma_3(\varphi) > 0$, three types of steady state are possible.

1. The homogeneous state: $z_1 = z_2 = 0$.

2. Rolls: $z_1 = \sqrt{1/\gamma_3(0)}e^{i\psi_1}$, $z_2 = 0$ or $z_1 = 0$, $z_2 = \sqrt{1/\gamma_3(0)}e^{i\psi_2}$.

3. Squares or rhombics: $z_j = \sqrt{1/[\gamma_3(0) + 2\gamma_3(\varphi)]}e^{i\psi_j}$, $j = 1, 2$.

for arbitrary phases $\psi_1, \psi_2$. A standard linear stability analysis shows that if $2\gamma_3(\varphi) > \gamma_3(0)$ then rolls are stable whereas the square or rhombic patterns are unstable. The opposite holds if $2\gamma_3(\theta) < \gamma_3(0)$. Note that here stability is defined with respect to perturbations with the same lattice structure.

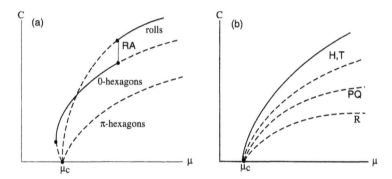

Fig. 24. Bifurcation diagram showing the variation of the amplitude $C$ with the parameter $\mu$ for patterns on a hexagonal lattice. Solid and dashed curves indicate stable and unstable solutions respectively. *(a) Even patterns*: Stable hexagonal patterns are the first to appear (subcritically) beyond the bifurcation point. Subsequently the stable hexagonal branch exchanges stability with an unstable branch of roll patterns due to a secondary bifurcation that generates rectangular patterns $RA$. Higher–order terms in the amplitude equation are needed to determine its stability. *(b) Odd patterns*: Either hexagons (H) or triangles (T) are stable (depending on higher–order terms in the amplitude equation) whereas patchwork quilts (PQ) and rolls ($R$) are unstable. Secondary bifurcations (not shown) may arise from higher–order terms.

*Hexagonal lattice*   Next consider planforms on a hexagonal lattice with $N = 3$, $\varphi_1 = 0$, $\varphi_2 = 2\pi/3$, $\varphi_3 = -2\pi/3$. The cubic amplitude equations take the form

$$\frac{dz_j}{d\tau} = z_j\left[1 - \gamma_3(0)|z_j|^2 - 2\gamma_3(2\pi/3)(|z_{j+1}|^2 + |z_{j-1}|^2)\right] + \gamma_2 z_{j+1}^* z_{j-1}^*$$
(5.56)

where $j = 1, 2, 3 \bmod 3$. Unfortunately, equation (5.56) is not sufficient to determine the selection and stability of the steady-state solutions bifurcating from the homogeneous state. One has to carry out an *unfolding* of the amplitude equation that includes higher-order terms (quartic and quintic) in $z, \bar{z}$. One can classify the bifurcating solutions by finding the axial subgroups of the symmetry group of the lattice (up to conjugacy) as explained in the previous section. Symmetry arguments can also be used to determine the general form of higher-order contributions to the amplitude equation (5.56), and this leads to the bifurcation diagrams shown in Fig. 24 [14, 15]. It turns out that stability depends crucially on the sign of the coefficient $2\gamma(2\pi/3) - \gamma(0)$, which is assumed to be positive in Fig. 24. The subcritical bifurcation to hexagonal patterns in the case of even patterns is a consequence of an additional quadratic term appearing on the right-hand side of equation (5.56).

### 5.7. From cortical patterns to geometric visual hallucinations

We have now identified the various planforms that are expected to occur as primary bifurcations from a homogeneous, low activity state of the coupled hypercolumn model (5.1). These planforms consist of certain linear combinations of the roll patterns shown in Fig. 22 and can thus be classified into non-contoured ($n = 0$) and contoured ($n = 1$ even or odd) patterns. Given a particular activity state in cortex, we can infer what the corresponding image in visual coordinates is like by applying the inverse of the retino-cortical map shown in Fig. 5. (In the case of contoured patterns, one actually has to specify the associated tangent map as detailed in [14]). Examples of non–contoured V1 planforms and their corresponding visual images have already been presented in Fig. 1 and Fig. 2. Two examples of contoured V1 planforms and their associated visual images are shown in Fig. 25; the first visual image is very similar to the hallucination shown in Fig. 3. Note that the original model of Ermentrout and Cowan [34] could only reproduce the non–contoured hallucinations.

The success of the model in reproducing the various hallucination form constants is quite striking. However, certain caution must be exercised since there is a degree of ambiguity in how the cortical patterns should be interpreted. A working assumption is that the basic visual hallucinations can be understood without the need to invoke higher-order processing from extrastriate visual areas. Given this assumption, the interpretation of non–contoured planforms is relatively straightforward, since to lowest order in $\beta$ the solutions are $\theta$–independent and can thus be directly treated as activity patterns $a(\mathbf{r})$ with $\mathbf{r} \in \mathbf{R}^2$. At the simplest level, such patterns can be represented as contrasting regions of high and low activity depending on whether $a(\mathbf{r})$ is above or below threshold. These regions form square, triangular, or rhombic cells that tile the cortical plane as illustrated in Fig. 1. When such patterns are mapped back to the visual field they generate alternating light and dark contrast images. The case of contoured planforms is more subtle. At a given location $\mathbf{r}$ in V1 we have a sum of two or three sinusoids with different phases and amplitudes (see Tables 3 and 4), which can be written as $a(\mathbf{r}, \phi) = A(\mathbf{r}) \cos[2\theta - 2\theta_0(\mathbf{r})]$ (to lowest order in $\beta$). The phase $\theta_0(\mathbf{r})$ de-

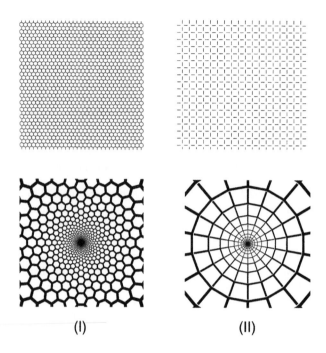

Fig. 25. *First row*: (I) Hexagonal and (II) Square even contoured V1 planforms. *Second row*: Corresponding visual field images.

termines the peak of the orientation tuning curve at $\mathbf{r}$ (see Fig. 18b). Hence the contoured solutions generally consist of iso–orientation regions or patches over which $\theta_0(\mathbf{r})$ is constant but the amplitude $A(\mathbf{r})$ varies. As in the non–contoured case these patches are either square, triangular, or rhombic in shape. The contoured patterns in Fig. 25 are constructed by representing each patch by a locally oriented contour centered at the point of maximal amplitude $A(\mathbf{r}_{max})$ within the patch. Thus we have made a particular choice in how to sample and interpret the activity patterns. For example, each patch contains a distribution of hypercolumns within the continuum model framework so that picking out a single orientation preference corresponds to a discrete sampling of the pattern. Additional difficulties arise when higher order terms in $\beta$ are included, since it is possible to have more than one preferred orientation within a patch [15].

## 6. Pattern formation in a planar model of V1

Part of the problem in sampling and interpreting the activity patterns of the coupled hypercolumn model is that the details of the underlying feature maps have been lost. An alternative approach is to consider the planar model introduced in section 3.1, in which correlations between the long-range connections and the feature maps are included explicitly without any artificial partitioning into hypercolumns. Any spontaneously formed activity pattern can then be overlayed on top of the feature maps thus providing a direct means to construct the corresponding visual image. We will consider a one–population version of equation (3.1) with weights decomposed into local and long–range parts according to equation (3.17):

$$
\frac{\partial a(\mathbf{r}, t)}{\partial t} = -a(\mathbf{r}, t) + \int_{\mathbf{R}^2} W(|\mathbf{r} - \mathbf{r}'|) f(a(\mathbf{r}', t)) d\mathbf{r}'
$$
$$
+ \beta \int_{\mathbf{R}^2} W^\Delta(\mathbf{r}|\mathbf{r}') f(a(\mathbf{r}', t)) d\mathbf{r}' + \bar{h}. \tag{6.1}
$$

For simplicity, we take the long–range horizontal connections to be of the form

$$
W^\Delta(\mathbf{r}|\mathbf{r}') = J(\mathbf{r} - \mathbf{r}')[1 + \kappa D(\mathbf{r})], \tag{6.2}
$$

where

$$
J(\mathbf{r}) = \sum_{\boldsymbol{\ell} \in \mathcal{L}} J_{\boldsymbol{\ell}} \Delta(\mathbf{r} - \boldsymbol{\ell}). \tag{6.3}
$$

Here $J_{\boldsymbol{\ell}}$ represents the decrease in the strength of long–range connections with cortical separation, $\Delta$ represents the patchiness of the connections (linking cells with similar feature preferences), and $D$ is a spatially periodic modulation of the patchy connections that correlates with the underlying periodic feature maps (see section 3.4).

### 6.1. Homogeneous weights

Let us first consider the case of zero horizontal interactions by setting $\beta = 0$ in equation (6.1) to obtain the homogeneous and isotropic network equation

$$
\frac{\partial a(\mathbf{r}, t)}{\partial t} = -a(\mathbf{r}, t) + \int_{\mathbf{R}^2} W(|\mathbf{r} - \mathbf{r}'|) f(a(\mathbf{r}', t)) d\mathbf{r}' + \bar{h}. \tag{6.4}
$$

Equation (6.4) is equivariant with respect to the Euclidean group $\mathbf{E}(2)$ of rigid body transformations in the plane (see section 3.4). Linearize the equation about an equilibrium solution by setting $a(\mathbf{r}, t) = \bar{a} + e^{\lambda t} a(\mathbf{r})$ to obtain the eigenvalue equation

$$\lambda a(\mathbf{r}) = -a(\mathbf{r}) + \mu \int_{\mathbf{R}^2} W(|\mathbf{r} - \mathbf{r}'|) a(\mathbf{r}') d^2\mathbf{r}', \tag{6.5}$$

where $\mu = f'(\bar{a})$. Since the local weight distribution $W$ is homogeneous, it follows from continuous translation symmetry that the eigenmodes are in the form of plane waves $a(\mathbf{r}) = e^{i\mathbf{k}\cdot\mathbf{r}}$ with wavenumber $\mathbf{k}$. Substitution into equation (6.5) shows that the corresponding eigenvalues satisfy the dispersion relation

$$\lambda = \lambda(k) \equiv -1 + \mu \widehat{W}(k), \tag{6.6}$$

where $k = |\mathbf{k}|$ and $\widehat{W}(k)$ is the Fourier transform of $W(|\mathbf{r}|)$,

$$\widehat{W}(k) = \int_{\mathbf{R}^2} W(|\mathbf{r}|) e^{-i\mathbf{k}\cdot\mathbf{r}} d^2\mathbf{r}. \tag{6.7}$$

Suppose that we impose periodic boundary conditions of the form $a(\mathbf{r} + \mathbf{L}_1) = a(\mathbf{r} + \mathbf{L}_2) = a(\mathbf{r})$ for all $\mathbf{r}$. Given the periodic boundary conditions, the wavevectors $\mathbf{k}$ are constrained to lie on a dual lattice, $\mathbf{k} = 2\pi(m_1\hat{\mathbf{L}}_1 + m_2\hat{\mathbf{L}}_2)$ for integers $m_1, m_2$ and $\hat{\mathbf{L}}_i \cdot \mathbf{L}_j = \delta_{i,j}$. It will be assumed that $L = |\mathbf{L}_j|$ is sufficiently large so that the wavevectors are quasi-continuous.

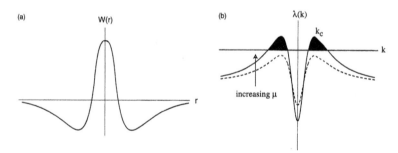

Fig. 26. Neural basis of the Turing mechanism. (a) Mexican hat interaction function showing short-range excitation and long-range inhibition. (b) Dispersion curves $\lambda(k)$ for Mexican hat function. If the excitability $\mu$ of the cortex is increased, the dispersion curve is shifted upwards leading to a Turing instability at a critical parameter $\mu_c = \widehat{W}(k_c)^{-1}$ where $\widehat{W}(k_c) = [\max_k \widehat{W}(k)\}]$. For $\mu_c < \mu < \infty$ the homogeneous fixed point is unstable.

We now use the dispersion relation (6.6) to determine conditions under which the homogeneous state loses stability leading to the formation of spatially periodic patterns. The standard mechanism for such an instability, which is the neural analog of the Turing instability in reaction-diffusion equations, is a combination of short-range excitation and long-range inhibition [34, 105]. This can be represented by the so-called "Mexican Hat" function (see Fig. 26a):

$$W(r) = \frac{W_+}{2\pi\sigma_+^2}e^{-r^2/2\sigma_+^2} - \frac{W_-}{2\pi\sigma_-^2}e^{-r^2/2\sigma_-^2}, \tag{6.8}$$

with $\sigma_+ < \sigma_-$ and $W_+\sigma_-^2 > W_-\sigma_+^2$. The corresponding Fourier transform

$$\widehat{W}(k) = W_+e^{-\frac{1}{2}\sigma_+^2k^2} - W_-e^{-\frac{1}{2}\sigma_-^2k^2} \tag{6.9}$$

has a maximum at $k = k_c \neq 0$ as shown in Fig. 26b. Since $\widehat{W}(k)$ is bounded, it follows that when the network is in a low activity state such that $\mu \approx 0$, any solution of equation (6.5) satisfies Re $\lambda < 0$ and the fixed point is linearly stable. However, when the excitability of the network is increased, either through the action of some hallucinogen or through external stimulation, $\mu$ increases and the fixed point becomes marginally stable at the critical value $\mu_c$ where $\mu_c^{-1} = \widehat{W}(k_c) \equiv \max_k\{\widehat{W}(k)\}$. For $\mu > \mu_c$, the fixed point is unstable and there exists a band of wavevectors $\mathbf{k}$ that have positive $\lambda$ and can thus grow to form spatially varying patterns. Sufficiently close to the bifurcation point these patterns can be represented as linear combinations of plane waves

$$a(\mathbf{r}) = \sum_{j=1}^{N}(z_je^{i\mathbf{k}_j\cdot\mathbf{r}} + z_j^*e^{-i\mathbf{k}_j\cdot\mathbf{r}}), \tag{6.10}$$

where the sum is over all wave vectors lying within some neighborhood of the critical circle $|\mathbf{k}_j| = k_c$. The number $N$ of excited nodes will rapidly grow beyond the bifurcation point due to two distinct factors. First, rotation symmetry implies that the spectrum only depends on $|\mathbf{k}|$ so that all modes related by a continuous rotation will be excited. Second, for large system size $L$ there will be a correspondingly large number of modes within a band $(k_c, k_c + \Delta k)$. The latter issue will be addressed in section 6.3.

The high–dimensionality arising from rotation symmetry can be eliminated by using continuous translation symmetry to restrict the space of solutions to functions that are doubly–periodic with respect to some lattice $\mathcal{L}$ as explained in section 5.3. However, we can now identify the lattice $\mathcal{L}$ with the physical lattice of orientation pinwheels or CO blobs that form organizing centers for the periodic feature maps. Moreover, suppose that we incorporate homogeneous horizontal connections by taking $\beta > 0$ in equation (6.1) and $\kappa = 0$ in equation (6.2). Setting

$$w(\mathbf{r}) = W(|\mathbf{r}|) + \beta J(\mathbf{r}), \tag{6.11}$$

and repeating the linear stability analysis about the corresponding homogeneous fixed point, the dispersion relation (6.6) becomes

$$\lambda = \lambda(\mathbf{k}) \equiv -1 + \mu \widehat{w}(\mathbf{k}). \tag{6.12}$$

Note that $\lambda(\mathbf{k})$ now depends on the direction as well as the magnitude of $\mathbf{k}$, since it is invariant with respect to the discrete rotation symmetry group of the dual lattice $\widehat{\mathcal{L}}$, rather than the continuous rotation group. It follows that a particular wavevector $\mathbf{k}^*$ will be excited, together with all modes generated by discrete rotations of the dual lattice (ignoring accidental degeneracies). The eigenmodes will naturally tend to be low-dimensional patterns such as rolls and hexagons, that is, they will satisfy equation (6.10) with $\mathbf{k}_n = R_{2\pi n/N}\mathbf{k}^*$ for $n = 2, \ldots N$ and $N = 2$ if $\mathcal{L}$ is a square lattice or $N = 3$ if $\mathcal{L}$ is a hexagonal lattice. Thus the existence of a lattice $\mathcal{L}$ underlying the distribution of patchy horizontal connections provides a physical mechanism for generating low-dimensional, doubly-periodic patterns. In the standard Euclidean symmetric model of cortical pattern formation discused above, the double–periodicity of the solutions was imposed by hand as a mathematical simplification rather than reflecting the existence of a real lattice [14, 34, 36]).

## 6.2. *Pinning of cortical patterns by spatially periodic horizontal connections*

We now wish to determine the effects of a spatially periodic modulation of the long–range horziontal connections. For simplicity, we consider a one–dimensional version of equations (6.1) and (6.2):

$$\frac{\partial a(x, t)}{\partial t} = -a(x, t) + h(x) \tag{6.13}$$

$$+ \int_{\mathbf{R}} \left[ w(x - x') + \kappa D(x) J(x - x') \right] f(a(x', t)) dx'$$

with

$$w(x) = W(x) + \beta J(x), \qquad J(x) = \sum_{n \neq 0} J_n \Delta(x - n d_0) \tag{6.14}$$

and we have absorbed a factor of $\beta$ into $\kappa$. The patch spacing is taken to be $d_0$ whereas $D(x)$ is taken to be $d$–periodic with $d_0 = 2d$. This is motivated by the idea that horizontal connections link cells with the same ocularity, whereas the periodic modulations are correlated with the locations of the CO blobs. The presence of the periodic term $D(x)$ means that the fixed points of the network are no longer homogeneous. However, linearization about such equilibria only introduce additional periodic contributions that do not alter the basic results detailed below. Therefore, for simplicity, we will assume that there is an additional inhomogeneous input such that there exists a homogeneous fixed point $\bar{a}$. Linearization then leads to the eigenvalue equation

$$\lambda a(x) = \mathcal{H} a(x) \equiv -a(x) + \mu \int_R [w(x - x') + \kappa D(x)J(x - x')]a(x')dx'.$$

$$(6.15)$$

The periodic term $D(x)$ breaks continuous translation symmetry so that the eigensolutions take the form of Bloch waves [3]

$$a(x) = e^{ikx}u_k(x) \tag{6.16}$$

with $u_k(x + nd) = u_k(x)$ for all $n \in \mathbf{Z}$. In order to prove this, let us introduce the translation operator $T_n$ such that $T_n f(x) = f(x + nd)$ for any function $f$. Then

$$
\begin{aligned}
T_n \mathcal{H} a(x) &= -a(x + nd) \\
&\quad + \mu \int_R [w(x + nd - x') \\
&\qquad + \kappa D(x + nd)J(x + nd - x')]a(x')dx' \\
&= -a(x + nd) \\
&\quad + \mu \int_R [w(x - x') + \kappa D(x)J(x - x')]a(x' + nd)dx' \\
&= \mathcal{H}[T_n a(x)]
\end{aligned}
\tag{6.17}
$$

so that $T_n \mathcal{H} = \mathcal{H} T_n$. Shur's lemma then implies that $\mathcal{H}$ and $T_n$ have simultaneous eigensolutions:

$$\mathcal{H} a = \lambda a, \quad T_n a = C(n)a \tag{6.18}$$

Since $T_n T_{n'} = T_{n+n'}$, we have

$$C(n)C(n') = C(n + n') \tag{6.19}$$

which implies that $C(n) = e^{ink}$ and the result follows.

Another way to establish the general form of the eigensolutions (6.16) is to introduce the Fourier series expansion

$$D(x) = \sum_q D_q e^{iqx}, \quad q = 2\pi m/d, \quad m \in \mathbf{Z} \tag{6.20}$$

We impose periodic boundary conditions $a(x+L) = a(x)$ where $L$ is the system size such that

$$a(x) = \frac{1}{L} \sum_k a_k e^{ikx}, \quad k = \frac{2\pi m}{L}, \quad m \in \mathbf{Z} \tag{6.21}$$

Equation (6.5) then reduces to

$$[\lambda + 1 - \mu \widehat{w}(k)] a_k = \kappa \sum_q \mathcal{V}_q(k) a_{k-q}, \tag{6.22}$$

where

$$\mathcal{V}_q(k) = \widehat{J}(k-q) \mathcal{D}_q. \tag{6.23}$$

and we have absorbed a factor $\mu$ into $\kappa$. Equation (6.22) implies that the lateral interactions only couple together those coefficients $a_k, a_{k-q_1}, a_{k-q_2}, \dots$ whose wavenumbers differ by $q_n = 2\pi n/d$. In other words, if we fix $k$ then

$$a(x) = \sum_q a_{k-q} e^{i(k-q)x} = e^{ikx} u_k(x) \tag{6.24}$$

which is of the form (6.16) with $u_k(x)$ given by the $d$–periodic function

$$u_k(x) = \sum_q a_{k-q} e^{-iqx}. \tag{6.25}$$

It is generally not possible to find exact solutions of the eigenvalue equation (6.22). Therefore, we proceed by carrying out a perturbation expansion in $\kappa$ [20,22]. Let $k_c$ be the critical wavenumber for a pattern forming instability when $\kappa = 0$. We then have to distinguish between two cases, based on whether or not the following degeneracy condition holds: there exists an integer $n$ such that

$$|\widehat{w}(k_c) - \widehat{w}(k_c - 2\pi n/d)| = \mathcal{O}(\kappa) \tag{6.26}$$

and $|\widehat{w}(k_c) - \widehat{w}(k_c - 2\pi m/d)| \gg \kappa$ for all $m \neq n$. Note that the exact degeneracy condition $\widehat{w}(k_c) = \widehat{w}(k_c - 2\pi n/d)$ is only satisfied if $k_c = \pi n/d$, since $\widehat{w}(k)$ is an even function of $k$. We now present an argument for the degeneracy condition $k_c = \pi/d$ to hold. For the sake of illustration suppose that $\Delta(x) = \delta(x)$. Taking the one-dimensional Fourier transform of $J(x)$ gives

$$\widehat{J}(k) = \sum_{n \neq 0} J_n e^{-ikn d_0} \tag{6.27}$$

In the case of the exponential function $J_n = e^{-|n|/\xi}/2\xi$, the sum over lattice sites may be performed explicitly to yield

$$\widehat{J}(k) = \frac{e^{d_0/\xi}\cos(kd_0) - 1}{[e^{d_0/\xi}\cos(kd_0) - 1]^2 + [e^{d_0/\xi}\sin(kd_0)]^2} \tag{6.28}$$

It follows that for infinitesimal patch size, the Fourier transform of the homogeneous part of the horizontal weight distribution is a $(2\pi/d_0)$-periodic function of $k$ and thus has an infinite set of global maxima and global minima. This infinite degeneracy is then broken by the local connections, which are described by a Mexican hat function (see Fig. 26a), resulting in dispersion curves such as shown in Fig. 27. Given the range of the local connections $W$, the maximum of $\widehat{w}(k)$ is likely to occur close to the peak $k_c \approx 2\pi/d_0 = \pi/d$. Note that the above argument is easily extended to the case of small but finite patch-size.

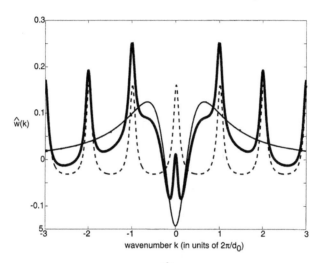

Fig. 27. *Thin solid curve*: Fourier transform $\widehat{W}(k)$ of a local Mexican hat weight distribution. *Dashed curve*: Fourier transform $\widehat{J}(k)$ of long-range weight distribution satisfying equation (6.28) with $\xi = 2d_0$. *Thick solid curve*: Fourier transform of the total weight distribution $\widehat{w}(k) = \widehat{W}(k) + \beta\widehat{J}(k)$ for $\beta = 0.25$.

*Non-degenerate case*   First, suppose that equation (6.26) does not hold. We then fix $k$ and look for solutions satisfying $a_k \to 1$ and $a_{k-q} \to 0$, $q \neq 0$, in the limit $\kappa \to 0$. This corresponds to a small perturbation of the solution $e^{ikx}$. Equation

(6.22) implies that (for $q \neq 0$)

$$\left[\lambda + 1 - \mu \widehat{w}(k - q)\right] a_{k-q} = \kappa \sum_{q'} \mathcal{V}_{q'}(k - q) a_{k-q-q'} \tag{6.29}$$

which can be rewritten in the form

$$a_{k-q} = \kappa \frac{\mathcal{V}_{-q}(k - q) a_k}{\lambda + 1 - \mu \widehat{w}(k - q)} + \kappa \sum_{q' \neq 0} \frac{\mathcal{V}_{q'-q}(k - q) a_{k-q'}}{\lambda + 1 - \mu \widehat{w}(k - q)} \tag{6.30}$$

The first term on the right-hand side will be an order of magnitude larger than the remaining terms provided that the degeneracy condition (6.26) does not hold. Therefore

$$a_{k-q} = \kappa \frac{\mathcal{V}_{-q}(k - q) a_k}{\lambda + 1 - \mu \widehat{w}(k - q)} + \mathcal{O}(\kappa^2) \tag{6.31}$$

Substituting this back into equation (6.22) gives to $\mathcal{O}(\kappa^2)$

$$\left[\lambda + 1 - \mu \widehat{w}(k)\right] a_k = \kappa \mathcal{V}_0(k) a_k + \kappa^2 \sum_{q \neq 0} \frac{\mathcal{V}_q(k) \mathcal{V}_{-q}(k - q) a_k}{\lambda + 1 - \mu \widehat{w}(k - q)} \tag{6.32}$$

Hence, replacing the denominator by the lowest order contribution yields

$$\lambda = -1 + \mu \widehat{w}(k) + \kappa \mathcal{V}_0(k) + \kappa^2 \sum_{q \neq 0} \frac{\mathcal{V}_q(k) \mathcal{V}_{-q}(k - q)}{\mu \left[\widehat{w}(k) - \widehat{w}(k - q)\right]} + \mathcal{O}(\kappa^3) \tag{6.33}$$

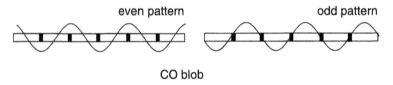

CO blob

Fig. 28. Even and odd eigenmodes localized around blob and inter-blob regions.

*Degenerate case*   Now suppose that the degeneracy condition (6.26) is satisfied for some integer $n$, that is, $k_c = m\pi/d$ for some integer $m$. It follows that the pair of coefficients $a_{\pm k}$ associated with the dominant wavenumber $k = k_c$ are coupled in equation (6.22), since $-k = k - Q$ for $Q = 2\pi m/d$. This implies that there is an approximate two-fold degeneracy, and we must treat equation (6.22) separately for the two cases $k$ and $k - Q$ where $Q = 2\pi n/d$. Thus, to first order in $\kappa$, equation (6.22) reduces to a pair of equations for the coefficients $a_k$ and $a_{k-Q}$:

$$\begin{pmatrix} \lambda - E(k) & -\kappa V_Q(k) \\ -\kappa V_{-Q}(k-Q) & \lambda - E(k-Q) \end{pmatrix} \begin{pmatrix} a_k \\ a_{k-Q} \end{pmatrix} = 0, \tag{6.34}$$

where $E(k) = -1 + \mu \widehat{w}(k) + \kappa V_0(k)$. Assume for simplicity that the exact degeneracy condition $\widehat{w}(k_c) = \widehat{w}(k_c - Q)$ holds so that $k_c = \pi n/d$. We also take $D(x)$ to be an even function of $x$ so that $\mathcal{D}_Q = \mathcal{D}_{-Q}$. Since $\widehat{w}'(k_c) = 0$, it follows that for $k = k_c + \mathcal{O}(\kappa)$ we have $E(k) = E(k - Q) + \mathcal{O}(\kappa^2)$ and $V_{-Q}(k-Q) = V_Q(k) + \mathcal{O}(\kappa^2)$. The above matrix equation then has solutions of the form (to first order in $\kappa$)

$$\lambda_\pm(k) = -1 + \mu \widehat{w}(k) + \kappa[V_0(k) \pm V_Q(k)]) \tag{6.35}$$

with $a_{k-Q} = \pm a_k$. Thus there is a splitting into even and odd eigenmodes of the form

$$a^+(x) = A\cos(n\pi x/d), \quad a^-(x) = A\sin(n\pi x/d). \tag{6.36}$$

where $A$ is an arbitray amplitude (at the linear level). The even and odd eigenmodes for $n = 1$ are shown in Fig. 28. Note that the even mode has extrema at sites corresponding to the CO blobs, whereas the odd mode has extrema at sites corresponding to inter–blobs. One might have expected the activity patterns in Fig. 28 to be have their *maxima* at the CO blobs, since the latter are supposed to be sites of higher metabolic activity. However, it should be remembered that the staining of the blobs to cytochrome oxidase occurs under conditions of normal vision. Thus the primary source of higher activity in this case is due to the external drive from the lateral geniculate nucleus induced by visual stimuli. Here, on the other hand, we are considering spontaneous cortical activity in the absence of any visual stimuli. Thus, in the case of homogeneous cortical interactions, there is no intrinsic mechanism for ensuring that the peaks of spontaneous activity coincide with the CO blobs. We have shown that inhomogeneities in the horizontal connections that correlate with the distribution of CO blobs can break translation symmetry in such a way as to pin the activity pattern to the blobs along the lines of Fig. 28.

*Two–dimensional pinning* The perturbation analysis of pinning in a one–dimensional network can be extended to the full planar model (6.1). In particular, it can be shown that the periodic modulation term $D(\mathbf{r})$ couples together eigenmodes with wavevectors $\mathbf{k}$ that differ by a dual lattice vector $\mathbf{q}$ [22]. This then leads to the following commensurability condition for the pinning of a two–dimensional activity pattern to a regular lattice of CO blobs, say: there exists a dual lattice vector $\mathbf{Q} \in \widehat{\mathcal{L}}$ such that $|\mathbf{k} - \mathbf{Q}| = |\mathbf{k}| = k_c$. Geometrically this means that $\mathbf{k}$ must lie on the perpendicular bisector of the line joining the origin of the reciprocal lattice $\widehat{\mathcal{L}}$ to the lattice point $\mathbf{Q}$. [In the theory of crystalline solids these perpendicular bisectors partition the reciprocal lattice into domains known as Brillouin zones. Commensurability then requires that $\mathbf{k}$ lie on the boundary of a Brillouin zone [3]]. Two–dimensional pinning is also expected to occur for other sources of spatially periodic modulations in the weights. For example, correlations between the distribution of patchy anisotropic horizontal connections and the orientation preference map could lead to a pinning of a spontaneously generated activity pattern that is commensurate with the orientation map. Interestingly, this provides a possible mechanism for the generation of the ordered lattice of edges seen in the hallucinatory image of Fig. 3. Without some form of pinning the resulting image would likely be a random distribution of edges. (Recall that in the coupled hypercolumn model the image was generated by a judicious sampling of the contoured activity pattern, see section 5.7).

### 6.3. Ginzburg–Landau equation

To what extent does the above pinning mechanism persist when nonlinearities are included? In order to address this issue we first need to take into account the existence of a large number of excited modes within a band of wavenumbers around $k_c$, see Fig. 26b. In the vicinity of the bifurcation point, one can exploit a separation of time and space scales along analogous lines to the analysis of fluid convection patterns [73, 81, 102]. Suppose for the moment that $\kappa = 0$. Expand the one–dimensional dispersion relation $\lambda(k) = -1 + \mu \widehat{w}(k)$ about the critical wavenumber $k_c$ using the fact that $\widehat{w}(k_c)$ is a maximum:

$$
\begin{aligned}
\lambda(k) &= -1 + (\mu_c + \Delta\mu)\left(\widehat{w}(k_c) + \frac{1}{2}(k - k_c)^2 \frac{d^2\widehat{w}}{dk^2}(k_c) + \dots\right) \\
&= \Delta\mu \widehat{w}(k_c) - \alpha(k - k_c)^2 + \dots,
\end{aligned}
\tag{6.37}
$$

where $\Delta\mu = \mu - \mu_c$ and $\alpha = -\mu_c \widehat{w}''(k_c)/2 > 0$. Introducing the small parameter $\varepsilon$ and setting $\Delta\mu = \varepsilon^2$ we see that the unstable modes occupy a band of width $\mathcal{O}(\varepsilon)$ and the rate of growth of the patterns is $\mathcal{O}(\varepsilon^2)$. The interaction of two or more modes within the unstable band gives rise to a spatial modulation of the periodic pattern on a length-scale that is $\mathcal{O}(1/\varepsilon)$ compared to the pattern wavelength $2\pi/k_c$. Within a multiscale perturbation expansion procedure, this corresponds to a slow spatial scale $X = \varepsilon x$. Similarly, the slow growth rate of the patterns implies that there is a slow time scale $T = \varepsilon^2 t$.

The band of modes around $k_c$ are taken into account by considering a wavepacket of the form

$$a(x, X, T) = \varepsilon \left[ A(X, T)e^{ik_c x} + A^*(X, T)e^{-ik_c x} \right], \tag{6.38}$$

where $A(X, T)$ represents a slowly varying envelope of the activity pattern. The linear part of the evolution equation for the amplitude $A(X, T)$ is determined by the Fourier-Laplace transform of the rescaled dispersion relation

$$\Lambda(K) = \omega - \alpha K^2 + \ldots, \tag{6.39}$$

where $K = (k - k_c)/\varepsilon$ and $\omega = \widehat{w}(k_c)$. Symmetry arguments then constrain the form of the leading-order nonlinear term so that we obtain the Ginzburg-Landau equation

$$\frac{\partial A}{\partial T} = \omega A + \alpha \frac{\partial^2 A}{\partial X^2} - \nu |A|^2 A. \tag{6.40}$$

The coefficient $\nu$ may be determined by carrying out an explicit multiple-scale expansion in time and space [22]. Note that the Ginzburg-Landau equation is equivariant with respect to the transformation $A \to Ae^{i\phi}$, which reflects the translation invariance of the full system in the absence of horizontal connections.

Most studies of cortical pattern formation neglect slow spatial modulations of the amplitude by dropping the diffusion term in equation (6.40) [17, 34, 36]. However, it is well known from fluid convection that diffusion-induced long-wavelength phase modulations can induce secondary instabilities away from the primary bifurcation point [28, 102]. A major reason for being interested in these phase instabilities is that they can couple to weak, spatially periodic variations in the strength of the horizontal connections as specified by the function $D(x)$. Such a coupling leads to additional terms in the amplitude equation. For example, consider the periodic function $D(x) = \mathcal{D}_0 \cos(2\pi x/d)$ with the lattice spacing $d$ satisfying the *near-resonance* condition

$$\frac{2\pi}{d} = \frac{n}{m}(k_c + \varepsilon q), \qquad n \neq m, \tag{6.41}$$

for integers $n, m$. If $q \equiv 0$ then the lattice spacing $d$ is rationally related to the critical wavelength of the pattern. The amplitude $\mathcal{D}_0$ of the spatial inhomogeneity can be absorbed into $\kappa$ so that, in terms of fast and slow variables,

$$D(x, X) = \left[ e^{ink_c x/m} e^{inq X/m} + c.c. \right]. \tag{6.42}$$

Assuming the near-resonance condition (6.41), the amplitude equation in the presence of weak spatially periodic modulated horizontal connections takes the general form

$$\frac{\partial A}{\partial T} = \omega A + \alpha \frac{\partial^2 A}{\partial X^2} - \nu |A|^2 A + \gamma A^{*n-1} e^{inqX}, \tag{6.43}$$

where $\gamma \sim \kappa^m$. The general form of the amplitude equation (6.43) can be derived using symmetry arguments [25]. In particular, the final term arises because the amplitude equation is now equivariant with respect to transformations $A \rightarrow A e^{ik_c p d}$ for integers $p$.

## 6.4. Commensurate-incommensurate transitions in cortex

The amplitude equation (6.43) was previously derived within the context of fluid convection under external periodic forcing [25], where it was used to investigate commensurate-incommensurate transitions and quasiperiodic structures associated with the mismatch between the periodicities of the forcing and the critical mode. Here we reinterpret these results in terms of the mismatch in the periodicities of cortical activity patterns and the underlying functional architecture of cortex. First, it is useful to rewrite equation (6.43) in the rescaled form

$$\frac{\partial A}{\partial T} = \omega A + \frac{\partial^2 A}{\partial X^2} - |A|^2 A + \gamma A^{*n-1} e^{inqX}, \tag{6.44}$$

where $X \rightarrow X/\sqrt{\alpha}$, $A \rightarrow \sqrt{\nu} A$, $q \rightarrow \sqrt{\alpha} q$ and $\gamma \rightarrow \gamma \nu^{(2-n)/2}$. Setting $A = R e^{i\Theta}$, we obtain the pair of equations

$$\frac{\partial R}{\partial T} = \omega R - R^3 + \frac{\partial^2 R}{\partial X^2} - \left(\frac{\partial \Theta}{\partial X}\right)^2 R + \gamma R^{n-1} \cos[n(\Theta - qX)] \tag{6.45}$$

and

$$R \frac{\partial \Theta}{\partial T} = R \frac{\partial^2 \Theta}{\partial X^2} + 2 \frac{\partial R}{\partial X} \frac{\partial \Theta}{\partial X} - \gamma R^{n-1} \sin[n(\Theta - qX)]. \tag{6.46}$$

These have stationary solutions of the form

$$\Theta = \Theta_0(X) = qX + \frac{p\pi}{n}, \qquad R = R_0, \tag{6.47}$$

with $R_0$ satisfying

$$\omega - q^2 - R_0^2 + \gamma(-1)^p R_0^{n-2} = 0 \tag{6.48}$$

for integers $p$. These correspond to the activity patterns $R_0 \cos(\pi[2mx/d + p]/n)$, which in the particular case $m = 1, n = 2$ become the even/odd modes

$$a^+(x) = R_0 \cos(k_c x), \quad a^-(x) = R_0 \sin(k_c x). \tag{6.49}$$

We can investigate the stability of the stationary solutions by substituting

$$R = R_0 + r(X, T), \quad \Theta = \Theta_0(X) + \phi(X, T) \tag{6.50}$$

into equations (6.45) and (6.46) and expanding to first order in $\phi, r$. This yields the pair of linear equations

$$\frac{\partial r}{\partial T} = -\Omega_r r + \frac{\partial^2 r}{\partial X^2} - 2R_0 q \frac{\partial \phi}{\partial X}, \tag{6.51}$$

and

$$\frac{\partial \phi}{\partial T} = -\Omega_\phi \phi + \frac{\partial^2 \phi}{\partial X^2} + \frac{2q}{R_0} \frac{\partial r}{\partial X}, \tag{6.52}$$

with

$$\Omega_r = -[\omega - q^2 - 3R_0^2 + (-1)^p (n-1) \gamma R_0^{n-2}] \tag{6.53}$$

and

$$\Omega_\phi = (-1)^p \gamma n R_0^{n-2}. \tag{6.54}$$

It immediately follows from equations (6.52) and (6.54) that the stationary solutions for odd $p$ are unstable with respect to phase perturbations, since $\Omega_\phi < 0$. Now suppose $\Omega_r \gg \Omega_\phi > 0$ for even $p$, so that the amplitude perturbations $r$ adiabatically follow the phase fluctuations $\phi$. We can then make the approximation

$$\Omega_r r = -2R_0 q \frac{\partial \phi}{\partial X}, \tag{6.55}$$

which leads to the following phase equation for $\Phi = \Theta - qX$ [25]:

$$\frac{\partial \Phi}{\partial T} = \Delta \frac{\partial^2 \Phi}{\partial X^2} - K \sin(n\Phi) = -\frac{\delta V}{\delta \Phi}, \tag{6.56}$$

where

$$\Delta = 1 - \frac{4q^2}{\Omega_r}, \quad K = \gamma R_0^{n-2} \tag{6.57}$$

and

$$V = \int \left[ \frac{\Delta}{2} (\partial_X \Phi + q)^2 - \frac{\mathcal{K}}{n} \cos(n\Phi) \right] dx. \tag{6.58}$$

Equation (6.56) is identical to the sine–Gordon equation used to describe commensurate-incommensurate transitions in solids, for which structural configurations correspond to minima of the effective potential $V$ [4]. For the purposes of the current discussion, we will neglect the $q$–dependence of $\Delta$ and assume that it is fixed. We note, however, that it is possible for $\Delta$ to become negative as $q$ increases, leading to the analog of an Eckhaus instability (see [28, 102] for more details). From well known properties of the sine-Gordon equation, the following results then hold:

1. There exists a critical value $q_0$ defined by

$$q_0^2 = \frac{16\mathcal{K}}{n\pi^2 \Delta} \tag{6.59}$$

such that when $q < q_0$ the minimum of the potential $V$ corresponds to a locked state $\Phi = 2\pi p/n$. Recall that $q$ is a measure of the mismatch between the spatial frequencies of the primary activity pattern and the cortical lattice as specified by the near-resonance condition (6.41). It follows that when this mismatch is sufficiently small, the activity pattern is pinned to the lattice with $a(x) \sim \cos(2\pi mx/nd)$. In the particular case $n = 2$ this corresponds to the even mode shown in figure 28. Thus our nonlinear analysis has established that the even (odd) mode is stable (unstable) to phase fluctuations and that the pinning of the stable mode persists over a range of values of the pattern wavenumber.

2. When the mismatch $q > q_0$, the minimum of the potential $V$ occurs for soliton–like solutions of the sine–Gordon equation. A single soliton solution is of the form

$$\Phi(X) = \frac{4}{n} \tan^{-1} \exp(\sqrt{n\mathcal{K}}X). \tag{6.60}$$

This describes a kink centered at $X = 0$ that separates two regions of the activity pattern each of which is commensurate with the cortical lattice, one with phase $\Phi = 0$ and one with phase $\Phi = 2\pi/n$. More generally, the solutions are regularly spaced solitons as illustrated in figure 29. The average phase-shift of the activity pattern per unit length is then

$$\rho = \frac{2\pi}{nl}, \tag{6.61}$$

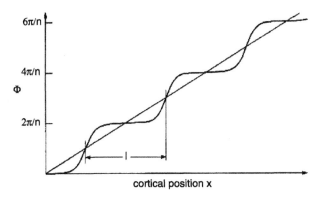

Fig. 29. Multi-soliton solution to the sine-Gordon equation that describes regions of the activity pattern that are commensurate with the cortical lattice separated by domain walls of mean separation $l$

where $l$ is the separation between neighboring kinks. Equivalently, $\rho$ is a measure of the soliton density. It can be shown that

$$\frac{\rho}{q} = \frac{4}{\pi^2} K(\eta) E(\eta), \tag{6.62}$$

where $K$ and $E$ are complete elliptic integrals of first and second kind, and $\eta$ satisfies the equation

$$\frac{q_0}{q} = \frac{\pi \eta}{2 E(\eta)}. \tag{6.63}$$

An asymptotic analysis of these equations establishes that $\rho \sim \log(q - q_0)$ for $q \to q_0$ [4]. We conclude that if the mismatch between the periodicities of the activity pattern and cortical lattice is increased such that $q > q_0$, then there is a commensurate-incommensurate transition to a state in which the activity pattern is no longer perfectly pinned to the lattice due to the formation of soliton-like phase defects.

## 7. Pattern formation in a model of cortical development

We end these lectures by briefly discussing another important example of pattern formation in visual cortex, namely, the development of the feedforward connectivity patterns that generate the various cortical feature maps described in section 2. Activity-based developmental models of retinotopy, ocular dominance and orientation maps typically involve some Hebbian-like competitive mechanism for the modification of feedforward afferents (see the review [90]). Intracortical interactions consisting of short-range excitation and longer-range inhibition mediate a pattern forming instability with respect to the feedforward synaptic weights that is sensitive to the two–point spatial correlation structure of the inputs to cortex. In order to point out some of the formal similarities with the continuum models discussed previously, we will review a recent model of ocular dominance (OD) column formation that takes into account the existence of an array of intrinsically defined CO blobs [23].

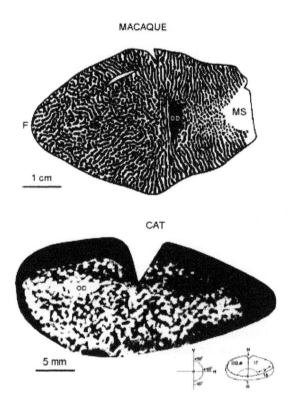

Fig. 30. Intra-ocular injection of proline revealing OD columns in layers 4 of macaque and cat

The OD columns in macaque form a stripe–like pattern in which the CO blobs are distributed at evenly spaced intervals along the center of the OD columns. In cat the OD columns form a spot–like pattern and the spatial relationship with the blobs is weaker, that is, although blobs avoid OD borders they are not strictly aligned along the center of the columns. Examples of OD patterns are shown in Fig. 30. In macaque both CO blobs and OD columns emerge prenatally, so that at birth the pattern of OD columns and their spatial relationship with blobs is adult-like. Moreover, this spatial relationship and the spacing of CO blobs is not influenced by visual experience [51,69]. In contrast, the cat's visual cortex is quite immature at birth. For example, supragranular layers of cat V1 differentiate postnatally and the blobs in these layers are normally first visible around 2 weeks of age. Nevertheless, altering visual experience by either monocular deprivation, binocular deprivation or dark rearing does not affect the pattern of blobs, establishing that visual experience is not necessary for the initial expression or early development of CO blobs in cat [70]. Thus, the lack of influence of visual experience on CO blob development in both cats and primates suggests that the blobs may reflect an innate columnar organization within superficial layers of cortex [69,70]. An intrinsic cortical specification of the blobs, which is distinct from the competition-driven development of cortical circuitry, could be generated by a periodic array of molecular markers. A number of anatomical markers are distributed in a patchy fashion early in development and are thought to play an important role in synaptic plasticity, including the NMDA receptor [95]. It is possible that these patchy distributions are themselves generated by a diffusion–driven Turing instability [98].

*Developmental model* We now describe a simple model of OD column formation that incorporates the effects of an array of CO blob markers [23]. Let $n_R(\mathbf{r}, t)$ and $n_L(\mathbf{r}, t)$ be, respectively, the right and left eye densities of synaptic connections to the visual cortex modeled as a two–dimensional sheet with $\mathbf{r} = (x, y)$ a point on the cortical sheet. The feedforward weights are taken to evolve according to a modified version of the competitive Hebbian learning rule introduced by Swindale [88]

$$\frac{\partial n_i}{\partial t} = \left[ \Gamma(n_i) + \sum_{j=R,L} \int W_{ij}(|\mathbf{r} - \mathbf{r}'|) n_j(\mathbf{r}', t) d^2\mathbf{r}' \right] F_N(n_i) \qquad (7.1)$$

for $i = R, L$. The logistic function $F_N(n) = n(N - n)$ ensures that growth terminates at a maximum density $N$ and that the densities remain positive, that is, $0 \leq n \leq N$. The function $\Gamma(n) = \mu(-n + M)$ for $\mu > 0$ is a term that is absent in Swindale's model, and represents a process that tend to stabilize the binocular state $n_L = n_R = M$ for some constant $M$, which is identified as the initial state prior to the onset of activity–driven synaptic modification. The transition to the activity–driven phase can then be modeled as a reduction in the parameter $\mu$. There are four types of interaction function $W_{LL}, W_{LR}, W_{RL}, W_{RR}$, which lump together distinct left/right eye input correlations with intracortical synaptic interactions. Each of the interaction functions is assumed to be be a difference of Gaussians of the form (6.8). We are assuming that early in development the patchy long–range connections and cortical feature maps are only crudely formed, so that the types of spatially periodic modulations considered in section 6.2 are not yet present. That is, we take the interaction functions to be homogeneous. Note that equation (7.1) reduces to the original Swindale model if $\mu = 0$ and the maximal density $N$ is spatially uniform.

Suppose there exists an array of instrinsically defined CO blobs that locally enhance experience–dependent plasticity. In the simplified developmental model given by equation (7.1) this can be interpreted as a local increase in $N$ at the CO blob centers. [In more detailed models of the competitive process underlying synaptic modification [99], the CO blob marker could modulate the local level of neurotrophic factors, for example]. Letting $u(\mathbf{r})$ denote the patchy distribution of the CO marker, we set $N(\mathbf{r}) = \overline{N} + \kappa u(\mathbf{r})$ where $\overline{N}$ is a baseline level for the maximal density of feedforward afferents and $\kappa$ is a positive constant that determines the strength of the modulation. The distribution $u(\mathbf{r})$ has maxima at the CO blob centers, which are located at the sites $\boldsymbol{\ell} \in \mathcal{L}$ of a two-dimensional lattice $\mathcal{L}$. In the case of a regular planar lattice we can write $\boldsymbol{\ell} = m_1\boldsymbol{\ell}_1 + m_2\boldsymbol{\ell}_2$ for integers $m_1, m_2$ with $\boldsymbol{\ell}_1, \boldsymbol{\ell}_2$ the generators of the lattice. Moreover, $u(\mathbf{r})$ becomes a doubly periodic function on the lattice, $u(\mathbf{r} + m_1\boldsymbol{\ell}_1 + m_2\boldsymbol{\ell}_2) = u(\mathbf{r})$ for all integers $m_1, m_2$, and can thus be expanded as a Fourier series. Note that the actual distribution of CO blobs in cortex is more accurately described in terms of a disordered lattice [69]. However, in order to highlight the basic alignment mechanism we will consider regular lattices here. Moreover, as we establish below, the proposed alignment mechanism is robust to disorder in the OD dominance pattern. Thus some additional disorder arising from the distribution of CO blobs will not significantly affect our results.

*Linear stability analysis*   Since $N$ is spatially varying, equation (7.1) has at most one homogeneous fixed point $(n_L^*, n_R^*)$ given by the solution of the pair of equations $n_i^* = M + \mu^{-1}\sum_j \widehat{W}_{ij}(0)n_j^*$ with $\widehat{W}_{ij}(0) = \int w_{ij}(|\mathbf{r}|)d\mathbf{r}$. Imposing the left-right eye symmetry constraints $W_{LL} = W_{RR} = -W_{LR} = -W_{RL} = W$ ensures that the equilibrium reduces to the binocular state $n_L^* = n_R^* = M$. The linear stability of the uniform binocular state can be determined by setting $n_j(\mathbf{r}, t) = M + a_j(\mathbf{r}, t)$ and expanding equation (7.1) to first order in $a_j$. In terms of the sum and difference densities $n_\pm = (n_L \pm n_R)/2$ and $a_\pm = (a_L \pm a_R)/2$ we find that

$$\frac{\partial a_+}{\partial t} = -\mu a_+ H(\mathbf{r}) \tag{7.2}$$

and

$$\frac{\partial a_-}{\partial t} = \left[ -\mu a_-(\mathbf{r}, t) + \int W|(\mathbf{r} - \mathbf{r}'|)a_-(\mathbf{r}', t)d\mathbf{r}' \right] H(\mathbf{r}) \qquad (7.3)$$

where $H(\mathbf{r}) = M(\overline{N} + \kappa u(\mathbf{r}) - M)$. We will assume that $\overline{N} > M$ so that $H(\mathbf{r})$ is always positive. [If $\mu = 0$ then the equilibrium is only marginally stable with respect to small perturbations $a_+$ of the total density and one has to impose an additional constraint in order to specify the equilibrium uniquely [88].]

*(a) Homogeneous case ($\kappa = 0$).* Let us first ignore the modulatory effect of the CO blobs by setting $\kappa = 0$ so that $H(\mathbf{r}) = M(\overline{N} - M)$. Equation (7.2) with $\mu > 0$ implies that $a_+(\mathbf{r}, t) \to 0$ as $t \to \infty$ so that the stability of the binocular state $n_j(\mathbf{r}) = M$ will depend on the asymptotic behavior of $a_-$. Since equation (7.3) is homogeneous when $\kappa = 0$, it has solutions of the form $a_-(\mathbf{r}, t) = e^{\lambda t}e^{i\mathbf{k}\cdot\mathbf{r}}$ where $\mathbf{k}$ denotes two-dimensional spatial frequency and the growth factor $\lambda$ satisfies the dispersion relation

$$\lambda = \lambda(k) \equiv [-\mu + \widehat{W}(k)]M(\overline{N} - M) \qquad (7.4)$$

with $\widehat{W}(k) = \int e^{-i\mathbf{k}\cdot\mathbf{r}}W(|\mathbf{r}|)d\mathbf{r}$ and $k = |\mathbf{k}|$. Since $W$ is given by a Mexican hat function of the form (6.8), it follows that $\lambda(k)$ is a difference of Gaussians with a maximum at a non-zero wavenumber $k_c$. This is similar to the dispersion curve shown in Fig. 26 except that now the maximum shifts downwards rather than upwards as $\mu$ increases. If $\widehat{W}(k_c) < \mu$ then the equilibrium is stable with respect to local perturbations. On the other hand, if $\widehat{W}(k_c) > \mu$ then the equilibrium is unstable due to the growth of a finite band of unstable Fourier modes lying in some interval containing $k_c$. Treating $\mu$ as a bifurcation parameter, we see that there is a primary pattern forming instability at $\mu = \mu_c \equiv \widehat{W}(k_c)$. Sufficiently close to the bifurcation point with $0 < \mu_c - \mu \ll \mu_c$, the instability leads to the formation of a spatially periodic stripe pattern of the form $n_-(\mathbf{r}) = A\cos(\mathbf{k}.\mathbf{r} + \phi)$ with $|\mathbf{k}| \approx k_c$. This represents alternating left/right eye ocular dominance columns of approximate width $\pi/k_c$, each of which is a stripe running orthogonal to the direction of $\mathbf{k}$. Ignoring boundary effects, the direction $\arg(\mathbf{k})$ and spatial phase $\phi$ of the stripe pattern are arbitrary when $\kappa = 0$, which reflects the rotational and translational symmetry of the effective lateral interactions $w$. In the case $\mu = 0$ considered by Swindale [88], the band of unstable modes extends to $\infty$, and the dynamics is far from the primary bifurcation point. The emerging pattern tends to have a more disordered stripe-like morphology, which is consistent with the pattern of OD columns found in macaque.

*(b) Inhomogeneous case ($\kappa > 0$).* Including the effects of the CO blobs by taking $\kappa > 0$ breaks continuous translation symmetry so that the spatial phase of

the pattern will no longer be arbitrary. For small $\kappa$ we can carry out a small–$\kappa$ perturbation analysis of equation (7.1) along identical lines to section 6.2, with $D \to u$ and $a \to n_-$. In the one–dimensional case with $k \approx \pi/d$ this leads to the following pair of dispersion branches (after rescaling $\lambda$ and $\kappa$)

$$\lambda_{\pm}(k) = -\mu + \widehat{W}(k) + \kappa[\mathcal{V}_0(k) \pm \mathcal{V}_Q(k)] \tag{7.5}$$

where $\mathcal{V}_1(k) = [-\mu + \widehat{W}(k)]\mathcal{U}_q$ and we have introduced the Fourier series expansion $u(x) = \sum_q \mathcal{U}_q e^{iqx}$. It follows that if the emerging pattern of OD columns has a critical wavenumber that is commensurate with the lattice of CO blobs then pinning will occur. Here, however, the pinning mechanism involves mode–locking between spatial frequency components of the emerging pattern of OD columns and the distribution of intrinsically defined CO blobs. In particular, we can reinterpret the even mode shown in Fig. 28 as showing alternating left and right ocular dominance columns of width $\pi/d$ whose centers are located at the CO blob centers. Interestingly, the same patchy molecular markers that initially generate the CO blobs could also play a role in seeding the formation of the im-mature horizontal connections, thus leading to commensurability. That is, the Mexican hat interaction functions could be replaced by a homogeneous weight distribution whose Fourier transform is similar to that shown in Fig. 27.

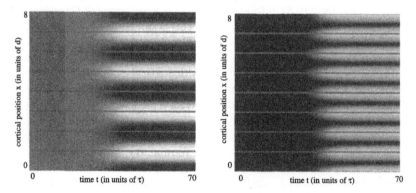

Fig. 31. One–dimensional pinning of OD columns to a periodic array of CO blobs. (a) Space–time plot of the density $n_-(x, t)$ obtained by numerically solving a one–dimensional version of equation (7.1) for $N = 1 + 0.5\kappa[1 + \cos(2\pi x/d)]$, $\kappa = 0.4$, $\mu = 0$ and periodic boundary conditions with system size $L = 8d$. The initial binocular state evolves into alternating left and right eye ocular dominance columns whose centers align with the centers of the CO blobs (indicated by horizontal lines). (b) Correspond-ing plot of $n_+(x, t)$ showing the emergence of a periodic variation in the total density of feedforward afferents.

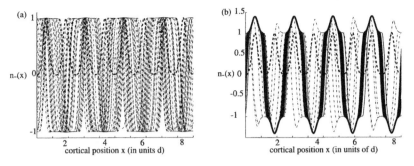

Fig. 32. Superposition of one–dimensional OD patterns obtained over repeated trials using the same parameter values as Fig. 31 but different choices of $\kappa$. For each trial the steady–state density $n_-(x)$ is plotted as a function of $x$. The CO blob density is shown as a solid dashed curve. (a) $\kappa = 0$: there is no pinning so that the spatial phase of the resulting pattern is arbitrary. (b) $\kappa = 0.4$: pinning occurs over most trials.

*Nonlinear effects*  As the amplitude of the selected pattern grows the dynamics will become dominated by nonlinearities arising from the logistic function $F_N(n)$. If a point $x$ lies within a left eye ocular dominance column we expect $n_L(x, t) \to \overline{N} + \kappa u(x)$ and $n_R(x, t) \to 0$ as $t \to \infty$, whereas we expect the opposite to occur if $x$ lies within a right eye column. This suggests that there will be a periodic variation in the total density of feedforward afferents such that $n_+(x) \approx (\overline{N} + \kappa u(x))/2$. Thus the pinning mechanism also provides a means for generating a higher density of feedforward afferents to the CO blobs so that they ultimately become defined extrinsically. To what extent does the pinning of the OD pattern to the lattice of CO blobs persist when these nonlinearities are included, and is the resulting pattern stable? In contrast to the planar model analyzed in section 6, it is not possible to use weakly nonlinear analysis to analyze the pattern forming instability of the developmental model (7.1). This is due to the fact that beyond the point of instability, a large–amplitude OD pattern forms. Therefore, we proceed by numerically solving a one–dimensional version of equation (7.1) for $N = 1 + 0.5\kappa(1 + \cos(2\pi x/d))$. We choose the parameters of the Mexican hat function $W$ such that $k_c \approx \pi/d$. Space–time plots of the densities $n_+(x, t)$ and $n_-(x, t)$ are shown in figure 31 for $\kappa = 0.4$. It can be seen that the initial binocular state evolves into alternating left and right eye ocular dominance columns whose centers align with the CO blobs. A periodic variation in the total density of feedforward afferents also emerges. We find that pinning to the CO blobs occurs over most trials, although occasionally the system evolves into either a misaligned state or an antipinned state corresponding to an odd eigenmode. This is illustrated in Fig. 32. Increasing the system size tends to make pinning more reliable through the establishment of phase defects, see figure 33.

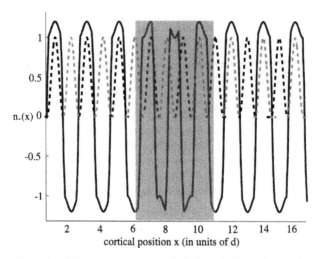

Fig. 33. One–dimensional OD pattern with a spatial phase defect. The steady–state density $n_-(x)$ is plotted as a function of $x$ for $\Lambda = 16d$. Other parameter values are as in Fig. 31. It can be seen that pinning occurs except within a transition region (shaded in gray) where the OD columns are gradually shifted by an additional spatial phase equal to a single blob spacing.

In Fig. 34 we show examples of two–dimensional OD patterns obtained by numerically solving equation (7.1) for CO blobs on a square lattice. We take $N(x, y) = 1 + 0.25\kappa(2 + \cos(2\pi x/d) + \cos(2\pi y/d))$, $\kappa = 0.4$ and $\mu = 0$. The parameters of the weight distribution $w(r)$ are chosen so that $k_c \approx \pi/d$. The resulting OD stripes tend to be oriented in directions corresponding to symmetries of the lattice and the centers of the OD columns coincide with CO blob centers. Interestingly, pinning occurs even though the patterns are quite irregular with a number of defects and branches.

## 8. Future directions

In these lectures we have described in some detail the theory of neural Turing instabilities in large-scale models of primary visual cortex (V1). We have emphasized the role of symmetries in determining the types of pattern that emerge. We have also argued that the standard assumption of isotropic and homogeneous

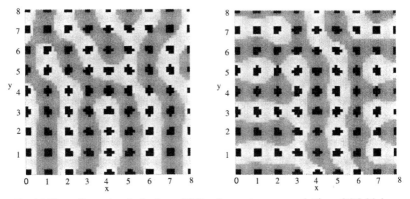

Fig. 34. Two–dimensional pinning of OD columns to a square lattice of CO blobs.

intracortical interactions breaks down when the detailed functional architecture of V1 is taken into account. That is, there are spatially periodic modulations of the interactions arising from correlations between the patchy horizontal (and feedback connections) and cortical feature preference maps (obtained using optical imaging for example). These modulations can induce a pinning of cortical activity patterns to the underlying feature maps. An analogous pinning mechanism may also account for the alignment of OD columns with the CO blobs during development. Here the pattern forming instability occurs with respect to the feedforward afferents from the two eyes, and a periodic modulation in synaptic plasticity arises from patchy molecular markers.

There are a number of directions in which the theory of cortical pattern formation could be extended. First, one could consider a more detailed model of cortex that takes the following aspects into account: (i) the various cortical layers, (ii) the different classes of interneuron and their distribution across layers, (iii) short–term synaptic plasticity and adaptation, (iv) more biophysically realistic conductance–based neuron models and (v) dendritic structure. Second, one could investigate to what extent spontaneously generated global activity patterns reflect properties of the cortex in response to visually generated stimuli. For example, how do the long–range connections provide a substrate for the extra–classical receptive field? Third, one could construct more detailed developmental models that investigate the possible role of patchy molecular markers in the formation of cortical feature maps through Hebbian learning. Finally, one could consider other forms of spatio–temporal coherent structures including localized bumps and traveling waves.

# References

[1] Angelucci A, Levitt J B and Lund J S, 2002. Anatomical origins of the classical receptive field and modulatory surround field of single neurons in macaque visual cortical area v1. *Progress in Brain Research 136* 373–388.

[2] Angelucci A, Levitt J B, Walton E J S, Hupe J-M, Bullier J and Lund J S, 2002. Circuits for local and global signal integration in primary visual cortex. *J. Neurosci 22* 8633–8646.

[3] Ashcroft N W and Mermin N D. *Solid state physics*. (Saunders College Publishing, 1976).

[4] Bak P, 1982 Commensurate phases, incommensurate phases and the devil's staircase. *Rep. Prog. Phys. 45* 587–629.

[5] Ben-Yishai R, Lev Bar-Or R and Sompolinsky H, 1995 Theory of orientation tuning in visual cortex. *Proc. Nat. Acad. Sci.* **92** 3844-3848

[6] Ben-Yishai R, Hansel D and Sompolinsky H, 1997 Traveling waves and the processing of weakly tuned inputs in a cortical network module. *J. Comput. Neurosci.* **4** 57-77

[7] Blasdel G G and Salama G, 1986 Voltage-sensitive dyes reveal a modular organization in monkey striate cortex. *Nature 321* 579-585

[8] Blasdel G G, 1992 Orientation selectivity, preference, and continuity in monkey striate cortex. *J. Neurosci.* **12** 3139-3161

[9] Bonhoeffer, T. & Grinvald, A., 1991. Orientation columns in cat are organized in pinwheel like patterns. *Nature* **364**: 166–146.

[10] Bonhoeffer T, Kim D S, Malonek D, Shoham D and Grinvald A, 1995 Optical imaging of the layout of functional domains in area 17/18 border in cat visual cortex. *European J. Neurosci.* **7** 1973-1988

[11] Bosch Vivancos I, Chossat P and Melbourne I, 1995 New planforms in systems of partial differential equations with Euclidean symmetry. *Arch. Rat. Mech.* **131**: 199–224.

[12] Bosking W H, Zhang Y, Schofield B and Fitzpatrick D, 1997 Orientation selectivity and the arrangement of horizontal connections in tree shrew striate cortex. *J. Neurosci.* **17** 2112-2127.

[13] Bressloff P C, Bressloff N W and Cowan J D, 2000 Dynamical mechanism for sharp orientation tuning in an integrate–and–fire model of a cortical hypercolumn. *Neural Comput.* **12** 2473-2511

[14] Bressloff P C, Cowan J D, Golubitsky M, Thomas P J and Wiener M, 2001 Geometric visual hallucinations, Euclidean symmetry and the functional architecture of striate cortex. *Phil. Trans. Roy. Soc. Lond. B* **356** 299-330

[15] Bressloff P C, Cowan J D, Golubitsky M and Thomas P J, 2001 Scalar and pseudoscalar bifurcations: pattern formation on the visual cortex. *Nonlinearity* **14** 739-775.

[16] Bressloff P C, Cowan J D, Golubitsky M, Thomas P J and Wiener M, 2002 What geometric visual hallucinations tell us about the visual cortex. *Neural Comput.* **14** 473-491

[17] Bressloff P C and Cowan J D, 2002 Spontaneous pattern formation in primary visual cortex. In: S. J. Hogan *et al* editors *Nonlinear Dynamics and chaos: where do we go from here?* Institute of Physics: Bristol.

[18] Bressloff P C and Cowan J D, 2002 SO(3) symmetry breaking mechanism for orientation and spatial frequency tuning in visual cortex *Phys. Rev. Lett.* **88** 078102

[19] Bressloff P C and Cowan J D, 2002 The visual cortex as a crystal. *Physica D* **173** 226–258.

[20] Bressloff P C, 2002 Bloch waves, periodic feature maps and cortical pattern formation. *Phys. Rev. Lett.* **89** 088101.

[21] Bressloff P C and Cowan J D, 2003 Spherical model of orientation and spatial frequency tuning in a cortical hypercolumn. *Phil. Trans. Roy Soc. B.* **358** 1643–1667

[22] Bressloff P C, 2003 Spatially periodic modulation of cortical patterns by long–range horizontal connections. *Physica D*. **185**, 131-157

[23] Bressloff P C and Oster A M, 2004 A theory for the alignment of cortical feature maps during development (submitted).

[24] Bullier J, Hupe' J-M, James A J and Girard P, 2001 The role of feedback connections in shaping the response of visual cortical neurons. *Prog. Brain Res.* **134** 193-204.

[25] Coullet P, 1986 Commensurate-incommensurate transition in nonequilibrium systems. *Phys. Rev. Lett.* **56** 724–727.

[26] Cowan J D, 1982 Spontaneous symmetry breaking in large scale nervous activity. *Intl. J. Quantum Chem.* **22** 1059–1082.

[27] Cowan J D, 1997 Neurodynamics and Brain Mechanisms. In Ito M, Miyashita Y and Rolls E T, editors, *Cognition, Computation, and Consciousness* (Oxford University Press) p. 205-233

[28] Cross M C and Hohenberg P C, 1993 Pattern formation outside of equilibrium. *Rev. Mod. Phys.* **65** 851–1111.

[29] Daugman J G, 1980 Two–dimensional Spectral analysis of Cortical Receptive Field Profiles. *Vision Res.*, 20:847–856.

[30] Dayan P and L F Abbott L F, 2001 *Theoretical Neuroscience*. MIT Press, Cambridge MA.

[31] De Valois, R L and De Valois, K K, 1988. *Spatial Vision*. Oxford: Oxford University Press.

[32] Dionne B, Golubitsky M, Silber M and Stewart I, 1995 Time–periodic spatially periodic planforms in Euclidean equivariant partial differential eqnarrays. *Phil. Trans. Roy. Soc. Lond. A* **325** 125-168

[33] Douglas R, Koch C, Mahowald M, Martin K and Suarez H, 1995 Recurrent excitation in neocortical circuits. *Science* **269** 981-985

[34] Ermentrout G B and Cowan J D, 1979 A mathematical theory of visual hallucination patterns. *Biol. Cybernetics* **34** 137-150

[35] Ermentrout G B and Cowan J D, 1979 Temporal oscillations in neuronal nets. *J. Math. Biol.* **7** 265–280

[36] Ermentrout G B, 1998 Neural networks as spatial pattern forming systems. *Rep. Prog. Phys.* **61** 353-430

[37] Eysel U, 1999. Turning a corner in vision research. *Nature* **399**: 641–644.

[38] Fitzpatrick D, 2000 Seeing beyond the receptive field in primary visual cortex. *Curr. Op. in Neurobiol.* **10** 438-443

[39] Ferster D, Chung S and Wheat H, 1997 Orientation selectivity of thalamic input to simple cells of cat visual cortex. *Nature* **380** 249-281

[40] D Ferster and K D Miller (2000). Neural mechanisms of orientation selectivity in the visual cortex. *Ann. Rev. Neurosci.* 23:441–471.

[41] Gerstner W, 1995 Time structure of the activity in neural network models. *Phys. Rev. E* **51** 738–758

[42] Gilbert C D and Wiesel T N, 1983 Clustered intrinsic connections in cat visual cortex. *J. Neurosci.* **3** 1116-1133

[43] Gilbert C D, 1992 Horizontal integration and cortical dynamics *Neuron* **9** 1-13

[44] Gilbert C D, Das A, Ito M, Kapadia M and Westheimer G, 1996 Spatial integration and cortical dynamics. *Proc. Nat. Acad. Sci.* **93** 615-622

[45] Golubitsky M, Stewart I and Schaeffer D G, 1988 *Singularities and Groups in Bifurcation Theory II*. (Berlin: Springer–Verlag)

[46] Grossberg S and Williamson J R, 2001   A neural model of how horizontal connections of visual cortex develop into adult circuits that carry out perceptual grouping and learning. *Cerebral Cortex* **11** 37–58.

[47] Hansel D and Sompolinsky H. Modeling feature selectivity in local cortical circuits. In C. Koch and I. Segev, editors, *Methods of Neuronal Modeling*, pages 499–567. Cambridge: MIT Press, 2nd edition, 1998.

[48] Haüssler A and von der Malsburg C 1983  Development of Retinotopic Projections: An Analytical Treatment *J. Theoret. Neurobiol.* **2** 47-73

[49] Hirsch J D and Gilbert C D, 1991  Synaptic physiology of horizontal connections in the cat's visual cortex. *J. Physiol. Lond.* **160** 106-154

[50] Horton J C, 1984.  Cytochrome oxidase patches: a new cytoarchitectonic feature of monkey visual cortex. *Phil. Trans. R. Soc. Lond. B* **304** 199–253.

[51] Horton J C and Hocking D R, 1997  *J. Neurosci.* **17** 3684-3709.

[52] Hubel D H and Wiesel T N, 1962  Receptive fields, binocular interaction and functional architecture in the cat's visual cortex. *J. Neurosci.* **3** 1116-1133

[53] Hubel D H and Wiesel T N, 1974a  Sequence regularity and geometry of orientation columns in the monkey striate cortex. *J. Comp. Neurol.* **158** 267-294.

[54] Hubel D H and Wiesel T N, 1974  Uniformity of monkey striate cortex: A parallel relationship between field size, scatter, and magnification factor. *J. Comp. Neurol.* **158** 295-306.

[55] Hubener M, Shoham D, Grinvald A and Bonhoeffer T, 1997  Spatial relationship among three columnar systems. *J. Neurosci.* **15** 9270-9284.

[56] Issa N P, Trepel C and Stryker M P, 2000.  Spatial frequency maps in cat visual cortex. *J. Neurosci.* **20** 8504-8514.

[57] Jones J P and Palmer L A, 1987  An evaluation of the two–dimensional Gabor filter model of simple receptie fields in cat striate cortex. *J. Neurophysiol.*, 58(6):1233–1258.

[58] Kang K, Shelley M and Sompolinsky H, 2003.  Mexican hats and pinwheels in visual cortex. *Proc. Natl. Acad. Sci. USA.*

[59] Klüver, H., 1966.  *Mescal and Mechanisms of Hallucinations.* University of Chicago Press.

[60] LeVay S and Nelson S B 1991.  Columnar organization of the visual cortex. *The neural basis of visual function* ed Leventhal A G (Boca Raton: CRC Press) pp 266-315

[61] Livingstone M S and Hubel D H, 1984  Anatomy and physiology of a color system in the primate visual cortex. *J. Neurosci.* **4** 309-356.

[62] Lowe M and Gollub J P, 1985  Solitons and the commensurate-incommensurate transition in a convective nematic fluid. *Phys. Rev. A*, **31** 3893–3896.

[63] Lund J S, Wu Q, Hadingham P T and Levitt J B, 1995  Cells and circuits contributing to functional properties in area V1 of macaque monkey cerebral cortex: Bases for neuroanatomically realistic models. *J. Anat. (Lond.)*, 187:563–581.

[64] Lund J S, Angelucci A and Bressloff P C, 2003.  Anatomical substrates for functional columns in macaque monkey primary visual cortex. *Cerebral Cortex* **12** 15-24.

[65] Malach R, Amir Y, Harel M and Grinvald A, 1993  Relationship between intrinsic connections and functional architecture revealed by optical imaging and in vivo targeted biocytin injections in primate striate cortex. *Proc. Natl. Acad. Sci.* **90** 10469-10473

[66] Maldonado P E, Gödecke I, Gray C M and Bonhoeffer T, 1997  Orientation selectivity in pinwheel centers in cat striate cortex. *Science* **276** 1551-1555.

[67] Marr D and Poggio T, 1976  Cooperative computation of stereo disparity *Science* **194** 283-287

[68] McLaughlin D R, Shapley R, Shelley M and Wielaard J, 2000.  A neuronal network model of macaque primary visual cortex (V1): orientation selectivity and dynamics in the input layer 4Ca, Proc. Natl. Acad. Sci. USA **97** 8087-8092.

[69] Murphy K M, Jones D G, Fenstemaker S B, Pegado V D, Kiorpes L and Movshon J A, 1998 *Cerebral Cortex* **8** 237-244.

[70] Murphy K M, Duffy K. R, Jones D G and Mitchell D E, 2001 *Cerebral Cortex* **11** 237-244.

[71] Murray J D, 2002 *Mathematical Biology* volumes I and II. Springer Verlag, Berlin.

[72] Nelson S, Toth L J, Seth B and Sur M, 1994. Orientation selectivity of cortical neurons during extra-cellular blockade of inhibition. Science **265** 774-777.

[73] Newell A C and Whitehead J A, 1969 Finite bandwidth, finite amplitude convection. *J. Fluid Mech.* **38** 279.

[74] Obermayer K and Blasdel G G, 1993 Geometry of orientation and ocular dominance columns in monkey striate cortex. *J. Neurosci.* **13** 4114-4129

[75] Phleger B and Bonds A B, 1995 Dynamic differentiation of gaba$_A$–sensitive influences on orientation selectivity of complex cells in the cat striate cortex. *Exp. Brain Res.* **104**

[76] Reid R C and Alonso J M, 1995 Specificity of momosynaptic connections from thalamus to visual cortex. *Nature* **387** 281–284.

[77] Ringach D L, Hawken M J and Shapley R, 1997. Dynamics of orientation tuning in macaque primary visual cortex. *Nature*, 387:281–284.

[78] Rockland K S and Lund J, 1983 Intrinsic laminar lattice connections in primate visual cortex. *J. Comp. Neurol.* **216** 303–318

[79] Salin P-A and Bullier J, 1995 Corticocortical connections in the visual system: structure and function *Physiological Reviews* **75** 107–154

[80] Schwartz E, 1977. Spatial mapping in the primate sensory projection: analytic structure and relevance to projection. *Biol. Cybernetics* **25**: 181–194.

[81] Segel L A, 1969 Distant side-walls cause slow amplitude modulation of cellular convection. *J. Fluid Mech.* **38** 203.

[82] Shelley M and McLaughlin D, 2002. A Neuronal Network Model of Macaque Primary Visual Cortex (V1): Orientation Tuning and Dynamics in the Input Layer 4C$\alpha$. *J. Comput. Neurosci.*, 12:97–122.

[83] Shmuel A, Korman M, Harel M, Grinvald A and Malach R, 1998 Relationship of feedback connections from area V2 to orientation domains in area V1 of the primate. *Soc Neurosci. Abstr.* **24** 767

[84] Sillito A M, 1975 The contribution of inhibitory mechanisms to the receptive–field properties of neurones in the striate cortex of the cat. *J. Physiol. Lond.* **250**: 305–329.

[85] Sincich L C and Blasdel G G, 2001. Oriented axon projections in primary visual cortex of the monkey. *J. Neurosci* **21**: 4416–4426.

[86] Somers D C, Nelson S and Sur M, 1995 An Emergent Model of Orientation Selectivity in Cat Visual Cortical Simple Cells. *J. Neurosci.* **15** 5448-5465

[87] Sompolinsky H and Shapley R, 1997. New perspectives on the mechanisms for orientation selectivity. *Curr. Opin. Neurobiol.*, 7:514–522.

[88] Swindale N V, 1980 A model for the formation of ocular dominance stripes. *Proc. Roy. Soc. Lond. B* **208** 243-264

[89] Swindale N V, 1982 A model for the formation of orientation columns *Proc. Roy. Soc. Lond. B* **215** 211-230

[90] Swindale N V, 1996 The development of topography in visual cortex: a review of models *Network* **7** 161-247

[91] Tass P, 1997 Oscillatory cortical activity during visual hallucinations. *J. Biol. Phys.* **23** 21-66

[92] Tootell R B H, Silverman M S and De Valois R L, 1981. Spatial frequency columns in primary visual cortex. *Science* **214**: 813–815.

[93]  Tootell R B, Silverman M S, Switkes E and DeValois R L , 1982. Deoxyglucose analysis of retinotopic organization in primate striate cortex. *Science*, **218** 902–904.

[94]  Toth L J, Rao S C, Kim D-S, Somers D C and Sur M, 1996 Subthreshold facilitation and suppression in primary visual cortex revealed by intrinsic signal imaging. *Proc. Natl. Acad. Sci.* **93** 9869-9874

[95]  Trepel C, Duffy K R, Pegado V D and Murphy K M, 1998 *J. Neurosci.* **18** 3404-3415.

[96]  Troyer T W, Krukowski A, Priebe N J and Miller K D, 1998. Contrast-invariant orientation tuning in cat visual cortex: feedforward tuning and correlation-based intracortical connectivity. *J. Neurosci.*, **18** 5908–5927.

[97]  Tsodyks M, Kenet T, Grinvald A and Arieli A, 1999. Linking spontaneous activity of single cortical neurons and the underlying functional architecture, *Science* **286** 1943-1946.

[98]  Turing A M, 1952 The Chemical Basis of Morphogenesis *Phil. Trans. Roy Soc. Lond. B* **237** 32-72

[99]  Van Ooyen A, 2003. In: *Modeling Neural Development* Van Ooyen A (editor) pp. 213–244. MIT Press, Cambridge MA.

[100]  Von der Malsburg C, 1973 Self-organization of Orientation Selective cells in the Striate Cortex. *Kybernetik* **14** 85-100

[101]  Von der Malsburg C and Willshaw D J, 1977 How to label nerve cells so that they can interconnect in an ordered fashion. *Proc.Natl.Acad.Sci. USA* **74** 5176-5178

[102]  Walgraef D, 1997 *Spatio-Temporal Pattern Formation*. Berlin: Springer–Verlag.

[103]  Wielaard D J, Shelley M J, McLaughlin D W and Shapley R, 2001 Lateral Inhibition Generates Simple cells in a Computational Model of Primary Visual Cortex.

[104]  Wilson H R and Cowan J D, 1972 Excitatory and inhibitory interactions in localized populations of model neurons. *Biophys. J.* **12** 1-24

[105]  Wilson H R and Cowan J D, 1973 A mathematical theory of the functional dynamics of cortical and thalamic nervous tissue. *Kybernetik* **13** 55-80

[106]  Yabuta N H and Callaway E M, 1998 Cytochrome–oxidase blobs and intrinsic horizontal connections of layer 2/3 pyramidal neurons in primate V1 *Visual Neurosci.* **15**: 1007–1027.

[107]  Yoshioka T, Blasdel G G, Levitt J B and Lund J S, 1996 Relation between patterns of intrinsic lateral connectivity, ocular dominance, and cytochrome oxidase–reactive regions in macaque monkey striate cortex. *Cerebral Cortex* **6** 297-310

Course 12

# SYMMETRY BREAKING AND PATTERN SELECTION IN VISUAL CORTICAL DEVELOPMENT

## F. Wolf

*Research Group Theoretical Neurophysics, Department of Nonlinear Dynamics,
Max-Planck-Institute for Dynamics and Selforganisation
and Institute for Nonlinear Dynamics, Fakultät für Physik, Universität Göttingen,
D-37073 Göttingen, Germany*

*C.C. Chow, B. Gutkin, D. Hansel, C. Meunier and J. Dalibard, eds.*
*Les Houches, Session LXXX, 2003*
*Methods and Models in Neurophysics*
*Méthodes et Modèles en Neurophysique*

# Contents

## 1. Introduction

The ontogenetic development of the cerebral cortex of the brain is a process of astonishing complexity. In every cubic millimeter of cortical tissue in the order of $10^6$ neurons must be wired appropriately for their respective functions such as the analysis of sensory inputs, the storage of skills and memory, or motor control [1]. In the brain of an adult animal, each neuron receives input via about $10^4$ synapses from neighboring and remote cortical neurons and from subcortical inputs [1]. At the outset of postnatal development, the network is formed only rudimentarily: For instance in the cat's visual cortex, most neurons have just finished the migration from their birth zone lining the cerebral ventricle to the cortical plate at the day of birth [2]. The number of synapses in the tissue is then only 10% and at the time of eye-opening, about two weeks later, only 25% of its adult value [3]. In the following 2-3 months the cortical circuitry is substantially expanded and reworked and the individual neurons aquire their final specificities in the processing of visual information [4].

It is a very attractive but still controversial hypothesis that in the ontogenetic development of the brain the emerging cortical organization is constructed by learning mechanisms which are similar to those that enable us to aquire skills and knowledge in later life [5–7]. Several lines of evidence strongly suggest that the brain in a very fundamental sense *learns* to see. First, visual experience is very important for the normal development of sight. If the use of the visual sense is prevented early in life vision becomes irreversibly impaired [4]. Since this is not due to a malformation of the eye or of peripheral stages of the visual pathway, it suggests that in development visual input it used to improve the processing capabilities of the visual cortical networks. In addition, the performance of the developing visual system responds very sensitively to visual experience. In human babies, for instance, already a few hours of visual experience lead to a marked improvement of visual acuity [8]. Second, the synaptic organization of the visual cortex is highly plastic and responds with profound and fast functional and structural reorganization to appropriate experimental manipulations of visual experience [9, 10]. These and similar observations suggest that the main origin of perceptual improvement in early development is due to an activity-dependent and thus use-dependent refinement of the cortical network. In the course of this neuronal activity patterns that arise in the processing of visual information in turn guide the further refinement of the cortical network until a mature configuration is reached. Whereas, theoretically, this hypothesis is very attractive, it is, experimentally, still controversial, whether neural activity actually plays such an *instructive* role (for discussion see [11–13]).

Viewed from a dynamical systems perspective, the activity-dependent remodeling of the of the cortical network described above is a process of dynamical pattern formation. In this picture, spontaneous symmetry breaking in the developmental dynamics of the cortical network underlies the emergence of cortical selectivities such as orientation preference [14]. The subsequent convergence of the cortical circuitry towards a mature pattern of selectivities can be viewed as the development towards an attractor of the developmental dynamics [15]. In this set of lectures, I will discuss universal dynamical properties of a paradigmatic process in visual cortical development: the development of orientation columns and the formation of so called orientation pinwheels.

In the visual cortex, as in most areas of the cerebral cortex information is processed in a 2-dimensional (2D) array of functional modules, called cortical columns [16, 17]. Individual columns are groups of neurons extending vertically throughout the entire cortical thickness that share many functional properties. Orientation columns in the visual cortex are composed of neurons preferentially responding to visual contours of a particular stimulus orientation [18]. In a plane parallel to the cortical surface, neuronal selectivities vary systematically, so that columns of similar functional properties form highly organized 2D patterns, known as functional cortical maps. In the case of orientation columns, this 2D organization is characterized by so called pinwheels, regions in which columns preferring all possible orientations are organized around a common center in a radial fashion [19, 20] (see Figure 1).

Experimental evidence suggests that the formation of orientation columns is a dynamical process guided by neuronal activity and sensitive to visual experience. In normal development, orientation columns first form at about the time of eye opening [21–23]. Comparison of this process to the development under conditions of modified visual experience demonstrates that adequate visual experience is essential for the complete maturation of orientation columns and that impaired visual experience, as with experimentally closed eye-lids can suppress or impair the formation of orientation columns [23] (see Figure 2). Most intriguingly, when visual inputs are experimentally redirected to drive what would normally become primary auditory cortex, orientation selective neurons and a pattern of orientation columns even forms in this brain region that would normally not at all be involved in the processing of visual information ( [24] see Figure 3). In particular the latter observation strongly suggests that the capability to form a system of orientation columns is intrinsic to the learning dynamics of the cerebral cortex given appropriate inputs. Taken together these lines of evidence mark the formation of orientation columns as a paradigmatic problem in the dynamics of cortical development and plasticity.

Owing to the large number of degrees of freedom of any realistic scale microscopic model of visual cortical development, the description of the development of the pattern of columns by equations for the synaptic connections between the LGN and cortex is very complicated. On the order of $10^6$ synaptic strengths would be required to realistically describe, for example, the pattern of orientation preference in a $4x4mm^2$ piece of the visual cortex. This complexity and the presently very incomplete knowledge about the nature of realistic equations for the dynamics of visual cortical development demand that theoretical analyzes concentrate on aspects that are relatively independent of the exact form of the equations and are representative for a large class of models. An appropriate framework for this is provided by models in which the emerging cortical architecture is described by order parameter fields and its development by a dynamics of such fields [15, 25–31]. A few years ago, Theo Geisel and the author discovered that experimentally accessible signatures of an activity-dependent refinement of the cortical network are predicted by universal properties of this very general class of models for the development of visual cortical orientation preference maps [15]. We could demonstrate that that if the pattern of orientation preferences is set up by learning mechanisms, then the number of pinwheels generated early in development exhibits a universal minimal value that depends only on general symmetry properties of the cortical network. This implies that in species exhibiting a lower number of pinwheels in the adult, pinwheels must move and annihilate in pairs during the refinement of the cortical circuitry. Verification of this intriguing prediction would provide striking evidence for the activity-dependent generation of the basic visual cortical processing architecture. In the initial sections of this chapter (Sections 2-5), I will present a self-contained treatment of the mathematical origin of this kind of universal behavior. In particular, I will discuss the description of the development of orientation preference columns in terms of a dynamics of abstract order parameter fields, connect this description to the theory of Gaussian random fields, and show how the theory of Gaussian random fields can be used to obtain quantitative information on the generation and motion of pinwheels, in the two dimensional pattern of visual cortical orientation columns. I will then extended the symmetry based approach used to derive this prediction to study also the kind of patterns to which the map will asymptotically converge and the interactions essential for the stabilization of different kinds of solutions (Section 6). In Section 7, I will provide an exposition of an appropriate perturbation method called weakly nonlinear analysis for the problem of orientation column formation. Using this method, a class of generalized Swift-Hohenberg models for the formation of patterns of contour detecting neurons during visual cortical development is constructed in Section 8. In this model class, a permutation symmetry of the model equations satisfies the requirement that the visual cortex develops selectivity for all contour orientations. By this symmetry a large number of dynamically degenerate solutions exist that quantitatively reproduce the experimentally observed patterns. Long-range interactions are found to be essential for the stability of realistic solutions.

## 2. The pattern of orientation preference columns

In the following, I will briefly introduce the mathematical description of the spatial layout of orientation columns in the visual cortex in terms of a complex valued order parameter field. Experimentally, the pattern of orientation preferences can be visualized using the optical imaging method [20, 32]. In such an experiment, the activity patterns $E_k(\mathbf{x})$ produced by stimulation with a grating of orientation $\theta_k$ are recorded. Here $\mathbf{x}$ represents the location of a column in the cortex. Using the activity patterns $E_k(\mathbf{x})$, a field of complex numbers $z(\mathbf{x})$ can be constructed that completely describes the pattern of orientation columns:

$$z(\mathbf{x}) = \sum_k e^{i\, 2\, \theta_k}\, E_k(\mathbf{x}) \qquad (2.1)$$

The pattern of orientation preferences $\vartheta(\mathbf{x})$ is then obtained from $z(\mathbf{x})$ as follows:

$$\vartheta(\mathbf{x}) = \frac{1}{2}\, arg(z) \; . \qquad (2.2)$$

Typical examples of such activity patterns $E_k(\mathbf{x})$ and the patterns of orientation preferences derived from them are shown in Figure 1. Numerous studies confirmed that the orientation preference of columns is a almost everywhere continuous function of their position in the cortex. Columns with similar orientation preferences occur next to each other in "iso-orientation domains" [33]. Neighboring iso-orientation domains preferring the same stimulus orientation exhibit a typical lateral spacing $\Lambda$ in the range of 1mm, rendering the pattern of preferred orientations roughly repetitive. Furthermore, it was found experimentally that the iso-orientation domains are often arranged radially around a common center [20]. Such an arrangement had been previously hypothesized on the basis of electrophysiological experiments [19, 34] and theoretical considerations [26]. The regions exhibiting this kind of radial arrangement were termed "pinwheels" (see Fig.1). The centers of pinwheels are point discontinuities of the field $\vartheta(\mathbf{x})$ where the mean orientation preference of nearby columns changes by 90 degrees. They can be characterized by a topological charge which indicates in particular whether the orientation preference increases clockwise around the center of the pinwheel or counterclockwise:

$$q_i = \frac{1}{2\pi} \oint_{C_j} \nabla\vartheta(\mathbf{x})d\mathbf{s} \tag{2.3}$$

where $C_j$ is a closed curve around a single pinwheel center at $\mathbf{x}_i$. Since $\vartheta$ is a cyclic variable within the interval $[0, \pi]$ and up to isolated points is a continuous function of $\mathbf{x}$, $q_i$ can in principle only have the values

$$q_i = \frac{n}{2} \tag{2.4}$$

where $n$ is an integer number [35]. If its absolute value $|q_i|$ is $1/2$, each orientation is represented exactly once in the vicinity of a pinwheel center. Pinwheel centers with a topological charge of $\pm 1/2$ are simple zeros of $z(\mathbf{x})$. In experiments only pinwheels that had the lowest possible topological charge $q_i = \pm 1/2$ are observed. This means there are only two types of pinwheels: those whose orientation preference increases clockwise and those whose orientation preference increases counterclockwise. This organization has been confirmed in a large number of species and is therefore believed to be a general feature of visual cortical orientation maps [36–43].

## 3. Symmetries in the development of orientation columns

Because on a phenomenological level the pattern of orientation columns can be represented by a order parameter field $z(\mathbf{x})$, the simplest conceivable models for the developmental formation the pattern are dynamic equations for the field of the form

$$\frac{\partial}{\partial t} z(\mathbf{x}) = F[z(\cdot)] + \eta(\mathbf{x}, t) \ . \tag{3.1}$$

Here $F[z(\cdot)]$ is a nonlinear operator and the random term $\eta(\mathbf{x}, t)$ describes intrinsic, e.g., activity-dependent fluctuations. In Equation (3.1), it is assumed that, except for random effects, changes in the pattern of orientation columns during development can be predicted on the basis of a knowledge of the current pattern. Swindale was the first to study models of this type, with the intent to show that roughly periodical patterns of columns can develop from a homogeneous initial state [25, 26, 44].

It is biologically plausible to assume that Equation (3.1) exhibits various symmetries. Considered anatomically, the cortical tissue appears rather homogeneous. If we look at the arrangement of the cortical neurons and their patterns

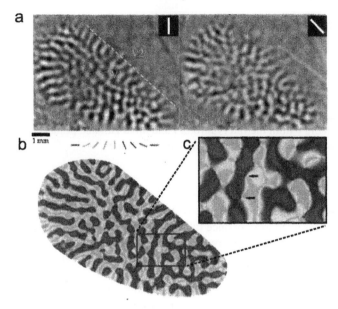

Fig. 1. Patterns of orientation columns in the primary visual cortex of a tree shrew visualized using optical imaging of intrinsic signals (figure modified from [42]). Activity patterns resulting from stimulation with vertically and obliquely oriented gratings are shown in (a). White bars depict the orientation of the visual stimulus. Activated columns are labeled dark grey. The used stimuli activate only columns in Area 17 (V1 in the lower left parts of (a)). The patterns thus end at the boundary between Areas 17 and 18 (V2). The pattern of orientation preferences calculated from such activity patterns is shown in (b). The orientation preferences of the columns are color coded as indicated by the bars. A part of the pattern of orientation preferences is shown at higher magnification in (c). Two pinwheel centers of opposite topological charge are marked by arrows.

of connections, there is no region of the cortical layers and no direction parallel to the layers that is distinguishable from other regions or directions [1]. If the development of the pattern of orientation preferences can be described by an equation in the form of Equation (3.1), it is thus very plausible to require that it is symmetric with respect to translations,

$$F[\hat{T}_{\mathbf{y}} z] = \hat{T}_{\mathbf{y}} F[z] \ \text{ with } \ \hat{T}_{\mathbf{y}} z(\mathbf{x}) = z(\mathbf{x} + \mathbf{y}) \tag{3.2}$$

and rotations,

$$F[\hat{R}_{\beta} z] = \hat{R}_{\beta} F[z] \ \text{ with } \ \hat{R}_{\beta} z(\mathbf{x}) = z \left( \begin{bmatrix} \cos(\beta) & \sin(\beta) \\ -\sin(\beta) & \cos(\beta) \end{bmatrix} \mathbf{x} \right) \tag{3.3}$$

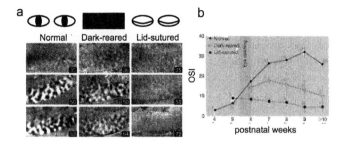

Fig. 2. Time-course and experience dependence of the development of orientation columns (figure modified from [23]). (a) Emergence of patterns of orientation columns in area 17 of ferret visual cortex under three different rearing conditions: normal visual experience (left column), rearing in total darkness (middle column), rearing with closed eye-lids (right column). The age of the animal at the time of the experiment (days postnatal) is indicated for each pattern. The panels show optically assessed activity patterns in response to horizontally oriented grating stimuli. (b) Orientation selectivity index (OSI) as a function of postnatal age for the three treatment groups (mean ± SEM). Numbers near symbols indicate sample size.

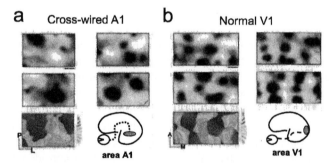

Fig. 3. Orientation maps in area A1 of 'cross-wired' ferrets (a) and V1 of control animals (b) (figure modified from [24]). In (a) and (b) the upper four panels show cortical activity patterns evoked by stimulation with horizontal, vertical, and left and right oblique grating stimuli. The maps of preferred orientations are displayed in the lower left panels. As indicated by the schematics (lower right panels) visual input that normally drives the primary visual cortex (V1) was experimentally redirected to the primary 'auditory' cortex (A1) in the cross-wired animals.

of the cortical layers. This means that patterns that can be converted to one another by translation or rotation of the cortical layers are equivalent solutions of
Equation (3.1). If the orientation preference of a column is determined by the
afferent connections from the lateral geniculate nucleus (LGN), it is also plausible to assume that the spatial arrangement of iso-orientation domains contains no
information about the orientation preferences of the columns. This is guaranteed
by a further symmetry. If Equation (3.1) is symmetric with respect to shifts in
orientation,

$$F[e^{i\phi} z] = e^{i\phi} F[z] \tag{3.4}$$

then patterns whose arrangement of iso-orientation domains is the same but whose
orientation preference values differ by a given amount, are equivalent solutions
of Equation (3.1).

The three symmetries Equations (3.2-3.4) imply a number of basic properties
of Equation (3.1). Owing to the symmetry under orientation shifts, $z(\mathbf{x}) = 0$ is a
stationary solution of Equation (3.1) because:

$$F[0] = e^{i\phi} F[0] \Rightarrow F[0] = 0. \tag{3.5}$$

Near this state, $z(\mathbf{x})$ develops approximately according to a linear equation:

$$\frac{\partial}{\partial t} z(\mathbf{x}) = \hat{L} z(\mathbf{x}) + \eta(\mathbf{x}, t) \ , \tag{3.6}$$

where $\hat{L}$ is a linear operator. Like $F[\cdot]$, the operator $\hat{L}$ must also commute with
rotation and translation of the cortical layer and with global shifts of orientation.
Thus, the Fourier representation of $\hat{L}$ is diagonal and its eigenvalues $\lambda(\mathbf{k})$ are
only a function of the absolute value of the wave vector $k = |\mathbf{k}|$.

A qualitative requirement placed on Equation (3.1) is that it be able to describe the spontaneous generation of a roughly repetitive pattern of orientation
preferences from an initially homogeneous state $z(\mathbf{x}) \approx 0$. This requirement further constrains the class of dynamic equations that are to be considered. Equation
(3.6), and thus also Equation (3.1), describes the generation of a repetitive pattern
of orientation preferences when the spectrum $\lambda(\mathbf{k})$ has positive eigenvalues, and
exhibits them only in one interval of wave numbers $(k_l, k_h)$ with $0 < k_l < k_h$.
Such a system is said to exhibit a finite wavelength instability of the homogenous
state $z(\mathbf{x}) \approx 0$ (for a review see [45]). Throughout the following considerations
we will assume that, in addition, the dynamics of $z(\mathbf{x})$ derives from a potential.
This guarantees that the process of visual development can be interpreted as a
optimization process.

## 4. From learning to dynamics

It is not difficult to construct models with the form of Equation (3.1) that represent the features of activity-dependent plasticity in an idealized fashion [15]. One instructive possibility is to start from an equation that describes how the pattern of orientation preferences $z(\mathbf{x})$ changes under the influence of a sequence of patterns of afferent activity $\mathcal{A}_i$:

$$z_i(\mathbf{x}) \xrightarrow{\mathcal{A}_i} z_{i+1}(\mathbf{x}) \tag{4.1}$$

In a minimal model, the changes

$$\delta z(\mathbf{x}) = z_{i+1}(\mathbf{x}) - z_i(\mathbf{x}) \tag{4.2}$$

in the pattern must be dependent on both the current pattern $z_i(\mathbf{x})$ and the patterns of activity $\mathcal{A}_i$:

$$\delta z(\mathbf{x}) = f(\mathbf{x}, z_i(\cdot), \mathcal{A}) \ . \tag{4.3}$$

If the maximum absolute value $\max_\mathcal{A}(|\delta z(\mathbf{x})|)$ of a modification induced by a single activity pattern is much smaller than the amplitude of the pattern and if the patterns of afferent activity $\mathcal{A}_i$ are random variables with a stationary probability distribution, then the changes in $z(\mathbf{x})$ on a long time scale are described by the following equation:

$$\frac{\partial}{\partial t} z(\mathbf{x}) = \left[ f(\mathbf{x}, z(\cdot), \mathcal{A}) \right]_\mathcal{A} \equiv F[z(\cdot)] \ , \tag{4.4}$$

where $[\ ]_\mathcal{A}$ denotes averaging over the ensemble of activity patterns [46]. This equation has the form of Equation (3.1).

A requirement that results directly from the activity-dependent nature of synaptic plasticity is that only the selectivity of the columns that are activated by $\mathcal{A}$ is changed. For cortical activity patterns $e(\mathbf{x}) = e(\mathbf{x}, z(\cdot), \mathcal{A})$, this requirement is fulfilled by making the modification proportional to the cortical activity:

$$f(\mathbf{x}, z(\cdot), \mathcal{A}) \ \propto \ e(\mathbf{x}, z(\cdot), \mathcal{A}) \ . \tag{4.5}$$

For simplicity, let us assume that a single activity pattern $\mathcal{A}$ forces the activated cortical neuron to take on a certain orientation preference $\theta$ and a certain orientation selectivity $|s|$, described by the complex number

$$s(\mathcal{A}) = |s(\mathcal{A})| \, e^{2i\theta(\mathcal{A})} \ , \tag{4.6}$$

then the simplest modification rule has the form

$$f(\mathbf{x}, z(\cdot), \mathcal{A}) \ \propto \ (s(\mathcal{A}) - z(\mathbf{x})) \, e(\mathbf{x}, z(\cdot), \mathcal{A}) \ . \tag{4.7}$$

Several models have been proposed that can be interpreted in the just described way as an order parameter dynamics [28–30, 47, 48]. These models are defined by modification formulas with the form of Equation (4.7). They differ mainly in how the pattern $e(\mathbf{x})$ of the activity of the cortex is modeled.

## 5.  Generation and motion of pinwheels

It is easily see that within this class of models pinwheel will typically form during the initial symmetry breaking phase of development. If the eigenvalues $\lambda(k)$ are real, which is expected when $z(\mathbf{x})$ develops to a stationary state, then beginning with a homogeneous state $z(\mathbf{x}) \approx 0$ the real and imaginary parts of $z(\mathbf{x})$ initially develop independently of each other. In particular, the zero lines of the real and imaginary parts will develop independently of each other and thus typically intersect at points $\mathbf{x}_i$. These points are simple zeros of $z(\mathbf{x})$ and therefore are the centers of pinwheels with a topological charge $q_i = \pm 1/2$.

It is important to note that the possible forms of a change in the pinwheel configuration over time $\{\mathbf{x}_i, q_i\}$ are already constrained by assuming a smooth dependence of the pattern on time as is implied by any equation for the developmental dynamics with the form of Equation (3.1). Since the field $z(\mathbf{x})$ can only be a continuous function of time, the entire topological charge of a given area $A$ with a boundary $(A)$,

$$ Q_A \equiv \frac{1}{2\pi} \oint_{(A)} \nabla \vartheta(\mathbf{x}) ds = \sum_{\mathbf{x}_i \in A} q_i \ , \tag{5.1} $$

is invariant as long as no pinwheel transgresses the boundary of the area [35]. If the pattern contains only pinwheels with $q_i = \pm 1/2$, then only three qualitatively different modifications of the pinwheel configuration are possible. First, movement of the pinwheel within the area; second, generation of a pair of pinwheels with opposite topological charges; third, the annihilation of two pinwheels with opposite topological charge when they collide. Only these transformations conserve the value of $Q_A$ and are therefore permitted.

### 5.1.  Random orientation maps

Since pinwheels are an important structural element of the spatial layout of orientation let us next consider how the pinwheel density which results from the initial breaking of symmetry can be analytically estimated. To do this in a model independent fashion, I will start by examining general ensembles of random fields

$z(\mathbf{x})$. Such an ensemble can be characterized by its spatial correlation functions:

$$C(\mathbf{x}, \mathbf{y}) \;=\; \langle z(\mathbf{x})\bar{z}(\mathbf{y})\rangle \tag{5.2}$$

$$C^*(\mathbf{x}, \mathbf{y}) \;=\; \langle z(\mathbf{x})z(\mathbf{y})\rangle \tag{5.3}$$

Here angular brackets, $\langle\ \rangle$, represent the expectation value for the ensemble. The form of these correlation functions can be constrained by symmetry assumptions. Because of the symmetries Equations (3.2-3.4), we assume that the ensemble is statistically invariant with respect to translations and rotations and that the patterns that can be transformed into each other by a global orientation shift $z(\mathbf{x}) \rightarrow e^{i\phi}\,z(\mathbf{x})$ occur with the same probability. The latter assumption implies that the expectation of $z(\mathbf{x})$ is equal to zero:

$$\langle z(\mathbf{x})\rangle = 0 \tag{5.4}$$

This means that any orientation preference can occur at any location $\mathbf{x}$ in the cortex. Moreover, invariance under orientation shifts implies that the correlation function (5.3) is also equal to zero, since only in this case can the following relation be fulfilled for any $\phi$:

$$\langle z(\mathbf{x})z(\mathbf{y})\rangle \;=\; \left\langle e^{i\phi}z(\mathbf{x})e^{i\phi}z(\mathbf{y})\right\rangle \tag{5.5}$$

$Re(z(\mathbf{x}))$ describes the patterns of columns that prefer horizontal and vertical stimuli. $Im(z(\mathbf{x}))$ describes the patterns of columns that prefer oblique stimuli. These two patterns are not correlated and both have the same correlation function because the correlation function (5.3) is zero:

$$
\begin{aligned}
C^*(\mathbf{x}, \mathbf{y}) \;&=\; \langle z(\mathbf{x})z(\mathbf{y})\rangle \\
&=\; \langle Re(z(\mathbf{x}))\,Re(z(\mathbf{y}))\rangle - \langle Im(z(\mathbf{x}))\,Im(z(\mathbf{y}))\rangle \\
&\quad +i\,(\langle Re(z(\mathbf{x}))Im(z(\mathbf{y}))\rangle + \langle Im(z(\mathbf{x}))Re(z(\mathbf{y}))\rangle) \\
&=\; 0
\end{aligned}
\tag{5.6}
$$

Invariance with respect to translations and rotations implies that the correlation function $C(r)$ is a function only of the distance $r = |\mathbf{x} - \mathbf{y}|$ of the respective pair of locations in the cortex:

$$C(\mathbf{x}, \mathbf{y}) = C(|\mathbf{x} - \mathbf{y}|) = C(r) \tag{5.7}$$

The correlation function $C(r)$ mainly provides information about the characteristic wavelength and the correlation length of the pattern. The characteristic wavelength $\Lambda$ of the pattern can be defined using the Fourier transform of $C(r)$:

$$P(|\mathbf{k}|) = \frac{1}{2\pi}\int d^2x\,C(\mathbf{x})\,e^{i\mathbf{kx}} \tag{5.8}$$

which is called the power spectral density. It is of advantage to use the mean wavenumber $\bar{k}$ to define the characteristic wavelength:

$$\Lambda = \frac{2\pi}{\bar{k}} = \frac{2\pi}{\int_0^\infty dk\, k\, P(k)} \quad . \tag{5.9}$$

Without loss of generality, the power spectral density is assumed to be normalized as follows: $\int_0^\infty dk\, P(k) = 1$.

There is an infinite number of ensembles of maps that have the same two-point correlation function. In the following considerations, we will assume that the ensembles considered consists of Gaussian random fields. In the next section, we will show that this assumption is fulfilled for a large class of dynamic models, owing to the central limit theorem.

The centers of the pinwheels are the zeros of the field $z(\mathbf{x})$. The pinwheel density is therefore obtained from the number of these zeros in an area $A$,

$$N = \int_A d^2x \, \delta(z(\mathbf{x})) \left| \frac{\partial(Re\, z(\mathbf{x}),\, Im\, z(\mathbf{x}))}{\partial(x_1, x_2)} \right| \tag{5.10}$$

where

$$\frac{\partial(Re\, z(\mathbf{x}),\, Im\, z(\mathbf{x}))}{\partial(x_1, x_2)} = \frac{\partial Re\, z(\mathbf{x})}{\partial x_1} \frac{\partial Im\, z(\mathbf{x})}{\partial x_2} - \frac{\partial Re\, z(\mathbf{x})}{\partial x_2} \frac{\partial Im\, z(\mathbf{x})}{\partial x_1} \tag{5.11}$$

is the Jacobian of $z(\mathbf{x})$. The expectation value of the number of zeros in an ensemble is

$$\langle N \rangle = \int_A d^2x \, \left\langle \delta(z(\mathbf{x})) \left| \frac{\partial(Re\, z(\mathbf{x}),\, Im\, z(\mathbf{x}))}{\partial(x_1, x_2)} \right| \right\rangle \tag{5.12}$$

from which follows that

$$\rho = \left\langle \delta(z(\mathbf{x})) \left| \frac{\partial(Re\, z(\mathbf{x}),\, Im\, z(\mathbf{x}))}{\partial(x_1, x_2)} \right| \right\rangle \tag{5.13}$$

is the density of the pinwheels. This expectation value will now be evaluated. Superficially, $\rho$ as defined in Equation (5.13) is the expectation value for an ensemble of fields $z(\mathbf{x})$ and thus is a functional integral. It is, however, important to note that this expectation value depends on locally defined parameters only, namely the values of the field $z(\mathbf{x})$ and its derivatives $\nabla z(\mathbf{x})$ at a given location $\mathbf{x}$. To integrate Equation (5.13) it is, therefore, sufficient to know the joint probability density $p(z(\mathbf{x}), \nabla z(\mathbf{x}))$. Since this is also Gaussian, it is given by the cross-correlations and autocorrelations of $z(\mathbf{x})$ and $\nabla z(\mathbf{x})$. Most of these correlations are equal to zero owing to the symmetry requirements:

$$\langle z(\mathbf{x}) \nabla z(\mathbf{x}) \rangle = 0 \tag{5.14}$$

$$\langle z(\mathbf{x}) z(\mathbf{x}) \rangle = 0 \tag{5.15}$$

$$\langle \nabla z(\mathbf{x}) \nabla z(\mathbf{x}) \rangle = 0 \tag{5.16}$$

$$\langle \bar{z}(\mathbf{x}) \nabla z(\mathbf{x}) \rangle = 0 \tag{5.17}$$

$$\langle z(\mathbf{x}) \bar{z}(\mathbf{x}) \rangle = c_a \tag{5.18}$$

$$\langle \nabla z(\mathbf{x}) \nabla \bar{z}(\mathbf{x}) \rangle = c_g \tag{5.19}$$

(5.14-5.16) result from the statistical invariance of the ensemble under an orientation shift. Equation (5.17) results from the rotation invariance of the ensemble. Equation. (5.18) defines the scale on which the order parameter is measured. Thus, $c_a$ and $c_g$ are the only nontrivial correlations of the distribution $p(z(\mathbf{x}), \nabla z(\mathbf{x}))$. Because there is no correlation between $z(\mathbf{x})$ and $\nabla z(\mathbf{x})$, the distribution is composed of two factors:

$$p(z, \nabla z) = \frac{1}{\pi^3 c_g^2 c_a} \exp\left(-2 \frac{\nabla z \nabla \bar{z}}{c_g}\right) \exp\left(-\frac{z\bar{z}}{c_a}\right). \tag{5.20}$$

Because this is true for any location $\mathbf{x}$, the argument of the field $z(\mathbf{x})$ is omitted here and in the following equations. Because Equation (5.20) is factored, the expectation in Equation (5.13) is also factored. Thus,

$$\rho = \frac{1}{\pi^3 c_g^2 c_a} \int d^4(\nabla z) \exp\left(-2 \frac{\nabla z \nabla \bar{z}}{c_g}\right) |(\nabla z + \nabla \bar{z}) \times (\nabla z - \nabla \bar{z})|$$

$$\int d^2 z \, \delta(z) \exp\left(-\frac{z\bar{z}}{c_a}\right) \tag{5.21}$$

where $|(\nabla z + \nabla \bar{z}) \times (\nabla z - \nabla \bar{z})|$ is the Jacobian of $z(\mathbf{x})$. This integral is easily evaluated by converting to the spherical coordinates $g \in [0, \infty)$, $\theta \in [0, \pi)$ and $\phi_1 \phi_2 \in [0, 2\pi)$ in gradient space with the volume element

$$d^4(\nabla z) = g^3 \, dg \, |\cos(\theta) \sin(\theta)| \, d\theta \, d\phi_1 \, d\phi_2 \tag{5.22}$$

Integration then yields

$$\rho = \frac{1}{\pi^3 c_g^2 c_a} \int_0^\infty dg \, g^5 \exp\left(-2 \frac{g^2}{c_g}\right) \int_0^\pi d\theta \, |\cos(\theta) \sin(\theta)|^2$$

$$\times \int_0^{2\pi} d\phi_1 \, d\phi_2 \, |\cos(\phi_1) \sin(\phi_2) - \cos(\phi_2) \sin(\phi_1)| \tag{5.23}$$

$$= \frac{1}{4\pi} \frac{c_g}{c_a}. \tag{5.24}$$

Since in the following discussion the dependence of this expression on the power spectral density $P(k)$ is of interest, we express $c_a$ and $c_g$ as functionals of $P(k)$:

$$c_a = \int d^2k\, P(k) \tag{5.25}$$

$$c_g = \int d^2k\, |\mathbf{k}|^2 P(k) \tag{5.26}$$

The pinwheel density is then given by

$$\rho = \frac{1}{4\pi} \frac{\int d^2k\, |k|^2 P(k)}{\int d^2k\, P(k)} \tag{5.27}$$

The exact form of the correlation function $C(r)$ and the structure function $P(k)$ at the beginning of development is not known. In particular, it is to be expected that these functions vary from species to species and from individual to individual. In spite of this, the following argument shows that Equation (5.27) implies a quantitative estimate of the initial pinwheel density. Since it may be assumed that the pinwheel density is inversely proportional to the square of the characteristic wavelength $\rho \propto \Lambda^{-2}$, we shall rewrite the expression for the density:

$$\begin{aligned}
\rho &= \frac{\bar{k}^2}{4\pi} \frac{\int_0^\infty dk\, k^3 P(k)}{\left(\int_0^\infty dk\, k P(k)\right)^3} \\
&= \frac{\pi}{\Lambda^2} \frac{\int_0^\infty dk\, k^3 P(k)}{\left(\int_0^\infty dk\, k P(k)\right)^3} .
\end{aligned} \tag{5.28}$$

Owing to Jensen's inequality [49],

$$\int_0^\infty dk\, k^3 P(k) \geq \left(\int_0^\infty dk\, k P(k)\right)^3 \tag{5.29}$$

it follows that $\rho$ has a lower bound:

$$\rho = \frac{\pi}{\Lambda^2} (1 + \alpha) \tag{5.30}$$

where $\alpha > 0$. Thus, the exact form of the power spectral density influences the expected pinwheel density only via the positive-definite functional

$$\alpha = 3 \int_0^\infty dk\, \frac{(k - \bar{k})^2}{\bar{k}^2} P(k) + \int_0^\infty dk\, \frac{(k - \bar{k})^3}{\bar{k}^3} P(k), \tag{5.31}$$

where $\alpha$ is zero only when the power spectral density is the Dirac delta distribution $P(k) = \delta(k - \bar{k})$, i.e., when the correlation length of the pattern diverges. Thus, it can be seen that a Gaussian random pattern of orientation columns has a minimum pinwheel density,

$$\rho_m = \frac{\pi}{\Lambda^2} , \tag{5.32}$$

independently of the exact form of its spatial correlation functions.

### 5.2. Gaussian fields from a dynamic instability

In the previous section, we made the assumption that the ensemble of possible initial patterns has Gaussian statistics. We will now show that this is actually the case for a large class of models. This establishes that the lower bound for the pinwheel density calculated in the previous section places a quantitative constraint on the dynamics of pinwheels in this class of models. For this purpose, we will use the class of models defined in Section 3 that can be represented by an equation for the dynamics of the order parameter field $z(\mathbf{x})$:

$$\frac{\partial}{\partial t} z(\mathbf{x}) = F[z(\cdot)] + \xi(\mathbf{x}, t) . \tag{5.33}$$

The assumed symmetries of this equation imply that $z(\mathbf{x}) = 0$ is a stationary solution. In the vicinity of this point, the dynamics of $z(\mathbf{x})$ is approximately linear:

$$\frac{\partial}{\partial t} z(\mathbf{x}) = \hat{L} z(\mathbf{x}) + \xi(\mathbf{x}, t) \tag{5.34}$$

where $\hat{L}$ is a linear operator. This equation generates an orientation map from an initially homogeneous state.

As discussed in Section (3), the class of operators $\hat{L}$ is further constrained by the requirement that Equation (5.33) describe the spontaneous generation of a pattern of orientation preferences starting with an initially homogeneous state. Equation (5.34), and thus also Equation (5.33), describes the generation of a repetitive pattern of orientation preferences if the spectrum $\lambda(\mathbf{k})$ has positive eigenvalues only within a single interval $(k_l, k_h)$ of $k$-values. Using the Green's function

$$G(\mathbf{x}, t) = \frac{1}{2\pi} \int d^2k \, e^{-i\mathbf{k}\mathbf{x} + \lambda(|\mathbf{k}|)t} , \tag{5.35}$$

Equation (5.34) with the initial condition $z_0(\mathbf{x}) \approx 0$ at time $t = 0$ is solved to yield

$$z_L(\mathbf{x}, t) = \int d^2y \int_0^t dt' \, G(\mathbf{y} - \mathbf{x}, t - t') \, (\xi(\mathbf{y}, t') + z_0(\mathbf{x})\delta(t')) . \tag{5.36}$$

Since $z_L(\mathbf{x}, t)$ is the sum of linear transforms of the random fields $z_0(\mathbf{x})$ and $\xi(\mathbf{x}, t)$, it will always have Gaussian statistics when $z_0(\mathbf{x})$ and $\xi(\mathbf{x}, t)$ are also Gaussian. This is independent of the form of their correlation functions:

$$C_{z_0}(\mathbf{r}) = \langle z_0(\mathbf{x})\bar{z}_0(\mathbf{x} + \mathbf{r})\rangle \qquad (5.37)$$

$$C_\xi(\mathbf{r}, t) = \langle \xi(\mathbf{x}, t')\bar{\xi}(\mathbf{x} + \mathbf{r}, t' + t)\rangle . \qquad (5.38)$$

In general, the statistical properties of $z_L(\mathbf{x}, t)$ are also Gaussian for a much larger class of random processes. The field $z_L(\mathbf{x}, t)$ is an integral over a number of random variables. When this integral consists of a large number of independent terms then the central limit theorem [49] gives the conditions under which the statistical properties of $z_L(\mathbf{x}, t)$ are Gaussian, even if $z_0(\mathbf{x})$ and $\xi(\mathbf{x}, t)$ are not Gaussian. We will now briefly present these conditions and discuss their biological significance. For simplicity we will consider only sufficiently smooth functions $z_0(\mathbf{x})$ and $\xi(\mathbf{x}, t)$ such that the right side of Equation (5.36) can be expressed approximately by a sum:

$$z_L(\mathbf{x}, t) \approx \sum_{i,j} \Delta y_i^2 \, \Delta t_j \, G(\mathbf{y_i} - \mathbf{x}, t - t_j) \left(\xi(\mathbf{y_i}, t_j) + z_0(\mathbf{x_i})\delta_{j,0}\right) . \quad (5.39)$$

The central limit theorem applies if the distributions of $z_0(\mathbf{x})$ and $\xi(\mathbf{x}, t)$ fulfill the Lindeberg criterion:

$$\lim_{\beta \to \infty} \int_{|w|>\beta} dw \, w^2 \, P(w) = 0 , \qquad (5.40)$$

where the variable $w$ represents either $z_0(\mathbf{x})$ or $\xi(\mathbf{x}, t)$ at an arbitrary location $\mathbf{x}$ at any time $t$ and $P(w)$ is the probability density. Equation (5.40) is fulfilled when $|z_0(\mathbf{x})|$ and $|\xi(\mathbf{x}, t)|$ are bounded or have a finite variance. Both appear to be plausible for the observed biological process. There is very little orientation selectivity, which is observed in the visual cortex considerably before the eyes open [21]. Thus, it can be taken as certain that the $|z_0(\mathbf{x})|$ field, which describes this early selectivity, is bounded. A similar argument can be made for the fluctuations $\xi(\mathbf{x}, t)$. All available evidence indicates that the process in which the initial orientation map is generated is continuous [21]. This speaks for the assumption that fluctuations in this process are bounded or at least have a finite variance.

In order to determine the conditions under which $z_L(\mathbf{x}, t)$ consists of a large number of independent terms, it is necessary to compare the correlation times and lengths of $z_0(\mathbf{x})$ and $\xi(\mathbf{x}, t)$ with the temporal and spatial scales of the Green's function $G(\mathbf{x}, t)$. The characteristic time scale of $G(\mathbf{x}, t)$ is

$$\tau = 1/\lambda(k_{max}) \qquad (5.41)$$

where $\lambda(k_{max})$ is the largest eigenvalue of $\hat{L}$. The amplitude $|G(0, \Delta t)|$ of $G(\Delta \mathbf{x}, \Delta t)$ increases with time with time constant $\tau$. The Green's function also has a characteristic spatial scale $L$ on which $G(\Delta \mathbf{x}, \Delta t)$ decays as a function of the distance $|\Delta \mathbf{x}|$ for a given time difference $\Delta t$. In order to estimate $L$, let us assume that $\lambda(k)$ has a quadratic maximum between $k_l$ and $k_h$. The spatial Fourier transform $\tilde{G}(\mathbf{k}, \Delta t)$ of the Green's function $G(\Delta \mathbf{x}, \Delta t)$ is then approximated in the vicinity of the maximum $k_{max}$ by a Gaussian function:

$$\tilde{G}(\mathbf{k}, \Delta t) = e^{\lambda(|\mathbf{k}|)t} \tag{5.42}$$

$$\approx e^{\lambda_{max}\left(1-\left(\frac{|\mathbf{k}|-k_{max}}{(k_h-k_l)/2}\right)^2\right)t}. \tag{5.43}$$

The bandwidth $\Delta k = |k_h - k_l|/2\sqrt{\Delta t/\tau}$ of this function is a function of time. Its inverse provides an estimate of $L$:

$$L \approx \frac{2\sqrt{\Delta t/\tau}}{|k_h - k_l|}. \tag{5.44}$$

Thus, the field $z_L(\mathbf{x}, t)$ Equation (5.36) integrates fluctuations $\xi(\mathbf{x}, t - \Delta t)$ that occurred in the past at time $t - \Delta t$ within a distance $L$. The size of this volume as given by Equation (5.44) increases diffusively: $L \propto \sqrt{\Delta t/\tau}$. If it is assumed that the correlation function $C_{z0}(\mathbf{r})$ decays on a spatial scale $L_{z0}$ and the same is assumed for $C_{z0}(\mathbf{r})$ on the spatial and temporal scales $L_\xi$ and $\tau_\xi$, so that

$$C_{z0}(\mathbf{r}) \le A_{z0}e^{-|x|/L_{z0}} \tag{5.45}$$

$$C_\xi(\mathbf{r}, t) \le A_\xi e^{-|x|/L_\xi - |t|/\tau} \tag{5.46}$$

are fulfilled with positive constants $A_{z0}$ and $A_\xi$, then the number of independent terms in $z_L(\mathbf{x}, t)$ is large if

$$\tau_\xi \ll \tau \tag{5.47}$$

and

$$\max\{L_{z0}, L_\xi\} \ll \frac{1}{|k_h - k_l|}. \tag{5.48}$$

If $z_L(\mathbf{x}, t)$ is determined mainly by $\xi(\mathbf{x}, t)$, then either $\tau_\xi \ll \tau$ or $L_\xi \ll \frac{1}{|k_h-k_l|}$ is a sufficient condition. If $z_L(\mathbf{x}, t)$ is determined by $z_0(\mathbf{x})$, then $L_{z0} \ll \frac{2\sqrt{t/\tau}}{|k_h-k_l|}$ is sufficient, where $t$ is the time at which the amplitude of $z(\mathbf{x}, t)$ saturates.

Especially the first of the two cases is compatible with the biological situation. Fluctuations caused by afferent activity patterns, like the activity patterns themselves, can be correlated only over time intervals of a few hundred milliseconds. Since the first pattern of orientation preferences is formed over a period of

several hours to several days, the number of independent activity-induced fluctuations is surely large. Therefore, the statistical properties of the initial pattern
can be assumed to be Gaussian.

## 5.3. Predicting pinwheel annihilation

The analyzes presented above demonstrate that the density of the pinwheels
in a Gaussian random pattern of orientation preferences is always larger than
$\rho_m = \pi/\Lambda^2$, where $\Lambda$ is the characteristic wavelength of the pattern. This feature is universal, meaning that it applies, no matter what the detailed structure
of the fields' statistical correlations are. In addition, we showed that orientation
preference patterns generated by a dynamic instability have Gaussian statistical
properties in a large class of models owing to the central limit theorem, and that
this property is not affected by interaction with two other patterns of neuronal
selectivities. These results imply that the dynamics of pinwheels is quantitatively
constrained in a large class of models for the development of orientation preference patterns. Independent of modeling details, pinwheel densities that are less
than the lower limit for the initial density $\rho_m = \pi/\Lambda^2$ can develop under the
given conditions only by pairwise annihilation of pinwheels. This implies that
pinwheels must move during development and must be pairwise annihilated in
those species in which low pinwheel densities are observed in adults (Fig. 4).

Because of this, it is of interest to compare densities in adult animals with
the calculated lower limit. The characteristic wavelength $\Lambda$ differs considerably
from species to species. Values of the relative pinwheel density $\hat{\rho} = \rho \Lambda^2$ that are
less than $\pi$ imply pinwheel annihilation during development. Observed relative
densities range from 2.0 to 4.0 (see [15] and refs therein)). Macaques appear to
have the highest relative density $\hat{\rho} \approx 3.7$. The lowest relative densities observed
so far are in the visual cortex of adult tree shrews which have many regions of
V1 that are relatively pinwheel sparse (see Fig.1). The relative densities of both
tree shrews and cats appear to be distinctly less than $\pi$. Thus, pinwheel movement and annihilation is implied for both species by the theory developed above.
Pinwheel movement cannot be excluded in ferrets and squirrel monkeys. This
would be especially the case when the initial pinwheel density is substantially
larger than the minimum value. The high pinwheel density observed in macaques
can occur as a result of a dynamic instability without any subsequent rearrangement of the pattern. Thus, tree shrews are the species in which the most extensive
rearrangement is expected during development.

Besides the density of pinwheels, there are many more layout properties of
Gaussian random maps, such as the probability of finding a pinwheel center of
given topological charge at a distance $r$ from another pinwheel center or the
average length of an iso-orientation domain that can be analytically characterized

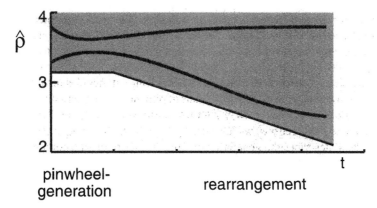

Fig. 4. Symmetries constrain the time course of the pinwheel density. The range of permissible trajectories of the scaled pinwheel density $\hat{\rho}(t)$ is shaded gray. Two permissible trajectories are shown. The final density of the lower trajectory is less than $\pi$ and can develop only by annihilation of pinwheels (modified from [15]).

in a similar, model independent fashion. Many useful expressions for this can be found in a recent paper by Berry and Dennis [50]. Analogous to the case of the pinwheel density considered above, one can ask for any such property whether the observed statistics is consistent with the assumption Gaussian statistics. If this assumption is rejected by the data substantial rearrangement of the orientation map after its initial emergence is predicted as in the example of low density maps above.

## 6. The problem of pinwheel stability

The question why there are pinwheels in the visual cortex of adult animals has not yet been answered. Are persistent pinwheels a component of genuinely stable solutions of the developmental dynamics, or are they to be interpreted as a residue of a random initial condition? Many concrete models for the developmental formation of orientation columns imply that the initially pinwheel rich random state decays towards a pinwheel sparse pattern of orientation stripes as the developmental dynamics converges towards its attractors [15, 31, 51–53]. A trivial explanation for the persistence of pinwheels in adult animals would therefore be that the changes in the pattern are terminated relatively early by external factors. The observed patterns would then be interpreted as a frozen intermediate stage. If in contrast to this we want to explain the observed pattern by the structure of the attractors of the pattern-forming dynamics, we must conclude that such models are unsuitable. From a theoretical perspective, the question thus arises whether there is another biologically plausible class of models that can explain the persistence of pinwheels via the structure of stable patterns. This is the problem of pinwheel stability. Mathematically, it includes the questions of whether intrinsically stable configurations of pinwheels exist, and if so, what they look like, as well as, what kind of interactions are necessary to stabilize them?

In the previous sections, we have used primarily the linear equation, Equation (3.6), which describes the generation of a pattern of orientation columns from an initially homogeneous state. Equation (3.1), which models the entire process of development must be nonlinear in order to describe the saturation of the orientation selectivity $|z(\mathbf{x})|$ and a possible subsequent reorganization of the pattern. The symmetry with respect to orientation shifts also constrains the nonlinear terms. It means that when $F[\cdot]$ is represented by a power series only terms with an odd power can occur. If the parameter values in Equation (3.1) are assumed to be in the vicinity of an instability of the homogeneous state $z(\mathbf{x}) = 0$ and the transition to inhomogeneous solutions is continuous, then also the stationary patterns that bifurcate from $z(\mathbf{x}) = 0$ can be studied perturbatively in a largely model independent fashion. In the following section 7, I will describe a mathematical formalism for examining this question in order parameter models. In the subsequent section 8, I will use this formalism to construct a simple model that has the major structural properties characteristic of the dynamics of cortical development. It will be seen that stable pinwheel configurations develop in a substantial region of parameter values. The pinwheel densities of these patterns are strikingly similar to the experimentally observed densities. Analysis of the model also shows that the stability of pinwheels requires the existence of long-range interactions within the cortex.

## 7. Weakly nonlinear analysis of pattern selection

In order to discuss the question of pinwheel stability, we will use a perturbation method that enables to analytically examine the structure and stability of inhomogeneous solutions in the vicinity of an instability. In this section, I present the concepts and methods needed for this, called weakly nonlinear stability analysis. Here the stability of so-called planforms is examined [45, 54]. Planforms are patterns that are composed of a finite number of Fourier components, such as

$$z(\mathbf{x}) = \sum_j A_j e^{i\mathbf{k}_j \mathbf{x}} \tag{7.1}$$

for a pattern of orientation columns. When the dynamics is close to a finite wavelength instability, the essential Fourier components of the emerging pattern are located on the critical circle. The dynamic equations for the amplitudes of these Fourier components are called amplitude equations. To discuss the emergence of such patterns in a model independent fashion, we will make use of the fact that the amplitude equations, depend mainly on the symmetries of the problem and on the qualitative nature of the bifurcation [45, 54]. The amplitude equations can be used to find the stable planforms of a pattern-forming system. These planforms represent the attractors and thus determine the properties of the patterns that eventually form. In Section 7.1, I will derive the general form of the amplitude equation for the dynamics of the complex order parameter field $z(\mathbf{x})$

$$\frac{\partial}{\partial t} z(\mathbf{x}) = F[z(\mathbf{x})] \tag{7.2}$$

symmetric under translations, rotations, and orientation shifts.

In Section 7.2, I will discuss stationary solutions of these amplitude equations and their stability. In Section 7.3, I will show how expressions for the parameters in the amplitude equations can be obtained for specific models and how solutions of these equations approximate their stationary solutions. The concepts presented in this section were originally developed to study problems involving instabilities in hydrodynamic convection [55]. In most such applications real order parameter fields were examined [45, 54]. In this chapter, the concepts of weakly nonlinear stability analysis are generalized for examining complex order parameter fields. As will become apparent, there are characteristic differences from the case(s) of real fields that have important consequences. For this reason, I will give here a complete derivation of all the essential relationships.

### 7.1. Amplitude equations

It is assumed that the dynamics of the field $z(\mathbf{x})$ are derived from an energy functional whose fourth order $z(\mathbf{x})$ terms are positive-definite. In this case, inhomogeneous solutions branch of continuously in the vicinity of a finite wavelength instability of the homogenous state. Since the amplitude of the solutions at the instability is arbitrarily small, it is possible for the dynamics of the amplitudes $A_i$ of a discrete number of $N$ Fourier components of $z(\mathbf{x})$ whose wave vectors lie equally spaced on the critical circle to be approximated by a system of third-order equations:

$$\frac{\partial}{\partial t} A_i = \lambda A_i + \sum d_{j-i,k-i,l-i} A_j \bar{A}_l A_m \tag{7.3}$$

The form of these amplitude equations is greatly constrained by the assumed symmetries. In Equation (7.3) it is assumed that the observed modes are not coupled linearly and that the system is symmetric with respect to rotation and orientation shifts. The former follows from the symmetry with respect to translation, which decouples the linearized equation in the Fourier representation. Owing to the latter properties, third-order terms must contain one complex conjugated amplitude and their coefficients $d_{j-i,k-i,l-i}$ can only be functions of the differences between the indices.

The system (7.3) should also be symmetric with respect to translations $\mathbf{x} \to \mathbf{x} + \mathbf{y}$. Since the amplitudes $A_i$ are transformed by translation as follows:

$$A_j \to A_j e^{i \mathbf{k}_j \mathbf{y}} \tag{7.4}$$

the coefficients $d_{j-i,k-i,l-i}$ can differ from zero only when

$$e^{i \mathbf{k}_{j'} \mathbf{y}} = e^{i(\mathbf{k}_j - \mathbf{k}_l + \mathbf{k}_m) \mathbf{y}} \tag{7.5}$$

This is possible for arbitrary $\mathbf{y}$ only if the wave vectors fulfill the following condition:

$$\mathbf{k}_{j'} = \mathbf{k}_j - \mathbf{k}_l + \mathbf{k}_m \tag{7.6}$$

This is a nonlinear resonance condition. Since all wave vectors have the same absolute value $|\mathbf{k}_i| = k_c$, this condition can be fulfilled only when the wave vectors form the sides of a parallelogram. The four wave vectors $-\mathbf{k}_{j'}, \mathbf{k}_j, \mathbf{k}_m$ and $-\mathbf{k}_l$ must, therefore, be two pairs of antiparallel vectors in order to fulfill the condition of Equation (7.6). For a given $\mathbf{k}_{j'}$ these can be

$$\mathbf{k}_{j'} = \mathbf{k}_j \tag{7.7}$$
$$\mathbf{k}_l = \mathbf{k}_m \tag{7.8}$$

or

$$\mathbf{k}_{j'} = \mathbf{k}_m \tag{7.9}$$
$$\mathbf{k}_j = \mathbf{k}_l \tag{7.10}$$

or

$$\mathbf{k}_{j'} = -\mathbf{k}_l \tag{7.11}$$
$$\mathbf{k}_j = -\mathbf{k}_m \tag{7.12}$$

Thus, the interaction of the modes comprises only two kinds of terms. The vector pairs of Equations (7.7-7.10) lead to terms with the form $A_j \bar{A}_j A_{j'}$. The vector pairs of Equations (7.11-7.12) lead to terms with the form $A_j A_{j-} \bar{A}_{j'-}$. where $j^-$ is the index of the wave vector that is antiparallel to $\mathbf{k}_j$

$$\mathbf{k}_{j-} = -\mathbf{k}_j \tag{7.13}$$

The equation for the dynamics of the amplitudes $A_i$, therefore, always have the form

$$\frac{\partial}{\partial t} A_i = \lambda A_i - \sum_j g_{ij} |A_j|^2 A_i - \sum_j f_{ij} A_j A_{j-} \bar{A}_{i-} \tag{7.14}$$

where $g_{ij}$ and $f_{ij}$ are coupling coefficients. If the dynamics are also equivariant with respect to rotations and reflections, these coupling coefficients may only be a function of the difference between the indices $|i - j|$. $g_{ij}$ and $f_{ij}$ are, therefore, elements of two symmetric, cyclic matrices. It is also always possible to choose $f_{ii} = 0$, since a nonzero value of $f_{ii}$ can always be absorbed into $g_{ii-}$.

The first characteristic difference from the case of real order parameter fields appears at this point. For real fields, the amplitudes of modes with antiparallel wave vectors would be complex conjugated: $A_{i-} = \bar{A}_i$. In this case, the second and third terms would not be constitutively different and the amplitude equations would form a system of Landau equations:

$$\frac{\partial}{\partial t} A_i = \lambda A_i - \sum_j (g_{ij} + f_{ij}) |A_j|^2 A_i \tag{7.15}$$

The dynamics of the amplitudes as given by Equation (7.14) are potential if $g_{ij}$ and $f_{ij}$ are real valued. The potential is

$$U_A = -\frac{\lambda}{2} \sum_i |A_i|^2 + \frac{1}{4} \sum_{i,j} g_{ij} |A_j|^2 |A_i|^2 + \frac{1}{4} \sum_{i,j} f_{ij} A_j A_{j-} \bar{A}_i \bar{A}_{i-} \tag{7.16}$$

Note that, the third term in this equation is real valued owing to the symmetry of $f_{ij}$.

### 7.2. Fixed points of the amplitude equations and their stability

Let us investigate the stationary solutions of Equation (7.14) and their stability. First, I will consider the case when $f_{ij} = 0$. This is fulfilled especially when $N$ is an odd number, because only one of the wave vectors $\mathbf{k}$ and $-\mathbf{k}$ are then in the set of wave vectors. In addition, the case $f_{ij} = 0$ will play a prominent role in the construction of model equations with stable pinwheel patterns. In a second step, I will then examine the general dynamics when $f_{ij} \neq 0$ to determine which of the relationships found for the $f_{ij} = 0$ case can be generalized for $f_{ij} \neq 0$.

If $f_{ij} = 0$, the amplitude equations form a system of Landau equations

$$\frac{\partial}{\partial t} A_i = \lambda A_i - \sum_j g_{ij} |A_j|^2 A_i \ . \tag{7.17}$$

The stationary solutions of this system are either trivial, $A_i = 0$, or fulfill the condition

$$0 = \lambda - \sum_j g_{ij} |A_j|^2 \tag{7.18}$$

for all modes $i$ with non-zero amplitudes. These solutions are called "active modes" in the following discussion. If $\lambda$ and $g_{ij}$ are positive, then this condition Equation (7.18) is fulfilled especially for solutions with $N$ active modes and a uniform absolute value $|A_j| = A_0$ for all $j$, where

$$A_0 = \sqrt{\frac{\lambda}{\sum_j g_{ij}}} \,. \tag{7.19}$$

Owing to rotation symmetry, the numerator in this equation is independent of $i$. Such an $N$-mode solution corresponds for $N \leq 4$ and $N = 6$ to a periodic pattern, and for $N = 5$ and $N > 6$ it corresponds to a quasi-periodic pattern. All stationary solutions of equation (7.17) are degenerate with respect to the phases $\phi_j$ of $A_j = |A_j| e^{\phi_j}$. These phases may take any arbitrary value and are conserved variables in Equation (7.17). In a third class of stationary solutions, all amplitudes disappear except one: $A_i \neq 0$ and $A_j = 0$ for $j \neq i$. In this case,

$$|A_i| = \sqrt{\frac{\lambda}{g_{ii}}}. \tag{7.20}$$

These solutions represent plane waves with different orientations. It should be noted that these two examples are not the only nontrivial solutions. If the matrix $\mathbf{G} = (g_{ij})$ is not singular, then there is a family of solutions for each number $M < N$ of active modes.

The stability of the trivial solution $A_i = 0$ is governed by $\lambda$. Only when $\lambda$ is negative is this solution stable. If $\lambda$ is positive, then the stability of the solutions $|A_j| = A_0$ is governed by the eigenvalues of the matrix $\mathbf{G}$. This is revealed by analysis of the dynamics of $|A_i|$, given by Equations (7.17) in the case of real valued $A_i$. The stability of a solution is determined by the linearized equations for the perturbation $a_j = |A_j| - A_0$ :

$$\frac{\partial}{\partial t} a_i = -\frac{2\lambda}{\sum_j g_{ij}} \sum_j g_{ij} a_j \tag{7.21}$$

from which it can be seen that solutions with $|A_j| = A_0$ are stable if the matrix $\mathbf{G}$ has only positive eigenvalues. Since the matrix $\mathbf{G}$ is cyclic, it has $N$ complex eigenvectors $\mathbf{V}_l$ with the form

$$\mathbf{V}_l = (1, \, e^{i 2\pi \frac{l}{N}}, \, e^{i 2\pi \frac{2l}{N}}, \, ..., \, e^{i 2\pi \frac{(N-1)l}{N}}) \tag{7.22}$$

where $l = 0, 1, ..., (N - 1)$. The eigenvalues $\omega_l$ of **G** are real and given by

$$\omega_l = \sum_j g_{0,j} \cos(2\pi \frac{jl}{N}) . \tag{7.23}$$

Equation (7.21) implies a simple criterion for the instability of a $N$-mode solution: If $g_{ij} > g_{ii}$ for one arbitrary pair $(i, j)$, then the $N$-mode solution is unstable. In order to see this, consider the quadratic form

$$Q(a_1, a_2, ...) = \sum_{i,j} a_i g_{ij} a_j \tag{7.24}$$

If all of the eigenvalues of the symmetrical matrix **G** are positive, then $Q > 0$ for arbitrary $a_i$. However, if

$$g_{ij} > g_{ii} . \tag{7.25}$$

for one pair $(i, j)$ of modes, then choosing $a_i = 1, a_j = -1$, and $a_k = 0$ for $k \neq i, j$ $Q = g_{ii} - g_{ij} < 0$ will be obtained. Thus if Equation (7.25) holds, the matrix **G** must have at least one negative eigenvalue. In this case, the $N$-mode solution cannot possibly be stable.

The stability of the plane waves $|A_i| = \sqrt{\frac{\lambda}{g_{ii}}}$, $A_j = 0$ for $j \neq i$ is governed by two conditions:
(1) The linearized dynamics of $a_i$ is

$$\frac{\partial}{\partial t} a_i = -2\lambda a_i . \tag{7.26}$$

This is always the case when $\lambda$ is positive.
(2) The active mode $A_i$ must be able to suppress the growth of the inactive modes $A_{j \neq i} = 0$. The linearized dynamics of these inactive modes is

$$\frac{\partial}{\partial t} A_j = \lambda \left(1 - \frac{g_{j,i}}{g_{ii}}\right) A_j \tag{7.27}$$

implying stability if $(1 - g_{ji}/g_{ii})$ is negative, i.e., when the condition

$$g_{ij} > g_{ii} \tag{7.28}$$

is fulfilled for all $j \neq i$. This implies that plane waves and $N$-mode solutions cannot be stable simultaneously.

It should be noted at this point that the condition

$$g_{ij} < g_{ii} \tag{7.29}$$

for all $j \neq i$ is only a necessary, but not sufficient condition for the stability of a $N$ - mode solution. This is demonstrated by the following example: Consider the following $4 \times 4$ matrix:

$$\mathbf{M} = \begin{pmatrix} 1 & g_1 & g_2 & g_1 \\ g_1 & 1 & g_1 & g_2 \\ g_2 & g_1 & 1 & g_1 \\ g_1 & g_2 & g_1 & 1 \end{pmatrix} . \tag{7.30}$$

This matrix has three different eigenvalues:

$$\omega_0 = 1 + 2g_1 + g_2 \tag{7.31}$$
$$\omega_1 = 1 - g_2 \tag{7.32}$$
$$\omega_2 = 1 - 2g_1 + g_2 \tag{7.33}$$

where $\omega_1$ is doubly degenerate. If $0 < g_1, g_2 < 1$, then $\omega_0$ and $\omega_1$ are positive. The third eigenvalue $\omega_2$ is positive, however, only when the condition $g_1 < \frac{1+g_2}{2}$ is also fulfilled.

For the case of $f_{ij} = 0$, these results show that depending on the coefficients $g_{ij}$ there are solutions of the amplitude equation (7.14) corresponding to very complicated, spatial non-periodic patterns. For a wide range of parameter values, defined by the existence of a pair $(i, j)$ of modes with $g_{ii} < g_{ji}$, these patterns are not stable, however. If $g_{ii} < g_{ji}$ is fulfilled for all $j \neq i$, then the attractors of Equation (7.14) correspond to plane waves. In the following discussion I will examine the general case in which $f_{ij} \neq 0$. The main interest here will be to clarify which of the above conclusions remains valid under the more general conditions.

In the general amplitude equation

$$\frac{\partial}{\partial t} A_i = \lambda A_i - \sum_j g_{ij} |A_j|^2 A_i - \sum_j f_{ij} A_j A_{j-} \bar{A}_{i-} \tag{7.34}$$

the interaction of the modes is not determined only by the absolute value of the amplitudes $|A_j|$. Their phases $\phi_j$ are also coupled by the third term in Equation (7.34). It is advantageous to split this equation (7.34) into equations

$$\frac{\partial}{\partial t} A_i = \lambda A_i - \sum_j g_{ij} A_j^2 A_i$$

$$- \sum_j f_{ij} A_j A_{j-} A_{i-} \cos \left( (\phi_j + \phi_{j-}) - (\phi_i + \phi_{i-}) \right) \tag{7.35}$$

$$\frac{\partial}{\partial t} \phi_i = - \sum_j f_{ij} A_j A_{j-} A_{i-} / A_i \sin \left( (\phi_j + \phi_{j-}) - (\phi_i + \phi_{i-}) \right) \tag{7.36}$$

for the amplitudes $|A_j| = A_j$ and the phases $\phi_j$. It can be easily seen that the linearized equations for the dynamics of the system for $A_i$ and $\phi_i$ are decoupled in the vicinity of a stationary solution. Hence, it is possible to discuss the stability of stationary solutions with respect to phase perturbations and amplitude perturbations separately.

A stationary configuration of the absolute values of the amplitudes $A_i$ fulfills

$$0 = \lambda A_i - \sum_j g_{ij} A_j^2 A_i - \sum_j f_{ij} A_j A_{j-} A_{i-} , \tag{7.37}$$

when the phases satisfy $\phi_j + \phi_{j-}) = \Phi_0 = $ const. . An $N$-mode solution with identical absolute values of the amplitudes $A_i = A_0$ is thus stationary for

$$A_0 = \sqrt{\frac{\lambda}{\sum_j (g_{ij} + f_{ij})}} \tag{7.38}$$

For $f_{ij} \geq 0$, the absolute value of the equilibrium amplitudes is thus reduced compared to the case of $f_{ij} = 0$. The linearized dynamics of the perturbation $a_i = A_i - A_0$ of such an equilibrium configuration is given by

$$\frac{\partial}{\partial t} a_i = -2A_0^2 \sum_j g_{ij} a_j - A_0^2 \sum_j f_{ij} \left( a_j + a_{j-} + a_{i-} \right) \tag{7.39}$$

$$= -2A_0^2 \sum_j g_{ij} a_j - A_0^2 \sum_j (f_{ij} + f_{ij-}) a_j - a_{i-} A_0^2 \sum_j f_{ij} \tag{7.40}$$

$$= -A_0^2 \sum_j \hat{g}_{i,j} a_j \tag{7.41}$$

The $N$-mode solution is stable when the matrix $\hat{\mathbf{G}} = (\hat{g}_{i,j})$ where

$$\hat{g}_{ij} = 2g_{ij} + f_{ij} + f_{ij-} + \delta_{j,i-} \sum_k f_{ik} \tag{7.42}$$

has only positive eigenvalues. In complete analogy to the case when $f_{ij} = 0$, this solution is unstable when a pair $(i, j)$ of modes exists that fulfills

$$\hat{g}_{ij} > \hat{g}_{ii} \tag{7.43}$$

When $f_{ij} > 0$ and $\hat{g}_{ii} = 2g_{ii} + f_{ii-}$, the existence of a pair $(i, j)$ that fulfills the condition

$$2g_{ij} + f_{ij} + f_{ij-} > 2g_{ii} + f_{ii-} \tag{7.44}$$

is a sufficient condition for the instability of the $N$-mode solution. For a pair $(i, i^-)$ of modes with antiparallel wave vectors this reduces to

$$g_{ii^-} > g_{ii} .  \tag{7.45}$$

As in the case of $f_{ij} = 0$, an $N$-mode solution is always unstable if $g_{ij} > g_{ii}$ for all $j \neq i$.

Again plane wave solutions defined by $\mathcal{A}_i \neq 0$ and $\mathcal{A}_j = 0$ for $j \neq i$ are stationary, where

$$\mathcal{A}_i = \sqrt{\frac{\lambda}{g_{ii}}}  \tag{7.46}$$

These solutions are stable when $\lambda > 0$ and $g_{ii} < g_{ij}$ for all $j \neq i$, since linearized Equations (7.26) and (7.27) still describe the dynamics in the vicinity of plane wave solutions. Thus, also in this case, the stability of planes wave excludes the stability of an $N$-mode solution.

## 7.3. Linking amplitude equations and field dynamics

Next, let us clarify the relationship of the system of amplitude equations (7.14) and the dynamics of the field $z(\mathbf{x})$

$$\frac{\partial}{\partial t} z(\mathbf{x}) = F[z(\mathbf{x})]  \tag{7.47}$$

Again, it is assumed that the dynamics described by Equation (7.47) occur in the vicinity of a critical point at which the homogeneous solution $z(\mathbf{x}) = 0$ becomes unstable and a nontrivial solution branches of continuously from the homogeneous solution. It is assumed that the inhomogeneous solutions of Equation (7.47) can be represented by a power series in the vicinity of the critical point. Equation (7.47) then leads to expressions for the first term of this series from which the parameters $g_{ij}$ and $f_{ij}$ of Equation (7.14) can be determined.

For a perturbative treatment, it is of advantage to expand the right hand side of Equation (7.47) into a power series:

$$\frac{\partial}{\partial t} z(\mathbf{x}) = \hat{L}z(\mathbf{x}) + N_3[z(\mathbf{x})] + N_5[z(\mathbf{x})] + \dots  \tag{7.48}$$

where $\hat{L}$ is the operator that determines the linearized equation in the vicinity of $z(\mathbf{x}) = 0$ and $N_i[z(\mathbf{x})]$ represents the respective terms of $F[z(\mathbf{x})]$ that are of order $i$. It is assumed that the spectrum of $\hat{L}$ is real and has only one maximum at the critical wavenumber $k_c$. In this case the operator $\hat{L}$ is selfadjoint. The largest eigenvalue of $\hat{L}$ is called $r$ in the following discussion. Owing to the symmetry of Equation (7.47) with respect to orientation shifts, the series Equation (7.48) only contains odd-order terms. The nonlinear operators $N_i[z(\mathbf{x})]$ can always be written as operators linear in $i$ arguments. For example, $N_3[z(\mathbf{x})]$ may be written

$$N_3[z(\mathbf{x})] = N_3\left(z(\mathbf{x}), z(\mathbf{x}), \bar{z}(\mathbf{x})\right) \tag{7.49}$$

with the three-argument operator $N_3\,(...)$ fulfilling

$$N_3\left(\sum_i \alpha_i u_i(\mathbf{x}), \sum_j \beta_j v_j(\mathbf{x}), \sum_k \gamma_k w_k(\mathbf{x})\right) =$$
$$\sum_{i,j,k} \alpha_i \beta_j \gamma_k N_3\left(u_i(\mathbf{x}), v_j(\mathbf{x}), w_k(\mathbf{x})\right) \tag{7.50}$$

for arbitrary functions $u_i(\mathbf{x})$, $v_j(\mathbf{x})$, and $w_k(\mathbf{x})$ and arbitrary coefficients $\alpha_i$, $\beta_j$, $\gamma_k$. I will denote this $i$ -argument notation for the operators $N_i$ by parentheses (Ě). It can be seen in Equation (7.50) that one of the three arguments in $N_3(...)$, like in Equation (7.49), must be the complex conjugate of $z(\mathbf{x})$. Were this not the case, then operator $N_3[z(\mathbf{x})]$ would not be symmetric with respect to orientation shifts $z(\mathbf{x}) \rightarrow e^{i\phi} z(\mathbf{x})$.

For the further derivations, it will be useful to separate the largest eigenvalue $r$ from the spectrum as follows:

$$\hat{L} = r + \hat{L}^0 \tag{7.51}$$

Since the spectrum $\lambda(k)$ of $\hat{L}$ reaches a maximum for $k_c$, the eigenvalues of $\hat{L}^0$ are zero for this wavenumber.

For the analysis of the dynamics (Equation (7.47)) in the vicinity of an instability ($r \approx 0$), it is assumed that the solution $z(\mathbf{x}, t)$ of Equation (7.47) and the largest eigenvalue $r$ above the point of instability ($r > 0$) can be represented as a power series with respect to a formal control parameter $\epsilon$:

$$r = r_1\epsilon + r_2\epsilon^2 + r_3\epsilon^3 + ... \tag{7.52}$$
$$z(\mathbf{x}, t) = \epsilon z_1(\mathbf{x}, t) + \epsilon^2 z_2(\mathbf{x}, t) + \epsilon^3 z_3(\mathbf{x}, t) + ... \tag{7.53}$$

This will generally be possible when the inhomogeneous solution $z(\mathbf{x}, t)$ for $r > 0$ branches of continuously from the homogeneous solution. In this case, the leading order term $z_1(\mathbf{x})$ will describe the shape of the pattern emerging at the instability.

It must also be taken into consideration that the intrinsic time scale $\tau = r^{-1}$ of the linear instability diverges for $r \rightarrow 0$, i.e., the dynamics of $z(\mathbf{x}, t)$ becomes arbitrarily slow. This can be compensated for by rescaling the time as follows:

$$T = r\,t \tag{7.54}$$

The dynamics of $z(\mathbf{x})$, as a function of the rescaled time variable $T$

$$r \frac{\partial}{\partial T} z(\mathbf{x}) = F[z(\mathbf{x})] \tag{7.55}$$

does not show critical slowing in the vicinity of the instability. Since the amplitude of $z(\mathbf{x})$ and the time $t$ are rescaled differently when the formal control parameter $\epsilon$ is changed, the subsequent expansion is called a multiscale expansion [54].

The rescaled Equation (7.55) leads to a system of equations for the fields $z_i(\mathbf{x}, T)$ of the power series Equation (7.53). To determine these equations, Equations (7.51,7.52,7.53) are inserted into Equation (7.55) and the resulting terms are ordered separating different powers of the control parameter $\epsilon$:

$$
\begin{aligned}
0 &= \hat{L}^0 z - r \frac{\partial}{\partial T} z + N_3[z] + N_5[z] + \dots \tag{7.56} \\
&= \epsilon \left( \hat{L}^0 z_1 \right) + \epsilon^2 \left( \hat{L}^0 z_2 + r_1 z_1 - r_1 \frac{\partial}{\partial T} z_1 \right) \\
&+ \epsilon^3 \left( \hat{L}^0 z_3 + r_1 z_2 + r_2 z_1 - r_1 \frac{\partial}{\partial T} z_2 - r_2 \frac{\partial}{\partial T} z_1 + N_3(z_1, z_1, \bar{z}_1) \right) \\
&+ \epsilon^4 \left( \hat{L}^0 z_4 + r_1 z_3 + r_2 z_2 + r_3 z_1 - r_1 \frac{\partial}{\partial T} z_3 - r_2 \frac{\partial}{\partial T} z_2 \right. \\
&\quad \left. - r_3 \frac{\partial}{\partial T} z_1 + N_3(z_1, z_1, \bar{z}_2) + N_3(z_1, z_2, \bar{z}_1) + N_3(z_2, z_1, \bar{z}_1) \right) \\
&+ \epsilon^5 \left( \hat{L}^0 z_5 + \dots \right) \\
&\quad \vdots \tag{7.57}
\end{aligned}
$$

Equation (7.57) can be fulfilled for $\epsilon > 0$ only if each individual term is zero. The conditions for this have the form:

$$\hat{L}^0 z_i = \dots \tag{7.58}$$

where $i$ is the order of the respective term. An equation with this form can be fulfilled only if the right side of the equation is orthogonal to the kernel of the operator $\hat{L}^0$. This requirement is often called a compatibility condition [54]. We will see that the desired equation for $z_1$ in the power series in $z$ is identical to the third-order compatibility condition. To reveal this, the equations are considered in increasing order. The condition

$$\hat{L}^0 z_1 = 0 \tag{7.59}$$

results from the first-order term. It implies that $z_1(\mathbf{x}, T)$ must be an element of the kernel of $\hat{L}^0$ and thus have only Fourier modes whose wave vectors lie on the critical circle. The second-order term leads to

$$\hat{L}^0 z_2 = r_1 \left( -z_1 + \frac{\partial}{\partial T} z_1 \right) . \tag{7.60}$$

Since $z_1(\mathbf{x}, T)$ must be element of the kernel of $\hat{L}^0$, the compatibility condition for this equation (7.60) is fulfilled for a selfadjoint operator $\hat{L}^0$ only by $r_1 = 0$. The third-order term leads to

$$\hat{L}^0 z_3 = r_2 \left( -z_1 + \frac{\partial}{\partial T} z_1 \right) - N_3(z_1, z_1, \bar{z}_1) . \tag{7.61}$$

If the projection operator onto the kernel of $\hat{L}^0$ is denoted $\hat{P}_c$, then

$$0 = \hat{P}_c \left( r_2 \left( -z_1 + \frac{\partial}{\partial T} z_1 \right) - N_3(z_1, z_1, \bar{z}_1) \right) \tag{7.62}$$

$$= r_2 \left( -z_1 + \frac{\partial}{\partial T} z_1 \right) - \hat{P}_c N_3[z_1] \tag{7.63}$$

is the compatibility condition for Equation (7.61). If this condition is fulfilled, then the value of $r_2$ defines the units in which the magnitude of $z_1$ is measured and can be chosen arbitrarily. Choosing $r_2 = 1$, Equation (7.63) defines the dynamics of the first term $z_1(\mathbf{x})$ in the power series for $z(\mathbf{x})$:

$$\frac{\partial}{\partial T} z_1(\mathbf{x}) = z_1(\mathbf{x}) - \hat{P}_c N_3(z_1(\mathbf{x}), z_1(\mathbf{x}), \bar{z}_1(\mathbf{x})) \tag{7.64}$$

The equation that describes the dynamics of one of the Fourier components $A_i e^{i\mathbf{k}_i \mathbf{x}}$ of $z_1(\mathbf{x})$ with $|\mathbf{k}_i| = k_c$ is obtained by projection of Equation (7.64) onto the $e^{i\mathbf{k}_i \mathbf{x}}$ subspace of the kernel of $\hat{L}^0$. If the operator of the projection onto this subspace is denoted $\hat{P}_i$, then

$$\hat{P}_i \frac{\partial}{\partial T} z_1(\mathbf{x}) = e^{i\mathbf{k}_i \mathbf{x}} \frac{\partial}{\partial T} A_i$$

$$= e^{i\mathbf{k}_i \mathbf{x}} A_i + \hat{P}_i \hat{P}_c N_3[z_1(\mathbf{x})] \tag{7.65}$$

$$= e^{i\mathbf{k}_i \mathbf{x}} A_i + \hat{P}_i N_3[z_1(\mathbf{x})] .$$

The dynamics of the amplitudes $A_i$ are therefore, given by

$$\frac{\partial}{\partial T} A_i = A_i + e^{-i\mathbf{k}_i \mathbf{x}} \hat{P}_i N_3[z_1(\mathbf{x})] \tag{7.66}$$

Taking

$$z_1(\mathbf{x}) = \sum_i A_i(T)e^{i\mathbf{k}_i \mathbf{x}} \tag{7.67}$$

with the wave vectors $\mathbf{k}_i$ equidistant on the critical circle, Equation (7.66) leads to the following system of amplitude equations:

$$\frac{\partial}{\partial T} A_i = A_i + \sum_{j,k,l} A_j A_k \bar{A}_l \, e^{-i\mathbf{k}_i \mathbf{x}} \hat{P}_i N_3(e^{i\mathbf{k}_j \mathbf{x}}, e^{i\mathbf{k}_k \mathbf{x}}, e^{-i\mathbf{k}_l \mathbf{x}}) \,. \tag{7.68}$$

As shown above, the number of nonzero terms in this equation (7.68) is restricted by the symmetries. Thus, the equation (7.68) must have the form:

$$\frac{\partial}{\partial T} A_i = A_i - \sum_j g_{ij}|A_j|^2 A_i - \sum_j f_{ij} A_j A_{j-} \bar{A}_{i-} \tag{7.69}$$

Thus the coupling constants $g_{ij}$ and $f_{ij}$ are identified by comparison of the coefficients in Equation (7.68) and Equation (7.69). It is found that

$$
\begin{aligned}
g_{ij} \;=\; & -e^{-i\mathbf{k}_i \mathbf{x}} N_3(e^{i\mathbf{k}_i \mathbf{x}}, e^{i\mathbf{k}_j \mathbf{x}}, e^{-i\mathbf{k}_j \mathbf{x}}) \\
& -e^{-i\mathbf{k}_i \mathbf{x}} N_3(e^{i\mathbf{k}_j \mathbf{x}}, e^{i\mathbf{k}_i \mathbf{x}}, e^{-i\mathbf{k}_j \mathbf{x}}) \tag{7.70}
\end{aligned}
$$

$$
\begin{aligned}
f_{ij} \;=\; & -\frac{1}{2} e^{-i\mathbf{k}_i \mathbf{x}} N_3(e^{i\mathbf{k}_j \mathbf{x}}, e^{-i\mathbf{k}_j \mathbf{x}}, e^{i\mathbf{k}_i \mathbf{x}}) \\
& -\frac{1}{2} e^{-i\mathbf{k}_i \mathbf{x}} N_3(e^{-i\mathbf{k}_j \mathbf{x}}, e^{i\mathbf{k}_j \mathbf{x}}, e^{i\mathbf{k}_i \mathbf{x}}) \tag{7.71}
\end{aligned}
$$

for $j \neq i$ and

$$g_{ii} \;=\; -e^{-i\mathbf{k}_i \mathbf{x}} N_3(e^{i\mathbf{k}_i \mathbf{x}}, e^{i\mathbf{k}_i \mathbf{x}}, e^{-i\mathbf{k}_i \mathbf{x}}) \tag{7.72}$$
$$f_{ii} \;=\; 0 \tag{7.73}$$

for the diagonal elements of the matrix.

If the directions of the wave vectors $\mathbf{k}_i = (\cos(\alpha_i), \sin(\alpha_i)) k_c$ are represented by the angles $\alpha_i$, then the coefficients $g_{ij}$ and $f_{ij}$ are functions only of the angle $\alpha = |\alpha_i - \alpha_j|$ between the wave vectors $\mathbf{k}_i$ and $\mathbf{k}_j$. Thus, one can obtain them from the continuous functions

$$
\begin{aligned}
g(\alpha) \;=\; & -e^{-i\mathbf{k}_0 \mathbf{x}} \Big( N_3(e^{i\mathbf{k}_0 \mathbf{x}}, e^{i\mathbf{h}(\alpha)\mathbf{x}}, e^{-i\mathbf{h}(\alpha)\mathbf{x}}) \\
& +N_3(e^{i\mathbf{h}(\alpha)\mathbf{x}}, e^{i\mathbf{k}_0 \mathbf{x}}, e^{-i\mathbf{h}(\alpha)\mathbf{x}}) \Big) \tag{7.74}
\end{aligned}
$$

$$
\begin{aligned}
f(\alpha) \;=\; & -\frac{1}{2} e^{-i\mathbf{k}_0 \mathbf{x}} \Big( N_3(e^{i\mathbf{h}(\alpha)\mathbf{x}}, e^{-i\mathbf{h}(\alpha)\mathbf{x}}, e^{i\mathbf{k}_0 \mathbf{x}}) \\
& +N_3(e^{-i\mathbf{h}(\alpha)\mathbf{x}}, e^{i\mathbf{h}(\alpha)\mathbf{x}}, e^{i\mathbf{k}_0 \mathbf{x}}) \Big) \tag{7.75}
\end{aligned}
$$

of the angle $\alpha$ where $\mathbf{k}_0 = (1, 0) k_c$ and $\mathbf{h}(\alpha) = (\cos(\alpha), \sin(\alpha)) k_c$. Both functions $\alpha$ are periodic with a period of $2\pi$. The function $f(\alpha)$ is also periodic with a period of $\pi$, since the right side of (7.75) is invariant with respect to the transformation $\mathbf{h}(\alpha) \rightarrow \mathbf{h}(\alpha + \pi) = -\mathbf{h}(\alpha)$. If the function $g(\alpha)$ and $f(\alpha)$ are known, then the coefficients of interaction are given by

$$
\begin{aligned}
g_{ij \neq i} &= g(|\alpha_i - \alpha_j|) \tag{7.76} \\
g_{ii} &= g(0)/2 \tag{7.77}
\end{aligned}
$$

and

$$
f_{ij \neq i} = (1 - \delta_{ij-}) f(|\alpha_i - \alpha_j|) \tag{7.78}
$$

The functions $g(\alpha)$ and $f(\alpha)$ are called the angle-dependent interaction functions.

Equations (7.74 - 7.78) determine the coefficients of the system of amplitude equations that describe the dynamics of a pattern with $N$ discrete Fourier components for any dynamics of $z(\mathbf{x})$ in the class considered here. In particular if $g(\alpha) > 0$ and $f(\alpha) > 0$, these amplitude equations imply a simple criterion for determining whether plane waves are the only stable solution: If $g(\alpha)$ fulfills the condition

$$
g(\alpha) > g(0)/2 \tag{7.79}
$$

for all angles $\alpha$, then plane wave are the only stable planform.

They also indicate that considering patterns with $N$ discrete Fourier components is sufficient to study the structure and stability of stationary solution of Equation (7.2) close to the instability.

Such patterns imply that independently of the exact form of $g(\alpha)$ and $f(\alpha)$ an $N_{max}$ always exists such that patterns with $N > N_{max}$ active modes are unstable. To see this, we examine the amplitude equations for a pattern with $N$ discrete Fourier components. The minimum angle between two wave vectors in such a pattern is $\alpha_{min} = 2\pi/N$. The $N$-mode solution is unstable when a single nondiagonal coupling coefficient $g_{ij}$ is larger than the diagonal element $g_{ii} = g(0)/2$. The coefficient that describes the interaction on the critical circle of neighboring modes is given by $g_{i(i+1)} = g(2\pi/N)$. Therefore,

$$
g_{i(i+1)} \xrightarrow[N \to \infty]{} g(0) \tag{7.80}
$$

for any continuous bounded function $g(\alpha)$. It follows that an finite number $N_{max}$ must exist such that

$$
g_{ii} = g(0)/2 < g(2\pi/N) = g_{i(i+1)} \tag{7.81}
$$

is fulfilled for all $N > N_{max}$. Solutions with a larger number of active modes are thus always unstable. In order to identify the stable patterns for a given dynamics, it is sufficient to examine the patterns with a number of active modes $N < N_{max}$.

### 7.4. Extrinsic stability

The conditions derived in the previous section for the stability of a planform are not sufficient to demonstrate that a stationary solution of the form

$$z(\mathbf{x}) = \sum_j A_j e^{i\mathbf{k}_j \mathbf{x}},\tag{7.82}$$

is also an attractor of the dynamics of Equation (7.2). If $A_j$ are a stable solution of the amplitude equations derived from Equation (7.2), then it is only guaranteed that such a solution is *intrinsically* stable: Such a solution would not decay as a result of interaction between the active modes. In addition to this, it must be clarified whether the planform is *extrinsically* stable, i.e whether or not it decays into a solution that contains wave vectors besides $\mathbf{k}_j$. Again we will study this in the vicinity of the instability implying that only modes on the critical circle are admitted.

In order to answer the question of extrinsic stability, we examine the dynamics of the amplitudes of one test mode interacting with the active modes of the planform. This interaction follows from the amplitude equations for a generalized planform

$$z(\mathbf{x}) = Be^{i\mathbf{h}\mathbf{x}} + \sum_j A_j e^{i\mathbf{k}_j \mathbf{x}},\tag{7.83}$$

with the wave vector $\mathbf{h} = k_c(\cos(\alpha), \sin(\alpha))$ and the amplitude $B$ of the test mode in the vicinity of a stationary solution with $|A_j| = |A_0|$ and $B = 0$. The dynamics of $B$ is then

$$\frac{\partial}{\partial t} B = \left(1 - \sum_j g(\alpha - \alpha_j)|A_j|^2\right) B\tag{7.84}$$

$$= \left(1 - |A_0|^2 \sum_j g(\alpha - \alpha_j)\right) B .\tag{7.85}$$

The solution $B = 0$ of this equation is stable when the condition

$$1 - |A_0|^2 \sum_j g(\alpha - \alpha_j) < 0\tag{7.86}$$

is fulfilled. In this case, the interaction with the active modes $A_j$ suppresses the growth of the test mode. If this condition is fulfilled for all $\alpha$, then the Planform is also extrinsically stable: Such a solution suppresses the growth of all Fourier components on the critical circle that it does not contain.

The condition for extrinsic stability of a stationary planform assumes a particularly simple form for $f_{ij} = 0$. The stationary value of $|A_0|$ for such a solution is

$$|A_0|^{-2} = \sum_j g_{ij} = -g(0)/2 + \sum_j g(\alpha_j) \tag{7.87}$$

and the stability condition is thus given by

$$-g(0)/2 + \sum_j g(\alpha_j) - \sum_j g(\alpha - \alpha_j) < 0 \tag{7.88}$$

equivalent to

$$g(0)/2 - g(\alpha - \alpha_0) + \sum_{j=1}^{N-1} \left( g(\alpha_j) - g(\alpha - \alpha_j) \right) < 0 . \tag{7.89}$$

Condition (7.89) must be fulfilled for all values of $\alpha$. Note that the Condition (7.89) represents a generalization of Condition (7.79), which guarantees that plane waves are the only intrinsically stable planforms of the dynamics. This results in the special case of Equation (7.89) for $N = 1$.

### 7.5. Swift–Hohenberg models

The analysis in Section 7.3 showed that in the vicinity of a supercritical bifurcation it is mainly the third-order terms that determine the behavior of the dynamics

$$\frac{\partial}{\partial t} z(\mathbf{x}) = F[z(\mathbf{x})] \tag{7.90}$$

The leading term $z_1(\mathbf{x})$ of the solution fulfills the following equation in the vicinity of the instability:

$$\frac{\partial}{\partial T} z_1(\mathbf{x}) = z_1(\mathbf{x}) - \hat{P}_c N_3(z_1(\mathbf{x}), z_1(\mathbf{x}), \bar{z}_1(\mathbf{x})) \tag{7.91}$$

This equation is independent of the precise form of the linear component of the dynamics. It is based only on the assumption that the critical modes have a fixed finite wavelength in the vicinity of the instability.

To answer questions of a qualitative nature, one therefore often studies model equations in which the linear component is as simple as possible. The simplest power series for a spectrum that has a maximum at a finite wavelength is

$$\lambda(k) = r - (k_0^2 - k^2)^2 \tag{7.92}$$

This spectrum has a quadratic minimum at $k = 0$ and quadratic maximum at $k = k_0$. The largest eigenvalue $r$ is assumed for $k = k_0$. The operator defined by this spectrum is given by

$$\hat{L} = \left( r - (k_0^2 + \Delta)^2 \right) \tag{7.93}$$

This is the simplest differential operator that has the assumed symmetries and whose spectrum has a simple maximum for the wavenumber $k_0$.

For this reason, model equations with the form

$$\frac{\partial}{\partial t} z(\mathbf{x}) = \left( r - (k_0^2 + \Delta)^2 \right) z(\mathbf{x}) - N_3[z(\cdot)] \tag{7.94}$$

are often examined in the study pattern formation processes. Swift and Hohenberg first used such an equation in 1977 in a discussion of hydrodynamic fluctuations in the vicinity of a convective instability [56]. Equations with the form of Equation (7.94) are therefore called Swift-Hohenberg models or generalized Swift-Hohenberg equations. In the following, we will use such models to examine the stability of patterns with many pinwheels.

## 8. A Swift–Hohenberg model with stable pinwheel patterns

In this section, I will identify the basic structural properties of a class of models exhibiting stable pinwheel patterns. For this purpose, I will first construct a simple model that fulfills two conditions required of the dynamics of neuronal development of orientation preference patterns: (1) Interactions can be nonlocal. (2) Patterns that do not exhibit all possible orientation preferences should not be stable solutions of the developmental dynamics. An examination of the possible solutions of this model and its stability will then show that there is a range of parameter values in which patterns of stable pinwheels are formed. Within this range of parameter values, the model exhibits a pronounced multistability of different solutions: The number of simultaneously stable planforms increases exponentially with increasing interaction-range. The set of solutions in this parameter range has the topology of a set of nested tori, which can be visualized as a graph in the space of possible patterns. Finally, I will show that as a result of the multistability of different patterns a spectrum of allowed pinwheel densities forms that is in good agreement with the range of experimentally observed pinwheel densities.

### 8.1. Construction of the model

In the dynamics of cortical development, interactions are generally not strictly local. The equations for neuronal patterns typically contain terms in the form of integral operators. This property has a simple structural cause. The activity patterns that drive changes in a neuronal pattern are spatially extended and are shaped by intracortical interactions. Therefore, changes in the pattern occurring some distance from one another are almost instantaneously coupled. Mathematically, this coupling is described by integral operators and, therefore, the equations for the dynamics of the system takes the form of integrodifferential equations. In the following discussion, I assume the dynamics of neuronal development have this structural mathematical property.

In order to analyze which structural properties make it possible for the dynamics of neuronal development to stabilized pinwheels, I will next construct a mathematically simple example of a nonlinear integrodifferential equation. As presented in Section 7, the attractors of such equations are determined in the vicinity of a continuous bifurcation by the third-order terms. The exact form of the linear component is not important in the vicinity of the instability as long as the spectrum has only a single maximum. Thus, the simplest form of such an equation is

$$\frac{\partial}{\partial t} z(\mathbf{x}) = \hat{L} z(\mathbf{x}) + N_3[z(\mathbf{x})] \tag{8.1}$$

This assumes the dynamics are symmetric with respect to translation and rotation and with respect to orientation shifts. The simplest essentially nonlocal nonlinear integral operators that have these symmetries are

$$N_3^1[z(\mathbf{x})] = -z(\mathbf{x}) \int d^2y \, K(\mathbf{y} - \mathbf{x}) \, |z(\mathbf{y})|^2 \tag{8.2}$$

$$N_3^2[z(\mathbf{x})] = -\bar{z}(\mathbf{x}) \int d^2y \, K(\mathbf{y} - \mathbf{x}) \, z(\mathbf{y})^2 \tag{8.3}$$

where $K(\mathbf{x})$ is a rotation symmetric, real integral kernel. For example,

$$K(\mathbf{x}) = \frac{1}{2\pi\sigma^2} e^{-\frac{x^2}{2\sigma^2}} \tag{8.4}$$

where $\sigma$ describes the spatial extent of the interactions. The operators $N_3^1[z(\mathbf{x})]$ and $N_3^2[z(\mathbf{x})]$ are especially simple, as they allow only two point interactions. They are essentially nonlocal, since the form of their angle-dependent interaction functions

$$g^1(\alpha) = 1 + e^{-\sigma^2 k_0^2 (1 - \cos(\alpha))} \tag{8.5}$$

$$f^1(\alpha) = \frac{1}{2}\left(e^{-\sigma^2 k_0^2 (1 - \cos(\alpha))} + e^{-\sigma^2 k_0^2 (1 + \cos(\alpha))}\right) \tag{8.6}$$

$$g^2(\alpha) = 2 e^{-\sigma^2 k_0^2 (1 + \cos(\alpha))} \tag{8.7}$$

$$f^2(\alpha) = 1 \tag{8.8}$$

and thus of the stable planforms, are functions of $\sigma$, the range of the interactions. In Equations (8.5-8.8), the critical wavenumber is $k_0 = 2\pi/\Lambda$.

Since it cannot be assumed that a typical equation for the dynamics of the system contains only nonlocal terms, let us include also the simplest local term:

$$N_3^0[z(\mathbf{x})] = -|z(\mathbf{x})|^2 z(\mathbf{x}) , \tag{8.9}$$

whose angle-dependent interaction functions are constant and given by

$$g^0(\alpha) = 2 \tag{8.10}$$

$$f^0(\alpha) = 1 \tag{8.11}$$

The general form of a nonlinearity that contains these three terms is

$$N_3[z(\mathbf{x})] = g_0 N_3^0[z(\mathbf{x})] + g_1 N_3^1[z(\mathbf{x})] + g_2 N_3^2[z(\mathbf{x})] . \tag{8.12}$$

The angle-dependent interaction functions assigned to $N_3[z(\mathbf{x})]$ are then linear combinations of the functions $g^i(\alpha)$ and $f^i(\alpha)$. If $g_1$ and $g_2$ are zero and if $g_0 > 0$ and, then (8.1) has only plane wave attractors, if the homogeneous state is unstable. In this case,

$$g(\alpha) = 2g_0 > g^0(0)/2 = g_0 \tag{8.13}$$

$$f(\alpha) = g_0 > 0 \tag{8.14}$$

for any $\alpha$.

The choice of $g_1$ and $g_2$ can be constrained by a second plausible requirement, resulting from the function of the column pattern. The column pattern shows the spatial distribution of neurons that can detect different stimulus orientations. If not all orientations are present in the pattern, then the system being examined is blind for a range of possible stimulus orientations. If it is to be excluded that the visual system resulting from the development process is blind for certain orientations, then the attractors of (8.1) must contain all possible orientation preferences. In particular, stationary solutions that do not contain all orientation preferences may not be stable.

However, such patterns are easy to construct. Consider an N planform

$$z(\mathbf{x}) = \sum_j A_j e^{i\mathbf{k}_j \mathbf{x}}, \tag{8.15}$$

with an even number of active modes. There is a stationary solution with

$$|A_j| = |A_0| \tag{8.16}$$

and the phases $\phi_j$ of $A_j = |A_j|e^{i\phi_j}$ fulfilling

$$\phi_j + \phi_{j^-} = \Phi_0 \tag{8.17}$$

where $\Phi_0$ is a constant. When $\Phi_0 = 0$, this means

$$\phi_j = -\phi_{j^-}, \tag{8.18}$$

which is equivalent to

$$\bar{A}_j = A_{j^-} \tag{8.19}$$

Such a planform is, therefore, real and contains only two orientation preferences: horizontal and vertical. When $\Phi_0 \neq 0$, then the resulting planform also contains only two orientation preferences: $\Phi_0/2$ and $(\Phi_0 + \pi)/2$. For this reason, I call such a planform "not essentially complex".

Since it is required that the resulting system is not blind for any orientation, then all not essentially complex planforms must be unstable. As shown before, this requirement is always fulfilled when $g_{ii^-} > g_{ii}$ and $f_{ij} > 0$. Thus, for the angle- dependent interaction function $g(\alpha)$

$$g(\pi) > g(0)/2. \tag{8.20}$$

This condition can be fulfilled by selecting $g_1$ and $g_2$ so that $g(\alpha)$ is periodic with a period of $\pi$

$$g(\alpha + \pi) = g(\alpha) \tag{8.21}$$

and is positive. The interaction function $g(\alpha)$ is periodic with a period of $\pi$ exactly when $g_1 = 2g_2$. The assumption of Equation (8.21) guarantees the instability of all not essentially complex planforms and reduces the number of independent parameters from four to three.

As we will see in the following, the symmetry property of Equation (8.21) also considerably simplifies the structure of the essentially complex stationary planforms and the investigation of their stability properties. In general, this symmetry will be fulfilled in every model in which the third order term of the dynamics exhibits the permutation symmetry

$$N_3(u, v, w) = N_3(w, u, v) \tag{8.22}$$

for arbitrary test functions $u(\mathbf{x})$, $v(\mathbf{x})$ and $w(\mathbf{x})$.

Finally, the number of independent coupling parameters can be further reduced since the stability of possible planforms does not depend on the absolute scale of the interaction functions $g(\alpha)$ and $f(\alpha)$. This scale can, therefore, be fixed without changing the set of solutions of the model. If $g_1 = 2g_2$, as above, then

$$g(\alpha) = 2g_0 + g_2 \left(1 + e^{-\sigma^2 k_0^2 (1-\cos(\alpha))} + e^{-\sigma^2 k_0^2 (1+\cos(\alpha))}\right) . \qquad (8.23)$$

For $g_0 > 0$, this function displays a maximum $g(0) \approx 2(g_0 + g_2)$ at $\alpha = 0$ and a minimum $g(\pi/2) > 2g_0 + g_2$ at $\alpha = \pi/2$. Its form depends only on the ratio of $g_0$ to $g_2$. If the absolute scale for $g(\alpha)$ is set to

$$2(g_0 + g_2) = 2 \qquad (8.24)$$

and the lower bound for $g(\alpha)$ is set to $\hat{g}_1$

$$2g_0 + g_2 = \hat{g}_1 \qquad (8.25)$$

then $N_3[z(\mathbf{x})]$, $g(\alpha)$, and $f(\alpha)$ has only two effective parameters, $\sigma$ and $\hat{g}_1$, and

$$
\begin{aligned}
N_3[z(\mathbf{x})] &= (1 - g_1)|z(\mathbf{x})|^2 z(\mathbf{x}) - \frac{2 - g_1}{2\pi\sigma^2} z(\mathbf{x}) \int d^2 y \, e^{-\frac{|y-x|^2}{2\sigma^2}} |z(\mathbf{y})|^2 \\
&\quad - \frac{2 - g_1}{4\pi\sigma^2} \bar{z}(\mathbf{x}) \int d^2 y \, e^{-\frac{|y-x|^2}{2\sigma^2}} z(\mathbf{y})^2 \qquad (8.26) \\
g(\alpha) &= g_1 + (2 - g_1)\left(e^{-\sigma^2 k_0^2 (1-\cos(\alpha))} + e^{-\sigma^2 k_0^2 (1+\cos(\alpha))}\right) \qquad (8.27) \\
f(\alpha) &= \frac{g_1}{2} + \frac{2 - g_1}{2}\left(e^{-\sigma^2 k_0^2 (1-\cos(\alpha))} + e^{-\sigma^2 k_0^2 (1+\cos(\alpha))}\right) . \qquad (8.28)
\end{aligned}
$$

In these equations and the further discussion, the hat is dropped from $g_1$. Equation (8.26) defines a Swift-Hohenberg model.

The interaction function $g(\alpha)$ is shown in Figure 5 for several values of $g_1$ and $\sigma k_0 = \pi$, i.e., $\sigma = \Lambda/2$. For $g_1 < 2$, $g(\alpha)$ displays maxima at $\alpha = 0$ and $\alpha = \pi$ and minima at $\alpha = \pi/2$ and $\alpha = 3\pi/2$. Moreover, owing to the symmetry $g(\alpha) = g(-\alpha)$ and the periodicity of $g(\alpha)$ with a period of $\pi$, this function also displays the following symmetry property:

$$g(\alpha) = g(\pi - \alpha) \qquad (8.29)$$

For $\sigma k_0 \gg 0.1$, the value of the maximum approaches 2 and the value of the minimum approaches $g_1$. For $g_1 = 2$, the nonlinear term is purely local and $g(\alpha)$ is a constant. For $g_1 > 0$, $g(\alpha) > 0 f(\alpha) > 0$. It follows from these properties

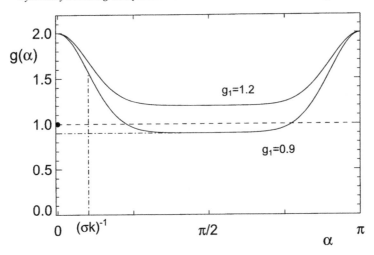

Fig. 5. Angle-dependent interaction function $g(\alpha)$ of the model defined by Equation (8.26): $g(\alpha)$ is shown for two values of the coupling parameter $g_1$: upper curve: $g_1 = 1.2$; lower curve: $g_1 = 0.9$. $g_1$ is the minimum value of the interaction function. The width of the maxima of $g(\alpha)$ at $\alpha = 0$ and $\alpha = \pi$ are given by $(\sigma k)^{-1}$. The dashed line indicates the stability condition of plane wave solutions: $g(\alpha) > g(0)/2$.

that for $g_1 > 1$, plane waves are the stable planform of this model. The condition for the stability of plane waves

$$g(\alpha_{min}) = g(\pi/2) > g(0)/2 \tag{8.30}$$

is fulfilled in the range of parameter values defined by

$$g_1 > 2 \frac{1 - 4e^{-4\pi^2(\frac{\sigma}{\Lambda})^2} + e^{-8\pi^2(\frac{\sigma}{\Lambda})^2}}{2 - 4e^{-2\pi^2(\frac{\sigma}{\Lambda})^2} + e^{-8\pi^2(\frac{\sigma}{\Lambda})^2}}$$

$$\text{for } \frac{\sigma}{\Lambda} > \sqrt{-\ln(2 - \sqrt{3})/2\pi} \tag{8.31}$$

$$g_1 > 0$$

$$\text{for } \frac{\sigma}{\Lambda} \leq \sqrt{-\ln(2 - \sqrt{3})/2\pi} \tag{8.32}$$

Thus, for large ranges of the nonlinearity, $\sigma/\Lambda > \sqrt{-\ln(2 - \sqrt{3})/2\pi} \approx 0.18$, an interval of $g_1$ values exists in which plane waves are unstable.

The structure of the developing pattern differs qualitatively in the two regions of parameter values mentioned. In Figure 6, this is demonstrated by numerically obtained solutions of the Swift-Hohenberg model

$$\frac{\partial}{\partial t} z(\mathbf{x}) = \left( r - (k_0^2 + \Delta)^2 \right) z(\mathbf{x}) + N_3[z(\cdot)] \tag{8.33}$$

defined by Equation (8.26). Three simulations with different values of $g_1$ and different initial conditions are compare in this Figure. In all three simulations, the value of the control parameter $r = 0.1$. Hence, the characteristic time scale $\tau = r^{-1}$ is the same in all three simulations. The range of nonlinear interactions $\sigma = 3\Lambda$ is also the same and is distinctly larger than the characteristic wavelength $\Lambda$ of the developing pattern.

If $g_1 > 1$, it is expected that the solution of the model converges to a pattern composed of parallel stripes. For such parameters and randomly chosen unselective initial conditions $z(\mathbf{x}) \approx 0$, the behavior is like that observed in the previously proposed models (see [15] and references therein). After the initial proliferation of a large number of pinwheels, a pattern with almost no pinwheels develops as a result of movement and pairwise annihilation. A representative example is shown in Figure 6a for $g_1 = 1.02$.

If $g_1 < 1$, a qualitatively different behavior is observed. In this region of parameter values, most of the initially generated pinwheels remain for long periods of time in configurations that change only slowly over time. A representative example is shown in Figure 6b for $g_1 = 0.98$. The simulation in 6c shows that parallel orientation stripes are unstable in this region of parameter values. All parameter values are the same in simulations b and c. Note that distinctly different pinwheel densities develop in simulations 6b and c. The final state in the simulation of Figure 6c has a lower pinwheel density than that of simulation of Figure 6b. This observation suggests that the model has several attractors in the region of parameter values in which plane waves are unstable and that these attractors have different pinwheel densities.

In order to obtain a transparent picture of the behavior observed for $g_1 < 1$, I will determine in the next section the stable planforms of the above model equation in this region of parameter space. First, I will construct a system of stationary, essentially complex planforms (Section 8.2) and then determine their stability regions within the model (Section 8.3).

## 8.2. Essentially complex planforms

The above model is designed so that it is not possible for regular n-planforms with an even number of active modes to be stable. However, N planforms with an odd number of active modes may possibly be stable. A system of potentially stable planforms can be obtained by generalization of such planforms. N planforms with an odd number of active modes can be considered to be a special case of a

(a)  $g_1 = 1.02$

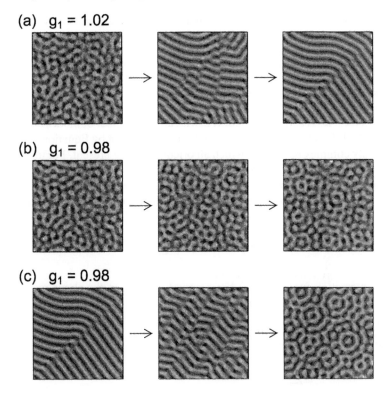

(b)  $g_1 = 0.98$

(c)  $g_1 = 0.98$

Fig. 6. Pinwheel annihilation and preservation in the dynamics of the Swift-Hohenberg model de-fined by Equation (8.26): The patterns in the three rows a to c are snapshots for different times in three numerical simulations of the model obtained for different parameter values and initial condi-tions. Equation (8.33) was integrated numerically for patterns with a $20\Lambda$ size and periodic boundary conditions on a square grid ($256 \times 256$). Only parts of the patterns are shown. In all three simula-tions, the control parameter $r = 0.1$. The range $\sigma$ of the nonlocal term is $3\Lambda$. Simulations a and b differ in the value of the parameter $g_1$: (a) $g_1 = 1.02$; (b) $g_1 = 0.98$. The initial conditions (left column) of simulations a and b are the same. The middle and right columns are for the times $t = 50\tau$ and $t = 500\tau$, where $\tau = r^{-1}$. The parameter values for simulations b and c are identical. The initial conditions for simulation c are the same as the final state of simulation a, the middle and right columns are for the times $t = 200\tau$ and $t = 2000\tau$.

more general system of irregular planforms:

$$z(\mathbf{x}) = \sum_j A_j e^{il_j \mathbf{k}_j \mathbf{x}} \tag{8.34}$$

The $n$ wavevectors $\mathbf{k}_i$ in (8.34) are equally spaced on the semicircle $|\mathbf{k}| = k_0$, $k_y > 0$, where $k_y$ is the $y$-component of the wavevector $\mathbf{k}$, i.e.,

$$\mathbf{k}_j = k_0 \left(\cos(j\pi/n), \sin(j\pi/n)\right) \tag{8.35}$$

where $j = 0, 1, 2, ..., n - 1$. The binary variable $l_j = \pm 1$ then indicates whether the mode with the wavevector $\mathbf{k}_j$ or the mode with the corresponding antiparallel wavevector $-\mathbf{k}_j$ is active. N planforms with an odd number of modes correspond to $l_j = (-1)^j$ and odd $n$. Planforms with the form given by Equation (8.34) can form only complex patterns in which all orientations are present. For this reason, I call such a planform "essentially complex".

The amplitude equations for a essentially complex planform with $n$ active modes are determined by specialization of the amplitude equations for an N planform with $2n$ modes

$$z(\mathbf{x}) = \sum_j A_j e^{il_j \mathbf{k}_j \mathbf{x}} + \sum_j B_j e^{-il_j \mathbf{k}_j \mathbf{x}}. \tag{8.36}$$

If the interaction function $g(\alpha)$ has a period of $\pi$, as in the above model, then these amplitude equations are

$$\frac{\partial}{\partial t} A_i = A_i - \sum_j g_{ij} |A_j|^2 A_i - \sum_j (1 + \delta_{ij}) g_{ij} |B_j|^2 A_i$$

$$-2 \sum_j f_{ij} A_j B_j \bar{B}_i \tag{8.37}$$

$$\frac{\partial}{\partial t} B_i = B_i - \sum_j g_{ij} |B_j|^2 B_i - \sum_j (1 + \delta_{ij}) g_{ij} |A_j|^2 B_i$$

$$-2 \sum_j f_{ij} A_j B_j \bar{A}_i \tag{8.38}$$

The form of these amplitude equations is independent of the configuration $(l_0, l_1, ..., l_{n-1})$. They have stationary solutions with uniform absolute values $|A_i|$ and $|B_i|$:

$$|A_i| = \left(\sum_j g_{ij}\right)^{-1/2} \tag{8.39}$$

$$B_i = 0 \tag{8.40}$$

To determine the stability of the essentially complex planforms, the stability of these fixed points must be determined.

It is to be observed that the number of different planforms increases exponentially with an increasing number of active modes, $n$. For a given $n$, there are $2^n$ different combinations of the binary parameters $l_j$. Not all of these combinations, however, lead to different spatial patterns. The $2^n$ possible planforms can be grouped in classes of equivalent planforms, i.e., those that converted into each other by rotation and/or reflection. The planforms in different classes are called constitutively different and the number of these classes is $N_{Pf}$.

Representative examples of such planforms are shown in Figure 7 for increasing numbers $n$ of active modes. If $n < 3$, then the patterns contain no pinwheels. In contrast, when $n \geq 3$, all possible patterns contain pinwheels. Moreover, only for $n \geq 3$ are there several constitutively different planforms for a given number of active modes. In the case of $n = 3$, there are two different hexagonal configurations of groups of two or six pinwheels. When $n > 3$, the pattern, and therefore also the configuration of pinwheels, is not periodic but instead is spatially quasiperiodic. Planforms with ten or more active modes are qualitatively very similar to the experimentally observed patterns.

The number $N_{Pf}$ of different planforms is plotted in Figure 8 as a function of the number of active modes $n$. If $n$ is small, then $N_{Pf}$ can be determined by counting. For example, when $n = 3, 4$, there are two constitutively different planforms; when $n = 5$, there are four. For large values of $n$, $N_{Pf}$ and a set of representative planforms can be determined by exhaustive search through all $2^n$ possible planforms, represented by $n$-tuples $(l_0, l_1, ..., l_{n-1})$, and counting all of the tuples that cannot be generated from one of the previous tuples by a symmetry operation. Figure 8 shows the number $N_{Pf}$ of constitutively different planforms obtained in this way for $n \leq 16$. $N_{Pf}$ increases exponentially with increasing $n$. For example with $n = 16$ there are more than 1000 constitutively different planforms.

To conclude, the discussion of the system of essentially complex planforms Equation (8.34), let us briefly characterize the structure of the set of patterns that can be represented by planforms with a given $n$. For these patterns, the spatial mean of the squared difference between two patterns $z_1(\mathbf{x})$ and $z_2(\mathbf{x})$ forms a metric:

$$d_{12}^2 = \lim_{R \to \infty} \frac{1}{\pi R^2} \int_{K_R} d^2x \, |z_1(\mathbf{x}) - z_2(\mathbf{x})|^2 \tag{8.41}$$

where $K_R$ is the area of the circle defined by $|\mathbf{x}| < R$. Since $d_{12}^2$ (i) is equal to zero if and only if the patterns $z_1(\mathbf{x})$ and $z_2(\mathbf{x})$ are identical, (ii) is symmetrical, $d_{12}^2 = d_{21}^2$, and (iii) fulfills $d_{13}^2 \leq d_{12}^2 + d_{23}^2$ it defines a metric. Note that $d_{12}^2$ is not a metric of the function space $L^2$, but is a metric in the topological space composed of essentially complex planforms. The global structure of this space may be represented by a graph. For this purpose, two constitutively different

*F. Wolf*

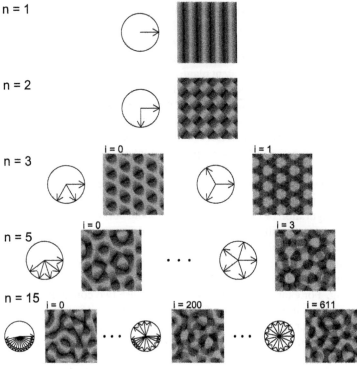

n = 1

n = 2

n = 3   i = 0   i = 1

n = 5   i = 0   i = 3

n = 15   i = 0   i = 200   i = 611

Fig. 7. Essentially complex planforms with different numbers $n = 1, 2, 3, 5, 15$ of active modes: The patterns of orientation preferences $\theta(\mathbf{x})$ are shown. The diagrams to the left of each pattern display the position of the wavevectors of active modes on the critical circle. For $n = 3$, there are two patterns; for $n = 5$, there are four; and for $n = 15$, there are 612 different patterns.

planforms are considered to be neighboring if two of their realizations are related by a single flip of a wave vector. The distance of the constitutively different planforms from each other on this graph in is then a measure of the overall similarity of the patterns that can be obtained from them. By this construction, each of the constitutively different planforms is assigned a coordination number, the number of neighbors it has in the graph. The graph structure can be extended to the entire set of essentially complex planforms in a similar way by an arbitrary rotation of obtainable patterns. Each planform then generates an $(n + 1)$ torus of patterns; the planforms that can be transformed into another by rotation lead to identical tori. These considerations provide a first glimpse of the very rich structure of the phase space of our model. They imply in particular that if a planform with n

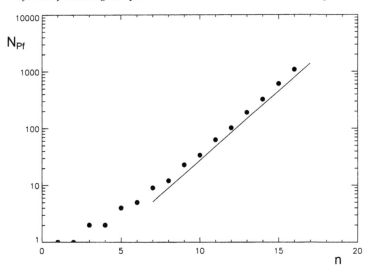

Fig. 8. The number $N_{Pf}$ of constitutively different planforms as a function of the number $n$ of active modes: The points give the number of constitutively different planforms determined by exhaustive search. For comparison, the line shows $N_{Pf} \propto (7/8)^n$.

active modes is stable, then a set of attractors of the Swift-Hohenberg model lies on $M \geq N_{Pf}$ nested $(n+1)$ tori in the space of the possible patterns.

### 8.3. Stability ranges and phase diagram

The stability properties of the planforms in the model are independent of the specific configuration $(l_0, l_1, ..., l_{n-1})$ of the active modes. This results from independence of $(l_0, l_1, ..., l_{n-1})$ of the amplitude equations (8.37 and 8.38). The intrinsic stability of the planforms can be analyzed using the amplitude equations in the vicinity of a fixed point that fulfills $|A_i| = |A_0|$, $B_i = 0$. To this end, the amplitude equations (8.37 and 8.38) are linearized about $B_i = 0$ with respect to the amplitudes $B_i$ of the inactive modes to obtain

$$\frac{\partial}{\partial t} A_i = A_i - \sum_j g_{ij} |A_j|^2 A_i \tag{8.42}$$

$$\frac{\partial}{\partial t} B_i = \left( 1 - \sum_j (1 + \delta_{ij}) g_{ij} |A_j|^2 \right) B_i - 2 \sum_j f_{ij} A_j \bar{A}_i B_j . \tag{8.43}$$

In the vicinity of such a fixed point, the dynamics of $A_i$ are decoupled from the dynamics of $B_i$ and form a system of Landau equations. According to the results of Section 7, a fixed point of the system of equations (8.42) with $|A_0| = (\sum_j g_{ij})^{-1/2}$ is stable if the matrix of $g_{ij}$ has only positive eigenvalues.

The fixed point $B_i = 0$ of the system of (8.43) is stable if the matrix

$$F_{ij} = \left(1 - \sum_j (1 + \delta_{ij}) g_{ij} |A_j|^2\right) - 2 f_{ij} A_j \bar{A}_i f_{ij} \tag{8.44}$$

$$= -\frac{g_{ii}}{\sum_j g_{ij}} - 2 A_j \bar{A}_i f_{ij} \tag{8.45}$$

has only negative eigenvalues. To evaluate whether these conditions are fulfilled, consider the eigenvalues of the matrix $\bar{F}_{ij} = A_j \bar{A}_i f_{ij}$. The matrix $\bar{F}_{ij}$ is self-adjoint and therefore has real eigenvalues. They are identical to those of the matrix $|A_0|^2 f_{ij}$, since $\bar{F}_{ij}$ is mapped onto it via

$$|A_0|^2 f_{ij} = \sum_{k,l} T_{ik} \bar{F}_{kl} T_{lj}^{-1} \tag{8.46}$$

where $T_{ij} = \delta_{ij} e^{i\phi_j}$ is a unitary matrix. Hence, it is sufficient to look at the eigenvalues of the matrix

$$\tilde{F}_{ij} = -\frac{g_{ii}}{\sum_j g_{ij}} - 2|A_0|^2 f_{ij} \tag{8.47}$$

$$= -\frac{1}{\sum_j g_{ij}} (1 + 2 f_{ij}) \tag{8.48}$$

The eigenvalues of this matrix are negative if the smallest eigenvalue of $f_{ij}$ is larger than $-1/2$. This condition is fulfilled for the range of parameter values $0 < g_1 < 1$, within which plane waves are unstable.

Next let us discuss the intrinsic and extrinsic stability of an essentially complex planform. When the minimum eigenvalue of the $n \times n$ matrix of $g_{ij}$ is called $\omega_{min}^n$, then

$$\omega_{min}^n > 0 \tag{8.49}$$

is necessary and sufficient for the intrinsic stability of a essentially complex planform with n active modes. The planform is extrinsically stable if the following condition

$$g(0)/2 - g(\alpha - \alpha_0) + \sum_{j=1}^{N-1} \left(g(\alpha_j) - g(\alpha - \alpha_j)\right) < 0 . \tag{8.50}$$

is fulfilled for all angles $0 \leq \alpha < 2\pi$ (see Section 7) where $\alpha_j = j\pi/n$ for all angles $0 \leq \alpha < 2\pi$. The function $\sum_j g(\alpha - \alpha_j)$ is periodic with respect to $\alpha$ with a period of $\pi/n$. Hence it is sufficient to examine it for $\alpha$ in the interval $[0, \pi/n]$. In this interval it has a single minimum at $\alpha = \pi/(2n)$. The condition for extrinsic stability is thus reduced to

$$g(0)/2 - g\left(\frac{\pi}{2n}\right) + \sum_{j=1}^{N-1} \left(g\left(\frac{j\pi}{n}\right) - g\left(\frac{\pi}{2n}(1 - 2j)\right)\right) < 0 . \tag{8.51}$$

The different essentially complex planforms with a given number of active modes are also degenerate with respect to their potential. The potential of such a planform is given by

$$U_n = -\frac{1}{4}\left(\frac{\sum_j g_{ij}}{n}\right)^{-1} \tag{8.52}$$

$$= -\frac{1}{4}\left(\frac{\sum_j g(j\pi/n) - g(0)/2}{n}\right)^{-1} \tag{8.53}$$

For large $n$, the potential $U_n$ increases algebraically as a function of $n$ as

$$U_n \approx \frac{1}{4}\left(\frac{g(0)}{2n} - \int_0^\pi d\alpha\, g(\alpha)\right)^{-1} \tag{8.54}$$

Because $U_n$ is finite, this implies that for fixed parameter values the minimum potential is obtained for a finite number of active modes $n_{min}$.

An understanding of the behavior of the model Equation (8.33) an be obtained with the help of the potential Equation (8.53) and the conditions for intrinsic and extrinsic stability of the planforms: (1) For a given number of active modes $n$, the potential $U_n = U_n(\sigma, g_1)$ defines a parameter domain in which the planforms with $n$ active modes have minimum potential. (2) For a given $n$, the conditions for intrinsic and extrinsic stability (8.49) and (8.51) define two stability boundaries $g_n^{*i}(\sigma k_0)$ and $g_n^{*e}(\sigma k_0)$.

$$0 = \omega_{min}^n\big|_{\sigma, g_1 = g_n^{*i}(\sigma k_0)} \tag{8.55}$$

$$0 = g(0)/2 - g\left(\frac{\pi}{2n}\right)$$

$$+ \sum_{j=1}^{N-1} \left(g\left(\frac{j\pi}{n}\right) - g\left(\frac{\pi}{2n}(1 - 2j)\right)\right)\Bigg|_{\sigma, g_1 = g_n^{*i}(\sigma k_0)} , \tag{8.56}$$

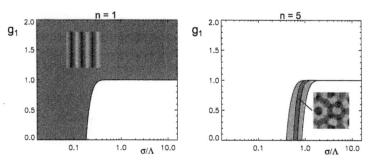

Fig. 9. The stability ranges of two solutions of the model (Equation 8.33): The graph on the left shows the stability range of plane waves ($n = 1$) as a function of the coupling parameter $g_1$ and the nonlinear interaction range $\sigma$. Plane waves are stable and have minimal energy within the shaded area (left). The graph on the right shows the stability range of quasi-periodic solutions with $n = 5$ active modes. These solutions are stable within entire shaded area. In the darkly shaded area, they have minimal energy.

which bound the parameter range in which a planform with $n$ active modes is stable.

These parameter ranges are shown in Figure 9 for $n = 1$ and $n = 5$. For plane wave solutions ($n = 1$), the domain in which this planform has minimum energy coincides with the domain of stability of the planform. This is not the case for $n > 1$. As shown on the right in Figure 9, a quasi-periodic planform with $n = 5$ active modes has a larger stability range than the parameter range in which it has a minimum potential (shaded dark). Since plane waves and other planforms cannot be stable simultaneously the stability range for large $\sigma$ the border of the stability domain asymptotes towards the line $g_1 = 1$. The lower (left) boundary of the stability range is formed by the boundary of stability to intrinsic perturbations. At the upper (right) boundary of the stability range, the planforms become extrinsically unstable, i.e., they decay to solutions with a number of active modes larger than n. With increasing $\sigma$, both stability boundaries converge to the line $g_1 = 1$.

Figure 9 shows that besides the degenerate stability of various planforms with the same number of active modes, there is a second origin of multistability in the model. At the edge of the stability range, the $n = 5$ planforms are still stable, whereas other planforms ($n = 4$ to the right, $n = 6$ to the left) have minimum potential and are thus also stable. The model thus also exhibits multistability of planforms with different numbers of active modes.

Let us examine the (non-overlapping) regions of parameter space in which planforms with a given number of active modes have minimal potential. A phase

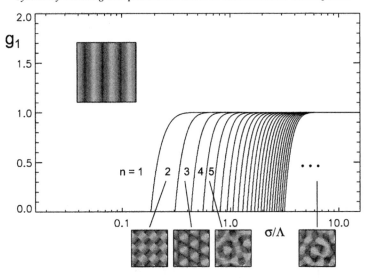

Fig. 10. Phase diagram of the model Equation (8.33): The graph shows the ranges in which patterns with $n$ active modes have minimum energy. For $g_1 > 1$ and for very small values of $\sigma/\Lambda \leq 0.18$, plane waves ($n = 1$) are the only stable solution (cf. Fig.9). For $g_1 < 1$, the number of active modes of stable solutions increases with increasing $\sigma/\Lambda$.

diagram that shows these areas is shown in Figure 10. For small values of $\sigma/\Lambda$ and $g_1 > 1$, only plane waves are stable planforms of the model and thus minimize the potential. For $\sigma/\Lambda > 0.2$ and $g_1 < 1$, planforms with an increasing number of active modes increasing with $\sigma$ minimize the potential. Since for $g_1 < 1$, the number of constitutively different planforms increases exponentially with increasing $n$, the number of different stable planforms also increases with increasing $\sigma$. Figure 10 suggests that the number of active modes increases without limit with increasing $\sigma/\Lambda$. Since the minimum distance between active modes in a stable planform decreases proportional to $(\sigma k_0)^{-1}$ in the model examined here, it is expected that the number of active modes $n$ increases with increasing $n \propto \sigma k_0 \propto \sigma/\Lambda$.

This supposition can be confirmed by calculating the stability boundaries for $n \gg 1$. Let the stability boundaries of a planform be given by $g_n^{*i}(\sigma k_0)$ and $g_n^{*e}(\sigma k_0)$ and assume that, except for a scale transformation, the shape of the stability ranges, $g_n^{*e}(\sigma k_0) < g_1 < g_n^{*i}(\sigma k_0)$, are similar, then it is expected that the stability boundaries for $n \gg 1$ show the following scaling behavior:

$$g_n^{*i/e}(\sigma k_0) = g^{*i/e}\left(\frac{\sigma k_0}{n}\right) \qquad (8.57)$$

I.e. that there are functions $g^{*i/e}\left(\frac{\sigma k_0}{n}\right)$ that describe the asymptotic form of the stability range.

The following calculation of the stability borders confirms this scaling behavior. Let us first examine the condition for extrinsic stability of a planform with $n$ active modes Equation (8.51). For $\sigma k_0/n > 1$, the left hand side of the inequality (8.51) is given approximately by

$$g(0)/2 + g(\pi/n) - 2g(\pi/(2n)) \tag{8.58}$$

because in this case,

$$\sum_{j=2}^{n-1}\left(g\left(\frac{j\pi}{n}\right) - g\left(\frac{\pi}{2n}(1-2j)\right)\right) \simeq \sum_{j=2}^{n-1}(g_1 - g_1) = 0. \tag{8.59}$$

The condition for extrinsic stability of such a planform is therefore given by

$$1 - g_1 + (2 - g_1)\left(e^{-\frac{\pi^2}{2}\left(\frac{\sigma k_0}{n}\right)^2} - 2e^{\frac{\pi^2}{8}\left(\frac{\sigma k_0}{n}\right)^2}\right) < 0 \tag{8.60}$$

Since for $\sigma k_0/n > 1$ the inequality

$$1 + e^{-\frac{\pi^2}{2}\left(\frac{\sigma k_0}{n}\right)^2} - 2e^{\frac{\pi^2}{8}\left(\frac{\sigma k_0}{n}\right)^2} > 0 \tag{8.61}$$

holds the stability condition inequality (8.60) can be rearranged to obtain the expression

$$g_1 > g_n^{*e}(\sigma k_0) = 1 - \frac{2e^{\frac{\pi^2}{8}\left(\frac{\sigma k_0}{n}\right)^2} - e^{-\frac{\pi^2}{2}\left(\frac{\sigma k_0}{n}\right)^2}}{1 + e^{-\frac{\pi^2}{2}\left(\frac{\sigma k_0}{n}\right)^2} - 2e^{\frac{\pi^2}{8}\left(\frac{\sigma k_0}{n}\right)^2}} \tag{8.62}$$

for the stability boundary. The boundary for extrinsic stability depends, therefore, only on $\sigma k_0/n$, and shows the expected scaling behavior.

This scaling behavior can also be verified for the intrinsic stability boundary. To show this, let us first calculate the eigenvalues $\omega_l^n$ of the matrix of $g_{ij}$ and then examine the dependence of the smallest eigenvalue on $g_1$ and $\sigma k_0$. For $\sigma k_0 \gg \pi^{-1}$, the matrix elements $g_{0j}$ are given approximately by

$$g_{0j} \simeq -\delta_{j0} + g_1$$
$$+(2 - g_1)\left(e^{-(\sigma k_0)^2\left(1-\cos\left(\frac{\pi j}{n}\right)\right)} + e^{-(\sigma k_0)^2\left(1+\cos\left(\frac{\pi j}{n}\right)\right)}\right) \tag{8.63}$$
$$\simeq -\delta_{j0} + g_1 + (2 - g_1)\left(e^{-\frac{(\sigma k_0)^2}{2}\left(\frac{\pi}{n}j\right)^2} + e^{-\frac{(\sigma k_0)^2}{2}\left(\frac{\pi}{n}(j-n)\right)^2}\right) \tag{8.64}$$

The eigenvalue $\omega_l^n$ are

$$\omega_l^n = \sum_{j=0}^{n-1} \cos\left(\frac{2\pi}{n} jl\right) g_{0j} \tag{8.65}$$

$$= -1 + n g_1 \delta_{l0} + (2 - g_1)\Omega_l^n \tag{8.66}$$

with

$$\Omega_l^n = \sum_{j=0}^{n-1} \cos\left(\frac{2\pi}{n} jl\right) \left(e^{-\frac{(\sigma k_0)^2}{2}\left(\frac{\pi}{n} j\right)^2} + e^{-\frac{(\sigma k_0)^2}{2}\left(\frac{\pi}{n}(j-n)\right)^2}\right). \tag{8.67}$$

where $\Omega_l^n$ can be approximated as

$$\Omega_l^n \simeq \sum_{j=-(n-1)}^{n-1} \cos\left(\frac{2\pi}{n} jl\right) e^{-\frac{(\sigma k_0)^2}{2}\left(\frac{\pi}{n} j\right)^2} \tag{8.68}$$

$$\simeq \sum_{j=-(n-1)}^{n-1} e^{i\frac{2\pi}{n} jl} e^{-\frac{(\sigma k_0)^2}{2}\left(\frac{\pi}{n} j\right)^2} \tag{8.69}$$

$$\simeq \sum_{j=-\infty}^{\infty} e^{i\frac{2\pi}{n} jl} e^{-\frac{(\sigma k_0)^2}{2}\left(\frac{\pi}{n} j\right)^2}. \tag{8.70}$$

This can be evaluated using the identity

$$\frac{n}{\sqrt{2\pi}} \sum_{j=-\infty}^{\infty} h(l - jn) = \sum_{j=-\infty}^{\infty} e^{i\frac{2\pi}{n} jl} \tilde{h}\left(\frac{2\pi}{n} j\right) \tag{8.71}$$

where $\tilde{h}(s)$ is the Fourier transform of

$$h(l) = \frac{1}{\sqrt{2\pi}} \int ds\, e^{isl}\, \tilde{h}(s) \tag{8.72}$$

Equation (8.71) is an elementary property of a function represented by a periodic superposition of one basic element $h(l)$. Choosing

$$\tilde{h}(s) = \frac{1}{2} e^{-\frac{(\sigma k_0)^2}{2}\left(\frac{\pi}{n} j\right)^2}, \tag{8.73}$$

i.e.,

$$h(l) = \frac{4}{\sigma k_0} e^{-\frac{2l^2}{(\sigma k_0)^2}} \tag{8.74}$$

the expression

$$\Omega_l^n \simeq \frac{n}{\sqrt{2\pi}} \frac{2}{\sigma k_0} \sum_{j=-\infty}^{\infty} e^{-\frac{2(l-jn)^2}{(\sigma k_0)^2}} . \tag{8.75}$$

is obtained. The minimum of $\Omega_l^n$,

$$\Omega_{n/2}^n \simeq \frac{n}{\sqrt{2\pi}} \frac{2}{\sigma k_0} \sum_{j=-\infty}^{\infty} e^{-2\left(\frac{n}{\sigma k_0}\right)^2 (j-1/2)^2} \tag{8.76}$$

is obtained for $l = n/2$.

Under the given conditions, the minimum eigenvalue $\omega_{min}^n$ is given by

$$\omega_{min}^n = -1 + (2 - g_1)\Omega_{n/2}^n \tag{8.77}$$

and because $\Omega_{n/2}^n = \Omega_{n/2}^n(\frac{\sigma k_0}{n})$ it is dependent only on $\frac{\sigma k_0}{n}$. The condition for intrinsic stability

$$\omega_{min}^n > 0 \tag{8.78}$$

leads, therefore, to

$$g_1 < g_n^{*i}(\sigma k_0) = 2 - \Omega_{n/2}^n \left(\frac{\sigma k_0}{n}\right)^{-1} . \tag{8.79}$$

This relation confirms that the boundary of intrinsic stability also has the expected scaling behavior.

The good quality of the approximations of Equations (8.62) and (8.79) is illustrated in Figure 11. The stability ranges determined approximately above and the exact, numerically calculated stability boundaries are shown next to each other in this figure. For $n > 2$, the actual stability boundaries rapidly converge to the asymptotic forms of Equations (8.62) and (8.79). The above results also imply that the number of active modes is approximated by

$$n \approx \sigma k_0 \tag{8.80}$$

Summarizing, it can be said that pinwheels in the model constructed here can be interpreted as elements of stable patterns. The structure of the attractors of the model in the parameter range $g_1 < 1$ is governed mainly by the range of nonlinear interactions $\sigma$. In this parameter range the number of active modes of the stable planforms increases without limit with increasing $\sigma$: $n \approx \sigma k_0$. Within this parameter range, in which pinwheels are stable elements of the pattern, the dynamics exhibits a distinct multistability, and different planforms with the same number of active modes and even planforms with different numbers of active modes are simultaneously stable.

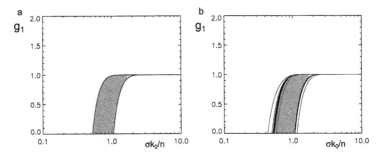

Fig. 11. Asymptotic form of the stability boundaries of the essentially complex planforms: The form of the stability range analytically obtained for $n \gg 1$ is shown on the left. The range of nonlinear interactions $\sigma$ is normalized by $n$. The numerically calculated boundaries of the exact stability ranges are shown on the right for $n = 2, 3, ..., 20$ overlaid on the analytically calculated range (shaded).

### 8.4. The spectrum of pinwheel densities

In the previous section it was shown that in the model described by Equation (8.26), pinwheels can occur as elements of a stable pattern. If the existence of such solutions is to explain the persistence of pinwheels in the visual cortex, then the density of the pinwheels in such solutions must be compatible with the experimentally observed densities. It was also shown that the stability of pinwheels is associated with a distinct multistability of various solutions of the model Equation (8.26). Hence, it is also of theoretical interest to quantitatively characterize the differences between simultaneously stable patterns. The following discussion will show that the simultaneously stable planforms differ especially with respect to pinwheel density. The multistability of the model (8.26) leads to a range of stable pinwheel densities that for $\sigma/\Lambda > 1$ is in good agreement with the experimentally observed pinwheel densities. In addition, the constitutively different planforms can be ordered according to their pinwheel densities.

Each essentially complex planform

$$z(\mathbf{x}) = \sum_j A_j e^{il_j \mathbf{k}_j \mathbf{x}} \qquad (8.81)$$

where $|A_j| = |A_0|$, defines a manifold of possible patterns. This manifold is parameterized by the phases $\phi_j$ of the amplitudes $A_j = |A_j| e^{i\phi_j}$. If $n \leq 3$, then a change in the phases $\phi_j$ does not lead to different patterns. Instead, a change in the phases $\phi_j$ only leads to a spatial translation and a shift in orientation of the pattern. If $n > 3$, then $n - 3$ of the $n$ phase variable parameterize the multiplicity of the spatial organization of the different patterns than can be formed

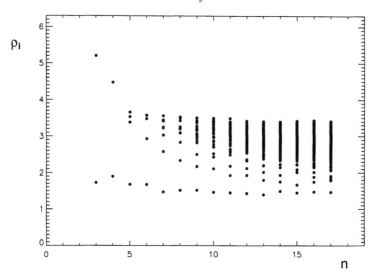

Fig. 12. Pinwheel densities (scaled, $\Lambda = 1$) of the constitutively different essentially complex planforms for $n = 3, 4, ..., 17$: In the case of $n = 3$, a periodic pattern is formed in which the pinwheel densities are exactly equal to $2\cos(\pi/6) \simeq 1.73$ and $6\cos(\pi/6) \simeq 5.2$. The points mark the numerically determined pinwheel densities $\rho_i$ of the constitutively different planforms for different numbers of active modes $n$. With increasing numbers of active modes the pinwheel densities occur within an interval of allowed pinwheel densities $1.4 < \rho < 3.5$.

by the planform. If it is assumed that the phases $\phi_j$ are selected at random, then Equation (8.81) defines a random ensemble of patterns. Therefore, a plausible way to assign a pinwheel density to a planform is to consider the mean value of the pinwheel density in this ensemble. Since it is expected that the phases of a pattern generated by spontaneous breaking of symmetry are really determined at random, the pinwheel densities defined in this way are also the densities that are expected to be observed experimentally.

Numerically, these pinwheel densities can be estimated with arbitrary precision. The pinwheel densities of all constitutively different planforms with $n \leq 17$ active modes were calculated to the second decimal place. The spectrum of pinwheel densities determined in this way is shown in Figure 12. The pinwheel densities of the different planforms differ considerably. If $n > 7$, all pinwheel densities lie within the interval [1.4, 3.6]. With increasing number of active modes, the pinwheel densities increasingly fill this interval. If $n \geq 14$, then the different planforms take values in the range of experimentally observed values [2, 3.7].

## 9. Discussion

Universality, the phenomenon that collective properties of very different systems exhibit identical quantitative laws, is of great importance for the mathematical modeling of complex systems. Originally, the phenomenon of universality gained widespread recognition when it was realized that the quantitative laws of phase transitions in physically widely different equilibrium thermodynamic systems were determined only by their dimensionalities and symmetries and were otherwise insensitive to the precise nature of physical interactions (see e.g. [57]). Subsequent research in nonlinear dynamics and statistical physics has uncovered that universal behavior extends far beyond equilibrium thermodynamics and is found for instance in pattern forming systems far from equilibrium (see e.g. [45]), in chaotic dynamics (see e.g. [58]), and in turbulence (see e.g. [59]).

It is for two reasons that universal behavior is particularly important for the mathematical modeling of complex systems such as the brain. First, in order to understand the universal properties of a system it is sufficient to study fairly simplified models as long as they are in the right universality class. Second, predictions for experiments that are derived from universal model properties are *critical*: Because universal properties are insensitive to changing microscopic interactions and numerical parameters or refining the level of detail in a model, verification or falsification of universal predictions can determine whether a certain modeling approach is appropriate or not. This is particularly important in theoretical neuroscience because for the neuronal networks of the brain a complete microscopic characterization of all interactions cannot be achieved experimentally and even if available would preclude comprehensive mathematical analysis.

In the present lectures we have made extensive use of the concept of universality. In the first sections we have discussed methods that allow to quantitatively analyze the generation of patterns of orientation columns and pinwheels in a largely model independent fashion. Subsequently, we have investigated the problem of pinwheel stability, starting from the general discussion of structure and properties of the systems of amplitude equations that will describe the emerging pattern in the vicinity of a supercritical finite wavelength bifurcation. As the theory of Gaussian random fields that we used in the analysis of pinwheel generation, the amplitude equations of the bifurcation problem aquire their structure and many of their properties from the symmetry assumption we made about the model and apply to any model with these symmetries. To answer the question of pinwheel stability, we finally constructed an analytically tractable model equation that exhibits stable pinwheel configurations in a transparent fashion and indicates that long-range interactions are essential for the stabilization of the pattern of orientation columns. I would like to conclude this lecture by discussing the reasons why also the detailed properties of this simple model equation are expected to be representative for a large class of models and are not critically dependent on the precise form of the model.

Let us briefly summarize the key properties of the constructed Swift-Hohenberg model. The model has two key ingredients: (1) Nonlocal interactions. (2) An inner symmetry that guarantees that the model system cannot become "blind" to a subset of orientations. Its stable planforms are selected from the system of stationary planforms constructed in Section 8.2. The planforms have different numbers, $n$, of active modes and different pinwheel densities, $\rho$. They represent periodic patterns for $n \leq 3$ and quasi-periodic for $n > 3$. When the model parameter $g_1 > 1$, all planforms with $n > 1$ are unstable. When the model parameter $g_1 < 1$, then all planforms exhibit a finite stability range. All planforms with the same number of active modes exhibit degenerate stability and potential. With increasing $n$, the range of stability shifts to larger interaction ranges $\sigma$. Thus, with increasing range $\sigma$, the number of active modes of stable solutions increases without limit. For a given $n$, the constitutively different planforms differ with respect to their pinwheel density. The number of different planforms increases exponentially with increasing $n$: for $n = 17$ there about 1000 different planforms. If $n > 5$, then the scaled pinwheel densities of these planforms lie within the interval $[1.5, 3.6]$ and increasingly fill this interval with increasing $n$. For large $n$, therefore, there is a range of stable pinwheel densities in the model that is in good agreement with the experimentally observed pinwheel densities.

It is obvious from the analysis that a similar behavior may be expected for every model in which the angle-dependent interaction function $g(\alpha)$ has a shape that is qualitatively similar to the model examined here, i.e., when it has maxima at $\alpha = 0$ and $\alpha = \pi$, so that $g(\pi) > g(0)/2$ is fulfilled and its minimum is less than $g(0)/2$, dependent on the control parameters. In such a model, as the width of the maxima at $\alpha = 0$ and $\alpha = \pi$ decreases, solutions with an increasing number of active modes are becoming stable. Also the width of the maxima $\Delta \alpha$ will be in general inversely proportional to some effective spatial range of nonlinear interactions $\sigma$ intrinsic to the dynamics: $\Delta \alpha \propto \sigma^{-1}$. This can be derived by dimensional analysis: The interaction function $g(\alpha)$ is defined by

$$
\begin{aligned}
g(\alpha) &= -e^{-i\mathbf{k}_0\mathbf{x}}\Big(N_3(e^{i\mathbf{k}_0\mathbf{x}}, e^{i\mathbf{h}(\alpha)\mathbf{x}}, e^{-i\mathbf{h}(\alpha)\mathbf{x}}) \\
&\quad + N_3(e^{i\mathbf{h}(\alpha)\mathbf{x}}, e^{i\mathbf{k}_0\mathbf{x}}, e^{-i\mathbf{h}(\alpha)\mathbf{x}})\Big) & (9.1) \\
&= \tilde{g}(\mathbf{k}_0, \mathbf{h}(\alpha) - \mathbf{k}_0) & (9.2)
\end{aligned}
$$

The width $\Delta \alpha$ of a maximum of $g(\alpha)$ thus corresponds to a characteristic scale $\Delta k$ of the nonlinear interactions in Fourier space. It is expected, therefore, that $\Delta \alpha \propto \Delta k \propto \sigma^{-1}$. In general, the number of active modes should thus increase with increasing interaction range $\sigma$.

The symmetry requirement $g(\alpha) = g(\pi - \alpha)$ is also not of critical importance for the observed qualitative behavior. We have seen, nevertheless, that this symmetry considerably facilitates the stability analysis and implies that potential and stability of the stationary planforms

$$z(\mathbf{x}) = \sum_j A_j e^{i l_j \mathbf{k}_j \mathbf{x}} \qquad (9.3)$$

are independent of the specific configuration of $l_j$. If the function $g(\alpha)$ deviates from the symmetry, then the values of the equilibrium amplitudes $A_j$ dependent on the specific choice of $l_j$. Moreover, these values would then, in general, no longer be identical. Also in this case, the stability of the planforms would no longer be degenerate. If the interaction function $g(\alpha)$, however, has a form that is qualitatively similar to that of the examined model, then it should be possible to analyze the stability of the planforms by decomposing the function

$$g(\alpha) = g_s(\alpha) + \epsilon g_a(\alpha) \qquad (9.4)$$

into a symmetric term $g_s(\alpha)$ and an asymmetric term $\epsilon g_a(\alpha)$. The stationary planforms and their stability can then be treated as a perturbation problem with the formal perturbation parameter $\epsilon$. Since the stability and form of the planforms depend continuously on $\epsilon$, it is obvious that a multitude of asymmetric interaction functions $g(\alpha)$, and thus models with qualitatively similar behavior, exist. Hence, their stable planforms can be mapped onto the system of planforms examined here and can be obtained from this system using a transform that is continuous with respect to $\epsilon$. These heuristic considerations lead to the conclusion that the observed multistability of the model is a robust property of the considered class of developmental dynamics.

What does the theory presented above teach us about visual cortical development? Most importantly, it demonstrates that the formation of orientation maps can be successfully modeled as a dynamic optimization process. In homogenous, spatially extended systems, relaxation towards the minima of an energy functional often implies the emergence of regular states exhibiting long range order. The seemingly disordered structure of orientation maps might thus be taken to indicate that frozen randomness rather than a functional optimization principle governs the layout of the maps. In fact, in many optimization models considered in the past pinwheel rich irregular states only occur transiently and decay with progressive relaxation [15, 31, 51–53]. The model class introduced above intrinsically possesses a multitude of spatially aperiodic solutions of realistic defect densities demonstrating that such states can be produced and sustained by optimization. The example analyzed in detail highlights that in models incorporating long-range interactions the emerging long-range order can indeed be of very subtle kind fully consistent with a seemingly irregular spatial structure.

Experimentally, it remains a challenging task to provide unequivocal experimental evidence for the occurrence of dynamic rearrangements of orientation columns in the developing or even in the adult brain (see however [60]). It is hoped that the transparence of the theoretical concepts and the analytical tractability of the models developed above will help to provide theoretical guidance for solving this key question of developmental neuroscience.

## References

[1] Braitenberg, V., Schüz, A., *Cortex: statistics and geometry of neuronal connectivity.* Springer, Berlin, 1998.

[2] Luskin, M.B., Shatz, C.J., *J. Comp. Neurol.*, 242 (1985) 611.

[3] Cragg, B.G., *J. Comp. Neurol.*, 160 (1975) 147.

[4] Daw, N.W., *Visual Development.* Plenum Press, New York, 1995.

[5] Stryker, M.P., Activity-dependent reorganization of afferents in the developing mammalian visual system. In Dominic. M. Lam and Carla J. Shatz, editors, *Development of the Visual System*, volume 3 of *Proceedings of the Retina Resarch Foundation Symposia*, chapter 16, pages 267–287. MIT Press, Cambridge, Mass, 1991.

[6] Singer, W., *Science*, 270 (1995) 758.

[7] Katz, L.C., Shatz, C.J., *Science*, 274 (1996) 1133.

[8] Maurer, D., Lewis, T.L., Brent, H.P., Levin, A.V., *Science*, 286 (1999) 108.

[9] Antonini, A., Stryker, M.P., *Science*, 260 (1993) 1819.

[10] Trachtenberg, J.T., Stryker, M.P., *J. Neurosci.*, 21 (2001) 3476.

[11] Crair, M.C., *Curr. Opin. Neurobiol.*, 9 (1999) 88.

[12] Kaschube, M., Wolf, F., Geisel, T., Löwel, S., *J Neurosci*, 22 (2002) 7206.

[13] Miller, K.D., Erwin, E., Kayser, A., *J Neurobiol.*, 41 (1999) 44.

[14] Miller, K.D., *J. Neurosci.*, 14 (1994) 409.

[15] Wolf, F., Geisel, T., *Nature*, 395 (1998) 73.

[16] Creutzfeldt, O.D., *Cortex Cerebri : performance, structural and functional organization of the cortex.* Oxford Univ. Press,, Oxford (UK), 1995.

[17] LeVay, S., Nelson, S.B., Columnar organization of the visual cortex. In *Vision and Visual Dysfunction*, chapter 11, pages 266–315. Macmillan, Houndsmill, 1991.

[18] Hubel, D.H., Wiesel, T.N., *J. Physiol.*, 160 (1962) 215.

[19] Swindale, N.V., Matsubara, J.A., Cynader, M.S., *J. Neursci.*, 7 (1987) 1414.

[20] Bonhoeffer, T., Grinvald, A., *Nature*, 353 (1991) 429.

[21] Chapman, B., Stryker, M.P., Bonhoeffer, T., *J. Neurosci.*, 16 (1996) 6443.

[22] Crair, M., Gillespie, D., Stryker, M., *Science* 279, (1998) 566.

[23] White L.E., Coppola D.M., Fitzpatrick D., *Nature*, 411 (2001) 1049.

[24] Sharma, J., Angelucci, A., Sur M., *Nature*, 404 (1998) 820.

[25] Swindale, N.V., *Proc. R. Soc. Lond.*, 208 (1980) 243.

[26] Swindale, N.V., *Proc. R. Soc. Lond.*, 215 (1982) 211.

[27] Swindale, N.V., *Network*, 7 (1996) 161.

[28] Durbin, R., Mitchinson, G., *Nature*, 343 (1990) 644.

[29] Obermayer, K., Blasdel, G.G., Schulten, K., *Phys. Rev. A*, 45 (1992) 7568.

[30] Erwin, E., Obermayer, K., K. Schulten, K., *Neural Comp.*, 7 (1995) 425.

[31] Wolf, F., Geisel, T., *Lecture notes in physics*, 527 (1999) 174.

[32] Blasdel, G.G., Salama, G., *Nature*, 321 (1986) 579.

[33] Swindale, N.V., Iso–orientation domains and their relationship with cytochrome oxidase patches. In D. Rose and V.G. Dobson, editors, *Models of the Visual Cortex*, chapter 47, pages 452–461. Wiley, New York, 1985.

[34] Albus, K., *Exp. Brain Res.*, 24 (1975) 181.

[35] Mermin, N.D., The topological theory of defects in ordered media. *Rev. Mod. Phys.*, 51 (1979) 591–648.

[36] Bonhoeffer, T., Kim, D.-S., Malonek, D., Shoham, D., Grinvald, A., *Europ. J. Neurosci.*, 7 (1995) 1973.

[37] Bartfeld, E., Grinvald, A., *Proc. Natl. Acad. Sci.*, 89 (1992) 11905.

[38] Blasdel, G.G., *J. Neurosci.*, 12 (1992) 3139.

[39] Blasdel, G.G., Livingstone, M., Hubel, D., *J. Neuroscience*, 12 (1993) 1500.

[40] Weliky, M., Katz, L.C., *J. Neurosci.*, 14 (1994) 7291.

[41] Rao, S.C., Toth, L.J., Sur, M., *J. Comp. Neurol.*, 387 (1997) 358.

[42] Bosking, W.H., Zhang, Y., Schofield, B.R., Fitzpatrick, D., *J. Neurosci.*, 17 (1997) 2112.

[43] Löwel, S., Schmidt, K.E., Kim, D.S., Wolf, F., Hoffsümmer, F., Singer, W., Bonhoeffer, T., *Eur. J. Neurosci.*, 10 (1998) 2629.

[44] Swindale, N.V., *Biol. Cybern.*, 66 (1992) 217.

[45] Cross, M.C., Hohenberg, P.C., *Rev. Mod. Phys.*, 65 (1993) 850.

[46] Geman, S., *SIAM J. Appl. Math.*, 36 (1979) 86.

[47] Goodhill, G.J., Willshaw, D., *Network*, 1 (1990) 41.

[48] Goodhill, G.J., Willshaw, D., *Neural Comp.*, 6 (1994) 615.

[49] Rényi, A., *Probability theory*. North-Holland Publishers, Amsterdam, 1970.

[50] Berry, M.V., and Dennis, M.R., *Proc. R. Soc. Lond. A*, 456 (2000) 2059.

[51] Koulakov, A.A., Chklovskii, D.B., *Neuron* 29 (2001) 519.

[52] Lee, H.Y., Yahyanejad, M., Kardar, M., *Proc. Natl. Acad. Sci. USA*. 100 (2003) 16036.

[53] Cho, M.W., Kim, S., *Phys Rev Lett.* 92 (2004) 018101.

[54] P. Manneville, *Dissipative Structures and Weak Turbulence* (Academic Press, San Diego, CA, 1990).

[55] F. H. Busse, Rep. Prog. Phys. **41** (1978) 1929.

[56] Swift, J. and Hohenberg, P., *Phys. Rev. A* 15, (1977) 319.

[57] Kadanoff, L.P., *Statistical Physics - Statics, Dynamics and Renormalization*. World Scientific, Singapore, 2000.

[58] Schuster, H.G., *Deterministic Chaos*. VCH, Weinheim, 3rd edition, 1995.

[59] Frisch, U., *Turbulence*. Cambridge University Press, Cambridge, UK, 1995.

[60] Godde, B., Leonhardt, R., Cords, S.M., Dinse, H.R., *Proc. Natl. Acad. Sci. USA* 99 (2002) 6352.

Course 13

# OF THE EVOLUTION OF THE BRAIN

Alessandro Treves[1,2] and Yasser Roudi[1]

[1] *SISSA, Cognitive Neuroscience sector, Trieste, Italy*
[2] *NTNU, Center for the Biology of Memory, Trondheim, Norway*

*C.C. Chow, B. Gutkin, D. Hansel, C. Meunier and J. Dalibard, eds.*
*Les Houches, Session LXXX, 2003*
*Methods and Models in Neurophysics*
*Méthodes et Modèles en Neurophysique*

# Contents

## 1. Introduction and summary

We review the common themes, the network models and the mathematical formalism underlying our recent studies about different stages in the evolution of the human brain. The first pair of studies both deal with radical changes in neuronal circuitry presumed to have occurred at the transition from early reptilians to mammals, introduced in sect. 2: the lamination of sensory cortex (sect. 4) and the differentiation into sub-fields of the mammalian hippocampus (sect. 5). In neither case the qualitative structural change seems to be accompanied by an equally dramatic functional change in the operation of those circuits. The last study, introduced in sect. 6, deals, instead, with the neuronal dynamics that might underlie the faculty for language in the human frontal lobes, a qualitatively new functional capacity that is not apparently associated with any new structural feature. These studies therefore all discuss the evolution of cortical networks in terms of their computations, quantified by simulating simplified formal models. All such models can be conceived as variants of a basic autoassociative neural network model, and their storage capacity, even when not formally analyzed, plays an important role in the results. We thus sketch, in sects. 3 and 7, the formalism that leads to storage capacity calculations, particularly in view of the fact that all three studies dwell on the interrelationship between qualitative and quantitative change, and all would benefit from more detailed mathematical analysis. Moreover, all studies include, as a necessary ingredient of the relevant computational mechanism, a simple feature of pyramidal cell biophysics: firing rate adaptation; a feature which to be treated properly requires extending the thermodynamics formalism into a full analysis of network dynamics. Overall, our approach is that while there is not necessarily a coupling between structural and functional phase transitions, understanding both at the mechanistic neural network level is a necessary step to understand the evolution of the organ of thought.

## 2. The phase transition that made us mammals

Mammals originate from the therapsids, one order among the first amniotes, or early reptiles, as they are commonly referred to. They are estimated to have radiated away from other early reptilian lineages, including the anapsids (the progenitors of modern turtles) and diapsids (out of which other modern reptilians, as well as birds, derive) some three hundred million years ago [25]. Perhaps mammals emerged as a fully differentiated class out of the third-to-last of the great extinctions, in the Triassic period. The changes in the organization of the nervous system, that mark the transition from proto-reptilian ancestors to early mammals, can be reconstructed only indirectly. Along with supporting arguments from the examination of endocasts (the inside of fossil skulls; [54]) and of presumed behavioural patterns [103], the main line of evidence is the comparative anatomy of present day species [32]. Among a variety of *quantitative* changes in the relative development of different structures, changes that have been extended, accelerated and diversified during the entire course of mammalian evolution [40], two major *qualitative* changes stand out in the forebrain, two new features that, once established, characterize the cortex of mammals as distinct from that of reptilians and birds. Both these changes involve the introduction of a new "input" layer of granule cells.

In the first case, it is the medial pallium (the medial part of the upper surface of each cerebral hemisphere, as it bulges out of the forebrain) that reorganizes into the modern-day mammalian hippocampus. The crucial step is the detachment of the most medial portion, that loses both its continuity with the rest of the cortex at the hippocampal sulcus, and its projections to dorso-lateral cortex [99]. The rest of the medial cortex becomes Ammon's horn, and retains the distinctly cortical pyramidal cells, while the detached cortex becomes the dentate gyrus, with its population of granule cells, that project now, as a sort of pre-processing stage, to the pyramidal cells of field CA3 [8]. In the second case, it is the dorsal pallium (the central part of the upper surface) that reorganizes internally, to become the cerebral neocortex. Aside from special cases, most mammalian neocortices display the characteristic isocortical pattern of lamination, or organization into distinct layers of cells (traditionally classified as 6, in some cases with sublayers). The crucial step, here, appears to be the emergence, particularly evident in primary sensory cortices, of a layer of non-pyramidal cells (called spiny stellate cells, or granules) inserted in between the pyramidal cells of the infragranular and supragranular layers. This is layer IV, where the main ascending inputs to cortex terminate [33].

## 2.1. An information-theoretical advantage in the hippocampus

What is the evolutionary advantage, for mammals, brought about by these changes? In the case of the hippocampus, attempts to account for its remarkable internal organization have been based, since the seminal paper by David Marr [70], on the computational analysis of the role of the hippocampus in memory. The hippocampus is important for spatial memory also in birds. A reasonable hypothesis is that the "invention" of the dentate gyrus enhances its capability, in mammals, to serve as a memory store. Following the approach outlined by David Marr, it was proposed 12 years ago [90] that the new input to CA3 pyramidal cells from the mossy fibers (the axons of the dentate granule cells) serves to create memory representations in CA3 richer in information content than they could have been otherwise. The crucial prediction of this proposal was that the inactivation of the mossy fiber synapses should impair the formation of new hippocampal dependent memories, but *not* the retrieval of previously stored ones. This prediction has recently been supported [61] at the behavioural level in mice, while neurophysiological experiments are in progress with rats. If validated, this hypothesis suggests that indeed a quantitative, information-theoretical advantage may have favored a qualitative change, such as the insertion of the dentate gyrus in the hippocampal circuitry. This possibility raises the issue of whether also the insertion of layer IV in the isocortex might be accounted for in quantitative, information-theoretical terms, an issue discussed in section 4. At the same time, the DG argument does not itself address the CA3-CA1 differentiation, which is equally prominent in the mammalian hippocampus. Section 5 will review a computational approach to this problem, and mention fresh experimental results that are shedding an entirely new light on it.

## 2.2. *An information-theoretical hypothesis about layers and maps*

It has long been hypothesized that isocortical lamination appeared together with fine topography in cortical sensory maps [6], pointing at a close relationship between the two phenomena. All of the cortex, which develops from the upper half of the neural tube of the embryo, has been proposed to have been, originally, sensory, with the motor cortex differentiating from the somatosensory portion [34, 65]. In early mammals, the main part of the cortex was devoted to the olfactory system, which is not topographic, and whose piriform cortex has never acquired isocortical lamination [45]. The rest of the cortex was largely allocated to the somatosensory, visual and auditory system, perhaps with just one topographic area, or map, each [32]. Each sensory map thus received its inputs directly from a corresponding portion of the thalamus, as opposed to the network of cortico-cortical connections which has been greatly expanded [2, 22] by the evolution of multiple, hierarchically organized cortical areas in each sensory system [56,59]. In the thalamus, a distinction has been drawn [55] between its matrix and core nuclei. The matrix, the originally prevalent system, projects diffusely to the upper cortical layers; while the core nuclei, which specialize and become dominant in more advanced species [38], project with topographic precision to layer IV, although their axons contact, there, also the dendrites of pyramidal cells whose somata lie in the upper and deep layers.

*A. Treves, Y. Roudi*

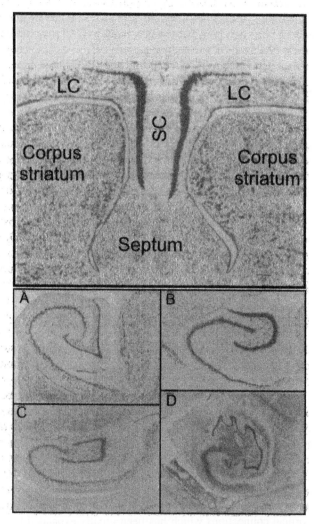

Fig. 1. The structural phase transition in the hippocampus. The medial pallium of a reptile (a lizard), top, with indicated the Large Cell (LC) and Small Cell (SC) subdivisions. Examples of the reorganized medial pallium in 4 highly divergent mammalian species, bottom: A – opossum; B – rat; C – cat; D – human. The homolog of the SC subdivision has become the detached dentate gyrus, which sends connections to the CA3 portion of the homolog of the LC subdivision, that has remained continuous with the rest of the cortex. Figures redrawn from those in P. Gloor, The Temporal Lobe and Limbic System (Oxford Univ. Press, New York, 1997).

The crucial aspect of fine topography in sensory cortices is the precise correspondence between the location of a cortical neuron and the location, on the array of sensory receptors, where a stimulus can best activate that neuron. Simple visual and somatosensory cortices thus comprise 2D maps of the retina and of the body surface, while auditory cortices map sound frequency in 1 dimension, and what is mapped in the other dimension is not quite clear [78]. Some of the parameters characterizing a stimulus, those reflected in the position of the receptors it activates, are therefore represented continuously on the cortical sheet. We define them as providing *positional* information. Other parameters, which contribute to identify the stimulus, are not explicitly mapped on the cortex. For example, the exact nature of a tactile stimulus at a fixed spot on the skin, whether it is punctuate or transient or vibrating, and to what extent, are reflected in the exact pattern of activated receptors, and of activated neurons in the cortex, but not directly in the position on the cortical sheet. We define these parameters as providing *identity* information. Advanced cortices, like the primary visual cortex of primates, include complications due to the attempt to map additional parameters on the sheet, like ocular dominance or orientation, in addition to position on the retina. This leads to the formation of so-called columns, or wrapped dimensions, and to the differentiation of layer IV in multiple sublayers. They should be regarded as specializations, which likely came much after the basic cortical lamination scheme had been laid out. The sensory cortices of early mammals therefore received from the thalamus, and had to analyse, information about sensory stimuli of two basic kinds: positional or *where* information, $I_p$, and identity or *what* information, $I_i$. These two kinds differ also in the extent to which cortex can contribute to the analysis of the stimulus. Positional information is already represented explicitly on the receptor array, and then in the thalamus, and each relay stage can only degrade it. At best, the cortex can try to maintain the spatial resolution with which the position of a stimulus is specified by the activation of thalamic neurons: if these code it inaccurately, there is no way the cortex can reconstruct it any better, because any other position would be just as plausible. The identity of a stimulus, however, may be coded inaccurately by the thalamus, with considerable noise, occlusion and variability, and the cortex can reconstruct it from such partial information. This is made possible by the storage of previous sensory events in terms of distributed efficacy modifications in synaptic systems, in particular on the recurrent collaterals connecting pyramidal cells in sensory cortex. Neural network models of autoassociative memories [53, 70] have demonstrated how simple "Hebbian" rules modelling associative synaptic plasticity can induce weight changes that lead to the formation of dynamical attractors [9]. Once an attractor has been formed, a partial cue corresponding *e.g.* to a noisy or occluded version of a stimulus can take the recurrent network within its basin of attraction, and hence lead to a pattern of activation of cortical neurons, which represents the stored identity of the original stimulus. Thus by exploiting dishomogeneities in the input statistics - some patterns of activity, those that have been stored, are more "plausible" than others - the cortex can reconstruct the identity of stimuli, over and beyond the partial information provided by the thalamus. This analysis of current sensory experience in the light of previous experience is hypothesized here to be the generic function of the cortex, which thus blends perception with memory [101]. Specialized to the olfactory sense, this function does not seem to require new cortical machinery to be carried out efficiently. A novel circuitry may instead be advantageous, when the generic function is specialized to topographic sensory systems, which have to relay both where and what information, $I_p$ and $I_i$. We take up the validation of this possibility after considering in the next section a fully defined model, which exemplifies the mathematical structures underlying our arguments.

## 3. Maps and patterns of threshold-linear units

The notion of autoassociative networks refers to a family of neuronal architectures, which in the simplest way can be thought of as one of the three main building blocks of cortical networks [82]. The two others are pattern associators and Competitive networks. By an autoassociative network we refer to a recurrent neuronal network with plastic connections. As briefly mentioned previously, associatively modifiable synapses, which might be modeled by a simple Hebbian plasticity mechanism, together with massive recurrent connections give a network of neurons the ability to function as a content addressable memory device.

In the past two decades, physicists have studied various models of autoassociative memory using different model neurons and different "learning rules" to implement Hebbian learning [12,53]. Mathematical methods have been adapted from statistical and spin glass physics for the purpose of analyzing these neuronal networks [7]. Although most of these investigation have been made on very abstract and simplified models, they have provided us with a good understanding of the general properties of associative memories, *e.g.* storage capacity and retrieval dynamics. Many of the methods borrowed from physics are based on the assumption of the existence of a Hamiltonian describing the dynamics of the system. The condition of the existence of a Hamiltonian imposes the important constraint of symmetric interactions on the network; this may be taken to be a good first approximation, but obviously it is not satisfied in the cortex. Actually, cortical networks belong to a subclass of asymmetrically wired networks, in which connections are not just asymmetric but, in addition, nearby neurons are more likely to make synapses with each other. This kind of more realistic models in terms of connectivity is what we want to briefly introduce in this section. To sketch the analytical treatment, we use an improved version of the self-consistent signal-to-noise analysis [85].

We thus introduce and analyze an autoassociative network which is comprised of threshold-linear units and includes a geometrical organization of neuronal connectivity, meant as a simplistic model of the type of organization of connections observed in the cortex.

### 3.1. A model with geometry in its connections

Consider a network of $N$ units. The level of activity of unit $i$ is a variable $v_i \geq 0$, which corresponds to the firing rate of the neuron. We assume that each unit receives $C$ inputs from the other units in the network. The specific covariance 'Hebbian' learning rule we consider prescribes that the synaptic weight between units $i$ and $j$ be given as:

$$J_{ij} = \frac{1}{Ca^2} \sum_{\mu=1}^{p} c_{ij} \left( \eta_i^{\mu} - a \right) \left( \eta_j^{\mu} - a \right),$$ (3.1)

where $\eta_i^{\mu}$ represents the activity of unit $i$ in pattern $\mu$, $c_{ij}$ is a binary variable and is equal to 1 if there is a connection running from neuron $j$ to the neuron $i$ and 0 otherwise. Each $\eta_i^{\mu}$ is taken to be a quenched variable, drawn independently from a distribution $p(\eta)$, with the constraints $\eta \geq 0$, $\langle \eta \rangle = \langle \eta^2 \rangle = a$, where $\langle \rangle$ stands for the average over the distribution of $\eta$. Here we concentrate on the binary coding scheme $p(\eta) = a\delta(\eta - 1) + (1 - a)\delta(\eta)$, but the calculation can be carried out for any probability distribution. As in one of the first extensions of the Hopfield model [98], we thus allow for the mean activity $a$ of the patterns to differ from the value $a = 1/2$ of the original model [87]. We further assume that the input (local field) to unit $i$ takes the form

$$h_i = \sum_{j \neq i} J_{ij} v_i + b \left( \frac{1}{N} \sum_j v_j \right),$$ (3.2)

where the first term enables the memories encoded in the weights to determine the dynamics; the second term is unrelated to the memory patterns, but is designed to regulate the activity of the network, so that at any moment in time $x \equiv \frac{1}{N} \sum_i v_i = a$. The activity of each unit is determined by its input through a threshold-linear function

$$v_i = F[v_i] = g(h_i - T_{thr})\Theta(h_i - T_{thr})$$ (3.3)

where $T_{thr}$ is a threshold below which the input elicits no output, $g$ is a gain parameter, and $\Theta(...)$ the Heaviside step function. The exact details of the updating rule are not specified further, here, because they do not affect the steady states of the dynamics, and we take "fast noise" levels to be vanishingly small, $T \to 0$. Discussions about the biological plausibility of this model for networks of pyramidal cells can be found in [11, 88], and will not be repeated here.

In order to analyze this network, we first define a set of order parameters $\{m_i^{\mu}\}$, with $\mu = 1 \ldots p; i = 1 \ldots N$, which we call the *local overlaps*, as follows:

$$m_i^{\mu} = \frac{1}{C} \sum_j c_{ij}(\eta_j^{\mu}/a - 1)v_j, \tag{3.4}$$

If we rewrite the local field $h_i$ defined above in terms of these order parameters we have:

$$h_i = \sum_{\mu} \left(\eta_i^{\mu}/a - 1\right) m_i^{\mu} - c_{ii}\alpha(1/a - 1)v_i + b(x) \tag{3.5}$$

in which $\alpha = p/C$ is the storage load. We will use this identity for the local field in the next section.

## 3.2. Retrieval states

A pattern $\mu$ is said to be retrieved if $\sum_i m_i^{\mu} = O(N)$. Without loss of generality, we suppose that the first pattern is the retrieved one and therefore $m_i^{\nu} \ll m_i^1$ for $\nu \neq 1$ and any $i$. When one pattern is retrieved, the local field to each unit can be decomposed into two terms. One is the *signal*, which is in the direction of keeping the network in a state with large overlap with the retrieved pattern. The second term, which we call *noise*, interferes with it. The idea is to calculate these terms as a function of the local overlap with the retrieved pattern. In other words we wish to express the r.h.s of 3.5 solely as a function of $m_i = m_i^1$ and $\eta_i^1$. If we are able to do so then we can calculate the activity of each unit as a function of $m_i$ and by using it in the definition of local overlaps, we will be able to find a self consistent equation for the local overlap with the first pattern.

To proceed further, we define two more order parameters $z_i^{\mu}$ and $\gamma$ through the equality below:

$$\sum_{\nu \neq 1, \mu} (\eta_i^{\nu}/a - 1)m_i^{\nu} = z_i^{\mu} + \gamma_i v_i \tag{3.6}$$

with this, we can write the activity of the network as:

$$
\begin{aligned}
v_i \;=\; & F[(\eta_i^1/a - 1)m_i^1 \\
& + (\eta_i^\mu/a - 1)m_i^\mu + z_i^\mu + \gamma_i v_i - c_{ii}\alpha(1/a - 1)v_i + b(x) - T_{thr}]
\end{aligned}
\tag{3.7}
$$

from which $v_i$ can be found self consistently:

$$
v_i = G[(\eta_i^1/a - 1)m_i^1 + (\eta_i^\mu/a - 1)m_i^\mu + z_i^\mu + b(x) - T_{thr}]
\tag{3.8}
$$

Assuming that $\Gamma_i = \gamma_i - c_{ii}\alpha(1/a - 1) < 1/g$, [1] the function $G[x]$ is of the following form for a threshold-linear unit:

$$
G[x] = \frac{g}{1 - g\Gamma} x\Theta(x)
\tag{3.9}
$$

now we expand the r.h.s. of the above equation for $v_i$ up to the linear term in $m_i^\mu$ and insert the result in 3.4, to get:

$$
m_i^\mu = L_i^\mu + \sum_j K_{ij}^\mu m_j^\mu
\tag{3.10}
$$

where:

$$
L_i^\mu \;=\; \frac{1}{C}\sum_j c_{ij}(\eta_j^\mu/a - 1)G[(\eta_j^1/a - 1)m_j^1 + z_j^\mu + b(x) - T_{thr}]
\tag{3.11}
$$

$$
K_{ij}^\mu \;=\; \frac{c_{ij}}{C}(\eta_j^\mu/a - 1)^2 G'[(\eta_j^1/a - 1)m_j^1 + z_j^\mu + b(x) - T_{thr}].
\tag{3.12}
$$

For the above equation, the solution for $m_i^\mu$ can be approximated as:

$$
m_i^\mu = \frac{1}{C}R_{ii}^\mu(\eta_i^\mu/a - 1)G^\mu[i] + \frac{1}{C}\sum_{j\neq i}R_{ij}^\mu(\eta_j^\mu/a - 1)G^\mu[j]
\tag{3.13}
$$

where $R_{ij}^\mu$ is defined as:

$$
R_{ij}^\mu = c_{ij} + \sum_l K_{il}^\mu c_{lj} + \sum_{lt} K_{il}^\mu K_{lt}^\mu c_{tj} + \ldots
\tag{3.14}
$$

in which we have used the notation $G^\mu[i] = G[(\eta_i^1/a - 1)m_i^1 + z_i^\mu + b(x) - T_{thr}]$.

---

[1] We shall see later that this is a correct assumption at least when one deals with diluted networks or very low storage loads.

Now that we have the local overlaps with the non-condensed patterns as a function of $m_i^1$, we can write the noise also as a function of it:

$$\sum_{v \neq \mu, 1} (\eta_{i/a}^v - 1) m_i^v = \frac{1}{C} \sum_{v \neq \mu, 1} R_{ii}^v (\eta_i^v / a - 1)^2 G^v[i] \tag{3.15}$$

$$+ \frac{1}{C} \sum_{j \neq i, v \neq \mu, 1} R_{ij}^v (\eta_i^v / a - 1)(\eta_j^v / a - 1) G^\mu[j]$$

for the first sum in the r.h.s of Eq.3.15 above, using the independence of different patterns and assuming that $z_i^\mu$ does not depend on $\eta_i^\mu$, one can write:

$$\frac{1}{C} \sum_{v \neq \mu, 1} R_{ii}^v (\eta_i^v / a - 1)^2 G^v[i] = \alpha \langle R_{ii}^v (\eta_i^v / a - 1)^2 G^v[i] \rangle \tag{3.16}$$

$$= \alpha \langle R_{ii}^v (\eta_i^v / a - 1)^2 \rangle v_i$$

and as a result of this we have:

$$\gamma_i = \alpha < R_{ii}^v (\eta_i^v / a - 1)^2 > = \alpha T_0 \langle R_{ii}^v \rangle, \quad T_0 \equiv 1/a - 1 \tag{3.17}$$

and therefore:

$$\Gamma_i = \frac{\alpha T_0^2}{C} \sum_j c_{ij} c_{ji} \langle G'[j] \rangle. \tag{3.18}$$

The second term is a bit tricky. For this term, by replacing the sum with the average we get zero mean, but for its deviation we have:

$$\rho^2 = \frac{\alpha}{C} (1/a - 1) \sum_j c_{ij} \langle R_{ij}^{\mu 2} (\eta_j^\mu / a - 1)^2 G^\mu[j]^2 \rangle \tag{3.19}$$

which is, actually, the standard deviation of the noise. Now we replace the second term, in the noise sum corresponding to $z_i^\mu$, with a Gaussian random variable with mean zero and standard deviation $\rho$, and take it into account in our fixed point equations by averaging the equations over this Gaussian measure.

Having done so, with some mathematical manipulations we derive the following fixed point equations ($\eta_i = \eta_i^2$):

$$\psi_{ij} = \frac{g T_0}{C} \sum_k c_{ik} c_{kj} \langle \int^+ Dz (1 - g\Gamma_k)^{-1} + \ldots \rangle$$

$$\Gamma_i = \alpha T_0 \psi_{ii} \tag{3.20}$$

$$\rho_i^2 = \frac{\alpha g^2 T_0^2}{C} \sum_j \left(c_{ij} + 2c_{ij}\psi_{ij} + \psi_{ij}^2\right) \times$$

$$\langle \int^+ Dz \left((\frac{\eta_j}{a} - 1)m_j + b(x) - T_{thr} - \rho_j z\right)^2 (1 - g\Gamma_j)^{-2} \rangle$$

where $Dz = dz\frac{e^{-z^2/2}}{\sqrt{2\pi}}$ and the superscript $+$ indicates that the integration has to be carried out in the range where $(\frac{\eta_i}{a} - 1)m_i + b(x) - T > \rho_i z$. Using the definitions of $m_i$ and $x$ we can get the following for their corresponding fixed point equations:

$$m_i = \frac{g}{C} \sum_j c_{ij}(\eta_j/a - 1) \times$$

$$\int^+ Dz \left((\frac{\eta_j}{a} - 1)m_j + b(x) - T_{thr} - \rho_j z\right)(1 - g\Gamma_j)^{-1} \quad (3.21)$$

$$x = \frac{g}{N} \sum_j \langle \int^+ Dz \left((\frac{\eta_j}{a} - 1)m_j + b(x) - T_{thr} - \rho_j z\right)(1 - g\Gamma_j)^{-1} \rangle.$$

### 3.3. The network without structure

Assume that the $c_{ij}$'s are randomly generated with probability $Pr\{c_{ij} = 1\} = C/N$. When $C/N \to 0$ the network is said to be in the highly diluted regime and the case of $C/N = 1$ corresponds to the fully connected network. Of course in these cases where the connectivity is randomly drawn from a non-geometric probability distribution, the order parameters become uniform in space and solutions have no spatial dependence. It can be shown that for the network without geometry the mean-field equations read:

$$\Omega = \frac{gT_0}{1 - g\Gamma} < \int^+ Dz >$$

$$\psi = \frac{C}{N}\left(\Omega + \Omega^2 + \Omega^3 + \ldots\right)$$

$$\Gamma = \alpha T_0 \psi$$

$$\rho^2 = \alpha \left(\frac{gT_0}{(1 - g\Gamma)}\right)^2 \left(1 + 2\psi + \frac{N}{C}\psi^2\right) \times \quad (3.22)$$

$$\langle \int^+ Dz \left((\frac{\eta}{a} - 1)m + b(x) - T_{thr} - \rho z\right)^2 \rangle$$

$$m = \frac{g}{1 - g\Gamma}\langle \int^+ Dz(\eta/a - 1)\left((\frac{\eta}{a} - 1)m + b(x) - T_{thr} - \rho z\right)\rangle$$

$$x = \frac{g}{1 - g\Gamma} \langle \int^{+} Dz \left( (\frac{\eta}{a} - 1)m + b(x) - T_{thr} - \rho z \right) \rangle.$$

It is worth noting that the contribution of the activity reverberating in the loops of the network is measured by the order parameter $\psi$. Also, $\Gamma$ essentially measures the effect of the activity of each unit on itself, after it has reverberated through the network. These order parameter disappear when $C/N = 0$, reflecting that when one considers a highly diluted network, the number of loops becomes negligible, and they do not contribute to network dynamics. This also makes the inequality $\Gamma_i < 1/g$ a valid assumption, as the renormalization of the gain and this effect becomes negligible when one deals with an extremely diluted network.

We can then define the new variables $r = m/\rho$ and $w = [b(x) - m - T_{thr}]/\rho$ and the following integrals, which are functions of $r$ and $w$, as in [87]:

$$A_2 = \frac{1}{rT_0} \langle (\frac{\eta}{a} - 1) \int^{+} Dz(w + \frac{r\eta}{a} - z) \rangle$$

$$A_1 = A_2 - \langle \int^{+} Dz \rangle \tag{3.23}$$

$$A_3 = \langle \int^{+} Dz(w + \frac{v\eta}{a} - z)^2 \rangle.$$

By using this notation the mean-field equations can be reduced to:

$$E_1(r, w) = A_2^2 - \left(1 + \frac{C}{N}\left(\frac{(2 - \Omega)\Omega}{(1 - \Omega)^2}\right)\right)\alpha A_3 = 0 \tag{3.24}$$

$$E_2(r, w) = (\frac{1}{gT_0} - \alpha\frac{C\Omega}{N(1 - \Omega)}) - A_2 = 0 \tag{3.25}$$

which extend and interpolate the results of [88] to finite values of $C/N$.

The first equation above appears as a closed curve in the $(w, r)$, plane, which shrinks in size when one increases $\alpha$ and then disappears; whereas the second equation is an almost straight curve, which for a certain range of $g$ intersects twice with the closed curve above. Since for a given value of $\alpha$ such that the first equation is satisfied, there always exists a value for $g$ that satisfies the second equation, the storage capacity is the value of $\alpha$ for which the closed curve shrinks to a point. We treat $g$ as a free parameter, because it can be easily changed in a network by mechanisms like multiplicative inhibition, if required in order to approach the optimal storage load.

In the limit of extreme dilution, i.e. $C/N \to 0$, $\Omega$ does not contribute to the equation for the storage capacity. The result of calculating the storage capacity as a function of the sparseness of the coding is shown in Fig.2 (the full curve). For other values of $C/N$ the contribution from $\Omega$ should be taken into account, which for small $C/N \neq 0$ results in deviations from the storage capacity of a highly diluted network. An example is illustrated in in Fig.2 for $C/N = 0.05$. It is clear that, at least for small $a$, a network with 5% connectivity can be considered as highly diluted, in the sense that for sparse patterns of activity, the effect of loops – what produces the difference between $A_2$ and $A_1$ – becomes unimportant.

An equivalent approach to study such a network is to use the replica method, from spin glass physics. If one considers a fully connected network with symmetric connections, then the dynamics can be described by a Hamiltonian. Using this Hamiltonian it is possible to calculate the partition function, and therefore the mean field equations, *e.g.* for the fully connected version of this model. Then one sends the order parameter corresponding to $\psi$ for the fully connected network to zero, to obtain the extremely diluted limit. This was basically the way the threshold linear network was first solved. One can look at [87, 89] for details of the calculations.

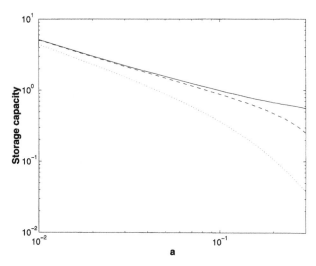

Fig. 2. Storage Capacity vs. $a$ for $C/N = 0$ (full curve), $C/N = 0.05$ (dashed line) and $C/N = 1$ (dotted line).

### 3.4. Appearance of bumps of activity

If we consider a network with a low connectivity level which is spatially organized, there can exist solutions of the fixed point equations that are spatially non-uniform. This is what one might call pattern formation. An interesting case, in one dimension, is a network with a Gaussian connectivity probability distribution:

$$Pr\{c_{ij} = 1\} = \frac{C}{\sqrt{2\pi\sigma^2}}e^{\frac{-(i-j)^2}{2\sigma^2}} + \text{Baseline.} \qquad (3.26)$$

The baseline is considered for $\sigma \propto N$. In this network it can be shown that there exists a critical $\sigma$ at which a second order phase transition occurs, to the appearance of spatially non-uniform solutions (more precisely, the first Fourier mode). Together with this appearance of non-uniform solutions, one can observe a sort of decrease in the storage capacity. Decreasing $\sigma$ further[2] results in the appearance of bumps of activity, *i.e.* fixed points of the dynamics that have large overlap with the stored pattern, and on the other hand are localized in space. An example of such bumps is shown in Fig.3. The dependence of the critical sigma and the properties of the bumps are beyond the scope of this paper and are being reported elsewhere [84], but what is important for us at this stage is the existence and stability of these spatially non-uniform retrieval states, which can be analyzed using the above formalism.

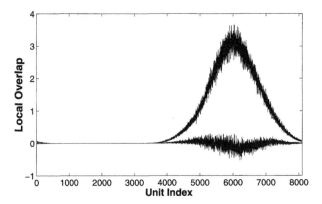

Fig. 3. The result of simulating a network of N=8100 units on a 1D ring, with C=405, p=5 and $\sigma$=300. The big bump is the local overlap with the retrieved pattern, and the curve fluctuating at low values is the overlap with one of the non retrieved patterns. Periodic boundary conditions were used.

---

[2]It should be noted that in the case of small $\sigma$ the approximations leading to disappearance of $\psi$ and $\Gamma$ from our equations are not applicable, since the loops become important again. One can use these equations in the case of small $\alpha$, where $\rho$ becomes zero and so does $\Gamma$, and the effects of the loops become unimportant. We will discuss this issue in more detail elsewhere.

## 3.5. The main points

Let us summarize the main results of the model discussed above, which are relevant to the forthcoming sections. The first point is the way the critical storage capacity scales with the relevant parameters of the model. As we stated before, this model shows that for a diluted network, which is close to a biologically plausible structure, we obtain a relation $p_c \propto C/a$. This expresses the computational advantage of sparse coding for memory storage, in structures like the hippocampus (see sect. 5). The second point is the appearance of bumps of retrieval activity, *i.e.* sustained activity localized both in physical space and in the space of stored pattern. This phenomenon, analysed in the more complicated situation of a partially recurrent and partially feedforward network comprised of multiple layers, and operating under the influence of sustained external input, is the basis of the results of [95], reported in the next section.

## 4. Validation of the lamination hypothesis

Does preserving accurate coding of position, in an isocortical patch, conflict with the analysis of stimulus identity? This is obviously a quantitative question, which has to be addressed with a suitable neural network model. An appropriate model can be designed with similar features as the one considered above, but with the additional option of differentiating multiple layers. In particular, the model of a cortical patch, receiving inputs from a thalamic array of units, can be investigated in its ability to generate localized retrieval states, that correspond to the stored patterns modulated by bumps, studied analytically in the section above. Unlike the analytical study, which is easier to conduct in a well defined limit case, e.g. looking at the existence of asymptotic attractor states after an afferent cue that has initialized activity in the network has been removed, with simulations one can also study the dynamics of localization and retrieval in time, with a cue that follows its own time course. Contrasting a model network with differentiated layers with one that has the same number of units and connections, but statistically equivalent layers, allows to approach the question of the role of lamination. The presence of several layers would anyway make an analytical treatment, while not impossible, very cumbersome, and computer simulations appear to be the method of choice. This is the approach taken in Ref. [95], the results of which are briefly summarized here.

A patch of cortex was modeled as a wafer of 3 arrays, each with $N \times N$ units. Each unit receives $C_{ff}$ feedforward connections from a further array of $N \times N$ "thalamic" units, and $C_{rc}$ recurrent connections from other units in the patch. Both sets of connections are assigned to each receiving unit at random, with a

Gaussian probability in register with the unit itself, and of width $S_{ff}$ and $S_{rc}$, respectively[3]. To model, initially, a *uniform*, non-laminated patch, the 3 arrays are identical in properties and connectivity, so the $C_{rc}$ recurrent connections each unit receives are drawn at random from all arrays. To model a laminated patch, later, different properties and connectivity are introduced among the arrays, but keeping the same number of units and connections, to provide for a correct comparison of performance. The 3 arrays will then model supragranular, granular and infragranular layers of the isocortex [95]. A local pattern of activation is applied to the thalamic units, fed forward to the cortical patch and circulated for $N_{iter}$ time steps along the recurrent connections, and then the activity of some of the units in the patch is read out. To separate out "what" and "where" information, the input activation is generated as the product of one of a set of $M$ predetermined global patterns, covering the entire $N \times N$ input array, by a local focus of activation, defined as a Gaussian tuning function of width $R$, centered at any one of the $N^2$ units. The network operates in successive *training* and *testing* phases. In a training phase, each of the possible $M \times N \times N$ activations is applied, in random sequence, to the input array; activity is circulated in the output arrays, and the resulting activation values are used to modify connections weights according to a model associative rule. In a testing phase, input activations are the product of a focus, as for training, by a *partial cue*, obtained by setting a fraction of the thalamic units at their activation in a pattern, and the rest at a random value, drawn from the same general distribution used to generate the patterns. The activity of a population of output units is then fed into a decoding algorithm - external to the cortical network - that attempts to predict the actual focus (its center, $p$) and, independently, the pattern $i$ used to derive the partial cue. $I_i$ is extracted from the frequency table $P(i, i_d)$ reporting how many times the cue belonged to pattern $i = 1, \ldots, M$ but was *decoded* as pattern $i_d$:

$$I_i = \sum_{i, i_d} P(i, i_d) \log_2 \frac{P(i, i_d)}{P(i) P(i_d)} \tag{4.1}$$

and a similar formula is used for $I_p$. The learning rule used to modify connection weights was

$$\Delta w_{ij} \propto r_j^{post} \cdot (r_i^{pre} - < r^{pre} >) \tag{4.2}$$

applied, at each presentation of each training phase, to weight $w_{ij}$. Weights are originally set at a constant value (normalized so that the total strength of afferents equals that of recurrent collaterals), to which is added a random component of similar but asymmetrical mean square amplitude, to generate an approximately

---

[3]Periodic boundary conditions are used, to limit finit size effects, so the patch is in fact a torus.

exponential distribution of initial weights onto each unit. $r$ denotes the firing rates of the pre- and postsynaptic units, and $< \ldots >$ an average over the corresponding array.

Among the several parameters that determine the performance of the network, $R << S_{rc}$ was fixed, while $S_{ff}$ was varied from $S_{ff} \simeq R$ up to $S_{ff} \simeq S_{rc}$. It is intuitive that if the feedforward connections are focused, $S_{ff} \simeq R$, "where" information can be substantially preserved, but the cortical patch is activated over a limited, almost point-like extent, and it may fail to use efficiently its recurrent collaterals to retrieve "what" information. If the other hand $S_{ff} \simeq S_{rc}$, the recurrent collaterals can better use their attractor dynamics, leading to higher $I_i$ values, but the spread of activity from thalamus to cortex means degrading $I_p$. This conflict between $I_p$ and $I_i$ is depicted in Fig. 4, which reports their joint values extracted from simulations, as they vary as a function of the spread of the afferents, at the end of the training phase (full curve). What is decoded is the activity of all units in the upper array of the patch. Since the patch is not differentiated, however, the other two arrays provide statistically identical information. Further, since information of both the what and where kinds is extracted from a number of units already well in the saturation regime [94], even decoding all units in all 3 arrays at the same time, or only, say, half of the units in any single array, does not alter the numbers significantly. $I_i$ is monotonically increasing with $S_{ff}$. $I_p$, instead, decreases with $S_{ff}$, and as a result one can vary $S_{ff}$ to select a compromise between what and where information, but *not optimise both* simultaneously. This conflict between what and where persists whatever the choice of all the other parameters of the network, although of course the exact position of the $I_p - I_i$ limiting boundary varies accordingly. Is it possible to go beyond such boundary?

### 4.1. Differentiation among isocortical layers

Several modifications of the "null hypothesis" uniform model were explored, as reported in [95]. Figure 4 illustrates, along with the results of the uniform model, results pertaining to slightly different versions of a 3-layer laminated model. Basically, the granular layer is differentiated by (i) focusing the thalamic afferents to the granular layer, while those to the two pyramidal layers are diffuse; (ii) restricting the recurrent collateral system of the granular units, by focusing the connections departing from granular units and decreasing the number of connections arriving at layer IV from the pyramidal layers; finally (iii) layer IV units follow a non-adaptive dynamics, and they do not operate during training, but only during testing. The non-adapting dynamics is effected, in the simulations, by making their effect on postsynaptic units, whatever their layer, scale up linearly with iteration cycle. Thus, compared to the model pyramidal units, whose firing rate would adapt over the first few interspike intervals, in reality (but is kept in constant ratio to the input activation, in the simulations), the firing rate of granule units, to model lack of adaptation, is taken to actually increase in time for a given input activation.

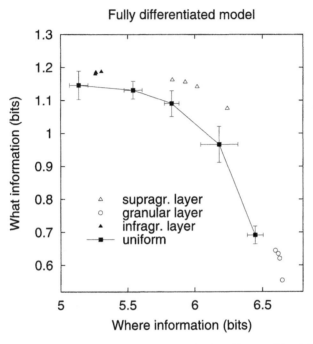

Fig. 4. $I_i$ and $I_p$ values obtained, after 3 training epochs, with the uniform model and, for 4 different parameter choices, for the fully differentiated model. 3 of the data points for the infragranular layer (black triangles) are nearly superimposed.

Differentiating infra- from supra-granular connections is effected by simply replacing the connections from layer IV to the infragranular pyramidal units with connections to the same units from supragranular units. In the real cortex, the supragranular layers project mainly onward, to the next stage of processing. The infragranular layers project mainly backward [14], or subcortically. Among their chief target structures are the very thalamic nuclei from which projections arise to layer IV. It is clear that having different preferential targets would in principle favour different mixes of what and where information. In particular, cortical units that project back to the thalamus would not need to repeat to the thalamus "where" a stimulus is, since this information is already coded, and more accurately, in the activity of thalamic units. They would rather report in its full glory the genuine contribution of cortical processing, that is, the retrieval of identity information. Units that project to further stages of cortical processing, on the other hand, should balance the "what" added value with the preservation of positional information. With this combination of modifications, layer III becomes the main source of recurrent collaterals [73, 104], which are spread out and synapse onto both supra- and infra-granular units and also, to a lesser degree, layer IV units.

The effect of the overall model of the differentiation can be appreciated by decoding the activity in the three layers, separately, as shown in Fig. 4 by the isolated symbols. From layer IV one can extract a large $I_p$ but limited $I_i$; from layer III one obtains a balanced mix. From layer V, on the other hand, one can extract predominantly "what" information, $I_i$, at the price of a rather reduced $I_p$ content. Thus, the last connectivity change, by effectively reducing the coupling between granular and infragranular layers, has made the latter optimize "what" information, while neglecting "where" information, of limited interest to their target structures. Can we understand the advantage brought about by lamination? The modifications required in the connectivity of layer IV are intuitive: they make granule units more focused in their activation, in register with the thalamic focus, while allowing the pyramidal units, that receive diffuse feedforward connections, to make full use of the recurrent collaterals. What is less intuitive is the requirement for non-adapting dynamics in the granule layer. It turns out that without this modification in the dynamics, the laminated network essentially averages linearly between the performances of uniform networks with focused and with diffuse connectivity, without improving at all on a case with, say, intermediate spread parameters for the connections. This is because the focusing of the activation and the retrieval of the correct identity interfere with each other, if carried out simultaneously, even if the main responsibility for each task is assigned to a different layer. Modifying the dynamics of the model granules, instead, enables the recurrent collaterals of the pyramidal layers to first better identify the attractor, i.e. the stored global pattern, to which the partial cue "belongs", and to start the dynamical convergence towards the bottom of the corresponding basin of attraction [7]. Only *later on*, once this process is – in most cases – safely underway, the granules make their focusing effect felt by the pyramidal units. The focusing action, by being effectively delayed after the critical choice of the attractor, interferes with it less – hence, the non-linear advantage of the laminated model.

## 5. What do we need DG and CA1 for?

If the synapses on the recurrent collaterals among pyramidal cells of the primitive cortex were endowed, as it is likely, with associative, "Hebbian", plasticity, such as that based on NMDA receptors [28], that cortex could have operated as an associative memory [22] – provided it had an effective way of distinguishing its operating modes. A generic problem with associative memories based on recurrent collaterals is to distinguish a storage mode from a retrieval mode. To be effective, recurrent collaterals should dominate the dynamics of the system when it is operating in retrieval mode; whereas while storing new information the dynamics should be primarily determined by afferent inputs, with limited interference from the memories already stored in the recurrent collaterals. The recurrent collaterals, instead, should modify their weights to store the new information [90]. In the model considered analytically in section 3, the learning phase is not explicitly considered. In the simulations of the laminated model, in section 4, the distinction is partially inserted by hand, by forcing the layer IV units to be silent during training.

## 5.1. Distinguishing storage from retrieval

The most phylogenetically primitive solution to achieve a similar effect is to use a modulator which acts differentially on the afferent inputs (originally, those arriving at the apical dendrites) and on the recurrent connections (predominantly lower on the dendritic tree). Acetylcholine (ACh) can achieve this effect, exploiting the orderly arrangement of pyramidal cells dendrites [47]. Acetylcholine is one of several very ancient neuromodulating systems, well conserved across vertebrates, and it is likely that it operated in this way already in the early reptilian cortex, throughout its subdivisions. In recent years, Mike Hasselmo has been emphasizing this role of ACh in memory, with a combination of slice work and neural network modeling [48,49]. This work has been focused on the hippocampus – originally, the medial wall – and on piriform cortex – originally, the lateral wall. In the hippocampus, however, it appears that mammals have devised a more refined trick to separate storage from retrieval, and perform both efficiently: operating the dentate gyrus preprocessor. It is illuminating, in fact, to contrast the avian and mammalian hippocampi. They are structurally very different, with birds having stayed close to their reptilian progenitors, and mammals having detached the dentate gyrus from Ammon's Horn, as mentioned above. Yet, at the behavioural level, the hippocampus of birds has been implicated in spatial memory in a role qualitatively similar to the prevailing description for the rodent hippocampus. Evidence comes from pigeons [18] and other species, and there is extensive literature to document it [26, 27].

Initially, the neural network approach, aiming at explaining structure from function, seemed to apply indiscriminately to hippocampal function in both birds and mammals, and therefore to be unable to say anything about the structural differences between the two. In his early paper, David Marr guessed the importance of recurrent collaterals, a prominent feature of the CA3 subfield [8], even though his own model was not really affected by the presence of such collaterals, as shown later [102]. Although the paper by Marr was nearly simultaneous with two of the most exciting experimental discoveries related to the hippocampus, that of place cells [75] and that of long term synaptic potentiation [19], for a long time it did not seem to inspire further theoretical analyses – with the exception of an interesting discussion of the collateral effect in a neural network model [44]. Marr himself become disillusioned with his youthful enthusiasm for unraveling brain circuits, and in his mature years took a much more sedate – and less neural – interest in vision. From 1987, however, McNaughton and Morris (1987) and then an increasing number of other investigators rediscovered the young Marr, and tried to elaborate those ideas in order to pin down the contribution of specific elements of the hippocampal circuitry. Edmund Rolls (1989) and several others have emphasized the crucial role probably played by the CA3 recurrent collaterals, that may form an autoassociator, a well studied network model of a content addressable memory. An autoassociator may subserve both the storage of episodic memories, e.g. in humans, and the storage of memory for space, e.g. in rats [15]). The emphasis on the essential role of the CA3 recurrent collaterals opened the way for attempting to understand the specialization of the dentate gyrus, in mammals [90]. A quantitative analysis of different network architectures (essentially, an autoassociator, CA3, operating with and without dentate gyrus to aid it in storing new memories) indicated an information theoretic advantage of one over the other in forming new representations. The models used were very abstract, and thus amenable to theoretical analysis [92] instead of just simulation, yet broadly consistent with generic cortical circuitry at all levels of details below the one being investigated. Conceptually, the function ascribed to the dentate is equivalent to the function ascribed to acetylcholine – to enhance the relation between hippocampal activity and afferent inputs during memory storage. The quantitative argument, however, allows a functional prediction at the neural level, which can be tested with suitable experiments. The prediction is that if mossy fibers are inactivated, the system is not able to acquire new hippocampal memories; or, more precisely, new memories rich in their information content. It should be able, nevertheless, to retrieve the memories already in store (and perhaps to form very impoverished representations of new memories). Somewhat surprisingly, the prediction has already been borne out of a purely behavioral experiment, in which mice were tested in a Morris water maze while transmission of dentate granule cell action potentials was reversibly blocked [61]. Another behavioural experiment has provided converging evidence [63]. Physiological experiments using similar techniques, in conjunction with measures of the information content of neural patterns of activity, will allow for more stringent tests of the argument. The mammalian 'invention' of the dentate gyrus, an ingenuity which likely took a long time to evolve from the simpler early reptilian organization, may thus represent a quantitative not qualitative improvement: qualitatively, we had acetylcholine already; but we managed to further improve on that.

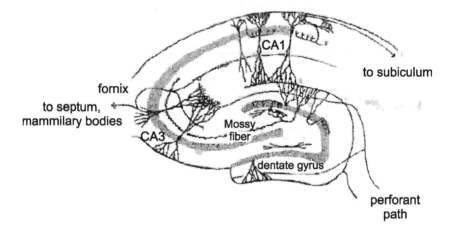

Fig. 5. Scheme of some of the main subfields and synaptic systems of the hippocampus proper. Redrawn from [82].

## 5.2. CA1 in search of a role

If DG can be understood as a CA3 preprocessor, perhaps CA1 should be understood as a CA3 postprocessor. Yet studies based solely on the notion of the usefulness of a further associative memory and recoding stage after CA3 (Treves, 1995) failed to illustrate impressive advantages to adding such a stage. More interesting hints come from neuropsychological studies in rats [58], that indicate a more salient role for CA1 along the temporal dimension. CA3 may specialize in associating information that was experienced strictly at the same time, whereas CA1 may link together, more than CA3, information across adjacent times. A way to formulate a qualitative implication of such a functional differentiation is to state that CA1 is important for *prediction*, i.e. for producing an output representation of what happened just after, at the time of storage, whatever is represented by the pattern of activity retrieved at the CA3 stage. Note, however, that reading the Kesner review in full indicates that the table at the end is a well-meaning simplification. Their Fig.31.2 suggests that CA3 may be involved in temporal pattern separation just as much as CA1. Moreover, the role of either DG or CA3 in temporal pattern association has never really been assessed. Further, available studies on the role of CA1 fail to make a clear distinction between tasks in which massive hippocampal outputs to the cortex are crucial, and tasks in which a more limited hippocampal influence on the cortex may be sufficient. In the first case, lesioning CA1 should have an effect independently of what CA1 specifically contributes to information processing, simply because one is severing the main hippocampo-cortical output pathway. In the second, CA3 outputs through the fimbria/fornix could enable hippocampal-mediated influences to be felt, deprived, though, of the specific CA1 contribution.

Structurally, CA3 and CA1 are contiguous portions of the dorsomedial cortex. When this reorganizes into the mammalian hippocampus, CA3 and CA1 differentiate in two important ways. First, only CA3 receives the projections from the dentate gyrus, the mossy fibers. Second, only CA3 is dominated by recurrent collaterals, while most of the inputs to CA1 cells are the projections from CA3, the Schaffer collaterals (Amaral *et al*, 1990). In [96] the hypothesis was explored that the differentiation between CA3 and CA1 may help solve precisely the computational conflict between pattern completion, or integrating current sensory information on the basis of memory, and prediction, or moving from one pattern to the next in a stored sequence. Neural network simulations, based on the same sort of model as those analyzed in section 3 and reviewed in section 4, were used to assess to what extent CA3 would take care of the former, while CA1 would concentrate on the latter. With the simulations, at the price of some necessary simplification, one can compare the performance of the differentiated circuit with a non-differentiated circuit of equal number and type of components (one in which CA3 and CA1 have identical properties, e.g. both receive mossy fibers and are interconnected with recurrent collaterals). Lesion studies, instead, can only compare the normal circuit with others with missing components, and it is thus difficult for them to say the last word on the meaning of a differentiation. The hypothesis was not really supported by neural network simulations. The conflict indeed exists, but the crucial parameter that regulates it appears to be simply the degree of firing frequency adaptation in pyramidal cells. The differentiation between the architectures of CA3 and CA1 has a minor effect on temporal prediction, while it does significantly increase the information content of hippocampal outputs.

After those simulations were completed, new experimental results from the labs of Edvard Moser [66] and James Knierim [62] have shed a completely new light on the significance of the CA3-CA1 differentiation. As explained in forthcoming papers [64, 67], activity in CA3 and CA1 differs remarkably when rats are asked to navigate in environments that some cues suggest are the same, and others indicate they are different. CA3 appears to take an all-or-none decision, usually allocating nearly orthogonal neural representations to even very similar environments, and switching to essentially identical representations only above a high threshold of physical similarity. Activity in CA1, instead, varies smoothly to reflect the degree of similarity. This functional differentiation, and the finding that new representations in CA3 emerge slowly, presumably through iterative processing, are entirely consistent with the recurrent character of the CA3 network, and the prevailing feedforward character of the CA1 network. Thanks to these experimental findings, therefore, we are beginning to finally 'understand' CA1, and to make complete sense of the events that drastically altered the structure of our medial pallium nearly 200 million years ago.


## 6. Infinite recursion and the origin of cognition

A cornerstone of the search for cortical network mechanisms of cognition is the observation, old but often neglected, that each small patch of neocortex is internally wired up in the same basic way. This holds across areas as well as across mammalian species, with relatively minor differentiations and specializations that do not alter the neocortical microcircuitry scheme [30, 35, 81] nor the basic overall cortical plan [59, 105]. It holds also in areas implicated in language functions in humans. This suggests that the local network mechanisms subserving the rich variety of cognitive functions are always essentially the same, and functional differentiation corresponds solely, to a first approximation, to differences in the long-range connections of different cortical areas. The local transaction, or elementary cortical network operation, is likely to be roughly the same everywhere [68], in sensory cortex as in association cortex.

Further, the long-range connections, denoted as the A system by Braitenberg [21] in contrast with the B system of local connections that do not leave the gray matter, follow indeed a specific wiring plan, which - when compared to simple mammalian species - is similar (although more complex) in elaborated species such as ours. However these connections do not seem to differ in other ways than in their overall wiring diagram: their layers of origin and termination, their synaptic mechanisms, their plasticity, their modulation by different neurotransmitters, all follow the same set of basic rules across cortical areas and across mammalian species.

One is led therefore to speculate that an understanding of the cortical operations underlying cognitive functions requires two main steps. First, the local network transaction has to be captured by a functional description, abstract enough to apply independently of areas and modalities yet accurate enough at the network level to be useful as a building block for system-level analyses. Second, global network operations have to be reduced to the combination of multiple instances of the universal local transaction, implemented along the wiring diagram relevant to each cognitive function.

## 6.1. Infinite recursion and its ambiguities

Are there clues about the nature of such global network operations that come from purely cognitive analyses? In a recent review, Marc Hauser, Noam Chomsky and Tecumseh Fitch [50] re-evaluate the requirements for the faculty of language. They state that language in the broad sense requires an adequate sensory-motor system and an adequate conceptual-intentional system, which however are both unlikely to be uniquely human attributes. They further propose that what may be uniquely human is a third necessary component of the faculty of language, that is a computational mechanism for recursion. Such a mechanism would provide the capacity to generate an infinite range of expressions from a finite set of elements. They also speculate that a capacity for infinite recursion may have evolved for reasons other than language, such as number processing, navigation, or social relations. In a related analysis, Daniele Amati (personal communication) wonders what could be a component that distinguishes uniquely human cognitive abilities, which he calls *H-abilities*, from the *1-H* abilities shared with other species; and he identifies this component with a capacity for producing arbitrarily long sequences that, at an abstract level, obey certain rules. This implies an ability to process cognitive states remote from those directly elicited by sensory inputs, and to generate such states recursively, *i.e.* a notion very close to the Chomskian one of infinite recursion in language, as manifested in a generative grammar. Thus a computational mechanism for infinite recursion may be utilized in other $H$-abilities, for example (as proposed by Amati) in the production of music (see [51], and other articles in the same issue).

Recursion, referred to the generation of infinite sequences of elements drawn from a finite alphabet, is an abstract and very loose notion. Computationally, at the most pedestrian level, it might simply mean that the transitions from one element to the next follow certain rules instead of being effectively random. A mathematical formulation of a grammar can be reduced in fact to the study of a system with certain forbidden transitions among its elements (see e.g. [72] and references therein). What are the elements, how they may be represented in the brain, and how restrictive are the rules they have to adhere to, remains to be clarified. Linguistically, and in other cognitive domains, recursion is often implied to mean something less pedestrian, like an embedding of clauses one inside the other in syntax, or the nesting of do-loops in Fortran codes. Recursion in this more sophisticated sense tends to be domain-specific, however, and is hardly ever infinite. An approach to explain infinite recursion mechanistically, in general, while taking into account domain-specific connotations is therefore likely to be ill-directed. More promising appears an almost opposite, nearly bottom-up approach, that considers the generic, pedestrian meaning of recursion, assumes the quality of its being infinite as critical, and focuses on the universal cortical transaction at the local network level.

## 6.2. Memory—statics and dynamics

In one of his ambitious and difficult papers discussing the organization of the brain, David Marr [69] proposed to regard the cerebral cortex in terms of its ability to decode the outside world using memory of previous experiences. In their 1991 book, Braitenberg and Schüz [22] summarized a series of insightful observations on the quantitative anatomy of the cortex, concluding that in general terms it operates as an associative memory machine. Over the last 15 years the interpretation of local cortical networks as attractor networks performing memory computations [9], which has informed our sect. 3, has diffused across the neuroscience community, leading to increased attention to the role of recurrent collateral processing even in early vision [80] and in slices [29, 86]. Memory computations take different forms [82], including self-organized recoding useful for categorization, pattern association (or directed memory in the terminology of Marr, [70]), and autoassociation or free memory. Common to all is the use of neurons as simple devices summating multiple inputs, of representations distributed over the activity of many neurons, and of associative plasticity mechanisms at their synaptic connections, as used in the earlier sections of this chapter.

With such minimal and neurally plausible ingredients, memory operations at the single neuron level can be depicted as simple analog operations on vectors of synaptic weights and on vectors of firing rates. These analog computations, widely accepted as the neural basis of memory in cortical networks, are seemingly far removed from the symbolic computations often subsumed as the logical basis of language and other higher cognitive faculties. Yet, apparent differences notwithstanding, analog computations at the single neuron level can implement symbolic computations at the local network level. The crucial element for this to occur is the discrete nature of local network attractors. The discreteness of local attractors can provide the error-correction capability and the robustness to noise that are often associated with the processing of discrete symbols.

In most simple models, local attractors are viewed as final states reached by a relaxation type of dynamics, that is, they coincide with the distribution of firing rates across neurons, that the local network would tend to reach in the absence of new perturbing inputs. This is however not necessarily the case. In his proposal of the notion of *synfire chains* [1], Moshe Abeles has envisioned inherently dynamical attractors, in which the identity of neurons firing in each attractor changes rapidly with time, along chains of links, each comprised of simultaneously firing neurons. The attractive nature of a chain expresses itself both in the convergence towards one of the sequences of links that are stored in memory, and in the progressive synchronization of the units comprising each link [17, 52].

Distinct sequences and distinct links within a sequence can share a number of participating units, and if this number does not exceed a value that can in principle be calculated, each chain continues to operate as a dynamical attractor. Further, distinct sequences can share the very same link or set of links, provided the activation of a link depends on previous links extending sufficiently into the past as to disambiguate each sequence. Although the original synfire chain model may be oversimplified, theoretically this notion has the merit of unleashing the computational capabilities of attractors, with their analog-to-symbolic transformation, into the temporal dimension. If provisions are made for the composition of individual chains, and for switch links with multiple possible outcomes, synfire chains can implement the structure of transition probabilities of a grammar. It has indeed been noted how synfire chains, or objects of a similar nature, could be at the basis of language [77]; interestingly, the concept derives from the experimental observation of neural activity recorded in the frontal cortex of monkeys [3]. In contrast, the *static* notion of fixed-point attractors finds its most salient experimental inspiration in data recorded in the temporal lobe [71].

In the temporal cortex, if a local static attractor may be taken to correspond to a feature represented over a limited patch of cortex, a global attractor extending over many patches may be taken to correspond to an item from semantic memory [31, 39]. In the past we have analyzed quantitatively simple multi-modular associative memory networks, to check whether they could serve as models of semantic memory. In line with the distinction between the A and B systems of connections among pyramidal cells [21], we considered models in which each module, including $N$ units, is densely connected through associatively modifiable weights (in fact each pair of units in the same module are pre- and post-synaptic to each other, so the number of local connections per unit $C_B$ equals $N - 1$) while different modules are sparsely connected (each unit receives $C_A$ connections, coming from other units widely distributed over $M$ modules). Anatomical evidence suggests that $C_B$ and $C_A$ are large numbers of similar magnitude, *e.g.* of order $10^4$ in primates [2]. $N$ determines the number of local attractor states, denoted here as $S$, which analytical studies show scales up with the number of local connections per units, *i.e.* is proportional to $C_B$.

In a first study of a multi-modular network, we concluded that the number $p$ of global attractor states cannot be much larger than $S$ for the system to retrieve each memory item correctly. Analytical results show that if $p$ is much larger than $S$, random combinations of local attractors, which do not correspond to any stored global pattern of activity, prevail as fixed points over the meaningful, stored combinations, which the $C_A$ long-range connections per unit try to enforce [76]. Thus a simple-minded multimodular network could not serve as an effective semantic memory, since it would be limited to storing a very low number of items, of the same order as that of local attractor states. In a subsequent study, we identified two modifications to the first model we had considered, which increase its storage capacity beyond such a limited value [42]. The first modification is a long-range connectivity that is not uniformly sparse across modules, but is concentrated between a module and a subset of other modules that strongly interact with it. The second modification is to consider global activity patterns, or semantic memory items, that are not defined across all modules, but only over a subset, different for each pattern, which tends to include strongly interacting modules. With these combined modifications, the storage capacity, as measured by $p$, can increase well beyond the local capacity $S$, although its exact value is difficult to calculate, and depends on the details of the model. The tentative conclusion of the second study, therefore, was that a viable model of semantic memory based on a collection of interacting local associative networks should include (a) nonuniformly distributed long-range connections and (b) activity patterns distributed over a sparse fraction of the modules [42].

### 6.3. Memory latching as a model of recursive dynamics

The analyses above refer to the operation of semantic memory retrieval, which has to be initiated by an input that conveys a partial cue. In the temporal cortex, the so-called stimulus specific delay activity, which is observed for up to a few seconds following the offset of the stimulus, is typically weak and disrupted by successively intervening stimuli [23]. In the frontal cortex, similar delay activity can instead be quite strong and persist in the face of intervening stimuli [41], reflecting the overall weaker influence that sensory inputs to the cortex have on frontal networks, compared to that on networks of the temporal lobe (as modeled in [79]). It becomes pertinent to ask, then, especially in the case of frontal cortex, what type of dynamics may follow semantic memory retrieval: what happens to a network comprised of multiple associative modules, once it has been activated by a cue and it has retrieved a given semantic memory. In the following, it is proposed that what can happen depends critically on the number of semantic memories stored, that is, on the number of global attractor states. While allowing for a special contribution of the frontal cortex to temporal integration, due to its position in the overall cortical plan [43], and while broadly compatible with the declarative/procedural model of Ullman [100], the proposal focuses on a network mechanism that is not restricted to frontal cortex, but that in human frontal cortex may have found a novel expression because of a purely quantitative feature: the abundance of its connections.

The hypothesis requires one additional ingredient, which however in the cortex comes for free, so to speak. This is a passive mechanism for moving a local network out of an attracting state, after some time. A combination of firing rate adaptation in pyramidal cells, short-term depression at excitatory synapses and slow rebound inhibition would produce such an effect, and in different proportions would tend exclusively to inactivate the local network or also to favour its transition to a different attractor state, or even to enable flip-flop switching between pairs of states, as in binocular rivalry [60]. Globally, under certain conditions the collection of modules will move continuously from global attractor to global attractor or, more precisely, it will hop from state to state, given the discrete nature of the attractor states. It may rapidly pass through intermediate states, but in a well behaving semantic system mixture states are unstable (see [83] for a simplified model) and the trajectory, in the absence of new inputs, will essentially include periods close to attracting states, which would be fixed points except for the adaptation/inhibition mechanism, and rapid transitions between them. The system *latches* between attractors.

We now focus on whether such transitions will continue to occur, one after the other in the absence of inputs, and, if they occur, on the degree to which they follow rules, or are effectively random. When relatively few global attractors exist, in the high-dimensional space in which they live, the attractors will tend to be orthogonal, or approximately equally distant from each other. This is a statistical tendency that follows simply from the high dimensionality of the space, without special assumptions. In such a regime, transitions will be nearly random, if they occur at all. This is because as the system moves out of the previous global attractor none of the other attractors will be strongly engaged to take over; small fluctuations in the instantaneous condition of the system may favour a particular hopping among many essentially equiprobable ones, or else selective activity in the system may simply die out. When more global patterns exist, they populate more densely their high dimensional space, and at some point each pattern will have a subset of other patterns that are closer to it, or more similar, than the rest. In such a regime transitions between states will tend to be structured, and the dynamics will appear to follow certain rules, *i.e.* a grammar. The critical density of global attractor states at which structured transitions begin to prevail depends markedly on how patterns are generated, and one has to make more concrete assumptions in order to proceed with more quantitative arguments. It is not fully clear at this stage whether the transition between the two regimes takes the sudden character of a phase transition, akin perhaps to a percolation transition [46]. In general, however, it should remain valid that such critical density does not depend on the long-range connectivity. The storage capacity for semantic memory, instead, does depend on the connectivity. The hypothesis, then, is that a connectivity increase may increase the storage capacity of a frontal semantic multi-modular network, until it can store enough patterns that, when left without inputs, it can follow structured dynamics, which express a sort of transition rules. This hypothesis can be formulated in more detail by considering a concrete model, amenable to computer simulations.

Before discussing the toy model, it is tempting to freely speculate on the relation, within this framework, between the universal grammar, posited to underlie all human languages, and the grammar constraining each particular language, characterized by its choice of parameters (see e.g. [13]). The universal grammar should reflect the associative nature of the semantic network, largely embodied in a time-independent matrix of similarities between global attractors, but also endowed with the restricted extent of time arrows characteristic of any action semantics [106]. Such time arrows, or directed associations in Marr's terms, can be realized by simple and biologically plausible mechanisms, e.g. by spike-timing dependent synaptic plasticity. The same mechanisms can operate, when learning a specific language, to resolve the residual temporal order ambiguities left by the fact that action semantics does not specify all the temporal relations necessary to produce (one-dimensional) speech. Thus, in this interpretation, language parameters are set (arbitrarily, from a formal point of view, that is according to one's mother tongue) when funneling the more loosely time-constrained action semantics into the strict order of sequential discourse. Also parameters that seemingly do not reflect simple temporal order, like the polysynthesis parameter, might be indirect by-products of such a funneling effort.

## 7. Reducing local networks to Potts units

Consider again a network comprised of $M$, modules each of which functions as an autoassociative network. Assume that each module stores $S$ patterns and that, together with the intra-modular connections, there are also connections running between units in different modules. The full analysis of such a system, when including in addition non-uniform connectivity like the one discussed in sect. 2, would be very hard; in order to proceed one should thus consider some simplified model. The first natural choice is to consider a network with all to all connectivity inside modules and dilute connectivity between any two of them. This was the model investigated by O'Kane and Treves [76]. The critical factor in the revised model considered by Fulvi Mari and Treves [42] is the existence of what a *null state* as a new attractor that a module can reach in addition to all the stored patterns in it. This null state differs form the normal attractors in the sense that if a module goes to its null-state, it would have no effect on the other modules. Basically this null-state is something like the zero activity state for a single neuron, generalized to the network level. The technical problem associated with this model is that even though it appears to have a larger storage capacity compared to the network without null state, a full analysis of storage capacity cannot be done analytically. To circumvent this problem one can first make a further drastic simplification, and consider a new reduced model based on Potts neural networks [20, 57]. Then one essentially neglects the internal structure inside each module and represents the state of each module by its correlation with the '0' (null) state or with one of the $S$ attractor states. At its simplest, this can be just one discrete variable, taking one of $S + 1$ values. Such a discrete variable simply indicates the closest stored pattern to the current state of the module. Then we model the interactions between two different modules, which in reality is the set of all weights associated to connections between them, with a $S(S + 1)/2$-dimensional weight vector.

### 7.1. A discrete-valued model

At any time we associate a Potts variable $s_i$ which takes one of the values $0 \ldots S$ to the $i^{th}$ module in the following way: $s_i$ takes the value $q, q \neq 0$ if and only if pattern $q$ is the closest pattern to the current activity of the network; and $s_i = 0$ if its closest pattern is the null state. Obviously being the closest one is nothing but having the largest overlap. Note that to facilitate comparisons with refs. [20, 57], one should convert to the notation $Q$ $(=S+1)$, for the number of Potts states. The interaction between modules $i$ and $j$ would be modeled through a set of weights $w_{ij}^{kl}, i, j = 1 \ldots M; k, l = 0 \ldots S$ symmetric in both $\{ij\}$ and $\{kl\}$. Now suppose that at time $t$ the configuration of the system is $\{s_i\}$. Then at time $t + 1$ we randomly choose one of the modules, say module $i$, and calculate a set of local fields $\{h_i^s\}, s = 0 \ldots S$ defined as:

$$h_i^s = \sum_{j=1, j \neq i}^{M} \sum_{k,l=0}^{S} w_{ij}^{kl} u_{s_i,k} u_{s_j,l} \tag{7.1}$$

where $u_{k,l} = (S + 1)\delta_{k,l} - 1$.

At time step $t + \Delta t$ the state variable $s_i$ is set equal to the value $s$ which maximizes $h_i^s$. The effect of Hebbian plasticity on the weights, which results in the formation of network attractors coinciding with, or near to the specified global patterns, can be described, for example, by the learning rule:

$$w_{ij}^{kl} = \frac{1}{(S + 1)^2 M} \sum_{\mu=1}^{p} u_{\xi_i^\mu, k} u_{\xi_j^\mu, l} (1 - \delta_{k0})(1 - \delta_{l0}) \tag{7.2}$$

in which $\xi_i^\mu$ is the local attractor in module $i$ which participates in the global pattern $\mu$. It is drawn from a uniform probability distribution, *i.e.* all local attractors are assumed equally likely to participate in a global pattern. With this weight matrix, global patterns defined by $\{\xi_i^\mu\}$ (or network states very close to them) become the global attractors of the network, provided their number does not exceed a critical value (when $M$ is large; in a small network the critical value is not well defined, as evident in the simulations below). Notice that we have considered the peculiar role of the null state in the dynamics of the network through the delta functions above. Also it should be noted that we have not yet considered whether the fraction of modules in the null state in each global memory pattern is the same or different as the fraction of modules in any other of the $S$ 'genuine' local attractors.

## 7.2. Storage capacity

In order to find the storage capacity of this network, we start by writing the Hamiltonian of the system. This is where we need the symmetry property of the weight matrix. If the weights are symmetric, as in 7.2, the dynamics of the network can be described by the following Hamiltonian:

$$H = -\frac{1}{2M} \sum_{i,j,j\neq i} \sum_{k,l} w_{ij}^{kl} u_{s_i,k} u_{s_j,l}. \tag{7.3}$$

One can then apply the classical methods of spin glasses to obtain the mean field equations of the system. The above formulation is basically nothing but a variation of the Potts-neural network first investigated by Kanter [57]. Kanter's model does not include the notion of the null state, and it treats all $S + 1$ local states in the same way. It also assumes full connectivity, so that the number of units providing input to any given unit, $C$, equals $M - 1$. For such a network Kanter found that the storage capacity for small values of $S$ scales like $MS(S+1)$. As noted by Kanter, this critical storage load scales up with the number $S$ of Potts states *squared* because, effectively, a connection weight between a pair of Potts units is comprised of $S(S+1)/2$ independently tunable synaptic variables. When the network is loaded close to its memory capacity each such variable ends up storing up to a fraction of a bit, as in the Hopfield model [57]. This result, it turns out, is valid only when $S$ is small, and cannot be generalized to large values of $S$, which is the case of interest for us. In the large $S$ limit we found that the critical load scales like $MS(S + 1)/\log(kS)$. The numerical factor $k$ is in practice quite large (of order $10^6$), and the correction term $\log S$ becomes important only for $S$ very large.

To apply a Potts model to our multimodular semantic network one needs to consider a number of extensions of the Kanter model. The first is incomplete connectivity between the Potts units. As for the analog extensions of the Hopfield model [82] the formula for the storage capacity is modified in that the number $C$ of connections each unit receives replaces $M$, the number of modules, and the numerical prefactor becomes larger (due to less reverberation of the noise along closed loops).

## 7.3. Sparse coding

In the above formulation, all local patterns have the same probability of appearing in a given global pattern. We are particularly interested, instead, in the case where this probability is much higher for the null state than for the others. In other words the fraction $a$ of modules in genuine local attractors (those different from the null state) should be small. This is equivalent to the notion of *sparse coding* in autoassociative memories. Adding the additional '0' state is in fact analog to considering 0-1 spin extensions of the Hopfield model with sparse coding [24, 98]. As in the associative networks with sparsely coded patterns, one expects that using sparse coding, the modular network will have a larger storage capacity. For sparsely coded global patterns we can rewrite the definition of the weights as:

$$w_{ij}^{kl} = \frac{1}{(S+1)^2 M} \sum_{\mu=1}^{p} (u_{\xi_i^\mu,k} - B_k)(u_{\xi_j^\mu,l} - B_l)(1 - \delta_{k0})(1 - \delta_{l0}) \qquad (7.4)$$

where the $\{B_k\}$'s, following [20], are defined through the equality:

$$Pr\{\xi_i^\mu = k\} = \frac{1 + B_k}{S+1}. \qquad (7.5)$$

Bolle *et al* [20], while not aiming to consider a null state, studied a generic Potts neural network with biased patterns, *i.e.* with non-zero $\{B_k\}$, although without considering optimal threshold setting, a bit like in [10]. Their formalism can be slightly modified and utilized to study a sparsely coded Potts neural network, with a null state. Optimal threshold-setting amounts, as in the transition from [10] to [98], to removing the constant coupling among non-null states, i.e. adding a term

$$\Delta w_{ij}^{kl} = \frac{-p}{(S+1)^2 M}(1 + B_k)(1 + B_l)(1 - \delta_{k0})(1 - \delta_{l0}) \qquad (7.6)$$

This is the form of the couplings used in the simulations reported below. A full analytical treatment is still to be carried out, but based on signal to noise analyses and computer simulations we expect a scaling behavior like $p_c \simeq CS^2/a \log(S/a)$ for large $S$ and small $a$. That is, the storage capacity benefits from sparser codings, unlike what happens without optimal threshold-setting.

### 7.4. A Potts model with graded response

In more realistic models of semantic storage, the stabilization into local attractors cannot be assumed to be an all-or-none phenomenon, and global attractor states cannot be assumed to be independent of one another and spatially uncorrelated. To deal with the first aspect, in the simulations we abandon the discrete Potts units used in the original storage capacity calculations, in favour of graded, analog variables representing the degree of overlap of local activity with a local attractor, and summing up to one:

$$\sum_{k=0}^{S} s_{i,k} = 1 \tag{7.7}$$

which reflect input variables $\sigma_{i,k}$ according to a standard sigmoidal activation transform:

$$s_{i,k} = \frac{\exp \beta \sigma_{i,k}}{\sum_{l=0}^{S} \beta \sigma_{i,l}} \tag{7.8}$$

where $\beta$ has the role of an inverse temperature, and the $\sigma_{i,k}$'s could simply be taken to reflect the weighted summation of inputs from other modules. To model somewhat more accurately the dynamics of entering and leaving a local attractor, however, it is convenient to assume the $\sigma_{i,k}$'s to integrate another set of variables, which themselves reflect summed inputs:

$$\tau_1 \dot{\sigma}_{i,k} = -\sigma_{i,k} + h_{i,k} - h_{i,k}^T \tag{7.9}$$

for $k \geq 1$, with the local fields $h_{i,k} = \sum_{j,l} w_{ij}^{kl} s_{j,l}$. Note the difference with the discrete-valued model in Eq. 7.1. The attractor-specific thresholds $h_{i,k}^T$'s evolve with a slower time constant to track recent correlation with the corresponding local attractor:

$$\tau_2 \dot{h}_{i,k}^T = s_{i,k} - h_{i,k}^T. \tag{7.10}$$

For $k = 0$, the 'activation' variable $\sigma_{i,0}$ acts as a general threshold for all local attractors, modulated on an even slower time scale, $\tau_3$, by the extent to which activity in the network is correlated to local attractors, as opposed to being in the null state:

$$\begin{aligned} \sigma_{i,0} &= r_0^T - h_{i,0} \\ \tau_3 \dot{h}_{i,0} &= \sum_{k=1}^{S} s_{i,k} - h_{i,0}. \end{aligned} \tag{7.11}$$

In the simulations below, the inverse temperature $\beta$ and the fixed threshold baseline $r_0^T$ were given values estimated to favour near optimal retrieval behaviour, while the time constants $\tau_1$, $\tau_2$ and $\tau_3$ were given values of *e.g.* 10, 33 and 100 basic integration time steps (a time step was indicatively taken to correspond to 1 msec of real neuronal dynamics. With such differential equations the graded variables describing local network behaviour evolve in time similarly to the collective variables describing an autoassociator network of integrate-and-fire units with adaptation [15].

## 7.5. Correlated patterns

Correlations among patterns can drastically reduce the storage capacity of an autoassociative network. However we hypothesize that in some models with correlations, one of which is adopted in the simulations sketched below, storage capacity is indeed reduced, but essentially by a prefactor dependent on the correlations, preserving the general dependence of $p_c$ on the connectivity per unit, $C$ – a linear dependence; and on the number $S$ of local attractors – roughly, a quadratic dependence.

Memory retrieval was simulated in a network of Potts units, in which global activity patterns to be stored as memory items were generated by a two-step algorithm, that could be parametrically varied from producing independent to highly correlated patterns. In the first step, a number of underlying *factors* were generated, defined simply as distinct random subsets of the entire set of Potts units. In the simulations, each subset included 50 units out of the total 300 units, and a total of 200 such factors were generated. The overlaps in the spatial distribution of different factors therefore are purely random, and clustered around their mean value $50^/300 = 8.33$.

In the second step, global patterns were generated from the factors, which had been indexed by $r$ in order of decreasing mean importance. For each global pattern, the specific importance of each factor was given by a coefficient $\gamma_r^\mu$ obtained by multiplying the overall factor $\exp(\zeta r)$ by a random number, taken to be 0 with probability $1 - a$, and otherwise drawn with a flat distribution between 0 and 1, specifically for pattern $\mu$. A value taken by factor $r$, $s_r$, was randomly drawn among the $S$ 'genuine' attractors, and a contribution $\gamma_r^\mu$ was added to the field onto each Potts unit over which factor $r$ was defined, in the direction $s_r$. After accumulating contributions from all factors, the direction in which each unit received the largest field was computed, and the $aM$ units receiving the largest maximal fields were assigned the corresponding direction $s_r$ in pattern $\mu$, while the remaining $(1 - a)M$ units were assigned the null state in pattern $\mu$.

With this procedure, pairs of Potts units have uncorrelated activity when averaged across patterns (because the different patterns that both engage the pair will span nearly evenly the different local states). Pairs of patterns, instead, can by highly correlated once averaged across units, particularly if they share one or a few most important factors; and positively correlated if these factors have been assigned the same direction in Potts space. Thus correlations among patterns will be higher if the importance of different factors decreases rapidly (e.g., in the simulations the value $\zeta = 0.02$ was used, equivalent to assuming of order 50 'important' factors); and they will tend to vanish if all factors are equally important, in general ($\zeta = 0$). When correlations are very high each pattern tends to be significantly correlated with a specific subset of the others, those sharing the main factor that influences them, and positively correlated with a fraction $1/S$ of this subset. In this scheme, the number of memory items significantly overlapping with one recently retrieved, and which can be the target of a non-random transition, scales up as $p/S$, and does not depend on $C$. By contrast, the storage capacity for retrieval, although severely reduced by correlations, should still scale as $p_c \simeq CS^2/a$. This leads to the two diagrams in Fig. 6, which indicate that conjoint semantic retrieval and structured transitions should be possible only above critical values for $C$ and $S$. Translated into the language of an underlying multimodular network, the expectation is that there should be critical values for both the short and the long-range connectivity, $C_A$ and $C_B$, beyond which a model which follows this factorial scheme would be able of both semantic retrieval and infinite recursion.

Before discussing the simulations of the Potts model, it is useful to clarify how its connectivity parameters could be mapped onto those of an underlying multimodular network model. In the *reduced* Potts model, each unit receives $C$ connections from other units, for a total of $CS(S-1)/2$ independently variable weights per unit. A storage capacity of $p_c \simeq CS^2/[a \log(kS)]$ patterns, each of which contains about $Ma \log_2(S)$ bits of information, implies that the total information that can be stored in the reduced network is of order $I_{tot} \simeq MCS^2$, that is, of order one bit per synaptic variable. In the full multimodular network, including $N$ units per module, each unit would receive $C_A$ single-variable weights from units in other modules. Note that one can further take $S$, the number of local attractors to be of order the number $C_B$ of short-range (local) connections per unit in the underlying model, that is of order $N$. If also the full network, like the reduced network, can store of order one bit per synaptic variable, in this case it would amount, even counting only long range connections, to $I_{tot} \simeq MNC_A \simeq MC_BC_A$. This implies that the bound on the number of global patterns, or semantic items, should scale up as $p_c \simeq C_AC_B/[a \log(kS)]$, that is not only it should increase with sparser modular coding (the $a$ factor), but it should also scale up with the *product* of the number of long- and short-range connections per unit in the underlying model, not with their *sum*. This is a possibility left open in the Fulvi Mari & Treves [42] calculation, which should be verified by further analysis and simulation. From a quantitative point of view, it would resurrect the idea [21, 22] that multimodular cortical networks can serve as efficient semantic memory storage devices, raising their capacity from several thousands to several millions of items.

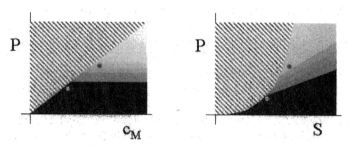

Fig. 6. Useful ranges for the number of global attractors. In the striped area above the critical line, which is linear in the $C$ (left) and almost quadratic in $S$ (right) semantic retrieval is not possible, because $p$ is above the maximum storage load. Below the critical line, there is expected to be a (dark) region of low $p$ values where long sequences of structured transitions are not possible. This region extends up to $p$ values that are independent of $C$ and proportional (in the multifactor model) to $S$. The allowed region for both semantic retrieval and infinite recursion, therefore, is close to the upper right corner of both the $p - C$ and the $p - S$ plane (uniform light area). The transition from dark to light should be sudden in a system with large $S$ and $C$ (akin to a percolation phase transition).

### 7.6. Scheme of the simulations

Whereas simulating the full multimodular network is a long-term project, the reduced Potts model requires only manageable CPU times and memory loads and can easily be simulated on a standard PC. Figure 7 shows a sample of the types of network dynamics which emerge in the simulation of the reduced Potts model.

When adaptation is turned off, typically the network remains in the retrieved attractor indefinitely. When adaptation is on, it gradually decreases the overlap between current network activity and the retrieved attractor. During this decay phase, other attractors see their overlaps increase. If one of them becomes sufficiently strong to pass an effective threshold (around 0.5 in the simulations), it manages to attract the entire network, and rapidly it reaches values close to 1, before decaying away in its turn. This transition can be repeated several times (bottom panel), reminiscent of the series of transitions seen in monkey frontal cortex [4]. The crucial ingredient for an indefinite repetition, and thus for infinite recursion to occur, is that any activated global pattern must have at least one *neighbour* that can reach an overlap above threshold before its predecessor has decayed away. Although this is a dynamical phenomenon, it is closely related to the (static) matrix of similarities among stored patterns. The more significantly

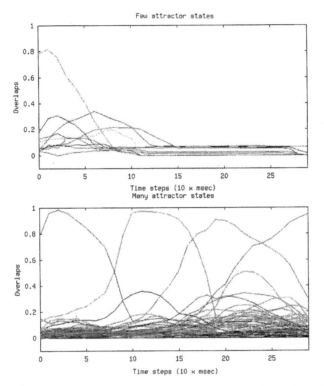

Fig. 7. Examples of global attractor retrieval with and without ensuing structured transitions among attractor states. Both examples were produced by simulating a Potts model with 300 units, $S = 10$, global patterns generated by a multi-factor model with $\zeta = 0.02$, and by applying a cue to 50% of the Potts units. Top panel, $C = 5$ and $p = 10$, and selective activity decays away after retrieval, as a second attractor is almost recruited, but it does not have a sufficient overlap with the first to emerge above an effective threshold. Bottom panel, $C = 25$ and $p = 50$, and a sequence of attractors dynamically replace each other, with the next one being recruited by its strong association with the previous one, thus generating structured transitions.

correlated global patterns exist to the one currently activated, the more likely is *latching* to proceed. For it to proceed indefinitely, each of the patterns activated in sequence must be able to activate the next, and this is more likely to occur when the density of patterns is higher, as posited in Fig. 6. To check more quantitatively the expectation expressed in those diagrams, we have run extensive simulations in which we have varied systematically $C$, $S$ and the storage load $p$, and kept other parameters constant.

Fig. 8 summarizes how these 3 parameters determine the network ability to combine the retrieval of the first, cued pattern with successive latching to dif-

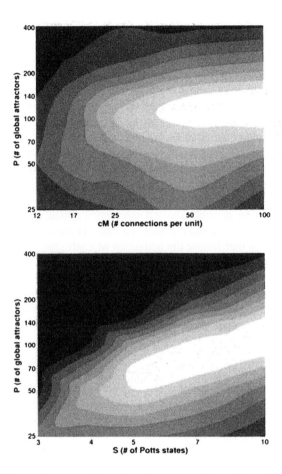

Fig. 8. Simulation results expressed as phase diagrams similar to Figure 6, but plotted on bi-logarithmic scales. In both the $p - C$ (top) and the $p - S$ plane (bottom), what is plotted in shades of gray is the product of a measure of retrieval ability (the degree to which activity is still best correlated with the cued pattern after 7 time steps) with a measure of latching ability (the degree with which after 30 time steps activity is still specifically correlated with one pattern, but not with the one cued). Both measures run from 0 to 1, and white corresponds to their product being higher than 0.3. Each diagram was obtained with 5x7 simulation datapoints, interpolated by Matlab. A datapoint reports the average of thousands of simulations with identical parameters.

ferent patterns. The light areas correspond to regions where both retrieval and latching occur frequently (averaging across thousands of independent runs). In the dark areas either retrieval tends to fail (towards the top of both plots) or latching tends to die out (towards the bottom of both plots). The simulations clearly demonstrate the existence of a limit $p_c$ on the storage load, beyond which retrieval of the pattern that best matches a partial cue is not possible (the striped regions of Fig. 6, and the top portions of Fig. 8). Below this limit (e.g. at the marked points on the $p - C$ and $p - S$ planes in Fig. 6) cued retrieval does occur, and latching can occur as well if $p$ is high enough, but still below $p_c$, hence inside the right *wedges* appearing in both plots of Fig. 6 and Fig. 8. Note that Fig. 8, represents the results of simulations limited in the size of the system but also in the time of each run (30 time steps), and this contributes to its smoother, more graded appearance than Fig. 6. It is expected, though, that comparing over longer runs behaviour corresponding to the marked points in Fig. 6, long series of structured transitions will prevail only in the higher $p$ regime (the upper marked point in each panel of Fig. 6), possibly above a *percolation* critical point. This could be assessed quantitatively even looking only at a limited time window, by measuring the entropy of states that follow the activation of each attractor. For structured transitions, this entropy is smaller than for random transitions, thereby quantifying the metric content [93] of the underlying grammar. We are currently working at the fully analytic approach sketched above, wich should allow clarification of these issues, beyond the limits of computer simulations and their dependence on specific choices of parameters (Kropff et al., in preparation).

## 7.7. Conclusions

The proposal [97] is that a generic capacity for infinite recursion (intended in its basic meaning) may have evolved as a consequence of the refinement of the semantic system. Such a refinement may have been triggered by the increase in connectivity among pyramidal cells in the cortex, particularly for some mammalian lineages including primates, and particularly in the temporal and frontal lobes [36, 37]. Such a development may have then been accelerated in the frontal cortex, relative to the temporal lobe and its sensory semantics, because action semantics invoked more structure along the time dimensions. This may have led to a capacity for syntax in communication in humans, favored by the further connectivity increase in their frontal cortex.

This proposal is still vague in several details, and it requires an analytical approach to be validated at least at the level of the self consistency of the mathematical model, even before implications for the evolution of cognition are explored in full. Its relation to a number of related approaches are discussed in [97], while here we note the distinction from the concept of phase transitions explored, in relation to language dynamics, in [74], and the potential relation to studies of the chaotic behaviour of analog systems which are close to neural networks [5]. Further work on Potts neural networks and their analog versions may pave the way to a better understanding of the rich dynamics of multi-modular neuronal networks, and indirectly contribute to illuminate the mysteries surrounding the sudden appearance, perhaps 40,000 years ago, of qualitatively new cognitive capabilities in our species.

## Acknowledgments

This chapter is based on three lectures given by AT at the Les Houches 2003 summer school, integrated by reports on analytical work in progress, written by YR. We are most grateful to Edvard and May-Britt Moser, Stefan Leutgeb, Jim Knierim and coworkers for making us appreciate the hippocampal experimental findings from their labs before publication. We are thankful to Baktash Babadi and Pooya Pakarian for the nice discussions and the hospitality. AT enjoyed the opportunity of giving lectures on these topics at the IPM, School of Cognitive Sciences in Tehran and also at Bar Ilan, Brain Research Center in Tel Aviv-Ramat Gan.

## References

[1] Abeles M (1982) Local Cortical Circuits. (Springer, Newyork)

[2] Abeles M (1991) Corticonics: Neural Circuits of the Cerebral Cortex (Cambridge Univ. Press, Cambridge)

[3] Abeles M et al (1993) Spatiotemporal firing patterns in the frontal cortex of behaving monkeys. *J Neurophysiol* 70:1629-1638

[4] Abeles M et al (1995) Cortical activity flips among quasi-stationary states. *Proc Natl Aca Sci* 92:8616-8620

[5] Afraimovich VS et al (2004) Heteroclinic contours in neural ensembles and the winnerless competition principle. *Int J Bifurc Chaos* 14:1145-1208.

[6] Allman J (1990) Evolution of neocortex. In *Cerebral Cortex*, vol 8A *Comparative Structure and Evolution of Cerebral Cortex* Jones, E.G. & Peters, A., eds. (Plenum Press, New York), 269-283

[7] Amit DJ (1989) Modelling Brain Function (Cambridge Univ. Press, New York).

[8] Amaral DG et al (1990) Neurons, numbers and the hippocampal network. *Progress in Brain Research* 83:1-11

[9] Amit DJ (1995) The Hebbian paadigm reintegrated: local reverberations as internal representations. *Behavioral and Brain Sciences* 18:617-657

[10] Amit DJ et al (1987) Information storage in neural networks at low levels of activity. *Phys Rev* A 35:2293

[11] Amit DJ, Tsodyks MV (1991) Quantitative study of attractor neural network retrieveing at low spike rates: I. Substrates – spikes, rates and neuronal gain. *Network: Comput Neural Syst* 2:259

[12]  Amit DJ & Brunel N (1997) Dynamics of a recurrent network of spiking neurons before and following learning. *Network: Comput Neural Syst* 1:381

[13]  Baker MC (2002) The Atoms of Language. (Oxford University Press, New York)

[14]  Batardiere A et al (1998) Area-specific laminar distribution of cortical feedback neurons projecting to cat area 17: Quantitative analysis in the adult and during ontogeny. *J Comp Neurol* 396:493-510

[15]  Battaglia FP & Treves A (1998a) Stable and rapid recurrent processing in realistic autoassociative memories. *Neural Computation* 10: 431-450

[16]  Battaglia FP & Treves A (1998b) Attractor neural networks storing multiple space representations: a model for hippocampal place fields. *Physical Review E* 58:7738-7753

[17]  Bienenstock E (1995) A model of neocortex. *Network: Comput. Neural Syst.* 6:179-224.

[18]  Bingman VP & Jones T-J (1994) Sun-compass based spatial learning impaired in homing pigeons with hippocampal lesions *Journal of Neuroscience* 14:6687-6694

[19]  Bliss TV & Lomo T (1973) Long-lasting potentiation of synaptic transmission in the dentate area of the anaesthetized rabbit following stimulation of the perforant path. *Journal of Physiology* 232:331-356

[20]  Bolle D et al (1993) Mean-field theory for the Q-state Potts-glass neural network with biased patterns. *J. Phys. A: Math. Gen.* 26:549

[21]  Braitenberg V (1978) Cortical architectonics: general and areal. In: Brazier MAB & Petsche H (eds) Architectonics of the cerebral cortex. (Raven, New York)

[22]  Braitenberg V & Schuz A (1991) Anatomy of the Cortex (Springer-Verlag, Berlin).

[23]  Brunel N (2003) Dynamics and plasticity of stimulus-selective persistent activity in cortical network models. *Cereb. Cortex* 13:1151-1161

[24]  Buhmann J et al (1989) Associative memory with high information content.*Phys. Rev. A* 39:2689-2692.

[25]  Carroll RL (1988) Vertebrate Paleontology and Evolution (W H Freeman & Co., New York).

[26]  Clayton N & Krebs JR (1995) Memory in food-storing birds: from behaviour to brain. *Current Opinion in Neurobiology* 5:149-154

[27]  Clayton NS, Griffiths DP, Emery NJ & Dickinson A (2001) Elements of episodic-like memory in animals. *Philosophical Transactions of the Royal Society of London* B 356:1483-1491

[28]  Collingridge GL & Bliss TV (1995) Memories of NMDA receptors and LTP. *Trends in Neuroscience* 18:54-56

[29]  Cossart R et al (2003) Attractor dynamics of network UP sates in the neocortex.*Nature* 423:283-288.

[30]  DeFelipe J et al (2002) Microstructure of the neocortex: comparative aspects. *J Neurocytol* 31:299-316

[31]  Devlin J et al (1998) Category specific semantic deficits in focal and widespread brain damage. A computational account. *J Cogn Neurosci* 10:77-94

[32]  Diamond IT & Hall WC (1969) Evolution of neocortex. *Science* 164:251-262

[33]  Diamond IT et al (1985) Laminar organization of geniculocortical projections in Galago senegalensis and Aotus trivirgatus. *J Comp Neurol* 242:610

[34]  Donoghue JP et al (1979) Evidence for two organizational plans in the somatic sensory-motor cortex in the rat. *J Comp Neurol* 183:647-666

[35]  Douglas RJ & Martin KAC (1991) A functional microcircuit for cat visual cortex. *J. Physiol.* 440:735-769

[36]  Elston GN (2000) Pyramidal cells of the frontal lobe: all the more spinous to think with. *J Neurosci* 20:RC95(1-4)

[37] Elston GN et al (2001) The pyramidal cell in cognition: a comparative study in human and monkey. *J Neurosci* 21:RC163(1-5)

[38] Erickson RP et al (1967) Organization of the posterior dorsal thalamus of the hedgehog.*J Comp Neurol* 131:103-130

[39] Farah M & McClelland J (1991) A computational model of semantic memory impairment: modality specificity and emergent category specificity. *J Exp Psychol: Gen* 120:339-357

[40] Finlay BL & Darlington RB (1995) Linked regularities in the development and evolution of mammalian brains. *Science* 268:1578-1584

[41] Freedman DJ et al (2003) A comparison of primate prefrontal and inferior temporal cortices during visual categorization. *J Neurosci* 23:5235-5246

[42] Fulvi Mari C & Treves A (1998) modeling neocortical areas with a modular neural network. *Biosystems* 48:47-55

[43] Fuster JM (2002) Frontal lobe and cognitive development. *J Neurocytol* 31:373-385

[44] Gardner-Medwin AR (1976) The recall of events through the learning of associations between their parts. *Proceedings of the Royal Society of London* B 194:375-402

[45] Haberly LB (1990) Comparative aspects of olfactory cortex. In Cerebral Cortex, vol. 8B: Comparative Structure and Evolution of Cerebral Cortex. (Jones EG, Peters A, eds.), pp.137-166. New York: Plenum Press

[46] Hammersley JM (1983) Origins of percolation theory. *Ann Israel Phys Soc* 5:47-57

[47] Hasselmo ME & Schnell E (1994) Laminar selectivity of the cholinergic suppression of synaptic transmission in rat hippocampal region CA1: Computational modeling and brain slice physiology. *Journal of Neuroscience* 14:3898-3914

[48] Hasselmo M et al (1995) Dynamics of learning and recall at excitatory recurrent synapses and cholinergic modulation in rat hippocampal region CA3. *Journal of Neuroscience* 15:5249-5262

[49] Hasselmo M et al (1996) Encoding and retrieval of episodic memories: role of cholinergic and GABAergic modulation in hippocampus. *Hippocampus* 6:693-708

[50] Hauser MD et al (2002) The faculty of language: what is it, who has it, and how did it evolve? *Science* 298:1569-1579

[51] Hauser MD & McDermott J (2003) The evolution of the music faculty: a comparative perspective. *Nature Neurosci* 6:663-668

[52] Hertz J & Prugel-Bennet A (1996) Learning synfire chains: turning noise into signal. *Int J Neural Syst* 7:445-450

[53] Hopfield JJ (1982) Neural networks and physical systems with emergent collective computational abilities. *Proc Natl Aca Sci USA* 79:2554-2558

[54] Jerison HJ (1990) In Cerebral Cortex, vol. 8A: Comparative Structure and Evolution of Cerebral Cortex, eds.Jones, EG & Peters, A (Plenum Press, New York), pp.285-309

[55] Jones EG (1998) Viewpoint: the core and matrix of thalamic organization. *Neuroscience* 85:331-45

[56] Kaas JH (1982) In: Contributions to sensory physiology, vol. 7 (Academic Press, New York) pp 201-240

[57] Kanter I (1988) Potts-glass models of neural networks, Phys. Rev. A. 37:2739

[58] Kesner RP et al (2002) Subregional analysis of hippocampal function in the rat. In *Neuropsychology of Memory*, LR Squire & DL Schacter (Eds.), 3rd Ed. (Guilford Press)

[59] Krubitzer L (1995) The organization of neocortex in mammals: are species differences really so different? *Trends Neurosci.* 18:408-417

[60] Laing CR & Chow, CC (2002) A spiking neuron model for binocular rivalry. *J. Comput. Neurosci.* 12:39-53

[61] Lassalle JM et al (2000) Reversible inactivation of the hippocampal mossy fiber synapses in mice impairs spatial learning, but neither consolidation nor memory retrieval, in the Morris navigation task *Neurobiol. Lear. Mem.* 73:243-257

[62] Lee I et al (2003) Differential coherence of CA1 vs CA3 place field ensembles in cue-conflict environments. *Soc Neurosci abs* 29:91.11

[63] Lee I & Kesner RP (2004) Encoding versus retrieval of spatial memory: double dissociation between the dentate gyrus and the perforant path inputs into CA3 in the dorsal hippocampus. *Hippocampus* 14:66-76.

[64] Lee I, Yoganarasimha D, Rao G & Knierim JJ (2004) Autoassociative network properties of the ensemble representation of environments in the CA3 field of the hippocampus, Nature 430:456-459

[65] Lende RA (1963) Cerebral cortex: a sensorimotor amalgam in the Marsupialia. *Science* 141:730-732

[66] Leutgeb S et al (2003) Differential representation of context in hippocampal areas CA3 and CA1. *Soc Neurosci abs* 29:91.5

[67] Leutgeb S, Leutgeb JK, Treves A, Moser M-B & Moser EI (2004) Distinct ensemble codes in hippocampal areas CA3 and CA1, Science 305:1295-1298

[68] Lorente de Nó R (1938) Architectonics and structure of the cerebral cortex. In Physiology of the Nervous System (Fulton JF, ed) pp. 291-330. (Oxford University Press, New York)

[69] Marr D (1970) A theory for cerebral neocortex. *Proc Roy Soc Lond* B 176:161-234

[70] Marr D (1971) Simple memory: a theory for archicortex. *Phil Trans Roy Soc (London)* B 262:23-81

[71] Miyashita Y & Chang HS (1988) Neuronal correlate of pictorial short-term memory in the primate temporal cortex. *Nature* 331:68-70

[72] Namikawa J & Hashimoto T (2003) Dynamics and computation in functional shifts, Nonlinearity 17:1317-1336.

[73] Nicoll A & Blakemore C (1993) Patterns of local connectivity in the neocortex. *Neural Comput* 5:665-68

[74] Nowak MA et al (2002) Computational and evolutionary aspects of language. *Nature* 417:611-617

[75] O'Keefe J & Dostrovsky J (1971) The hippocampus as a spatial map: preliminary evidence from unit activity in the freely moving rat. *Brain Research* 34:171-175

[76] O'Kane D & Treves A (1992) Why the simplest notion of neocortex as an autoassociative memory would not work. *Network: Comp Neural Syst* 3:379-384

[77] Pulvermuller F (2002) A brain perspective on language mechanisms: from discrete neuronal ensembles to serial order. *Progr Neurobiol* 67:85-111

[78] Rauschecker JP et al (1995) Processing of complex sounds in the macaque nonprimary auditory cortex. *Science* 268:111-114

[79] Renart A et al (1999) Associative memory properties of multiple cortical modules. *Network: Comp Neural Syst* 10:237-255

[80] Ringach DL et al (2003) Dynamics of orientation tuning in macaque V1: the role of global and tuned suppression. *J. Neurophysiol.* 90:342-352

[81] Rockel AJ et al (1980) The basic uniformity in structure of the neocortex. *Brain* 103:221-24

[82] Rolls ET & Treves A (1998) Neural Networks and Brain, (Oxford University Press: Oxford)

[83] Roudi Y & Treves A (2003) Disappearance of spurious states in analog associative memories. *Phys Rev E* 67:041906

[84] Roudi Y & Treves A (2004) An associative network with spatially organized connectivity, JStat, P07010.

[85] Shiino M & Fukai T (1993) Self-consistent signal-to-noise analysis of the statistical behavior of analog neural networks and enhancement of the stoarge capacity. *Phys. Rev. E* 48:867

[86] Shu Y et al (2003) Turning on and off recurrent balanced cortical activity. *Nature* 423:288-293

[87] Treves A (1990) Graded-response neurons and information encoding. *Phys Rev A* 42:2418

[88] Treves A & Rolls ET (1991) *Network: Comp. Neural. Syst.* 2:371

[89] Treves A (1991) Dilution and sparse coding in theshold-linear nets. *J Phys A: Math Gen* 24:327

[90] Treves A & Rolls ET (1992) Computational constraints suggest the need for two distinct input systems to the hippocampal CA3 network. *Hippocampus* 2:189-199

[91] Treves A (1995) Quantitative estimate of the information relayed by the Schaffer collaterals. *J Comput Neurosci* 2:259-272

[92] Treves A et al (1996) How much of the hippocampus can be explained by functional constraints? *Hippocampus* 6:666-674

[93] Treves A (1997) On the perceptual structure of face space. *Biosystems* 40:189-196

[94] Treves A (2001) In Handbook of Biological Physics, vol. 4: Neuro-Informatics and Neural Modelling, eds Moss F & Gielen S (Elsevier, Amsterdam) pp. 825-852

[95] Treves A (2003) Computational constraints that may have favoured the lamination of sensory cortex, *J Comput Neurosci* 14:271-282

[96] Treves A (2004) Computational constraints between retrieving the past and predicting the future, and the CA3-CA1 differentiation. *Hippocampus* 14:539-556.

[97] Treves A (2005) Frontal latching networks: a possible neural basis for infinite recursion. *Cogn Neuropsy:* in press

[98] Tsodyks MV & Feigel'man MV (1988) The enhanced storage capacity in neural networks with low activity level. *Europhysics Lett* 6:101-105

[99] Ulinski PS (1990) The cerebral cortex of reptiles, In Cerebral Cortex, vol. 8A: Comparative Structure and Evolution of Cerebral Cortex, eds EG Jones & A Peters (Plenum Press, New York) pp 139-215

[100] Ullman MT (2001) A neurocognitive perspective on language: the declarative/procedural model. *Nat Rev Neurosci* 2:717-726

[101] Whitfield IC (1979) The object of the sensory cortex. *Brain Behav Evol* 16:129-154

[102] Willshaw D & Buckingham J (1990) An assessment of Marr's theory of the hippocampus as a temporary memory store. *Philosophical Transaction of the Royal Society of London* B 329:205-215

[103] Wilson EO (1975) Sociobiology. The New Synthesis (Harvard Univ. Press, Cambridge, MA)

[104] Yoshioka T et al (1992) Intrinsic lattice connections of macaque monkey visual cortical area V4. *J. Neurosci.* 12:2785-2802

[105] Young MP et al(1994) Analysis of connectivity: neural systems in the cerebral cortex. *Rev Neurosci* 5:227-250

[106] Zanini S et al (2002) Action sequencing deficit following frontal lobe lesions. *Neurocase* 8:88-99

Course 14

# THEORY OF POINT PROCESSES FOR NEURAL SYSTEMS

Emery N. Brown

*Department of Anesthesia and Critical Care*
*Massachusetts General Hospital*
*Division of Health Sciences and Technology*
*Harvard Medical School*
*Massachusetts Institute of Technology*

*C.C. Chow, B. Gutkin, D. Hansel, C. Meunier and J. Dalibard, eds.*
*Les Houches, Session LXXX, 2003*
*Methods and Models in Neurophysics*
*Méthodes et Modèles en Neurophysique*

# Contents

## 1. Neural spike trains as point processes

Modeling analyses of neural systems are typically performed with Hodgkin and Huxley, integrate-and-fire and neural network models. In general, these models treat the process of action potential production as deterministic. Much insight into the behavior of neural systems has been obtained from these kinds of modeling analyses. However, for actual neurons, the deterministic representation is never completely true as many factors which these models assume are rarely known with certainty, even in controlled experiments. Indeed, the non-deterministic nature of neural processes is readily apparent from the plot of any neural spike train recorded in an electrophysiological study. In general, the deterministic models cannot suggest strategies or methods to analyze the non-deterministic properties of neural spike trains. Therefore, it is important to have a stochastic framework in which to model neural processes and to analyze neural spike train data.

While action potentials are not instantaneous, it is typical to assign them occurrence times such as the time at which the membrane voltage crosses threshold. The discrete, all-or-nothing nature of a sequence of action potentials together with their stochastic structure suggests that a neuronal spike train can be viewed as a point process. A point process is a stochastic process composed of a sequence of binary events that occur in continuous time. The theory of point processes is a highly developed subdiscipline in the field of stochastic processes [1]. There has recently been extensive theoretical study of point processes as well as application of this theory in biostatistics, geophysics and stochastic control. In many cases, these results have been presented in a highly theoretical framework that is not accessible to most students of computational neuroscience. Often, the more accessible, applied work is not related to the types of point process models that are relevant for neural systems.

These notes offer an introduction to the theory of univariate point processes for neural systems. We focus on results from the theory of univariate point processes that are relevant for modeling and data analysis in neuroscience. Section 2 reviews the derivation of interspike interval models from elementary stochastic dynamical systems models. Section 3 defines the conditional intensity function for a point process and discusses its properties. Section 4 derives the joint probability density of a point process. Section 5 discusses some special point process models including the Poisson process, renewal processes and stationary processes. Section 6 derives the time-rescaling theorem and discusses its implications for neural spike train data analysis. Section 7 discusses methods for simulating point processes. Section 8 mentions briefly some Poisson limit results.

## 2. Integrate and fire models and interspike interval distributions

There are two primary ways to characterize a point process. The first is in terms of the interevent probability model and the second is in terms of the conditional intensity function. The intervent probability model is the interspike interval probability model whereas the conditional intensity function, in its most general sense, is a history dependent rate function. Defining one defines the other. It is important to develop facility with using the two characterizations and to appreciate the interrelation between them. In this section, we derive the relation between elementary integrate-and-fire neuron models and the interspike interval distributions. We begin our discussion with these elementary stochastic models. This will make explicit the relation between the stochastic versions of elementary dynamical systems, neural models and statistical models used in the analysis of neural data. In Section 3, we develop the characterization of a point process in terms of its conditional intensity function and we relate it to the interspike interval probability model. This section follows closely [2].

### 2.1. Non-leaky integrator with excitatory Poisson inputs

Consider a neuron whose membrane voltage time course is defined by

$$dV(t) = \alpha_E dN(t), \tag{2.1}$$

where $N(t)$ is a Poisson process with constant rate parameter $\lambda$, and $\alpha_E$ is the magnitude of each excitatory input (Fig. 1A). The solution to Eq. 2.1 is

$$V(t) = \int_0^t \alpha_E dN(u) = \alpha_E N(t). \tag{2.2}$$

Suppose that the resting membrane potential at time 0 is $V(0) = 0$ and the neuron discharges an action potential when $V(t) \geq \theta$, where $\theta$ is a constant threshold voltage. Notice that for $V(t) \geq 0$, we must have $\alpha_E N(t) \geq \theta$ or $N(t) \geq \theta \alpha_E^{-1}$. If we let $[x]$ denote the greatest integer $\leq x$, then we require $N(t) > 1 + [\theta \alpha_E^{-1}]$, to observe an action potential.

What is the probability density of the times between the action potentials? To evaluate this probability we need to compute the probability density of the waiting time until the $k^{th}$ event for a Poisson process, beginning at an arbitrary time point $s$. This is

$$
\begin{aligned}
p_k(t)\Delta &= \Pr(k^{th} \text{ event in } (s+t, s+t+\Delta]) \\
&= \Pr(k-1st \text{ event in } (s, s+t] \cap 1 \text{ event}(s+t, s+t+\Delta]) \\
&= \Pr(k-1st \text{ event in } (s, s+t])
\end{aligned}
$$

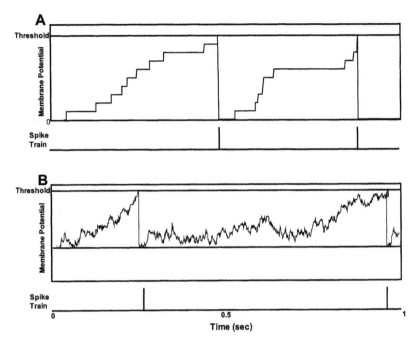

Fig. 1. Voltage time course for A, a non-leaky integrate and fire neuron with inputs given as a Poisson point process with rate $\lambda$ and input amplitude $\alpha_E$. (Eq. 2.1) and B, a non-leaky integrate-and-fire neuron following a Wiener process with drift model (Eq. 2.12). For both models, when the membrane potential crosses threshold an action potential is discharged and the membrane voltage resets immediately to the resting potential.

$$\times \text{Pr}(1 \ event(s+t, s+t+\Delta]|k-1st \ event \ in \ (s, s+t])$$
$$= \ \text{Pr}(k-1st \ event \ in \ (s, s+t]) \, \text{Pr}(1 \ event(s+t, s+t+\Delta])$$
$$= \ \frac{e^{-\lambda t}(\lambda t)^{k-1}}{(k-1)!}\lambda\Delta. \tag{2.3}$$

Hence,

$$p_k(t)\Delta = \frac{e^{-\lambda t}(\lambda\Delta)^{k-1}}{(k-1)!}\lambda\Delta,$$

which gives

$$p_k(t) = \frac{e^{-\lambda t}(\lambda t)^k}{(k-1)!} = \frac{e^{-\lambda t}\lambda t^{(k-1)-1}}{\Gamma(k)}. \tag{2.4}$$

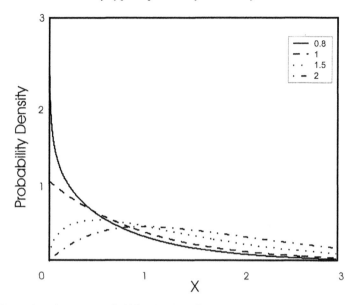

Fig. 2. Examples of gamma probability densities for $\lambda = 1$, and $k = 0.8$, $k = 1.0$, $k = 1.5$ and $k = 2.0$.

We see that $p_k(t)$ is a gamma probability density with parameters $k$ and $\lambda$. We have $E(t) = k\lambda^{-1}$ and $Var(t) = k\lambda^{-2}$ (Fig. 2).

For the primitive neuron model, we need $k = 1 + [\theta\alpha_E^{-1}]$ to generate a spike. Hence, the interspike interval probability density is the gamma probability density with parameters $1 + \theta\alpha_E^{-1}$ and $\lambda$. The interspike probability density is exponential if and only if $k = 1$. Hence, Poisson models are not the point process model associated with even elementary spike train models. The shape of the probability density is right-skewed.

### 2.2. Non-leaky integrator with excitatory and inhibitory Poisson inputs

We next consider a non-leaky integrator neuron with both excitatory and inhibitory Poisson inputs. For this neuron the time course of the membrane voltage is

$$dV(t) = \alpha_E dN_E(t) - \alpha_I dN_I(t), \tag{2.5}$$

where $N_E(t) \sim P(\lambda_E)$ and $N_I(t) \sim P(\lambda_I)$ are independent Poisson processes that govern the respective times of the excitatory and inhibitory inputs, and $\alpha_E$

and $\alpha_I$ are respectively the magnitudes of the inhibitory and excitatory inputs. We have

$$V(t) = \alpha_E \int_0^t dN_E(u) - \alpha_I \int_0^t dN_I(u) = \alpha_E N(t) - \alpha_I N_I(t). \tag{2.6}$$

For this model what is the interspike interval probability density? The interspike interval probability density for this model when $\alpha_E = \alpha_I = 1$ is given by [2]

$$p_\theta(t) = \theta \left(\frac{\lambda_E}{\lambda_I}\right)^{\theta/2} \frac{e^{-(\lambda_E + \lambda_I)t}}{t} I_\theta \left(2t(\lambda_E \lambda_I)^{\frac{1}{2}}\right), \tag{2.7}$$

for $t > 0$ where

$$I_\rho(x) = \sum_{k=0}^\infty \frac{1}{k!\Gamma(k+\rho+1)} \left(\frac{x}{2}\right)^{2k+\rho},$$

is a modified Bessel function. We also have that

$$\Pr(t_\theta < \infty) = \begin{cases} 1 & \lambda_I \leq \lambda_E \\ \lambda_E^\theta \lambda_I^{-1} & \lambda_E < \lambda_I. \end{cases} \tag{2.8}$$

Equation 2.8 means that if $\lambda_E \geq \lambda_I$ the time to threshold is finite with probability 1, whereas if $\lambda_E < \lambda_I$, the threshold may not be reached. This model is analytically tractable and leads to the Wiener process model we discuss next as the limiting case.

### 2.3. Non-leaky integrator with random walk inputs

Finally, we consider a non-leaky integrator neuron with Gaussian random walk inputs. That is, we define the membrane voltage equation as (Fig. 1B)

$$dV_\alpha(t) = \alpha dN_E(t) - \alpha dN_I(t), \tag{2.9}$$

or

$$V_\alpha(t) = \alpha \int_0^t dN_E(u) - \int_0^t dN_I(u) = \alpha[N_E(t) - N_I(t)], \tag{2.10}$$

where $\alpha_E = \alpha_I = \alpha$ and $\lambda_E = \lambda_I$ and $E[V(t)] = 0$ and $Var[V(t)] = 2\alpha^2 \lambda t$. We have as $V_\alpha(t) \to W(t)$, i.e., $V_\alpha(t)$ converges to $W(t)$ in distribution, where $W(t)$ is a Wiener process. The Wiener process (Brownian motion), $W(t)$, $t \geq 0$, is defined by the following three properties

i. $W(0) = 0$.

ii. If $(t_j, t_{j+1}]$ and $(t_k, t_{k+1}]$ are non-overlapping intervals, then $W(t_{j+1}) - W(t_j)$ and $W(t_{k-1}) - W(t_k)$ are independent.

iii. $W(t_{k+1}) - W(t_k) \sim N(0, \sigma^2(t_{k+1} - t_k))$.

Its probability density function is given as

$$p_w(w|t) = (2\pi t)^{-\frac{1}{2}} \exp\{-(2t)^{-1}w^2\}. \tag{2.11}$$

From this model we can define the Wiener process with drift by making the transformation

$$V(t) = V_0 + \sigma W(t) + \beta t, \tag{2.12}$$

and we find that this is a Gaussian process with mean and variance defined as

$$E[V(t)] = V_0 + \beta t \tag{2.13}$$

$$Var[V(t)] = \sigma^2 t. \tag{2.14}$$

If this primitive neuron receives stochastic inputs from a Wiener process or a Wiener process with drift, then what is the interspike interval probability density? We define the first passage time as the condition $t_\theta = \inf\{u|V(u) = \theta\}$, $V(0) = V_0 < \theta$. We consider two cases [2, 3]. First, we start with the driftless Wiener process, that is with $\beta = 0$. For the driftless Wiener process model, the first passage time probability density is

$$p_\theta(t) = \frac{\theta - V_0}{(2\pi\sigma^2 t^3)^{\frac{1}{2}}} \exp\left\{-\frac{(\theta - V_0)^2}{2\sigma^2 t}\right\}, \tag{2.15}$$

$t > 0$, $\theta > x_0$ (Fig. 3). For the Wiener process model with drift the first passage time probability density is given by the inverse Gaussian probability density [3,4] defined as

$$p_\theta(t) = \frac{\theta - V_0}{(2\pi\sigma^2 t^3)^{\frac{1}{2}}} \exp\left\{\frac{(\theta - V_0 - \beta t)^2}{2\sigma^2 t}\right\}, \tag{2.16}$$

$t > 0$ $\theta > x_0$. The probability of reaching threshold in finite time is defined by

$$\Pr\{t_\theta < \infty\} = \begin{cases} 1 & \beta \geq 0 \\ \exp\left[-\frac{2|\beta|(\theta - V_0)}{\sigma^2}\right] & \beta < 0. \end{cases} \tag{2.17}$$

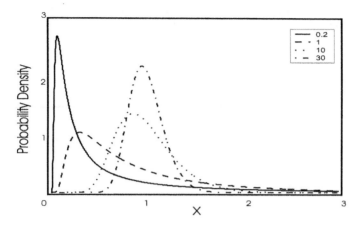

Fig. 3. Examples of inverse Gaussian probability densities for $\lambda = 1$, and $\mu = 0.8$, $\mu = 1.0$, $\mu = 1.5$ and $\mu = 2.0$.

For the inverse Gaussian probability density, the mean and variance are

$$
E[t_\theta] = \frac{(\theta - V_0)}{\beta} = \frac{\theta}{\alpha_E \lambda_E - \alpha_I \lambda_I}
$$

$$
V[t_\theta] = \frac{(\theta - V_0)\sigma^2}{\beta^3} = \frac{\theta(\alpha_E^2 \lambda_E + \alpha_I^2 \lambda_I)}{(\alpha_E \lambda_E - \alpha_I \lambda_I)^3}, \tag{2.18}
$$

$\beta > 0$, $\theta \geq x_0$ and $\alpha_E \lambda_E > \alpha_I \lambda_I$. The coefficient of variation is

$$
CV = \frac{Var[t_\theta]^{\frac{1}{2}}}{E[t_\theta]} = \left[ \frac{\alpha_E \lambda_E + \alpha_I \lambda_I}{\alpha_E \lambda_E - \alpha_{I\lambda_I}} \right]^{\frac{1}{2}}. \tag{2.19}
$$

### 2.4. Remarks

1. If we let the parameter be defined as $\mu = E[t_\theta]$ and $V[t_\theta] = \mu^3 \lambda^{-1}$, then under this change of parameters the inverse Gaussian probability density can be expressed as

$$
p(t \mid \mu, \lambda) = \left( \frac{\lambda}{2\pi t^3} \right)^{\frac{1}{2}} \exp\left\{ -\frac{1}{2} \frac{\lambda(t - \mu)^2}{\mu^2 t} \right\}. \tag{2.20}
$$

2. Schrödinger in (1915) gave one of the first derivations of this probability model [3].

3. The probability density is right-skewed and was first suggested by Gerstein and Mandelbrot [4] as a model for a neuron's ISI probability density.

4. If $(\mu/\lambda)^{-1}$ is large then, the inverse Gaussian probability density is well approximated by a Gaussian probability density (Fig. 3).

5. If we record only the spikes, then we are only able to recover the two parameters $\mu$ and $\lambda$, whereas in the original random walk model, there are four parameters $V_0, \beta, \theta$ and $\sigma^2$.

6. The results of this section show that if we construct even the simplest stochastic dynamical model a neuron, then the resulting interspike interval probability model is right-skewed and non-exponential (Figs. 2, 3). Hence, exponential interspike interval or Poisson models are not the elementary models most likely to agree with neural spike train data.

## 3. The conditional intensity function and interevent time probability density

Neural spike trains are characterized by their interspike interval probability models. In Section 2, we showed how elementary interspike interval probability models can be derived from elementary stochastic dynamical systems models of neurons. By viewing the neural spike trains as a point process, we can present characterization of the spike train in terms of its conditional intensity function. We develop this characterization in this section and we relate the conditional intensity function to the interspike interval probability models in Section 2. The presentation here follows closely [5].

Let $(0, T]$ denote the observation interval and let $0 < u_1 < u_2 <, \ldots, < u_{J-1} < u_J \leq T$ be a set of $J$ spike time measurements. For $t \in (0, T]$ let $N_{0:t}$ be the sample path of the point process over $(0, t]$. It is defined as the event $N_{0:t} = \{0 < u_1 < u_2, \ldots, u_j \leq t \cap N(t) = j\}$, where $N(t)$ is the number of spikes in $(0, t]$ and $j \leq J$. The sample path is a right continuous function that jumps 1 at the spike times and is constant otherwise [1, 5–8]. The function $N_{0:t}$ tracks the location and number of spikes in $(0, t]$ and hence, contains all the information in the sequence of spike times (Fig. 4A). The counting process $N(t)$ gives the total number of events that have occurred up through time $t$. The counting process satisfies

i) $N(t) \geq 0$.

ii) $N(t)$ is an integer-valued function.

iii) If $s < t$, then $N(s) \leq N(t)$.

iv) For $s < t$, $N(t) - N(s)$ is the number of events in $(s, t)$.

We define the conditional intensity function for $t \in (0, T]$ as

$$\lambda(t|H_t) = \lim_{\Delta \to 0} \frac{\Pr(N(t + \Delta) - N(t) = 1|H_t)}{\Delta}, \tag{3.1}$$

where $H_t$ is the history of the sample path and of any covariates up to time $t$. In general $\lambda(t|H_t)$ depends on the history of the spike train and therefore, it is also termed the stochastic intensity. In survival analysis the conditional intensity function is called the hazard function [9, 10]. This is because the hazard function can be used to define the probability of an event in the interval $[t, t + \Delta)$ given that there has not been an event up to $t$. For example, it might represent the probability that a piece of equipment fails in $[t, t + \Delta)$ given that it was worked up to time $t$ [9]. As another example, it might define the probability that a patient receiving a new therapy dies in the interval $[t, t + \Delta)$ given that he/she has survived up to time $t$ [10]. It follows that $\lambda(t|H_t)$ can be defined in terms of the interspike interval probability density at time $t$, $p(t|H_t)$, as

$$\lambda(t|H_t) = \frac{p(t|H_t)}{1 - \int_0^t p(u|H_t)du}. \tag{3.2}$$

We gain insight into the definition of the conditional intensity function in Eq. 3.1 by considering the following heuristic derivation of Eq. 3.2 based on the definition of the hazard function. We compute explicitly the probability of the event, a spike in $[t, t + \Delta)$ given $H_t$ and that there has been no spike in $(0, t)$. That is,

$$
\begin{aligned}
\Pr(u \in [t, t + \Delta)|u > t, H_t) &= \frac{\Pr(u \in [t, t + \Delta) \cap u > t|H_t)}{\Pr(u > t|H_t)} \\
&= \frac{\Pr(u \in [t, t + \Delta)|H_t)}{\Pr(u > t|H_t)} \\
&= \frac{\int_t^{t+\Delta} p(u|H_u)du}{1 - \int_0^t p(u|H_u)du} \\
&= \frac{p(t|H_t)\Delta}{1 - \int_o^t p(u|H_u)du} + o(\Delta) \\
&= \lambda(t|H_t)\Delta + o(\Delta), \tag{3.3}
\end{aligned}
$$

where $o(\Delta)$ refers to all events of order smaller than $\Delta$, such as two or more spikes occurring in an arbitrarily small interval. This establishes Eq. 3.2.

The power of the conditional intensity function is that if it can be defined, as Eq. 3.3 suggests, then it completely characterizes the stochastic structure of the spike train. In any time interval $[t, t + \Delta)$, $\lambda(t|H_t)\Delta$ defines the probability of

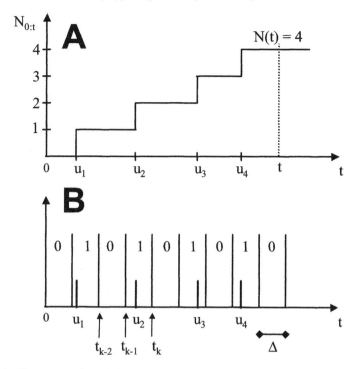

Fig. 4. A. The construction of the sample path $N_{0:t}$ from the spike times $u_1, u_2, u_3, u_4$. At time $t$, $N_{0:t} = \{u_1, u_2, u_3, u_4 \cap N(t) = 4\}$ B. The discretization of the time axis allows us to evaluate the probability of each spike occurrence or non-occurrence as a local Bernoulli process. By Eq. 3.3 the probability of the event $u_2$, i.e. a 1 between $t_{k-1}$ and $t_k$, is $\lambda(t_k|H_k)\Delta$, whereas the probability of the immediately prior to event, a 0 between $t_{k-2}$ and $t_{k-1}$, is $1 - \lambda(t_{k-1}|H_{k-1})\Delta$. In this plot, we have taken $\Delta_k = \Delta$ for all $k = 1, \ldots, K$ (reprinted and used with permission of CRC Press).

a spike given the history up to time $t$. If the spike train is an inhomogeneous Poisson process, then $\lambda(t|H_t) = \lambda(t)$ becomes the Poisson rate function. Thus, the conditional intensity function (Eq. 3.1) is a history-dependent rate function that generalizes the definition of the Poisson rate function. Similarly, Eq. 3.1 is also a generalization of the hazard function for renewal processes [9, 10].

**Example 3.1. Conditional intensity function of the Gamma probability density**. The gamma probability density for the integrate and fire model in Eq. 2.4

is

$$p_k(t) = \frac{e^{-\lambda t} \lambda t^{(k-1)-1}}{\Gamma(k)}. \tag{3.4}$$

From Eq. 3.2, it follows that the conditional intensity function is

$$\lambda(t) = \frac{e^{-\lambda t} \lambda t^{(k-1)-1}}{\Gamma(k)[1 - \int_0^t p_k(u)du]}. \tag{3.5}$$

**Example 3.2. Conditional intensity function of the inverse Gaussian probability density.** The inverse Gaussian probability density for the Wiener process integrate and fire model in Eq. 2.20 is

$$p(t|\mu, \rho) = \left(\frac{\rho}{2\pi t^3}\right)^{\frac{1}{2}} \exp\left\{-\frac{1}{2}\frac{\rho(t-\mu)^2}{\mu^2 t}\right\}. \tag{3.6}$$

From Eq. 3.2, the conditional intensity function for this model is

$$\lambda(t|\mu, \rho) = \frac{(\frac{\rho}{2\pi t^3})^{\frac{1}{2}} \exp\{-\frac{1}{2}\frac{\rho(t-\mu)^2}{\mu^2 t}\}}{1 - \int_0^t (\frac{\rho}{2\pi u^3})^{\frac{1}{2}} \exp\{-\frac{1}{2}\frac{\rho(u-\mu)^2}{\mu^2 u}\}du}. \tag{3.7}$$

## 4. Joint probability density of a point process [5]

The likelihood function is a primary tool used in constructing statistical models [5]. The likelihood function of a neural spike train, like that of any statistical model, is defined by finding the joint probability density of the data. We show in the next proposition that the joint probability of any point process and hence, spike train is easy to derive from the conditional intensity function. We begin with a lemma.

### 4.1. Derivation of the joint probability density

**Lemma 1.** Given $n$ events $E_1, E_2, \ldots, E_n$ in a probability space, then

$$\Pr\left(\bigcap_{i=1}^{n} E_i\right) = \prod_{i=2}^{n} \Pr\left(E_i \middle| \bigcap_{j=1}^{i-1} E_j\right) \Pr(E_1). \tag{4.1}$$

**Proof:** By the definition of conditional probability for $n = 2$, $\Pr(E_1 \cap E_2) = \Pr(E_2 | E_1) \Pr(E_1)$. By induction

$$\Pr\left(\bigcap_{i=1}^{n-1} E_i\right) = \prod_{i=2}^{n-1} \Pr\left(E_i \left| \bigcap_{j=1}^{i-1} E_j\right.\right) \Pr(E_1). \tag{4.2}$$

Then

$$\Pr\left(\bigcap_{i=1}^{n} E_i\right) = \Pr\left(E_n \left| \bigcap_{i=1}^{n-1} E_i\right.\right) \Pr\left(\bigcap_{i=1}^{n-1} E_i\right)$$

$$= \Pr\left(E_n \left| \bigcap_{i=1}^{n-1} E_i\right.\right) \prod_{i=2}^{n-1} \Pr\left(E_i \left| \bigcap_{j=1}^{i-1} E_j\right.\right) \Pr(E_1)$$

$$= \prod_{i=1}^{n} \Pr\left(E_i \left| \bigcap_{j=1}^{i-1} E_j\right.\right) \Pr(E_1). \tag{4.3}$$

$\square$

**Theorem 1.** Given $0 < u_1, < u_2, \ldots, u_J < T$, a set of neural spike train measurements, the sample path probability density of this neural spike train, i.e. the probability density of exactly these $J$ events in $(0, T]$ is

$$p(N_{0:T}) = \prod_{j=1}^{J} \lambda(u_j | H_{u_j}) \exp\left\{-\int_0^T \lambda(u | H_u) du\right\}$$

$$= \exp\left\{\int_0^T \log \lambda(u | H_u) dN(u) - \int_0^T \lambda(u | H_u) du\right\}. \tag{4.4}$$

**Proof:** Let $\{t_k\}_{k=1}^{K}$ be a partition of the observation interval $(0, T]$. Take $\Delta_k = t_k - t_{k-1}$, where $t_0 = 0$. Assume that the partition is sufficiently fine so that there is at most one spike in any $(t_{k-1}, t_k]$. For most neural spike trains, choosing $\Delta_k < 1$ msec would suffice. We define $dN(k) = 1$ if there is a spike in $(t_{k-1}, t_k]$ and 0 otherwise, and the events

$$A_k = \{\text{spike in } (t_{k-1}, t_k] | H_k\}$$
$$E_k = \{A_k\}^{dN(k)} \{A_k^c\}^{1-dN(k)} \tag{4.5}$$
$$H_k = \left\{\bigcap_{j=1}^{k-1} E_j\right\},$$

for $k = 1, \ldots, K$. In any interval $(t_{k-1}, t_k]$ we have (Fig. 3B)

$$\Pr(E_k) = \begin{cases} \lambda(t_k|H_k)\Delta_k + o(\Delta_k) & \text{if } dN(k) = 1 \\ 1 - \lambda(t_k|H_k)\Delta_k + o(\Delta_k) & \text{if } dN(k) = 0. \end{cases} \tag{4.6}$$

By construction of the partition, we must have $u_j \in (t_{k_j-1}, t_{k_j}]$, $j = 1, \ldots, J$ for a subset of the intervals satisfying $k_1 < k_2 < \ldots < k_J$. The remaining $K - J$ intervals have no spikes. The spike events form a sequence of correlated Bernoulli trials. It follows from Eq. 4.6 and Lemma 1 that given the partition, the probability of exactly $J$ events in $(0, T]$ may be computed as

$$p(N_{0:T}) \prod_{j=1}^{J} \Delta_{k_j} = p(u_j \in (t_{k_j-1}, t_{k_j}], j = 1, \ldots, J \cap N(T) = J) \prod_{j=1}^{J} \Delta_{k_j}$$

$$= \Pr\left(\bigcap_{k=1}^{K} E_k\right)$$

$$= \prod_{k=2}^{K} \Pr\left(E_k \middle| \bigcap_{j=1}^{k-1} E_j\right) \Pr(E_1)$$

$$= \prod_{k=1}^{K} [\lambda(t_k|H_k)\Delta_k]^{dN(t_k)} [1 - \lambda(t_k|H_k)\Delta_k]^{1-dN(t_k)} + o(\Delta_*)$$

$$= \prod_{j=1}^{J} [\lambda(t_{k_j}|H_{k_j})\Delta_{k_j}]^{dN(t_{k_j})} \prod_{\ell \neq k_j} [1 - \lambda(t_\ell|H_\ell)\Delta_\ell]^{1-dN(t_\ell)}$$

$$\quad + o(\Delta_*)$$

$$= \prod_{j=1}^{J} \lambda(t_{k_j}|H_{k_j})\Delta_{k_j}^{dN(k_j)} \prod_{\ell \neq k_j} \exp\{-\lambda(t_\ell|H_\ell)\Delta_\ell\} + o(\Delta_*)$$

$$= \exp\left\{\sum_{j=1}^{J} \log \lambda(t_{k_j}|H_{k_j}) dN(t_{k_j}) - \sum_{\ell \neq k_j} \lambda(t_\ell|H_\ell)\Delta_\ell\right\}$$

$$\quad \times \exp\left\{\sum_{j=1}^{J} \log \Delta_{k_j}\right\} + o(\Delta_*), \tag{4.7}$$

where because the $\Delta_k$ are small, we have used the approximation $[1 - \lambda(k)\Delta_k] \approx \exp\{-\lambda(k)\Delta_k\}$ and $\Delta_* = \Delta^J$. If follows that the probability density of exactly these $J$ spikes in $(0, T]$ is

$$
\begin{aligned}
p(N_{0:T}) &= \lim_{\Delta_* \to 0} \left\{ \exp\left\{ \sum_{j=1}^{J} \log \lambda(t_{k_j} | H_{k_j}) dN(t_{k_j}) \right. \right. \\
&\quad \left. - \sum_{\ell \neq k_j} \lambda(t_\ell | H_\ell) \Delta_\ell \right\} \exp\left\{ \sum_{j=1}^{J} \log \Delta_{k_j} \right\} + o(\Delta_*) \right\} \Big/ \prod_{j=1}^{J} \Delta_j \\
&= \exp\left\{ \int_0^T \log \lambda(u | H_u) dN(u) - \int_0^T \lambda(u | H_u) du \right\}. \quad (4.8)
\end{aligned}
$$

$\square$

Theorem 1 shows that the joint probability density of a spike train can be written in a canonical form in terms of the conditional intensity function [1, 5–8, 10]. That is, when formulated in terms of the conditional intensity function, all point process likelihoods have the form given in Eq. 4.8. The approximate probability density expressed in terms of the conditional intensity function (Eq. 4.7e) was given in [11]. The insight provided by this proof is that correct discretization for computing the local probability of a spike event is given by the conditional intensity function.

### 4.2. An alternative derivation of the joint probability density of a spike train

This derivation is based on [1, 12]. The sample path probability density is

$$
\begin{aligned}
p(u_1, u_2, \ldots, u_n \cap u_{n+1} > T) &= \prod_{k=1}^{n} p(u_k | u_1, \ldots, u_{k-1}) \\
&\quad \times \Pr(no \ spikes \ in \ (u_n, T] | u_1, \ldots, u_n) \\
&= \prod_{k=1}^{n} p(t_k | H_k)[1 - P(T | H_{n+1})], \quad (4.9)
\end{aligned}
$$

where $[1 - P(T | H_{n+1})] = \int_{t_n}^{T} p(u | H_{n+1}) du$. From Eq. 3.2 we have for $t > t_{k-1}$,

$$
r(t | H_k) = \frac{p(t | H_k)}{1 - P(t | H_k)}, \quad (4.10)
$$

and

$$
\lambda(t | H_k) = \begin{cases} r(t) & 0 < t < t_1 \\ r(t | H_k) & t_{k-1} < t < t_k \end{cases}. \quad (4.11)
$$

Integrating both sides of Eq. 4.10 from $t_{k-1}$ to $t$ gives

$$-\log[1 - P(t|H_k)] = \int_{t_{k-1}}^{t} \lambda(u)du, \tag{4.12}$$

and after exponentiating both sides we obtain

$$[1 - P(t|H_k)] = \exp\left\{-\int_{t_{k-1}}^{t} \lambda(u|H_k)du\right\}. \tag{4.13}$$

Rearranging Eq. 4.10 and using Eq. 4.13 we have

$$
\begin{aligned}
p(t_k|H_k) &= r(t|H_k)[1 - P(t|H_k)] \\
&= \lambda(t|H_k)\exp\left\{-\int_{t_{k-1}}^{t} \lambda(u|H_u)du\right\},
\end{aligned} \tag{4.14}
$$

which is the first part of the desired result. Returning to Eq. 4.9 and using Eq. 4.14, we find that the joint probability density of the spike train is

$$
\begin{aligned}
&p(t_1, t_2, \ldots, t_n \cap t_{n+1} > T) \\
&= \prod_{k=1}^{n} p(t_k|H_k)[1 - P(T|H_{n+1})] \\
&= \prod_{k=1}^{n} \lambda(t_k|H_k)\exp\left\{-\int_{t_{k-1}}^{t_k} \lambda(u|H_u)du\right\}[1 - P(T|H_{n+1})] \\
&= \prod_{k=1}^{n} \lambda(t_k)\exp\left\{-\int_{0}^{T} \lambda(u)du\right\} \\
&= \exp\left\{\int \log \lambda(u|H_u)dN(u) - \int_{0}^{T} \lambda(u|H_u)du\right\}.
\end{aligned} \tag{4.15}
$$

## 5. Special point process models

Thus far, we have described how elementary stochastic dynamical systems models lead to interspike interval models (Section 2), developed a characterization of point processes in terms of the conditional intensity function (Section 3) and derived the joint probability density of a point process in terms of the conditional intensity function (Section 4). In this section we describe some special point process models. These are the Poisson process, renewal processes and stationary point process models. This material follows [13–15].

A counting process has *independent increments* if the number of events which occur in disjoint time intervals are independent. A counting process has *stationary increments* if the distribution of the number of events which occur in any time interval depends only on the length of the interval. That is, the process has stationary increments if the number of events in the interval $(t_1 + s, t_2 + s)$ (i.e. $N(t_2 + s) - N(t_1 + s)$) has the same distribution as the number of events in the interval $(t_1, t_2)$ (i.e. $N(t_2) - N(t_1)$) for all $t_1 < t_2$ and $s > 0$. Independent increments and stationary increments are very strong conditions that are rarely satisfied exactly for neural processes since history dependence of the spiking activity is the rule rather than the exception.

## 5.1. Poisson processes

### 5.1.1. Axioms for a Poisson process
A counting process $N(t)$ $t \geq 0$ is a Poisson process with rate $\lambda$, $\lambda > 0$, if

i) $N(0) = 0$.

ii) The process has independent increments.

iii) $\Pr(N(t + \Delta) - N(t) = 1) = \lambda \Delta + o(\Delta)$.

iv) $\Pr(N(t + \Delta) - N(t) \geq 2) = o(\Delta)$.

The probability mass function is

$$p((N(t) = k) = \frac{e^{-\lambda t}(\lambda t)^k}{k!},$$ (5.1)

for $k = 0, 1, 2, 3, \ldots$. $E[N(t)] = \lambda t$ and $Var[N(t)] = \lambda t$. There are many approaches to deriving this probability mass function. The probability mass function can be derived from assumptions i)–iv) by using the Poisson approximation to the binomial (See Section 8).

### 5.1.2. Interevent and waiting time probability densities for a Poisson process
To construct the interspike interval probability density for the Poisson process, we note that

$$\Pr(T > t) = \Pr(N(t) = 0) = e^{-\lambda t},$$ (5.2)

or

$$\Pr(t \leq T) = 1 - e^{-\lambda t}.$$ (5.3)

If we differentiate Eq. 5.3, we find that the probability density is the exponential probability density

$$p(t) = \lambda e^{-\lambda t}, \tag{5.4}$$

for $t > 0$ and $\lambda > 0$. Given the assumption that the interspike interval probability density of a counting process is exponential, it is possible to show that this implies that the counting process is a Poisson process.

The waiting time probability density is the probability density of the time until the occurrence of the kth event. To construct the waiting time probability density, we consider $s_k = \sum_{i=1}^{k} t_i$. Using the properties of a Poisson process we obtain

$$\Pr(t < s_k < t + \Delta)$$
$$= \Pr(N(t) = k - 1 \cap \text{ an event in } (t, t + \Delta)) + o(\Delta)$$
$$= \Pr(N(t) = k - 1) \Pr(\textit{one event in } (t, t + \Delta) | N(t) = 1) + o(\Delta)$$
$$= \Pr(N(t) = k - 1)\lambda\Delta + o(\Delta)$$
$$= \frac{(\lambda t)^{k-1}}{(k-1)!} e^{-\lambda t} \lambda \Delta. \tag{5.5}$$

Hence,

$$p_{s_k}(t) = \lim_{\Delta \to 0} \frac{\Pr(t < s_k < t + \Delta)}{\Delta} = \frac{\lambda^k t^{k-1} e^{-\lambda t}}{(k-1)!}. \tag{5.6}$$

The waiting time probability density is, as we saw in Section 2, the gamma distribution with parameters $k$ and $\lambda$. The conditional intensity function is

$$\lambda(t) = \frac{p(t)}{1 - \int_0^t p(u)du}$$
$$= \frac{\lambda e^{-\lambda t}}{1 - \int_0^t \lambda e^{-\lambda u}du}$$
$$= \lambda. \tag{5.7}$$

The joint probability density of the spike train (Eq. 4.15)

$$p(u_1, \ldots, u_n) \Pr(u_{n+1} > T - u_n)$$
$$= \prod_{i=1}^{n} p(u_i | \lambda) \exp(-\lambda(T - u_n))$$
$$= \lambda^n \exp\left(-\lambda\left(\sum_{i=1}^{n}(u_i - u_{i-1}) - (T - u_n)\right)\right)$$
$$= \lambda^n \exp(-\lambda T). \tag{5.8}$$

There are three ways to generalize a simple Poisson process in order to construct more complex point process models. These are: 1) use a Poisson process with non-stationary increments, which leads to an inhomogeneous Poisson process; 2) chose an interevent probability model other than the exponential probability density which leads to a renewal process; and 3) allow the conditional intensity function to have history dependence which leads to the general class of point process models discussed in Sections 3 and 4. We examine these first two extensions next.

### 5.1.3. Inhomogeneous Poisson process
The following axioms define an inhomogeneous Poisson process.

i) $N(0) = 0$.

ii) $\{N(t), t \geq 0\}$ is a counting process with independent increments.

iii) $\Pr\{N(t + \Delta) - N(t) = 1\} = \lambda(t)\Delta + o(\Delta)$.

iv) $\Pr\{N(t + \Delta) - N(t) \geq 2\} = o(\Delta)$.

Given these axioms, it is straightforward to show that the probability mass function of the inhomogeneous Poisson process is

$$\Pr(N(t) = k) = \frac{[\Lambda(t)]^k e^{-\Lambda(t)}}{k!}, \tag{5.9}$$

where $\Lambda(t) = \int_0^t \lambda(u)du$. The interevent time probability density of the inhomogeneous Poisson can be derived by noting that

$$\Pr(T > t) = \Pr(N(t) = 0) = \exp\left\{-\int_0^t \lambda(u)du\right\}, \tag{5.10}$$

or

$$\Pr(T \leq t) = 1 - \exp\left\{-\int_0^t \lambda(u)du\right\}, \tag{5.11}$$

and hence, by differentiating with respect to $t$ we obtain

$$p(t) = \lambda(t)\exp\left\{-\int_0^t \lambda(u)du\right\}. \tag{5.12}$$

For an alternative derivation of this probability density we take $p(z) = e^{-z}$ and let $z = \int_0^t \lambda(u)du$. We have $p(t)dt = f(z)dz$ and hence $p(t) = p(z)\frac{dz}{dt}$ which is the standard change-of-variables formula. Now $\frac{dz}{dt} = \lambda(t)$ and hence $p(t) =$

$\lambda(t)e^{-z} = \lambda(t)\exp\{-\int_0^t \lambda(u)du\}$. We have inverted the computation in the time-rescaling theorem (see Section 6).

From the arguments used to construct the waiting time probability density for the simple Poisson process, it is easy to show that the waiting time probability density for the inhomogeneous Poisson process is

$$p_{s_k}(t) = \frac{\lambda(t)\Lambda(t)^{k-1}\exp(-\Lambda(t))}{\Gamma(k)}. \tag{5.13}$$

From Theorem 1, it follows that the joint probability density of the spike times is

$$
\begin{aligned}
p(u_1, &\ldots, u_n \cap u_{n+1} > T) \\
&= p(u_1, \ldots, u_n)\Pr(u_{n+1} > T | u_1, \ldots, u_n) \\
&= \prod_{i=1}^{n} \lambda(u_i)\exp\left\{-\int_{u_{i-1}}^{u_i} \lambda(u)du\right\}\exp\left\{-\int_{u_n}^{T} \lambda(u)du\right\} \\
&= \prod_{i=1}^{n} \lambda(u_i)\exp\left\{-\int_0^T \lambda(u)du\right\},
\end{aligned} \tag{5.14}
$$

because

$$
\begin{aligned}
\Pr(u_{n+1} > T | u_1, \ldots, u_n) &= \Pr(u_{n+1} > T) = \Pr(\text{No spike in } (u_n, T] \\
&= \exp\left\{-\int_{s_n}^{t} \lambda(u)du\right\}.
\end{aligned} \tag{5.15}
$$

To show that the rate function and the conditional intensity function are the same for an inhomogeneous Poisson model we substitute Eq. 5.12 in Eq. 3.2 to obtain

$$
\begin{aligned}
\lambda(t|H_k) &= \frac{\lambda(t)\exp\{-\int_{t_{k-1}}^{t} \lambda(u)du\}}{1 - P(t|H_k)} \\
&= \frac{\lambda(t)\exp\{-\int_{t_{k-1}}^{t} \lambda(u)du\}}{\exp\{-\int_{t_{k-1}}^{t} \lambda(u)du\}} \\
&= \lambda(t).
\end{aligned} \tag{5.16}
$$

Therefore, the conditional intensity function for the inhomogeneous Poisson process is the rate function.

Given an observation interval $(0, T]$ and a Poisson process with intensity parameter $\lambda(t)$, we consider the distribution of the Poisson events conditional on

$N(T) = n$. That is

$$\Pr(u_1, u_2, \ldots, u_n | N(T) = n)$$
$$= \frac{\Pr(u_1, \ldots, u_n \cap N(T) = n)}{\Pr(N(T) = n)}$$
$$= \frac{\prod_{k=1}^n \lambda(u_k) \exp\{-\int_0^T \lambda(u)du\}}{\frac{[\int_0^T \lambda(u)du]^n}{n!}} \exp\left\{-\int_0^T \lambda(u)du\right\}$$
$$= \frac{n! \prod_{k=1}^n \lambda(u_k)}{[\int_0^T \lambda(u)du]^n}. \tag{5.17}$$

This is the joint distribution of the ordered observation from the probability density

$$p(t) = \frac{\lambda(t)}{\int_0^T \lambda(u)du}, \tag{5.18}$$

where $t \varepsilon (0, T]$. This result is very relevant for neuroscience data analysis because a common practice is to use the normalized peristimulus time histogram (PSTH) as a probability density. The above result gives conditions under which this is reasonable.

## 5.2. Renewal processes

A renewal process is a point process in which the interevent intervals are independent and drawn from the same probability density. More specifically, let $T_i$ be independent, identically distributed interevent times from the probability density $p(t)$. Let $s_r = \sum_{i=1}^r T_i$. The process $s_r$ is a *renewal process*. The gamma and inverse Gaussian interspike interval probability models derived in Section 2 are renewal processes. Since we have defined the interevent (interspike) interval distribution we go back and find the counting process probability mass function.

### 5.2.1. Counting process associated with a renewal process
The counting process $N(t)$, is the number of events in $(0, t]$. Now $N(t) < r$ if and only if $s_r > t$ or, $\Pr(N(t) < r) = \Pr(s_r > t) = 1 - P_r(t)$. Therefore,

$$\Pr(N(t) \geq r) = \Pr(N(t) = r \cup N(t) > r)$$
$$= \Pr(N(t) = r \cup N(t) \geq r + 1)$$
$$= \Pr(N(t) = r) + \Pr(N(t) \geq r + 1).$$

Therefore, we have

$$\Pr(N(t) = r) = \Pr(N(t) \geq r) - \Pr(N(t) \geq r + 1)$$
$$= P_r(t) - P_{r+1}(t). \tag{5.19}$$

where $P_r(t)$ is the probability distribution function associated with the $r$-fold convolution of $p(t)$. While Eq. 5.19 gives the probability mass function for the counting process associated with any renewal process, it can only be evaluated explicitly in specific cases.

### 5.2.2. Asymptotic distribution of $N(t)$ for large $t$

Because Eq. 5.19 is a challenge to evaluate, asymptotic approximations are frequently used. It can be shown [14, 15] that for large $t$ the distribution of $N(t)$ may be approximated as a Gaussian random variable with mean and variance defined by

$$E[N(t)] = t\mu^{-1} \quad Var[N(t)] = \sigma^2 \mu^{-3} t, \tag{5.20}$$

where

$$\mu = E[t_i] \quad \sigma^2 = Var[t_i]. \tag{5.21}$$

### 5.3. Stationary process

We mention briefly stationary point process models [14, 15] and show how their properties can be analyzed in terms of the conditional intensity function. Under the assumption that the point process is stationary, we have that

$$E[dN(t)] = \lambda(t|H_t)$$
$$= \lambda dt, \tag{5.22}$$

and that the variance is given by

$$Var[dN(t)] = \sigma^2 dt. \tag{5.23}$$

### 5.3.1. Covariance function

To find the covariance function for a stationary point process we consider

$$E(dN(t+v), dN(t)) = \Pr(dN(t) = 1 \cap dN(t+v) = 1)$$
$$= \Pr(dN(t) = 1)\Pr(dN(t+v) = 1|dN(t) = 1)$$
$$= \lambda \Delta t \lambda(v) \Delta t = \lambda \lambda(v)(\Delta t)^2, \tag{5.24}$$

where

$$\lambda(v) = \lim_{\Delta t \to 0} \frac{\Pr(\text{event } (t+v, t+v+\Delta t] | \text{ event at } t)}{\Delta t}, \tag{5.25}$$

is the conditional intensity function of a stationary point process model. Hence, we have

$$Cov(dN(t+v), dN(t)) = \gamma(v)(\Delta t)^2, \tag{5.26}$$

where $\gamma(v) = \lambda(\lambda(v) - \lambda)$ is the covariance function of the process or more generally

$$\gamma(v) = \lambda(\lambda(v) - \lambda) + \sigma^2 \delta(v), \tag{5.27}$$

where $\delta(v)$ is the Dirac function and allows us to consider the variance of the process as part of the same expression.

### 5.3.2. Spectral density function
Taking the Fourier transform of the covariance function we find that the spectral density function is

$$\phi(\omega) = \frac{\lambda}{2\pi} \int_{-\infty}^{\infty} (\lambda(v) - \lambda) e^{-i\omega v} dv + \frac{\sigma^2 \delta(v)}{2\pi}. \tag{5.28}$$

Unlike with stationary Gaussian processes, it is possible to have two different stationary point process models with the same spectral density.

## 6. The time-rescaling theorem

This result, originally due to [16] and [17], states that every point process with a conditional intensity function maps into a Poisson process with unit rate. In addition to being an interesting theoretical result, it has important implications for assessing goodness-of-fit for point process models of neural spike trains.

We begin by restating the result established Section 4. The joint probability density of exactly $n$ event times in $(0, T]$ is

$$p(u_1, u_2, \ldots, u_n \cap N(T) = n)$$
$$= p(u_1, u_2, \ldots, u_n \cap t_{n+1} > T)$$
$$= p(u_1, u_2, \ldots, u_n) \Pr(u_{n+1} > T | u_1, u_2, \ldots, u_n)$$
$$= \prod_{k=1}^{n} \lambda(u_k | H_{t_k}) \exp \left\{ -\int_{u_{k-1}}^{u_k} \lambda(u | H_u) du \right\}$$

$$\times \exp\left\{-\int_{u_n}^{T} \lambda(u|H_u)du\right\},\tag{6.1}$$

where

$$p(u_1, u_2, \ldots, u_n) = \prod_{k=1}^{n} \lambda(u_k|H_{t_k}) \exp\left\{-\int_{u_{k-1}}^{u_k} \lambda(u|H_u)du\right\}\tag{6.2}$$

$$\Pr(t_{n+1} > T|t_1, t_2, \ldots, t_n) = \exp\left\{-\int_{t_n}^{T} \lambda(u|H_u)du\right\}.\tag{6.3}$$

and $t_0 = 0$. We can now state and prove the time-rescaling theorem.

### 6.1. An elementary proof of the time-rescaling theorem

This proof follows closely [18].

**Theorem 2.** Let $0 < u_1 < u_2 <, \ldots, < u_n < T$ be a realization from a point process with a conditional intensity function $\lambda(t|H_t)$ satisfying $0 < \lambda(t|H_t)$ for all $t \in (0, T]$. Define the transformation

$$\Lambda(u_k) = \int_{0}^{u_k} \lambda(u|H_u)du\tag{6.4}$$

for $k = 1, \ldots, n$, and assume $\Lambda(t) < \infty$ with probability one for all $t \in (0, T]$. Then the $\Lambda(u_k)$s are a Poisson process with unit rate.

**Proof:** Let $\tau_k = \Lambda(u_k) - \Lambda(u_{k-1})$ for $k = 1, \ldots, n$ and set $\tau_T = \int_{u_n}^{T} \lambda(u|H_u)du$. To establish the result it suffices to show that the $\tau_k$s are independent, identically distributed exponential random variables with mean one. Because the $\tau_k$ transformation is one-to-one and $\tau_{n+1} > \tau_T$ if and only if $u_{n+1} > T$, the joint density of the $\tau_k$'s is

$$p(\tau_1, \tau_2, \ldots, \tau_n \cap \tau_{n+1} > \tau_T) = p(\tau_1, \ldots, \tau_n)\Pr(\tau_{n+1} > \tau_T|\tau_1, \ldots, \tau_n).\tag{6.5}$$

We evaluate each of the two terms on the right side of Eq. 6.5. The following two events are equivalent

$$\{\tau_{n+1} > \tau_T|\tau_1, \ldots, \tau_n\} = \{u_{n+1} > T|u_1, u_2, \ldots, u_n\}.\tag{6.6}$$

Hence

$$\Pr(\tau_{n+1} > \tau_T | \tau_1, \tau_2, \ldots, \tau_n) = \Pr(u_{n+1} > T | u_1, u_2, \ldots, u_n)$$

$$= \exp\left\{ -\int_{u_n}^{T} \lambda(u | H_{u_n}) du \right\}$$

$$= \exp\{-\tau_T\}, \tag{6.7}$$

where the last equality follows from the definition of $\tau_T$. By the multivariate change-of-variable formula [19]

$$p(\tau_1, \tau_2, \ldots, \tau_n) = |J| p(u_1, u_2, \ldots, u_n \cap N(u_n) = n), \tag{6.8}$$

where $J$ is the Jacobian of the transformation between $u_j$, $j = 1, \ldots, n$ and $\tau_k$, $k = 1, \ldots, n$. Because $\tau_k$ is a function of $u_1, \ldots, u_k$, $J$ is a lower triangular matrix and its determinant is the product of its diagonal elements defined as $|J| = |\prod_{k=1}^{n} J_{kk}|$. By assumption $0 < \lambda(t | H_t)$, and by Eq. 6.4 and the definition of $\tau_k$, the mapping of $u$ into $\tau$ is one-to-one. Therefore, by the inverse differentiation theorem [20] the diagonal elements of $J$ are

$$J_{kk} = \frac{\partial u_k}{\partial \tau_k} = \lambda(u_k | H_{u_k})^{-1}. \tag{6.9}$$

Substituting $|J|$ and Eq. 6.2 into Eq. 6.8 yields

$$p(\tau_1, \tau_2, \ldots, \tau_n) = \prod_{k=1}^{n} \lambda(u_k | H_{u_k})^{-1} \prod_{k=1}^{n} \lambda(u_k | H_{u_k}) \exp\left\{ -\int_{u_{k-1}}^{u_k} \lambda(u | H_u) du \right\}$$

$$= \prod_{k=1}^{n} \exp\{-[\Lambda(u_k) - \Lambda(u_{k-1})]\}$$

$$= \prod_{k=1}^{n} \exp\{-\tau_k\}. \tag{6.10}$$

Substituting Eq. 6.10 and Eq. 6.7 into Eq. 6.5 yields

$$p(\tau_1, \tau_2, \ldots, \tau_n \cap \tau_{n+1} > \tau_T) = p(\tau_1, \ldots, \tau_n) \Pr(\tau_{n+1} > \tau_T | \tau_1, \ldots, \tau_n)$$

$$= \left( \prod_{k=1}^{n} \exp\{-\tau_k\} \right) \exp\{-\tau_T\}, \tag{6.11}$$

which establishes the result. □

The time-rescaling theorem generates a history-dependent rescaling of the time axis that converts a point process into a Poisson process with a unit rate (Fig. 5).

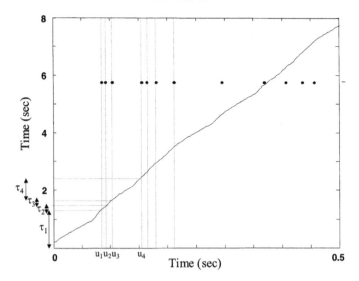

Fig. 5. Illustration of the time-rescaling theorem. The interspike intervals (x-axis) $u_k -$
$u_{k-1}$ are transformed into the $\tau_k s$ (y-axis) using Eqs. 6.12. The integrated conditional
intensity function is the solid line. By the time rescaling theorem the $\tau_k s$ are independent,
identically distributed exponential random variables.

### 6.2. The time-rescaling theorem: assessing model goodness-of-fit

We may use the time-rescaling theorem to construct goodness-of-fit tests for a
spike data model. Once a model has been fit to a spike train data series we can
compute from its estimated conditional intensity the rescaled times

$$\tau_k = \Lambda(u_k) - \Lambda(u_{k-1}). \tag{6.12}$$

If the model is correct, then according to the theorem, the $\tau_k s$ are independent
exponential random variables with mean 1. If we make the further transformation

$$z_k = 1 - e^{-\tau_k}, \tag{6.13}$$

then $z_k s$ are independent uniform random variables on the interval [0, 1). Because the transformations in Eqs. 6.12 and 6.13 are both one-to-one, any statistical assessment that measures agreement between the $z_k s$ and a uniform distribution directly evaluates how well the original model agrees with the spike
train data [21]. Here we present two methods: Kolmogorov-Smirnov tests and
quantile-quantile plots.

### 6.2.1. Kolmogorov-Smirnov test

To construct the Kolmogorov-Smirnov test we first order the $z_k$s from smallest to largest, denoting the ordered values as $z_{(k)}$s. We then plot the values of the cumulative distribution function of the uniform density defined as $b_k = \frac{k-\frac{1}{2}}{n}$ for $k = 1, \dots, n$ against the $z_{(k)}$s. If the model is correct, then the points should lie on a 45° line [22]. Confidence bounds for the degree of agreement between the models and the data may be constructed using the distribution of the Kolmogorov-Smirnov statistic [22]. For moderate to large sample sizes the 95% (99%) confidence bounds are well approximated as $b_k \pm 1.36/n^{1/2}(b_k \pm 1.63/n^{1/2})$ [22]. We term such a plot a Kolmogorov-Smirnov (KS) plot.

### 6.2.2. Quantile-quantile plot

Another approach to measuring agreement between the uniform probability density and the $z_k$s is to construct a quantile-quantile (Q-Q) plot [22]. In this display we plot the quantiles of the uniform distribution, denoted here also as the $b_k$s, against the $z_{(k)}$s. As in the case of the KS plots, exact agreement occurs between the point process model and the experimental data if the points lie on a 45° line. Pointwise confidence bounds can be constructed to measure the magnitude of the departure the plot from the 45° line relative to chance. To construct pointwise bounds we note that if the $\tau_k$s are independent exponential random variables with mean 1 and the $z_k$s are thus uniform on the interval (0, 1], then each $z_{(k)}$ has a beta probability density with parameters $k$ and $n - k + 1$ defined as

$$p(z|k, n - k + 1) = \frac{n!}{(n-k)!(k-1)!} z^{k-1}(1-z)^{n-k}, \tag{6.14}$$

for $0 < z \leq 1$ [22]. We set the 95% confidence bounds by finding the 2.5[th] and 97.5[th] quantiles of the cumulative distribution associated with Eq. 6.14 for $k = 1, \dots, n$. These exact quantiles are readily available in many statistical software packages. For moderate to large spike train data series, a reasonable approximation to the 95% (99%) confidence bounds is given by the Gaussian approximation to the binomial probability distribution as $z_{(k)} \pm 1.96[z_{(k)}(1 - z_{(k)})/n]^{1/2}(z_{(k)} \pm 2.575[z_{(k)}(1 - z_{(k)})/n]^{1/2})$.

In general, the KS confidence intervals will be wider than the corresponding Q-Q plot intervals. From the Gaussian approximation to the binomial, the maximum width of the 95% confidence interval for the Q-Q plots occurs at the median, i.e., $z_{(k)} = 0.50$, and is $2[1.96/(4n)^{1/2}] = 1.96n^{-1/2}$. For $n$ large, the width of the 95% confidence intervals for the KS plots is $2.72n^{-1/2}$ at all quantiles. The KS confidence bounds consider the maximum discrepancy from the 45° line along all quantiles; the bands show the discrepancy that would be exceeded 5%

of the time by chance if the plotted data were truly uniformly distributed. The Q-Q plot confidence bounds consider the maximum discrepancy from the 45° line for each quantile separately. These pointwise 95% confidence bounds mark the amount by which each $z_{(k)}$ would deviate from the true quantile 5% of the time purely by chance. The KS bounds are broad because they consider the distribution of the largest of all deviations. The Q-Q plot bounds are narrower because they measure the deviation at each quantile separately. Used together, the two plots can help approximate upper and lower limits on the discrepancy between a proposed model and a spike train data series.

### 6.2.3. Normalized point process residuals

Let $0 < t_1^* < \cdots < t_K^* < T$ be a coarse partition of the observation interval $(0, T]$. Let $N_k^*$ denote the number of spikes in the interval $(t_{k-1}^*, t_k^*]$. Define the point process residuals

$$r_k = \frac{N_k^* - \Lambda(\Delta_k^*)}{[\Lambda(\Delta_k^*)]^{\frac{1}{2}}},$$

for $k = 1, \ldots, K$. The point process residuals have zero expectation and variance 1. A plot of the point process residuals against time can reveal systematic temporal structures in model lack of fit.

As a second application of the time-rescaling theorem we discuss in Section 7 an approach to simulating point processes based on the time-rescaling theorem.

## 7. Simulation of point processes

Lewis and Shedler [23] developed an efficient algorithm for simulating a Poisson process termed thinning. Ogata [24] developed extension to general point processes using the conditional intensity function. The algorithms are as follows.

### 7.1. Thinning algorithm 1

Given a regular point process $N(t)$ on $(0, T]$ with a bounded intensity function $\lambda(t)$. That is, for all $t\varepsilon(0, T]$ there exists $\lambda$ such that $\lambda(t) \le \lambda$. To simulate a random sample from $N(t)$ use the following two stage algorithm:

A. Draw a Poisson Sample

    1. Draw $u_i$ from $u(0, 1)$

    2. Compute $w_i = -\frac{\log(1-u_i)}{\lambda}$

    3. Compute $t_i = t_{i-1} + w_1$

4. If $t_i \geq T$ stop else $i = i + 1$ go to 1        (7.1)

5. $N(T) = i$

B. Thinning Algorithm

   0. Set $i = 1$ and $j = 1$

   1. Draw $v_i$ from $U(0, 1)$

   2. Compute $\lambda(t_i)/\lambda$

   3. If $\lambda(t_i)/\lambda \leq v_i$ accept $t_j = t_i$; $j = j + 1$

   4. If $i = N(T)$ stop else $i = i + 1$; go to 1.

The $t_j$'s are a random sample from $N(t)$.

The proof is straightforward. It suffices to note that

$$\Pr(\text{accepting } t_i \text{ in } [t, t + \Delta t))$$
$$= \Pr(t_i \text{ in } [t, t + \Delta t)) \Pr(\text{accepting } t_i | t_i \varepsilon [t, t + \Delta t))$$
$$= \lambda \Delta t \frac{\lambda(t_i)}{\lambda}$$
$$= \lambda(t_i) \Delta t. \tag{7.2}$$

Hence $t_i$ is a sample from the point process with conditional intensity function $\lambda(t)$.

### 7.2. Simulating a multivariate point process model

Although we have not formally defined a multivariate point process, it is roughly speaking a set of univariate point processes whose probability structure is characterized by a vector valued joint intensity function. Such models will be necessary to accurately describe multiple single unit neural spike train data. See [25] for a discussion of these models.

**Theorem 3** [24]. Consider an m-dimensional multivariate point process model on an interval $(0, T]$ with joint intensity function $\lambda_t = (\lambda_t^1, \ldots, \lambda_t^m)$. Suppose that there is a homogeneous (univariate) Poisson process with rate $\lambda^*$ such that for all $t$

$$\sum_{k=1}^{m} \lambda_t^k \leq \lambda^* \tag{7.3}$$

and

$$\lambda_t^0 = \lambda^* - \sum_{k=1}^{m} \lambda_t^k. \tag{7.4}$$

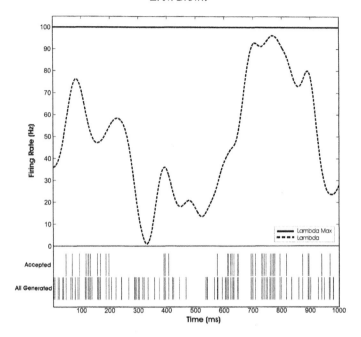

Fig. 6. Illustration of the thinning algorithm. Initially, a set of candidate spikes (All Generated) is simulated from the simple Poisson process with parameter $\lambda$ (——). The candidate spikes are thinned using the algorithm in Eq. 7.1 to obtain a sample (Accepted) from the point process with conditional intensity function $\lambda(t|H_t)$ (- - - -).

Let $0 < t_{1*}, < t_{2*}, \ldots, < t_{n*} \leq T$ be the points of the Poisson process with intensity $\lambda^*$. For each of the points accept $t_{j*}$ to be a sample point from the $k^{th}$ component of $\lambda^*$ with probability $\lambda_t^k/\lambda^*$. Then the accepted points are a sample from a multivariate point process with parameter $\lambda_t$.

**Proof:** The proof is again straightforward. It suffices to note that

$$\Pr(\text{accepting } t_{j*} \text{ in } [t, t + \Delta t))$$
$$= \Pr(t_{j*} \text{ from } \lambda_t^k \text{ in } [t, t + \Delta t)) \Pr(\text{accepting } t_{j*}|t_{j*} \in [t, t + \Delta t))$$
$$= \lambda^* \Delta t \frac{\lambda_t^k(t_{j*})}{\lambda^*}$$
$$= \lambda_t^k(t_{j*}) \Delta t. \tag{7.5}$$

Hence $t_{j*}$ is a sample from the point process with conditional intensity function $\lambda_t^k(t_{j*})$. $\qquad\square$

Notice that this probability is a multinomial probability, for each $t_{j*}$. That is, there is probability that $\lambda_t^k(t_{j*})$ that $t_{j*}$ is assigned neuron $k$ for $k = 1, \ldots, m$ and probability $\lambda_t^0$ that it is assigned to no neuron. Notice that as a consequence, this procedure does not admit simultaneous spikes. If in the proof above, the parent homogeneous Poisson process from which the candidate spike times are drawn is chosen to be piecewise continuous so that the bound can be chosen locally, then the algorithm can be made more efficient [24].

### 7.3. Simulating a univariate point process by time-rescaling

As a second application of the time-rescaling theorem we describe how the theorem may be used to simulate a general point process [18]. The time-rescaling theorem provides a standard approach for simulating an inhomogeneous Poisson process from a simple Poisson process. The general form of the time-rescaling theorem suggests that any point process may be simulated from a Poisson process with unit rate by rescaling time with respect to the conditional intensity (rate) function of the process. Given an interval $(0, T]$ the simulation algorithm proceeds as follows:

1. Set $u_0 = 0$; Set $k = 1$.
2. Draw $\tau_k$ an exponential random variable with mean 1.
3. Find $u_k$ as the solution to $\tau_k = \int_{u_{k-1}}^{u_k} \lambda(u|u_0, u_1, \ldots, u_{k-1}) du$. $\qquad$ (7.6)
4. If $u_k > T$ then stop.
5. $k = k + 1$
6. Go to 2.

By using Eq. 7.6, a discrete version of the time-rescaling algorithm can be constructed as follows. Choose $J$ large, and divide the interval $(0, T]$ into $J$ bins each of width $\Delta = T/J$. For $k = 1, \ldots, J$ draw a Bernoulli random variable $u_k^*$ with probability $\lambda(k\Delta|u_1^*, \ldots, u_{k-1}^*)\Delta$ and assign a spike to bin $k$ if $u_k^* = 1$, and no spike if $u_k^* = 0$.

While in many instances there will be faster, more computationally efficient algorithms for simulating a point process, such as model based methods for specific renewal processes [26] and thinning algorithms, the algorithm in Eq. 7.6 is simple to implement. For an example of where this algorithm is crucial for point process simulation, we consider the inhomogeneous inverse gamma model in [12]. Its conditional intensity function is infinite immediately following a spike if its location parameter $\psi < 1$. If in addition, $\psi$ is time-varying, i.e.,

$\psi = \psi(t) < 1$ for all $t$, then neither thinning nor standard algorithms for making draws from a gamma probability distribution may be used to simulate data from this model. Thinning fails because the conditional intensity function is not bounded and the standard algorithms for simulating a gamma model cannot be applied because $\psi$ is time-varying. In this case, the time-rescaling simulation algorithm may be applied as long as the conditional intensity function remains integrable as $\psi$ varies temporally.

For algorithms to simulate specific renewal models see [26].

## 8. Poisson limit theorems

One method for deriving a Poisson process is as a limit of a binomial process when the number of attempts or trials, $n$, is large and the probability of success $p$ is small. A second approach is a superposition of non-Poisson processes. We discuss both briefly.

### 8.1. Poisson approximation to the binomial probability mass function

Suppose $x$ has a binomial distribution with parameters $n$ and $p$, where $n$ is large and $p$ is small. Let $\lambda = np$

$$
\begin{aligned}
\Pr(x = k) &= \frac{n!}{(n-k)!k!} p^k (1-p)^{n-k} \\
&= \frac{n!}{(n-k)!k!} \left(\frac{\lambda}{n}\right)^k \left(1 - \frac{\lambda}{n}\right)^{n-k} \\
&= \frac{n(n-1)\ldots(n-k+1)}{n^k} \frac{\lambda^k}{k!} \frac{(1 - \frac{\lambda}{n})^n}{(1 - \frac{\lambda}{n})^k}
\end{aligned}
\tag{8.1}
$$

Now for $n$ large and $p$ small

$$
\left(1 - \frac{\lambda}{n}\right)^n \approx e^{-\lambda}, \qquad \frac{n(n-1)\ldots(n-k+1)}{n^k} \approx 1, \qquad \left(1 - \frac{\lambda}{n}\right) \approx 1.
$$

Hence for $n$ large and $p$ small we have

$$
\Pr(x = k) \approx \frac{\lambda^k}{k!} e^{-\lambda}.
\tag{8.2}
$$

## 8.2. Superposition of point processes

Another way to construct a point process is to superpose a set of point processes. The question then arises as to what is the distribution of the superposed process. A case in neuroscience data analysis where this is relevant is in the analysis of the PSTH. The PSTH is constructed by choosing a fixed reference point and superposing the spike trains across multiple run in an experiment with respect to this point. It is often assumed that the resultant distribution of spikes is Poisson. In general superposition of point processes will not lead to a composite process that is a Poisson process [14]. This assumption however, can be easily verified empirically by using the goodness-of-fit analysis based on the time-rescaling theorem discussed in Section 6. One case in which the resultant process is likely to be Poisson is if the spiking activity is sparse as the number of component processes increases, in the spirit of the Poisson approximation to the binomial.

## 9. Problems

1. Let $w_1, \ldots, w_n$ be independent, identically distributed gamma random variables where

$$f(w_i | \alpha, \beta) = \frac{\beta^\alpha}{\Gamma(\alpha)} w_i^{\alpha-1} e^{-\beta w_i},$$

$w_i > 0$, $\alpha > 0$, $\beta > 0$ and $\Gamma(\alpha) = \int_0^\infty x^{\alpha-1} e^{-x} dx$. Let $s_n = \sum_{i=1}^n w_i$. Show that the probability density of

$$f(s_n | \alpha^*, \beta) = \frac{\beta^{\alpha^*}}{\Gamma(\alpha^*)} s_n^{\alpha^*-1} e^{-\beta s_n},$$

where $\alpha^* = n\alpha$.

2. Show that if $w$ has an inverse Gaussian probability density with parameters $\mu'$ and $\lambda'$. Verify that $E(w) = \mu'$ and $Var(w) = \mu'^3 \lambda'^{-1}$.

3. Let $w_1, \ldots, w_n$ be independent, identically distributed gamma random variables. Write their joint probability density in terms of the interevent time probability density and in terms of the conditional intensity function. Show that while the interevent times are independent, the counting process increments are correlated.

4. If $z$ has an exponential probability density with parameter $\lambda$, show that $u = 1 - e^{-z}$ is a uniform random variable on the interval $(0, 1)$.

5. Consider a point process whose conditional intensity function is

$$\lambda(t|H_t) = \mu + \alpha \int_0^t e^{-\beta(t-u)} dN(u),$$

defined an interval $(0, T]$, where $\mu > 0$, $\alpha > 0$, and $\beta > 0$. Write a thinning algorithm to simulate this point process model. How can you use the time-rescaling theorem and the $KS$ plots to check the accuracy of your simulation? Write a continuous time and a discrete time algorithm to simulate this model based on the time-rescaling theorem.

## Acknowledgments

Support for this work was provided in part by NIH grants MH66410, MH59733, MH61637 and DAO15644. We thank Uri Eden and Riccardo Barbieri for help preparing the figures.

## References

[1] D. Daley and D. Vere-Jones, *An Introduction to the Theory of Point Process (2nd ed.)* (Springer-Verlag, New York, 2003).

[2] H.C. Tuckwell, *Introduction to Theoretical Neurobiology: Nonlinear and Stochastic Theories* (Cambridge, New York, 1988).

[3] R.S. Chhikara and J.L. Folks, *The Inverse Gaussian Distribution: Theory, Methodology, and Applications* (Marcel Dekker, New York, 1989).

[4] G.L. Gerstein and B. Mandelbrot, J. Biophys. **4** (1964) 41-68.

[5] E.N. Brown, R. Barbieri, U.T. Eden, and L.M. Frank, in: *Computational Neuroscience: A Comprehensive Approach* (CRC, London, 2003).

[6] D. Snyder and M. Miller, *Random Point Processes in Time and Space (2nd ed.)* (Springer-Verlag, 1991).

[7] M. Jacobsen, *Statistical Analysis of Counting Processes* (Springer-Verlag, New York, 1982).

[8] P. Guttorp, *Stochastic Modeling of Scientific Data* (Chapman & Hall, London, 1995).

[9] J. Kalbfleisch and R. Prentice, *The Statistical Analysis of Failure Time Data* (Wiley, New York, 1980).

[10] P.K. Andersen, O. Borgan, R.D. Gill, and N. Keiding, *Statistical Models Based on Counting Processes* (Springer-Verlag, New York, 1993).

[11] D.R. Brillinger, Biol. Cyber. **59** (1988) 189-200.

[12] R. Barbieri, M.C. Quirk, L.M. Frank, M.A. Wilson, and E.N. Brown, J. Neurosci. Meth. **105** (2001) 25-37.

[13] S.M. Ross, *Introduction to Probability Models* (6th ed.) (Academic Press, San Diego, 1997).

[14] D.R. Cox and V. Isham, *Point Processes* (Chapman and Hall, London, 1980).

[15] D.R. Cox and H.D. Miller, *The Theory of Stochastic Processes* (Methuen, London, 1965).

[16] P. Meyer, *Démonstration Simplifiée d'un Théorème de Knight, Séminaire Probabilitié V. Lecture Notes in Mathematics* (Springer-Verlag, New York, 1969) 191-195.

[17] F. Panangelou, Trans. Amer. Math. Soc. **165** (1972) 483-506.

[18] E.N. Brown, R. Barbieri, V. Ventura, R.E. Kass, and L.M. Frank, Neural Computation **14(2)** (2002) 325-46.

[19] S. Port, *Theoretical Probability for Applications* (Wiley, New York, 1994).

[20] M.H. Porter and C.B. Morrey, *A First Course in Real Analysis (2nd ed.)* (Springer-Verlag, New York, 1991).

[21] Y. Ogata, Journal of the American Statistical Association **83** (1988) 9-27.

[22] A. Johnson, S. Kotz, *Distributions in Statistics: Continuous Univariate Distributions-2* (Wiley, New York, 1970).

[23] P.A.W. Lewis and G.S. Shedler, Naval Res. Logistics Quart. **26** (1978) 403-413.

[24] Y. Ogata, IEEE Transactions on Information Theory **27** (1981) 23-31.

[25] E.N. Brown, R.E. Kass, and P.M. Mitra, Nature Neuroscience 7(5) (2004) 456-461.

[26] B. Ripley, *Stochastic Simulation* (Wiley, New York, 1987).

Course 15

# TECHNIQUE(S) FOR SPIKE-SORTING

## C. Pouzat

*Laboratoire de Physiologie Cérébrale, CNRS UMR 8118*
*UFR Biomédicale de l'Université René Descartes (Paris V)*
*45, rue des Saints Pères*
*75006 Paris*
*France*
*http://www.biomedicale.univ-paris5.fr/physcerv/C_Pouzat.html*
*e-mail: christophe.pouzat@univ-paris5.fr*

*C.C. Chow, B. Gutkin, D. Hansel, C. Meunier and J. Dalibard, eds.*
*Les Houches, Session LXXX, 2003*
*Methods and Models in Neurophysics*
*Méthodes et Modèles en Neurophysique*

# Contents

# 1. Introduction

Most of this book is dedicated to the presentation of models of neuronal networks and of methods developed to work with them. Within the frameworks set by these models, the activity of populations of neurons is derived based on the intrinsic properties of "simplified" neurons and of their connectivity pattern. The experimentalist trying to test these models must start by collecting data for which model predictions can be made, which means he or she must record the activity of several neurons at once. If the model does in fact predict the compound activity of a local neuronal population (*e.g.*, a cortical column or hyper-column) on a relatively slow time scale (100 msec or more), techniques like intrinsic optical imaging [9] are perfectly adapted. If the model makes "more precise" predictions on the activity of individual neurons on a short time scale (~ msec) then appropriate techniques must be used like fast optical recording with voltage sensitive dyes [36] or the more classical extracellular recording technique [22].

We will focus in this chapter on the problems associated with the extracellular recording technique which is still a very popular investigation method. This popularity is partly due to its relative ease of implementation and to its low cost. The author is moreover clearly biased toward this technique being one of its users. We hope nevertheless that users of other techniques will learn something from what follows. We will explain how to make inferences about values of the parameters of a rather complex model from noisy data. The approach we will develop can be adapted to any problem of this type. What will change from an experimental context to another is the relevant data generation model and the noise model. That does not mean that the adaptation of the method to other contexts is trivial, but it is still doable.

The data generation model with which we will work in this chapter is qualitatively different from the ones which have been considered in the past [21, 30, 31, 34]. The reader unfamilliar with the spike-sorting problem and willing to get some background on it can consult with profit Lewicki's review [22]. To get an introduction on the actual statistical methods adapted to the models previously considered, we recommend the consultation of the statistical literature (*eg*, [3]) rather than the spike-sorting one, which tends, in our opinion, to suffer from a strong taste for ad-hoc methods.

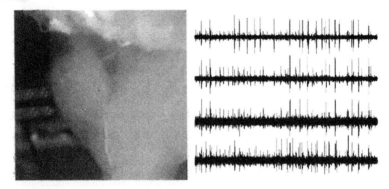

Fig. 1. Example of a tetrode recording from the locust (*Schistocerca americana*) antennal lobe. Left, recording setting. The probe, a silicon substrate made of two shanks with 16 iridium deposits (the actual recording sites), can be seen on the left side of the picture. Each group of 4 recording sites is called a *tetrode* (there are therefore 4 tetrodes by probe). The recording sites are the bright spots. The width of the shanks is 80 $\mu m$ , the side length of each recording site is 13 $\mu m$ , the diagonal length of each tetrode is 50 $\mu m$ , the center to center distance between neighboring tetrodes is 150 $\mu m$ . The structure right beside the probe tip is the *antennal lobe* (the first olfactory relay of the insect), its diameter is approximately 400 $\mu m$ . Once the probe has been gently pushed into the antennal lobe such that the lowest two tetrodes are roughly 100 $\mu m$ below the surface one gets on these lowest tetrodes data looking typically as shown on the right part of the figure. Right, 1s of data from a single tetrode filtered between 300 and 5kHz.

## 2. The problem to solve

Extracellular recordings are typically a mixture of spike waveforms originating from a generally unknown number of neurons to which a background noise is superposed as illustrated on Fig 1. Several features can be used to distinguish spikes from different neurons [22] like the peak amplitude on a single of several recording sites (Fig 2), the spike width, a bi - or tri - phasic waveform, etc.

*What do we want?*

- Find the number of neurons contributing to the data.
- Find the value of a set of parameters characterizing the signal generated by each neuron (*e.g.*, the spike waveform of each neuron on each recording site).
- Acknowledging the classification ambiguity which can arise from waveform similarity and/or signal corruption due to noise, the probability for each neuron to have generated each event (spike) in the data set.
- A method as automatic as possible.
- A non ad - hoc method to answer the above questions. By non ad - hoc we mean a method based on an *explicit probabilistic* model for data generation.

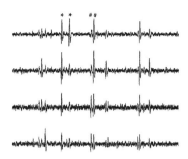

Fig. 2. The last 200 ms of Fig. 1. Considering two pairs of spikes (** and ##) the interest of the tetrodes becomes clear. On the first recording site (top) the two spikes of the pair (**) look very similar and it would therefore be hard for an analyst having only the top recording site information to say if these two spikes originate from the same neuron or from two different ones. If now, one looks at the same spikes on the three other recording sites the difference is obvious. The same holds for the two spikes of the pair (##). They look similar on sites 3 and 4 but very dissimilar on sites 1 and 2.

## 3. Two features of single neuron data we would like to include in the spike-sorting procedure

### 3.1. *Spike waveforms* from a single neuron *are usually not stationary on a short time–scale*

#### 3.1.1. *An experimental illustration with cerebellar Purkinje cells*
One commonly observes that, when principal cells fire bursts of action potentials, the spike amplitude decreases[1] during the burst as illustrated on Fig. 3.

#### 3.1.2. *A phenomenological description by an exponential relaxation*
Following [6] we will use an exponential relaxation to describe the spike waveform dependence upon the inter - spike interval:

$$\mathbf{a}\,(isi) = \mathbf{p} \cdot (1 - \delta \cdot \exp\,(-\lambda \cdot isi)), \tag{3.1}$$

where **p** is the vector of maximal amplitudes (or full waveforms) on the different recording sites, $\delta \in [0, 1]$, $\lambda$, measured in 1/s, is the inverse of the relaxation time constant.

---

[1]More generally the spike shape changes and basically slows down [13]. This is mainly due to sodium channels inactivation.

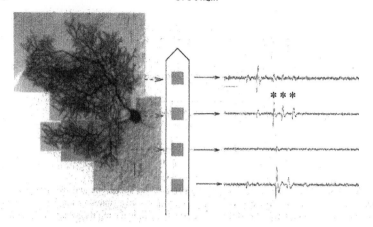

Fig. 3. An example of a multi - electrode recording from a rat cerebellar slice. Left, picture of a slice (from a 38 days old rat) with a biocytin filled Purkinje cell. Middle, the tip of a Michigan probe drawn to scale. Right, 100 ms of data filtered between 300 and 5K Hz, Scale bar: 10 ms, Stars: a triplet of spikes coming from a single Purkinje cell. (Delescluse and Pouzat, unpublished)

## 3.2. Neurons inter–spike interval probability density functions carry a lot of information we would like to exploit

### 3.2.1. An experimental illustration from projection neurons in the locust antennal lobe

It is well known and understood since the squid giant axon study by Hodgkin and Huxley that once a neuron (or a piece of it like its axon) has fired an action potential, we must wait "a while" before the next action potential can be fired. This delay is dubbed the *refractory period* and is mainly due to inactivation of sodium channels and to strong activation of potassium channels at the end of the spike, meaning that we must wait for the potassium channels to de - activate and for sodium channel to de - inactivate. Phenomenologically it means that we should observe on the inter - spike interval (*ISI*) histogram from a single neuron a period without spikes (*i.e.*, the *ISI* histogram should start at zero and stay at zero for a finite time). In addition we can often find on *ISI* histograms some other features like a single mode[2] a "fast" rise and a "slower" decay as illustrated on Fig 4. The knowledge of the *ISI* histogram can in principle be used to improve spike - sorting because it will induce correlations between the labeling of successive spikes. Indeed, if in one way or another we can be sure of the labeling of a given spike to a given neuron, we can be sure as well that the probability to have an other spike from the same neuron within say the next 10

---

[2]That's the statistician's terminology for local maximum.

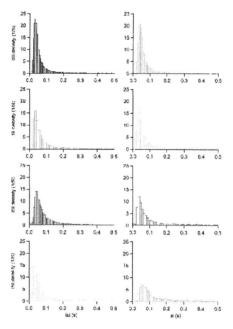

Fig. 4. An example of *ISI pdf* estimates for 8 projection neurons simultaneously recorded in the locust antennal lobe (Pouzat, Mazor and Laurent, unpublished).

ms is zero, that this probability is high between 30 and 50 ms and high as well between 60 and 100 (you just need to integrate the *ISI* histogram to see that).

### 3.2.2. A phenomenological description by a log-normal pdf

We will use a log - Normal approximation for our empirical *ISI* probability density functions (*pdf*s):

$$
\pi \left( ISI = isi \mid S = s, F = f \right) =
$$

$$
\frac{1}{isi \cdot f \cdot \sqrt{2\pi}} \cdot \exp \left[ -\frac{1}{2} \cdot \left( \frac{\ln isi - \ln s}{f} \right)^{2} \right], \tag{3.2}
$$

where S is a scale parameter (measured in seconds, like the *ISI*) and *F* is a shape parameter (dimensionless). Fig. 5 shows three log - Normal densities for different values of *S* and *F*. The lowest part of the figure shows that when we look at the density of the logarithm of the *ISI* we get a Normal distribution which explains the name of this density.

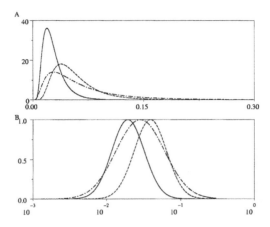

Fig. 5. A, examples of log - Normal densities. plain: $S = 0.025$, $F = 0.5$; dashed: $S = 0.05$, $F = 0.5$; dot-dashed: $S = 0.05$, $F = 0.75$. B, peak normalized densities displayed with a logarithmic abscissa.

In the sequel, we will in general use $\pi$ to designate proper probability density functions (*pdf* for continuous random variables) or probability mass functions (*pmf* for discrete random variables). By proper we mean the integral (or the sum) over the appropriate space is 1. We will (try to) use uppercases to designate random variables (*e.g.*, *ISI, S, F*) and lowercases to designate their realizations (*e.g.*, *isi, s, f*). We will use a Bayesian approach, which means that the model parameters (like $S$ and $F$ in Eq. 3.2) will be considered as random variables.

## 4. Noise properties

It's better to start with a reasonably accurate noise model and Fig. 6 illustrates how the noise statistical properties can be obtained from actual data. Once the empirical noise auto - and cross - correlation functions have been obtained, assuming the noise is stationary, the noise covariance matrix ($\Sigma$) is constructed as a block - Toeplitz matrix. There are as many blocks as the square of the number of recording sites. The first row of each block is the corresponding auto- or cross - correlation function, see [31]. Then, if $\Sigma$ is a complete noise description, the *pdf* of a noise vector **N** is multivariate Gaussian:

$$\pi\,(\mathbf{N} = \mathbf{n}) = \frac{1}{(2\pi)^{\frac{D}{2}}} \cdot \left|\Sigma^{-1}\right|^{\frac{1}{2}} \cdot \exp\left(-\frac{1}{2} \cdot \mathbf{n}^T \Sigma^{-1} \mathbf{n}\right), \qquad (4.1)$$

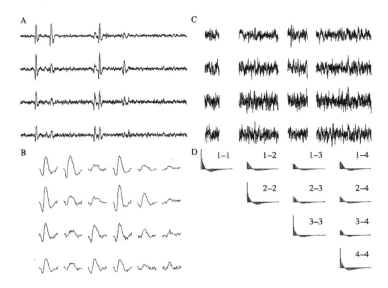

Fig. 6. Procedure used to obtain the second order statistical properties of the recording noise illustrated with a tetrode recording from the locust antennal lobe. Putative "events" (B, the sweeps are 3 ms long) and "noise" (C) are separated from the complete data (A) after a threshold crossing and template matching detection of the events. The noise auto - and cross - correlation functions (D, 3 ms are shown for each function, the autocorrelations functions are on the diagonal, the crosscorrelation functions on the upper part) are computed from the reconstructed noise traces (C). Adapted from [31].

where $D$ is the dimension of the space used to represent the events (and the noise), $|\Sigma^{-1}|$ stands for the determinant of the inverse of $\Sigma$, $\mathbf{n}^T$ stands for the transpose of $\mathbf{n}$.

## 4.1. Noise whitening

If the noise covariance matrix is known, it is useful to know about a transformation, *noise whitening*, which is easy to perform. It makes the equations easier to write and computations faster. If $\Sigma$ is indeed a covariance matrix, that implies:

$$\mathbf{v}^T \Sigma^{-1} \mathbf{v} \geq 0, \forall \mathbf{v} \in \Re^D. \tag{4.2}$$

That is, the *Mahalanobis distance* ($\mathbf{v}^T \Sigma^{-1} \mathbf{v}$) is a proper distance which is stated differently by saying that the inverse of $\Sigma$ is positive definite. Then one can show (Exercise 1 below) that there exists a unique lower triangular matrix $A$

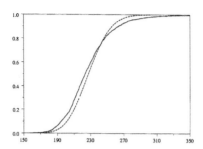

Fig. 7. How good is a noise description based on a multivariate Gaussian *pdf*? A typical example (from locust antennal lobe) where $\Sigma$ turns out to be a reasonable noise description. A sample of 2000 noise events $\{\mathbf{n}_1, \ldots, \mathbf{n}_{2000}\}$ was randomly selected from the reconstructed noise trace (Fig. 6C), the *Mahalanobis distance* to the origin: $\mathbf{n}_i^T \Sigma^{-1} \mathbf{n}_i$, was computed for each of them and the cumulative distribution of these distances (plain) was compared with the theoretical expectation (dashed) a $\chi^2$ distribution (with in that case 238 degrees of freedom). See [31] for details.

such that $AA^T = \Sigma^{-1}$ and transform vector $\mathbf{v}$ of Eq. 4.2 into $\mathbf{w} = A^T\mathbf{v}$. We then have:

$$\mathbf{w}^T\mathbf{w} = \mathbf{v}^T\Sigma^{-1}\mathbf{v}. \tag{4.3}$$

That is, we found a new coordinate system in which the *Mahalanobis distance* is the *Euclidean distance*. We say we perform *noise whitening* when we go from $\mathbf{v}$ to $\mathbf{w}$.

Assuming the noise is stationary and fully described by its second order statistical properties is not enough. One can check the validity of this assumption after whitening an empirical noise sample as illustrated in Fig. 7.

*Exercise 1*

Assume that $\Gamma = \Sigma^{-1}$ is a symmetric positive definite matrix. Show there exists a unique (in fact 2, one being obtained from the other by a multiplication by - 1) lower triangular matrix $A$ (the *Cholesky* factor) such that:

$$AA^T = \Gamma. \tag{4.4}$$

In order to do that you will simply get the algorithm which computes the elements of $A$ from the elements of $\Gamma$. See Sec. 13.1 for solution.

## 5. Probabilistic data generation model

### 5.1. Model assumptions

We will make the following assumptions:

1. The firing statistics of each neuron is fully described by its time independent inter - spike interval density. That is, the sequence of spike times from a given neuron is a realization of *a homogeneous renewal point process* [18]. More specifically, we will assume that the *ISI pdf* is log - Normal (Eq. 3.2).

2. The amplitude of the spikes generated by each neuron depends on the elapsed time since the last spike of this neuron. More specifically, we will assume that this dependence is well described by Eq. 3.1.

3. The measured amplitude of the spikes is corrupted by a Gaussian white noise which sums linearly with the spikes and is statistically independent of them. That is, we assume that noise whitening (Eq. 4.3) has been performed.

### 5.2. Likelihood computation for single neuron data

With a data set $\mathcal{D}' = \{(t_0, \mathbf{a}_0), (t_1, \mathbf{a}_1), \ldots, (t_N, \mathbf{a}_N)\}$, the likelihood is readily obtained (Fig. 8 ). One must first get the *isi* values: $i_j = t_j - t_{j-1}$, which implies one has to discard the first spike of the data set[3] to get the "effective" data set: $\mathcal{D} = \{(i_1, \mathbf{a}_1), \ldots, (i_N, \mathbf{a}_N)\}$. The likelihood is then[4]:

$$L(\mathcal{D} \mid \mathbf{p}, \delta, \lambda, s, f) = \prod_{j=1}^{N} \pi_{isi}(i_j \mid s, f) \cdot \pi_{amplitude}(\mathbf{a}_j \mid i_j, \mathbf{p}, \delta, \lambda), (5.1)$$

---

[3]To avoid discarding the first spike we can assume periodic boundary conditions, meaning the last spike (*N*) "precedes" the first one (*0*), we then have: $i_0 = T - t_N + t_0$, where $T$ is the duration of the recording which started at time 0.

[4]Purists from both sides (frequentist and Bayesian) would kill us for writing what follows (Eq. 5.1)... A frequentist would rather write:

$L(\mathbf{p}, \delta, \lambda, s, f; \mathcal{D})$

because for him the likelihood function is a random function (it depends on the sample through $\mathcal{D}$) and its arguments are the model parameters. The Bayesian, probably to show the frequentist that he perfectly understood the meaning of the likelihood function, would introduce a new symbol, say $h$ and write:

$h(\mathcal{D} \mid \mathbf{p}, \delta, \lambda, s, f) = L(\mathbf{p}, \delta, \lambda, s, f; \mathcal{D})$.

But the likelihood is, within a normalizing constant, the probability of the data for given values of the model parameters. We will therefore use the heretic notation of Eq. 5.1.

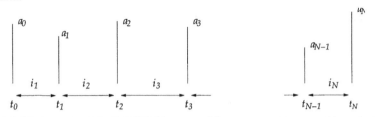

Fig. 8. Illustration of the likelihood computation for data from a single neuron recorded on a single site. For simplicity we assume that the spike peak amplitude only is used.

where $\pi_{isi}\left(i_j \mid s, f\right)$ is given by Eq. 3.2 and:

$$\pi_{amplitude}\left(\mathbf{a}_j \mid i_j, \mathbf{p}, \delta, \lambda\right) =$$
$$\frac{1}{(2\pi)^{\frac{D}{2}}} \cdot \exp\left\{-\frac{1}{2}\left\|\mathbf{a}_j - \mathbf{p} \cdot \left(1 - \delta \cdot \exp\left(-\lambda \cdot i_j\right)\right)\right\|^2\right\}. \qquad (5.2)$$

This last equation is obtained by combining Eq. 3.1 with our third model assumption (Gaussian white noise). For convenience we write $\pi_{isi}\left(i_j \mid s, f\right)$ for $\pi_{isi}\left(ISI_j = i_j \mid S = s, F = f\right)$ and $\pi_{amplitude}\left(\mathbf{a}_j \mid i_j, \mathbf{p}, \delta, \lambda\right)$ for $\pi_{amplitude}\left(\mathbf{A}_j = \mathbf{a}_j \mid ISI_j = i_j, \mathbf{P} = \mathbf{p}, \Delta = \delta, \Lambda = \lambda\right)$, where the $ISI_j$ are considered independent and identically distributed (*iid*) random variables (conditioned on $S$ and $F$) as well as the $\mathbf{A}_j$ (conditioned on $ISI_j$, $\mathbf{P}$, $\Delta$ and $\Lambda$).

The log - likelihood ($\mathcal{L} = \ln L$) can be written as a sum of two terms:

$$\mathcal{L}\left(\mathcal{D} \mid \mathbf{p}, \delta, \lambda, s, f\right) = \mathcal{L}_{isi}\left(\mathcal{D} \mid s, f\right) + \mathcal{L}_{amplitude}\left(\mathcal{D} \mid \mathbf{p}, \delta, \lambda\right), \qquad (5.3)$$

where:

$$\mathcal{L}_{isi}\left(\mathcal{D} \mid s, f\right) = -N \cdot \ln f - \sum_{j=1}^{N}\left\{\ln i_j + \frac{1}{2}\left[\frac{\ln\left(\frac{i_j}{s}\right)}{f}\right]^2\right\} + Cst, \qquad (5.4)$$

and:

$$\mathcal{L}_{amplitude}\left(\mathcal{D} \mid \mathbf{p}, \delta, \lambda\right) =$$
$$-\frac{1}{2}\sum_{j=1}^{N}\left\|\mathbf{a}_j - \mathbf{p} \cdot \left(1 - \delta \cdot \exp\left(-\lambda \cdot i_j\right)\right)\right\|^2 + Cst. \qquad (5.5)$$

The term $\mathcal{L}_{isi}$ is a function of the *ISI pdf* parameters only and the term $\mathcal{L}_{amplitude}$ is a function of the "amplitude dynamics parameters" only.

*Exercise 2*

Starting with the expression of the log - likelihood given by Eq. 5.4 , show that the values $\tilde{s}$ and $\tilde{f}$ of $s$ and $f$ maximizing the likelihood are given by:

$$\ln \tilde{s} = \frac{1}{N} \sum_{j=1}^{N} \ln i_j, \tag{5.6}$$

and:

$$\tilde{f} = \sqrt{\frac{1}{N} \sum_{j=1}^{N} \left[ \ln \left( \frac{i_j}{\tilde{s}} \right) \right]^2}. \tag{5.7}$$

*Exercise 3*

Remember that in a Bayesian setting your start with a prior density for your model parameters: $\pi_{prior}(s, f)$, you observe some data, $\mathcal{D}$, you have an explicit data generation model (*i.e.*, you know how to write the likelihood function), $L(\mathcal{D} \mid s, f)$ and you can therefore write using Bayes rule:

$$\pi(s, f \mid \mathcal{D}) = \frac{L(\mathcal{D} \mid s, f) \cdot \pi_{prior}(s, f)}{\int_{u,v} du \, dv \, L(\mathcal{D} \mid u, v) \cdot \pi_{prior}(u, v)} \tag{5.8}$$

The "Bayesian game" in simple cases is always the same, you take the numerator of Eq. 5.8, you get rid of everything which does not involve the model parameters (because that will be absorbed in the normalizing constant) and you try to recognize in what's left a known *pdf* like a Normal, a Gamma, a Beta, etc.

In the following we will assume that our prior density is uniform on a rectangle, that is:

$$\pi_{prior}(s, f) =$$
$$\frac{1}{s_{max} - s_{min}} \cdot \mathcal{I}_{[s_{min}, s_{max}]}(s) \cdot \frac{1}{f_{max} - f_{min}} \cdot \mathcal{I}_{[f_{min}, f_{max}]}(f), \tag{5.9}$$

where $\mathcal{I}_{[x,y]}$ is the indicator function:

$$\mathcal{I}_{[x,y]}(u) = \begin{cases} 1, & if \ x \leq u \leq y \\ 0, & otherwise \end{cases} \tag{5.10}$$

Detailed answers to the following questions can be found in Sec. 13.2.

*Question 1*   Using Eq. 8 and Eq. 3.2 show that the posterior density of ln *S* conditioned on $F = f$ and $\mathcal{D}$ is a truncated Gaussian with mean:

$$\overline{\ln i} = \frac{1}{N} \sum_{j=1}^{N} \ln i_j, \tag{5.11}$$

and un-truncated variance (that is, the variance it would have if it was not truncated):

$$\sigma^2 = \frac{f^2}{N}. \tag{5.12}$$

And therefore the posterior conditional *pdf* of the scale parameter is given by:

$$\pi(s \mid f, \mathcal{D}) \, \alpha$$
$$\exp\left[-\frac{1}{2}\frac{N}{f^2}\left(\overline{\ln i} - \ln s\right)^2\right] \cdot \frac{1}{s_{max} - s_{min}} \cdot \mathcal{I}_{[s_{min}, s_{max}]}(s). \tag{5.13}$$

*Question 2*   Write an algorithm which generates a realization of *S* according to Eq. 5.13.

*Question 3*   We say that a random variable $\Theta$ has an *Inverse Gamma* distribution and we write: $\Theta \sim Inv - Gamma(\alpha, \beta)$ if the *pdf* of $\Theta$ is given by:

$$\pi(\Theta = \theta) = \frac{\beta^\alpha}{\Gamma(\alpha)} \, \theta^{-(\alpha+1)} \, \exp\left(-\frac{\beta}{\theta}\right). \tag{5.14}$$

Using Eq. 8 and Eq. 3.2 show that the posterior density of $F^2$ conditioned on $S = s$ and $\mathcal{D}$ is a truncated *Inverse Gamma* with parameters:

$$\alpha = \frac{N}{2} - 1$$

and

$$\beta = \frac{1}{2} \sum_{j=1}^{N} \left(\ln i_j - \ln s\right)^2.$$

Therefore the posterior conditional *pdf* of the shape parameter is:

$$\pi(f \mid s, \mathcal{D}) \, \alpha$$
$$\frac{1}{f^N} \exp\left[-\frac{1}{2}\frac{\sum_{j=1}^{N}(\ln i_j - \ln s)^2}{f^2}\right] \cdot \frac{1}{f_{max} - f_{min}} \cdot \mathcal{I}_{[f_{min}, f_{max}]}(f). \tag{5.15}$$

*Question 4* Assuming that your favorite analysis software (*e.g.*, Scilab[5]or Mat-lab) or your favorite C library (*e.g.*, GSL[6]) has a *Gamma* random number generator, write an algorithm which generates a realization of $F$ from its posterior density Eq. 5.15. The *pdf* of a *Gamma* random variable $\Omega$ is given by:

$$\pi\left(\Omega = \omega\right) = \frac{\beta^{\alpha}}{\Gamma\left(\alpha\right)}\,\omega^{\alpha-1}\,\exp\left(-\beta \cdot \omega\right). \tag{5.16}$$

### 5.3. Complications with multi–neuron data

#### 5.3.1. Notations for the multi–neuron model parameters
In the following we will consider a model with $K$ different neurons, which means we will have to estimate the value of a maximal peak amplitude parameter ($\mathbf{P}$), an attenuation parameter ($\Delta$), the inverse of a relaxation time constant ($\Lambda$), a scale parameter ($S$) and a shape parameter ($F$) for each of the $K$ neurons of our model. We will write $\Theta$ the random vector lumping all these individual parameters:

$$\Theta = \left(\mathbf{P}_1, \Delta_1, \Lambda_1, S_1, F_1, \ldots, \mathbf{P}_K, \Delta_K, \Lambda_K, S_K, F_K\right). \tag{5.17}$$

#### 5.3.2. Configuration and data augmentation
When we deal with multi - neuron data we need to formalize our ignorance about the origin of each individual spike. In order to do that, assuming we work with a model with $K$ neurons, we will "attach" to each spike, $j$, in our data set ($\mathcal{D}$) a *label*, $l_j \in \{1, \ldots, K\}$ whose value corresponds to the number of the neuron which generated it. From this view point, a spike - sorting procedure is a method to set the labels values. But we don't know, at least until we have estimated the model parameters, what is the value of each $l_j$. In such an uncertain situation the best we can do is to consider $l_j$ as a realization of a random variable $L_j$ taking values in $\{1, \ldots, K\}$, then the best we can expect from a spike - sorting procedure is the *distribution* (or the *pmf*) of $L_j$ for $j \in \{1, \ldots, N\}$.

Our data generation model will necessarily induce a dependence among the $L_j$ because we take into account both the *ISI* densities and the spike waveform dynamics, we will therefore introduce an other random variable, the *configuration*, $C$ defined as a vector of $L_j$ that is:

$$C = \left(L_1, \ldots, L_N\right)^{T}. \tag{5.18}$$

The introduction of this random variable $C$ is a very common procedure in the statistical literature where it is called *data augmentation*. The key idea being that we are given an *incomplete* data set ($\mathcal{D}$), which basically does not allow us

---

[5]http://www.scilab.org
[6]The Gnu Scientific Library: http://sources.redhat.com/gsl

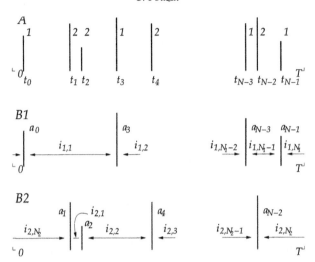

Fig. 9. A simple example of sub-trains extraction. We consider a case with 2 neurons in the model and a single recording site. A, Spikes are shown with their labels in the present configuration and their occurrence times. The recording goes from $0$ to $T$. B1, the sub-train attributed to neuron 1 has been isolated. The relevant *ISIs* are shown, as well as the amplitudes of the events. B2, same as B1 for the sub-train attributed to neuron 2.

to write the likelihood easily, and if we knew some extra properties of the data (like the *configuration*), we could write down the likelihood of the *augmented* (or *completed*) data in a straightforward way[7]. Indeed, with $C$ introduced the likelihood of the augmented data ($L\,(\mathcal{D}, c \mid \theta)$) is obtained in two steps:

1. Using the configuration realization, $c$, we can separate the "complete" data set into sub‑trains corresponding to the different neurons of the model. This is illustrated with a simple case, $K = 2$ and a single recording site on Fig. 9. Here we have: $i_{1,1} = t_3 - t_0$, $i_{1,N_1-1} = t_{N-3} - t_{N-1}$, $i_{1,N_1} = T - t_{N-1} + t_0$, where periodic boundary conditions are assumed and where $N_1$ is the number of spikes attributed to neuron 1 in $c$. For neuron 2 we get: $i_{2,1} = t_2 - t_1$, $i_{2,2} = t_4 - t_2$, $i_{2,N_2} = T - t_{N-2} + t_1$, where $N_2$ is the number of spikes attributed to neuron 2 in $c$. Using Eq. 8 we can compute a "sub‑likelihood" for each sub‑train: $L_{l_j=1}\,(\mathcal{D}, c \mid \theta)$, $L_{l_j=2}\,(\mathcal{D}, c \mid \theta)$.

2. Because our data generation model does not include interactions between neu-

---

[7] $C$ is also called a *latent* variable in the statistical literature. Statisticians, moreover, commonly use the symbol $Z$ for our $C$. We have change this convention for reasons which will soon become clear.

rons, the "complete" likelihood is simply[8]:

$$L(\mathcal{D}, c \mid \theta) = \prod_{q=1}^{K} L_{l_j=q}(\mathcal{D}, c \mid \theta).$$ (5.19)

### 5.3.3. Posterior density

Now that we know how to write the likelihood of our *augmented* data set, we can obtain the posterior density of all our unknowns $\Theta$ *and* $C$ applying Bayes rule:

$$\pi_{posterior}(\theta, c \mid \mathcal{D}) = \frac{L(\mathcal{D}, c \mid \theta) \cdot \pi_{prior}(\theta)}{Z},$$ (5.20)

with the normalizing constant $Z$ given by:

$$Z = \sum_{c \in C} \int_{\theta} d\theta \, L(\mathcal{D}, c \mid \theta) \cdot \pi_{prior}(\theta),$$ (5.21)

where $C$ is the set of all configurations.

### 5.4. Remarks on the use of the posterior

It should be clear that if we manage to get the posterior: $\pi_{posterior}(\theta, c \mid \mathcal{D})$, then we can answer any quantitative question about the data. For instance if we want the probability of a specific configuration $c$ we simply need to integrate out $\theta$:

$$\pi(c \mid \mathcal{D}) = \int_{\theta} d\theta \, \pi_{posterior}(\theta, c \mid \mathcal{D}).$$ (5.22)

If we are interested in the cross - correlogram between to neurons, say 2 and 4, of the model and if we formally write $Cross_{2,4}(c)$ the function which takes as argument a specific configuration ($c$) and returns the desired cross - correlogram, then the *best* estimate we can make of this quantity is:

$$\langle Cross_{2,4} \rangle = \sum_{c \in C} \int_{\theta} d\theta \, \pi_{posterior}(\theta, c \mid \mathcal{D}) \cdot Cross_{2,4}(c),$$ (5.23)

where we use the symbol $\langle \rangle$ to designate averaging with respect to a proper *pdf*.

---

[8] Strictly speaking, this is true only if we can ignore spike waveforms superpositions.

## 5.5. The normalizing constant problem and its solution

### 5.5.1. The problem

If we take a look at our normalizing constant Z (Eq. 5.21) we see a practical problem emerging. To compute Z we need to carry out a rather high dimensional integration (on $\theta$) *and* a summation over every possible configuration in $\mathcal{C}$ . That's where the serious problem arises because $\mathcal{C}$ contains $K^N$ elements! Blankly stated, we have no way, in real situations where $K$ varies from 2 to 20 and where $N$ is of the order of 1000, to compute Z.

Fine, if we can't compute directly Z, let us look for a way around it. First, we introduce an "energy function":

$$E\left(\theta, c \mid \mathcal{D}\right) = -\ln\left[L\left(\mathcal{D}, c \mid \theta\right) \cdot \pi_{prior}\left(\theta\right)\right]. \tag{5.24}$$

$E$ is defined in such a way that given the data ($\mathcal{D}$ ), a value for the model parameters ($\theta$) and a configuration ($c$) we can compute it.

Second, we rewrite our posterior expression as follows:

$$\pi_{posterior}\left(\theta, c \mid \mathcal{D}\right) = \frac{\exp\left[-\beta\, E\left(\theta, c \mid \mathcal{D}\right)\right]}{Z}, \tag{5.25}$$

where $\beta = 1$.

Third, we make an analogy with statistical physics: $E$ is an energy, $Z$ a *partition function*, $\beta$ is proportional to the inverse of the temperature ($(kT)^{-1}$) and $\pi_{posterior}\left(\theta, c \mid \mathcal{D}\right)$ is the canonical distribution [19, 29].

Fourth, we look at the way physicists cope with the combinatorial explosion in the partition function...

### 5.5.2. Its solution

Physicists are interested in computing expected values (*i.e.*, things that can be measured experimentally) like the expected internal energy of a system:

$$\langle E \rangle = \sum_{c \in \mathcal{C}} \int_{\theta} d\theta \, \frac{\exp\left[-\beta\, E\left(\theta, c \mid \mathcal{D}\right)\right]}{Z} \cdot E\left(\theta, c \mid \mathcal{D}\right), \tag{5.26}$$

or like the pair - wise correlation function between two molecules in a gas or in a fluid which will give rise to equations like Eq. 5.23. Of course they are very rarely able to compute Z explicitly so the trick they found fifty years ago [24] is to *estimate* quantities like $\langle E \rangle$ with *Monte Carlo* integration.

At this point to make notation lighter we will call *state* of our "system" the pair $(\theta, c)$ and we will write it as: $x = (\theta, c)$. We will moreover drop the explicit dependence on $\mathcal{D}$ in $E$. That is, we will write $E(x)$ for $E(\theta, c \mid \mathcal{D})$.

Now by *Monte Carlo* integration we mean that we "create" a sequence of states: $x^{(1)}, x^{(2)}, \ldots, x^{(M)}$ such that:

$$\lim_{M \to \infty} \overline{E} = \langle E \rangle, \tag{5.27}$$

where:

$$\overline{E} = \frac{1}{M} \sum_{t=1}^{M} E\left(x^{(t)}\right), \tag{5.28}$$

is the empirical average of the energy. Showing how to create the sequence of states $\{x^{(t)}\}$ and proving Eq. 5.27 will keep us busy in the next section.

## 6. Markov chains

Before embarking with the details of Monte Carlo integration let's remember the warning of one of its specialists, Alan Sokal [35]:

> Monte Carlo is an extremely bad method; it should be used only when all alternative methods are worse.

That being said it seems we are in a situation with no alternative method, so let's go for it.

Our first problem is that we can't directly (or independently) generate the $x^{(t)}$ from the posterior density because this density is too complicated. What we will do instead is generate the sequence as a realization of a *Markov Chain*. We will moreover build the transition matrix of the Markov chain such that regardless of the initial state $x^{(0)}$, we will have for any bounded function $H$ of $x$ the following limit:

$$\lim_{M \to \infty} \frac{1}{M} \sum_{t=1}^{M} H\left(x^{(t)}\right) = \langle H \rangle. \tag{6.1}$$

In principle we should consider here Markov chains in a (partly) continuous space because $x = (\theta, c)$ and $\theta$ is a continuous vector. But we will take the easy solution, which is fundamentally correct anyway, of saying that because we use a computer to simulate the Markov chain, the best we can do is approximate the continuous space in which $\theta$ lives by a *discrete and finite* space. We therefore split the study of the conditions under which Eq. 6.1 is verified in the general state space (a product of a discrete and a continuous space) into two questions[9]:

---

[9]The results which will be presented in the next section have of course equivalents in the general case. The reference which is always cited by the serious guys is [25].

1. Under what conditions is Eq. 6.1 verified for a Markov chain defined on a *discrete and finite* state space?

2. What kind of error do we produce by approximating our general state space (a product of a discrete and of a continuous state space) by a discrete one?

We will answer here the first of these questions. For the second we will hope (and check) that our discrete approximation is not too bad.

### 6.1. Some notations and definitions

Now that we are dealing with a discrete and finite state space we label all our states $x$ with an integer and our state space $\mathcal{X}$ can be written as:

$$\mathcal{X} = \{x_1, \ldots, x_v\}. \tag{6.2}$$

We can, moreover, make a unique correspondence between a random variable (rv), $X$, and its *probability mass function* $\pi_X$ that we can view as a row vector:

$$\pi_X = \left(\pi_{X,1}, \pi_{X,2}, \ldots, \pi_{X,v}\right), \tag{6.3}$$

where:

$$\pi_{X,i} = Pr\left(X = x_i\right), \ \forall x_i \in \mathcal{X}. \tag{6.4}$$

Strictly speaking a *rv* would not be defined on a set like $\mathcal{X}$ but on a sample space $\mathcal{S}$ which would include every possible outcome, that is: all elements of $\mathcal{X}$ as well as every possible combination of them (like $x_i \cup x_j$, that we write $x_i + x_j$ if $i \neq j$ because then $x_i$ and $x_j$ are mutually exclusive) and the empty set $\emptyset$ (corresponding to the impossible outcome). It is nevertheless clear that for a finite state space, knowing the *pmf* of a *rv* is everything one needs to compute the probability of any outcome (e.g., $Pr\left(x_i + x_j\right) = \pi_{X,i} + \pi_{X,j}$). We will therefore in the sequel say that *rvs* are defined on the state space $\mathcal{X}$ acknowledging it is not the most rigorous way to speak.

*Formal definition of a Markov chain* A sequence of random variables $X^{(0)}$, $X^{(1)}, \ldots$, defined on a finite state space $\mathcal{X}$ is called a *Markov chain* if it satisfies the *Markov property*:

$$Pr\left(X^{(t+1)} = y \mid X^{(t)} = x, \ldots, X^{(0)} = z\right)$$
$$= Pr\left(X^{(t+1)} = y \mid X^{(t)} = x\right). \tag{6.5}$$

If $Pr\left(X^{(t+1)} = y \mid X^{(t)} = x\right)$ does not explicitly depend on $t$ the chain is said to be *homogeneous*.

An homogeneous Markov chain on a finite state space is clearly fully described by its initial random variable $X^{(0)}$ and by its *transition matrix*[10] $T(x, y) = Pr(X^{(t+1)} = y \mid X^{(t)} = x)$. This transition matrix must be *stochastic* which means it must satisfy (Sec 13.3):

$$\sum_{y \in \mathcal{X}} T(x, y) = 1, \ \forall x \in \mathcal{X}. \tag{6.6}$$

If $m$ is the *pmf* of a *rv* defined on $\mathcal{X}$, we can easily show that the row vector $n$ defined as:

$$n_j = \sum_{i=1}^{\nu} m_i T(x_i, x_j),$$

is itself a *pmf* (*i.e.*, $0 \le n_i \le 1$, $\forall i \in \{1, \dots, \nu\}$ and $\sum_{i=1}^{\nu} n_i = 1$) and must therefore correspond to a *rv*. That justifies the following vector notation:

$$\pi_{X^{(t+1)}} = \pi_{X^{(t)}} T, \tag{6.7}$$

where $\pi_{X^{(t)}}$ and $\pi_{X^{(t+1)}}$ are the *pmf* associated to two successive *rv*s of a Markov chain.

*Stationary probability mass function*  A *pmf* $m$ is said to be stationary with respect to the transition matrix $T$ if: $mT = m$.

Applied to our Markov chain that means that if the *pmf* of our initial *rv* ($\pi_{X^{(0)}}$) is stationary with respect to $T$, then the chain does not move (*i.e.*, $X^{(t+1)} \equiv X^{(t)}$, $t \ge 0$).

The notion of stationary *pmf* introduced we can restate our convergence problem (Eq 6.1). What we want in practice is a Markov transition matrix, $T$, which admits a single stationary *pmf* (which would be in our case $\pi_{posterior}$) and such that for any $\pi_{X^{(0)}}$ we have:

$$\lim_{t \to \infty} \pi_{X^{(t)}} = \pi_{posterior},$$

which means:

$$\forall x \in \mathcal{X}, \ \forall \epsilon > 0, \ \exists t \ge 0 : \left| \pi_{posterior}(x) - \pi_{X^{(t)}}(x) \right| \le \epsilon. \tag{6.8}$$

---

[10]Be careful here, the notations are, for historical reasons, misleading. We write down the matrix: $T(x, y)$ where $x$ is the state from which we start and $y$ the state in which we end. When we use the probability transition kernel notation $Pr(y \mid x)$ the starting state is on the right side, the end site is on the left side! In this section where we will discuss theoretical properties of Markov chains, we will use the matrix notation which can be recognized by the the "," separating the start end end states. When later on, we will use the kernel notation, the 2 states will be separated by the "|" (and the end will be on the left side while the start will be on the right).

It is clear that if we knew explicitly a candidate transition matrix, $T$, we could test that the posterior *pmf* is stationary and then using eigenvalue decomposition we could check if the largest eigenvalue is 1, etc... See for instance: [2,7,23,27,29,33,35]. The problem is that we won't in general be able to write down explicitly $T$. We need therefore conditions we can check in practice and which ensure the above convergence. These conditions are *irreducibility* and *aperiodicity*.

*Irreducible and aperiodic states*   (After Liu [23]) A state $x \in \mathcal{X}$ is said to be irreducible if under the transition rule one has nonzero probability of moving from $x$ to any state and then coming back in a finite number of steps. A state $x \in \mathcal{X}$ is said to be aperiodic if the greatest common divisor of $\{t : T^t(x,x) > 0\}$ is 1.

It should not be to hard to see at this point that if one state $x \in \mathcal{X}$ is aperiodic, then all states in $\mathcal{X}$ must also be. Also, if one state $y \in \mathcal{X}$ is aperiodic and the chain is irreducible, then every state in $\mathcal{X}$ must be aperiodic.

## Exercise 4
Prove the validity of Eq. 6.6 for a Markov chain on a finite state space $\mathcal{X}$ (Eq. 6.2). See Sec. 13.3 for a solution.

### 6.2. The fundamental theorem and the ergodic theorem

### Lemma
Let $\left(X^{(0)}, X^{(1)}, \ldots\right)$ be an irreducible and aperiodic Markov chain with state space $\mathcal{X} = \{x_1, x_2, \ldots, x_v\}$ and transition matrix $T$. Then there exists an $M < \infty$ such that $T^n(x_i, x_j) > 0$ for all $i, j \in \{1, \ldots, v\}$ and $n \geq M$.

### Proof
(After Häggström [11]) By definition of irreducibility and aperiodicity, there exist an integer $N < \infty$ such that $T^n(x_i, x_i) > 0$ for all $i \in \{1, \ldots, v\}$ and all $n \geq N$. Fix two states $x_i, x_j \in \mathcal{X}$. By the assumed irreducibility, we can find some $n_{i,j}$ such that $T^{n_{i,j}}(x_i, x_j) > 0$. Let $M_{i,j} = N + n_{i,j}$. For any $m \geq M_{i,j}$, we have:

$$Pr\left(X^{(m)} = x_j \mid X^{(0)} = x_i\right) \geq$$
$$Pr\left(X^{(m-n_{i,j})} = x_i, X^{(m)} = x_j \mid X^{(0)} = x_i\right),$$

and:

$$Pr\left(X^{(m-n_{i,j})} = x_i, X^{(m)} = x_j \mid X^{(0)} = x_i\right) =$$
$$Pr\left(X^{(m-n_{i,j})} = x_i \mid X^{(0)} = x_i\right) Pr\left(X^{(m)} = x_j \mid X^{(0)} = x_i\right)$$

but $m - n_{i,j} \geq 0$ implies that: $Pr\left(X^{(m-n_{i,j})} = x_i \mid X^{(0)} = x_i\right) > 0$ and our choice of $n_{i,j}$ implies that: $Pr\left(X^{(m)} = x_j \mid X^{(0)} = x_i\right) > 0$ therefore:

$$Pr\left(X^{(m)} = x_j \mid X^{(0)} = x_i\right) > 0.$$

We have shown that $T^m\left(x_i, x_j\right) > 0$ for all $m \geq M_{i,j}$. The lemma follows with:

$$M = \max\left\{M_{1,1}, M_{1,2}, \ldots, M_{1,\nu}, M_{2,1}, \ldots, M_{\nu,\nu}\right\}.$$

*Remark* It should be clear that the reciprocal of the lemma is true, that is, if there exists an $M < 0$ such that $T^n(x_i, x_j) > 0$ for all $i, j \in \{1, \ldots, \nu\}$ and $n \geq M$ then the chain is irreducible and aperiodic. A transition matrix which has all its elements strictly positive, or such that one of its powers has all its elements strictly positive, will therefore give rise to irreducible and aperiodic Markov chains.

We can now state and prove our first "fundamental" theorem which shows that irreducible and aperiodic Markov chains on finite state spaces enjoy *geometric convergence* to their *unique* stationary *pmf*.

### Fundamental theorem

If an irreducible and aperiodic homogeneous Markov chain on a finite state space $\mathcal{X} = \{x_1, x_2, \ldots, x_\nu\}$ with a transition matrix $T(x, y)$ has $\pi_{stationary}$ as a stationary distribution then regardless of the initial *pmf* $\pi_{X^{(0)}}$ we have:

$$\lim_{t \to \infty} \pi_{X^{(t)}} = \pi_{stationary} \tag{6.9}$$

and there exists an $1 \geq \epsilon > 0$ such that for all $x \in \mathcal{X}$ and all $t > 1$ we have:

$$\left|\pi_{stationary}(x) - \pi_{X^{(t)}}(x)\right| \leq (1 - \epsilon)^t \tag{6.10}$$

### Proof

Adapted from Neal [27]. We can without loss of generality assume that $T(x, y) > 0$, $\forall x, y \in \mathcal{X}$. If such is not the case for our initial transition matrix, we can always use the preceding lemma to find an $M$ such that: $T'(x, y) = T^M(x, y) > 0$ and work with the Markov chain whose transition matrix is $T'$. Then the uniqueness of the stationary *pmf* regardless of the initial *pmf* will imply that the convergence of the "new" Markov chain holds for the initial Markov chain as well.

So let's define:

$$\epsilon = \min_{x,y \in \mathcal{X}} \frac{T(x, y)}{\pi_{stationary}(y)} > 0,$$

where we are assuming that $\pi_{stationary}(y) > 0, \forall y \in \mathcal{X}$ (if not we just have to redefine the state space to get rid of $y$ and to take out row and column $y$ from $T$). Then from the transition matrix definition and the stationarity assumption we have: $\pi_{stationary}(y) = \sum_x \pi_{stationary}(x) T(x, y), \forall y \in \mathcal{X}$ which with the above assumption gives:

$$1 = \sum_x \pi_{stationary}(x) \frac{T(x, y)}{\pi_{stationary}(y)} \geq \epsilon \sum_x \pi_{stationary}(x),$$

it follows that: $\epsilon \leq 1$. We will now show by induction that the following equality holds:

$$\pi_{X^{(t)}}(x) = \left[1 - (1 - \epsilon)^t\right] \pi_{stationary}(x) + (1 - \epsilon)^t r_t(x), \forall x \in \mathcal{X}, (6.11)$$

where $r_t$ is a *pmf*.

This equality holds for $t = 0$, with $r_0 = \pi_{X^{(0)}}$. Let us assume it is correct for $t$ and see what happens for $t + 1$. We have:

$$
\begin{aligned}
\pi_{X^{(t+1)}}(y) &= \sum_x \pi_{X^{(t)}}(x) T(x, y) \\
&= \left[1 - (1 - \epsilon)^t\right] \sum_x \pi_{stationary}(x) T(x, y) \\
&\quad + (1 - \epsilon)^t \sum_x r_t(x) T(x, y) \\
&= \left[1 - (1 - \epsilon)^t\right] \pi_{stationary}(y) \\
&\quad + (1 - \epsilon)^t \sum_x r_t(x) \cdot \\
&\quad \left[T(x, y) - \epsilon \pi_{stationary}(y) + \epsilon \pi_{stationary}(y)\right] \\
&= \left[1 - (1 - \epsilon)^{t+1}\right] \pi_{stationary}(y) \\
&\quad + (1 - \epsilon)^t \sum_x r_t(x) \left[T(x, y) - \epsilon \pi_{stationary}(y)\right] \\
&= \left[1 - (1 - \epsilon)^{t+1}\right] \pi_{stationary}(y) \\
&\quad + (1 - \epsilon)^{t+1} \sum_x r_t(x) \frac{T(x, y) - \epsilon \pi_{stationary}(y)}{(1 - \epsilon)}
\end{aligned}
$$

$$= \left[1 - (1 - \epsilon)^{t+1}\right] \pi_{stationary}(y)$$
$$+ (1 - \epsilon)^{t+1} r_{t+1}(y)$$

where:

$$r_{t+1}(y) = \left[\sum_x r_t(x) \frac{T(x, y)}{(1 - \epsilon)}\right] - \frac{\epsilon}{1 - \epsilon} \pi_{stationary}(y)$$

One can easily check that: $r_{t+1}(y) \geq 0$, $\forall y \in \mathcal{X}$ and that: $\sum_y r_{t+1}(y) = 1$. That is, $r_{t+1}$ is a proper *pmf* on $\mathcal{X}$.

With Eq 6.11 we can now show that Eq 6.9 holds (using $\pi_s$ for $\pi_{stationary}$):

$$
\begin{aligned}
\left|\pi_s(x) - \pi_{X^{(t)}}(x)\right| &= \left|\pi_s(x) - \left[1 - (1 - \epsilon)^t\right]\pi_s(x) - (1 - \epsilon)^t r_t(x)\right| \\
&= \left|(1 - \epsilon)^t \pi_s(x) - (1 - \epsilon)^t r_t(x)\right| \\
&= (1 - \epsilon)^t \left|\pi_s(x) - r_t(x)\right| \\
&\leq (1 - \epsilon)^t
\end{aligned}
$$

*Ergodic theorem*
Let $\left(X^{(0)}, X^{(1)}, \ldots\right)$ be an irreducible and aperiodic homogeneous Markov chain on a finite state space $\mathcal{X} = \{x_1, x_2, \ldots, x_\nu\}$ with a transition matrix $T(x, y)$ and a stationary distribution $\pi_{stationary}$. Let $a$ be a real valued function defined on $\mathcal{X}$ and let $\overline{a}_N$ be the empirical average of $a$ computed from a realization $\left(x^{(0)}, x^{(1)}, \ldots\right)$ of $\left(X^{(0)}, X^{(1)}, \ldots\right)$, that is: $\overline{a}_N = \frac{1}{N} \sum_{t=0}^{N} a\left(x^{(t)}\right)$. Then we have:

$$\lim_{N \to \infty} \langle \overline{a}_N \rangle = \langle a \rangle_{\pi_{stationary}},$$

where:

$$\langle a \rangle_{\pi_{stationary}} = \sum_{x \in \mathcal{X}} \pi_{stationary}(x) a(x),$$

and:

$$\langle \overline{a}_N \rangle = \frac{1}{N} \sum_{t=0}^{N} \sum_{x \in \mathcal{X}} \pi_{X^{(t)}}(x) a(x).$$

*Proof*
Using Eq. 6.11 in the proof of the fundamental theorem we can write:

$$\langle \bar{a}_N \rangle = \frac{1}{N} \sum_{t=0}^{N} \sum_{x \in \mathcal{X}} \pi_{X^{(t)}}(x)\, a(x)$$

$$= \sum_{x \in \mathcal{X}} a(x)\, \frac{1}{N} \sum_{t=0}^{N} \pi_{X^{(t)}}(x)$$

$$= \sum_{x \in \mathcal{X}} a(x) \left[ \pi_{stationary}(x) \frac{1}{N} \sum_{t=0}^{N} (1 - \zeta^t) + \frac{1}{N} \sum_{t=0}^{N} \zeta^t r_t(x) \right]$$

$$= \left[ \sum_{x \in \mathcal{X}} \pi_{stationary}(x)\, a(x) \right]$$
$$+ \left[ \sum_{x \in \mathcal{X}} a(x) \left( \frac{1}{N} \sum_{t=0}^{N} \zeta^t (r_t(x) - \pi_{stationary}(x)) \right) \right]$$

where $\zeta$ corresponds to $1 - \epsilon$ in Eq. 6.11 and therefore: $0 \le \zeta < 1$. To prove the ergodic theorem we just need to show that the second term of the right hand side of the above equation converges to 0 as $N \to \infty$. For that it is enough to show that its modulus goes to 0:

$$\left| \sum_{x \in \mathcal{X}} a(x) \left( \frac{1}{N} \sum_{t=0}^{N} \zeta^t (r_t(x) - \pi_{stationary}(x)) \right) \right|$$

$$\le \sum_{x \in \mathcal{X}} |a(x)| \left( \frac{1}{N} \sum_{t=0}^{N} \zeta^t |r_t(x) - \pi_{stationary}(x)| \right)$$

$$\le \sum_{x \in \mathcal{X}} |a(x)| \frac{1}{N} \sum_{t=0}^{N} \zeta^t$$

We just need to recognize here the geometrical series $\sum_{t=0}^{N} \zeta^t$ which converges to $\frac{1}{1-\zeta}$ when $N \to \infty$. That completes the proof.

*What did we learn in this section?*   For a given *pmf* $\pi$ defined on a finite state space $\mathcal{X}$ if we can find an irreducible aperiodic Markov matrix $T$ which admits $\pi$ as its (necessarily only) stationary distribution, then *regardless* of our initial choice of *pmf* $\pi_0$ defined on $\mathcal{X}$ and regardless of the function $a : \mathcal{X} \to \Re$, an *asymptotically unbiased* estimate of $\langle a \rangle = \sum_{x \in \mathcal{X}} a(x)\, \pi(x)$ can be obtained by simulating the Markov chain to get a sequence of states: $\left(x^{(0)}, x^{(1)}, \ldots, x^{(N)}\right)$ and computing the empirical average: $\bar{a} = \frac{1}{N} \sum_{t=0}^{N} a\left(x^{(t)}\right)$.

## 7. The Metropolis–Hastings algorithm and its relatives

What we need now is a way to construct the transition matrix $T$ and more specifically a method which works with an *unnormalized* version of the stationary density. We first introduce the notion of *detailed balance*.

*Detailed balance definition*    We say that a *pmf* $\pi$ defined on a (finite) state space $\mathcal{X}$ and a Markov transition matrix $T$ satisfy the *detailed balance* if: $\forall x, y \in \mathcal{X}$, $\pi(x) T(x, y) = \pi(y) T(y, x)$.

*Detailed balance theorem*    If the *pmf* $\pi$ defined on a (finite) state space $\mathcal{X}$ and a Markov transition matrix $T$ satisfy the *detailed balance* then $\pi$ is stationary for $T$.

*Exercise 5*
Prove the theorem. See Sec. 13.4 for solution.

### 7.1. The Metropolis–Hastings algorithm

#### 7.1.1. Second fundamental theorem
Let $\pi$ be a *pmf* defined on a (finite) state space $\mathcal{X}$ and $T$ and $G$ two Markov transition matrices defined on $\mathcal{X}$ satisfying:

$$
\begin{aligned}
T(x, y) &= A(x, y) G(x, y) \ if \ x \neq y \\
T(x, x) &= 1 - \sum_{y \in \mathcal{X}, y \neq x} T(x, y)
\end{aligned}
$$

where $G(y, x) = 0$ if $G(x, y) = 0$ and

$$
\begin{aligned}
A(x, y) &= \min\left(1, \frac{\pi(y) G(y, x)}{\pi(x) G(x, y)}\right), \ if \ G(x, y) > 0 \\
&= 1, \quad otherwise
\end{aligned}
\tag{7.1}
$$

then
- $\pi$ and $T$ satisfy the detailed balance condition
- $\pi T = \pi$
- if $G$ is irreducible and aperiodic so is $T$

*Exercise 6*
Prove the theorem.

## THE MOST IMPORTANT COMMENT OF THESE LECTURES NOTES

This theorem is exactly what we were looking for. It tells us how to modify an irreducible and aperiodic Markov transition $(G)$ such that a *pmf* $\pi$ of our choice will be the stationary distribution and it does that by requiring a knowledge of the desired stationary $\pi$ only up to a normalizing constant, because Eq. 7.1 involves the ratio of two values of $\pi$.

$G$ is often called the *proposal* transition and $A$ the *acceptance* probability. The *Metropolis algorithm* [24] (sometimes called the $M (RT)^2$ algorithm because of its authors names) is obtained with a symmetrical $G$ (*i.e.*, $G (x, y) = G (y, x)$). The above formulation (in fact a more general one) is due to Hastings [14]. The set of techniques which involves the construction of a Markov chain (with the Metropolis–Hastings algorithm) to perform Monte Carlo integration is called *Dynamic Monte Carlo* by physicists [19, 29, 35] and *Markov Chain Monte Carlo* (*MCMC*) by statisticians [10, 23, 33].

### 7.2. *Metropolis–Hastings and Gibbs algorithms for multi–dimensional spaces*

Talking of multi - dimensional spaces when we started by saying we were working with a discrete and finite one can seem a bit strange. It is nevertheless useful not to say necessary to keep a trace of the multi - dimensionality of our "initial" space (ideally all our model parameters: $P_k$, $\Delta_k$, $\Lambda_k$, $S_k$, $F_k$ "live" in continuous spaces). If not it would be extremely hard to find our way on the map between the discrete approximations of all these continuous spaces and the genuine discrete and finite space on which our simulated Markov chain evolves.

We consider therefore now that our random variables are "vectors" and $X^{(t)}$ becomes $\mathbf{X}^{(t)}$. We can think of it as follows: $\mathbf{X}_1^{(t)}$ corresponds to $P_{1,1}^{(t)}$ the maximal peak amplitude of the first neuron in the model on the first recording site after $t$ steps,..., $\mathbf{X}_4^{(t)}$ corresponds to $P_{1,4}^{(t)}$ the maximal peak amplitude of the first neuron in the model on the fourth recording site (assuming we are using tetrodes), $\mathbf{X}_5^{(t)}$ corresponds to the parameter $\Delta^{(t)}$ of the first neuron, $\mathbf{X}_6^{(t)}$ corresponds to the parameter $\Lambda^{(t)}$ of the first neuron, $\mathbf{X}_7^{(t)}$ corresponds to the parameter $S^{(t)}$ of the first neuron, $\mathbf{X}_8^{(t)}$ corresponds the parameter $F^{(t)}$ of the first neuron, $\mathbf{X}_9^{(t)}$ corresponds to $P_{2,1}^{(t)}$ the maximal peak amplitude of the second neuron in the model on the first recording site, etc. The problem is that it turns out to be very hard to find a transition matrix $G$ acting on object like these "new" *pmf*s: $\pi_{\mathbf{X}^{(t)}}$ such that the acceptance probabilities (Eq. 7.1) are not negligible. The way around this difficulty is to build the transition matrix $T$ as a combination of component - wise transition matrices like $T_1$ which acts only on the first component of $\pi_{\mathbf{X}^{(t)}}$, $T_2$ which acts only on the second, etc. We just need to make sure that our combination is irreducible, and aperiodic (we assume here that we build each individual $T_j$ such that $\pi_{posterior}$ is its stationary *pmf*). A way to do that is to construct each $T_j$ such that it is irreducible and aperiodic on its "own" sub - space which is obtained practically by building the matrix such that it has a strictly positive probability to produce a move anywhere on its sub - space. Then two combinations of these $T_j$s are typically used, the first one being:

$$T = w_1 T_1 + w_2 T_2 + \ldots + w_m T_m,$$

where $m$ is the number of components of the random vectors and were the $w_j$s are components of a *pmf* defined on the the set of coordinates $\{1, \ldots, m\}$. It should be clear that if each $T_j$ is irreducible and aperiodic on its own sub - space, then $T$ will be irreducible and aperiodic on the "complete" state space. Because each $T_j$ is built such that $\pi_{posterior}$ is its stationary *pmf*, $\pi_{posterior}$ will be the stationary *pmf* of $T$ as well. The concrete implementation of this scheme would go as follows: at each "sub - step" a coordinate $j$ is randomly drawn from $w$, then a random move is performed using $T_j$. It is customary to call "Monte Carlo step" a sequence of $m$ successive such "sub - steps" (that means that on average each model parameter will be "changed" during a Monte Carlo step).

The second combination is:

$$T = T_1 \times T_2 \times \ldots \times T_m, \tag{7.2}$$

that is, each model parameter is successively "changed". In the same way as above the irreducibility and aperiodicity of the $T_j$s in their sub - spaces will give rise to an irreducible and aperiodic $T$ on the "full" parameter space. The main difference is that detailed balance condition which be imposed by construction to the pairs $T_j$, $\pi_{posterior}$ is not kept by the pair $T$, $\pi_{posterior}$. We only have the stationarity property (which is enough to ensure convergence). Of course variations on that scheme can be used like using random permutations of the $T_j$s (which would restore the detailed balance condition for the pair $T$, $\pi_{posterior}$). A "Monte Carlo step" for those schemes is obtained after the application of the complete series of $T_j$s. See [7] for details.

### 7.2.1. An example: the Gibbs sampler for the parameters of the ISI density

The previous discussion seemed probably a bit abstract for most of the readers. In order to be more precise about what we meant we will start by considering the following "simple" situation. Let's assume that we are given a sample of 25 *isis*: $\mathcal{D} = \{i_1, \ldots, i_{25}\}$ drawn independently from a log-Normal density with parameters: $s_{actual}, f_{actual}$. We are asked for a Bayesian estimation of values of $s$ and $f$ (assuming flat priors for these parameters like in Exercise 3):

$$\pi_{isi,posterior}(s, f \mid \mathcal{D}) \propto L_{isi}(\mathcal{D} \mid s, f) \cdot \pi_{isi,prior}(s, f),$$

where $L_{isi}$ is easily obtained from Eq. 5.4:

$$L_{isi}(\mathcal{D} \mid s, f) = \frac{1}{f^N} \exp\left[-\frac{1}{2}\frac{1}{f^2}\sum_{j=1}^{N}(\ln i_j - \ln s)^2\right] \qquad (7.3)$$

We do not recognize any classical *pdf* in Eq. 7.3 and we choose to use a MCMC approach. Following the previous discussion we will try to build our transition matrix $T$ as a "product" $T_S \times T_F$. Where $T_S$ does only change $s$ and $T_F$ does only change $f$ and where both are irreducible and aperiodic on their own sub-space. According to the second fundamental theorem we first need to find proposal transitions: $G_s(s_{now}, s_{proposed} \mid f, \mathcal{D})$ and $G_f(f_{now}, f_{proposed} \mid s, \mathcal{D})$. But questions 1 and 3 of Exercise 3 provide us with such proposal transitions. In fact these transitions are a bit special because they do not depend on the present value of the parameter we will try to change and because they indeed correspond to the posterior conditional density of this parameter. A direct consequence of the latter fact is that the acceptance probability (Eq. 7.1) is 1. An algorithm where the proposal transition for a parameter is the posterior conditional of this parameter is called a *Gibbs sampler* by statisticians and a *heat bath algorithm* by physicists.

We therefore end up with the following algorithm (using the results of Exercise 3):

1. Chose randomly $s^{(0)} \in [s_{min}, s_{max}]$ and $f^{(0)} \in [f_{min}, f_{max}]$.

2. Given $f^{(t)}$ draw:

$$s^{(t+1)} \sim \pi\left(\ \mid f^{(t)}, \mathcal{D}\right)$$

where $\pi(\ \mid f^{(t)}, \mathcal{D})$ is defined by Eq. 5.13 (remark that $s^{(t+1)}$ is independent of $s^{(t)}$).

3. Given $s^{(t+1)}$ draw:

$$f^{(t+1)} \sim \pi\left(\ \mid s^{(t+1)}, \mathcal{D}\right)$$

where $\pi(\ \mid s^{(t+1)}, \mathcal{D})$ is defined by Eq. 5.15.

*Exercise 7*

Simulate 25 *isi* following a log-Normal density with values of your choice for the pair $s_{actual}$, $f_{actual}$. Implement the algorithm and compare its output with the maximum likelihood based inference. That is with the maximal likelihood estimates for $s$ and $f$ (given by Eq. 5.6 & 5.7). You should compute as well the Hessian of the log-likelihood function at its maximum to get confidence intervals (see the chapter of E. Brown).

### 7.2.2. Generation of the amplitude parameters of our model

By "amplitude parameters" we mean here the following parameters: $\mathbf{P}$, $\Delta$, $\Lambda$. Given a data set from a single neuron: $\mathcal{D} = \{(i_1, \mathbf{a}_1), \ldots, (i_N, \mathbf{a}_N)\}$ (see Sec. 5.2) we now try to perform Bayesian inference on all its parameters: $\mathbf{P}$, $\Delta$, $\Lambda$, $S$, $F$. We again split our transition matrix $T$ into parameter specific transitions: $T = T_{P_1} \times T_{P_2} \times T_{P_3} \times T_{P_4} \times T_\Delta \times T_\Lambda \times T_S \times T_F$. We have seen in the previous example how to get $T_S$ and $T_F$. Following the same line we could try to build a Gibbs sampler for the other parameters as well. The problem is that the part of the Likelihood function which depends on the amplitude parameters (obtained from Eq. 5.5):

$$L_{amp}\left(\mathcal{D} \mid \mathbf{p}, \delta, \lambda\right) = \exp\left[-\frac{1}{2}\sum_{j=1}^{N} \left\|\mathbf{a}_j - \mathbf{p} \cdot \left(1 - \delta \cdot \exp\left(-\lambda \cdot i_j\right)\right)\right\|^2\right]$$

does not correspond to any know *pdf* even when considered as a function of a single parameter, say $\delta$. The reader can notice that such would not be the case if we had $\delta = 0$ and if we knew it, see [30]. A robust solution to this problem is to use a piece-wise linear approximation of the posterior conditional as a proposal transition for each parameter (*e.g.*, $G_\Delta (\ \mid \mathbf{p}, \lambda, s, f, \mathcal{D})$) and then an acceptance probability as defined by Eq. 7.1. More specifically we can start with 101 regularly spaced "sampling values" of $\delta$:

$$\delta \in \{\delta_0 = 0, \delta_1 = 0.01, \ldots, \delta_{99} = 0.99, \delta_{100} = 1\},$$

compute 101 values of:

$$L_{amp}\left(\mathcal{D} \mid \mathbf{p}, \delta_i, \lambda\right) \tag{7.4}$$

and define:

$$G_\Delta\left(\delta \mid \mathbf{p}, \lambda, s, f, \mathcal{D}\right) = \mathcal{N}_\Delta \cdot \left[L_{amp}\left(\mathcal{D} \mid \mathbf{p}, \delta_i, \lambda\right) + \frac{L_{amp}\left(\mathcal{D} \mid \mathbf{p}, \delta_{i+1}, \lambda\right) - L_{amp}\left(\mathcal{D} \mid \mathbf{p}, \delta_i, \lambda\right)}{\delta_{i+1} - \delta_i}\left(\delta - \delta_i\right)\right]$$

where $\mathcal{N}_\Delta$ ensures that $G_\Delta$ is properly normalized and where $\delta \in \left[\delta_i, \delta_{i+1}\right]$. The obvious problem with this approach is that we need to have a reasonably good piece-wise linear approximation of the corresponding posterior conditional *pdf* in order to get reasonable values for our acceptance probability. That means that when we start our Markov chain and we do not have a precise idea of the actual shape of this posterior conditional density we have to use a lot of sampling points. We spend therefore a lot of time computing terms like Eq. 7.4. A very significant increase of the algorithm speed can thus be obtained by using a first run with a lot of sampling points (say 101), then reposition fewer sampling point (say 13)[11] using the output of the first run. That's explained in details in [32] and illustrated on Fig. 10. When dealing with multiple neurons data, Eq. 5.19 shows that the same approach can be immediately applied after the introduction of the configuration.

### 7.2.3. Generation of the configuration
This can be done (almost) exactly as for a "normal" Potts model [29, 35] and is left as an exercise to the reader (the answer can be found in [32]).

### 7.2.4. Our complete MC step
In Sec. 11 we will give an example of our algorithm at work with the following sequential MC step:

$$T_{l_1} \times \ldots \times T_{l_N} \times T_{P_{1,1}} \times \ldots \times T_{f_1} \times \ldots \times T_{P_{K,1}} \times \ldots \times T_{f_K} \qquad (7.5)$$

## 8. Priors choice

We will assume here that we know "little" *a priori* about our parameters values and that the joint prior density $\pi_{prior}(\theta)$ can be written as a product of the densities for each component of $\theta$, that is:

$$\pi_{prior}(\theta) =$$
$$\prod_{q=1}^{K} \pi(f_q) \cdot \pi(s_q) \cdot \pi(\delta_q) \cdot \pi(\lambda_q) \cdot \pi(P_{q,1}) \cdot \pi(P_{q,2}) \cdot \pi(P_{q,3}) \cdot \pi(P_{q,4})$$

where we are assuming that four recording sites have been used. We will further assume that our signal to noise ratio is not better 20 (a rather optimistic value), that our spikes are positive and therefore the $\pi(P_{q,1...4})$ are null below 0

---

[11]We typically use 13 points because we want to have the 10 quantiles, the 2 extrema (here, $\delta_{min}$ and $\delta_{max}$) allowed by our priors and an extra point corresponding to the smallest value obtained in our sample. The reader can "show" as an exercise the usefulness of this extra sample point. The easiest way to see the use of this point is to try without it and to remember that the piece-wise linear approximation *must be normalized*.

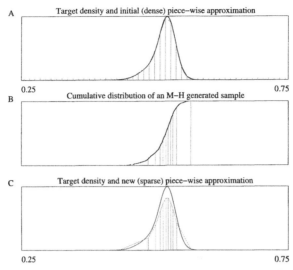

Fig. 10. A, a simple one-dimensional target density (black curve, a Gaussian mixture: $0.3\,\mathcal{N}\left(0.5, 0.025^2\right) + 0.7\,\mathcal{N}\left(0.525, 0.015^2\right)$) together with its linear piece-wise approximation (gray curve) based on 100 discrete samples (the prior density is supposed to be flat between 0 and 1 and zero everywhere else). B, A sample of size 1000 is drawn using the piece-wise linear approximation as a proposal and a MH acceptance rule. The cumulative distribution of the sample is shown (black) together with the location of the 10 quantiles. C, Using the location of the 10 quantiles, the boundaries of the prior and the location of the smallest generated value, a sparser piece-wise linear approximation of the target density is built.

and above +20 (remember we are working with normalized amplitudes). We will reflect our absence of prior knowledge about the amplitudes by taking a uniform distribution between 0 and +20. The $\lambda$ value reported by Fee et al [6] is 45.5 $s^{-1}$. $\lambda$ must, moreover be smaller than $\infty$, we adopt a prior density uniform between 10 and 200 $s^{-1}$. $\delta$ must be $\leq 1$ (the amplitude of a spike from a given neuron on a given recording site *does not* change sign) and $\geq 0$ (spikes do not become larger upon short ISI), we therefore use a uniform density between 0 and 1 for $\delta$. An inspection of the effect of the shape parameter $f$ on the ISI density is enough to convince an experienced neurophysiologist that empirical unimodal ISI densities from well isolated neurons will have $f \in [0.1, 2]$. We therefore take a prior density uniform between 0.1 and 2 for $f$. The same considerations leads us to take a uniform prior density between 0.005 and 0.5 for $s$.

## 9. The good use of the ergodic theorem. A warning

The ergodic theorem is the key theoretical result justifying the use of Monte Carlo integration to solve tough problems. When using these methods we should nevertheless be aware that the theorem applies only when the number of Monte Carlo steps of our algorithms go to infinity[12] and because such is never practically the case we will commit errors. We would commit errors even if our draws where directly generated from $\pi_{post}$. The difference between the MCMC/Dynamic MC based estimates and estimates based on direct samples ("plain MC") is that the variance of the estimators of the former have to be corrected to take into account the correlation of the states generated by the Markov chain. We explain now how to do that in practice, for a theoretical justification see Sokal [35] and Janke [17].

### 9.1. Autocorrelation functions and confidence intervals

We have to compute for each parameter, $\theta_i$, of the model the normalized autocorrelation function (ACF), $\rho_{norm}(l; \theta_i)$, defined by:

$$\rho(l; \theta_i) = \frac{1}{N_T - N_D - l} \cdot \sum_{t=N_D}^{N_T - l} (\theta_i^{(t)} - \bar{\theta}_i) \cdot (\theta_i^{(t+l)} - \bar{\theta}_i)$$

$$\rho_{norm}(l; \theta_i) = \frac{\rho(l; \theta_i)}{\rho(0; \theta_i)} \tag{9.1}$$

Where $N_T$ is the total number of MC steps performed and $N_D$ is the number of steps required to reach "equilibrium" (see Sec. 9.2, 11.2 & 11.3). Then we compute the *integrated autocorrelation time*, $\tau_{autoco}(\theta_i)$:

$$\tau_{autoco}(\theta_i) = \frac{1}{2} + \sum_{l=1}^{L} \rho(l; \theta_i) \tag{9.2}$$

where $L$ is the lag at which $\rho$ starts oscillating around 0. Using an empirical variance, $\sigma^2(\theta_i)$ of parameter $\theta_i$, defined in the usual way:

$$\sigma^2(\theta_i) = \frac{1}{N_T - N_D - 1} \sum_{t=N_D}^{N_T} (\theta_i^{(t)} - \bar{\theta}_i)^2 \tag{9.3}$$

---

[12]This is not even true because our algorithm use *pseudo - random - number* generators which among many shortcomings have a *finite* period. See Chap 7 of [7], Chap 2 of [33] and [20].

where $\bar{\theta}_i$ is defined like in Eq. 5.28. Our estimate of the variance, $Var\left[\bar{\theta}_i\right]$ of $\bar{\theta}_i$ becomes:

$$Var\left[\bar{\theta}_i\right] = \frac{2\,\tau_{autoco}(\theta_i)}{N_T - N_D - 1} \cdot \sigma^2(\theta_i) \qquad (9.4)$$

In the sequel, the confidence intervals on our parameters estimates are given by the square root of the above defined variance (Table 2 & 3).

We can view the effect of the autocorrelation of the states of the chain as a reduction of our effective sample size by a factor: $2\,\tau_{autoco}(\theta_i)$. This gives us a first quantitative element on which different algorithms can be compared (remember that the MH algorithm gives us a lot of freedom on the choice of proposal transition kernels). It is clear that the faster the autocorrelation functions of the parameters fall to zero, the greater the statistical efficiency of the algorithm. The other quantitative element we want to consider is the computational time, $\tau_{cpu}$, required to perform one MC step of the algorithm. One could for instance imagine that a new sophisticated proposal transition kernel allows us to reduce the largest $\tau_{autoco}$ of our standard algorithm by a factor of 10, but at the expense of an increase of $\tau_{cpu}$ by a factor of 100. Globally the new algorithm would be 10 times less efficient than the original one. What we want to keep as small as possible is therefore the product: $\tau_{autoco} \cdot \tau_{cpu}$.

## 9.2. Initialization bias

The second source of error in our (finite size) MC estimates is a bias induced by the state $(\theta^{(0)}, C^{(0)})$ with which the chain is initialized [7,35]. The bad news concerning this source of error is that there is no general theoretical result providing guidance on the way to handle it, but the booming activity in the Markov chain field already produced encouraging results in particular cases [23]. The common wisdom in the field is to monitor parameters (and labels) evolution, and/or functions of them like the energy (Eq. 5.24). Based on examination of evolution plots (*eg*, Fig. 12 & 13) and/or on application of time-series analysis tools, the user will decide that "equilibrium" has been reached and discard the parameters values before equilibrium. More sophisticated tests do exist [33] but they wont be used in this chapter. These first two sources of error, finite sample size and initialization bias, are clearly common to all MCMC approaches.

## 10. Slow relaxation and the replica exchange method

A third source of error appears only when the energy function exhibits several local minima. In the latter case, the Markov chain realization can get trapped in a local minimum which could be a poor representation of the whole energy function. This sensitivity to local minima arises from the local nature of the transitions generated by the MH algorithm. That is, if we use a sequential scheme like Eq. 7.5, at each MC time step, we first attempt to change the label of spike 1, then the one of spike 2, ..., then the one of spike N, then we try to change the first component of $\theta$ ($P_{1,1}$), and so on until the last component ($\sigma_K$). That implies that if we start in a local minimum and if we need to change, say, the labels of 10 spikes to reach another lower local minimum, we could have a situation in which the first 3 label changes are energetically unfavorable (giving, for instance, an acceptance probability, Eq. 7.1, of 0.1 per label change) which would make the probability to accept the succession of changes very low ( $0.1^3$ )... meaning that our Markov chain would spend a long time in the initial local minimum before "escaping" to the neighboring one. Stated more quantitatively, the average time the chain will take to escape from a local minimum with energy $E_{min}$ grows as the exponential of the energy difference between the energy, $E^*$, of the highest energy state the chain has to go through to escape and $E_{min}$:

$$\tau_{escape} \propto \exp\left[\beta\left(E^* - E_{min}\right)\right]$$

Our chains will therefore exhibit an Arrhenius behavior. To sample more efficiently such state spaces, the Replica Exchange Method (REM) [15, 26], also known as the Parallel Tempering Method [5, 12, 37], considers $R$ replicas of the system with an increasing sequence of temperatures (or a decreasing sequence of $\beta$) and a dynamic defined by two types of transitions : usual MH transitions performed independently on each replica according to the rule defined by Eq. 7.5 and a replica exchange transition. The key idea is that the high temperature (low $\beta$) replicas will be able to easily cross the energy barriers separating local minima (in the example above, if we had a probability of $0.1^3$ to accept a sequence of labels switch for the replica at $\beta = 1$, the replica at $\beta = 0.2$ will have a probability $0.1^{3 \cdot 0.2} \approx 0.25$ to accept the same sequence). What is needed is a way to generate replica exchange transitions such that the replica at $\beta = 1$ generates a sample from $\pi_{post}$ defined by Eq. 5.25. Formally the REM consists in simulating, on an "extended ensemble", a Markov chain whose unique stationary density is given by:

$$\pi_{ee}\left(\theta_1, C_1, \ldots, \theta_R, C_R\right) = \pi_{post,\beta_1}\left(\theta_1, C_1\right)\ldots\pi_{post,\beta_R}\left(\theta_R, C_R\right) \qquad (10.1)$$

where "$ee$" in $\pi_{ee}$ stands for "extended ensemble" [16], $R$ is the number of simulated replicas, $\beta_1 > \ldots > \beta_R$ for convenience and:

$$\pi_{post,\beta_i}\left(\theta_i, C_i\right) = \frac{\exp\left[-\beta_i E\left(\theta_i, C_i\right)\right]}{Z\left(\beta_i\right)} \qquad (10.2)$$

That is, compared to Eq. 5.25, we now explicitly allow $\beta$ to be different from 1. To construct our "complete" transition kernel we apply our previous procedure. That is, we construct it as a sequence of parameter, label and inter-replica specific MH transitions. We already know how to get the parameter and label specific transitions for each replica. What we really need is a transition to exchange replicas, say the replicas at inverse temperature $\beta_i$ and $\beta_{i+1}$, such that the detailed balance is preserved (Sec. 7):

$$\pi_{ee} (\theta_1, C_1, \ldots, \theta_i, C_i, \theta_{i+1}, C_{i+1}, \ldots, \theta_R, C_R)$$
$$\cdot T_{i,i+1} (\theta_{i+1}, C_{i+1}, \theta_i, C_i \mid \theta_i, C_i, \theta_{i+1}, C_{i+1}) =$$
$$\pi_{ee} (\theta_1, C_1, \ldots, \theta_{i+1}, C_{i+1}, \theta_i, C_i, \ldots, \theta_R, C_R)$$
$$\cdot T_{i,i+1} (\theta_i, C_i, \theta_{i+1}, C_{i+1} \mid \theta_{i+1}, C_{i+1}, \theta_i, C_i)$$

which leads to:

$$\frac{T_{i,i+1} (\theta_i, C_i, \theta_{i+1}, C_{i+1} \mid \theta_{i+1}, C_{i+1}, \theta_i, C_i)}{T_{i,i+1} (\theta_{i+1}, C_{i+1}, \theta_i, C_i \mid \theta_i, C_i, \theta_{i+1}, C_{i+1})}$$
$$= \frac{\pi_{ee} (\theta_1, C_1, \ldots, \theta_i, C_i, \theta_{i+1}, C_{i+1}, \ldots, \theta_R, C_R)}{\pi_{ee} (\theta_1, C_1, \ldots, \theta_{i+1}, C_{i+1}, \theta_i, C_i, \ldots, \theta_R, C_R)}$$
$$= \frac{\pi_{post,\beta_i} (\theta_i, C_i) \cdot \pi_{post,\beta_{i+1}} (\theta_{i+1}, C_{i+1})}{\pi_{post,\beta_i} (\theta_{i+1}, C_{i+1}) \cdot \pi_{post,\beta_{i+1}} (\theta_i, C_i)}$$
$$= \exp \left[ - (\beta_i - \beta_{i+1}) \cdot (E (\theta_i, C_i) - E (\theta_{i+1}, C_{i+1})) \right]$$

Again we write $T_{i,i+1}$ as a product of a proposal transition kernel and an acceptance probability. Here we have already explicitly chosen a deterministic proposal (we only propose transitions between replicas at neighboring inverse temperatures) which gives us:

$$\frac{A_{i,i+1} (\theta_i, C_i, \theta_{i+1}, C_{i+1} \mid \theta_{i+1}, C_{i+1}, \theta_i, C_i)}{A_{i,i+1} (\theta_{i+1}, C_{i+1}, \theta_i, C_i \mid \theta_i, C_i, \theta_{i+1}, C_{i+1})} =$$
$$\exp \left[ - (\beta_i - \beta_{i+1}) \cdot (E (\theta_i, C_i) - E (\theta_{i+1}, C_{i+1})) \right]$$

It is therefore enough to take:

$$A_{i,i+1} (\theta_i, C_i, \theta_{i+1}, C_{i+1} \mid \theta_{i+1}, C_{i+1}, \theta_i, C_i) =$$
$$\min \left\{ 1, \exp \left[ - (\beta_i - \beta_{i+1}) \cdot (E (\theta_i, C_i) - E (\theta_{i+1}, C_{i+1})) \right] \right\} \tag{10.3}$$

The reader sees that if the state of the "hot" replica ($\beta_{i+1}$) has a lower energy ($E (\theta_i, C_i)$) than the "cold" one, the proposed exchange is always accepted. The exchange can pictorially be seen as cooling down the hot replica and warming up the cold one. Fundamentally this process amounts to make the replica which is at the beginning *and* at the end of the replica exchange transition at the cold temperature to jump from one local minimum $(\theta_{i+1}, C_{i+1})$ to another one $(\theta_i, C_i)$. That is precisely what we were looking for. The fact that we can as well accept unfavorable exchanges (*i.e.*, raising the energy of the "cold" replica and decreasing the one of the "hot" replica) is the price we have to pay for our algorithm to generate samples from the proper posterior (we are not doing optimization here).

In order for the replica exchange transition to work well we need to be careful with our choice of inverse temperatures. The typical energy of a replica (*i.e.*, its expected energy) increases when $\beta$ decreases (Fig. 16A). We will therefore typically have a positive energy difference: $\Delta E = E_{hot} - E_{cold} > 0$ between the replicas at low and high $\beta$ before the exchange. That implies that the acceptance ratio (Eq. 10.3) for the replica exchange will be typically smaller than 1. Obviously, if it becomes too small, exchanges will practically never be accepted. To avoid this situation we need to choose our inverse temperatures such that the typical product: $\Delta\beta \cdot \Delta E$, where $\Delta\beta = \beta_{cold} - \beta_{hot}$, is close enough to zero [15, 16, 28]. In practice we used pre-runs with an a priori too large number of $\beta$s, checked the resulting energy histograms and kept enough inverse temperatures to have some overlap between successive histograms (Fig. 16B).

In Sec. 11 we will perform replica exchange transitions between each pair $\beta_i$, $\beta_{i+1}$ with an even, respectively odd, $i$ at the end of each even, respectively odd, MC step. With this replica exchange scheme, each MC time step will therefore be composed of a complete parameter and label transition for each replica, followed by a replica exchange transition. This scheme corresponds to the one described by Hukushima and Nemoto [15]. A rather detailed account of the REM can be found in Mitsutake et al [26]. Variations on this scheme do exist [4, 28].

## 11. An Example from a simulated data set

We will illustrate the performances of the algorithm with a simple simulated data set. The data are supposed to come from 3 neurons which are exactly described by our underlying model. Such an illustration has in our opinion several advantages. Being simple it helps the reader to concentrate on the inner working of the algorithm. Because the data correspond to the underlying model hypothesis, our implementation of the MCMC method should give us back the parameters used to simulate the data, we are therefore performing here a simple and *necessary* test of our code. The data do moreover exhibit features (strong cluster overlap, Fig. 11B,C) which would make them unusable by other algorithms. A presentation of the algorithm performances with a much worse data set can be found in [32].

Table 1

Parameters used to simulate the neurons. The maximal peak amplitude values ($P_i$) are given in units of noise SD. The scale parameters ($s$) and mean *isi* ($\langle isi \rangle$) are given in ms. The bottom row indicates the number of events from each neuron. The correspondence between neuron number and color on Fig. 11: 1, light gray, 2, dark gray, 3, black.

|  | neuron 1 | neuron 2 | neuron 3 |
| --- | --- | --- | --- |
| $P_1, P_2$ | 15,9 | 8,8 | 6,12 |
| $\delta$ | 0.7 | 0.8 | 0.6 |
| $\lambda$ | 33.33 | 40 | 50 |
| $s$ | 25 | 30 | 18 |
| $f$ | 0.5 | 0.4 | 1.0 |
| $\langle isi \rangle$ | 28.3 | 32.5 | 29.7 |
| # | 1060 | 923 | 983 |

### 11.1. Data properties

The parameters used to simulate the three neurons are given in Table 1. 30 seconds of data were simulated giving a total of 2966 spikes. The raw data, spike amplitude *vs* time on the two recording sites are illustrated on Fig. 11A1 & 11A2. Fig. 11B is a "Wilson plot" of the entire sample. Notice the *strong* overlap of points (spikes) arising from the 3 different neurons. Fig. 11C shows the theoretical iso-density contours for the clusters generated by each of the 3 neurons. Neuron 1 in Table 1 is light gray on the figure, neuron 2 is dark gray and neuron 3 is black. The reader can see that roughly 50% of the spikes generated by neuron 2 should fall in regions were neuron 1 or neuron 3 will also generate spikes. The theoretical densities associated with each neuron (cluster) *are not* 2-dimensional Gaussian. This can be most clearly seen for neuron 3 (black iso-density contours) which has a "flat" summit. None of these densities is symmetrical with respect to its maximum along its principal axis (which is the axis going through the graph's origin and the neuron density maximum). Fig. 11D1 represents the amplitude dynamics of each of the three neurons, while Fig. 11D2 displays their respective ISI densities.

### 11.2. Algorithm dynamics and parameters estimates without the REM

#### 11.2.1. Initialization

The algorithm being iterative we have to start it somewhere and we will use here a somewhat "brute force" initialization. We choose randomly with a uniform probability $\frac{1}{N}$ as many actual events as neurons in the model ($K=3$). That gives us our initial guesses for the $P_{q,i}$. $\delta$ is set to $\delta_{min} = 0$ for each neuron. All the other parameters are randomly drawn from their prior distribution (Sec. 8). The

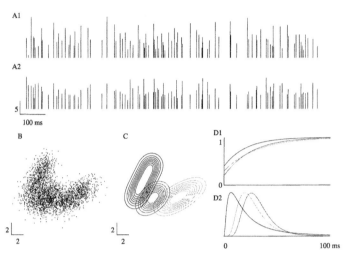

Fig. 11. Simulated data with 3 active neurons recorded on a stereode (2 recording sites). 30 s of data were simulated resulting in 2966 spikes. A, An epoch with 100 spikes on site 1 (A1) and 2 (A2). Vertical scale bar in units of noise SD. B, Wilson plot of the 2966 spikes: the amplitude of each spike on site 2 is plotted against its amplitude on site 1. The scale bars corner is at the origin (0,0). C, Expected iso-density plots for each of the three neurons: neuron 1 (light gray), neuron 2 (dark gray) and neuron 3 (black). D1, Normalized spike amplitude *vs* ISI for each neuron (normalization done with respect to the maximal amplitude). Horizontal scale, same as D2. D2, ISI density of each of the 3 neurons (peak value of black density: 35 Hz).

initial configuration is generated by labeling each individual spike with one of the $K$ possible labels with a probability $1/K$ for each label (this is the $\beta = 0$ initial condition used in statistical physics).

### 11.2.2. Energy evolution

At each MC step, a new label is proposed and accepted or rejected for each spike and a new value is proposed and accepted or rejected for each of the (18) model parameters (Eq. 7.5). Our initial state is very likely to have a very low posterior probability. We therefore expect our system to take some time to relax from the highly disordered state in which we have initialized it to the (hopefully) ordered typical states. As explained in Sec. 9.2, we can (must) monitor the evolution of "system features". The most reliable we have found until now is the system's energy as shown on Fig. 12. Here one sees an "early" evolution (first 60000 MC steps) followed by a much slower one which *looks like* a stationary regime during the last 100000 steps[13].

---

[13]The time required to perform such simulations depends on the number of sampling points used for the piece-wise linear proposals of the amplitude parameters (Sec. 7.2.2). During the first 2000

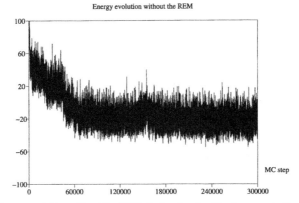

Fig. 12. Energy evolution without the REM. Notice the step like decreases in the early part.

### 11.2.3. Model parameters evolution

We can monitor as well the evolution of models parameters like the maximal peak amplitudes of the neurons[14] as shown on Fig. 13. The interesting, and typical, observation here is that the model parameters reach (apparent) stationarity much earlier than the energy (compare the first 60000 steps on Fig. 12 & 13). That means that most of the slow relaxation of the energy has a configurational origin. In other words, it comes from spike labels re-arrangements.

### 11.2.4. Model parameters estimates

Based on the energy trace (Fig. 12) and on the evolution of the model parameters (*e.g.*, Fig. 13) we could reasonably decide to keep the last 100000 MC steps for "measurements" and estimate the posterior (marginal) densities of our model parameters from these steps. We would then get for neuron 2 what's shown on Fig. 14. The striking feature of this figure is the time taken by the ACFs of the amplitude parameters ($P_1$, $P_2$, $\delta$, $\lambda$) to reach 0: 4000 steps. With the present data set we observe a similar behavior for the amplitude parameters of the first neuron but not for the third. That explains the un-acceptably low precision on the parameters estimates reported in Table 2 for neurons 1 & 2.

---

steps we use 101 regularly spaced sampling points. We then need less than 5' to perform 1000 steps. After that, we reposition the sampling points and use only 13 of them and the time required to perform 1000 steps falls to 50 s (Pentium IV computer at 3.06 GHz).

[14]We monitor in practice the evolution of every model parameter. We show here only the evolution of the maximal peak amplitude to keep this chapter length within bounds.

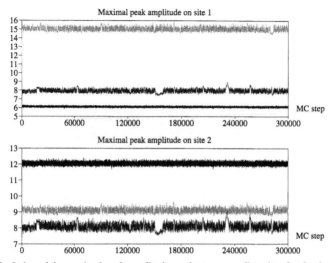

Fig. 13. Evolution of the maximal peak amplitude on the two recording sites for the three neurons. Notice that an apparently stationary regime is reached quickly for those parameters, while the energy (Fig. 12) is still slowly decreasing.

### 11.3. *Algorithm dynamics and parameters estimates with the REM*

The different behaviors of the model parameters, which relax to (apparent) equilibrium relatively fast ($2 \cdot 10^4$ to $3 \cdot 10^4$ MC steps) and of the energy which relaxes more slowly ($6 \cdot 10^4$ MC steps) suggests that the configuration dynamics is slower than the model parameters dynamics. It could therefore be worth considering a modification of our basic algorithm which could speed up the configuration (and if we are lucky the model parameters) dynamics like the Replica Exchange Method (Sec. 10). To illustrate the effect of the REM on our simple data set we have taken the first 2000 MC steps of the previous section and restarted[15] the algorithm augmenting our inverse temperature space from $\beta \in \{1\}$ to $\beta \in \{1, 0.95, 0.9, 0.85, 0.8, 0.75, 0.7, 0.65, 0.6, 0.55, 0.5\}$. The resulting energy trace for $\beta = 1$ and 30000 additional steps is shown on Fig. 15. The striking feature here is that the energy reaches after roughly $1.2 \cdot 10^4$ steps a *lower* energy than after $30 \cdot 10^4$ steps without the REM. Our REM scheme with 11 inverse temperatures implies that each step requires 11 times more CPU time than one

---

[15]When we introduce new $\beta$ values from one run to the next, the initial state of the replica with the new $\beta$ values is set to the final state of the replica with the closest $\beta$ in the previous run. That is, in our example, at step 2001 all the replicas are in the same state (same parameters values, same configuration).

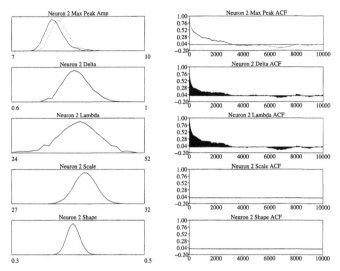

Fig. 14. Marginal posterior estimates for the parameters of neuron 2. The left part shown the esti-mated density of each of the six parameters (for the peak amplitudes, the first site corresponds to the black curve and the second to the gray curve). The right side shows the corresponding ACFs (Sec. 9.1 and Eq. 9.1), the abscissa is the lag in MC steps.

step with a single inverse temperature[16]. From the energy evolution view point the REM is clearly more efficient than a single long simulation at $\beta = 1$. A more striking illustration of the REM effect on energy relaxation (on an other data set) can be found in [32].

### 11.3.1. Making sure the REM "works"
As explained in Sec. 10 the REM will work efficiently only if the energies ex-plored by replicas at adjacent $\beta$ exhibit enough overlap. Fig. 16 illustrates how the efficiency of the "replica exchange" can be empirically assessed [5,12,26,37]. Energy traces at different $\beta$ values are displayed on Fig.16A. The overlap be-tween adjacent energy traces is already clear. The lowest trace ($\beta = 1$) is the trace of Fig. 15. Fig. 16B shows the energy histograms for the different $\beta$. Adjacent histograms clearly exhibit a significant overlap. A pictorial way to think of the REM is to imagine that several "boxes" at different pre-set inverse-temperatures are used and that there is one and only one replica per box at each step. After each MC step, the algorithm proposes to exchange the replicas located in neighboring boxes (neighboring in the sense of their inverse-temperatures) and this proposi-

---

[16]The computational overhead required accept or reject the proposed replica exchanges (Eq. 10.3) is negligible compared to the time required to update every spike label and every model parameter.

Table 2

Estimated parameters values using the last $10^5$ MC steps (among $3 \cdot 10^5$). The SDs are autocorrelation corrected (Sec. 9.1). The integrated autocorrelation time (Eq. 9.2) is given below each parameter between "()".

|  | neuron 1 | neuron 2 | neuron 3 |
|---|---|---|---|
| $\overline{P_1}$ | $15 \pm 7$ | $8 \pm 7$ | $6.1 \pm 0.2$ |
|  | (1071) | (1052) | (16) |
| $\overline{P_2}$ | $9 \pm 4$ | $8 \pm 6$ | $12.1 \pm 0.5$ |
|  | (945) | (930) | (27) |
| $\overline{\delta}$ | $0.7 \pm 0.3$ | $0.8 \pm 1.2$ | $0.58 \pm 0.05$ |
|  | (586) | (941) | (32) |
| $\overline{\lambda}$ | $33 \pm 64$ | $39 \pm 137$ | $47 \pm 13$ |
|  | (1250) | (1043) | (45) |
| $\overline{s}$ | $24.9 \pm 0.1$ | $29.8 \pm 0.1$ | $18.3 \pm 0.5$ |
|  | (6.5) | (7) | (1) |
| $\overline{f}$ | $0.51 \pm 0.05$ | $0.40 \pm 0.05$ | $1.01 \pm 0.03$ |
|  | (6.5) | (7) | (1) |

tion can be accepted or rejected (Eq. 10.3). Then if the exchange dynamics works properly one should see each replica visit all the boxes during the MC run. More precisely each replica should perform a random walk on the available boxes. Fig. 16C shows the random walk performed by the replica which start at $\beta = 1$ at step 2001. The ordinate corresponds to the box index (see legend). Between steps 2001 and 32000, the replica travels through the entire inverse temperature range.

One of the shortcomings of the REM is that it requires more $\beta$ to be used (for a given range) as the number of spikes in the data set increases because the width of the energy histograms (Fig. 16B) is inversely proportional to the square root of the number $N$ of events in the sample [12, 15, 16]. The necessary number of $\beta$ grows therefore as $\sqrt{N}$. The computation time per replica grows, moreover, linearly with $N$. We therefore end up with a computation time of the REM growing like: $N^{1.5}$.

### 11.3.2. Posterior estimates with the REM

We do not present here the Equivalent of Fig. 13 with the REM because the two figures would look too similar. The best way to see the effect of the REM on the model parameters dynamics is to look at the empirical *integrated autocorrelation time* (IAT) for each parameter as show in Table 3. When we compare these values with the ones obtained with our basic algorithm (without the REM), we see for instance that longest IAT is now 110 steps (parameter $\lambda$ of the second

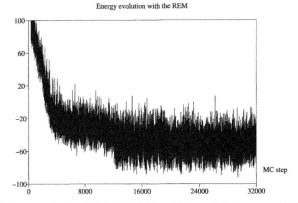

Fig. 15. Energy evolution with the REM. The ordinate scale is the same as in Fig. 12.

neuron) while the longest IAT without the REM was 1250 steps (parameter $\lambda$ of the first neuron, Table 2). We therefore get *much better parameters estimates although the absolute sample size we use* $10^4$ *is 10 times smaller than the one used previously.* This means that in this simple setting we were able without any fine tunning (we could have in fact used fewer inverse temperatures, we have not optimized the number of steps between replica exchange attempts) to get a much more efficient algorithm (in term of statistical efficiency and of relaxation to equilibrium) without extra computational cost.

### 11.3.3. Configuration estimate

As a quick way to compare estimated and actual configurations we can introduce what we will (abusively) call the "most likely" configuration estimate. First, we estimate the probability for each spike to originate from each neuron. For instance, if we discard the first $N_D = 22000$ on a total of $N_T = 32000$ steps we have for the probability of the 100th spike to have been generated by the second neuron:

$$Pr(l_{100} = 2 \mid Y) \approx \frac{1}{10000} \cdot \sum_{t=22000}^{32000} \mathcal{I}_2 \left[ l_{100}^{(t)} \right] \tag{11.1}$$

where $\mathcal{I}_q$ is the *indicator* function defined Eq. 5.10.

We then "force" the spike label to its most likely value. That is, if we have: $Pr(l_{100} = 1 \mid Y) = 0.05$, $Pr(l_{100} = 2 \mid Y) = 0.95$ and $Pr(l_{100} = 3 \mid Y) = 0$, we force $l_{100}$ to 2. We proceed in that way with each spike to get our "most likely" configuration. Fig. 17 shows what we get using the last $10 \cdot 10^3$ MC steps of our run using the REM. The actual configuration is shown too, as

*C. Pouzat*

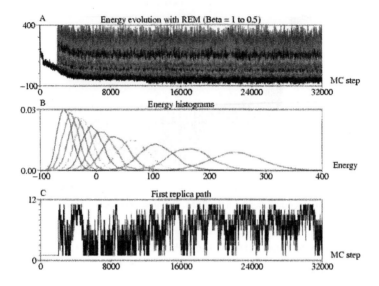

Fig. 16. Test of the REM. A, Energy evolution for several $\beta$ values (see text) during successive runs. 1 replica is used between steps 1 and 2000, 11 replicas are used between steps 2001 and 32000. B, Energy histograms at each $\beta$ value (histograms computed from the 10000 last steps with 25 bins): the left histogram correspond to $\beta = 1$, the right one to $\beta = 0.5$. C, Path of the first replica in the temperature space. The ordinate corresponds to the $\beta$ index (1 corresponds to $\beta = 1$ and 11 to $\beta = 0.5$ ).

well as the **50** misclassified events. We therefore get *1.7% misclassified spikes with our procedure.* We leave as an excercise to the reader to check that using an exact knowledge of the model parameters and the theoretical 2-dimensional Wilson plot densities one can get from them (Fig. 11C) we would get roughly 9% of the spikes misclassified. That would be the optimal classification we could generate using only the amplitude information. Our capacity to go beyond these 9% clearly stems from our inclusion of the ISI density in our data generation model. Using the last $10 \cdot 10^3$ MC steps of our run without the REM we generate 63 misclassifications, while we get "only" 59 of them with the last $100 \cdot 10^3$ steps[17].

---

[17]This difference in the number of misclassification explains the energy difference observed in the two runs (Fig. 12 & 15). It means clearly that the run without the REM has not reached equilibrium.

Table 3

Estimated parameters values using the last $10^4$ MC steps (among $6 \cdot 10^4$). The SDs are autocorrelation corrected (Sec. 9.1). The integrated autocorrelation time (Eq. 9.2) is given below each parameter between "()".

|  | **neuron 1** | **neuron 2** | **neuron 3** |
|---|---|---|---|
| $\overline{P_1}$ | $15 \pm 2$ | $8 \pm 2$ | $6.1 \pm 0.1$ |
|  | (62) | (91) | (12) |
| $\overline{P_2}$ | $9 \pm 1$ | $8 \pm 2$ | $12.1 \pm 0.3$ |
|  | (55) | (93) | (14) |
| $\overline{\delta}$ | $0.70 \pm 0.07$ | $0.8 \pm 0.3$ | $0.58 \pm 0.02$ |
|  | (25) | (93) | (9) |
| $\overline{\lambda}$ | $34 \pm 19$ | $40 \pm 38$ | $46 \pm 7$ |
|  | (60) | (110) | (20) |
| $\overline{s}$ | $24.9 \pm 0.6$ | $30.0 \pm 0.4$ | $18.3 \pm 0.7$ |
|  | (2) | (1.5) | (2.5) |
| $\overline{f}$ | $0.51 \pm 0.02$ | $0.40 \pm 0.01$ | $1.01 \pm 0.03$ |
|  | (2) | (1.5) | (2.5) |

### 11.3.4. A more detailed illustration of the REM dynamics

When the posterior density one wants to explore is non isotropic a problem arise with the use of an MH algorithm based on component specific transitions. In such cases, the transitions proposed by the MH algorithm are not optimal in the sense that only "small" changes of the present parameters values are likely to be accepted. These "small" changes will lead to a "slow" exploration of the posterior density and therefore to "long" autocorrelation times. This is illustrated for our MH algorithm without the REM on the left side of Fig. 18. Here we show only the situation in the plane defined by the maximal peak amplitude on site 1 and the parameter $\lambda$ of neuron 1 ($P_{2,1}, \lambda_1$) but the reader should imagine that the same holds in the 4 dimensional spaces defined by the 4 amplitude parameters of each neurons. The ACFs (Eq. 9.1) of the two parameters are shown. The last 1000 values of the parameters generated by the MH algorithm without the REM with the last 50 of them linked together by the broken line are shown. The movement of the "system" in its state space has clear Brownian motion features: small steps, random direction change from one step to the next, resulting in a "slow" exploration of the posterior density. If we now look at the system's dynamics using the REM, we see that the ACFs fall to zero much faster (right side of Fig. 18). This is easily understood by observing successive steps of the systems in the corresponding plane ($\lambda$ vs $P_{2,1}$ graph, bottom). Large steps going almost from one end to the other of the marginal posterior density are now observed, meaning that the resulting Markov chain explores very efficiently the posterior density we

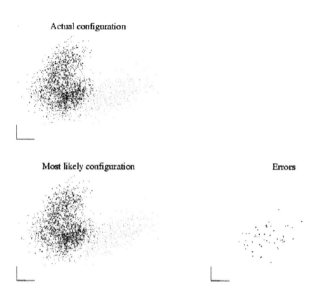

Fig. 17. Comparison between the actual and estimated configurations. Top, the Wilson plot of Fig. 11B color coded to show the neuron of origin of each spike. Bottom left, the most likely configuration (see text) estimated from the last 10000 steps with the REM. Bottom right, the 50 errors (among 2966 events).

want to study. This is clearly due to the fact that the high temperature replicas can make much larger steps with a still reasonably high acceptance probability. The replica exchange dynamics (cooling down "hot" replicas and warming up "cold" ones) does the rest of the job. This increased statistical efficiency ($\tau_{autoco}$ gets smaller) does in fact compensate for the increased computational cost ($\tau_{cpu}$ gets larger) of the REM (compare the integrated autocorrelation times in Table 2 & 3).

## 12. Conclusions

We have given a detailed presentation of a simple data generation model for extracellular recording. Although simple this model is already much more sophisticated than the ones previously used [21, 30, 31, 34], but this sophistication comes at a price: the usual algorithms presented in the spike-sorting literature cannot be used. To work with our model we had to find an analogy between our problem

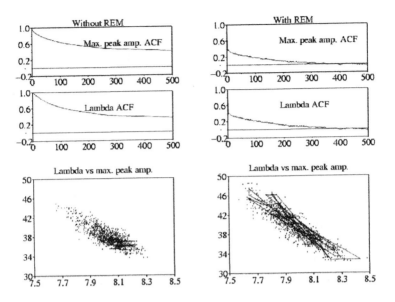

Fig. 18. REM and parameters auto-correlation functions. The ACFs of the maximal peak amplitude of neuron 2 on site 1 ($P_{2,1}$) are shown together with the ACFs of the parameter $\lambda$ of neuron 2. The ACFs without the REM (left side) were computed from the 100000 last iterations (Fig. 12). The ACFs with the REM (right side) were computed from the 10000 last iterations (Fig. 15). The ACFs with REM clearly fall faster than the ones without. The two graphs at the bottom give a qualitative explanation of why it is so. On each graph the last 1000 generated values for $P_{2,1}$ (abscissa) and $\lambda_1$ (ordinate) are shown (black dots). 50 values generated *successively* are linked together by the broken lines (black). The size of the jumps is clearly larger with REM than without.

and the Potts spin glass model studied in statistical physics. This analogy allowed us to get a Dynamic MC/MCMC algorithm with which model parameters and configurations could be estimated *with confidence intervals*.

We have not explained here how one of the critical questions of spike-sorting, the number of neuron contributing to the data, can be tackled. We have recently described in [32] how this can be done with an ad-hoc method. The next step is clearly a Bayesian solution to this problem which fundamentally implies estimating the normalizing constant $Z$ of Eq. 5.20 and 5.21 (which is the probability of the data). Nguyen et al [30] have already given a solution for a simple data generation model. This task of estimating $Z$ could seem daunting but is luckily not. Dynamic MC and MCMC methods have been used for a long time which means that other people in other fields already had that problem. Among the proposed solutions, what physicists call thermodynamic integration [8, 19] seems very attractive because it is known to be robust and it requires simulations to be performed between $\beta = 1$ and $\beta = 0$. And that's precisely what we are already (partially) doing with the REM. Of course the estimation of $Z$ is only half of the problem. The prior distribution on the number of neurons in the model has to be set properly as well. Increasing the number of neurons will always lead to a decrease in energy which will give larger $Z$ values (the derivative of the log of $Z$ being proportional to the opposite of the energy [8]), we will therefore have to compensate for this systematic $Z$ increase with the prior distribution.

Finally, although our method requires fairly long computation to be carried out, we hope we have convinced our reader that the presence of strong overlaps on Wilson plots does not necessarily precludes good sorting.

## 13. Exercises solutions

### 13.1. Cholesky decomposition (or factorization)

Write $a_{i,j}$ the elements of $A$ and $\sigma_{i,j}$ the elements of $\Gamma$, compute:

$$AA^T = B$$

with: $b_{i,j} = \sum_{k=1}^{D} a_{i,k} a_{j,k}$ and identify individual terms, that is:

$$a_{1,1} = \sqrt{\sigma_{1,1}},$$

where the fact that $A$ is lower triangular has been used.

$$a_{,j,1} = \frac{\sigma_{j,1}}{a_{1,1}} , \quad 2 \le j \le D.$$

$$a_{2,2} = \sqrt{\sigma_{2,2} - a_{2,1}^2}.$$

And keep going like that and you will find:

$$
\begin{aligned}
for\ i\ &=\ 1\ to\ D \\
for\ j\ &=\ i+1\ to\ D \\
a_{i,i}\ &=\ \sqrt{\sigma_{i,i} - \sum_{k=1}^{i-1} a_{i,k}^2} \\
a_{j,i}\ &=\ \frac{\sigma_{j,i} - \sum_{k=1}^{i-1} a_{j,k} a_{i,k}}{a_{i,i}}
\end{aligned}
$$

For details see [1].

*13.2. Bayesian posterior densities for the parameters of a log-normal* pdf

*Answer to question 1* We have:

$$\pi(s \mid f, \mathcal{D}) \propto$$

$$\exp\left[-\frac{1}{2}\frac{1}{f^2}\sum_{j=1}^{N}\left(\ln i_j - \ln s\right)^2\right] \cdot \frac{1}{s_{max} - s_{min}} \cdot \mathcal{I}_{[s_{min}, s_{max}]}(s).$$

If we introduce $\overline{\ln i}$ defined by Eq. 5.11 in the summed term we get:

$$\sum_{j=1}^{N}\left(\ln i_j - \ln s\right)^2 = \sum_{j=1}^{N}\left(\ln i_j - \overline{\ln i} + \overline{\ln i} - \ln s\right)^2$$

$$= \sum_{j=1}^{N}\left(\ln i_j - \overline{\ln i}\right)^2 + 2\left(\overline{\ln i} - \ln s\right)$$

$$\cdot \sum_{j=1}^{N}\left(\ln i_j - \overline{\ln i}\right) + N\left(\overline{\ln i} - \ln s\right)$$

The first term does not depend on $s$, the second is zero, we therefore have:

$$\pi(s \mid f, \mathcal{D}) \propto$$

$$\exp\left[-\frac{1}{2}\frac{N}{f^2}\left(\overline{\ln i} - \ln s\right)^2\right] \cdot \frac{1}{s_{max} - s_{min}} \cdot \mathcal{I}_{[s_{min}, s_{max}]}(s).$$

*Answer to question 2*

1. Generate $U \sim Norm\left(\overline{\ln i}, \frac{f^2}{N}\right)$[18]

2. if $s_{min} \le \exp u \le s_{max}$, then $S = \exp u$
   otherwise, go back to 1

*Answer to question 3* Using the first line of Answer 1 we get:

$$\pi(f \mid s, \mathcal{D}) \propto$$

$$\frac{1}{f^N}\exp\left[-\frac{1}{2}\frac{\sum_{j=1}^{N}\left(\ln i_j - \ln s\right)^2}{f^2}\right] \cdot \frac{1}{f_{max} - f_{min}} \cdot \mathcal{I}_{[f_{min}, f_{max}]}(f).$$

Identification does the rest of the job.

---

[18]This notation means that the random variable $U$ has a normal distribution with mean: $\overline{\ln i}$ and variance: $\frac{f^2}{N}$.

*Answer to question 4*   Using $\theta = \frac{1}{\omega}$ and the Jacobian of the change of variable, you can easily show that if $\Omega \sim Gamma\,(\alpha, \beta)$ then $\Theta = \frac{1}{\Omega} \sim Inv - Gamma\,(\alpha, \beta)$. The algorithm is simply:

1. Generate $\Omega \sim Gamma\left(\frac{N}{2} - 1, \frac{1}{2}\sum_{j=1}^{N}\left(\ln i_j - \ln s\right)^2\right)$

2. if $f_{min} \le \sqrt{\frac{1}{\omega}} \le f_{max}$ , then $F = \sqrt{\frac{1}{\omega}}$
   otherwise go back to 1.

### 13.3.  Stochastic property of a Markov matrix

$X^{(t+1)}$ being a *rv* defined on $\mathcal{X}$ we must have by definition:

$$Pr\left(X^{(t+1)} \subset \mathcal{X}\right) = 1,$$

where $X^{(t+1)} \subset \mathcal{X}$ should be read as: $X^{(t+1)} = x_1 + x_2 + \ldots + x_\nu$ (remember that the $x_i$ are mutually exclusive events). We have therefore:

$$
\begin{aligned}
Pr&\left(X^{(t+1)} \subset \mathcal{X}\right) \\
&= \sum_{i=1}^{\nu} Pr\left(X^{(t+1)} = x_i\right) \\
&= \sum_{i=1}^{\nu} Pr\left(X^{(t+1)} = x_i \mid X^{(t)} \subset \mathcal{X}\right) Pr\left(X^{(t)} \subset \mathcal{X}\right) \\
&= \sum_{i=1}^{\nu}\sum_{j=1}^{\nu} Pr\left(X^{(t+1)} = x_i \mid X^{(t)} = x_j\right) Pr\left(X^{(t)} = x_j\right) \\
&= \sum_{j=1}^{\nu} Pr\left(X^{(t)} = x_j\right)\left[\sum_{i=1}^{\nu} Pr\left(X^{(t+1)} = x_i \mid X^{(t)} = x_j\right)\right]
\end{aligned}
$$

Which implies:

$$\sum_{j=1}^{\nu} Pr\left(X^{(t)} = x_j\right)\left[\sum_{i=1}^{\nu} Pr\left(X^{(t+1)} = x_i \mid X^{(t)} = x_j\right)\right] = 1$$

And that must be true for any $X^{(t)}$ which implies that[19]:

$$\sum_{i=1}^{\nu} Pr\left(X^{(t+1)} = x_i \mid X^{(t)} = x_j\right) = 1$$

## 13.4. Detailed balance

If $\forall x, y \in \mathcal{X}$, $\pi(x) T(x, y) = \pi(y) T(y, x)$ then:

$$\begin{aligned}
\sum_{x \in \mathcal{X}} \pi(x) T(x, y) &= \sum_{x \in \mathcal{X}} \pi(y) T(y, x) \\
&= \pi(y) \sum_{x \in \mathcal{X}} T(y, x) \\
&= \pi(y)
\end{aligned}$$

Remember that $T$ is stochastic.

## References

[1] Bock RK and Krischer W (1998) *The Data Analysis BriefBook.* Springer - Verlag. http://physics.web.cern.ch/Physics/DataAnalysis/BriefBook/

[2] Brémaud P (1998) *Markov chains: Gibbs fields, Monte Carlo simulations and Queues.* New York: Springer - Verlag.

[3] Celeux G and Govaert G (1995) Gaussian parsimonious clustering models. *Pattern Recognition* **28**:781-793.

[4] Celeux G, Hurn M and Robert C (2000) Computational and inferential difficulties with mixture posterior distributions. *J American Statist Assoc* **95**:957-970.

[5] Falcioni M and Deem MW (1999) A biased Monte Carlo scheme for zeolite structure solution. *J Chem Phys* **110**:1754-1766.

[6] Fee MS, Mitra PP, and Kleinfeld D (1996) Variability of extracellular spike waveforms of cortical neurons. *J Neurophys* **76**: 3823 - 3833.

[7] Fishman GS (1996) *Monte Carlo. Concepts, Algorithms and Applications.* New York: Springer - Verlag.

[8] Frenkel D and Smit B (2002) *Understanding molecular simulation. From algorithms to applications.* San Diego: Academic Press.

[9] Frostig RD, Lieke EE, Tso DY and Grinvald A (1990) Cortical functional architecture and local coupling between neuronal activity and the microcirculation revealed by in vivo high resolution optical imaging of intrinsic signals. *PNAS* **87**: 6082 - 6086.

[10] Geman S, and Geman D (1984) Stochastic relaxation, Gibbs distributions and the Bayesian restoration of images. *IEEE Transactions on Pattern Analysis and Machine Intelligence* **6**: 721 - 741.

---

[19]If you're not convinced you can do it by *absurdum* assuming it's not the case and then find a $X^{(t)}$ which violate the equality.

[11] Häggström O (2002) *Finite Markov Chains and Algorithmic Applications*. Cambridge: Cambridge University Press.

[12] Hansmann UHE (1997) Parallel tempering algorithm for conformational studies of biological molecules. *Chem Phys Lett* **281**:140-150.

[13] Harris KD, Henze DA, Csicsvari J, Hirase H, and Buzsaki G (2000) Accuracy of tetrode spike separation as determined by simultaneous intracellular and extracellular measurements. *J Neurophys* **84**: 401 - 414.

[14] Hastings WK (1970) Monte Carlo sampling methods using Markov chains and their applications. *Biometrika* **57**: 92 - 109.

[15] Hukushima K and Nemoto K (1996) Exchange Monte Carlo and Application to Spin Glass Simulations. *J Phys Soc Japan* **65**:1604-1608.

[16] Iba Y (2001) Extended ensemble Monte Carlo. *Int J Mod Phys C* **12**:623-656.

[17] Janke W (2002) Statistical Analysis of Simulations: Data Correlations and Error Estimation. http://www.fz-juelich.de/nic-series/volume10

[18] Johnson DH (1996) Point process models of single - neuron discharges. *J Comput Neurosci* 3: 275 - 299.

[19] Landau D, and Binder K (2000) *A Guide to Monte Carlo Simulations in Statistical Physics*. Cambridge: Cambridge University Press.

[20] L'Ecuyer P (1994) Uniform random number generation. *Annals of Operations Research* **53**: 77 - 120.

[21] Lewicki MS (1994) Bayesian modeling and classification of neural signals. *Neural Comput* **6**:1005-1030.

[22] Lewicki MS (1998) A review of methods for spike-sorting: the detection and classification of neural action potentials. *Network: Comput Neural Syst* **9**: R53 - R78.

[23] Liu JS (2001) *Monte Carlo Strategies in Scientific Computing*. New York: Springer - Verlag.

[24] Metropolis N, Rosenbluth AW, Rosenbluth MN, Teller AH and Teller E (1953) Equations of state calculations by fast computing machines. *J Chem Phys* **21**: 1087 - 1092.

[25] Meyn SP and Tweedie RL (1996) *Markov Chains and Stochastic Stability*. New York: Springer - Verlag. A pdf version of the book (without figures) is available at the following address (Sean Meyn's web site): http://black.csl.uiuc.edu/ meyn/pages/TOC.html

[26] Mitsutake A, Sugita Y and Okamoto Y (2001) Generalized-ensemble algorithms for molecular simulations of biopolymers. *Biopolymers (Peptide Science)* **60**:96-123.

[27] Neal RM (1993) Probabilistic Inference Using Markov Chain Monte Carlo Methods. Technical Report CRG-TR-93-1, Department of Computer Science, Univ. of Toronto. http://www.cs.toronto.edu/ radford/papers-online.html

[28] Neal RM (1994) Sampling from multimodal distributions using tempered transitions. Technical Report No. 9421, Department of Statistics, Univ. of Toronto. http://www.cs.toronto.edu/ radford/papers-online.html

[29] Newman MEJ and Barkema GT (1999) *Monte Carlo Methods in Statistical Physics*. Oxford: Oxford University Press.

[30] Nguyen DP, Frank LM and Brown EN (2003) An application of reversible-jump Markov chain Monte Carlo to spike classification of multi-unit extracellular recordings. *Network: Comput Neural Syst* **14**: 61 - 82.

[31] Pouzat C, Mazor O, and Laurent G (2002) Using noise signature to optimize spike - sorting and to asses neuronal classification quality. *J Neurosci Methods* **122**: 43 - 57.

[32] Pouzat C, Delescluse M, Viot P and Diebolt J (2004) Improved spike-sorting by modeling firing statistics and burst-dependent spike amplitude attenuation: a Markov chain Monte Carlo approach. *J Neurophys* **91**: 2910 - 2928.

[33] Robert CP, and Casella G (1999) *Monte Carlo Statistical Methods*. New York: Springer - Verlag.

[34] Sahani M (1999) Latent variable models for neural data analysis. PhD Thesis, California Institute of Technology: Pasadena.

[35] Sokal AD (1996) Monte Carlo methods in statistical mechanics: foundations and new algorithms. Cours de Troisième cycle de la Physique en Suisse Romande. http://citeseer.nj.nec.com/sokal96monte.html

[36] Wu JY, Cohen LB and Falk CX (1994) Neuronal activity during different behaviors in Aplysia: a distributed organization? *Science* **263**: 820-823.

[37] Yan Q and de Pablo JJ (1999) Hyper-parallel tempering Monte Carlo: Application to the Lennard-Jones fluid and the restricted primitive model. *J Chem Phys* **111**:9509-9516.

Course 16

# THE EMERGENCE OF RELEVANT DATA REPRESENTATIONS: AN INFORMATION THEORETIC APPROACH

## Naftali Tishby

*School of Computer Science and Engineering*
*The Interdisciplinary Center for Neural Computation*
*The Hebrew University, Jerusalem 91904, Israel*
*www.cs.huji.ac.il/~tishby*

*C.C. Chow, B. Gutkin, D. Hansel, C. Meunier and J. Dalibard, eds.*
*Les Houches, Session LXXX, 2003*
*Methods and Models in Neurophysics*
*Méthodes et Modèles en Neurophysique*
© *2005 Elsevier B.V. All rights reserved*

787

# Contents

# Abstract

Information theory has a special place among theoretical approaches to neurobiology. While it is the framework that can provide general model independent bounds on information processing in biological systems, it is also one of the most elusive, misunderstood and abused conceptual theories. Most introductory texts to the theory follow Shannon's historical bottom-up construction. They begin with the axiomatic definition of Entropy - still falsely portrayed as the "measure of information" - then dive into involved technical arguments on "typicality" and "asymptotics". For the non-mathematician most of the intuition is lost at that point. It takes much more efforts to reveal the real beautify and integrity of Shannon's theory, that emerge only with the introduction of lossy communication and general cost and distortion tradeoffs. As a result many scientists treat applications of Information Theory with great suspicion. At best they are willing to apply its functions - like mutual information - to quantify independence or redundancies, but rarely take seriously attempts to apply the theory at the "whole system level".

Those notes is an attempt to do it differently. In the first part I try to introduce Shannon's theory in a non-standard top-down perspective. I begin with a general discussion of optimal tradeoffs, such as cost versus distortion or complexity versus accuracy. I then argue that Shannon in fact showed us - within the context his point-to-point communication model - that optimal cost-distortion tradeoff can be decomposed into two fundamental and dual relationships, the Rate-Distortion and the Capacity-Cost functions. To achieved this Shannon used the notion of mutual information to quantify an "internal representation" in "bits" and carefully balanced the source (past) information with the channel (future) information. In the second part such matched balance of information between compression and prediction is extended beyond the communication model to quantify optimal relevant representation of data. We achieve this through an "information bottleneck principle" (IB) that provide a general computational framework for "reverse engineering" of complex systems, merely from their input-output statistics. It also quantifies the efficiency of systems through their optimal complexity-accuracy tradeoff. The third part is a review of some of the application of IB to clustering, dimensionality reduction, and multivariate data analysis using the language of graphical models.

The style of the notes is informal and evokes technical arguments only when truly needed, to keep the flow and the intuitive nature of the exposition.

791

## 1. Part I: the fundamental dilemma

The volume of available data in a variety of different domains has exploded over the past years. Examples include the consistent growth in the amount of on-line text due to the expansion of the World Wide Web, the dramatic increase in the available genomic information due to the development of new technologies for gathering such data, and the vast complex data accumulated in neuroscience in multiple modalities and experimental technologies. As a result, there is a crucial practical need for exploratory data analysis and filtering methods.

While this is the main driving force behind much of machine learning and modern statistical techniques, filtering relevant information is also the primary objective of biological information processing and above all of the nervous system. In a well defined sense, extracting relevant information is the central unified task at all levels of life, as it accounts for noise reduction, feature extraction, adaptation and learning, on multiple time scales. It is at the heart of the dynamic and delicate balance between efficiency and survivability, which characterize life. This is also our motivation for developing principled techniques for extracting relevant information and - if possible - to obtain general quantitative limitations on efficient data representation.

### 1.1.  Fitting models to data—what is the goal?

Much of quantitative science in general and machine learning in particular is about fitting models to empirical data. This rudimentary task is illustrated in the regression of "a model function" to data points in two-dimensions, as shown in Figure 1.1. Given a few $(x, y)$ points we often imagine that they represent an underlying unknown functional relation of the form $y = f(x)$ for some function $f$. There are various mathematical techniques for fitting the "best" function to this data. We could fit - for instance - an interpolation polynomial that goes precisely through all the points, but surely non of you consider this a very useful "model". A simpler linear regression seems to makes much better sense in this case. But why? Do we know anything about the noise in this data or about the way it is sampled? The common learning theoretic argument is that simpler models are "better" since they provide better generalization beyond the given data (a principle also known as parsimonious, Occam's razor, etc.), but making this argument concrete requires various assumptions about the noise, sampling and the model class - all implicit at this point. What we intuitively do here is trading between the *complexity* of our possible fit and the (generalization) *accuracy* that it can provide us. We normally settle with the simplest point where additional complexity does not improve generalization accuracy significantly. Such a point may indeed exist for very simple functional relationships, but except for 'toy problems' most

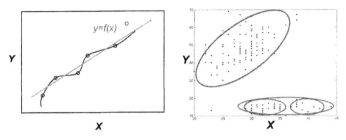

Fig. 1. Fitting a model to data always involve assumptions about the nature of the noise and the class of the possible models. This is true for both supervised learning, like regression (left), or unsupervised learning like clustering (right).

"natural problems" reveal more structure when adding more data. A much better procedure then is not to settle with "one correct model" but rather to understand how is the accuracy depends on the complexity of our model description.

Similar tradeoffs exist in unsupervised classification (like clustering) (see Figure 1.1 left) where adding more clusters obviously gives better fit - but often less "insight". Here again there a no meaningful "correct" number of clusters, except for toy problems. The standard statistical answer that it is the size of the data that determines the complexity of our model, through one or another complexity control principles - while epistemologically correct - is clearly unsatisfactory from an ontological perspective. Is there a meaning then to the "true" complexity of the data? Here again, our approach is that it is the way structure is revealed when adding more complexity, more clusters, which we would like to quantitatively understand.

### 1.2. Accuracy versus complexity

In both supervised learning, such as regression or function fitting, and unsupervised learning as clustering and feature extraction, we trade between the *complexity* and *accuracy* of our description of the data. Generally, more complex models provide better fit, by adapting more parameters for instance, and thus obtain more accurate description of the *given* (training) dataset. If we ignore for a moment the crucial concern about finite samples, over-fitting and generalization - the central issue in computational learning and statistics - we expect a monotonically increasing relationship between complexity and accuracy. Higher model complexity should allow us to obtain at least as accurate description as obtained with lower complexity (when considering all models up to some given complexity). When, presumingly, the underlying phenomena has finite complexity (much weaker then assuming a "correct" model) adding more complexity should eventually reach a point of diminishing return. We thus expect a monotonically increasing function with a diminishing slope - a concave function, whose precise characteristics depend on the our precise measures of accuracy and complexity and - of course - on the structure of the data itself. Figure 2 depicts this general expected tradeoff. We expect it to rise from the origin - zero complexity enables zero accuracy - and eventually saturate, as we exhaust the complexity of the data and reach the point of diminishing return. Can we make this intuition more quantitative and precise?

## 1.3. Quantifying complexity and accuracy

Before one can quantify such a general relationship it is better to define it properly through an optimization problem. Accuracy is usually quantified through a measure of distance or distortion to our target, such as the square distance. The problem however, is that for natural data finding the "right" measure of accuracy can become very tricky, as we see later on. But once the point-wise distance measure is established the overall accuracy of a model is usually taken as the *average* distance (or distortion) over the data. We adopt this convention as the total model accuracy measure.

Defining a general - yet practical - measure of complexity is one of the deepest predicaments of mathematical science. One natural and generally accepted measure of model complexity, following Kolomgorov and others, is its *minimal description length*, measured in bits. One can think of it as the minimal number of *yes/no* questions that are needed to specify the model, or as its "optimal" binary coding. We will return and elaborate on this measure and make it more precise and measurable (under some assumptions), but at this point we assume that we know how to assign complexity to models and that we measure it in bits.

Consider then the accuracy of *the most* accurate model achievable[1] for a given upper bound on the complexity, or alternately, the least complex model for a given lower bound on the accuracy level. Those two constrained optimization problems are equivalent and dual to each other, similar to the area-circumference duality in the isoperimetric problem, or to energy and entropy in equilibrium thermodynamics. Increasing the complexity, say - by one bit, does not necessarily mean that we can obtain the same improvement in accuracy for every additional bit of complexity. This is certainly the case when there is a fundamental quantization or noise level that determines the maximum achievable accuracy, as happens with most experimental data. We thus expect a monotonic downward concave accuracy-complexity relationship under very wide circumstances. Two important caveats are in order:

---

[1] Here again there are deeper computational issues which we ignore at this point.

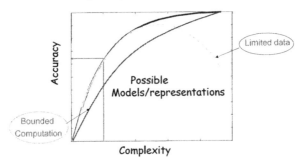

Fig. 2. Schematic plots of the Accuracy-Complexity tradeoff. The maximum achievable accuracy for a given model complexity. We expect a monotonically increasing concave curve, with "diminishing slope" for models that are "too complex". With limited data we see the effect of "over-fitting" that reduces accuracy for too high complexity. Bounded computational power on the other hand can limit the ability to achieve the most accurate model.

1. **Limited data.**

   With finite samples there is an optimal level of complexity beyond which we begin to *over-fit* the given sample and reduce performance and accuracy. Ways of avoiding it are the essence of statistical learning theory, but this is a different and basically solved issue that lies beyond the scope of these lectures.

2. **Bounded computation.**

   Bounded computational power can also reduce the ability to achieve the optimal relationship in various and highly complicated ways (shown here schematically) which are also beyond our scope.

   Tradeoffs of that nature are ubiquitous in science and engineering, but perhaps the deepest theoretical formulation of such tradeoffs and the one most relevant for our goals in this course is in Shannon's information theory.

## 2. Part II: Shannon's information theory—a new perspective

Information Theory is a truly remarkable conceptual achievement of one man, Claude Elwood Shannon. It was conceived, at least partly, during the war as an attempt to provide the fundamental mathematical limitations on point-to-point communication, independently of its specific physical realization. This general model independent nature of Shannon's theory makes it appealing when studying neural systems, for which there is really no single generally accepted computational paradigm. It is historically and conceptually interesting to note that Shannon was partly motivated by "practical" problems in military communication and specifically by cryptography. In fact his 1945 paper on the mathematical theory of cryptography [26], originally classified and still widely unknown, clearly hints at some of the key ideas of his seminal 1948 paper on the mathematical theory of communication [27]. Most noticeably, the ideas about redundancy reduction and entropy, the notion of a quantized/digital channels and the fundamental separation between source-coding and channel-coding - are directly linked with his ideas about cryptography. The source-channel separation, which enabled the treatment of compression and error correction as separated problems, was very much taken for granted during more than 50 years of information theory. It is also a departure point between the engineering of communication and problems of biological information processing, as has been pointed out recently by Toby Berger [3] and others [21]. This is also a starting point for our Information Bottleneck framework as we see later.

Shannon's theory is about the basic limitations on sending "messages" from one point - the "sender" - to another point - the "receiver" - efficiently and reliably. It does not appear to deal with the *contents* (or meaning) of those messages - a point that caused many confusions - nor does it deal with the physical nature of the communication. However, information theory - in its very heart - is about optimal tradeoffs. Similar to its older predecessor - statistical physics - information theory is formulated, in its ultimate form, through a competition between partially attainable "desires", very much like minimum energy and maximum entropy in physics.

A general and common optimization problem (not just in communication) is the tradeoff between distortion and cost. This "external" tradeoff is between the (average) cost required for communication (representation) through the channel and the minimal achievable data distortion that this transmission cost allows (see Figure 3). A communication system is considered optimal if it is operated along the optimal tradeoff between distortion and cost.

Shannon showed us that the cost-distortion tradeoff, in the point-to-point communication setting, can be broken into two - dual and complementary - basic tradeoffs. The first is between compression (short description) and distortion and is known as Rate-Distortion theory or source coding, while the second is between reliable error-correction and the cost of the transmission, known as the Capacity-Cost tradeoff or channel coding [6]. This remarkable decomposition of the cost-distortion tradeoff is made possible through the introduction of an internal, symbolic representation of messages in the "channel", and through matching of an *emergent* new property of the system - its (internal) mutual-information.

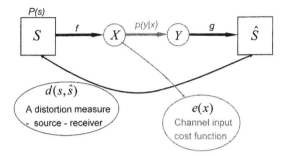

Fig. 3. Shannon's source-channel communication model, with distortion and channel cost. The source messages **S** are encoded through a map $f$ into channel symbols $X$. The channel output symbols $Y$ are stochastically related to $X$ through $p(y|x)$ due to possible channel noise. The output $Y$ is then decoded through the function $g$ into the received messages $\hat{S}$. A cost $e(x)$ is associated with every channel symbol $x$ and the distortion $d(s, \hat{s})$ is measured between the input and output messages.

### 2.1. Formalizing the problem: distortion versus cost

Source messages are stochastic variables that can be specified through their probability distribution[2] Denoting the source message variable by **S** with its probability distribution $P(\mathbf{S})$ and the received messages by $\hat{\mathbf{S}}$. We need to specify the external **distortion measure** between **S** and $\hat{\mathbf{S}}$, through a non-negative function $d(\mathbf{s}, \hat{\mathbf{s}})$ whose expected value $\Delta = \langle d(\mathbf{s}, \hat{\mathbf{s}}) \rangle_{p(s, \hat{s})}$ is to be minimized. To communicate, the messages must go through a *physical channel* between the source and the receiver. This can be chemical, acoustic, or electromagnetic or any other media that can be modulated. We denote the entry point to this channel by $X$ and the exit by $Y$, as the standard convention. Shannon realized (long before the digital age) that such physical channels, even when continuous in nature, must be quantized (digitized) into *channel symbols* which are discrete entities. While the precise nature of this quantization depends on the physical properties of the channel, such discretization is inevitable for the existence of distinguishable symbols and reliable communication. It also allows us to treat all channels in an abstract and unified way, as long as we can ignore "interactions" between the symbols (memoryless channels), as we do here. Hence we can assume for simplicity that the channel symbols are binary: 0 and 1.[3]

Since the source messages are not necessarily binary, some translation, or *encoding* is needed between the source and the channel, which transform the arbi-

---

[2] If they are not stochastic - there is no point communicating them at all.

[3] Any other set of countable symbols can be considered as finite strings of bits, even if for some channels it is easier to have other representations. True, this requires encoding as well...

trary messages into strings of channel-symbols. Moreover, for the communication to work this encoding mechanism must be *decodeable* by some decoder at the end of the channel - back into messages - without distorting them beyond the distortion (on average).

Transmitting a symbol through a physical channels has a **cost** in energy or other resources, which may depend on the transmitted symbol, denoted by $e(x)$ and assumed non-negative. One would like to minimize the expected cost, $\Gamma = \langle e(x) \rangle_{p(x)}$, which can be the transmission (or storage) cost, under a constraint on the expected distortion $\Delta$. If the channel cost is uniform (the bits have the same transmission cost) then minimum cost is equivalent to minimum average code length, for a given average distortion. This is the formulation of the familiar *source coding* problem: what is the minimum number of channel symbols per source message (hence, rate), that are needed to enable reconstruction (decoding) with at most $D$ expected distortion. This problem is known as the *Rate-Distortion tradeoff* and the optimal tradeoff function denoted by $R(D)$.

The problem with physical channels is that they are generally noisy - the symbols sent at the entry $X$ are not necessarily identical to those received at $Y$. In the simplest (memoryless and stationary) channel one can quantify this noise through the probability that the symbol $y$ is received when $x$ is transmitted, $p(y|x)$. Noisy channels introduce inherent stochasticity into the problem and it appears intuitively that the resulting communication must be stochastic as well. One of Shannon's brilliant observations was that noisy channels can be "repaired" and the error probability can be reduced arbitrarily, if one can encode arbitrarily long *blocks* of symbols together and use some fraction of the symbols for *error-correction* of the possible transmission errors. The maximum possible *ratio* of *useful* channel symbols to the total block length, at a given expected channel cost $E$, is called the *capacity-cost* function, $C(E)$. Shannon realized that this channel capacity (at a given cost) - depends only on the channel properties ($p(y|x)$) only and that the error-correcting coding can be *separated* from the source representation. This second fundamental problem, source coding or the *capacity-cost* tradeoff, and the function $C(E)$ can now be formulated as the maximum achievable fraction (rate) of reliably transmitted symbols through the noisy channel, at a given expected cost $E$.

## 2.2. Solving the cost–distortion tradeoff: the emergence of mutual information

Shannon's "external" problem can be formulated as follows: What is the minimal achievable expected cost, $E(D)$, for a given bound on the expected distortion $D$, given the source statistics $P(\mathbf{S})$, the non-negative distortion function $d(\mathbf{s}, \hat{\mathbf{s}})$, the channel noise distribution, $p(y|x)$, and the channel input symbol non-negative

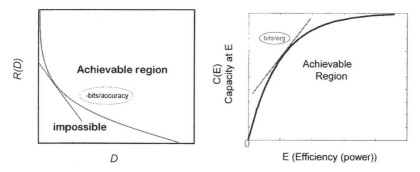

Fig. 4. The Rate-Distortion - convex downward - function (left) and the Capacity-Cost - concave upward - function (right). Both are boundaries of the achievable representation phases. The first is obtained vis minimization of mutual information between the source and its representation, subject to average distortion constarint. The second by maximization of mutual information across the channel, subject to the average cost constarint.

cost $e(x)$?

The complete solution to this problem was provided in Shannon's 1959 Rate-Distortion-Theory paper [28], though the formulation and parts of the proof appeared earlier. It amounts to matching of two values of a new, internal, quantity - mutual information - for the source compression (Rate-Distortion function) on the one hand and for the channel prediction (Cost-Capacity function) on the other hand. This remarkable decomposition remains true even without the assumption that the source and channel coding are separated. The transformation of the "physical" measures of cost and distortion into abstract values of (mutual) information serves as the starting point to our information-based analysis of complex systems.

Shannon's fundamental results can finally be summarized as follows (for details and proofs see e.g. [6]):

1. **Mutual information**:
   There exist a functional of any two distributions $p(x)$, and $q(y|x)$, called *the mutual information* between $X$ and $Y$, defined as:

$$I[p(x), q(y|x)] \equiv I(X; Y) \equiv \sum_{x,y} p(x, y) \log \frac{p(x, y)}{p(x)p(y)}, \qquad (2.1)$$

where $p(x, y) = p(x)q(y|x)$ and $p(y) = \sum_x p(x)q(y|x)$.[4] As can be easily verified $I[p, q]$ is a non-negative and symmetric function of $X$ and $Y$, and a

---

[4]As usual, sums should be interpreted as integrals for continuous variables throughout.

convex functional in both distributions $p$ and $q$ (due to the log-sum inequality, [6]).

2. The achievable **Rate-Distortion function** $R(D)$ is given by the optimization,

$$R(D) = \min_{\{q(x|s):\langle d(x,s)\rangle_{p(s)q(x|s)} \leq D\}} I[p, q] . \qquad (2.2)$$

This is a decreasing convex function whose value in the limit $D \to 0$ is the familiar source *Entropy*, denoted by $H(S)$.

3. The **Capacity-Cost function** $C(E)$ is obtained by the dual optimization problem,

$$C(E) = \max_{\{p(x):\langle e(x)\rangle_{p(x)} \leq E\}} I[p, q] . \qquad (2.3)$$

This function is monotonically increasing and concave and its asymptotic (unconstrained) value is known as *the* Capacity of the channel. For proper cost functions $e(x)$, the function $C(E)$ is strictly convex over the range $0 \leq E \leq e_{max}$ and its inverse function, $E(C)$, exists in the range $0 \leq C \leq Capacity$.

4. Given the source and channel specifications the **optimal cost-distortion trade-off** is decomposed as:

$$E(D) = E(C = R(D)) . \qquad (2.4)$$

Thus the *necessary and sufficient* condition for optimality is $R(D) = C(E)$, where $C(E)$ is the inverse function of $E(C)$.

The mutual information *emerges* here as the quantifier of the internal representation complexity (in bits) at the channel input, and as the interface between the source redundancy reduction (compression) problem and the channel error-correction (prediction) problem. It can be written as differences of the entropies $I(X; T) = H(X) - H(X|T)$ as shown in Figure 5.

Shannon's mutual information has several other interesting properties. It is the only functional (up to a multiplicative constant) of the joint distribution of any two variables, $X$ and $Y$, that quantifies our intuitive notion of information in $X$ about $Y$ (additive with respect to new information and symmetric). Since information is always "in something, about something else", this functional captures the statistical notion of information better than the entropy, which in fact measures disorder not information.

More importantly, mutual information provide the measure of the "volume" of jointly-typical sequences of the two variables, among all possible independent sequences. Minimizing mutual information, subject to our knowledge on the variables, thus provide the "least committed" joint distribution, similarly to the

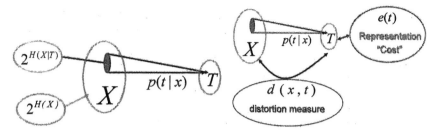

Fig. 5. The mutual information $I(X; T) = H(X) - H(X|T)$ is the number of bits need to specify (asymptotically) the compressed representation of $X$ by $T$ (left). The function $I(X; T)$ characterizes the measure of "jointly-typical" sequences of $X$ and $T$, and this is its key technical property. The Cost and Distortion associated with the general representation problem: how to minimize the cost of the representation of $X$ by $T$, subject to the reconstruction distortion as a constarint. It is the mutual information that links the two functions.

maximum entropy distribution for a single variable [15]. It can also serves as a natural measure of independence for two variable, but this is clearly its less interesting property.

### 2.3. The rate–distortion function

Using the form of the mutual information and the given distortion measure, one can derive the optimal tradeoff in the rate distortion problem. To simplify notation, we henceforth denote the representation variable by $T$ and a general input variable by $X$. We hence seek for the optimal representation of a variable $X$ by the variable $T$ (don't be confused: in the source representation problem our input variable $X$ is the message $S$, not the channel input...). The representation is through a conditional distribution, $q(t|x)$, for $t$ that represents $x$, such that the expected distortion $\sum_{x,t} p(x)q(t|x)d(x, t) = \langle d(x, t) \rangle_{p(x)q(t|x)} \leq D$. Introducing a Lagrange multiplier, $\beta$, for the average distortion constraint (taken as equality) we have to minimize the Lagrangian,

$$\mathcal{L}[q(t|x)] = I(T; X) + \beta \langle d(x, t) \rangle_{p(x)q(t|x)} . \tag{2.5}$$

among all normalized maps $q(t|x)$. It is not hard to verify that stationarity conditions for this optimization are given by

$$\begin{cases} q(t|x) = \frac{q(t)}{Z(x,\beta)} \exp(-\beta d(x, t)) \\ \\ q(t) = \sum_x q(t|x)p(x) , \end{cases} \tag{2.6}$$

where $Z$ is the normalization (partition) function. The above conditions must be satisfied self consistently for any positive value of $\beta$ and a solution can be found by iterating them alternating until convergence (this procedure is known as the Blahut-Arimoto algorithm for the Rate-Distortion function, see [6]). The Lagrange multiplier $\beta$ is clearly (minus) the slope of the Rate-Distortion function $R(D)$ at the convergence point. It is worth noting that the nature of the variable $T$ is implicit in this derivation and one has to verify (independently) that the chosen $T$ is indeed sufficient for a (global) optimum at a given $\beta$ (or $D$). This can be rather tricky.

A similar derivation applies to the Cost-Capacity problem, but there the optimization is over the channel input distribution $q(x)$ given the channel stochastic map $p(y|x)$. A similar iterative algorithm works there as well.

It is interesting to mention here the intriguing temporal aspects of this source-channel duality, so clearly pointed out by Shannon himself in an inspiring comment at the end of his 1959 Rate-Distortion theory paper:

*"There is a curious and provocative duality between the properties of a source with a distortion measure and those of a channel... if we consider channels in which there is a "cost" associated with the different input letters... The solution to this problem amounts, mathematically, to maximizing a mutual information under linear inequality constraint... which leads to a capacity-cost function $C(E)$ for the channel... In a somewhat dual way, evaluating the rate-distortion function $R(D)$ for source amounts, mathematically, to minimizing a mutual Information again under a linear inequality constraint.*

*This duality can be pursued further and is related to the duality between past and future and the notions of control and knowledge. Thus we may have knowledge of the past but cannot control it; we may control the future but have no knowledge of it."*

### 2.3.1. Example: the Gaussian source and channel

An illuminating concrete example to these notions is the "Gaussian source". This source is a real random variable with a normal distribution, $\mathbf{S} \sim N(0, \sigma^2)$ with the quadratic distortion, $d(\mathbf{s}, \hat{\mathbf{s}}) = (\mathbf{s} - \hat{\mathbf{s}})^2$ (see [6] Ch. 10,13 for details). In this case the Rate-Distortion function can be directly calculated and is given by:

$$R(D) = \frac{1}{2} \log \frac{\sigma^2}{D}$$

for $0 \leq D \leq \sigma^2$ and zero otherwise.

Matched with this source is the "Gaussian channel", $X$, which is also real val-

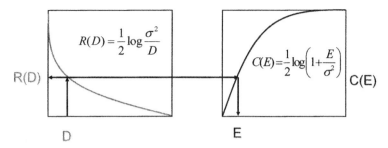

Fig. 6. The Rate-Distortion function $R(D)$ (left) and Capacity Cost function $C(E)$ (right). By equating the rate to the capacity $R(D) = C(E)$ we obtain the optimal Cost-Distortion tradeoff.

ued [5] with the Gaussian noise distribution $X \sim N(0, \eta^2)$ and power, $e(x) = |x|^2$, cost.

For the Gaussian channel the Capacity-Cost function is given by:

$$C(E) = \frac{1}{2} \log \left( 1 + \frac{E}{\eta^2} \right) .$$

The Gaussian $R(D)$ and $C(E)$ functions are plotted in Figure 6. The optimal Cost-Distortion tradeoff in this case is obtained for $R(D) = C(E)$, yielding the relation:

$$E = \eta^2 \left( \frac{\sigma^2}{D} - 1 \right) .$$

### 2.4. Perfect matching of source and channel

The duality of the Rate-Distortion and Capacity-Cost functions and the way they applied together to solve the communication problem is shown in Figure 6. The two coding problems are usually addressed independently (justified by Shannon's source-channel separation theorem). Much of the efforts in IT over the past 50 years thus went into finding efficient and ingenious algorithms for source coding (compression) and channel-coding (error-correction), which are notoriously difficult and in general require asymptotic (long blocks) coding and unbounded delays. However, those coding algorithms have probably little to do with biological and cognitive applications of IT and the focus on specific coding schemes in neuroscience is largely misleading.

But even without solving the above optimal coding problems independently the above tradeoffs are still very relevant biologically due to the notion of *matched*

---

[5]Q: What are the discrete channel symbols in this case?

*source and channel* [8] and the possibility of optimal cost-distortion operation without any delays and instantaneously by a single symbol blocks (see T. Berger, Living Information Theory, Shannon Lecture, 2003 [3]).

The example of the Gaussian source with square distortion and channel with a power constraint illustrates this issue. The optimal cost-distortion tradeoff can be obtained in this case by a simple rescaling of the channel variable by the noise ratio $(\eta/\sigma)^2$ and without any other coding! In other words, there is no need in any of the asymptotic block-coding associated with both channel and source coding when they are considered separately. To understand it, notice that there is a "match" between the distribution of the noise in the channel and the optimal representation map, which should be exponential in the distortion function, as we have seen. There is another match between the induced channel input distribution (also normal in this case) and the power constraint, which is a simple constraint on the input variance. This "conspiracy" of the distortion and cost functions with the source channel distributions essentially means that the allowed distortion $D$ is precisely the one generated by the noise in the channel, and thus no (block) coding is required to achieve optimality. In this case the optimality condition, $R(D) = C(E)$, is achievable with *a single letter code* as the source optimal (stochastic!) mapping induces the distribution $p(x)$ at the channel input which precisely achieves the channel capacity $C(E)$. This matching is possible, of course, only for lossy communication - when the channel error must not be corrected, but rather taken as the allowed distortion.

Recently, general necessary and sufficient conditions for this matching were formulated directly in terms of the distortion and cost and the channel distributions in [18]. These conditions can be summarized as,

$$
\begin{cases}
d_{GRV}(s, \hat{s}) = -d_1 \log p(s|\hat{s}) + d_0(s) \\
\\
e_{GRV}(x) = c_1 D_{KL}[p(y|x)\|p(y)] + c_0 ,
\end{cases}
\tag{2.7}
$$

where $c_0$, $c_1$ and $d_0$ are constants. In addition all loss of information in this system should be due to the noisy channel. It is an easy exercise to verify these conditions for the Gaussian source-channel and we will apply them again for the Information Bottleneck.

## 3. Part III: relevant data representation

Shannon's cost-distortion tradeoff can be simplified considerably, beyond his communication model, as shown in Figure 5-right.

Consider the source (input) variable $X$, with its distribution $p(x)$ and its representation $T$. We seek a stochastic map from $X$ to $T$, denoted by $q(t|x)$. Given the representation $t$ of $x$, there is a distortion measure associated with it, $d(x, t)$ that can be *any* non-negative function that quantifies the quality of $t$ as a representative of $x$. This is where our "understanding" of the usage of $t$ comes in. As before, would like to minimize the expected distortion $\langle d \rangle_{pq}$, under a constraint on the expect cost of the representation $\langle e(t) \rangle_{q(t)}$. When the cost is uniform (in $t$) the problem is clearly equivalent to maximum compression, under distortion constraints, i.e. minimum expected representation length.

The key observation is that the representation length, as shown in Figure 5-left, is determined by the ratio of two volumes. The first is the size of the space of $X$ and the other is the (average) size of a region of $X$ that is mapped to a point $t$ in $T$. When we consider long *typical n-long* sequences of $x$'s these volumes are completely characterized by the probability distributions $p(x)$ and $q(x|t)$ and are asymptotically given by $2^{nH(X)}$ and $2^{nH(X|T)}$ respectively. The entropy functions: $H(X) = -\sum_x p(x) \log_2 p(x)$ and $H(X|T) = -\sum_{x,t} p(t)q(x|t) \log_2 q(x|t)$, appear as usual for combinatorial counting. The typical size of the representation, or number of bits needed to specify $t$ given $x$, is given by the mutual information $I(X; T)$ through the ratio of these volumes. Namely:

$$2^{nI(X;T)} = 2^{nH(X)} / 2^{nH(X|T)} \, ,$$

which is the basic way mutual information emerges. Minimizing the (expected) representation size (cost in this case) amounts to minimizing the mutual information, just as in Section 2.3. It is easy to see that with a non-uniform cost $e(t)$ one obtains the same problem with a "shifted" distortion function $\tilde{d}(x, t) = d(x, t) + \gamma e(t)$, with $\gamma$ a Lagrange multiplier that determines the tradeoff between the cost and distortion.

Finally, we consider another observed variable - say - $Y$, that depends (causally and stochastically) on $X$. We are interested in revealing the most efficient connection between the two using the Information Theoretic framework. Such a "system-identification" problem is fundamental in neuroscience, as we often measure at two points which are connected through a complex network of system and would like to discover the simplest possible connection between the observed variables.

### 3.1. The "inverse Shannon problem"

We are now ready to connect Shannon's theory with neuroscience. As physicists, imagine that we are facing a complex information processing network that has external input and output signals and many (many) internal nodes connected in an intricate way. Imagine also that we can insert various probes and measure signals simultaneously at different nodes in the system (I am a theorists...). The system is complicated and operates via multiple chemical and electrical processes at every node. To characterize the system we may want to specify "costs" for the activity in some nodes or, possibly "distortions" between pairs of nodes... and try to evaluate the efficiency or other properties that characterize information processing in the system...

Well, even if we could know how to assign such costs and distortions this is clearly a hopeless task. The system is not only partially and extremely sparsely observed, but there is really no general way to translate our knowledge about the external (sensory?) inputs and (behavior?) outputs to meaningful constraints on the internal activities and representations.[6]

Imagine on the other hand, that we have an imaginary device, that I call "Shannon-meter", which can provide us with estimates of the mutual information between pairs of nodes in the system (told you I am a theorist), or even better - with estimates of the nodes joint activity distributions. What could we do with such information?

In principle, all the relevant physical and chemical costs and distortion constraints are manifested through those distribution and information values. This is really the essence of Shannon's decomposition. What we are facing is in fact the "inverse" of Shannon's engineering problem. We are given joint statistics of activities and can *assume* that it reflects optimal (e.g. due to evolution and learning) hidden internal channels and representations. Our goal is to reveal this (or an equivalent) internal representation solely from the input-output statistics. We may then be able to test those predictions with the Shannon-meter, avoiding - in theory - the need to handle the messy physical aspects of the system.

In its simplest form, this problem involves only three variables: input $X$, output $Y$, and an internal (black-box) representation variable $T$ which is assumed to depend on $X$. Give the external statistics, $p(x, y)$, the goal is an optimal representation $T$ that can account for, or approximate, this statistics. Solving this inverse-Shannon problem is the essence of the *Information Bottleneck* method.

### 3.2. *Looking into the black box: the information bottleneck method*

Can we reveal (identify) the system structure simply from the statistics of its input and output? In a sense this is the inverse of Shannon's source-channel problem. We would like to reveal the most efficient representation $T$ that could evolve between the observed system input $X$ and its output $Y$. As before, we seek for the optimal map $q(t|x)$ that captures precisely the components of $X$ that are required for making good predictions, or explaining the observed values of $Y$. In analogy with the source-channel picture, we imagine an unknown "channel", $q(y|t)$,

---

[6]Trying to construct everything bottom-up - from the biochemistry of the synapses - is at least as difficult as understanding computer algorithms from measurements of voltages across some gates in its memory... with all my respect to those who try to do that.

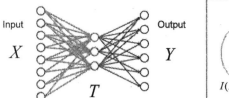 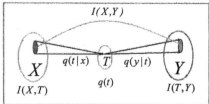

Fig. 7. The "Information Dynamics" of IB: find efficient representations of the input $X$, $T$ that maintain as much information on the output $Y$, given the joint statistics $p(x, y)$. It can be viewed as perfectly matching the source compression $q(t|x)$ with the channel prediction $q(y|t)$ by minimizing $I(X; T)$ and maximizing $I(T; Y)$ simultaneously, without any additional distortion or cost functions.

between the representation $T$ and $Y$, such that both the representation $q(t|x)$ and the channel $q(y|t)$ are optimized (through evolution and learning), the first by minimizing the mutual information $I(X; T)$ and the second by maximizing the "channel" information $I(T; Y)$ which can be also be thought as minimizing an expected "information distortion" $I(T; Y) = I(X; Y) - \langle d_{IB}(x, t) \rangle$. Furthermore, the representation distribution, $p(t)$, is such that perfectly match this source-channel combination and enable optimal block-less instantaneous operation, as required in biology. We show next that this optimization problem can be solved in a model independent way, without evoking any "arbitrary" distortion or cost measure, solely from the joint statistics of $X$ and $Y$.

This problem was formulated and termed the *Information Bottleneck* (IB) by Tishby, Pereira, and Bialek [37]. It turned out to be related to the a classical problem in information theory known as source coding with side-information([38, 12]).

The aim of IB is to find efficient (short) representation of $X$'s values that are maximally informative about $Y$. In the case of clustering it amounts to finding (possibly soft) partitions, or quantization, of $X$ that preserve as much mutual information as possible on $Y$. This requires balancing two goals: we want to lose "irrelevant" distinctions made by $X$, and at the same time maintain "relevant" ones.

In [37] we considered two variables, $X$ and $Y$, with their (assumed given) joint distribution $p(x, y)$. Here $X$ is the "source" variable we try to compress, with respect to the "relevant" variable $Y$. Namely, we seek a (possibly soft) partition of $X$ values through an auxiliary representation variable $T$ and the probabilistic mapping $q(t \mid x)$, such that the mutual information $I(T; X)$ is minimized (maximum compression) while the relevant information $I(T; Y)$ is maximized. Since $T$ is a compressed (efficient) representation of $X$ its distribution should be completely defined given $X$ alone. That is, $q(t \mid x, y) = q(t \mid x)$ which implies

$$q(x, y, t) = p(x, y)q(t \mid x) . \qquad (3.1)$$

An equivalent formulation is to require the following Markovian independence relation, known as the IB Markovian relation:

$$T \leftrightarrow X \leftrightarrow Y .^7 \qquad (3.2)$$

By introducing a positive Lagrange multiplier $\beta$ for the constrained relevant information, we have formulated this problem as minimizing the following IB-Lagrangian:

$$\mathcal{L}[q(t|x)] = I(T; X) - \beta I(T; Y) , \qquad (3.3)$$

over *all* normalized $q(t|x)$. The distributions $q(t)$ and $q(y|t)$ that are involved in this functional must satisfy the probabilistic consistency conditions:

$$\begin{cases} q(t) = \sum_x q(x, t) = \sum_x p(x)p(t \mid x) \\ q(y \mid t) = \frac{1}{q(t)} \sum_x q(x, y, t) = \frac{1}{q(t)} \sum_x p(x, y)q(t \mid x) , \end{cases} \qquad (3.4)$$

where the IB Markovian relation is used in the last equality.

As shown in [37], the stationary (minimum) points of the IB-Lagrangian must satisfy in addition the following relation between the unknown distributions:

$$q(t \mid x) = \frac{q(t)}{Z(x, \beta)} \exp\left(-\beta D_{KL}[p(y \mid x)\|q(y \mid t)]\right) , \qquad (3.5)$$

where $D_{KL}[p\|q] = \langle \log \frac{p}{q} \rangle_p$ is the familiar Kullback-Leibler (KL) divergence [6] and $Z(x, \beta)$ is a normalization (partition) function. It can be shown (using Bayes formula) that $Z$ is in fact a function of $\beta$ only *at the stationary points*. The three equations in Eq. (3.4) and Eq. (3.5) must be satisfied self-consistently at all the stationary points, for *any* choice of the representation variable $T$, and $\beta$. It is worth noting that the *effective IB distortion*

$$d_{IB}(x, t) = D_{KL}[p(y \mid x)\|q(y \mid t)] \qquad (3.6)$$

---

[7] As noted in [37], it is important to emphasize that this is *not* a modeling assumption about the quantization in $T$. In fact, this is not an assumption but rather a definition of the problem setup, hence the marginal over $q(x, y, t)$ with respect to $X$ and $Y$ is always consistent with the given (input) distribution, $p(x, y)$. In contrast, the standard mixture modeling approach defines $T$ as a hidden/latent variable in a model of the data, where in this case one *assumes* the Markov independence relation $X \leftrightarrow T \leftrightarrow Y$, which is typically not consistent with the input data. See [20, 35] for further discussion of this point.

*emerges* from the tradeoff principle and is related to the relevant information through

$$\langle d_{IB}(x,t)\rangle_{q(x,t)} = I(X; Y \mid T) = I(X; Y) - I(T; Y) \qquad (3.7)$$

which explains why the minimization of this (averaged) distortion is equivalent to the maximization of the relevant information term.

It was also shown in [37, 12] that the optimal tradeoff in the IB-Lagrangian $\mathcal{L}$ is quantified by one (concave) function, called the *complexity-accuracy tradeoff*, or *the information-curve*, which bounds $I(T; Y)$ for any given compression level $I(T; X)$. This curve, which depends solely on the joint distribution $p(x, y)$, provides information on the "structure" of the relationship between $X$ and $Y$. The simpler this structure - the steeper this curve becomes. The curve is in fact both the Rate-Distortion function and the Capacity-Cost functions and in a sense unifies them, as discussed in the introduction.

### 3.2.1. IB as perfectly matched source-channel

The IB problem can be viewed as a nonlinear Rate-Distortion problem, with the effective "information-distortion" given by Eq. (3.6), which depends on the optimal map $q(t|x)$, in contrast to normal RDT. It can also be viewed as Capacity-Cost problem, with the "information-cost" given by

$$e_{IB}(t) = D_{KL}[q(x|t)\|p(x)], \qquad (3.8)$$

whose expectation $\langle e_{IB}(t)\rangle_{q(t)} = I(X; T)$ is the compression information.

Interestingly, those IB distortion and cost functions obey the perfect matching conditions, given in Eq. (2.7), if we take into account that the cost and distortion are functions of the channel and map distributions, and thus we can add functions of the distribution that are constant "on the average". To see this, notice first that the exponential form obtained for the optimal map in IB, Eq. (3.5), immediately satisfy the first condition in Eq. (2.7), if we consider the distortion between $x$

and $t$. The condition on the cost function is also satisfied since,

$$
\begin{aligned}
e_{IB}(t) &= D_{KL}[q(x|t)\|p(x)] = \sum_x q(x|t) \log \frac{q(x|t)}{p(x)} \\
&= -\beta \sum_x q(x|t) \left(D_{KL}[p(y|x)\|q(y|t)] - \log Z(\beta, x)\right) \\
&= -\beta \sum_x q(x|t) \\
&\quad \times \sum_y p(y|x) \left(\log \frac{p(y)}{q(y|t)} + \log \frac{p(y|x)}{p(y)} - \log Z(\beta, x)\right) \\
&= -\beta \sum_y \sum_x q(x|t) p(y|x) \log \frac{p(y)}{q(y|t)} \\
&\quad - \beta \sum_{x,y} q(x|t) p(y|x) \log \frac{p(y|x)}{p(y)} - \sum_x q(x|t) \log Z(\beta, t) .
\end{aligned}
$$

$$(3.9)$$

If we now plug the Markov condition (with a Bayes inversion), Eq. (3.4), for the "channel", $q(y|t) = \sum_x p(y|x)q(x|t)$, and calculate the expected cost by averaging over the representation distribution $q(t)$, we obtain,

$$
\begin{aligned}
\langle e_{IB}(t) \rangle_{q(t)} &= \beta \sum_y q(y|t) \log \frac{q(y|t)}{p(y)} - \sum_{x,y} q(x|t) p(y|x) \log \frac{p(y|x)}{p(y)} \\
&\quad - \sum_x q(x|t) \log Z(\beta, t) \\
&= \beta \langle e_{GRV}(t) \rangle_{q(t)} - \beta I(X; Y) - \langle \log Z(\beta, x) \rangle_{p(x)} .
\end{aligned}
$$

$$(3.10)$$

Hence the IB information-cost acts like an optimal cost with respect to the effective IB-channel distribution $q(y|t)$, thus IB can be implemented optimally without using blocks or delays. This may have interesting biological implication, as we expect to observe with our "Shannon-meter" precisely the mutual information values predicted by IB, even if the system operated instantaneously without the usual asymptotic blocks of coding theory.

### 3.3. Alternating projections and the IB algorithm

The above self-consistent conditions on the unknown distributions can be turned into an iterative, alternating projection algorithm, in the spirit of [9], for finding the unknown distributions $q$, by iterating over those three equations, while holding the other variable distributions fixed at each step, similar to EM and the

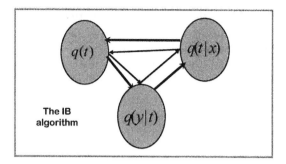

Fig. 8. The Alternating-Projection IB-algorithm. At each step two of the three unknown distributions are fixed and the free-energy is minimized with respect to the third. The free-energy is strictly reduced in each projection.

Blahut-Arimoto algorithm in information theory [6]. Each of the three equations becomes a projection on a convex set of distributions, yielding the following iterative algorithm,

$$
\begin{cases}
q_{k+1}(t) = \sum_x p(x) q_k(t|x) \\[2mm]
q_{k+1}(y|t) = \frac{1}{q_{k+1}(t)} \sum_x p(x, y) q_k(t|x) \\[2mm]
q_{k+1}(t|x) = \frac{q_{k+1}(t)}{Z_{k+1}(x,\beta)} e^{-\beta D_{KL}[p(y|x)\|q_{k+1}(y|t)]}
\end{cases}
\tag{3.11}
$$

whose general (local) convergence for every positive $\beta$ was proved in [37].

Such projections play an important role in our formulation and deserve a short clarification. Csiszár and Tusnády [9] introduced two types of projections of a distribution on (convex) sets of distributions that involve minimization of the KL-divergence. The first, called *I-projection* by Csiszár [7] (and e-projection by Amari [1]), is defined as,

$$
q^I =_{q \in Q} D_{KL}[q\|p]
\tag{3.12}
$$

which is the closest point in the set of distributions $Q$ to the distribution $p$, in an information geodesic sense. It is unique when the set $Q$ is convex. Here the varied distribution $q$ appears in the left argument of the divergence and hence the projection distribution typically has an exponential form.

The second projection, which we call *M-projection* (following Amari [1]), closely related to Maximum Likelihood approximation by a model class and the M-step in the EM algorithm, is defined as,

$$
q^M =_{q \in M} D_{KL}[p\|q] ,
\tag{3.13}
$$

and is also unique if the model-class $M$ is convex. Here the variational distribution $q$ is on the right argument of the divergence and the resulting projection has typically an expectation (weighted mixture) form.

Many alternating minimization algorithms, including the Blahut-Arimoto and the EM algorithm, involve the two type of projections and can be formulated as double minimization over two (or more) sets ,

$$\min_{q \in A} \min_{p \in B} D_{KL}[p\|q] \,. \tag{3.14}$$

Similarly, the above IB-algorithm involves two M-projections (to the sets of all $q(t)$ and all $q(y|t)$) and one I-projection to the set of all $q(t|x)$ that yields the exponential form. It is important to emphasize that all 3 variable distribution change through the iterations and their roles are equally important for the optimal solution. The M-projections are analogous to the centroids-update step and the I-projection to the data-membership re-estimation in typical central clustering algorithms (like K-means) and one determines the other. Since our variables are also distributions the M-projections are also consistent with the marginalization and known Markov composition rules, as was shown already in [9]. This picture is further generalized in the multivariate case.

Unfortunately, unlike the BA algorithm but like EM, there are generally many local solutions to those equations, but their structure is simpler for small values of $\beta$ (high "temperature"). This suggests a deterministic annealing scheme, by varying $\beta$ from lower to higher values, to obtain higher complexity representations, as implemented in [20, 37] and discussed in Section 4.7.

## 4. Part IV: applications and extensions

### 4.1. IB clustering

The simplest application of IB is as a clustering method, where the variable $T$ has a finite cardinality that represent discrete clusters of $X$ that are informative on $Y$. In that case Eq. (3.11) are very similar to distributional clustering of the conditional distributions $p(y|x)$ with respect to the KL-divergence as a distance measure. They should remind us other central-clustering algorithms (like K-means), where we iterate between estimation of the membership probability, $q(t|x)$ (of conditional distributions in this case) re-estimation of the centroids (also distributions) $q(y|t)$.

Typical application of this algorithm is to co-occurrence data, such as word-document statistics. In this case, the initial joint distribution of words in documents is transformed into clusters of words/documents, where the distinctions

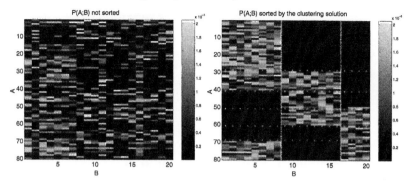

Fig. 9. IB as a distributional clustering method. Co-occurrence count matrix (left) is transformed into lower dimensional cluster-blocks matrix (right) with minimal loss of its mutual information. In the ideal case, the distinctions of $X$ within the clusters should not provide information about $Y$. .

within the clusters is less informative (irrelevant) with respect to the topic, or the identity, of the document (see Figure 4.1).

When applied to natural language corpus, such as Lang's 20-newsgroup database [39], we obtain an intriguing scaling phenomena that demonstrates the importance of the full complexity-accuracy tradeoff. In Figure 4.1 the complexity-accuracy tradeoff for English is plotted, based on word-topic IB-clustering. It reveals a remarkably accurate power-law scaling of meaningful information, measured by the cluster-conditional word-topic mutual-information, with the word complexity measured by the cluster-conditional word-entropy. This can be explained by a universal cognitive mechanism for generating new words in human language.

Similar analysis was done for neuroscience applications, where the "words" can be configurations of spike patterns and the "document" correspond to input stimuli, for instance. In that case the clusters will identify the spike-patterns distinctions that are irrelevant for the identity of the stimulus. Alternately, one could cluster the stimuli and reveal the "neural code" - what are the components of the stimulus that are distinguishable by the spikes. Examples can be found in [36, 23, 4].

### 4.2. IB dimensionality reduction: the Gaussian IB

A rather different application of the IB principle is to relevant dimension reduction, also known as feature extraction. An illuminating example for dimensionality reduction using IB has been recently provided in [10]. Here we make the assumption that the joint distribution $p(x, y)$ is a multivariate Gaussian, which

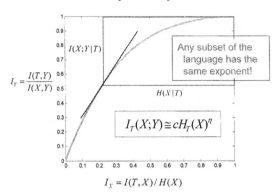

$I_x = I(T,X)/H(X)$

Fig. 10. Complexity-Accuracy relation ( the "information-curve") for English Documents as revealed by IB clustering. Clustering the word-topic joint distribution and calculating the optimal complexity-accuracy curve, we obtain a very accurate power-law, which express scaling of meaningful information $I(Word; Topic|T)$ with Word-entropy $H(Word|T)$, with an exponent $\eta \approx 1.92$, independent of vocabulary size or language subset. It may reflect a universal cognitive mechanism for "generating" new words at a fixed level of ambiguity. (Data taken from Slonim's PhD thesis, 2002).

allows us to obtain a complete analytical description of the IB solution space. In this case we thus assume that the joint distribution of $(X, Y)$ is multivariate Gaussian with dimensions $n_x$ and $n_y$. We denote by $\Sigma_x, \Sigma_y$ the covariance matrices of $X$ and $Y$ and by $\Sigma_{xy}$ their cross-covariance matrix, which carry the information between $X$ and $Y$. We seek a random variable $T \in R^{n_x}$ that compress $X$, while preserve information about $Y$, in a way that is controlled by the tradeoff parameter $\beta$. We show in [14] that the optimal $T$ for the Gaussian case is a (stochastic) linear projection of $X$, $T = AX + \xi$, where $\xi$ is a regularization (vector) white noise with covariance matrix $\Sigma_\xi$.

It is shown in [10] that all three unknown distributions, $q(t|x)$, $q(t)$ and $q(y|t)$ are Gaussian in this case and the problem reduce to the specification of the optimal projection matrix $A$ as function of $\beta$.

All the mutual information terms are Gaussian integrals in this case and the IB Lagrangian can be explicitly written as,

$$
\begin{aligned}
\mathcal{L} &= \log(|\Sigma_t|) - \log(|\Sigma_{t|x}|) - \beta \log(|\Sigma_t|) + \beta \log(|\Sigma_{t|y}|) \\
&= (1-\beta)\log(|A\Sigma_x A^T + \Sigma_\xi|) - \log(|\Sigma_\xi|) + \beta \log(|A\Sigma_{x|y} A^T + \Sigma_\xi|),
\end{aligned}
$$

$$(4.1)$$

where,

$$\Sigma_{t|y} = \Sigma_t - \Sigma_{ty}\Sigma_y^{-1}\Sigma_{yt} = A\Sigma_{x|y}A^T + \Sigma_\xi \ . \tag{4.2}$$

The optimal $A$ can thus be found by direct variation of this matrix equation, yielding the interesting equation for $A$,

$$\frac{\beta - 1}{\beta}\left[(A\Sigma_{x|y}A^T + I_d)(A\Sigma_x A^T + I_d)^{-1}\right]A = A\left[\Sigma_{x|y}\Sigma_x^{-1}\right] . \tag{4.3}$$

Equation 4.3 shows that the multiplication of $\Sigma_{x|y}\Sigma_x^{-1}$ by $A$ must reside in the span of the rows of $A$. This means that $A$ should be spanned by up to $d$ eigenvectors of $\Sigma_{x|y}\Sigma_x^{-1}$. We can therefore represent the projection $A$ as a mixture $A = WV$ where the rows of $V$ are left normalized eigenvectors of $\Sigma_{x|y}\Sigma_x^{-1}$ and $W$ is a mixing matrix that weights these eigenvectors. The eigenvalues of the $\Sigma_{x|y}\Sigma_x^{-1}$ matrix, $\{\lambda_1, \ldots, \lambda_k\}$ and the corresponding eigenvectors, $\{v_1^T, \ldots, v_k^T\}$, provide us with the full characterization of the projection. The global minimum of $\mathcal{L}$ is obtained with all $\lambda_i$ that satisfy $\beta > \frac{1}{1-\lambda_i}$. We can thus write the Gaussian information curve, in terms of the spectrum of $\Sigma_{x|y}\Sigma_x^{-1}$. To this end, we substitute the optimal projection $A(\beta)$ into $I(T; X)$ and $I(T; Y)$ and rewrite them as a function of $\beta$,

$$
\begin{aligned}
I_\beta(T; X) &= \frac{1}{2}\log\left(|A\Sigma_x A^T + I_d|\right) \\
&= \frac{1}{2}\log\left(|(\beta(I - D) - I)D^{-1}|\right) \\
&= \frac{1}{2}\sum_{i=1}^{n(\beta)}\log\left((\beta - 1)\frac{1-\lambda_i}{\lambda_i}\right) \\
I_\beta(T; Y) &= I(T; X) - \frac{1}{2}\sum_{i=1}^{n(\beta)}\log\beta(1-\lambda_i) \quad ,
\end{aligned}
\tag{4.4}
$$

where $n(\beta)$ is the maximal index $i$ such that $\beta \geq \frac{1}{1-\lambda_i}$. Isolating $\beta$ as a function of $I_\beta(T; X)$ in the correct range of $n_\beta$ and then $I_\beta(T; Y)$ as a function of $I_\beta(T; X)$ we obtain,

$$I(T; Y) = I(T; X) - \frac{n_I}{2}\log\left(\prod_{i=1}^{n_I}(1-\lambda_i)^{\frac{1}{n_I}} + e^{\frac{2I(T;X)}{n_I}}\prod_{i=1}^{n_I}\lambda_i^{\frac{1}{n_I}}\right) \tag{4.5}$$

where the products are over the *first* $n_I = n_{\beta(I(T;X))}$ eigenvalues, since these

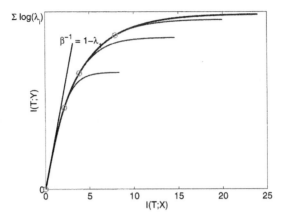

Fig. 11. GIB information curve obtained with four eigenvalues $\lambda_i = 0.1, 0.5, 0.7, 0.9$. The information at the critical points are designated by circles. For comparison, information curves calculated with smaller number of eigenvectors are also depicted (all curves calculated for $\beta < 1000$) The slope of the un-normalized curve at each point is the corresponding $\beta^{-1}$. The tangent at zero, with slope $\beta^{-1} = 1 - \lambda_1$, is super imposed on the information curve. (Taken from [10].)

obey the critical $\beta$ condition, with

$$c_{n_I} \leq I(T; X) \leq c_{n_I+1} \quad \text{and} \quad c_{n_I} = \sum_{i=1}^{n_I-1} \log \frac{\lambda_{n_I}}{\lambda_i} \frac{1 - \lambda_i}{1 - \lambda_{n_I}}.$$

The Gaussian information curve, illustrated in Figure 11, is continuous and smooth, but is built of several segments: as $I(T; X)$ increases additional eigenvectors are used in the projection. The derivative of the curve, which is equal to $\beta^{-1}$, can be easily shown to be continuous and decreasing, therefore the information curve is concave everywhere.

At each value of $I(T; X)$ the curve is bounded by a tangent with a slope $\beta^{-1}(I(T; X))$. Generally in IB, the data processing inequality yields an upper bound on the slope at the origin, $\beta^{-1}(0) < 1$, in GIB we obtain a tighter bound: $\beta^{-1}(0) < 1 - \lambda_1$. The asymptotic slope of the curve is always zero, as $\beta \to \infty$, reflecting the law of diminishing return: adding more bits to the description of $X$ does not provide higher accuracy about $T$.

A more general (non-Gaussian) application of IB to feature extraction and dimensionality reduction is closely related to another algorithm, called Sufficient Dimensionality Reduction (SDR) [13] and to association analysis in statistics, but is beyond the scope of these lectures.

### 4.3. Towards multivariate network IB

We would like to conclude this review with an exciting extension of IB to the multivariate case, which goes way beyond Shannon's original point-to-point communication model. In our neuro-biological "gedanken experiment" with the Shannon-meter, one would like to extend the observations of the system beyond just two probes - two measured variables $X$ and $Y$ - and make predictions on information values among multiple nodes in a complex network. A significant step in this direction has been made with Nir Friedman and Noam Slonim, in several recent publications.

The motivation is to provide a conceptual and practical framework for "looking inside the black box" with higher dimensional observations and with different probabilistic dependencies. In principle, it calls for an elevation of the point-to-point communication problem to a network communication setting - a notoriously difficult and largely unsolved problem. Surprisingly, such a network extension of IB, which does not require the complete solution of "network information theory", is viable and very natural. This extension combines the theory of probabilistic graphical models, such as Bayesian Networks, and a multivariate extension of mutual information known as *multi-information*.

Here we only introduce the basic components and the general principles. Details can be found in [36, 11, 31].

### 4.4. Bayesian networks and multi-information

General multivariate distributions over $n$ variables get (super) exponentially complicated, not only in the number of possible entries of probability values but also in the possible interdependencies structures among the variables. This complexity explosion turns multivariate distributions futile for most practical purposes without some simplifying assumptions. A powerful methodology for dealing and simplifying multivariate distributions, which has become tremendously important in recent years, are graphical probabilistic models and their predecessors Bayesian networks [19]. We find the graphical models language very instrumental for our goal.

A *Bayesian network* over a set of random variables $\mathbf{X} \equiv \{X_1, \ldots, X_n\}$ is a Directed A-cyclic Graph (DAG) $G$ in which vertices are annotated by names of random variables. For each variable $X_i$, we denote by $\mathbf{Pa}^G_{X_i}$ the (potentially empty) set of parents of $X_i$ in $G$. We say that a distribution $p(\mathbf{x})$ is *consistent* with $G$, if $p$ can be factored in the form:

$$p(x_1, \ldots, x_n) = \prod_i p(x_i \mid \mathbf{pa}^G_{X_i})$$

and use the notation $p \models G$ to denote that.

The other primary concept that we need for the extension is the amount of "information" that random variables $(X_1, \ldots, X_n) \sim p(x_1, \ldots, x_n)$ share, or contain, about each other. This measure should generalize the mutual information between two variables in a natural way. While there are other possible generalizations of mutual-information, the one that captures all the statistical dependencies, as well as the multivariate notion of joint-typicality, and is most suitable here is the *multi-information*, given by

$$
\begin{aligned}
\mathcal{I}[p(\mathbf{x})] \equiv \mathcal{I}(X_1, \ldots, X_n) & = D_{KL}[p(x_1, \ldots, x_n) \| p(x_1) \cdots p(x_n)] \\
& = \langle \log \frac{p(x_1, \ldots, x_n)}{p(x_1) \cdots p(x_n)} \rangle_p .
\end{aligned}
$$

The multi-information is the Kullback-Leibler divergence between the joint distribution and the product of all its single variable marginals, and hence measures how close the distribution $p(x_1, \ldots, x_n)$ is to the factored, independent variables, distribution. More precisely, it measures the average number of bits that can be gained by a joint compression of all the variables versus independent compression. Like the mutual information, it is non-negative, and equals to zero if and only if all the variables are all independent. Furthermore, as for the mutual information, in [17] it is shown that this is the *only* functional (up to a constant) that holds under simple and natural axiomatic requirements and it obeys a multivariate analog of the joint typicality theorem of mutual information (i.e. characterizes the "volume" of long typical sequences of all the variables).

Multi-information and Bayesian networks go hand-in-hand. When the variables have known independence relations, as in Bayesian networks, one can rewrite the multi-information in terms of the dependencies among the variables:

**Proposition 4.1 :** *Let* $\mathbf{X} = \{X_1, \ldots, X_n\} \sim p(\mathbf{x})$, *and let* $G$ *be a Bayesian network structure over* $\mathbf{X}$ *such that* $p \models G$. *Then*

$$
\mathcal{I}[p(\mathbf{x})] = \mathcal{I}(\mathbf{X}) = \sum_i I(X_i; \mathbf{Pa}_{X_i}^G) .
$$

That is, the multi-information is the sum of *local* mutual information terms between each variable and its parents. Notice that in general, even if $p(\mathbf{x})$ is *not* consistent with the graph $G$ the above sum is well defined, but it may not capture all the multi-information of the distribution. Hence, we introduce the following definition:

**Definition 4.2:** Let $\mathbf{X} = \{X_1, \ldots, X_n\} \sim p(\mathbf{x})$, and let $G$ be a Bayesian network structure over $\mathbf{X}$. The multi-information in $p(\mathbf{x})$ with respect to $G$ is defined as:

$$
\mathcal{I}^G[p(\mathbf{x})] = \sum_i I(X_i; \mathbf{Pa}_{X_i}^G) , \tag{4.6}
$$

where each of the local mutual information terms is calculated using the marginal distributions of $p(\mathbf{x})$.

When $p \models G$ then $\mathcal{I}[p(\mathbf{x})] = \mathcal{I}^G[p(\mathbf{x})]$. However, if $p$ is *not* consistent with $G$ then in general $\mathcal{I}[p(\mathbf{x})] \geq \mathcal{I}^G[p(\mathbf{x})]$ as we now show. In this case we want to know how close p *is* to the distribution class which is consistent with $G$. That is, what is the "distance" (or distortion) of $p$ from its "projection" onto the subspace of distributions consistent with $G$. We define this distortion through the *M-projection* of $p$ on that class, as

$$D_{KL}[p\|G] = \min_{q \models G} D_{KL}[p\|q] \,, \tag{4.7}$$

as defined by Eq. (3.13).

The following specifies the form of the M-projection $q$, for which the minimum is attained, as a product distribution.

**Proposition 4.3:** *Let $p(\mathbf{x})$ be a distribution and let $G$ be a DAG then*

$$D_{KL}[p\|G] = D_{KL}[p\|q^*] \,, \tag{4.8}$$

*where the M-projection $q^*$ is given by*

$$q^*(\mathbf{x}) = \Pi_{i=1}^n p(x_i \mid \mathbf{pa}_{X_i}^G) \,. \tag{4.9}$$

Expressed in words, $q^*$ is equivalent to the factorization of $p$ using the conditional independence implied by the graph of $G$.

Next, we generalizes Eq. (3.6), provides two possible interpretations of the M-projection distance $D_{KL}[p\|G]$, in terms of the structure of $G$.

**Proposition 4.4:** *Let $\mathbf{X} = \{X_1, \dots, X_n\} \sim p(\mathbf{x})$, and let $G$ be a Bayesian network structure over $\mathbf{X}$. Assume that the order $X_1, \dots, X_n$ is consistent with the DAG $G$ (i.e., $\mathbf{Pa}_{X_i}^G \subseteq \{X_1, \dots, X_{i-1}\}$). Then*

$$
\begin{aligned}
D_{KL}[p\|G] &= \sum_i I(X_i; \{X_1, \dots, X_{i-1}\} \setminus \{\mathbf{Pa}_{X_i}^G\} \mid \mathbf{Pa}_{X_i}^G) \\
&= \mathcal{I}[p(\mathbf{x})] - \mathcal{I}^G[p(\mathbf{x})] \,.
\end{aligned}
$$

Thus, we see that $D_{KL}[p\|G]$ can be expressed as a sum of local conditional mutual information terms, where each term corresponds to a (possible violation of a) Markov independence assumption with respect to the structure of $G$. If every $X_i$ is independent of $\{X_1, \dots, X_{i-1}\} \setminus \mathbf{Pa}_{X_i}^G\}$ given $\{\mathbf{Pa}_{X_i}^G\}$ (as implied by $G$) then $D_{KL}[p\|G]$ becomes zero. As these (conditional) independence assumptions are more extremely violated in $p$, the corresponding $D_{KL}[p\|G]$ will increase. Recall that the Markov independence assumptions (with respect to a given order) are necessary and sufficient to require the factored form of distributions consistent with $G$ [19]. Therefore, we see that $D_{KL}[p\|G] = 0$ if and only if $p$ is consistent with $G$.

An alternative interpretation of this measure is given in terms of multi-information terms. Specifically, we see that $D_{KL}[p\|G]$ can be written as the difference between the real multi-information of the variables, $\mathcal{I}[p(\mathbf{x})] = \mathcal{I}(\mathbf{X})$, and the multi-information that is captured by the mode class *as though* $p \models G$, denoted by $\mathcal{I}^G[p(\mathbf{x})]$, which in particular can not be larger. Hence, we can think of $D_{KL}[p\|G]$ as the residual information between the variables that is *not* captured by the dependencies that are implied in the structure of $G$. This value of information can not be captured by the *structure* of $G$ and requires more dependencies among the variables.

### 4.5. Network information bottleneck principle

The concept of multi-information allows us to introduce a natural "lift-up" of the original IB variational principle to the multivariate case, using the semantics of Bayesian networks. Given a set of observed variables, $\mathbf{X} = \{X_1, \ldots, X_n\}$, instead of one partition variable $T$, we now specify a set of random variables $\mathbf{T} = \{T_1, \ldots, T_k\}$, which corresponds to different possible partitions of various subsets of the observed variables. This specification should address two issues. First, loosely speaking, we need to specify "what represents what". More formally stated, for each subset of $\mathbf{X}$ that we would like to compress, we specify a corresponding subset of the compression variables $\mathbf{T}$. Second, analogous to the original IB problem, we define the solution space in terms of the independence we require between the observed $\mathbf{X}$ variables and the compression $\mathbf{T}$ variables. Recall that for the original IB problem this is achieved through the IB Markovian relation $T \leftrightarrow X \leftrightarrow Y$. As a result, the solution space consists of all the distributions over $X, Y, T$, such that $q(x, y, t) = p(x, y)q(t|x)$, where the free parameters correspond to the stochastic mapping between $X$ and $T$[8] In the multivariate case, the analogous situation would be to define the solution space through a *set* of IB Markovian independence relations, which imply that each compression variable, $T_j \in \mathbf{T}$, is completely defined given the variables it represents, denoted here as $\mathbf{U}_j \subset \mathbf{X}$.

We achieve these two goals by first introducing a DAG $G_{in}$ over $\mathbf{X} \cup \mathbf{T}$ where the variables in $\mathbf{T}$ are leafs. Given a joint distribution over the observed variables,

---

[8]They could be factorized in other - less "convenient" - ways, such as $q(x, y, t) = q(t)q(y|t)q(x|y, t)$ or $q(x, y, t) = q(t)q(x|t)p(y|x)$ to emphasize that all three unknown distributions play a role.

$p(\mathbf{x})$, $G_{in}$ is defined such that $p(\mathbf{x})$ is consistent with its structure restricted to $\mathbf{X}$. The edges from $\mathbf{X}$ to $\mathbf{T}$ define "what compresses what" and the independence implied by $G_{in}$ correspond to the required set of IB Markovian independence relations. In particular this implies that every $T_j$ is independent of all the other variables, given the variables it compresses, $\mathbf{U}_j = \mathbf{Pa}_{T_j}^{G_{in}} \subset \mathbf{X}$. Hence, the multivariate IB solution space consists of all the distributions over $\mathbf{X} \cup \mathbf{T}$ that satisfy $G_{in}$. Specifically, the form of these distributions is given by

$$q(\mathbf{x}, \mathbf{t}) = p(\mathbf{x}) \prod_{j=1}^{k} q(t_j \mid \mathbf{pa}_{T_j}^{G_{in}}), \qquad (4.10)$$

where the free parameters correspond to the stochastic mappings $q(t_j \mid \mathbf{pa}_{T_j}^{G_{in}})$, and the other unknown distributions are determined by the Markovian structure (or through M-projections on $G_{in}$). [9] Analogously to the original IB formulation, the information that we would like to minimize is now given by $\mathcal{I}^{G_{in}}$. Since $q(\mathbf{x}, \mathbf{t}) \models G_{in}$ then $\mathcal{I}^{G_{in}} = \mathcal{I}(\mathbf{X}, \mathbf{T})$, i.e., this is the real multi-information in $q(\mathbf{x}, \mathbf{t})$. Minimizing this quantity attempts to make the $\mathbf{T}$ variables as independent of the $\mathbf{X}$ variables as possible. Note that we only modify conditional distributions that refer to variables in $\mathbf{T}$, and we do not modify the dependencies among the original observed $\mathbf{X}$ variables.

Once $G_{in}$ is defined we need to specify the other part of the tradeoff - the relevant information that we want to preserve. We do that by specifying another DAG , $G_{out}$. Roughly speaking, $G_{out}$ determines "what predicts what". More formally stated, for each $T_j$, we define in $G_{out}$ which variables it predicts and preserve information about. These variables are simply its children in $G_{out}$. Thus, using Definition 4.2, we may think of $\mathcal{I}^{G_{out}}$ as a measure of how much information the variables in $\mathbf{T}$ maintain about their target variables. This suggests that we should maximize $\mathcal{I}^{G_{out}}$.

The *Network IB-Lagrangian* can now be written as a tradeoff between these two terms,

$$\mathcal{L}^{(network)}[q(\mathbf{x}, \mathbf{t})] = \mathcal{I}^{G_{in}}[q(\mathbf{x}, \mathbf{t})] - \beta \mathcal{I}^{G_{out}}[q(\mathbf{x}, \mathbf{t})], \qquad (4.11)$$

where the variation is done subject to the normalization constraints on the partition distributions, and $\beta$ is the positive inverse-temperature Lagrange multiplier controlling the information tradeoff. [10]

---

[9]For simplicity, we restrict attention to cases where the input distribution $p(\mathbf{x})$ is consistent only with the complete graph over $\mathbf{X}$. Hence, $G_{in}$ restricted to $\mathbf{X}$ must form a complete graph.

[10]Since $\mathcal{I}^{G_{out}}$ typically consists of several mutual information terms (Eq. (4.6)), in principle it is also possible to define a separate Lagrange multiplier for each of these terms. In some situations this

It leads to a tractable formal solution that involves as before a combination of I-projections, leading to exponential forms, and M-projections that lead to expectations and marginalization, as we show in the next section. Note that this functional is a direct generalization of the original IB-functional, Eq. (3.3). Again, we are interested in the competition between the complexity of the representation, measured through the compression (multi) information, $\mathcal{I}^{G_{in}}$, and the accuracy of the relevant predictions this representation provide, quantified by the (multi) information $\mathcal{I}^{G_{out}}$. Notice also that the "separation of source and channel" that exists in the two variable case no longer holds nor needed for this formulation. In a sense, the $G_{in}$ and $G_{out}$ replace the source and channel structures of the original IB, respectively.

As for the original IB principle the possible range of $\beta$ for the multivariate formulation is between 0 to $\infty$. consider the two-dimensional optimal curve of $\mathcal{I}^{G_{out}}$ versus $\mathcal{I}^{G_{in}}$ as the *generalized information curve*, or the generalized "complexity-accuracy tradeoff curve. From the variational Lagrangian it is clear that the slope of the curve is $\beta^{-1}$ and thus the curve - if smooth - is downward concave. As in the bivariate case, there is almost always a lower critical $\beta$ (upper critical temperature) beyond which there are no non-trivial solutions due to data-processing inequalities. We can always avoid that by modifying the $G_{out}$ to include the $G_{in}$ edges between the variables $X$ and their representation $T$, but in the opposite direction (this is a directed graph).

For $\beta \rightarrow 0$ the focus is on the compression term only, which yields a trivial solution in which the $T_j$'s are independent of their parents[11]. In other words, in this case each $T_j$ can consists of one value to which all the values of $\mathbf{Pa}_{T_j}^{G_{in}}$ are mapped and all the distinctions among these values (relevant or not) are lost. For $\beta \rightarrow \infty$ we ignore the need for compression and concentrate on maintaining the relevant (prediction) information terms as high as possible. This, in turn, yields a trivial solution at the opposite extreme, in which each $T_j$ is a 1-1 map of $\mathbf{Pa}_{T_j}^{G_{in}}$ with no loss of information. In this limit adding complexity to the representation does not enable any better predictions of the relevant variables - a "point of diminishing returns". The interesting cases are in between, where $\beta$ takes positive finite values.

---

option might be useful, for example if for some reason the preservation of one information term is of greater importance than the preservation of the others. Another such case can be information terms we would like to "ignore" or treat as "irrelevant information", as in [5].

[11] In fact there is always a lower critical positive value of $\beta$ below which there are no non-trivial solutions to the problem, due to (generalized) information processing inequalities.

### 4.6. *Characterization of the IB fix-points*

As in the bivariate IB it is possible to implicitly characterize the form of the optimal solutions, or the fix points of the IB-functional. Here we provide a similar characterization to the more complicated multivariate IB functionals, based on a similar analysis. The solution to the multivariate IB principle inherently, requires a higher level of abstraction, as it should apply to *any* specification of the DAG's $G_{in}$ and $G_{out}$. We expect to use such a solution as a *recipe*, whose concrete specification is provided only when the model classes are known.

We start by assuming that $G_{in}$, $G_{out}$, and $\beta$ are given. We now want to describe the properties of the representing distributions $q(t_j \mid \mathbf{pa}_{T_j}^{G_{in}})$, as well as the other unknown distributions that involve the variables $T$, which optimize the tradeoff defined by each of the two alternative principles. The characterization of the optimal solution provides a general extension to the self-consistent equations of the original IB problem.

In the presentation of this characterization, we need some additional notational shorthands, given by $\mathbf{U}_j = \mathbf{Pa}_{T_j}^{G_{in}}$, $\mathbf{V}_{T_j} = \mathbf{Pa}_{T_j}^{G_{out}}$, and $\mathbf{V}_{X_i} = \mathbf{Pa}_{X_i}^{G_{out}}$. We also denote $\mathbf{V}_{T_\ell}^{-j} = \mathbf{V}_{T_\ell} \setminus \{T_j\}$ and similarly for $\mathbf{V}_{X_i}^{-j} = \mathbf{V}_{X_i} \setminus \{T_j\}$.

In addition, we use the notation

$$\langle D_{KL}[p(y \mid \mathbf{z}, \mathbf{u}_j) \| p(y \mid \mathbf{z}, t_j)] \rangle_{p(\cdot \mid \mathbf{u}_j)}$$

$$= \sum_{\mathbf{z}} p(\mathbf{z} \mid \mathbf{u}_j) D_{KL}[p(y \mid \mathbf{z}, \mathbf{u}_j) \| p(y \mid \mathbf{z}, t_j)]$$

$$= \langle \log \frac{p(y \mid \mathbf{z}, \mathbf{u}_j)}{p(y \mid \mathbf{z}, t_j)} \rangle_{p(y, \mathbf{z} \mid \mathbf{u}_j)}$$

where $Y$ is a random variable and $\mathbf{Z}$ is a set of random variables. Note that this term implies averaging over all values of $Y$ and $\mathbf{Z}$ using the conditional distribution. In particular, if $Y$ or $\mathbf{Z}$ intersects with $\mathbf{U}_j$, then only the values consistent with $\mathbf{u}_j$ have positive weights in this averaging. Also note that if $\mathbf{Z}$ is empty, then this term reduces to the standard $KL$ divergence between $p(y \mid \mathbf{u}_j)$ and $p(y \mid t_j)$.

The main result, proved in [11, 31], is stated in the following theorem.

**Theorem 4.5:** *Assume that $p(\mathbf{x})$, $G_{in}$, $G_{out}$, and $\beta$ are given and that $q(\mathbf{x}, \mathbf{t}) \models G_{in}$. The conditional distributions $\{q(t_j \mid \mathbf{u}_j)\}_{j=1}^k$ are a stationary point of $\mathcal{L}^{(network)}[q(\mathbf{x}, \mathbf{t})] = \mathcal{I}^{G_{in}}[q(\mathbf{x}, \mathbf{t})] - \beta \mathcal{I}^{G_{out}}[q(\mathbf{x}, \mathbf{t})]$ if and only if*

$$q(t_j \mid \mathbf{u}_j) = \frac{q(t_j)}{Z_{T_j}(\mathbf{u}_j, \beta)} e^{-\beta d(t_j, \mathbf{u}_j)}, \ \forall \, t_j \in \mathcal{T}_j, \ \forall \, \mathbf{u}_j \in \mathcal{U}_j , \qquad (4.12)$$

where $Z_{T_j}(\mathbf{u}_j, \beta)$ is a normalization function, and

$$d(t_j, \mathbf{u}_j) \equiv \sum_{i:T_j \in \mathbf{V}_{X_i}} \langle D_{KL}[q(x_i \mid \mathbf{v}_{X_i}^{-j}, \mathbf{u}_j) \| q(x_i \mid \mathbf{v}_{X_i}^{-j}, t_j)] \rangle_{q(\cdot \mid \mathbf{u}_j)}$$

$$+ \sum_{\ell:T_j \in \mathbf{V}_{T_\ell}} \langle D_{KL}[q(t_\ell \mid \mathbf{v}_{T_\ell}^{-j}, \mathbf{u}_j) \| q(t_\ell \mid \mathbf{v}_{T_\ell}^{-j}, t_j)] \rangle_{q(\cdot \mid \mathbf{u}_j)}$$

$$+ D_{KL}[q(\mathbf{v}_{T_j} \mid \mathbf{u}_j) \| q(\mathbf{v}_{T_j} \mid t_j)] . \qquad (4.13)$$

The essence of this theorem is that it provides an *implicit* set of equations for $q(t_j \mid \mathbf{u}_j)$ through the *multivariate relevant-distortion* $d(t_j, \mathbf{u}_j)$, which in turn depend on those unknown distributions. This distortion measures the degree of proximity of the conditional distributions in which $\mathbf{u}_j$ is involved to those where we replace $\mathbf{u}_j$ with $t_j$. In other words, we can understand this as a measure of how well $t_j$ performs as a "representative" of the particular assignment $\mathbf{u}_j$. As the representatives behave more similarly to $\mathbf{u}_j$, $d(t_j, \mathbf{u}_j)$ become smaller, which in turn increase the membership probabilities, $q(t_j \mid \mathbf{u}_j)$.

The obtained exponential form is a direct result of the I-projections on $G_{in}$ that minimize the compression multi-information term. As in the original IB problem, those equations must be solved *self-consistently* with the equations for the other variable distributions that involve $Tj$, which emerge through the M-projections on $G_{out}$ and $G_{in}$ (or equivalently from the laws of probability decomposition, conditioning, and marginalization). Note that as in the original IB problem, the effective information-distortion measure, $d(t_j, \mathbf{u}_j)$, emerges directly from the principles, rather then being assumed in advance. In other words, it generalizes the Shannon's reduction of the physical cost and distortion properties to multiple tradeoffs between values of mutual information between different nodes of the network.

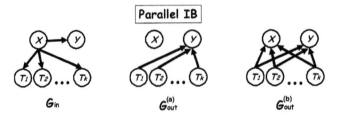

Fig. 12. Examples of input and output graphs that provide parallel clustering into multiple categories. The $G_{in}$ describe the compression dependencies (and the correct dependencies among the original variables), while $G_{out}$ give different types of prediction networks. ( Taken from [36].)

## 4.7. Deterministic annealing algorithm

The temperature parameter, $\beta$, which controls the complexity of the obtained representation, calls for a procedure that can take us from simple to complex representations, by modifying $\beta$ continuously. Such an algorithm mimics the physical process of slow cooling, or "annealing" of the system. When it is done without actually simulating the stochastic fluctuations it is called "deterministic annealing" [22]. In general, a deterministic annealing procedure works by iteratively increasing the parameter $\beta$ and then adapting the solution for the previous value of $\beta$ to the new one . It starts with low values of $\beta$ that correspond to simpler solutions and gradually increase the solution complexity. [12] It is computationally much simpler than stochastic (or simulated) annealing but has no guarantees to converge to a global optimum, though it is known to work well for many clustering and quantization problems.

In our context, the existence of the Lagrange multiplier $\beta$ makes deterministic annealing a natural choice. This allows the algorithm to "track" the changes in the solution as the tradeoff moves from high compression to high preservation of relevant (multi) information.

Recall that when $\beta \rightarrow 0$, the optimization problem tends to make each $T_j$ independent of its parents. At this point the solution consists of essentially one cluster for each $T_j$ which is not predictive about any other variable. As we increase $\beta$, at some (critical) point the values of some $T_j$ diverge and show two different behaviors. This phenomenon is correspond to a (usually second order) phase-transition, or a bifurcation point of the system. Successive increases of $\beta$ will reach additional phase transitions in which additional splits of some values of the $T_j$'s emerge. Our goal is to identify these cluster bifurcations and eventually record for each $T_j$ a bifurcating tree that traces the sequence of solutions at different values of $\beta$ (see, for example, Figure 13).

To detect these bifurcations we adopt the method suggested in [37] to multiple variables. At each step, we take the solution from the previous $\beta$ value we considered and construct an initial problem in which we *duplicate* each value of every $T_j$. Thus, we need to specify the conditional membership probabilities of these duplicated values. Suppose that $t_j^\ell$ and $t_j^r$ are two such duplications of some value $t_j \in T_j$. Then we set $q^*(t_j^\ell \mid \mathbf{u}_j) = q(t_j \mid \mathbf{u}_j) \left( \frac{1}{2} + \alpha \hat{\epsilon}(t_j, \mathbf{u}_j) \right)$ and $q^*(t_j^r \mid \mathbf{u}_j) = q(t_j \mid \mathbf{u}_j) \left( \frac{1}{2} - \alpha \hat{\epsilon}(t_j, \mathbf{u}_j) \right)$, where $\hat{\epsilon}(t_j, \mathbf{u}_j)$ is a (stochastic) noise term randomly drawn out of $U[-\frac{1}{2}, \frac{1}{2}]$ and $\alpha > 0$ is a (small) scale parameter. Thus, each copy $t_j^\ell$ and $t_j^r$ is a slightly perturbed version of $t_j$. If $\beta$

---

[12]In deterministic annealing terminology, $\frac{1}{\beta}$ is the "temperature" of the system, and thus increasing $\beta$ corresponds to "cooling" the system.

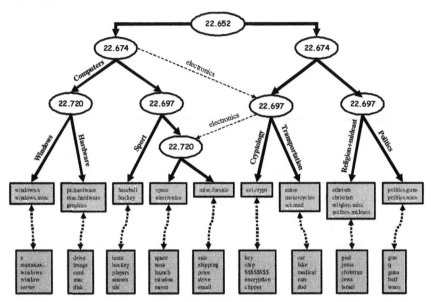

Fig. 13. Cluster-tree obtained by Noam Slonim for the 20-News-group word-category data, using a symmetric version of the multi-IB algorithm. The categories are hierarchically clustered by deterministic annealing (where the critical value of $\beta$ written at each node). The corresponding word clusters are associated to the topic clusters at the bottom. As can be seen, many of the transition splits occur at similar values of $\beta$, indicating the emergence of a relevant "semantic dimension" along which several clusters split together. A similar phenomena was observed in [20] for noun-verb clustering. (Taken from [36].)

is high enough, this random perturbation suffices to allow the two copies of $t_j$ to diverge. If $\beta$ is too small to support such bifurcation, both perturbed versions will collapse to the same solution.

Given this initial point, we simply apply the (asynchronous) multivariate iIB algorithm. After convergence is attained, if the behavior of $t_j^\ell$ and $t_j^r$ is "sufficiently different" [13] then we declare that the value $t_j$ has split, and incorporate $t_j^\ell$ and $t_j^r$ into the hierarchy we construct for $T_j$. Finally, we increase $\beta$ and repeat the whole process.

---

[13] Specifically, we denote by $\mathbf{Ne}_{T_j}^{G_{out}}$ the set of $T_j$'s neighbors in $G_{out}$, and consider the (JS) divergence between $p(\mathbf{ne}_{T_j}^{G_{out}} \mid t_j^\ell)$ and $p(\mathbf{ne}_{T_j}^{G_{out}} \mid t_j^r)$. Other techniques are also plausible.

## Acknowledgments

Those lecture notes are based on many years of fruitful interaction with Fernando Pereira and Bill Bialek and on insightful ideas and collaboration of Nir Friedman. Special thanks are due to Amir Globerson for many useful discussions during the writing of these notes. Most of the work described in the lectures has culminated and elaborated in the PhD theses and in numerous joint publications with Noam Slonim, Gal Chechik, Amir Globerson, Elad Schneidman, and Ilya Nemenamn. Our work has been supported in part by grants from the Israel Science Foundation (ISF-Center of Excellence), the Human Frontier Science Foundation (HFSP), The Israeli Ministry of Science, and by the US-Israel Bi-national Science Foundation (BSF). I thank the organizers of the Les-Houches summer school on "Methods and Models of Neurophysics" for their invitation and support and to the University of Pennsylvania for the kind hospitality.

## References

[1] S. Amari and N. Nagaoka. *Methods of Information Geometry*. Oxford University Press, 2000.

[2] L. D. Baker and A. K. McCallum. Distributional clustering of words for text classification. In ACM SIGIR 98, 1998.

[3] T. Berger. Living information theory. Shannon Lecture, ISIT, July 2002. IEEE ISIT, Lusanne.

[4] Gal Chechik. *An Information Theoretic Approach to the Study of Auditory Coding*. PhD thesis, Hebrew University of Jerusalem, 2004.

[5] G. Chechik and N. Tishby. Extracting relevant information with side information. *Advances in Neural Information Processing Systems (NIPS) 15*, 2002.

[6] T.M. Cover and J.A. Thomas. *The elements of information theory*. Plenum Press, New York, 1991.

[7] I. Csiszár. I-divergence geometry of probability distributions and minimization problems. *Annals of Probability*, 3(1):146–158, 1975.

[8] I. Csiszár and J. Körner . *Information Theory*. Akadémiai Kiadó, Budapest, 1981.

[9] I. Csiszár. A geometric interpretation of darroch and ratcliff's generalized iterative scaling. *Annals of Statistics*, 17(3):1409–1413, 1989.

[10] G. Chechik, A. Globerson N. Tishby and Y. Weiss. Information bottleneck for gaussian variables. *Journal of Machine Learning Research (submitted)*, 2004.

[11] N. Friedman, O. Mosenzon, N. Slonim, and N. Tishby. Multivariate information bottleneck. In *Proceedings of the Seventeenth Conference on Uncertainty in Artificial Intelligence (UAI)*, 2001.

[12] R. Gilad-Bachrach, A. Navot, and N. Tishby. An information theoretic tradeoff between complexity and accuracy. In *Proceedings of the 16'th Conference on Computational Theory (COLT)*, pages 595–609, Washington, 2003.

[13] A. Globerson and N. Tishby. Sufficient dimensionality reduction. *Journal of Machine Learning Research*, 3:1307–1331, 2003.

[14] A. Globerson and N. Tishby. On the optimality of the gaussian information bottleneck curve. Technical report, Hebrew University, February 2004.

[15]  A. Globerson and N. Tishby. The minimum information principle for discriminative learning. In *Submitted to "The Twentieth Conference on Uncertainty in Artificial Intelligence (UAI)"*, 2004.

[16]  J. Lin. Divergence Measures Based on the Shannon Entropy. In IEEE Transactions on Information Theory, 37(1):145–151, 1991.

[17]  I. Nemenman and N. Tishby. Network information theory. Preprint.

[18]  M. Gastpar, B. Rimoldi and M. Vetterli. To code or not to code: Lossy source-channel communictaion revistsed. *IEEE Trans. Info. Theory*, 49:1147–1158, 2002.

[19]  J. Pearl. Probabilistic Reasoning in Intelligent Systems. Morgan Kauffman, San Francisco, 1988.

[20]  Fernando C. Pereira, Naftali Tishby and Lillian Lee. Distributional similarity of english words. In *30th Annual Meeting of the Association for Computational Linguistics*, pages 183–190, 1993.

[21]  B. Rimoldi. Beyond the separtion principle: A broader approach to source-channel coding. In *4th Int. ITG Conf. on Source and Channel coding*, Berlin, January 2002. IEEE.

[22]  K. Rose. Deterministic annealing for clustering, compression, classification, regression, and related optimization problems. In Proceedings of the IEEE, 86:2210–2239, 1998.

[23]  Elad Schneidman. *Noise and Information in Neural Codes*. PhD thesis, Hebrew University of Jerusalem, 2002.

[24]  E. Schneidman, W. Bialek, and M. J. Berry. An information theoretic approach to the functional classification of neurons. *Advances in Neural Information Processing Systems (NIPS) 15*, 2002, to appear.

[25]  E. Schneidman, N. Slonim, R. R. de Ruyter van Steveninck, N. Tishby and W. Bialek Analyzing neural codes using the information bottleneck method. In preparation.

[26]  C.E. Shannon. A mathematical theory of cryptography. *Bell Labs Tech. MM 45-110-02*, 1945.

[27]  C. E. Shannon. A mathematical theory of communication. *Bell Sys. Tech. Journal*, 27:379–423 and 623–656, 1948.

[28]  C. E. Shannon. Coding theorems for a discrete source with fidelity criterion. *Institute of Radio Engineers, International Convention Record*, 7:4:142–163, 1959.

[29]  N. Slonim, N. Friedman, and N. Tishby. Agglomerative Multivariate Information Bottleneck. In Advances in Neural Information Processing Systems (NIPS) 14, 2001.

[30]  N. Slonim, N. Friedman, and N. Tishby. Unsupervised Document Classification using Sequential Information Maximization. In 25th Ann. Int. ACM SIGIR Conf. on Research and Development in Information Retrieval, 2002.

[31]  N. Slonim N. Friedman, , and N. Tishby. Multivariate information bottleneck. *Preprint*, 2004.

[32]  N. Slonim & N. Tishby. Agglomerative Information Bottleneck. In Advances in Neural Information Processing Systems (NIPS) 12, pp. 617-623, 1999.

[33]  N. Slonim and N. Tishby. Document clustering using word clusters via the information bottleneck method. In ACM SIGIR 2000, pages 208–215. 2000.

[34]  N. Slonim and N. Tishby. The power of word clusters for text classification. In 23rd European Colloquium on Information Retrieval Research, 2001.

[35]  N. Slonim and Y. Weiss. Maximum likelihood and the information bottleneck. In S. Becker, S. Thrun, and K. Obermayer, editors, *Advances in Neural Information Processing Systems 15*, Vancouver, Canada, 2002.

[36]  Noam Slonim. *Information Bottlneck theory and applications*. PhD thesis, Hebrew University of Jerusalem, 2003.

[37]  N. Tishby F.C. Periera and W. Bialek. The information bottleneck method. In *The 37th Allerton Conference on Communication, Control and Computing*, volume 37, pages 368–379. University of Illinios, 1999.

[38] A.D. Wyner. On source coding with side information at the decoder. *IEEE Trans. on Info Theory*, IT-21:294–300, 1975.

[39] The 20 newsgroups collection:
*http://kdd.ics.uci.edu/databases/20newsgroups/20newsgroups.html"*

Printed and bound by CPI Group (UK) Ltd, Croydon, CR0 4YY

03/10/2024

01040430-0008